H.-E. Reineck · I. B. Singh

Depositional Sedimentary Environments

With Reference to Terrigenous Clastics

Second, Revised and Updated Edition
Corrected Second Printing

With 683 Figures

Springer-Verlag
Berlin Heidelberg New York

Prof. Dr. HANS-ERICH REINECK
Institut für Meeresgeologie und Meeresbiologie
Senckenberg-Institut, Schleusenstraße 39 A
2940 Wilhelmshaven, FRG

Dr. INDRA BIR SINGH
Department of Geology, Lucknow University
Lucknow, India

ISBN 3-540-10189-6 2. Aufl. Springer-Verlag Berlin Heidelberg New York
ISBN 0-387-10189-6 2nd ed. Springer-Verlag New York Berlin Heidelberg

ISBN 3-540-07377-9 1. Aufl. Springer-Verlag Berlin Heidelberg New York
ISBN 0-387-07377-9 1st ed. Springer-Verlag New York Heidelberg Berlin

Library of Congress Cataloging in Publication Data. Reineck, Hans-Erich. Depositional sedimentary environments, with reference to terrigenous clastics. Bibliography: p. Includes index. 1. Rocks, Sedimentary. 2. Sedimentary structures. 3. Sediments (Geology). I. Singh, Indra Bir, joint author. II. Title. QE471.R425 1980 551.3′03 80-20429.

Cover design by W. Eisenschink, Heidelberg.

Typesetting: Bauer & Bökeler, Filmsatz KG, Denkendorf.
Printing and Bookbinding: Konrad Triltsch, Graphischer Betrieb, Würzburg.
32/3145-5 4 3 2 – Printed on acid-free paper

To
Harold N. Fisk and
Rudolf Richter

Preface to the Second Edition

Interest in the interpretation of depositional environment of sediments has increased enormously since the first edition of this book was published in 1973. There were innumerable comments and critical reviews of our book, some of them through talks with both professionals and students. Our colleagues working in modern sediments found that we have not dealt enough with sedimentation processes. On the contrary, field geologists were happy to have generalized information on modern environments in one work. We are thankful to all who gave us their comments.

Keeping the comments in mind, we decided to stick to the same basic pattern of the book which we also had in the first edition: to provide the wealth of information from Recent and experimental studies to the geologists working in the ancient sediments. No doubt, specialized and sophisticated methods of study are of immense value. However, in our book we emphasize mainly the methods which a geologist can use in the field, in the outcrops, and in the cores, and we are confident, that even by these simple methods much information relevant to environmental interpretation can be obtained.

We have benefited tremendously from information obtained by the environmental analysis of the ancient sediments. However, in our book we have not directly utilized the data obtained from the ancient sediments. Even today, the sequences which for years were considered to be typically aeolian deposits are being reinterpreted as fluvial deposits, and controversies between shallow versus deep-sea deposits of certain sequences are quite common. Nevertheless, we have mentioned a few selected examples of each environment from the ancient sediments.

For the revision of the first edition, information gathered in the last few years has been added, although basically the old text has been retained to avoid unnecessary increase in the price of the book.

While working on the second edition, a few points struck us; namely (a) much additional information is now available for certain environments, (b) some sedimentation models can be better represented today, (c) in some environments existing models have only been fortified with new, additional investigations. As in the first edition, we have been selective in citation. Although many new studies are cited, some significant references may have been missed. We hope that our colleagues will excuse us for this unavoidable mistake.

Although quite hesitant in the beginning to undertake the revision of our book, we gained courage mainly due to numerous reprints which our colleagues sent to us. At the same time, we also felt the responsibility of providing additional up-to-date information for our young colleagues and students. We would welcome any comments or critiscism offered to us on our modest compilation.

Many colleagues and publishers permitted us to reproduce their drawings or photographs in this book, to them we are extremely thankful.

Thanks are expressed to Professor J. D. Howard, Dr. J. Hülsemann, Dr. A. Schäfer, and Professor A. Seilacher for reviewing parts of the manuscript.

One of us (I.B.S.) is grateful to the Lucknow University, India, for granting leave of absence and to the Alexander von Humboldt Stiftung for a research scholarship during which part of the writing was done, and to Professor P. Wurster for providing the working facilities of the Geologisches Institut, Bonn University.

We wish to thank Mrs. Renate Flügel for making the line drawings; and Miss Astrid Grünert for the tedious job of typing the manuscript for the second edition.

We acknowledge with thanks the support of the Senckenbergische Naturforschende Gesellschaft, Frankfurt am Main, during preparation of this edition.

A final word of thank goes to Springer-Verlag who encouraged us, and cooperated in every aspect for completion of this second edition.

July 1980   H.-E. REINECK
                                                    I. B. SINGH

Preface to the First Edition

This book has been written with the aim of compiling from modern environments information that can be useful in the reconstruction of ancient environments. It is intended for all those interested in recognizing depositional environments. The study of sediments includes investigations of various aspects of sediments. This needs a study by standard methods. Methods of study have not been included, as many textbooks exist on the subject. However, the importance of various results obtained from such investigations has been discussed, as far as these results can be helpful in environmental reconstruction. Special attention is given to information that has accumulated during the last decades on the mode of genesis of various sedimentary features and their distribution in present-day environments.

As far as possible, existing terminology has been used. However, in several cases new simple groupings and classifications have been proposed. In making classification, generally, the form and shape of the features have been considered, so that they can be applied easily to ancient sediments. At the same time, the genesis of such features has been noted, and genetic names and their characteristics have been given for detailed work. The subject is so vast that several primary sedimentary features that have no direct bearing on environmental interpretation have been omitted.

The subject is vast, and numerous publications are available. Most of the publications were consulted, but only selected works have been cited. While no attempt is made to give a complete bibliography, most of the important, modern, and comprehensive studies have been cited. Preference has been given to articles and books that provide useful reviews of the subject and which themselves contain large, well-selected lists of references. Thus, this book provides a working bibliography, which hints at additional reading. We are aware, that many important references have been omitted in this way. But this has been almost unavoidable while compiling information from diverse fields of research.

Readers might note a significant difference in emphasis and interpretation of various features between this and other books. This is because of our different approach and experience.

On several points an opinion formulated by the authors is preferred over the other existing points of view. Some critical readers more deeply involved than us in various fields may find our opinions biased and may find themselves in disagreement with some of the opinions expressed herein. We have tried to be objective, but conflicting views are unavoidable.

Any textbook is essentially a compilation of previously published studies. The existence of this book is due mainly to the eminent research workers who have contributed to the better understanding of processes of sedimentation.

Furthermore, many collegues and publishers permitted us to reproduce in this book their original photographs and drawings.

The completion of this book is due to major efforts of Springer-Verlag, who cooperated in every way with the authors.

We also wish to acknowledge the criticism of those who read sections of the manuscript.

We wish to thank Mrs. Renate Flügel for most of the drawings; Mr. Hermann Schäfer for drawings of Figs. 25, 33, 36, 38 and 44; Mrs. Margot Bialdiga for drawings of Figs. 512 and 526; and Miss Elisabeth Beruda and Mrs. Gabriele Holzner for the final typing of the manuscript.

Last but not least we are greatly indebted to the Senckenbergische Naturforschende Gesellschaft, Frankfurt am Main, which enabled us to write this book and to the Deutsche Forschungsgemeinschaft for financial supporting most of our own researches.

Wilhelmshaven, September 1973   H.-E. REINECK
                                                 I. B. SINGH

Contents

Part I. Primary Structures and Textures

Part. II. Modern Environments

Part I. Primary Structures and Textures

Introduction

Geologists are mainly concerned with the study of the lithosphere. The sedimentary and metasedimentary rocks, although they constitute only 5 % of the lithosphere by volume, they occupy 75 % of the exposed land area (Pettijohn 1957). Thus, the study of sedimentary rocks is the major problem of geology. An obvious question arises in the study of sedimentary rocks: Under what conditions and by which processes were the sedimentary rocks in question deposited.

Pettijohn (1957) pointed out that geology is primarily history – a history of the earth. The study of sedimentary rocks concerns the history of sedimentary rocks. Pettijohn adds, further, that the history of any given sedimentary rock unit involves determining the source area and source rocks, determining the environment of deposition, the physical, chemical, and biological milieu in which the sediment accumulated, and finally, lithification the study of post-depositional changes.

Excluding the process of lithification, the sediment can be regarded as a product of source and environment30Pettijohn 1957). Discovering the origins and reconstructing the environment that produced the sedimentary rock is a difficult task. A thorough knowledge of sedimentary processes that cause the production, transportation, and deposition, is most important in environmental reconstruction. Most important for such purposes are primary structures and textures, together with the biological and chemical aspects of the sediment. In this book, mainly primary structures and textures have been considered. However, biological factors, as far as they are important, have been covered. A short account of mineralogical and chemical factors is also given.

Of all the sedimentary rocks, clastic detrital rocks – sandstones and shales – are most abundant. In the present work the focus is on the depositional environment of detrital sediments.

Interest in the study of recent sediments has increased tremendously, especially because an understanding of the processes of sedimentation and the resulting sedimentary features contributes to a better understanding of various features encountered in ancient sediments. With insight into the processes of sedimentation and the various features resulting therefrom, environmental reconstructions can be made with much more certainty. In the last few decades a tremendous amount of work has been done on processes of sedimentation and sedimentary structures in modern environments. Detailed laboratory and theoretical work on sedimentary structures have contributed much to the understanding of the process of formation of various structures.

In the present work the genetic aspects of various sedimentary structures – their mode of origin and distribution in modern environments – have been given greatest emphasis. No attempt was made to make rigid classifications, but rather broad groupings are proposed, keeping in mind their applicability to ancient sediments. The interrelationship of various features has been particularly emphasized. Features more likely to be preserved and those important in diagnosis of ancient environments have been given special attention. Various similar-looking structures may originate in altogether different ways. Wherever possible, criteria are given to differentiate such structures.

There are many excellent books, monographs, and symposia available on various aspects of sediments, sedimentary structures, and environmental analysis. The following is a list of supplementary reading: Walther (1894), Andrée (1920), Twenhofel (1932, 1950), Krumbein and Pettijohn (1938), Trask (1939), Sverdrup et al. (1942), Shrock (1948), Kuenen (1950a), Bagnold (1954a), Lombard (1956), Strackov (1956), Pettijohn (1957), Bruns (1958), Ruchin (1958), Termier and Termier (1960), Weller (1960), Milner (1962), Hill (1963), Krumbein and Sloss (1963), Potter and Pettijohn (1963), Pettijohn and Potter (1964), Dzulynski and Walton (1965), Middleton (1965a), Gubler et al. (1966), Duff et al. (1967), Hallam (1967), Allen (1968a), Conybeare and Crook (1968), Laporte (1968), Henningsen (1969), Allen (1970), Füchtbauer and Müller (1970), Glennie (1970), Ricci Lucchi (1970), Rigby and Hamblin (1972), Pettijohn et. al. (1972).

In recent years a number of books, monographs have been published which give information on various aspects of sedimentation. The important ones are Niggli (1948, 1952), Lombard (1959),

Dmitrieva et al. (1962), Gubler et al. (1966), Wei-mer (1970), Blatt et al. (1972), Fisher and Brown (1972), LeBlanc (1972), Rigby and Hamblin (1972), AAPG (1973 a, b, 1974), Amstutz et al. (1973), Wolf (1973), Busch (1974), Lützner et al. (1974), Spearing (1974), Dickinson (1975 b), Grumbt (1975, 1976), Klein (1975), SEPM Short Course No. 2 (1975), Andrews (1976), George (1976), Hayes and Kana (1976), Lützner (1976), Friedman and Sanders (1978), Reading (1978), Ricci Lucchi (1978).

Allen, J. R. L. (1982) gave an excellent review of se-dimentary structures, their character and physical basis.

Depositional Environments

The word "environment" is constantly used in geological studies. There is a wide range of geological environments one can study and interpret. One can study the environment of igneous rocks i.e., the physico-chemical environment of their formation, or the environment of metamorphic rocks and minerals which deals with the study of stability of various minerals under different pressure-temperature conditions, or the environment of sedimentary rocks which deals with the study of erosional, and depositional processes. What we are concerned with here are the depositional sedimentary environments – i.e. under what hydrodynamic, biological, and chemical conditions a given rock was deposited. A sedimentary environment can be an erosional or a depositional environment. An erosional environment is characterized by certain denudation processes and denudation geomorphology, resulting in definite denudation products. Associated with erosional environments are depositional environments, with their characteristic depositional processes and products of deposition (Walther 1894). In the geological record it is the depositional environment which leaves behind its imprints in sediment and produces sedimentary sequences. Thus, as geologists we are more interested in the depositional environment than the erosional environment of sediments.

Depositional sedimentary environment has been variously defined. A depositional environment can be defined in terms of physical, biological, chemical or geomorphic variables. Thus, a depositional sedimentary environment is a geomorphic unit in which deposition takes place. Such a place of deposition is characterized by an unique set of physical, biological, and chemical processes operating at a specified rate and intensity which impart sufficient imprint on the sediment, so that a characteristic deposit is produced. The character of a sediment so produced is determined both by the intensity of the formative processes operating on it and by the duration through which such action is continued (Pettijohn 1957).

Some of the geomorphic units are rather complex in character. In such geomorphic units physicals, biological, and chemical variables vary strongly from place to place, and as a result so does the type of sediment deposited. These are the subenvironments within a broad environment. For example, intertidal flats can be differentiated into channels, mud flats, sandy flats, etc. A fluvial environment can be differentiated into channel, levee, and flood basin deposits. These are the subenvironments. Moreover, variation can be produced as a result of changing weather conditions and hydrodynamic conditions.

The study of depositional environments in ancient sediments is, thus, essentially the study of geomorphology, i.e., recognition of geomorphic units. Geomorphic units are recognized by features preserved in ancient sediments. Only the processes which leave a lasting record in ancient sediments are useful in the recognition of environment. Such environmental parameters can be grouped into three groups – physical, biological, and chemical. Out of these three groups the study of physical processes is most important, as it provides the most basic information for geomorphic interpretation.

Here, we would like to add few words about facies. A sedimentary facies is defined as a sum of all the primary characteristics of a sedimentary unit. However, the term "facies" has been used in many different senses.

Detailed discussions on the concept of facies can be obtained in Dunbar and Rodgers (1957), Moore (1949, 1957), Teichert (1958), Weller (1960), Krumbein and Sloss (1963).

Some workers consider only one or two aspects of a sediment in defining facies –such as lithofacies, biofacies, geochemical facies, etc. In any case a sedimentary facies is a result of deposition in a given environment.

Quite often the term facies is used to distinguish various units of a succession on the basis of single criteria, like color – red facies, green facies, bedding characteristics – parallel-bedded facies, ripple-bedded facies, or fossils – shell limestone facies. In such a case the term facies is used to define the small identifiable units within a succession. Further, the term facies has also found usage in expressing the processes of sedimentation, e.g., tidal facies, fluvial facies, or even to define the large-scale tectonic setting of the sedimentation basin, e.g. flysch facies, molasse facies.

In the study of ancient sediments it is advisable to study various characteristics of a stratigraphic unit and describe it as a facies. Only after such a basic study should the facies in question be assigned an environment. Terms such as "beach facies" and "lagoon facies" should be avoided. A sandy facies of a given stratigraphic unit can be assigned to a beach environment only after a summing up of all the characteristics and a comparison with models based on the study of modern sediments.

Study of facies includes changes in horizontal direction, as well as changes in vertical direction. If genetically related facies (in the sense of Walther 1894) overlie each other, they can be grouped into sequences.

Among the most important physical factors are the deposition medium, current and wave intensity and velocity, water depth, etc. In other words, study of physical environments means determination of hydrodynamic conditions under which certain sediments were deposited. Information about hydrodynamic conditions is best obtained by detailed and careful study of primary sedimentary structures, both inorganic and organic, and of size distribution parameters. During the last few decades especially much work has been done both experimentally and in the field, regarding the genesis, stability, and distribution of sedimentary structures. This information helps us to understand and interpret environments.

Deposits of a geomorphic unit are also strongly influenced by the climate and the size and shape of the basin of deposition. In ancient rocks, information regarding the latter can be obtained by studying the geometry of units and their inter-relationships. Clues about general climatic conditions may be provided by faunal and floral studies, and to a certain extent by mineralogical and chemical parameters.

Sedimentary tectonics is another important factor which should be considered in reconstruction of ancient environments. Study of lateral and vertical facies changes, knowledge of rate of deposition, and palaeogeography provides important hints for sedimentary tectonics.

Sedimentary tectonics has a strong control over the thickness of sedimentary deposits, and the nature of juxtaposition of various facies in space and time. The characteristics of sedimentary deposits of a given environment can change with changing tectonic setting. Recently, attempts have been made to characterize the sedimentary deposits of different tectonic settings, especially in the light of the plate tectonic theory. Mitchell and Reading (1978) provide a useful text on the relationship of sedimentary deposits and plate tectonics. Other useful sources of information of sedimentary tectonics are Dickinson (1974) and Bott (1976).

Syn-depositional tectonics can be very significant in affecting the nature of sediments in a given environment. Thiede (1978) discussed this problem with respect to sedimentation in young, immature ocean basins. The sedimentation in young ocean basins begins after the end of initial rifting and associated volcanism. Most of such young ocean basins are separated from large oceans by shallow sills. In areas where evaporation exceeds precipitation and runoff from the land, mainly evaporites are deposited. Similarly, euxinic black shales can be deposited if circulation is poor. The Red Sea, Gulf of Aden, Gulf of California, and the Arctic ocean are such young ocean basins (Thiede 1978).

Recognition of minerals originally precipitated in the environment of deposition provide important clues to the Eh, pH, and salinity of deposition medium. Biological criteria, i.e., palaeoecological information, can provide conclusive information about water depth, salinity, temperature, turbulence, rate of sedimentation and resedimentation, etc. The study of the physical factors, if combined with the study of biological and chemical factors provides a more complete picture of the sedimentary depositional environment.

In this book we deal mainly with the physical criteria that are helpful in environmental reconstruction. Biological and chemical criteria are dealt only briefly. Important references dealing with geochemical and biological parameters have been listed.

The presence or absence of individual features cannot generally be used as a positive sign in environmental interpretation. Most of the structures occur in several environments. It is generally the spectrum of sedimentary structures and their presence in certain combinations that provide direct clues in environmental interpretation. Certain associations of sedimentary structures and certain lithological associations are typical for specific environments. Study of present-day environments has provided vast information, on the basis of which environmental models can be constructed. Such models can be based on actual information of the sedimentary structures, grain size etc. in the present-day environment of a given area; or process-response models can be constructed emphasizing the sequence of sedimentary structures and relating them to the processes responsible for their formation. Information of investigations in different geographic areas of a given environment may be put together to build up a generalized model for a given environment. However, such generalized models should be considered only as a working tool and a first step in understanding the environment in question. These models should be considered no more than "models", because many variations in the deposits of similar environments are possible. Climatic and geomorphological relief play very important roles, both in source area and in the basin of deposition. For example, tidal flat environments in a temperate climate show many variations from those in arid and warm humid climates. The problem of identifying environments becomes much more difficult in ancient sediments, where sufficient information about the geomorphology and the climate of a region is rarely available. Moreover, some of the conditions may have been unique in the geological past; and, we may say that no one sedimentary sequence is exactly identical to another. However, despite their deposition in varying conditions, certain basic patterns are always similar in deposits of a given environment. This makes the recognition of depositional environments, even in geological record, possible with some certainty.

In certain environments identification is rather difficult because some environments do not posses enough typical characteristics. Such is the case for recognizing an intertidal flat subenvironment within a tidal environment. For definite recognition of an intertidal flat subenvironment, definite indications of intermittent emergence of the depositing surface are necessary. But the probability that such features in these sediments will be preserved is very low. Other environments need both a detailed study and a regional study at the same time. Such is the case in recognition of delta environments, as various subenvironments of a delta can also occur independently. It is the specific association of various subenvironments and their geographic distribution which makes the recognition of an ancient delta possible.

Environmental analysis of sedimentary sequences is commonly a part of a broader study in a part or whole of a sedimentary basin.

For basin analysis, besides the knowledge of depositional environments, a detailed knowledge of nature and pattern of current systems, dispersal patterns of sediments, etc., is necessary. Potter and Pettijohn (1963) discuss in detail the methods of study of sedimentary basin. Detailed environmental analyses of selected profiles in a sedimentary basin provide a good background for basin analysis.

Thus, we think that the best way to study ancient sediments is to record all the primary sedimentary structures, grain features and parameters, and bioturbation structures of individual units of a sedimentary sequence, and try to interpret the hydrodynamic conditions under which the unit was probably deposited. A descriptive guide to rock types (occurring in the Cretaceous deposits of western North America) is given by Ruby et al. (1981).

The next step is the study of lateral and vertical relationship of various units and the determination of their geometry. Then these can be compared with models of comparable environments based on information from the study of present-day environments. The special features of the ancient sediments should be discussed, and an attempt should be made to find out under which special conditions such features were formed. Erection of such detailed sections within a sedimentary basin would help in reconstruction of basin configuration, sedimentary tectonics, and palaeogeography.

The importance of environmental reconstruction has been recognized from the very beginning of the study of sediments. Leonardo da Vinci undertook environmental interpretation in fifteenth century, when he recognized fossils as preserved animals. And he concluded that the presence of corals and oysters on the rocks of Monte Ferrato/ Lombardei were indicators of marine deposits and the presence of sea at that place (Salomon 1928). Since then, geologists have been engaged in interpretation of depositional environments of ancient sediments.

Recognition of ancient depositional environments is not only of academic interest, i.e., in the proper understanding of sedimentary processes in the history of the earth. Such information is very important in exploration for natural resources such as petroleum, coal, limestone, phosphate, placer deposits, etc., as such deposits are found associated with certain specific environments.

Physical Parameters

General Information

Earlier (see p. 5) the importance of physical parameters in the reconstruction of ancient environments was discussed. Primary sedimentary structures and textures of a sediment are the major features that provide information about the medium and mode of transport (by ice, water, or wind) and energy conditions at the time of deposition. Energy conditions include depth of water, velocity of flow, turbulence, wind velocity, etc. The study of physical parameters can be divided into two groups.

1. The study of inorganic primary sedimentary structures. It includes the study of all the bedding features and surface features produced at the time of deposition which are the direct manifestation of depositing medium and energy conditions prevalent at the time of deposition.

2. The study of sedimentary textures. It includes granulometric study of sediments. Grain size, grain size parameters, shape and roundness, surface texture, and primary fabric are the important aspects of such a study. These features are controlled mainly by the medium and mode of transport, and to a lesser degree by the depositing medium. Thus, they provide more accurate information about the transporting medium and only limited information about depositional conditions. This leads to the fact that the study of primary sedimentary structures is most important in the reconstruction of ancient physical environments.

Primary Sedimentary Structures

Pettijohn and Potter (1964) have defined primary sedimentary structures as those formed at the time of deposition or shortly thereafter and before consolidation of the sediments in which they are found.

Thus, they include various kinds of surface markings, bedforms, and bedding. Also included are the structures produced by the activity of organisms, i. e., bioturbation structures. Even penecontemporaneous deformation structures, produced after deposition but before consolidation of sediments, are included herein.

Inorganic primary sedimentary structures are produced as a result of interactions between gravity, the physical and chemical characteristics of the sediment and the fluid, as well as the hydraulic environment (Brush 1965). Brush adds, further, that the recognizable structures develop as a result of sorting of sediment with respect to size, shape, and specific gravity. Such sorting is, in turn, the result of variable settling rates of sediment grains, turbulent diffusion, gravitational avalanching, and boundary shear stress. In short, inorganic primary sedimentary structures provide useful information about hydrodynamic conditions of the environment of deposition.

Organic sedimentary structures, especially those produced by benthonic communities, can also provide useful information regarding conditions of sedimentation (see p. 158).

Many authorities have made attempts to classify primary sedimentary structures. Some deal only with a particular type of structures; others have tried to recognize all the sedimentary structures. Recent attempts in this direction are those by Pettijohn (1957). Dmitrieva et al. (1962), Pettijohn and Potter (1964), Conybeare and Crook (1968), and Ricci Lucchi (1978).

In general, inorganic primary structures can be organized into two groups:

1. Surface features. Includes all the features observed on the sediment surface or a bedding plane surface. In ancient sediments some of them may be found better preserved on the lower surface of a sediment layer.

2. Bedding features. Includes all the forms observed in a section cut normal to the sediment surface. Various types of bedding and deformation structures, (i. e. convolute bedding, ball-and-pillow structures) can be grouped here.

However, many of the surface features are closely related to bedding structures, and some of the bedding is simply the internal structure of surface features. For example, ripple marks, if superimposed one above the other and if viewed in section, show cross-bedding. However, for the sake of classification surface features and bedding structures have been described under different headings.

In the present discussion no attempt has been made to classify the sedimentary structures. For the sake of clarity in description they have been organized into various groups that do not represent any sort of strict classification, but just the grouping. Following groups have been organized: Current and wave ripples, wind ripples, surface markings and imprints, scour marks, tool marks, penecontemporaneous deformation structures, and bedding.

Sediment Grain Movement

The process and type of sediment grain motion in a fluid flow is of fundamental importance to sedimentologists. The problem of movement of sediment grains under the influence of fluid (mostly water) has been studied primarily by civil engineers. Leliavsky (1955) reviews the mechanisms of sediment grain movement. Sundborg (1956), Church and Gilbert (1975), and Middleton and Southard (1978) give useful reviews on the processes involved in the sediment grain motion, expecially from a sedimentologist's standpoint.

A grain starts to move on a sediment bed when the combined lift and drag forces produced by the fluid exceed the gravitational and cohesive forces of the sediment grain. Cohesive forces are important only in the finer-grained sediments.

The process of entrainment of grain (beginning of grain movement) on a sediment bottom is mainly determined by the water flow velocity and grain size of the sediments. However, shape of sediment grains, position of the grain, composition of sediment, and turbulences of the flow and type of packing of the sediments are also important in effecting the entrainment velocity of sediment grains.

Determination of entrainment velocity for a sediment is significant in that it gives the flow conditions necessary to just move the largest grain present in a given deposit. Hjulström (1935, 1939) plotted critical tractive velocity at which grain movement begins against the grain size and found that for coarser grains, size of material moved is proportional to the velocity. However, in finer sediments (< 0.1mm), due to cohesive forces, this relationship breaks out, and the energy needed for setting grains in motion increases with decreasing grain size. However, once in motion (especially in suspension), the behaviour of sediment grain is mainly the function of settling velocities. For example, it requires greater energy to set clay particles than sand particles into suspension;

but the sand grains settle rapidly, while the clay particles remain in suspension because of their smaller settling velocities (Fig. 1).

Entrainment velocity for cohesive sediments is strongly dependent upon the degree of consolidation. With increasing consolidation (reduction in porosity) in the cohesive sediments a much higher entrainment velocity is necessary to put the sediment in motion (Postma 1967). Further, studies of Migniot (1968), Terwindt and Breusers (1972) indicate that freshly deposited mud exhibits a rather quick initial consolidation, and a time of about 3–4 hrs is enough to cause a considerable increase in the entrainment velocity.

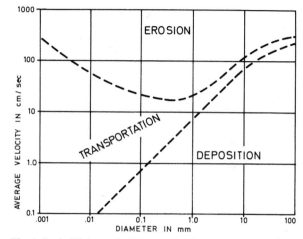

Fig. 1. Basic Hjulström's diagram showing relationship among erosion, transportation and deposition of sedimentary particles. (Modified after Hjulström 1939)

Later, Sundborg (1956, 1967) elaborated and supplemented Hjulström's diagram by including effects of specific gravity, suspension load concentration, and deposition velocity (Fig. 2). The grains once put into motion continue to move even if the flow velocity falls below the critical value of erosion, and according to Sundborg (1967), the cessation of bed movement occurs at a flow velocity which is about ⅔ of the critical erosion velocity. Thus, for a sediment of a given grain size and specific gravity, there is a minimum erosional velocity (entrainment velocity), and a velocity of cessation of movement (depositional velocity). Further, there is no well-marked limit of velocity sufficient to suspend the sediment. There is a continuous transition between the state where no material is in suspension and that of successive increase in concentration of suspension. In the velocity of flow versus grain size diagram (Fig. 2) four different areas are distinguished:

1. Upper left side of diagram, no distinct limit to area II, below. The area is characterized by entrainment of grains from the bottom, transport in suspension with resulting net erosion.

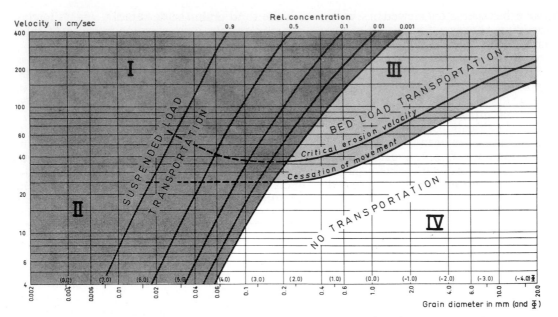

Fig. 2. Diagram showing the relationship between flow velocity, grain size, and state of sediment movement for uniform material of density 2.65 (quartz and feldspar). Velocity is measured
100 cm above the sediment bottom. The curves showing relative
concentration of suspended material indicate the ratio between
the concentration at half the water depth and the concentration
at a reference level close to the bottom. For sediments of different density, distribution of curves is different. (After Sundborg
1967)

2. Lower left side of diagram. The area is characterized by transport in suspension and net deposition of suspended load.

3. Upper right side of diagram. The area is characterized by entrainment of grains from the bottom
and transport of bed load, with resulting net erosion or net accumulation.

4. Lower right side of diagram. Absence of any
significant transport with resulting deposition of
bed load or suspended load.

The boundary between bed load and suspension
load lies at about 0.2 mm. However, the boundary
is diffuse, some suspension transport occurs for
coarser grains, and also traction transport for finer
grains. This diagram is empirical to semi-empirical
and must be considered as qualitative. In natural
material it is difficult to obtain or apply these values directly. One of the main factors is that the material is non-uniform, and flow is rather turbulent.
Moreover, bed roughness related to grain and bedforms has strong influence on the grain movement
velocities, and it has not been accounted for in the
diagram.

However, for the coarse-grained material, especially in the gravels, an approximate estimation of
the minimum current velocities can be obtained by
determining the critical erosion velocity for a maximum gravel size, if an estimate of water depth can
be obtained (Fig. 3) (see Sundborg 1956). Gustavson (1978) used this curve to estimate current ve-

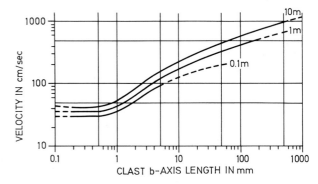

Fig. 3. Diagram showing relationship between mean particle
size of large particles and critical erosion velocity for material of
density 2.65. Curves for three different water depths are plotted
(0.1, 1.0, and 10.0 m depths). (Modified after Sundborg 1956;
Gustavson 1978)

locity in a gravelly river. An analysis of Helley
(1969) also indicates that an estimation of entrainment velocity can be made for the clasts of 15 to
60 cm in diameter. Novak (1973) discusses the significance of the Hjulström curve in coarse-grained
sediments.

However, the distribution of velocity in a flow at
various depths is highly variable and strongly depends upon the depth of flow. This mostly led the
engineers to use shear stress as an expression of energy of flow rather than the current velocity. The

use of shear stress has the advantage as it is the actual force operating on sediment bed. However, for geologists flow velocity is more easily to visualize.

Shields (1936) approached the problem of sediment grain movement by dimensional methods, using shear stress, specific weights of fluid and sediment, particle size, and dynamic viscosity. For practical purposes, Shields's (1936) data indicates that the largest grain that can be moved in a flow is determined by the value of the shear stress at the bed. Detailed analysis of Shields's criterion and shear stress is done by Gessler (1971), Graf (1971), and Yalin (1972).

Shields's curve is extensively used in predicting the competence of a flow or entrainment velocity for a given grain size. Miller et al. (1977) provide an up-to-date review of the literature of the threshold of sediment motion (entrainment velocity) under unidirectional currents, and calculated new threshold curves, modified Shields-type threshold diagrams.

Baker (1974), Baker and Ritter (1975), and Church (1978) have used Shields's criterion with slight modifications to determine the palaeoflow conditions. Baker and Ritter (1975) suggest that in grains larger than 10 mm in diameter, entrainment velocity (shear stress) can be determined using a modified Shields curve. They pointed out that shallow streams tend to initiate particle motion at shear stresses lower than those predicted by the Shields theory, while in deeper streams shear stress needed to entrain sediment is often larger than theoretically expected. Baker and Ritter (1975) use shear stress as $\tau = \nu\ RS$. τ is critical mean shear stress for initiating particle movement, ν-sp. wt. of the transporting fluid, R is hydraulic radius (cross-sectional Area/wetted perimeter), and S is the energy slope for alluvial reach. For rivers D (depth of flow) is approximately equal to R. ν is approximately the specific gravity of water, S is taken as channel bottom slope, or the slope of stratified deposit.

Church (1978) also discusses the use of the Shields curve in determining entrainment energy in natural environments. He argues that, out of many hydraulic parameters, the two, i.e., tractive force and sediment size, show direct relationship. The entrainment of a sediment grain is strongly influenced by the geometrical environment of the grain, e.g., packing, and in general terms a non-cohesive bed can be considered to exist in one of the three stages:
1. *Normal Boundary* – Sediment grains show random arrangement.
2. *Overloose Boundary* – Sediments are in a dilated state due to presence of a large volume of water within the sediment and flow toward the surface exerting dispersive stress. This includes also the sediments showing open-framework packing.
3. *Underloose Boundary* – Sediments show close packing or imbrication.

The experimental data for the Shields curve has been obtained from sediments with normal boundary, while in nature overloose and underloose boundary type sediments are more common. The overloose boundary sediments would need much lower shear stress for entrainment of particles than calculated from Shields's criteria, while the underloose boundary sediments would need much higher shear stress than calculated from Shields's criteria. That is one of the reasons why plots of natural sediments show wider scatter than predictable from Shields's criteria.

However, even the use of shear stress to determine the entrainment energy of a sediment is strongly criticized, as it is strongly affected by grain size distribution of the sediment, bedforms, slope etc., factors which are not accounted for in Shields's criteria. Further, high concentration of sediment in a flow causes deposition at much lower energy than the critical entrainment energy. Nevertheless, the use of Shields's criteria or Hjulström-type diagrams may provide some information about the minimum energy for flow, especially in the gravelly sediments. However, bedforms provide more reliable information on the hydrodynamic conditions of a flow.

In fluvial sediments shear velocities greater than about 20 cm/sec rapidly remove all sand, leaving only coarser materials, and a minimum shear velocity of about 1.4 cm/sec is required to move sand at all (Middleton 1976).

Mode of Sediment Transport

Once sediment grains are put into motion (above the critical erosional velocity) they are transported downstream with the flow. The mode in which grains are transported depends upon the energy of flow and the grain size. In the engineering literature three modes of transport are distinguished:

1. *Bed load* – The sediment grains are in contact with the sediment beds and move by traction, etc. The grains move by traction if their settling velocity is larger than the shear velocity of the flow. It is considered to be about a two-grain-thick bed layer.

2. *Suspended load* – The sediment grains move above the sediment bed, but can be intermittently changed with the bed load. The transport is either free moving in suspension or by saltation. A sediment grain is moved in suspension when the shear velocity is equal to or more than the settling velocity. The suspended load includes the saltation population and coarser suspension population. Middleton and Southard (1977) prefer to call this mode

of transport *intermittent suspension load*. Upward components of the turbulent currents helps in keeping sediments in suspension.

3. *Wash load* – They are very fine-grained particles, and once taken into suspension, remain in suspension until deposited by decellerating flows. They are mostly silt and clay size particles.

The suspended load (intermittent suspended load) is the most important process, which affects the sorting and grain size characteristics of a sediment deposit.

Hydrodynamic Factors and Bedforms in Water

Ripples and some other bedforms are produced on surfaces made up of incoherent material by moving fluids. In nature two important fluids with which geologists are concerned are water and air. The former is a liquid and the latter a gas, but both possess the properties of a fluid. Thus, interaction of a fluid with a noncohesive material, such as sand, produces such features as ripples, antidunes, etc. Important information on fluid dynamics can be found in the works of Prandtl (1952, 1956), Goldstein (1938, 1952), Truckenbrodt (1968), Sellin (1969), and Prandtl et al. (1969). Allen (1968a) gives a useful summary of the principles of fluid motion, mainly from a geologist's point of view. Middleton (1977) reviews the development and progress in the hydrodynamic interpretation of bedforms, and Middleton and Southard (1978) summarize the flow patterns and sediment movement patterns useful for geologists.

Since the investigations by Darcy and Bazin in 1865 (cited in Simons et al., 1965), a number of workers (mainly hydraulic engineers) have tried to interpret bedforms hydrodynamically. Gilbert (1914) provided the first extensive experimental data on the relationship between bedforms and the velocity of flow. Some later works worth mentioning are those by Anderson (1953), Sundborg (1956), Bagnold (1956), Liu (1957), Dillo (1960), Simons and Richardson (1961), Raudkivi (1963), Yalin (1964), Kennedy (1964), and Bogardi (1965). They explained the various bedforms and their relationship to intensity of flow. They observed, for example, that the simultaneous occurrence of small ripples and megaripples is due to mutual existence of two different modes of instability and that the bedforms and the bed roughness changes with the changing flow conditions. Engelund and Hansen (1966) and Engelund (1966) provide a thorough stability analysis of bedforms on the basis of the linearized theory. Engelund and Fredsøe (1974) give additional information about the stability of bedforms such as megaripples and antidunes.

It is beyond the scope of this book to consider these works in detail. Those interested in such details are referred to Allen (1968a). In the following pages we shall consider in detail only the concept of Simons and Richardson and their coworkers. Their concept of flow regime gives a useful hydrodynamic treatment of bedforms with which geologists are primarily concerned. This concept is based mainly on the results of work done in flume experiments from 1956 to 1961. Guy, Simons and Richardson (1966) summarize the data of their flume experiments. A useful summary of the flow regime concept is provided by Simons, Richardson and Nordin, Jr. (1965). Miller et al. (1977) provide an up-to-date review on the literature of the threshold of sediment motion under unidirectional currents and calculated new threshold curves.

The following short description of flow regimes and individual bedforms is based essentially on the paper by Simons et al. (1965).

Flow Regime

The flow in alluvial channels can be classified into a lower flow regime and an upper flow regime, with a transition between. This classification is based on the form of the bed configuration, the mode of sediment transport, the process of energy dissipation, and the phase relation between the bed and water surface, as shown in Table 1.

Lower Flow Regime

In the lower flow regime resistance to flow is large and sediment transport is relatively small. Water surface undulations are out-of-phase with the bed undulations. Considerable segregation of bed material occurs. The bedform is either small ripples or megaripples or a combination of both. The usual mode of bed material transport is for the individual grains to move up the back of the small ripples or megaripples, and avalanche down the lee face. The Froude number $\left(F = \dfrac{V}{\sqrt{g \cdot h}} \right)$ is < 1. The pattern of flow is tranquil flow (see p. 19).

Transition Regime

The bed configuration in the transition from the megaripples of the lower flow regime to the plane bed and antidunes of the upper flow regime is erratic. In the transition the bedform may range from

Table 1. Classification of flow regime and their characteristics. (After Simons et al. 1965)

Flow regime	Bed form	Bed material concentrations (ppm)	Mode of sediment transport	Type of roughness	Phase relation between bed and water surface
Lower regime	Small ripples	10–200	Discrete steps	Form roughness predominates	Out of phase
	Small ripples on megaripples	100–1200			
	Megaripples	200–2000			
Transition	Washed-out megaripples	1000–3000		Variable	
Upper regime	Plane beds	2000–6000	Continuous	Grain roughness predominates	In phase
	Antidunes	2000			
	Chutes and pools	2000			

the one typical of the lower flow regime to that typical of the upper flow regime. The Froude number is more or less 1.

Upper Flow Regime

In the upper flow regime resistance of flow is small and sediment transport is large. Water surface undulations and bed undulations are in-phase, and segregation of bed material is negligible. The usual bedforms are plane beds and antidunes. The main mode of sediment transport is for the individual grains to roll almost continuously downstream in sheets a few grain diamcter thick. Thc saltation movement of the grains is less dominant than in the lower flow regime. The Froude number is > 1. The pattern is rapid flow. Occasionally, transition from lower flow regime to upper flow regime takes place at much lower values than the Froude number of 1. In natural environments upper flow regime conditions can sometimes be achieved at Froude's number as low as 0.6 or 0.7.

Bedforms

Simons and Richardson recognize the following sequence of bedforms with increasing intensity of flow, expressed in terms of increasing stream power ($V\tau_o$).

V = velocity,
τ_o = γDS – shear stress,
γ = specific weight of the fluid and sediment,
D = stream depth,
S = slope of the energy gradient.

Figure 4 shows the various bedforms as a function of stream power and grain size. Results of all the runs of Guy et al. (1966) have been plotted, as are the stability fields of various bedforms.

Plane Bed Without Movement

A plane bed is a horizontal bed without any bedforms. The plane bed without movement has been studied to determine the shear stress for the beginning of motion and the bedform that would result after commencement of motion. After the beginning of motion the plane bed changes to small ripples for sand sizes smaller than 0.6 mm and to megaripples for sand larger than 0.6 mm. In a few runs, sand of 0.93 mm diameter gave a plane bed with movement at small stream powers before formation of megaripples (Guy et al., 1966). Liu (1957) and Bogardi (1961, 1965) also report plane beds with movement at small flow velocities. Bogardi's data refer to sediments of much smaller grain size than that of Guy et al. (1966). As also pointed out by Allen (1968a), at least for coarser sands, existence of plane bed with movement at low velocities is quite convincing. However, Simons et al. (1965) do not mention it in their paper.

Southard and Boguchwal (1973) found in the flume studies that at flow depths of 7 cm, in the sediment grain size range of 0.6 to 0.7 mm, a lower plane bed phase develops between small ripple and megaripple phases. The ripple field pinches out at about 0.7 mm and the sediment movement starts directly with a plane bed phase, followed by a megaripple phase. Southard (1975) points out that in sediment sizes finer than 0.1 mm, small ripple directly changes to plane bed of upper flow regime, and there is no development of larger bedforms such as megaripples or sand waves.

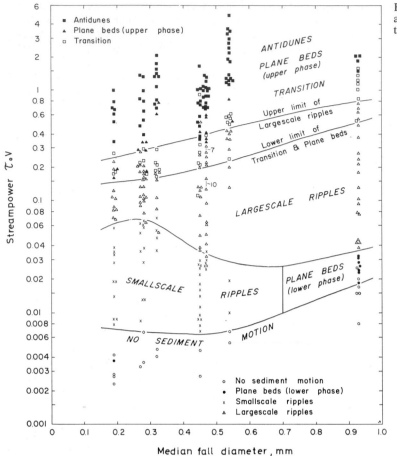

Fig. 4. Bedform in relation to stream power and grain size. Data of Guy et al. (1966). (After Allen 1968a)

Small Ripples

Small ripples are small bedforms with gentle upstream slopes and steep downstream slopes. They are usually less than 30 cm in length from trough to trough, and never exceed 60 cm in length. When initially formed, they are parallel, long-crested, and of small amplitude. But, with increased flow intensity, larger irregular ripples are formed. The length of ripples is independent of sand size. The lowest velocity capable of producing small ripples in fine sand is about 20 cm/sec. Rees (1966) gives current velocities of 8 to 15 cm/sec for producing ripples in clay-free silt.

Mantz (1978) reports development of ripples in flume experiments up to 2 mm height in cohesionless, granular sediments of 15 micron size. However, in natural environments, sediments of this size (15 μ) are normally cohesive and may not develop small ripples. Reineck (1974e) observed that in muddy sediments ripples are erosive, and suggests that sediments with less than 40% sand produce erosive ripples of the longitudinal ripple type. Southard and Harms (1972) observed

in silt-size sediments (30 μ) that ripples are quickly produced, becoming flat ripples with increase in velocity and ultimately changing to flat-bed phase. Little-Gadow and Reineck (1974) observed in a flume that in fine sand (Md ~ 0.09 mm), small ripples are active down to a current velocity value of 9 cm/sec. Stanley (1974) found that in the Pleistocene glacial silts ripples of two distinct sizes are developed, the larger ones ranging in length from 15–60 cm and height from 2-6 cm corresponding to the small ripple and probably megaripple phases. The smaller ones are "minute ripples", mostly 1–3 mm height and 3–6 cm in length. The "minute ripples" move upon the stoss side of the larger ones.

Megaripples

If by some means or other, one gradually increases the boundary shear stress and stream power associated with small ripples (or the plane beds in the case of coarse sand), a certain rate of bed material transport, magnitude of velocity, and degree of turbulence are soon achieved that cause the formation of larger sand waves called megaripples. The transition to megaripples is rather abrupt. With smaller shear stress values the megaripples have small ripples superimposed on their backs. These small ripples tend to disappear at larger shear values. Megaripples are similar in shape to small ripples. Their length ranges from 60 cm to several meters. In contrast to small ripples, the length of the megaripples can increase with increasing depth. However, in natural environment, with water depth more than a few metres, dimensions of megaripples are independent of flow depth. Furthermore, the length and shape of the megaripples are the function of the grain size of the bed material. Bagnold (1956) states that small ripples cannot exist in quartz sands of mean diameter greater than about 0.67 to 0.71 mm. Simons et al. (1965) accept a mean diameter of 0.6 mm for the upper limit of small ripples. Fine sand produces megaripples at much higher stream power than medium and coarser sand.

Southard (1975) considers dunes (megaripples) to be 1-5 m in length showing high L/H ratios and sinuous to cuspate forms developed at flow velocities of 70-150 cm/sec. In more recent literature, e. g., Southard (1975), Middleton and Southard (1978), the term dune denotes the three-dimensional forms only (sinuous to lunate megaripples), showing strongly sinuous crests. The two-dimensional forms (straight–crested megaripples) are mostly described as sand waves.

Plots of shear velocity and grain size, along with the suspension criteria, demonstrate that the change from sand wave (straight-crested megaripples) to dune (three-dimensional megaripples) takes place well above the suspension criteria (Fig.

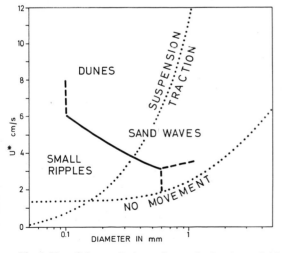

Fig. 5. Plot of shear velocity against grain size shows fields of no movement, traction movement, and suspension transport (a modified form of the Hjulström diagram). The stability fields of ripples and dunes are also shown. (Modified after Middleton 1977; Middleton and Southard 1978)

5). In other words, during dune phase (three-dimensional megaripples) suspension transport is significant and may be responsible for making scouring on the lee-side, and helping in the development of three-dimensional forms.

In the literature, especially that dealing with bedforms, the term dune is more commonly used than megaripples. Sometimes, the term sand wave is also used as a synonym for dunes. But the term sand waves is generally used to describe independent larger bed forms other than megaripples. Megaripples are also known as three-dimensional dunes.

Plane Bed with Sediment Movement

A plane bed is a bed without elevations or depressions larger than the maximum size of the bed material. The resistance to flow with the plane bed is relatively small, resulting mainly from grain roughness, which is related to the size of bed material. Grains move by rolling, hopping, and sliding along the bed. The magnitude of the stream power ($V\tau_o$) at which the megaripples or transition roughness changes to the plane bed depends mainly on the grain size of the bed material. With fine sand (low fall velocity), the megaripples disappear at relatively low stream power in comparison to the coarser sands.

Some bed load segregation manifested in the form of longitudinal striations does take place. As shown by Jopling (1967), some of the striations are slightly coarser grained and others slightly finer grained than the average material. Their random superposition probably provides the size contrast necessary to impart a horizontally laminated structure in section. The lamination is rather indistinct.

Antidunes

Antidunes are the bedforms which are in-phase with and strongly interact with water surface gravity waves. The height and length of these waves depend on the scale of the flow system and the characteristics of the fluid and bed material. Antidunes do not exist as a continuous train of waves that never change; the trains of waves are gradually built up with time from a plane bed and plane water surface. The water surface waves may grow in height until they become unstable and break or they may gradually subside. The former are known as breaking antidunes and the latter, standing waves. As the antidunes form and increase in height, they may move upstream or downstream, or remain stationary. Because of their possible upstream movement, Gilbert (1914) named them antidunes.

Chutes and Pools

These are sediment bed configurations, which develop at relatively large slopes and sediment discharges, consisting of large mounds of sediment forming chutes with supercritical flow (upper flow regime). They are connected by pools, in which flow can be supercritical to subcritical (lower flow regime). The chutes move slowly upstream. They are observed in the field only under exceptional conditions.

Although the pattern of flow over ripples and flow regimes is more or less established, Folk (1976) gives an interesting account where he explains ripples due to systematically spaced transverse roller vortex system (lower flow regime) changing to longitudinal (helicoidal) vortices in upper flow regime.

"Sand Waves"

Later flume experiments and observations in the field led some workers to identify another distinct bedform, namely "sand waves", which are formed at lower current velocities than the megaripples (dunes).

Experiments of several workers (for example, Chabert and Chauvin 1963; Pratt and Smith 1972; Pratt 1973; Costello 1974) produced certain low, straight-crested bedforms at velocities between small ripple and megaripple (dune) phases, and are named bars, bancs, or intermediate flattened dunes. Boothroyd and Hubbard (1971, 1974) and Costello and Southard (1974) deduced that intermediate to small ripples and megaripples, another distinct ripple-like bedform exists, at energies lower than megaripples (dunes). These bedforms are commonly referred to as "sand waves" (also bar) and have been plotted in the later flow regime and bedform diagrams (Fig. 6) (Southard 1975; Middleton and Southard 1978).

The "sand waves" are characterized by straight, continuous crests, produced at lower energies than the megaripples. They are low forms with high L/H ratios, showing poor correlation between height and length. Southard (1975) considers that the sand waves are 5–100 m in length, developed at moderate velocities of 30–80 cm/sec.

Actually, there are two types of ripple-like form which are described as sand waves:

a) small, straight-crested sand waves of low energy, corresponding to the forms produced in the flumes,
b) large straight-crested sand waves produced at higher energies, showing superimposed megaripples on the stoss side, and developed in sufficiently deep waters and coarse sediments.

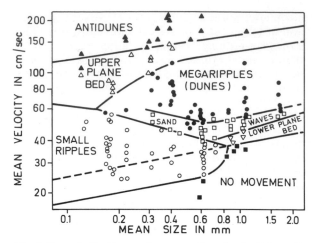

Fig. 6. Size velocity diagram for a mean flow depth of about 40 cm, and 10°C. Stability fields of various bedforms in the flume experiments are shown. (Originally from Boguchwal; modified after Middleton and Southard 1978)

Boothroyd (1978) considers two stability fields of sand waves, low-energy sand waves, having small ripples on their stoss side, and high-energy sand waves with megaripples on their stoss side.

As also shown by the study of Dalrymple et al. (1978), the low-energy sand waves correspond to the straight-crested megaripples or megaripples type I (see p. 41). In other words, "sand waves" of flume experiments correspond to the straight-crested megaripples formed at somewhat lower energies, while the "dunes" correspond to the lunate, three-dimensional megaripples, formed at higher energies.

Middleton and Southard (1978) also consider that the relationship between dunes and sand waves is not yet completely worked out. The stability field plot of sand waves in the velocity/grain-size diagram shows that in both fine and coarse sand, sand wave phase is absent; in fine sand ripples change to dunes, while in coarse sand lower flat bed changes to dunes, without an intervening sand wave phase. The development of straight-crested megaripples (sand waves) seems to be related to the grain size.

Jackson (1976b) describes large solitary forms of transverse bars from the Wabash River, and considers them to correspond to the stability field of sand waves, Cant (1978) considers rhomboid- or lingoid-type small bars as sand waves. Thus, there seems to be a tendency to group various types of bedform, which are produced at lower velocities than the three-dimensional dunes under the category of sand wave.

The term straight-crested megaripples is probably better suited for the sand waves of flume studies, as the term sand wave is also used to describe very large bedforms. Usually, in the oceanographic literature the term sand wave is used for megaripples and giant ripples. Sometimes, a differentiation between small sand waves (megaripples or dunes) and large sand waves (giant ripples) is made.

Megaripples and Sand Waves

As already discussed above, the low-energy sand waves can be better referred to as straight-crested megaripples. With inreasing energy these forms transform into three-dimensional megaripples.

Besides, there are high-energy sand waves, which are the largest bedforms and occur in water depths of more than 4–5 m, and coarser-grained sediments. They are mostly referred to as sand waves or giant ripples, and invariably show superimposed megaripples on the stoss side. The giant ripples are produced at somewhat higher energies than the megaripples. Thus, these larger sand waves of deeper flows are genetically different from the low-energy sand waves, and should not be grouped together as done by some workers, e. g., Southard (1975). However, Middleton and Southard (1978) also point out that these two types of sand wave are different kinds of features, and may not be genetically related.

Newton and Werner (1969) used the term sand waves ("Sandwellen") for the flat bedforms larger than 10 m on the tidal flats of the North Sea, and regarded them as different from megaripples. These sand waves are a few centimetres to a few decimetres high, and unlike megaripples, always consist of a form-discordant internal structure. We feel that reworking processes in the tidal flats are important in producing such sand waves.

Many workers do not differentiate between several dune-like bedforms and class them as dunes or sand waves.

Later, Newton and Werner (1972) discussed that there are no breaks in the size of ripples. According to them the smaller bedforms are single-phase structures (form-concordant internal structures), which, as they grow in size, acquire multiphase structure (internally form-discordant). After a certain time, and when they have acquired a certain size, the large form itself moves with lee-face deposition and the form is restructured.

In several areas, up to three to four morphologically distinct populations of sand waves can be observed. For example, Coleman (1969) reports four distinct types of sand wave.

Many of these large ripple-like bedforms may be related to temporal variations in flow conditions. On the present-day tidal flats, if the same megaripple field is visited under different conditions, such as spring tide, neap tide, storm tide, or fair weather, it may be developed as straight-crested megaripples, sinuous megaripples or even a strongly three-dimensional type with deep scours.

Further, development of bedforms in natural environment is more complex than observed in the flume studies (Dalrymple et al. 1978). Most of the recent studies in the bedforms of a natural environment (Dalrymple et al. 1978; Cant 1978) indicate that mostly small ripples, two types of megaripples (small sand waves, and dunes), and giant ripples (large sand waves) are developed. All these bedforms are definite morphological forms which are also in equilibrium with flow. The straight-crested megaripples (sand waves) are stable in current velocities of less than 70 cm/sec, while the three-dimensional megaripples (dunes) are common in current velocities of more than 60 cm/sec. The giant ripples are stable in current velocities of more than 70 cm/sec, and water depth of 4–5 m. It is, however, still not possible to say whether straight-crested megaripples (small sand waves) are hydrodynamically distinct forms or just low-energy forms of three-dimensional megaripples (dunes).

Allen (1978) discusses the significance of various types of dunes and sand waves and feels that polymodal assemblages of transverse dune-like bedforms may not represent hydrodynamically distinct subpopulations. As dune wavelength is directly related to water depth, larger dunes may develop at high-water stages, smaller dunes at low water stages. From experimental dune analysis, Allen (1978) demonstrates that in a flow depth–flow velocity diagram dunes of various sizes belong to a single hydrodynamically defined class of bed features. Polymodality of dunes is more prominent in channels undergoing manifold sudden changes in the discharge.

Geologically speaking, all these dune-like bedforms on migration produce large-scale cross-bedding, and at the present state of our knowledge it may not be possible to differentiate between the cross-bedding made by various types of dunes and sand waves, excepting that straight-crested megaripples (sand waves) produce planar cross-bedding; while three-dimensional dunes make trough-shaped cross-bedding.

Study of Bedload Transport

Moss (1972) made an extensive study of bed load transport in both the natural environment and in the flumes. He emphasized that texture is of primary importance in defining a bed stage and not the bedforms. Development of certain bedforms can be prevented by some mechanism associated with the flow unsteadiness. For example, plane bed can develop under conditions of small ripple and megaripple stages; however the grain-size characteristics of such plane bed deposits are similar to those of the ripple stage.

Moss (1972) refers to the transition and plane bed stage of upper flow regime as rheologic bed-stage and is characterized by the development of a rheologic layer, formed at intense bed-load motion, where sediment load proceeds as dense mass of colliding particles behaving as viscous fluid. This layer shows a sharp boundary with the flow above and maintains a constant thickness over the bed. If all grain sizes are readily available, the presence of pebbles larger than 30 mm is a good indicator of rheologic bed stage. In the sediments of rheologic bed stage, saltation population is quantitatively less dominant over suspension and traction populations, as compared to the sediments of small ripple and megaripple stages.

Rate of Migration of Bedforms

Very little information is available on the rate of migration of bedforms. Dillo (1960) made some detailed measurements on the rate of migration of various ripples. In the flume experiments he found that migration of small ripples is an irregular process. In general, phases of quick movement of ripples are followed by rather slow phases of movement. Moreover, it is not possible to follow the movement of ripples over longer distances, because they are continuously destroyed and rebuilt. Dillo (1960) measured the average rate of migration photographically (Table 2).

Reineck (1961) measured the rate of migration of megaripples to be 0.029 cm/sec. These megaripples are formed by swell on the longshore bars.

Table 2. Average rate of migration of small ripples and megaripples in three different kinds of sand at different current velocities. (After Dillo 1960)

Current velocity (cm/sec)	Average rate of migration in mm/sec in sand of different values of Md.					
	Md = 0.085		Md = 0.19		Md = 0.28	
	small ripples	mega-ripples	small ripples	mega-ripples	small ripples	mega-ripples
30	–	–	–	–	–	–
35	–	–	–	–	0.018	–
40	–	–	0.02	–	0.035	–
50	0.03	–	0.1	–	0.15	–
60	0.07	–	0.35	–	0.6	–
70	0.18	–	0.9	–	1.5	–
80	0.40	–	2.0	–	–	0.45
90	0.90	–	3.5	0.4	–	0.6
100	1.50	–	–	0.55	–	0.8
105	–	–	–	0.9	–	–
110	–	–	–	–	–	–
120	–	–	–	–	–	–

If other factors are kept constant, the rate of migration of ripples is directly dependent on the current velocity and the grain size. With increasing current velocity, average rate of migration of ripples also increases. Moreover, at the same current velocity the rate of movement of ripples is higher in coarser sand than in finer sand (Table 2). Small ripples move at a much quicker rate than megaripples. At lower values of current velocities where small ripples and megaripples co-exist, small ripples move much rapidly than the megaripples (Table 2). Giant ripples with heights up to 16 m are known to move at a rate of 700 m/24 hrs. On the average, giant ripples migrate at rates of 30 to 100 m/24 hrs (Coleman 1969).

Bedforms, Stream Power, and Water Depth

Figure 7 summarizes the above discussion and depicts graphically various bedforms and their relationship to grain size and stream power. Under the influence of flowing water, sediment movement starts on a noncohesive bed. At low energy conditions a few grains start rolling and may produce horizontal laminations if sufficient sediment is available and the process continues for a certain length of time. This plane bed phase at low-energy conditions is probably more prominent and stable in coarser sediments than in finer sediments. In sediments finer than 0.6 mm small ripples are produced. The first-formed small ripples are straight-crested. With increasing stream power, they become undulatory and lingoid small ripples. Small ripples at still higher energies form megaripples. (In sand coarser than 0.6 mm the plane bed phase changes directly into megaripples.) The first megaripples are straight-crested; with increasing stream power they change into undulatory and finally lunate ripples.

With $F \cong 1$, megaripples become unstable and form the plane bed phase of the upper flow regime. At still higher energies antidunes are formed.

The drawback of the graphic shown in Fig. 7 is that geologists would like to know more about water depth and current velocities and their relationships to various bedforms and resulting sedimentary structures which become preserved. However, the relationship between water depth, current velocity, and bedforms is not an easy one. Figure 8 shows the relationship among water depth, current velocity (expressed in terms of the Froude number), and the bedforms.

It clearly demonstrates that the relationship between current velocity and small ripples is strongly influenced by the depth of water, as long as water depth is a few centimeters to a few decimeters. Within this range increasing water depth demands higher current velocities for the generation of small

Fig. 7. Schematic representation of various bedforms and their relationship to grain size and stream power. Based on Simons et al. (1965) and Allen (1968a). a straight-crested ripples; b undulatory ripples; c lingoid ripples; d lunate ripples. For both small ripples and megacurrent ripples, ripple crests tend to become discontinuous (three dimensional) with increasing stream power. Recent flume experiments show that megaripple field pinches out at 0.1 mm grain size

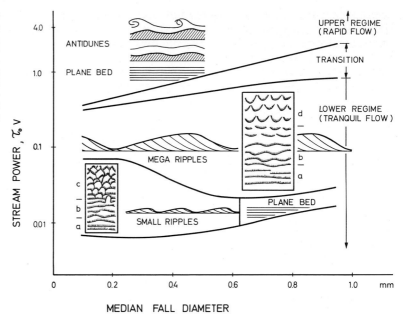

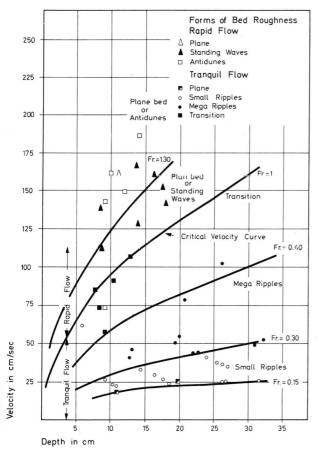

Fig. 8. Relationship between current velocity and water depth, in relation to various bedforms. Stability fields of various bedforms is indicated by the Froude number. (Modified after Simons and Richardson 1962)

ripples. However, above this range the curve is more or less horizontal. This means that after a certain critical water depth, water depth does not play any important role. The curve for megaripples shows a very similar trend.

However, curves for bedforms of the upper flow regime ($F = 1$ and $F > 1$) show a different trend. Here, with increasing water depths, increased current velocities are required to produce the same bedforms. The curves are rather steep.

To conclude, bedforms of tranquil flow after a critical water depth are more or less independent of water depth; however, the bedforms of rapid flow are strongly controlled by water depth.

The Froude number (F) is a rather useful and simple factor for determining the hydrodynamic conditions of a flow. The Froude number (F) =

$$\frac{V}{\sqrt{gh}}.$$

V = flow velocity,
g = acceleration due to gravity,
h = depth of flow.

$F < 1$ characterizes tranquil flow in which the bedforms of lower regime are stable. $F > 1$ characterizes rapid flow in which the bedforms of the upper flow regime are stable. If we look at the Froude equation, we see that $F > 1$ can be achieved either by increasing velocity and keeping the water depth constant, or by reducing the water depth and keeping the current velocity constant. However, in nature, conditions of rapid flow can be realized only in shallow-water conditions. A rough calculation shows that to obtain $F = 1$ in water depths of 10 m, one needs a current velocity of 9.90 m/sec. Such high velocities are very exceptional, and may occur in unusually rapid river flows. In shallow-water marine environments, current velocities of more than 3 m/sec are extremely rare; generally, they reach values up to only 2 m/sec. Table 3 gives the

Table 3. Relationship among water depth, current velocity, and Froude number

Water depth	Current velocity (m/sec)	Froude number
1 cm	0.31	1
10 cm	0.99	1
1 m	3.12	1
10 m	9.90	1
100 m	31.32	1

water depths and the corresponding current velocity needed to obtain $F = 1$. It suggests that rapid flow ($F = 1$ and $F = > 1$) are found only locally in water depths ranging from a few millimeters to a few meters.

However, upper flow regime sedimentation in deep-water environment is quite common and is related to turbidity currents, whose mechanism of transport and deposition are different from the normal currents in shallow-water environments. Middleton and Hampton (1973) give a good account of mechanism of sediment gravity flows, including turbidity currents.

Depth–Velocity–Size Diagram

Southard (1971) discusses that there are seven important variables controlling the bedforms in an open channel, namely depth of flow, mean velocity of flow, density of fluid, viscosity of fluid, mean size of sediment, density of sediment, and acceleration due to gravity. If the consideration is restricted to quartz sand, then a three-dimensional diagram can be plotted to characterize the bedforms with dimensionless measures of the depth, mean velocity, and sediment size as coordinates. Such diagrams are known as depth–velocity–size diagrams. Two-dimensional slices of such a diagram, namely depth–velocity diagram, and size–velocity diagram have found wide use in sedimentology (Figs. 9 and 10). Plots of Southard (1971, 1975) and Middleton and Southard (1978) demonstrate that the stability fields of various bedforms are non-overlapping. These diagrams clearly demonstrate that the bedforms are strongly controlled by the flow velocity, and the flow depth has little influence; except in the case of upper flow regime at extremely shallow water depths. Vanoni (1974) also demonstrates that if the depth is large

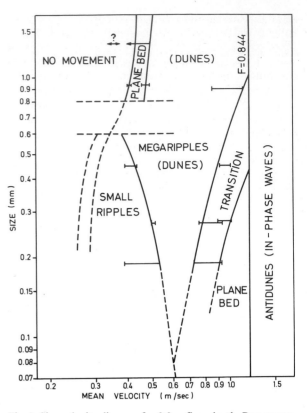

Fig. 9. Size velocity diagram for 0.2 m flow depth. Bars represent latitude in placing boundaries in the depth-velocity diagrams. Dashed curves are hypothetical extension of the boundaries. Lower part of megaripple (dunes) stability field corresponds to sand wave field, and the upper part corresponds to the dunes of Fig. 2. (Redrawn after Southard 1971)

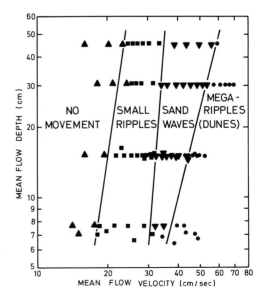

Fig. 10. Depth-velocity diagram for about 0.50 mm, very well-sorted sand. Triangles – no movement; squares – ripples; inverted triangles – sand waves; circles – megaripples. (Redrawn after Southard 1975; data from various sources)

enough, the stability fields of ripples and dunes (megaripples) can be conveniently shown in a diagram of shear stress vs. grain size. Figure 10 shows a depth–velocity diagram depicting stability fields of small ripples, sandwaves and megaripples (dunes). The schematic size–velocity diagram of Southard (1975) and Middleton and Southard (1978) clearly demonstrates the control of grain size upon the sequence of bedforms (Fig. 9). This diagram is based on data of many studies and constructed for a water depth of 20 cm. In general, in the very fine sand size, grain movement starts in the form of small ripples and remains stable over a very wide range of velocity, and ultimately changes to plane bed phase of upper flow regime. In the case of fine to medium sand small ripples change with increasing velocity to three-dimensional megaripples (dunes) and ultimately to plane bed phase. For very fine to medium sand the velocity at which first ripples develop is highly variable. In the case of medium to coarse sand a lower plane bed phase develops, changing to sand waves (straight-crested megaripples) to three-dimensional megaripples (dunes) and to plane bed phase of upper flow regime with increasing current velocity. In the size range of 0.5–0.6 mm, small ripples first change

to plane bed and then to megaripples. Figure 11 shows flow depth versus flow velocity plot for medium sand, showing stability fields of various bedforms as observed in flume studies and natural environments.

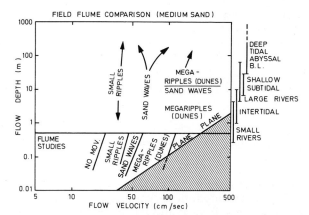

Fig. 11. Depth-velocity diagram to show bed phases developed in medium sand in flume experiments and in natural flows. In natural flows sand waves at higher energy than the dunes (megaripples) are also present (Modified after Middleton and Southard 1978)

Current and Wave Ripples

General Information

Ripple marks are present as undulations on a non-cohesive surface, though they may also be found infrequently in muddy sediments as well. They are produced as a result of the interaction of waves or currents on a sediment surface. Ripple marks are one of the commonest features of sedimentary rocks, both in Recent and ancient sediments. The shape and size of ripples vary considerably. The crests usually run parallel to each other or may anastomose partially. In transverse section they may be symmetrical or asymmetrical in shape. The crest may be sharp, rounded, or flattened. Only through later modification by change in water depth, etc., they may become rounded or flattened. Most of the classifications of ripple marks are based on their mode of origin, and their shape and size.

"The apparent contradictions of writers on ripple marks are so surprising that one fails to see how the student or even the textbook writer can find his way through the mist." These words were written by A. R. Hunt in 1904 (cited in Kindle 1917), almost 70 years ago. A vast amount of new information and knowledge about ripple marks, especially based on the experimental studies and observations in modern sediments, has accumulated. But the statement of Hunt bears some truth even today. Based on our present state of knowledge, we have attempted here to give as clear a picture of ripple marks as possible, avoiding clichés and complicated terminology, while at the same time exercising caution about oversimplification.

Description of Ripple Marks

A ripple is conventionally described in terms of its size and shape. Traditionally, ripple marks are represented and described in terms of vertical profile parallel to flow, at right angles to the ripple crest. A ripple is composed of a crest and a trough. Thus, each ripple crest has a trough shared by the adjoining ripple crest. In the following summary we shall give some important terms currently used in describing ripple marks (Figs. 12 and 13).

Ripple length is the horizontal distance, at right angles to the crest, between the troughpoints of both sides of the ripple crest.

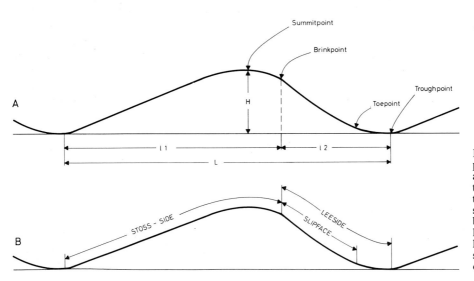

Figs. 12, 13. A ripple profile parallel to flow, at right angles to the elongation of the ripple crest. Various terms commonly used to describe a ripple mark have been depicted. L–ripple length, H–ripple height, l_1–horizontal projection of stoss side, l_2–horizontal projection of lee side

Ripple height is the vertical distance between the troughpoint and the summitpoint of a ripple

Summitpoint is the point of maximum elevation on the vertical profile through a ripple.

Toepoint is that point on the vertical profile of a ripple that separates the slipface from the bottomset.

Brinkpoint is the point on a vertical profile of a ripple that separates the slipface from the crestal shoulder.

Troughpoint is that point of minimum elevation on the vertical profile of a ripple which separates the lee side of the one ripple from the stoss side of the adjacent ripple.

Lee side is the steeply inclined part of a ripple extending downstream from the brinkpoint as far as the troughpoint.

Stoss side is the gently sloping upcurrent side of a ripple extending from the troughpoint as far as the brinkpoint.

Slipface is the steeply sloping portion of the lee side of a ripple, situated between the brinkpoint and the toepoint.

Bottomset is the gently sloping distal part of a lee side located between the toepoint and the troughpoint.

Vertical form index (ripple index) is the ratio of length to height (L/H). It was first used by Bucher (1919). It is the most commonly used ripple index.

Symmetry index is defined as the ratio of the horizontal projection of the stoss side (l_1) and the horizontal projection of the lee side (l_2). This terminology has been expressed graphically in Figs. 12 and 13.

Ripples normally occur in groups called ripple trains, in which ripple crests run almost parallel to each other. In one ripple train, usually one type of ripple is present.

Ripples can be further described depending upon the nature of the crest. Crests may be straight, undulatory, lingoid, cuspate, lunate, or rhomboid shaped. Undulatory crests may be further distinguished as being in-phase or out-of-phase. Crests are said to be in-phase when the lobes and the saddles of undulatory crests are in the same line. They arc out of phase when lobes and saddles of undulatory crests alternate with each other. These two can be regarded as end members of a continuous series. Commonly, forms intermediate to in-phase and out-of-phase crests are most abundant. Within a

single ripple train they may change from in-phase forms to out-of-phase and intermediate forms. In general terms, three major patterns of ripple crests can be recognized: Straight-crest, undulatory crest, and discontinuous crest. The last pattern may occur in varied forms, e.g., lingoid, lunate, cuspate, rhomboid. Straight crests: Ripple crests are straight and can be followed for considerable distances. Undulatory crests: Ripple crest are wavy in character but continuous over long distances. Discontinuous crests: Crestline of the ripples is broken, and cannot be followed. A broken crest line may be curved forward, backward, etc. Allen (1968a) provides a detailed description of all these terms, and several more terms not currently much in use.

Two terms commonly used by hydrodynamists, but almost unknown in geological literature, are two-dimensional and three-dimensional ripples. Ripples are called two-dimensional if they are almost uniform in height and possess straight crests at right angles to flow. Crests of three-dimensional ripples vary substantially in height along the crestline, and the crestline is locally skewed to the flow. In extreme cases the crest line is discontinuous and curved to the flow.

To show the geometry and internal structure of ripple marks, block diagrams are constructed. The geometry of ripples may be defined in terms of 3 axes, in which the *a*-axis runs across the ripple crests and is parallel to the direction of flow. The *b*-axis is also horizontal lying in the plane of the bed, but at right angles to the *a*-axis and parallel to the crest line. The *c*-axis is vertical and perpendicular to the plane containing the *a*- and *b*-axes.

Internal Structure of a Ripple

A well-developed current ripple is made up of a few or a single stoss side laminae, many foreset laminae, and a few or a single horizontal bottomset laminae (Fig. 14). Foreset laminae are the main constituents of the ripple body. They may acquire different shapes and angles. Their shapes may vary between angular, tangential, and sigmoidal. Maximum measured angles are up to 35°. Shrock (1948) gives an angle of 33°, while McKee (1953) reports

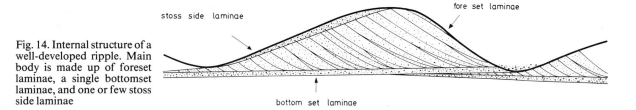

Fig. 14. Internal structure of a well-developed ripple. Main body is made up of foreset laminae, a single bottomset laminae, and one or few stoss side laminae

angles between 20 and 35°. The angle of the foreset laminae near its base is relatively lower than near the top, and can be used as an up-down criteria. Many of the foreset laminae pinch out upward, without reaching the full height of the ripple. Bottomset laminae cover the ripple base and run across the ripple body.

Except for the size, there is no difference in the internal structure of a small-current ripple and a megacurrent ripple. The internal structure of a wave ripple (especially an asymmetrical wave ripple) can be just the same as the internal structure of a current ripple. However, some of the symmetrical wave ripples show a rather different internal structure. The internal structure of wave ripples has been dealt with on p. 33.

A small or megaripple possessing an internal structure such as the one described above may be said to be form-concordant (simple), i. e., the internal structure of the ripple corresponds to and is concordant with the outer form. In other words, internal structure of the ripple is genetically related to the outer shape of the ripple. This can be regarded as primary form. The ripple consists internally of a single unit formed due to the migration of the ripple outer form in question. However, not all the ripples possess form-concordant or simple internal structure.

A ripple can also be form-discordant, or composite structure. In this case, ripples do not show the typical internal structure described above, but possess some "foreign" structures. Here the internal structure does not conform to the outer shape of the ripple. In other words, the outer shape of the ripple does not correspond to the internal structure, and the outer form of the ripple is genetically not related to the internal structure. Form-discordant ripples showing composite structure can be divided into two major subgroups:

1. Primarily Form-Discordant. Some of the ripples are primarily and genetically made up of composite internal structures. For example, giant ripples are known to be composed of composite megaripple bedding. Such forms are produced originally with a composite structure (Fig. 15).

2. Secondarily Form-Discordant. In several cases, especially in areas undergoing continuous reworking of bottom sediments, small ripples and megaripples are shown to possess composite internal structure or even completely foreign structures. They can be further subdivided in two types.

a) Reworked Forms. These are made up of more than one unit of reworked rests of ripple bedding. However, in this case a few laminae of the last phase of ripple building in which the ripple form is generated, are always present. These laminae usually cover the foreset laminae of previous ripples (Fig. 16).

b) Erosive Forms. Normally during the process of rippling, both erosion and deposition take place. In certain exceptional cases ripples or ripple-like forms are produced purely by erosion; however, on the crests no deposition takes place. Thus, ripple crests represent only the uneroded older surface and are composed internally completely of older structures. In other words, ripple forms are produced from scouring out of the ripple troughs. Sometimes such erosive forms may show structures like laminated sand, sand/mud rhythmic bedding, etc. Occasionally, one or two foreset laminae may be present on the ripple slopes (Fig. 17).

Reineck (1974a) demonstrates that in muddy sediments sometimes current ripples are generated, but they are always erosive; few foreset laminae may develop if sand content is more than 70 %. In tidal flat sediments, the silt and clay is mostly present as faecal pellets and helps in generation of

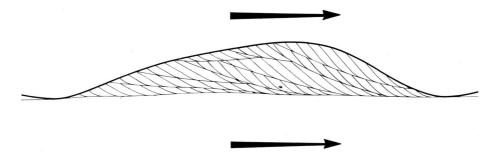

Fig. 15. A giant ripple showing primarily form-discordant structure. The giant ripple is originally formed with composite structure. Each unit is a megaripple bedding

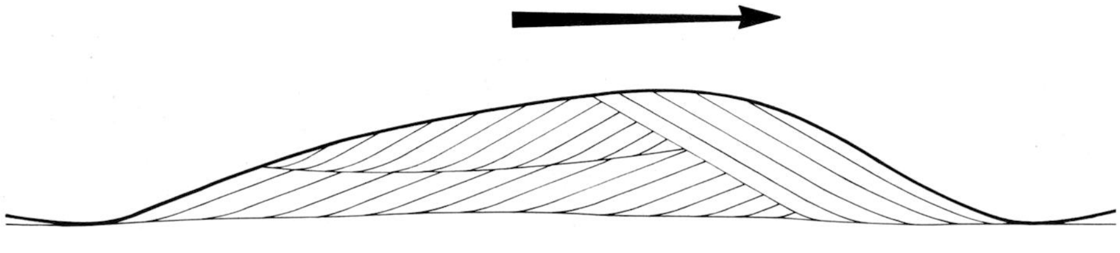

Fig. 16. Reworked form of a ripple. Foreset laminae on the present slipface are produced during the last phase of ripple migration

Fig. 17. Erosive forms of ripples. Ripple troughs are produced from erosion. Internally ripples are made up of different types of bedding; only a few laminae of ripple origin are present near the ripple crests

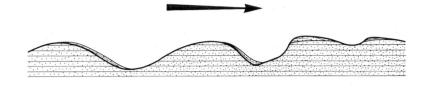

Fig. 17. Erosive forms of ripples. Ripple troughs are produced from erosion. Internally ripples are made up of different types of bedding; only a few laminae of ripple origin are present near the ripple crests

current ripples even at high silt and clay content. Longitudinal ripples and small wave-generated ripples in mud are also erosive.

Another type of ripple internal structure is the climbing ripple structure. In areas where much sediment is available in suspension the rippled surface is quickly buried under the new sediment layer, which in turn is rippled. Thus a series of superimposed, almost completely preserved ripple laminae is produced. This structure is known as climbing ripple lamination (for details see p. 109 ff.).

Pattern of Flow over a Lee Face of a Small Ripple, a Megaripple, and "Microdeltas"

In general, it can be stated that the flow of water over a noncohesive material due to the instability of the water-sediment interface produces small ripples and megaripples, which have a well-developed lee slope. A "microdelta" also possesses a well-defined lee slope. However, unlike ripples, it lacks a well-developed stoss side. Recent researches have shown that small current ripples[1] and megacurrent ripples[1] are the most important bedforms. They occur in large numbers in trains. Microdeltas are very insignificant, and locally may be present as isolated bodies. Even so, most of the hydrodynamic studies dealing with mode of transportation and deposition have been made on the lee side of deltalike bodies (Allen 1965a; Jopling 1963, 1965a). The only detailed observations of the lee side of ripples are those by Reineck (1961) and Jopling (1967).

The following discussion on lee side deposition is based on the foregoing contributions.

At the sediment-water interface exists a heavy-fluid layer, with a relatively high concentration of sediment. In this layer grains move by rolling, sliding, saltation, and suspension. Upward, away from the sediment surface, the concentration of coarser particles decreases enormously; however, the concentration of finer sediments in suspension does

not alter significantly. Important in the present discussion is the heavy-fluid layer, usually only a few millimeters thick. This heavy-fluid layer, generated along the back or stoss side of a ripple, is dissipated at the crest. This phenomena is known as boundary layer separation (cf. Allen 1965a). Particles are dispersed at the crest, where support from the bed for the particles is lost. Particles tend to settle on the lee face and the region beyond. The path of a particle depends upon its size and its position in the bed-load–heavy-fluid layer–at the time it leaves the crest.

Figures 18 and 19 show ripples in a laboratory tank. The heavy-fluid layer is clearly visible, as are the eddies in the lee faces of the ripples. The lee side can be divided into three major hydrodynamically different zones, as shown in Figs. 20A and B.

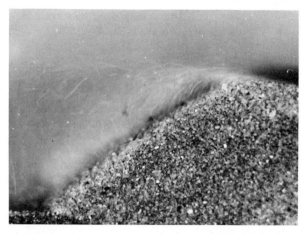

Fig. 18. Protuberance at the ripple crest and grain path on the upper lee face of a ripple in a flume experiment. (After Reineck 1961)

Fig. 19. Flow pattern over ripples in a flume experiment. Heavy-fluid layer and lee side eddies are clearly visible. (After Reineck 1961)

---

1 For the sake of brevity, in the text the terms "small ripples" and "megaripples" have been used for small current ripples and megacurrent ripples, respectively (the word "current" has been dropped). But we still mean current ripples. For wave ripples the prefix wave has always been used so that no confusion arises

Zone of No Diffusion

As described by Jopling (1963), this zone may be regarded as a "vestigal" part of the stream flow that carries sediment in suspension over and behind the foreset slope (lee slope). Its velocity distribution is similar to that of the upstream flow, but with a progressive truncation in the downstream direction.

Zone of Mixing

It is characterized by macroturbulence. The fluid layer which occupies the mixing region is unstable and rolls up into vortices. The fluid in the zone of mixing displays a rapidly changing distribution of longitudinal velocity, and there is an inflection point in the velocity profile (see Fig. 20A).

The separation of the flow from the foreset boundary results in the generation of a reverse circulation which is directed toward the toe of the foreset and up the foreset slope. The zone of mixing also works as a "sediment vorticity trap" for catching some of the sediment settling out from the zone of no diffusion. At the beginning the vortices are spatially periodic and represent a pseudoturbulence. They are small at the start of the rolling-up process, but gradually increase in size and strength as they are convected down the mixing region. At a certain critical distance from separation ($\frac{2}{3}$ of the distance of the next ripple stoss side), the vortices themselves become unstable and begin to break down into random turbulence.

Zone of Backflow

This zone shows a reversal of direction of flow. When vortices originating in the zone of mixing reach the zone of back flow, they develop a counter current flowing along the bottomset and up the lee slope. The back flow may attain velocities as large as 20 to 25 % of the average down-current velocity. Some of the grains moving in the backflow (in suspension) may be caught back in the eddies of the zone of mixing (Fig. 20B).

At about $\frac{2}{3}$ distance of a stoss side, transport is in the downstream direction only. A heavy-fluid layer is formed. In this layer grains move by sliding, rolling, saltation, and suspension. Sand grains move in an irregular pulsating movement along and up the stoss side. The bigger grains have a larger area exposed to running water, and so are taken away easily. However, the smaller grains are also transported away. Even then a sorting of the sediment takes place: Finer grains settle and percolate down the intergranular pore spaces. The net result is that on the stoss side a relative enrichment of finer grains takes place. Thus, the stoss side laminae are composed of relatively finer material (Reineck 1961; Jopling 1967). However, the processes causing this sorting on the ripple stoss side are not well understood.

At the crest, particles accumulate, and in lower current velocities they may form a protuberance there, before avalanching downslope the lee side (Figs. 18 and 19). According to Jopling (1963), the marginal diffusion and deceleration of the expanding flow results in a reduction of flow com-

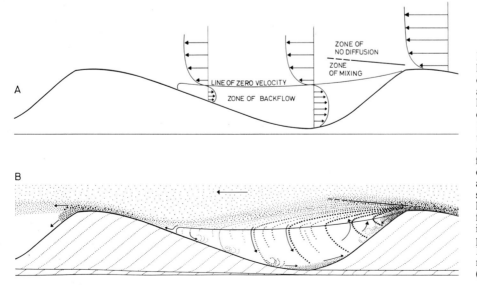

Fig. 20A. Flow pattern over a lee face of a ripple. Velocity distribution, flow separation, and three major zones on the lee side are depicted. (Modified after Jopling 1963, 1967) B. Flow pattern and sedimentation processes on the lee side of a ripple. The heavy-fluid layer accumulates particles in the form of a protuberance at the crest, from where sediment avalanches at the lee face. In the zone of backflow some sediment is caught into backflow eddy and is deposited at the toe of lee slope. Idealized path lines of sediment grains are shown. (Modified after Jopling 1967)

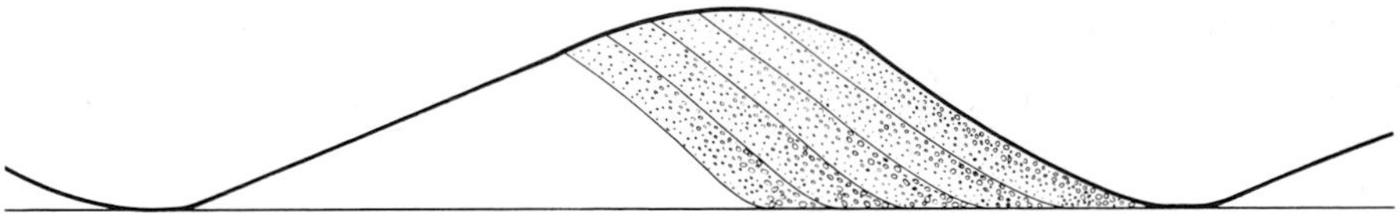

Fig. 21. Effect of sorting in the foreset laminae. Two sorting processes are active: Larger particles are concentrated toward the outer side of a lamina under the influence of shear; under the influence of gravity, larger particles are concentrated near the base of a lamina. (Based on Brush 1965)

petency and capacity, and the bulk of sediment finds a resting place on the upper part of the foreset slope, preceding the downslope avalanching.

The heavy-fluid layer is disrupted by the eddies of free turbulence shear flow, and most of the granular products are dumped on the upper part of the lee face. Most of the deposition from suspension also takes place at the ripple crest, i. e., the upper lee face. Larger grains, in particular, are deposited near the crest, whereas finer grains i. e., clay particles settle mainly on the lower part of the lee slope.

The lee face is a mobile surface shaped partly by the turbulent eddies of the transitional flow, but controlled primarily by downslope transfer of bedload material under the influence of gravity. Bagnold (1954) studied gravity sliding in air, and as also stated by Brush (1965), the same may apply to subaqueous sliding. Grain diameter is inversely proportional to the stress as given by the velocity gradient. Thus, larger particles tend to move in the zone of lesser shear, whereas the smaller particles tend to accumulate in areas of maximum shear. Thus, in avalanches down the lee face the particles would tend to become sorted normal to the plane of sliding. Larger particles tend to accumulate toward the outer side. An equally important aspect of sorting is that under the influence of gravity, by virtue of their size and higher settling velocities, the larger particles tend to reach the bottom of lee face in larger concentrations. This sorting effect is shown in Fig. 21.

Efficiency of the sorting process is strongly controlled by the length of the lee face. Sorting on the lee face of small ripples is rather indistinct, whereas in the case of megaripples and microdeltas, such sorting can be very well developed.

Development of Lamination in the Lee Face

Avalanching down the lee face is a result of instability created by more intense deposition of grains near the crest of the lee face than near the toe. After each avalanche, equilibrium is again established. As will be discussed later, the angle of repose at which avalanching takes place is controlled by several factors.

Below a certain critical velocity, episodes of avalanching alternate with periods when the lee face is built forward solely by the deposition of fine grains from suspension. The result is that thicker, coarser-grained laminae alternate with extremely thin laminae composed of fine-grained and light material, for example, mica flakes, clay floccules, etc. Above a certain critical velocity, avalanching is continuous, and grains in every part of the lee face are in sliding motion. Distinction between coarser- and finer-grained laminae is obscured.

When the avalanching is intermittent, the coarser-grained foreset laminae are longer, parallel, and continuous along the whole height of the lee face, but are less persistent laterally, and there is more distinct sorting along the foreset laminae. In the case of continuous avalanching, laminae are shorter but more persistent laterally and there is less distinct sorting along the foreset laminae (Allen 1965a). As will be discussed later, even the shape of the foreset laminae is controlled by the velocity of flow.

Foreset material is derived from the heavy-fluid layer, mostly as bedload. In relatively deeper waters, foreset may trap an appreciable amount of suspended load settling through the zone of mixing.

Unusually fine-grained material, or hydrodynamically lighter material (such as broken shells, faecal pellets, plant detritus, commonly known as "coffee ground," clay aggregates) accumulate in the trough of the ripple due to settling of particles from suspension passing over the crest and subsequently settling through the zone of mixing.

The finer material settling through the zone of mixing is caught in the backflow. It is transported back toward the foreset slope, and a part of it may be lodged near the base in the form of toeset deposit, which in terms of Jopling (1963), is a depositional continuum linking basal foresets and continuous bottomset.

If the backflow current is strong enough, it may reform the sediments of the ripple trough into small ripples, commonly known as backflow ripples, which move in the upstream direction of the main flow. Sometimes if abundant sediment is available they may develop in the form of climbing ripples. Boersma et al. (1968) found that sense of backflow ripples differ variably in direction. Backflow ripples have been shown in Figs. 22 and 23. These ripples develop only if the lee face is large enough, as in the case of megaripples.

In the trough of a ripple, few laminae or a set of bottomset laminae develop. Sediment of the bottomset is derived mainly from suspended load and the residual material left behind as the ripple migrates. Because bottomset deposition represents mainly suspension transport, the grain-size distribution, i. e., the maximum size of grains found in the bottomset is a good indicator of paleo-velocity (Jopling 1966).

Reineck (1961) suggested that foreset laminae are formed as a result of four processes which contribute to varying degrees: Avalanching contributes 75 to 90 %; coarser sediments falling out from suspension, 10 %; deposition of fine sediments from suspension, 5 %; backflow transport, 10 %.

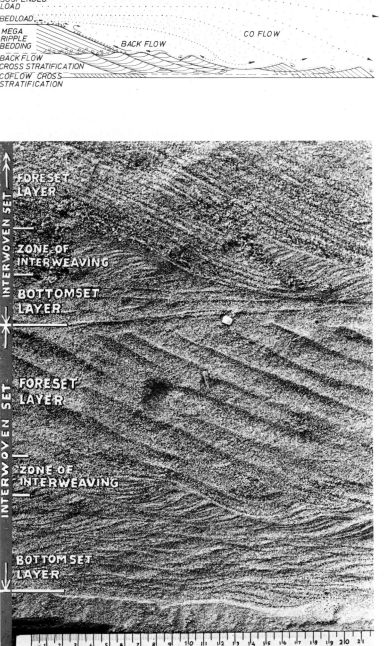

Fig. 22. Schematic illustration of flow pattern and resulting deposition in the lee face of a megaripple with a well-developed backflow system. Various sedimentation units are laid down simultaneously. Backflow beds are developed in the form of climbing ripples. (Modified after Boersma et al., 1968)

Fig. 23. Megaripple bedding showing well-developed backflow ripple bedding at the base of foreset laminae of megaripple bedding. (After Boersma et al., 1968)

Variables Controlling the Shape and Slope of Foreset Laminae

Important variables controlling the slope of the lee face, and thus the slope of foreset laminae are, according to Jopling (1963, 1965b):

1. Velocity and bed shear stress.
2. Ratio of the depth of stream flow to the depth of the basin of deposition (depth ratio).
3. Sediment type.

1. Velocity and Bed Shear Stress. To a geologist engaged in environmental studies, velocity is a useful parameter because it is a measure of the dynamics of transport and deposition. From a quantitative hydraulic point of view, however, bed shear stress is a more useful parameter.

The relationship between velocity and depth of stream flow is complex because of the changing roughness of the stream bed. In the same way, the relationship between velocity and bed shear is rather complex.

At low velocities of flow hardly exceeding the threshold value for particle movement, the sediment grains move along the bed and are deposited on the upper part of the lee face, from where it moves downslope under gravitational force, producing essentially a planar slip face. Sediment transport is by bedload movement.

With increasing velocity, a greater proportion of particles is taken in suspension, carried beyond the lee face, and deposited in the form of bottomset and toeset. Angular contact is replaced by a tangential contact, and, at the same time, the angle of repose of the lee face is also reduced substantially. In the case of megaripples, backflow ripples may develop in the bottomset and toeset.

Changes in the shape of foreset laminae with increasing velocity are: Angular contact → incipiently tangential → strongly tangential (concave) → sigmoidal (Fig. 24). With a sigmoidal profile it is no longer possible to differentiate between toeset and foreset, because eddy action has strongly reworked and scoured the face of the deposit.

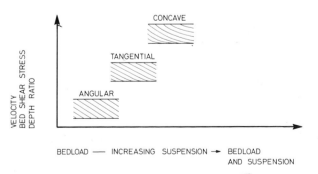

Fig. 24. Scheme showing factors controlling the shape of foreset laminae. (Modified after Jopling 1965b)

The effects of an increasing shear are more or less parallel to those of an increasing velocity. However, no direct relationship exists.

2. Influence of Depth Ratio. Low depth ratio (deeper water) favors deposition of angular foresets with steeper angles, whereas a high depth ratio (shallow water) favors the development of tangential units with gentle dips.

3. Sediment Type. The chances of developing a tangential foreset are improved as the sediment becomes finer, other factors remaining the same (Jopling 1965b). McKee (1957) found that the foreset slope becomes steeper when the sand grains are coarser and angular, sorting is poor, and clay is absent.

McKee (1957) states further that the presence of clay particles and the degree of sorting also controls the development of bottomset laminae. Moreover, a high precentage of clay in the sediment helps in reducing the angle of repose on lee face. In contrast, in subaerial sediments e. g., sand dunes the presence of clay causes an increase in the angle of repose.

Causes for the Production of Foreset Laminae of Varying Composition

Easy recognition of foreset laminae of the ripples is the result of the alternating coarser material laminae separated by a thin clay or fine-grained lamina. This is also a reason why in well-sorted clay-poor sands recognition or visibility of cross-laminae is so difficult.

As discussed by Jopling (1966), three processes seem to be important in the development of foreset laminae:

1. Selective transport related to differential settling velocity of grains.
2. Rapidly changing composition of the sediment mix discharged over the lee slope.
3. Pulses in sediment transport related to the moving bedform along the stream bed.

1. The avalanching down the lee face tends to be intermittent rather than continuous. This is especially so with megaripples and microdeltas. This process of avalanching produces layers composed of coarse material of bedload origin, i. e., sediment derived primarily from the lower part of the heavy-fluid layer. On the other hand, the material comprising the thin fine-grained laminae is derived from the suspended material of the upper part of the heavy-fluid layer. Thus, the fine laminae is derived mainly from the suspension load. However, effective sorting into fine-grained and coarse-grained laminae is prevented because of high sediment con-

centration and short distance and time of travel.

Many of the laminae of the lee face can be traced from the foreset through the toeset into the distal bottomset. This continuity affords important evidence of suspension transport, and lends support to the idea that selective transport is the dominant factor in the genesis of foreset laminae. Thus, a relative proportion of fine-grained laminae to coarse-grained laminae would provide a rough quantitative index to the relative importance of bedload and suspended load.

2. As mentioned earlier, avalanching along the lee face is intermittent. The material sliding down the slope is not of uniform composition. The composition of sediment mix avalanching shows great variation. These variations are mainly the result of segregation in the bedload and variability in the composition of the heavy-fluid layer. This would result in the formation of laminae of slightly differing composition. This process, combined with the sorting taking place within the laminae (see p. 27), can be sufficient to produce textural differences between adjoining foreset laminae, making their recognition possible.

3. Velocity pulsations have sometimes been considered the primary cause of laminae and cross-laminae formation. High-frequency pulsations do occur in streams, with a periodicity of a few seconds (7 to 14 sec). The time required for the deposition of a foreset lamina is about 1 to 2 min. Thus, periodicity of velocity pulsations is too rapid to produce the laminae. Pulsations may explain the gradational character of individual laminae. It could be argued, however, that the laminae are correlated with velocity pulsations of intermediate frequency, but no direct evidence is available. Anyway, the first two processes are probably most important and effective in the development of foreset laminae.

Jopling (1966) gives a list of useful indices of foreset laminae which may serve as a qualitative guide in determining the current strength in modern or ancient sediments.

1. *Maximum Angle of Dip of Foreset Laminae.* At low velocities the angle of dip may slightly exceed the static angle of repose (angle about 30°), whereas at high velocities the angle is less than the static angle (up to 20°).

2. *Character of Contact between Foreset and Bottomset.* With increasing velocity the character of contact between foreset and bottomset changes from an angular, through a tangential, into a sigmoidal shape.

3. *Laminae Frequency, or the Number of Laminae per Unit Area Measured at Right Angles to Bedding.* With increasing velocity there are more laminae per unit area.

4. *Sharpness of Foreset Laminae, i.e., Textural Contrast between Adjacent Laminae.* Sharply defined laminae of the foreset suggest a moderate amount of suspension transport. At higher velocities, laminae become less distinct.

5. *Occurrence of Regressive Ripples in the Toeset and Bottomset of a Megaripple and Equally Large Microdelta.* Their presence indicates relatively higher current velocities. The presence of a well-developed toeset deposit indicates that the velocity of backflow is somewhat in excess of the threshold for general movement of the sediment mix.

Sediment Movement in Symmetrical Wave Ripples

The motion in a fluid by progressive waves makes fluid particles move in a more or less circular path at the surface. If the fluid near the sediment bottom is also set into motion, the fluid particles do not follow a circular path but rather orbits that are much flattened, and the particles move only back and forth. Thus an oscillatory motion is produced. The intensity of motion of the sediment depends upon this oscillatory velocity near the bed. This velocity is a function of the height, period, wavelength of the surface wave, and of the depth of fluid (cf. Manohar 1955). Gilbert (1899) was probably the first to describe the process of wave oscillation and wave ripple formation.

At certain critical velocities a few particles from the sediment bottom are dislodged and sediment movement is initiated. With an increase in velocity, more and more grains start moving, essentially by rolling and creeping, and ultimately ripples are produced. Sediment particles are moved to the crest from the adjacent trough, and a steep downcurrent face (lee face) corresponding to the angle of repose is produced. At the same time a vortice exists on the lee side of the ripple. During the succeeding reverse flow, a similar profile is established in the reverse direction. The vortice of the previous lee side is moved up from the bottom, breaks up, and drifts away; a new vortice is formed in the new lee side (Bagnold 1946; Manohar 1955; Kolp 1958). Figure 25 demonstrates the movement of water particles during the generation of symmetrical wave ripples.

As more and more foreset laminae are added to the ridge, it increases in height. This increase in height of crests is only relative, for at the same time material from the trough is continuously removed and the troughs grow deeper.

With an increase in velocity, ripple height increases. Above a certain critical velocity, however, the height of the wave ripples begins to decrease,

Fig. 25. Schematic representation of movement of water particles during generation of symmetrical wave ripples. The orbit of water particles gradually flattens out with nearness to the sediment surface. Depending upon the direction of movement of water particles at a given time and point, lee-face vortices are moving toward the right or left, or at the point of breaking, while moving upward. (Modified after Reineck 1961)

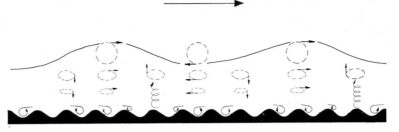

Sediment Movement in Asymmetrical Wave Ripples

and their length increases. At such values, instead of the rolling and creeping movement of grains, there is an intense motion of the top layer of the sediment as if most of the grains move in suspension. This was demonstrated by Manohar (1955) in a flume experiment. Inman (1957), however, in his study of ripples in nature could not find such cases in which above certain velocity the height of the ripples begins to decrease and the length increases.

Komar and Miller (1975) discuss that in the field wave ripples start developing at much lower threshold values than those obtained in the tank experiments. With increasing stress, the ripple heights progressively decrease and ultimately disappear at a critical stress value when sheet sand transport begins.

In other words, the mechanics of oscillatory wave motion producing symmetrical wave-formed ripples can be compared to those of a current being constantly reversed in direction, so that the lee side vortex is first on one side of the crest and then on the other.

As observed by Manohar (1955), no ripples are formed unless the flow is turbulent in the boundary layer. No ripples are produced in laminar flow.

Asymmetrical wave-formed ripples, also called half-stationary oscillation ripples by Reineck (1961), are produced because of a difference in velocity of forward and backward motion.

Theoretical and experimental studies of oscillatory waves have shown that the water particles in a wave do not move in a closed orbit, and therefore there exists a resultant mass transport of water in the direction of wave propagation. Moreover, with such waves, there exists a difference in the velocity of the forward and backward motion of the water particles (Fig. 26). These factors are responsible for the generation of asymmetrical wave ripples. This is especially the case in the surf zone and in shallow water. Forward velocity is greater than backward velocity near the shore and causes a landward sediment transport. When the waves are long and low, the velocity difference is great, while under short steep waves, the velocity differential is small (Manohar 1955).

Because the velocity is greater in the forward direction, more particles are rolled to the crest and over the crest from the adjacent trough. The lee face of the ripples approximates the angle of repose of the sediment, whereas the upstream

Fig. 26. Schematic representation of asymmetrical wave movement responsible for the asymmetrical wave ripples. Short periods of higher velocity toward land alternate with longer periods of lower velocity toward the sea. The velocity in the landward direction increases the critical velocity necessary to move the sand and sand is moved toward the land in form of asymmetrical wave ripples. Based on various measurements on North Sea tidal flats. (Modified after Reineck 1961)

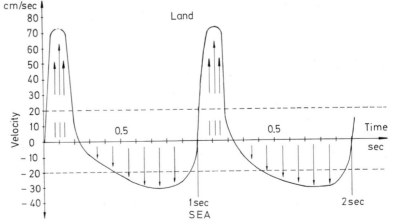

slope has a very flat slope. During the backward flow, due to the low velocity of the flow, the movement of particles is less and is mainly limited to the crest, from where particles are moved partly into the adjacent trough. But there is no reversal of ripple symmetry. Slowly, ripples move in the direction of greater velocity, the direction of wave propagation. Each new stroke produces a foreset lamina in a rather similar fashion to the ripples of unidirectional current flow.

Wave Ripples

Wave ripples are symmetrical or slightly asymmetrical undulations produced by the action of waves on a noncohesive surface. They are usually straight-crested, frequently showing bifurcation. Such bifurcations are never observed in the case of current-formed small ripples and megaripples.

The most complete study dealing with the various factors controlling size and shape of wave ripples is that by Inman (1957). He carried out detailed measurements in the coastal waters of California. His studies showed that ripples were always present on the sandy bottom when the velocity of wave propagation exceeded about 9 cm/sec. They disappeared when the velocity exceeded about 90 cm/sec, and sediment is moved along a plane bed (Fig. 27). Evans (1942, 1943, 1949) provides useful data on wave-generated ripples.

Besides velocity, the grain size is one of the most important factor in determining the size of the wave ripple. In general, larger ripples occur in coarser sand, and smaller ripples in finer sand. Moreover, for equivalent sizes of sand, wave ripples from deeper water along exposed coasts tend to be larger than the wave ripples in shallow water. This is due to the fact that in the deeper water of the open sea, wavelengths are longer and consequently possess larger orbital diameters, producing larger ripples.

In general, if sufficient sand is available, the material of the largest size and least density is typically found on the ripple crests, and the finest and heaviest material is found in the trough.

The ripple index (L/H) can attain much larger values for finer sand than for coarser sand. Ripples in fine sand near the surf zone can attain very high values of ripple index.

Tietze (1978) studied development of wave ripples in bimodal sands, both in nature and in the laboratory. He found that addition of coarse-sand particles in fine sand leads to the increase in the asymmetry of the asymmetrical wave ripples, and there is some tendency that ripple height is reduced by higher contents of coarse sand.

Draper (1967), based on long-term observations in northwestern Europe, concludes that larger waves in certain cases can move bottom sediments at 100 or 200-m depths on the open shelves.

Komar et al. (1972) observed in bottom photographs symmetrical wave ripples up to a depth of 204 m on the Oregon continental shelf and calculated that during storms velocities are sufficient to produce ripples up to 100 m water depth and occasionally to 200 m water depth.

Wave ripples, based on the symmetry of their crests, can be distinguished into two groups: symmetrical wave ripples and asymmetrical wave ripples.

Symmetrical Wave Ripples

Symmetrical wave ripples are marked by the symmetrical shape of their crests (Fig. 28). The crests are usually pointed and the troughs rounded. Occasionally crests may be rounded in shape. However, rounding of crests is a result of reworking of ripples during the process of emergence. Some-

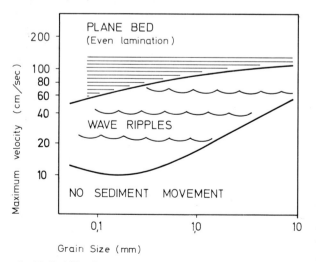

Fig. 27. Stability fields of wave ripples in relation to maximum wave-induced current velocity near the sediment bottom, and grain size. Wave ripples disappear at velocities exceeding 90 cm/sec. (Modified after Allen 1970)

Fig. 28. Symmetrical wave ripples. Tidal flats of the North Sea

Fig. 29. Various kinds of chevron structures developed in wave-generated ripple marks. The opposedly dipping laminae join in a central crotch zone, which may show variants. a–d chevron structure in ripple crests; e–f chevron structure in ripple troughs. (After Boersma 1970)

times a secondary ridge of small magnitude may be present along the axis of the trough. Symmetrical wave ripples are essentially straight-crested, partly showing bifurcation. Available data shows that the length of the symmetrical wave ripples varies from 0.9 to 200 cm, and height from 0.3 to 23 cm. The ripple index (L/H) varies from 4 to 13, mostly from 6 to 7.

A typical symmetrical wave ripple shows a distinctive internal structure characterized by superimposed chevron-like laminations. The laminae join the central crotch zone in an imbricate, often overlapping fashion. These chevron structures can develop in several variants. Boersma (1970) in his study of wave ripples found that in some cases deposition in ripple troughs may also produce laminae arranged in a chevron-like fashion, but open upward (Fig. 29).

However, as shown by the investigations of Newton (1968), most of the symmetrical wave ripples in the nearshore zone show an internal structure in which foreset laminae are only in one direction (Fig. 30).

Such ripples may be regarded as forms transitional between symmetrical and asymmetrical wave ripples. In this case the forward motion of a wave is somewhat stronger than the backward motion, and produces foreset laminae. On the other hand, the backward motion of a wave is only strong enough to maintain the symmetry of the ripple, but too weak to produce foreset laminae. The net result is that, although the symmetry of the ripple is maintained, foreset laminae are produced only in one direction–in the direction of wave propagation.

However, we disagree with Newton (1968) that wave ripples always possess unidirectional foreset laminae. In many cases a chevron-like structure is formed.

Moreover, if sufficient sediment supply is available, wave-formed ripple laminae in-phase (climbing ripples) may be produced. Thus, a wave ripple may be internally composed of slightly curved laminae lying one above the other with convexity upward – a wave-formed ripple laminae in-phase (Fig. 31).

Singh and Wunderlich (1978) propose the name mini-ripples for the very small wave ripples, 0.5–3 cm in length. The ripple crests are symmetrical or asymmetrical, mostly straight or slightly curved, and often show tuning-fork-like bifurca-

Fig. 30. Symmetrical wave ripple with foreset laminae dipping in a single direction (shoreward). Relief cast. North Sea. (After Newton 1968)

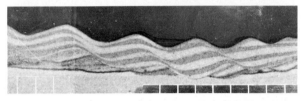

Fig. 31. Symmetrical wave ripples showing climbing-ripple lamination structure. Ripples are internally made up of slightly curved laminae lying one above the other with convexity upward. Ripples produced in a tank experiment, by providing sediment simultaneously to the ripple propagation. The scale is in centimeters. (After Reineck 1961)

tion. Such ripples are produced by slow-moving waves on the margins of a water body in water a few centimetres deep and do not need a fetch. These ripples possess a faint internal lamination and mostly show modified crests. They are good indicators of subaerial emergence or quasi-subaerial emergence of a sedimentation surface.

Quite often symmetrical wave ripples may show internal structure that is form-discordant. Within a wave ripple there are present foreset laminae of older and earlier ripples, which are not genetically related to the outer ripple form (Fig. 32). Newton (1968) found that the thickness of individual laminae within the wave ripples depends upon the grain size.

Asymmetrical Wave Ripples

Asymmetrical wave ripples show much similarity to straight-crested current ripples, in possessing a steep lee side and a gentle stoss side (Fig. 33). Pub-

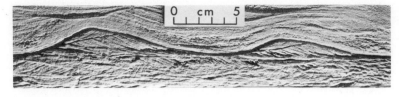

Fig. 32. Lacquer peel showing wave ripples with a form-discordant internal structure. A younger set of foreset laminae dipping to the left has truncated the older set of foreset laminae and caused a reversal in ripple symmetry. (After Newton 1968)

Fig. 33. Asymmetrical wave ripples. Land toward the right. North Sea tidal flats

lished information shows that asymmetrical wave ripples range in length between 1.5 and 105 cm, and height varies from 0.3 to 20 cm. The ripple index (L/H) varies from 5 to 16, mostly between 6 and 8. Asymmetry index ranges from 1.1 to 3.8.

The internal structure of normal asymmetrical wave ripples is almost identical to the internal structure of current ripples, composed of a single bottomset lamina and a few stoss side laminae. The main body is composed of foreset laminae dipping in a single direction. This is the form-concordant internal structure of asymmetrical wave ripples.

However, they may also possess internal structure that is form-discordant. In such cases ripples show a composite structure, made up of laminae of earlier formed ripples. In other words, the outer form of the ripples is genetically not related to the internal structure.

Asymmetrical wave ripples may also develop as climbing ripples or ripple laminae in-phase, if sufficient sediment is available in suspension (Reineck 1961; McKee 1965).

It is sometimes rather difficult to distinguish between asymmetrical wave ripples and current-formed small ripples with straight crests.

Recently, Boersma (1970) carried out a detailed study of wave ripples, especially of their internal structure. According to him, characteristic features of the internal structure of wave ripples are the irregular lower bounding surface, bundle-wise arrangement of foreset laminae, and foreset laminae with off-shoots (Fig. 34). Moreover, the same wave ripple, when cut at different intervals, shows a rather varied arrangement of internal laminae (see also wave-ripple bedding, p. 103).

Tanner (1967) and Reineck and Wunderlich (1968a) have carried out investigations in recent sediments and give some parameters to distinguish

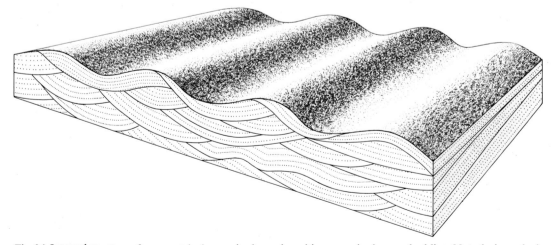

Fig. 34. Internal structure of asymmetrical wave ripples and resulting wave ripple cross-bedding. Note the irregular lower bounding surface, bundle-wise arrangement of foreset laminae, and foreset laminae with off-shoots. (Based on Boersma 1970)

between asymmetrical wave ripples and small-current ripples. Both papers deal with the outer form and shape of the ripples. From the geometrical properties of the ripples, several indices were calculated.

Along with the two widely used indices, ripple index (R.I.) and ripple symmetry index (R.S.I.), Tanner (1967) calculates several other indices as well (Fig. 35). All of them show that there is an area where characteristics of both types of ripples overlap.

Reineck and Wunderlich (1968a) also calculated R.I. and R.S.I. (Fig. 36). In addition, they calculated such indices as ripple cross-sectional area index, ripple stoss side angle, surface index, etc. They also found that characteristics of both types of ripples show a region of overlapping, and also suggest that R.I. and R.S.I. are the most important indices. Data of Tanner (1967) and Reineck and Wunderlich (1968a) are in rather close agreement. However, the method of presentation of data of Reineck and Wunderlich (1968a) in the form of relative frequency distribution has the advantage that one is able to interpret the degree of correctness of interpretation in the region of overlap.

Boersma (1970) adds that in ancient sediments, ripple size and with that the R.I. would be affected by the compaction processes. The same objections were put forward by Reineck and Wunderlich (1968a), who prefer the use of R.S.I. in the ancient sediments. The maximum value of R.S.I. acquired by asymmetrical wave ripples is 3.8.

Another criterion for differentiating the two types of ripples is that the ripples smaller than 4.5 mm are known only as asymmetrical wave ripples. Moreover, symmetry of successive ripples of a ripple train of wave origin can vary in direction. The crests of asymmetrical wave ripples (and symmetrical wave ripples) show repeated bifurcations in the shape of a tuning fork (Reineck and Wunderlich 1968a) and the profile of crests is rather regular (Fig. 37).

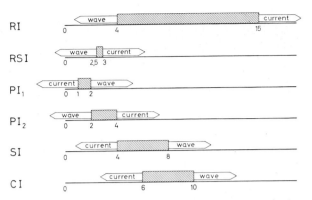

Fig. 35. Simplified scheme to distinguish between wave ripples and current ripples by means of various indices (dimensionless ratios). Based on data from Tanner[1967]. Hatched areas indicate where values of both wave and current ripples overlap. (Modified after Boersma 1970)

RI–Ripple index
$$= \frac{\text{ripple length}}{\text{ripple height}}$$

RSI–Ripple symmetry index =

$$= \frac{\text{length of horizontal projection of stoss side}}{\text{length of horizontal projection of lee side}}$$

PI₁–Parallelism index
$$= \frac{\text{length of curved part of crestline} \times \text{min. ripple length}}{\text{mean ripple length} \times \text{max. ripple length}}$$

PI₂–Parallelism index =
$$= \frac{\text{max. ripple length} - \text{min. ripple length}}{\text{mean ripple length}}$$

SI–Straightness index =

$$= \frac{\text{length of curved part of crestline}}{\text{departure of curvature from straight (crest) line}}$$

CI–Continuity index =
$$= \frac{\text{crest length between bifurcations}}{\text{mean ripple length}}$$

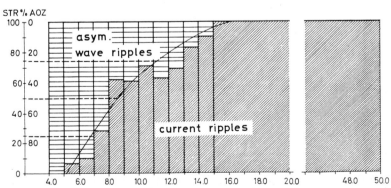

Fig. 36. Ripple index (RI = $\frac{L}{H}$ of asymmetrical wave ripples and current ripples. Frequency of occurrence of wave and current ripples at various values of ripple index has been plotted. At RI > 15, only current ripples are present. (After Reineck and Wunderlich 1968a)

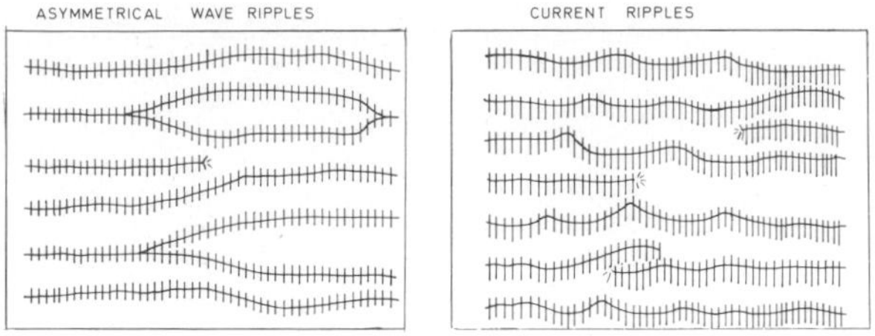

ASYMMETRICAL WAVE RIPPLES   CURRENT RIPPLES

Fig. 37. Schematic diagram showing asymmetrical wave ripples and current ripples in a surface view. Asymmetrical wave ripples show well-developed bifurcation of crests, and crests are rather regular. Crests of current ripples, on the other hand, are not so regular. Bifurcation is absent, also crests terminate and are replaced by other crests

Small ripples of current origin do not show such bifurcations. However, one does find crests which terminate and are replaced by other crests, and this may give a false impression of bifurcation. Moreover, the straight crests of small-current ripples are not so regular, but show many small tongue-like projections in the downstream direction (Fig. 38).

Fig. 38. Straight-crested current ripples showing small tongue-like projections in the down-current direction. (Photograph by Wunderlich)

Current Ripples

Ripples produced on a noncohesive bed surface due to an unidirectional current or flow constitute current ripples. They are elongated transversely to the flow, with regularly spaced crests alternating with troughs. Based on their size and morphology they can be distinguished into 4 groups:

1. small-current ripples
2. megacurrent ripples
3. giant-current ripples
4. antidunes

For the sake of brevity the prefix "current" will be dropped. Thus, small ripples, megaripples, and giant ripples all refer to current ripples. Wave-formed ripples will always be used with the prefix "wave."

Small-Current Ripples

These are the smaller symmetrical undulations on a bed surface produced by a current flowing over it. Normally, they are no more than about 30 cm in length. However, a generally accepted size limit is 60 cm (cf. Allen 1968a; Reineck et al. 1971); although purely current-formed ripples between 30 and 60 cm seem to be rather rare. Small ripples are not produced in sand having a Md > 0.60 mm (Simons et al. 1965). Existing literature and our experience seem to show that they range in length between 4 and 60 cm, whereas the height varies between 0.3 and 6 cm. The ripple index (L/H) is always more than 5, with most values falling between 8 and 15.

Some generalizations about the size of small ripples with respect to grain size and shear stress, as can be read from the graphs plotted by Allen (1968a) based on the data of Guy et al. (1966), are discussed below. To begin with, at the same shear stress coarser sands give larger ripples than finer sands. With increasing shear stress the length of small ripples increases, in coarser sand the small ripples acquire maximum sizes earlier than in the finer sand.

The height of small ripples in fine sand is less than in coarse sand. With increasing shear stress, the height of small ripples increases. For a given shear stress value, the height of small ripples in coarser sand is always more than for the corresponding small ripples in finer sand (Fig. 39).

The ripple index (L/H) of small ripples tends to decrease with higher values of ripple length. In other words, the smaller small ripples are relatively flatter than the larger small ripples.

Based on the nature of their crests, small ripples can be distinguished into following types:

Straight-Crested Small Ripples. These are ripples showing more or less straight and parallel crests (Figs. 38 and 40). They are formed at relatively low velocities, and Harms (1969) calls them low-energy ripples. Sometimes they may be difficult to distinguish from asymmetrical wave ripples. How-

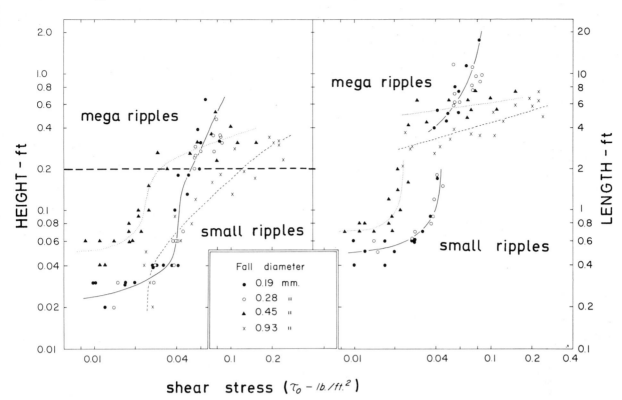

Fig. 39. Ripple height and ripple length as a function of grain size and boundary shear stress. Both small ripples and megaripples are plotted. Based on data of Guy et al. (1966). (After Allen, 1968a)

Fig. 40. Small ripples with almost straight crests. Small rill marks at the crests are formed during emergence of tidal flats. North Sea tidal flats. Knife as scale

and the bedding produced by their migration. They are small-scale units of planar cross-bedding. Individual units are usually less than 3 or 4 cm thick, never exceeding the maximum height of the small ripples.

Undulatory Small Ripples. Undulatory small ripples represent a transition form between low-energy straight-crested small ripples and higher-energy lingoid small ripples. The crest of the ripples can be traced for long distances, and it is wavy or undulatory (Figs. 42 and 43). These wavy crests are usually out-of-phase, although sometimes these undulations can also be in-phase. However, often in a single ripple train, both out-of-phase and in-phase undulations occur together, and all transitional stages may be present. Undulatory small ripples are very common. Allen's (1968a) ripples, both sinuous and catenary, fall in this group. Figure 44 demonstrates the internal structure of undulatory small ripples and the resulting cross-bedding produced from their migration. The units are weakly festoon-shaped.

Lingoid Small Ripples. In lingoid small ripples the crest of the ripples is discontinuous and broken with forward closures and thus cannot be traced over long distances. Small curved crests extend for-

ever, in general, straight-crested small ripples are flatter than asymmetrical wave ripples. Sometimes they may acquire extremely high ripple indices. Figure 41 shows the straight-crested small ripples

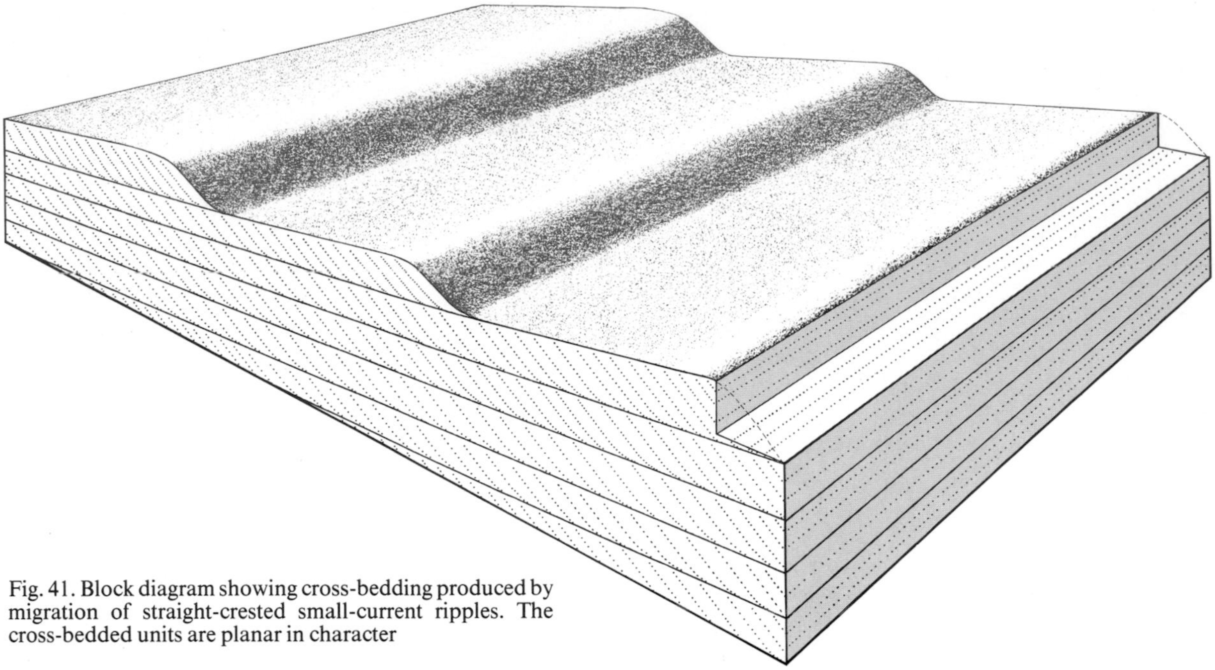

Fig. 41. Block diagram showing cross-bedding produced by migration of straight-crested small-current ripples. The cross-bedded units are planar in character

Fig. 42. A ripple train with weakly undulatory small-current ripples. Forms are transitional between straight-crested and undulatory ripples. North Sea tidal flats. Ripple length: 8 cm

Fig. 43. Strongly undulatory small-current ripples, almost tending to the lingoid type. North Sea tidal flats

ward in a tongue-like or lobe-like form (Fig. 45). The tongues are out-of-phase. These small ripples are produced at velocities or stream power higher than that needed to produce the earlier described small ripples. Figure 46 shows the internal structure and the resulting crossbedding from migration

of lingoid small ripples. The cross-bedded units are strongly festoon-shaped.

To conclude: In small ripples there is a tendency for the crests to become discontinuous or lobate with an increase in the energy of the environment. A somewhat rarer variant of the lingoid small ripples are cuspate small ripples. They differ from lingoid small ripples in that their lobes or tongues are in-phase. However, it seems that both the lingoid and the cuspate small ripples and other variants

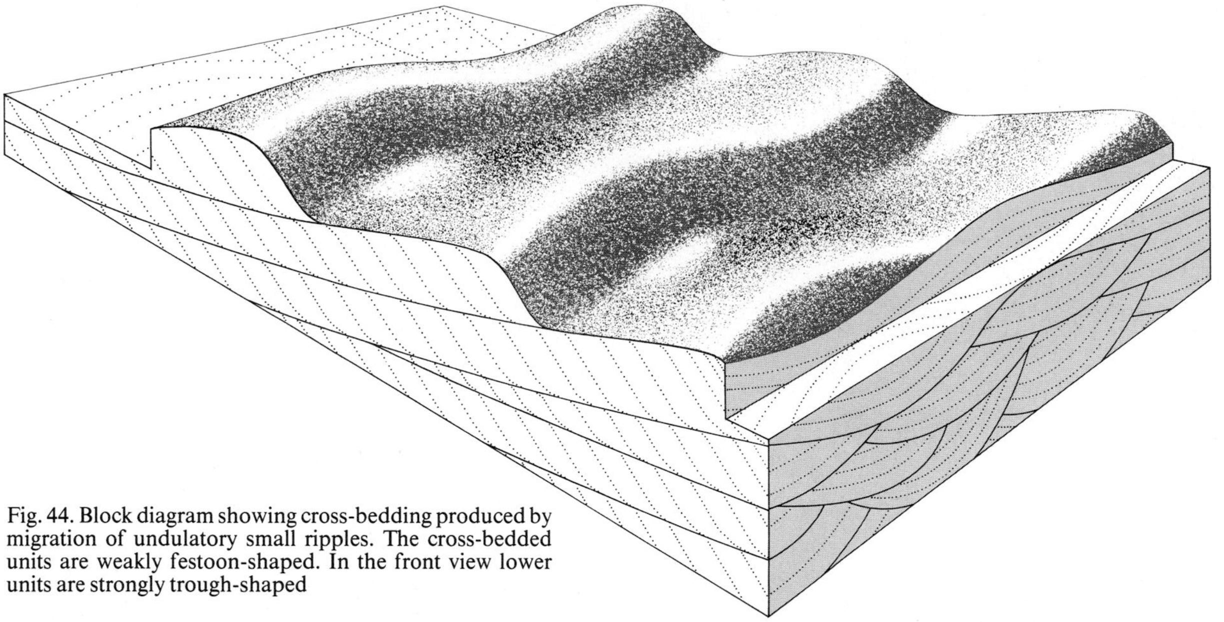

Fig. 44. Block diagram showing cross-bedding produced by migration of undulatory small ripples. The cross-bedded units are weakly festoon-shaped. In the front view lower units are strongly trough-shaped

Fig. 45. Lingoid-shaped small-current ripples. The direction of flow is from left to right. North Sea tidal flats

with discontinuous crests are produced under very similar hydrodynamic conditions. Thus, we would suggest classifying them as lingoid small ripples or small ripples with discontinuous crests. There is a gradual transition from straight-crested to undulatory, and undulatory to lingoid small ripples, and all transition forms exist.

Rhomboid Small Ripples. Crests of rhomboid small ripples show a reticulate, scale-like pattern on the bed surface, or in other words, the crests are in the form of small rhomboidal, scale-like tongues (Figs. 47 and 48).

These ripples develop under a very thin layer of water, usually on the seaward slopes of the beaches, by backwash, and may be common on the landward slopes of beach bars, produced by washovers. Water depth never exceeds 1 to 2 cm. Rhomboid small ripples may even develop under a film of water a few millimeters thick. Flow velocity is probably rather high. Thus, rhomboid small ripples are forms that develop under conditions of higher velocity and extremely shallow water. It seems that rhomboid ripples develop in flow conditions, where Froude's number is about 1 or slightly more. As they originate under such exceptional conditions, they constitute rather rare forms of small ripples. Because they are commonly produced by the backwash on the beaches, McKee (1953) calls them backwash marks. Trusheim (1935a) and Hoyt and Henry (1963) give some useful information about rhomboid ripple marks.

Small rhomboid ripples may occasionally show rather conspicuous crests, especially when they are produced by the washing away of earlier small ripples. Each scale-like tongue has an acute angle pointing downstream, formed by two steep lee sides, while the other upstream half generally smaller than the downstream half is formed by the gentle slope extending into the angle of re-entrance of the lee sides of the two tongues of the following alternating row (Twenhofel 1932). Scales of adjoining rows are strongly out-of-phase. In other cases rhomboid small ripples may be present only as a pattern on a more or less smooth, plane surface.

Because of the extremely low crests of rhomboid small ripples, it is not possible to see their internal structure. They can be, thus, considered in most

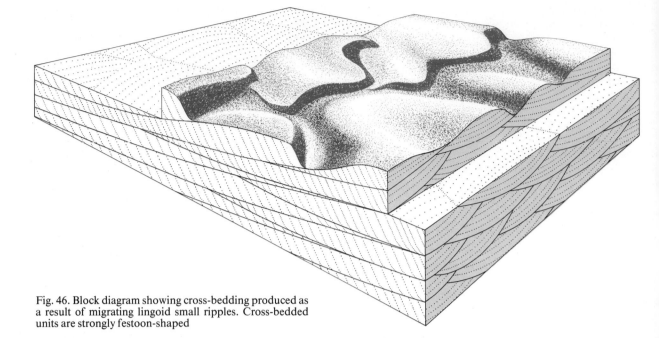

Fig. 46. Block diagram showing cross-bedding produced as a result of migrating lingoid small ripples. Cross-bedded units are strongly festoon-shaped

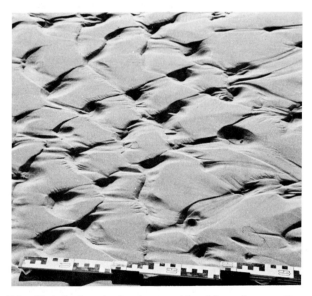

Fig. 47. A ripple train showing transitional forms between lingoid and rhomboid small ripples. Flow is from left to right. North Sea tidal flats

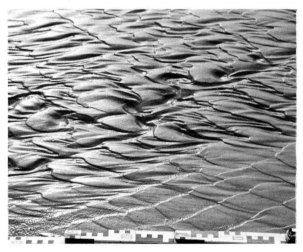

Fig. 48. Well-developed small rhomboid ripples. Flow is from right to left. North Sea tidal flats

cases merely surface forms. In several cases, rhomboid small ripples seem to grade into forms where only small scours are present, arranged in rhomboidal patterns. In such cases, scours migrate upcurrent, depositing laminae dipping upcurrent, generating a sort of backset bedding (Wunderlich 1973).

Megaripples

The fact that small ripples can be found superimposed on large ripples and that both types show the same general relationship to flow has led many workers to the view that the small ripples differ from the larger ones in origin. It has come to the force that there are at least two distinct populations or classes of transverse ripples which can be generated on a loose grain bed by flowing water.

Megaripples are also described under the terms dunes and sand waves. Southard (1975) and Boothroyd (1978) consider megaripples (dunes) to be 1–5 m in length developing at relatively high energies; while sand waves are 5–100 m in length, rather low forms, formed at lower energies. Their sand waves overlap with the dimensions of the giant ripples (30–100 m).

Based on his studies in the Klarälvan River, Sundborg (1956) states that there are no definite forms intermediate between small ripples and large ripples, and he further adds that the mode of origin in the two cases is quite different. Simons and Richardson (1962), Raudkivi (1963), Yalin (1964), Allen (1968a), and several other workers believe in the essential separateness of the two classes of ripples. Bagnold (1956) regards small ripples as distinct from the larger forms megaripples, but believes that the two classes may overlap.

Our experience in tidal environments shows that two distinct groups of ripples are present. Small ripples are almost always less than 30 cm in length. Megaripples are always more than 60 cm in length, ranging up to 30 m.

The presence of two distinct groups of ripples is also reflected in the occurrence of two distinct types of ripple bedding: Small-ripple bedding, originated through the migration of small ripples, and megaripple bedding, produced through the migration of megaripples (see p. 97 ff.).

The lowest limit of 60 cm for megaripples is agreed upon by Allen (1968a), Simons et al. (1965), Reineck et al. (1971), and most of the other workers. As already discussed on p. 18 in the coarser sand, migrating megaripples with superimposed active small ripples are produced at relatively lower energy. (This should not be confused with features where first megaripples are produced, and at a certain time, because of a decrease in current velocity or water depth they cease to migrate. Now on the top of inactive megaripples, small ripples are formed and migrate.) Megaripples without superimposed small ripples are produced at a relatively higher energy. Existing data indicate that megaripples may range in length from 0.6 to 30 m, and in height from 6 cm to 1.5 m. The ripple index (L/H) is generally above 15.

As mentioned previously, in sand with a grain size > 0.6 mm, no small ripples are produced and only megaripples are formed. This leads to the fact that at certain current velocities fine sand is reformed into small ripples, whereas the medium-to-coarse sand exposed to the same current velocities produces megaripples. Cases are known where shell and mud pebbles accumulations on the inter-

tidal flats are reformed in the shape of megaripples.

Megaripples show an increase in length with increasing shear stress of the bed. In finer sand, megaripples achieve greater lengths at relatively lower shear stress values than in coarser sand.

With increasing shear stress, the height of megaripples gradually increases. However, its dependence on grain size is rather obscure. Even so, in fine sand megaripples acquire, as a rule, greater height at lower shear stress values than in coarse sand.

The ripple index (L/H) of the megaripples increases with increasing size of ripples. In other words, smaller megaripples show lower ripple indices and are relatively higher than the larger megaripples, which show higher ripple indices. These relationships can be seen in Fig. 39. Singh and Kumar (1974) observed in the megaripples of the Son River that for smaller megaripples, height shows great variation, but for larger megaripples relationship between height and length is linear.

In ancient sediments well-preserved megaripples are not so commonly reported as small ripples. This is due in all probability to the larger size of the megaripples, which makes their identification in outcrops rather difficult. Other important factors are that the megaripples are more susceptible to reworking of the surface, which results in the capping-off or complete erosion of the megaripple surface, the result being that megaripples more commonly leave a record of their presence in the form of large-scale cross-bedding and only rarely in the form of surface features.

Dalrymple et al. (1978) in their study of bedforms in the Bay of Fundy distinguish two types of megaripple:

a) Type 1 Megaripples – Characterized by straight to smoothly sinuous crests, showing flat profiles.
b) Type 2 Megaripples – Sinuous to lunate in form and possessing well-developed scour-pits.

Based on the character of the crests of megaripples, they can be separated into several types.

Straight-Crested Megaripples. Trains of straight-crested megaripples are quite well known. The crest lines are straight and parallel over considerable distances (Fig. 49). Like other megaripple types, they may show small ripples on the stoss side, and sometimes regressive ripples and climbing ripples moving in the upstream direction on the lee side of the megaripple. They are produced at somewhat lower velocities than the undulatory and lunate megaripples. The type 1 megaripples described by Dalrymple et al. (1978) correspond to our straight-crested megaripples. They show straight to smoothly sinuous crests and flat pro-

Fig. 49. Straight-crested megaripples. Ripple troughs are still partly filled with water. On the lee face of the megaripples water-level marks are present. Megaripples are covered with small ripples, which were generated after megaripples ceased to migrate. Flow is toward the observer. North Sea tidal flats

files. The L/H ratio is mostly greater than 20, reaching values up to 150, and there is poor correlation between length and height. They are stable up to maximum current velocities of about 70 cm/sec (in water depth of more than 2 m). It seems that straight-crested megaripples correspond to the bar phase of Costello (1974), and the sand wave phase of Southard (1975) and Middleton and Southard (1978). The internal structure and resulting bedding is identical to the straight-crested small ripples, differing only in the scale (see Fig. 41). The cross-bedding produced by straight-crested megaripples is essentially large-scale planar cross-bedding.

Undulatory Megaripples. These ripples possess long continuous wavy or undulating crests (Fig. 50). Such undulations can be arranged in-phase or out-of-phase. However, transitional forms seem to predominate. Occasionally all types may be present in a single ripple train. Allen's (1968a) catenary and sinuous megaripples fall in this group. For the internal structure see Fig. 44.

Lunate Megaripples. Lunate megaripples do not show continuous crests. The crestline is broken, producing sickle-shaped or crescent-shaped lobes showing a backward closure (Fig. 51). Häntzschel (1938) calls them D-shaped megaripples. Usually in a ripple train they are arranged in en-echelon fashion or out-of-phase; rarely are some of them in-phase or partly so.

In some cases crests seem to be planed out or ill-developed; thus, a ripple train appears as if made up of regularly, out-of-phase arranged D-shaped troughs. Harms and Fahnestock (1965) call such troughs "spoon-shaped scours." Lunate megaripples are formed at higher velocities than

Fig. 50. Undulatory megaripples. Crests are continuous but projected fore-ward into tongue-like extensions. Smaller ripples on megaripples were produced during the emergence of the surface. Flow is from left to right. North Sea tidal flats

Fig. 51. Lunate megaripples

straight-crested and undulatory megaripples. Figure 52 shows the internal structure and the bedding resulting from the migration of lunate megaripples.

The type 2 megaripples of Dalrymple et al. (1978) correspond to lunate megaripples, which possess well-developed scour pits. They show steeper profiles than the straight-crested megaripples, L/H ratio is mostly less than 20, and there is good correlation between length and height. They are mostly produced at velocities higher than those required for straight-crested megaripples, mostly more than 70 cm/sec (in water depth of more than 2 m). These forms are also described as dunes, three-dimensional dunes, or megaripples.

All transitions from straight-crested to undulatory, and undulatory to lunate megaripples are known. They constitute a continuous series from straight-crested to lunate megaripples. Straight-

crested and lunate megaripples represent the end members of the series, whereas undulatory megaripples represent arbitrary intermediate transitional forms.

Lingoid Megaripples. Lingoid megaripples are rather rare, and except for their size, they are similar to lingoid small ripples. Normally lunate megaripples are produced. Van Straaten (1953a) described them from tidal channels of the Wadden Sea (Fig. 53) and Coleman (1969) from Brahmaputra River. It is not clear under what conditions the lunate megaripples are replaced by lingoid megaripples. They are not very common, and seem to be generated under exceptional conditions. Allen (1968a) prefers to call them lingoid bars instead of ripples. Many examples of supposed lingoid megaripples in the literature are actually lingoid giant ripples.

Some of the lingoid to rhomboid bars of fluvial braided streams may be actually lingoid megaripples (see under fluvial bar, p. 262). Some of the rhomb-shaped forms, described as sand waves (low-energy type) from the South Saskatchewan River, Canada (Cant 1978) may actually be lingoid megaripples. If this is true, then lingoid megaripples are produced at lower energy flow conditions (corresponding to the stability field of sand wave = straight-crested megaripples) than the lunate megaripples. Upon migration lingoid megaripples produce planar cross-bedding with strongly wedge-shaped units.

Rhomboid Megaripples. The shape of rhomboid megaripples is similar to rhomboid small ripples.

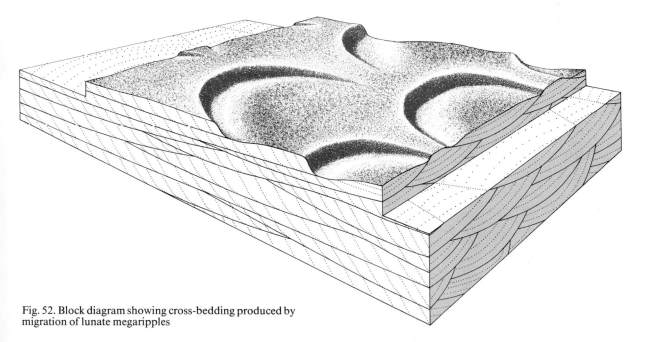

Fig. 52. Block diagram showing cross-bedding produced by migration of lunate megaripples

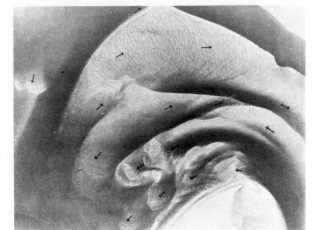

Fig. 53. Aerial photograph of Pinkegat Netherland, tidal inlet between Ameland and Engelsmanplaat at low tide. Large shoals are covered with megaripples. Some of the small tongue-shaped extensions in the lower half of the picture are probably lingoid giant ripples. Van Straaten (1953) groups them among lingoid megaripples. (After van Straaten 1953a)

They are present as large-scale reticulate patterns on sandy surfaces (Fig. 54). Rhombs may be several decimeters to more than a meter in length. At the same time, relief is rather low. They are few millimeters to 1 to 2 cm above the sediment surface. Reineck (1963a) described them from the beaches of the Frisian Islands. Superimposed on the rhomboid megaripples, rhomboid small ripples may be present. Rhomboid megaripples seem to develop under similar hydrodynamic conditions as rhomboid small ripples, i.e., conditions of high velocity and extremely shallow water (maximum of a few centimeters), in flows with Froude's number around 1 or more. Rhomboid megaripples, if preserved, are good indicators of very shallow water (water depth less than 1 m). They are known to occur on the seaward slopes of beaches and the land-

ward part of emerged longshore bars, where backwash or swash may attain high velocities due to a sloping surface.

In general, they seem to be merely surface features. Very little is known about their internal structure. Wunderlich (1973) found that active rhomboid megaripples may produce a set of laminae, dipping upstream, the scours migrating upstream. However, Reineck and Cheng (1978) observed that rhomboid megaripples upon migration produce distinct downcurrent dipping foreset laminae.

Giant Ripples

A ripple is defined as a giant ripple when it is more than 30 m in length; it usually shows superimposed megaripples. The relationship between a giant ripple and a megaripple seems to be similar to that between megaripples and small ripples. However, there does not seem to be a very clear-cut boundary between megaripples and giant ripples. Transitional forms are often recorded, and many workers do not make any distinction between the two. Giant ripples are produced only when water is several meters deep.

Nasner (1974) made measurements on the giant ripple fields of the River Weser over several years to study the changes in their pattern under differ-

Fig. 55. A single giant ripple exposed after retreating water level of a flood. The height of this giant ripple during active migration was 9 m. Brahmaputra River. (After Coleman 1969)

Fig. 54. Rhomboid megaripples. Superimposed on a large-scale reticulate pattern are rhomboid small ripples. Generally, a well-developed scour is present near the downcurrent end of a rhomb. Under certain conditions this scour migrates upstream and produces gently up-current-dipping laminae. Norderney Barrier Island, North Sea

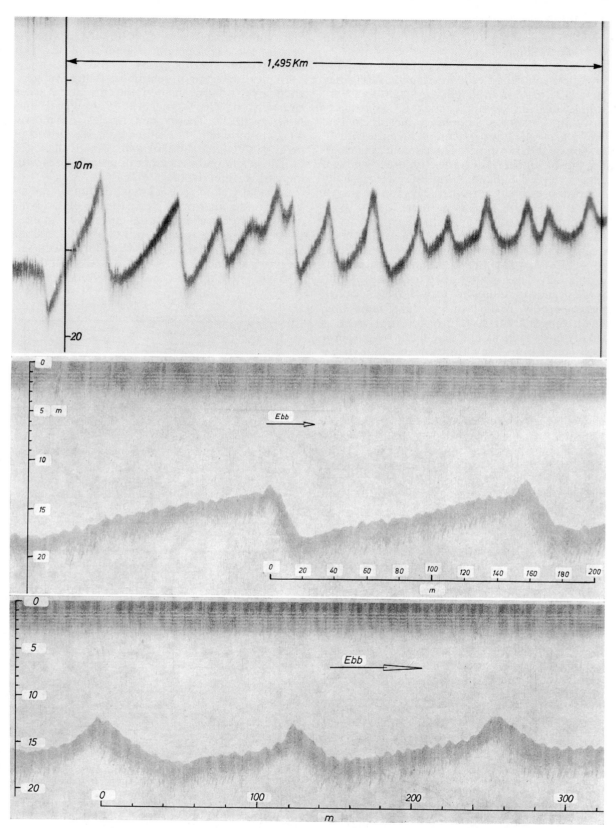

Fig. 56. Giant ripples as seen on an echo sounder profile. Giant ripples are themselves covered by smaller megaripples. Outer Jade, North Sea. (After Reineck 1963a)

ent conditions and with time. Ulrich (1973) gives a general review of giant ripples and megaripples in the German Bight.

Available data shows that giant ripples usually range between 30 and 1000 m in length; rarely they may be 20 to 30 m in length. In height they vary from 1.5 to 15 m. The ripple index (L/H) is usually above 30, sometimes attaining values up to 100. The index of giant ripples tends to increase with increasing length of ripples. Thus, smaller giant ripples show a higher ripple index. Figure 55 shows a giant ripple. Giant ripples are usually superimposed by megaripples (Fig. 56).

Dalrymple et al. (1978) in their study of bedforms in the Bay of Fundy, Canada, consider sand waves to be the largest scale bedforms with average heights and wave lengths more than those of megaripples. This corresponds to the giant ripples of our terminology. These giant ripples in the Bay of Fundy are straight to sinuous in form, the lee-face inclination is only 10–20° due to tidal reworking, length ranges from 10–215 m, the L/H ratio is usually more than 20, and height and length show good correlation. They always bear active megaripples on the stosside. Dalrymple et al. (1978) point out that megaripples (sinuous and lunate type = dunes) and giant ripples (sand waves) occur under the same water depth and current velocity conditions: the giant ripples in sand coarser than 0.3 mm, the megaripples in somewhat finer sand. Thus, the megaripples (sinuous and lunate type) and sand waves (giant ripples) seem to be distinct bedforms developed under similar hydrodynamic conditions, separated and controlled by the grain size.

In the study of bedforms of the Wabash River, Jackson (1976b) observed that the sand waves (giant ripples) are mostly 20–60 m in length, always with superimposed megaripples (dunes). The giant ripples occur in coarser sediment and higher water depths than the megaripples; however, the stability fields of giant ripples and megaripples overlap (Jackson 1976b).

Because of their large size, they are known only from the relatively deeper waters of shallow seas and larger rivers. They have never been recorded on intertidal flats. In existing literature they are usually grouped with megaripples.

Quite often, in the literature, giant ripples are referred to as dunes or sand waves. Sometimes the bedform sand waves, observed in the flumes and on the intertidal flats are also grouped together with giant ripples (see Southard 1975). However, we feel that giant ripples are different from the sandwaves of the intertidal flats and flume experiments. In the North Sea, giant ripples occur only in the deeper waters, and never on the intertidal flats. Within a channel, giant ripples die out laterally if the water depth decreases. Studies of Jackson

(1976b) and Dalrymple et al. (1978) also indicate that a minimum water depth of about 4 m is necessary for the development of giant ripples, which are then controlled mainly by the grain size.

Giant ripples are present in the Baltic Sea, generated by wind–driven aperiodic currents (Werner and Newton 1970; Werner et al. 1974; Werner and Newton 1975). These giant ripples are mostly 40–70 m in length, 1–2 m in height, developed in medium to coarse sand, in water depths of 12–22 m. The height of these giant ripples decreases with increasing water depth.

Lonsdale et al. (1972) record dunes (megaripples and giant ripples) from the top of a Guyot, produced by accelerated tidal currents of 15 cm/sec velocities in a water depth of about 2,000 m. Lonsdale and Malfait (1974) reports deep-sea megaripples and giant ripples, in water depths of 2.65 km. These ripples occur in transverse trains or

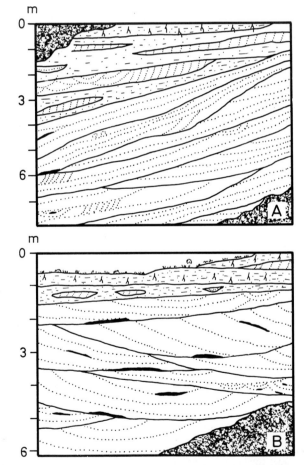

Fig. 57. Internal structure of a giant ripple in two sections, one parallel to the current (A) and one normal to the current (B). The giant ripple is made up of large-scale cross-bedded units of megaripple origin. The units are trough-shaped when seen in a section normal to the current. Intercalated to megaripple bedding are small-ripple bedding, penecontemporaneous deformation structures, scour-and-fill, layers of organic debris, etc. Brahmaputra River. (Modified after Coleman 1969)

Fig. 58. Antidunes on a sandy beach. Note
the gentle relief of antidunes. Sapelo Island,
Georgia, U.S.A.

Fig. 58. Antidunes on a sandy beach. Note
the gentle relief of antidunes. Sapelo Island,
Georgia, U.S.A.

as barchans, showing average length of 20 m,
height about 0.6 m with lee-faces possessing
25–30° dip. These large ripple forms are related to
episodic currents of about 30 cm/sec.

Giant ripples are commonly straight-crested.
However, undulatory and bifurcating forms are also known (Fig. 538). The crests of giant ripples
may be symmetrical or asymmetrical. As mentioned earlier, their internal structure is mostly
form-discordant, composed mainly of megaripple
bedding (Fig. 57) (Reineck 1963a; Coleman 1969).
Figure 537 shows schematically the internal structure of a giant ripple.

Werner et al. (1974) found that giant ripples of
the Baltic Sea mostly show lee-face dipping at
15–20°, often up to 34°. As revealed by the study
of vibro-corer samples, these giant ripples show
form-concordant internal structure. They show a
giant ripple about 2 m high, made up of one-phase
foreset laminae dipping in one direction. It suggests that the leeface is active and serves as an avalanche face. Only the topmost 20 cm of the giant
ripple sediments are made up of small ripple bedding, shell debris, and bioturbation structures produced during inactive periods of giant ripples.

Antidunes

Antidunes are trains of ripple-like bedforms produced by the free-surface flow in or near the supercritical state. They are bedforms which are in phase
with the water surface, and they are produced in
rapid flow conditions ($F > 1$).

Antidunes are more or less long crested, with
generally low relief. Slopes of antidunes are gentle
(Fig. 58). Antidunes may remain stationary, or
may move downward or even upstream when they
break. Thus, antidunes are scoured away, and the
bed becomes flat prior to the building up of a new
train of antidunes.

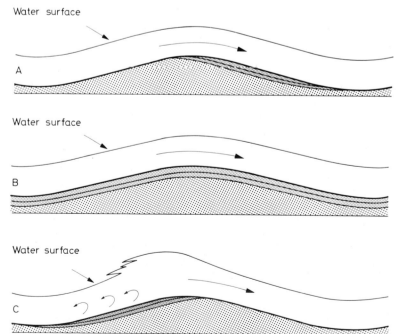

Fig. 59. Scheme showing three modes of deposition in antidunes. A poorly defined low-
angled laminae on the down-stream slope; B
laminae draping over the complete antidune; C low-angled inclined laminae, dipping upstream. Type C is most common and
originates when antidunes move upstream
and break. (Based on information from Kennedy 1961).

As deposition is a continuous process in antidunes, the internal structure is not prominent or
well developed. Deposition in antidunes takes
place in three ways (Kennedy 1961; Allen 1966).
Rather crudely developed laminae either drape
over the crest of the antidunes, or build up on the
downstream side (rare), or form a kind of low-
angle inclined bedding on the upstream face (common). Figure 59 demonstrates these three modes of
deposition on antidunes. The internal structure of
antidunes has been described in detail by Reineck
(1963a), Middleton (1965b), Jopling and Richardson (1966), and Panin and Panin (1967).

Available data shows that antidunes may vary in
length from 1 cm to 6 m, whereas their height
ranges from 1 mm to 45 cm. They are more or less
symmetrical in form.

Hand et al. (1972) discuss that antidunes formed
by turbidity currents possess large wave lengths
(more than 10 m) and the laminations formed by
such antidunes look like horizontal bedding.

Shaw and Kellerhals (1977) note that antidune
phase is common in a fluvial environment; regularly spaced gravel mounds can be interpreted as antidunes, spaced at 2–3 m and can be used in palaeohydraulic reconstructions. Foley (1977) discusses
that gravel can be trapped on the upstream side of
antidunes during breaking of stationary waves.
From the maximum size of the clasts of the gravel,
maximum bed shear stress can be calculated.

Combined Current / Wave Ripples

In shallow-water regions, where both current and
waves are present, ripple patterns under a combined influence are produced. In general, they can
be placed in two groups.

Longitudinal Combined Current / Wave Ripples

These ripples are the longitudinal ripples of van
Straaten (1951). They are straight-crested, usually
showing no bifurcation of crests. Ripple crests run
parallel to the current, and wave propagation is at
right angles to the current (Fig. 60). The genesis of
longitudinal ripples is not very well understood.
However, it seems that wave action is the major
force in producing them; but at the same time a
weak current flowing parallel to the crest is important in controlling their form. The current seems to
cause erosion in the ripple troughs and at the same
time helps to maintain the form of the crests. Some
sediment transport does take place down- current
along the troughs.

These ripples are also found on somewhat muddy bottoms–cohesive sediments. They seem to be

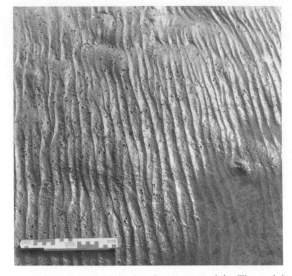

Fig. 60. Longitudinal ripples. Crests are straight. They originate
under the combined action of waves and currents. Ripple crests
run parallel to the current. The direction of wave propagation is
at right angles to it. North Sea tidal flats

mainly erosive forms, i.e., troughs have been
eroded in the sediments, but there is almost no deposition on the crests. In some cases, rather ill-defined occasional laminae are encountered along
the crests (Fig. 61).

The crests of the ripples are usually symmetrical
or slightly asymmetrical in form. The internal
structure of longitudinal ripples seems always to be
form-discordant (cf. van Straaten, 1951). Length
of the ripples may vary between 2.5 and 5 cm. Investigations of Reineck (1974a) also indicate that
longitudinal ripples are erosive forms, mainly due
to wave action, and can originate in sediments with
very low sand content (ca. 40%), an observation
earlier also mentioned by van Straaten (1951).

Fig. 61. Internal structure of longitudinal ripples. They are
mainly erosive in nature, showing form-discordant structure. A
few ill-defined laminae are present near the crest

Transverse Combined Current / Wave Ripples

Here, the direction of the wave motion is parallel to
the axis of the current flow. Quite often such ripple
marks develop near the shore line, where both current and waves are active and are important factors. Harms (1969) produced some of them in
flume experiments. The crests of the ripples run
transverse to the current direction.

When the direction of wave propagation is the same as the current axis, asymmetrical current ripples are produced. However, crests are often more rounded in comparison to the crests of ripples made purely by current action. The internal structure of such ripples can be either form-concordant or form-discordant.

The simultaneous presence of waves may influence the current ripples in several ways. Wave action may help in the formation and modification of current ripples. For example, our observations in the nearshore areas show that with other factors remaining the same, straight-crested small ripples may change over to higher-energy forms (e. g., lingoid ripples) if aided by suitable waves.

Thus wave action may influence the current action to produce another type of current ripple. However, wave action does not seem to take any direct part. Such ripples can never be distinguished from ripples formed purely from current action and should, for all practical purpose, be considered current ripples. But wave ripples may not show their typical bifurcation, if at the same time a weak current is at work.

Another interesting common observation is that if a weak current runs against the direction of wave propagation on intertidal flats, then in some cases, instead of asymmetrical wave ripples, symmetrical wave ripples are formed. The weak current helps to maintain the symmetry of wave ripples. Moreover, it may not be possible to distinguish them from purely wave-formed symmetrical ripples.

Isolated Ripples

Isolated ripples are more commonly referred to as incomplete ripples (Shrock 1948; Reineck 1961), and are sometimes even known as barchan ripples. However, we prefer the term isolated ripples. Allen (1968a) also calls them isolated ripples.

Isolated ripples originate when there is an insufficient supply of sand to cover the entire surface. Currents or waves concentrate this meager supply of sand into incompletely developed ripple crests in the form of disconnected ridges. In the troughs a different older substrate is usually visible (Fig. 62).

Isolated ripples are very similar to "normal ripples"–which are formed where sufficient sediment is available–except that the former are isolated and incompletely developed, and usually acquire lower heights, thus possessing relatively higher ripple indices (L/H). Both current ripples (small, mega, and giant) and wave ripples (symmetrical and asymmetrical) may develop as isolated ripples.

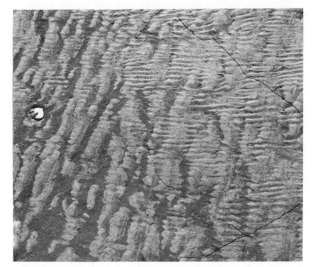

Fig. 62. Isolated current ripples originated on a firm bottom covered with a thin layer of sand. North Sea tidal flats. (Photography by F. Wunderlich)

On firm bottoms covered with a thin layer of sand, wave motion or current action tends to heap the grains or particles into isolated small and symmetrical or slightly asymmetrical lenticular bodies. It may be a case of sand ripples on large pebbly or cobbly bottoms, or even a case of sand on sand differing in grain size or in some other characteristics.

Inman (1957) reports isolated wave ripples in which sand is heaped up in the form of isolated crests on a rocky bottom. On muddy intertidal flats, under certain suitable conditions, isolated small ripples of sand can be observed.

The internal structure of isolated ripples can be form-concordant as well as form-discordant. When isolated ripples are found embedded in muddy sediments, they are designated as lenticular bedding. Sand lenses of lenticular bedding represent isolated ripples (see lenticular and flaser bedding).

Wind Ripples

Wind ripples are produced by the action of wind on noncohesive material. Before we go into the description of wind ripples, we should provide a background of the mechanism of the movement of sand through wind.

Movement of Sand in a Wind Regime

Movement of sand in a wind regime has been dealt with in detail by Bagnold (1954a), who gives a clear insight into the mechanism of sand movement in a wind regime. According to Bagnold (1954a), grain motion of three types is possible: Suspension, saltation, and surface creep. He adds that for the production of ripples and other bedforms in a wind regime, saltation and surface creep are the most important. When a grain flying in the air strikes a bigger grain too large to move, the smaller grain bounces off with very little loss of momentum. If the grain strikes the surface of loose sand consisting of smaller grains, it disperses most of its energy among the grains it strikes, ejecting them into the air. These in turn again strike the surface and either rebound or eject other grains. For any given power of wind, there exists a definite average or characteristic path (Fig. 63). This saltation process, or grains moving through saltation, partly transfers momentum to other grains that are not lifted but are merely pressed forward, so that a continuous surface creep of grains results.

In a sand admixture finer grains move by saltation and coarser ones move chiefly by creep, as they are too big to be lifted.

In their outer form, ripples produced by wind flowing over a dry sand surface may resemble ripples produced under water. In the air, however, the mechanism of sand grain transport is mainly saltation and surface creep, and therefore imparts certain characteristics to wind ripples which are absent or different in water-formed ripples. Bagnold (1954a) and Sharp (1963) give several useful characteristics of wind ripples, which may enable us to distinguish them from water-formed ripples. Sharp (1963) distinguishes two types of wind ripples:

1. wind sand ripples
2. wind granule ripples.

Wind sand ripples are formed of typical wind-blown sand, while wind granule ripples are formed in lag deposits of grains mostly greater than 1 mm in diameter.

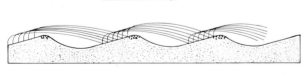

Fig. 63. Relation between characteristic path of saltating grains and sand ripples in wind regime. Coarser grains are concentrated near the crest. (Modified after Bagnold 1954a)

Wind Sand Ripples

Bagnold (1954a) defines sand ripples as wavy surface forms on sandy surfaces whose wavelength depends on the wind strength, and remains con-

stant as time goes on. Ripple wavelengths increase with the wind gradient, but ripples flatten out and disappear when the wind rises above a value of about three times the threshold value for the particular sand.

Wind ripples generally have straight long, parallel crests, and are asymmetrical, with high ripple indices (Fig. 64). They may show well-developed bifurcating of crests. There is a tendency for curved ripples to straighten out until they lie at right angles to the wind direction (Bagnold 1954a).

Fig. 64. Wind sand ripples on a wind-blown sandy surface. (Gripp 1963)

Bagnold (1954a) considers that the spacing between crests is equal to the length of the characteristic path of saltating grains. The latter is in turn a function of wind velocity. Sharp (1963), however, feels that the wavelength of a ripple is strongly influenced by the size of the grains composing a ripple and also by wind velocity, and that no relationship between the characteristic path of saltating grains and ripple wavelength seems to be required. Investigations by Bagnold (1954a) and Sharp (1963) indicate that a lee-side eddy, although present, has no significant influence on the ripple form and the development of wind ripples. The eddy is present when there is no sand transport.

Ripple indices range from 30 to 70. In fine-grained, well-sorted sands only very low ripples are possible. In coarser, poorly sorted sands, wind ripples are steeper, with ripple indices of 10 to 15 (Bagnold 1954a; Sharp 1963). The length of sand ripples in Kelso Sand, California, commonly ranges between 7.5 and 15 cm. The largest measured ripples are 25 cm and the smallest, about

2.5 cm. Heights are mostly between 0.5 and 1.0 cm (Sharp 1963). According to Sharp, the ripple index varies inversely with grain size and directly with the wind velocity. By contrast, the degree of asymmetry of individual sand ripples varies directly with the grain size and inversely with wind velocity. With increase in wind velocity sand ripples flatten out, and at high wind velocities, sand ripples disappear.

Sand ripples develop on surfaces undergoing neither marked erosion nor heavy deposition. Therefore, there may be no erosional or climbing ripples in the wind regime, in contrast to water-formed ripples.

Sharp (1963) studied the internal structure of wind ripples by impregnating them with resins. He writes that wind ripples do not represent an average sample of the immediately underlying material, but rather a concentrate of coarser particles moving mostly by creep through saltation impact.

The coarsest grains are commonly seen to collect on the crests of sand ripples (Cornish 1914; Twenhofel 1932; Bagnold 1954a). Bagnold (1954a) showed that the concentration of the largest grains on ripple crests is due to the fact that they are more difficult to move by the saltation and impact process. On the other hand, in water-formed ripples larger grains are carried away from the crests because they reach upward in higher velocities thus avalanching down the lee slope, and because of gravity, sorting processes are concentrated near the base of lee slopes.

Sharp (1963) expresses the opinion that the bulk of the ripple is made up of sand of essentially uniform coarseness. It is the concentration of fine grains in the trough of the ripples and the enrichment of fine grains as a thin veneer on the lower part of the windward slope that attracts attention more than the distribution of coarser grains.

In any case, in contrast to water-formed ripples, wind-formed ripples have coarser grains in the crest region and finer grains in the troughs. Figure 65 shows the internal structure of wind ripples.

According to Sharp (1963), sand ripples represent asymmetrical piles of relatively coarse, homogeneous, essentially structureless sand resting on a smooth platform of much finer, thinly laminated sand. Occasionally one or two foresets may stand out. Such foresets are unusually rich in fine grains, suggesting that they were laid down during a short interval of reduced wind velocity, when movement of coarser grains was minimal. The bedding produced from migration of wind ripples is horizontally laminated sand. No foreset laminae are present.

Hunter (1977 a,b) demonstrates that migrating wind ripples produce stratification due to net deposition and climb of the following wind ripples on the preceding ones, and names it aeolian climbing ripple structure. A general lack of foreset laminae in wind ripples makes their identification rather difficult. Each layer represents product of deposition of a single wind ripple, which migrated laterally and climb gently. The resulting strata look mostly like low-angle dipping parallel bedding. Hunter (1977b) discusses the genesis of climbing ripple structure due to migration and climbing of wind ripples during tractional deposition (see Fig. 323). He distinguishes four types of these structures:

1. Subcritically climbing translatent stratification – Angle of dip of strata and the depositional surface is low. Laminae are mostly thin with sharp, erosional contacts, mostly showing inverse grading.

2. Supercritically climbing translatent stratification – Angle of dip of strata variable, dip of depositional surface moderate, laminations several millimetres thick with gradational contacts, showing inverse grading.

3. Ripple-foreset cross-lamination – Relative to translatent stratification moderate dip, lamination thin, sharp, or gradational non-erosional contacts.

4. Rippleform lamination – Moderate angle of dip, thin laminations.

Wind Granule Ripples

Granule ripples may be designated as "large wind ripples" on the surface plane of sands, partly composed of granule-sized grains (Fig. 66). Bagnold (1954a) calls them ridges. However, Sharp (1963) prefers to them as ripples. Weir (1962) describes large granule and pebble ripples from the Coyte Lake region, United States.

Granule ripples are usually produced in areas undergoing excessive erosion, where excessive deflation produces a lag concentrate of grains between 1 and 3 mm in diameter, because they are too big to be transported through saltation in existing wind conditions.

Like wind sand ripples, surface creep through saltation is the main mechanism responsible for

Fig. 65. Internal structure of a wind ripple as seen in an X-ray radiograph. Internally, sediments are faintly laminated. Note the concentration of larger sand grains near the crests. The natural surface discovered by red lead powder (in the photo dark)

Fig. 66. Wind granule ripples. They are produced in areas of deflation. Granule-size sediment is concentrated at regular intervals in form of crests. Kelso Dunes, California. (After Sharp 1963)

the formation and growth of granule ripples. As sand ripples and granule ripples are formed under the same wind regime, grain size seems to be the major controlling factor (Sharp 1963).

In the surface plan, granule ripples are much more irregular than sand ripples, with a distinct tendency to produce cuspate forms, particularly in linear chains of crescent-shaped barchan-like ripples.

In cross-section they consist of a crown, with typical ripple asymmetry capping a larger, more symmetrical ridge. The length and inclination of both leeward and windward slopes below the crown tend to be comparable, a relationship attributed to the work of opposing winds. Most of the coarse grains are concentrated in the asymmetrical crown. The crest area is composed of 50 to 80 % coarser grains, whereas in surface material in the troughs large grains make up 10 to 20 % of the sediment.

Granule ripples are much larger than sand ripples. Sharp (1963) reports that in Kelso Dunes, California, granule ripples range in length from 25 cm to 2.3 m, in height from 2.5 to 13 cm, and in ripple index from 15 to 20. Bagnold (1954a) reports granule ripples with a wavelength of up to 20 meters, and heights of over 60 cm in the Libyan Desert.

The wavelength limitation of sand ripples is no more effective and granule ripples may grow to large sizes with time. Very large granule ripples can develop, if there are strong winds. In contrast to sand ripples, at least the upper part of a granule ripple has a well-developed internal structure of well-developed foreset layers (Fig. 67). This layering results primarily from variations in the ratio of average dune sand and granules. These variations reflect changes in the wind velocity. Foreset laminae dipping in opposition to surface symmetry are

also present, and are a result of different wind direction.

Ellwood et al. (1977) studied distribution and genesis of wind sand ripples and wind granule ripples and conclude that transitional forms between the two types also exist. They explain the existence of these transitional forms due to improved rebound of the fine fraction off the coarse fraction in bimodal sand, and suggest that saltation impact mechanism can produce large granule ripples.

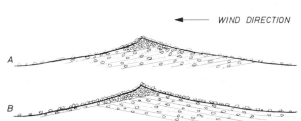

Fig. 67. Internal structure of granule ripples. Large grains are concentrated near the crest as an asymmetrical crown. Internally foreset layers are made up of variable grain size. Oppositely dipping layers are due to changing wind directions. (Modified after Sharp 1963)

Tables for the Identification of Ripples

The foregoing discussion has shown that the genesis and identification of various ripple types is a rather difficult problem. Reineck et al. (1971) have tabulated the existing information on the ripples and some other marine sand bodies. Table 4 (based mainly on Reineck et al. 1971) give the information on various types of ripples; some additional information has been added.

In Table 4 information about ripples of differing genesis has been tabulated, and their characteristic features have been pointed out. In most cases, information overlaps significantly, making the identification of ripples difficult. The data on the sizes indicates only the lower and upper limits of the size of a particular ripple type known up to the present day. They should in no case be regarded as valuves for definition. These values may change or become better established as more data from newer investigations becomes available.

Distribution of Ripples in Various Depositional Environments

Genesis of ripples is mainly controlled by the hydraulic conditions which dominate in an area, or exceptional hydraulic conditions which occur only

Table 4 a–e. Genetic classification of ripples. (After Reineck, Singh and Wunderlich 1971)

a) Current ripples (transverse)

Name	Nature of crest	Size parameters	Ripple Index L/H	Symmetry	Internal structure
Small-current ripples	Straight Undulatory Lingoid Rhomboid	L = 4–60 cm (11, 13) H = upto 6 cm (11, 13)	>5 (11, 13) Mostly 8–15	Asymmetrical	Form-concordant Form-discordant Climbing
Megacurrent ripples[a]	Straight Undulatory Lunate Lingoid Rhomboid	L = 0.6–30 m (9, 13) H = 0.06–1.5 m (9, 13)	Mostly > 15	Asymmetrical	Form-concordant Form-discordant
Giant-current ripples[b]	Straight Undulatory Bifurcating	L = 30–1 000 m (rarely 20–30 m) H = 1.5–15 m (2, 9, 10)	Mostly > 30 Upto about 100 (2, 9)	Asymmetrical and Symmetrical	Form-concordant Form-discordant (2, 9)
Antidunes	Straight	L = 0.01–6 m (7, 10) H = 0.01–0.45 m (7, 10)		Almost symmetrical	Form-concordant (8) Form-discordant (9)

[a] An active megaripple field can be covered by small ripples.
[b] An active giant ripple field can be covered by megaripples.

b) Wave ripples

Name	Nature of crest	Size parameters	Ripple index L/H	Symmetry	Internal structure
Symmetrical wave ripples	Straight, partly bifurcating	L = 0.5–200 cm[a] (4, 5, 11, 15, 17) H = 0.3–22.5 cm (4, 11)	4–13 Mostly 6–7 (4, 11)	Symmetrical	Form-concordant Form-discordant Climbing
Asymmetrical wave ripples	Straight, partly bifurcating	L = 1.5–105 cm H = 0.3–19.5 cm (4, 11)	5–16 Mostly 6–8 (4, 11)	Asymmetrical R.S.I. = 1.1–3.8 (11)	Form-concordant Form-discordant Climbing

[a] Singh and Wunderlich (1978)

c) Isolated (incomplete) ripples. (Formed on the foreign substratum by sediment paucity)

Name	Nature of crest	Size parameters	Symmetry	Internal structure
Isolated small-current ripples	Like small-current ripples	Like small-current ripples, but lesser in height	Asymmetrical	Form-concordant Form-discordant
Isolated mega-current ripples	Straight (6, 7) Curved (6) Sichel-shaped (10)	Like megacurrent ripples, but lesser in height	Asymmetrical	Form-concordant Form-discordant (6, 10)
Isolated giant-current ripples (16)	Like current ripples	Similar to giant-current ripples	Asymmetrical Symmetrical	Form-discordant
Isolated wave ripples	Straight Curved	Like wave ripples, but lesser in height	Symmetrical Asymmetrical	Form-concordant Form-discordant

d) Combined current / wave ripples

Name	Nature of crest	Size parameters	Symmetry	Internal structure
Longitudinal current/ wave ripples (direction of wave propagation at right angles to current direction	Straight, un- branched crests parallel to the current direction; also known to occur in mud	L = 2.6–5 cm (16)	Symmetrical and asymmetrical	Form-discordant (10, 14)
Transverse current/ wave small ripples (wave direction parallel to current direction)	Mostly rounded crests, transverse to current direction (3)		Asymmetrical	Form-concordant Form-discordant

e) Wind ripples

Name	Nature of crest	Size parameters	Ripple index L/H	Symmetry	Internal structure
Wind sand ripples	Straight, partly bifurcating	L = 2.5–25 cm H = 0.5–1.0 cm (1, 12)	10–70 and more (1, 12)	Asymmetrical	Laminated sand, rarely few foreset laminae. Concentrat- ion of coarse sand near the crest.
Wind granule ripples	Straight, cuspate, barchan-like	L = 25 cm–20 m H = 2.5–60 cm (1, 12)	12–20 (12)	Asymmetrical	Foreset laminae in opposing directions. On the crest enrichment of granules.

Legend: Longitudinal = ripple crest parallel to current direction; transverse = ripple crest at right angles to current direction. L = Ripple length; H = Ripple height.
(1) Bagnold (1954b), (2) Coleman (1969), (3) Harms (1969), (4) Inman (1957), (5) Inman (1958), (6) Newton and Werner (1969), (7) Nordin (1964), (8) Panin and Panin (1967), (9) Reineck (1963a), (10) Reineck and Singh (unpublished), (11) Reineck and Wunder- lich (1968a), (12) Sharp (1963), (13) Simons et al. (1965), (14) van Straaten (1951), (15) Werner (1963), (16) Wunderlich (1969); Singh and Wunderlich (1978).

occasionally, e. g., storms, sheet floods, etc. Other important factors controlling genesis of ripples are sediment-type, and other parameters of minor importance. The dominant hydrodynamic conditions also characterize an environment of deposition. This means that although many types of ripples can be found in several varied depositional environments, their distribution (especially their relative abundance) can give a parameter for the recognition of depositional environment. Table 5 gives information about the relative distribution of various types of ripples in different depositional environments. A rather simple scale of abundance has been selected. Four groups are differentiated: Abundant, common, rare, and absent. This scale gives only a general idea of the relative distribution of ripples. Moreover, some of the broad environments are made up of rather varied subenviron- ments. For example, flood basins of a fluvial environment rarely have well-formed ripples. In the bottom of muddy tidal channels, ripples are absent; however, they are abundant in sandy tidal channels.

For a lake environment wave ripples are given as common; however, this is true only on the shallow, coastal zone of a lake. Such details have been discussed in the individual environments of deposition.

However, it can be said that ripples are most abundant in shallow-water sandy environments, and certain special patterns also seem to be specific to individual environments. For example, the intertidal zone, because of its varied hydrodynamic conditions, wave energy, current energy, and waterdepth changes shows the most varied spectrum of ripple-mark types.

Table 5. Occurrence of ripples in various depositional environments

	Mega-current ripples	Small-current ripples	Wave ripples	Climbing ripple lamination	Rhombo-hedral ripples	Anti-dunes
River	++	++	0	++	0	0
Lake	–	0	+	–	–	–
Lake beach	–	+	++	–	0	0
Lagoon	–	0	+	0[a]	–	–
Intertidal flat	+	++	++	0	0	0
Tidal-channel and inlet	++	++	–	0	–	–
Backshore and foreshore	+	+	++	0	+	+
Upper shoreface	+	+	+	–	–	–

	Mega-current ripples	Small-current ripples	Wave ripples	Climbing ripple lamination	Rhombo-hedral ripples	Anti-dunes
Lower shoreface	–	0	+	–	–	–
Transition zone	–	0	+	–	–	–
Muddy shelf	–	–	0	–	–	–
Sandy shelf	+	+	0	–	–	–
Continental slope and rise	–	0	–	–	–	–
Deep sea	–	–	–	–	–	–
Sandy deep sea	–	+	?	–	–	–
Turbidite	0	+	–	+	–	0
Seamounts	–	+	0	–	–	–

[a] Climbing ripple lamination of only wave ripple origin.

++ abundant, + common, 0 rare, – absent.

Surface Markings and Imprints

Mud and Sand Volcanoes and Other Similar Features

Mud and sand volcanoes are small volcano-like features. They range in diameter from a few centimeters to several meters, and are more or less circular with a well-formed central crater made up of sand and silt ejected from the central vent. They form by springs welling up through quicksand, quickmud (Fig. 68), or slumped material (Gill and Kuenen 1958). Sediment volcanoes are associated with the expulsion of water from sediment as a result of slumping, very rapid sedimentation, or agitation of freshly deposited sediment. These processes cause a change in the packing of the sediment resulting in expulsion of water. Sediment in these rising columns of water becomes mobile, acting as quicksand or quickmud, and moves up to the sediment surface in form of a cone. The sediment columns formed on top of the compaction zone and the adjoinig cones or vents on the sediment surface may be preserved.

Dionne (1973) provides a good review on mud volcanoes and uses the term monroes for mud volcanoes on the tidal flats of cold regions. Neumann-Mahlkau (1976) describes sand volcanoes from a construction site, produced as a result of development of overpressure due to dewatering (Figs. 69 and 70).

Fig. 68. Mud volcano in modern sediment as seen from above. (After Ricci Lucchi 1970)

Fig. 69. Sand volcano development on a construction site. Diameter of the cone is 2 m, and the height 0.15 m. (After Neumann-Mahlkau 1976)

Fig. 70. Sand volcano seen in a cross-section. Note the well-developed vent. (After Neumann-Mahlkau 1976)

Sediment volcanoes are formed in regions undergoing intermittent, exceptionally high rates of sedimentation. They may occur in almost any environment–from deep-sea sediments (flysch sediments) to shallow-water marine or nonmarine sediment.

Pit and mound structures (Shrock 1948) are structures related to the above phenomena, but on a smaller scale. Gas bubbles and water currents moving vertically upward through the sediment produce at the point of emergence a shallow pit or less commonly a blister-like mound with a tiny central pit (cf. Häntzschel 1941). They are a few millimeters to 1 centimeter in diameter (Fig. 71). The pit leads downward to the vertical tube along which the bubble moved, but the tube itself is rarely apparent (Shrock 1948). Pits are generally better preserved as mounds on the lower surface of the overlying layer.

The escaping gas may simply be entrapped air, or gas formed by decomposition of organic matter in underlying sediments: in areas of volcanic activity it may be volcanic gas.

Spring pits (Quirke 1930) are features somewhat larger in size than pit and mound structures, and are produced on sandy beaches by ascending waters (Fig. 72).

A feature worth mentioning at this point is *mudlumps* of the Mississippi River delta (Fig. 453a). Mudlumps are regarded as the near-surface expressions of the diapiric intrusion of older shelf

Fig. 71. Pit and mound structure. A small blister-like mound with a central cone is visible

and prodelta clays into and through overlying sand bar deposits. Such intrusions are associated with reverse faulting and displacement of layers.

They are interpreted to be a result of rapid deposition of thick localized masses of heavier bar sediments directly upon lighter, plastic clay, causing instability which is relieved by diapiric intrusion of the clay. This results in the formation of

Fig. 72. Spring pits in beach sediments. Some mounds show a central pit

jection can be produced by the load of overlying sediments, accumulated gas pressure, or even by hydrostatic pressure. Fractures are generated by shrinkage, shock waves, slumping, etc. Clastic material in fluid form invades these cracks and becomes solidified.

Other types of clastic dykes result from filling of cracks from the surface. Material carried down into the cracks generally shows a considerable range in composition and texture. The walls of such dykes do not show any deformation commonly observed along dykes where material is injected from below. Surface cracks can be of varied origin.

Jenkins (1925) summarizes the genesis of clastic dykes. He states that in every case fissures are formed first and then clastic material is dropped, washed, or pressed into them from above, from below, or from the sides. This action takes place both in open fissures on the surface and in cracks in underwater sediments. The filling from below is asso-

mudlumps (Morgan et al., 1968). Such mudlumps appear in the form of low islands or shoals just offshore, and in general, seem to be associated with sites of maximum sedimentation at the mouths of major distributaries of the Mississippi River.

Clastic Dykes

Many sedimentary formations are cut across by tabular bodies of clastic material at various angles. These dyke-like bodies are composed of extraneous materials that have penetrated the host formations along cracks either from below or above. Clastic dykes are composed of different types of material: sand, gravel, silt, mud, asphalt, bituminous sediment. It seems that under favorable conditions, almost any unconsolidated or easily deformed material could invade a crack and solidify to form a clastic dyke (Shrock 1948).

Genetically, one can distinguish between clastic dykes originating from the injection of material from below and those formed by infilling of cracks by material from above. Clastic material in a liquefied form can be easily injected from below into fractures if there is sufficient pressure. They are related to the phenomena of liquefaction and fluidization (Lowe 1975). Sometimes, clastic dyke sediments show preferred orientation of tabular grains parallel to the wall, layering parallel to the wall with grading or inclined laminae, markings along the layering (Peterson 1968), and may be related to flow in fluidized state. The pressure needed for in-

Fig. 73. Clastic dykes in the tidal sediments of Alaska produced as a result of an earthquake. a Pattern of dykes seen on the exposed surface. b dyke seen in a section cutting across the sediments. (Photograph by E. Reimnitz)

ciated with some kind of pressure. Strauch (1966) reports clastic dykes of Plio-Pleistocene age from Iceland, most of them filled from above. He provides a review on the genesis of fissures and filling up to produce clastic dykes. Dionne and Shilts (1974) describe a vertical dyke, 40 cm in width, 200 cm in height, cutting glacio-lacustrine sand, and filled by till sediments from above. The filling of the dyke occurred when an ice-sheet overrode and fractured the underlying sediments.

Reimnitz and Marshall (1965) report clastic dykes in Recent sediments, produced in the tidal sediments of Alaska after an earthquake (Fig. 73). Oomkens (1966) describes clastic dykes from the Libyan Desert, where shrinkage cracks of the surface are filled mainly by water-saturated sediment from below.

Clastic dykes cannot be assigned to any particular environment. They are known to occur in deep-sea sediments (flysch sediments), shallow-water marine and nonmarine sediments, and even desert (inland sebkhas).

Mud Cracks

Desiccation and compaction of water-saturated muddy sediments produce a system of shrinkage cracks, which, in their typical development, form a network and divide the surface into polygonal areas (Fig. 74).

Mud cracks are straight or variously curved in plan view, and the polygons bounded by them pos-

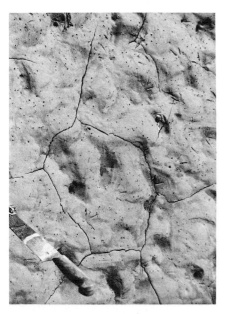

Fig. 74. Mud cracks produced on a muddy surface as a result of desiccation

sess rectilinear or curvilinear sides. The polygons generally have three to six sides sometimes more. The shape of the polygons may be equiradial, much elongated, or some other shape. Differently shaped polygons commonly occur together within a small area of a single surface.

Well-developed mud cracks generally show the development of cracks of several orders–first, second, third, etc. The principal fissures, which are formed first, are rather wide and deep. Secondary and tertiary cracks run across the area demarcated by the principal larger cracks. Thus, on a single surface, cracks and polygons of different magnitudes are present.

In the case of partial shrinkage or desiccation of a sediment surface, incomplete mud cracks are formed (Shrock 1948). They are single, bifurcating or trifurcating cracks marking the initial stages in the development of polygons. Schäfer (1954) demonstrated that the presence of organic traces, such as bird footprints on a muddy surface, may initiate and control the form and direction of mud-cracking.

Karcz and Goldberg (1967) report that in a wadi sediment, ripple marks control the pattern of mud cracks. The cracking starts along the ripple crests and edges to produce well-developed polygonal cracks.

Baldwin (1974) discusses control of mud cracks by groove-like trails of the gastropod Hydrobia ulvae, and Kues and Siemers (1977) discuss that bivalves control the development of mud cracks in supratidal areas.

In a transverse section mud cracks appear as fingerlike extensions of the overlying layer into underlying sediments. They normally possess V shaped profiles, but sometimes can be even parallel-sided.

The polygons of mud cracks are generally a few centimeters across. The cracks themselves are usually a few millimeters in cross-section and a few centimeters in depth. But sometimes cracks may be a few centimeters wide and many centimeters deep. In exceptional cases, cracks may be a few meters in depth. However, cracks are sometimes very fine–only a fraction of a millimeter wide.

Polygonal areas are generally concave upward; however, they may be concave downward. In extreme cases of desiccation the muddy layer becomes curved up and broken into smaller pieces. Such mud plates may be eroded away and rounded during transportation to produce flat discoidal bodies–mud pebbles. However, mud pebbles are abundantly and commonly formed by erosion of the mud layer into pieces which are reshaped into discoidal bodies during transport.

The commonest sites for the development of mud cracks are water-saturated sediment surfaces which are exposed subaerially, such as surfaces of

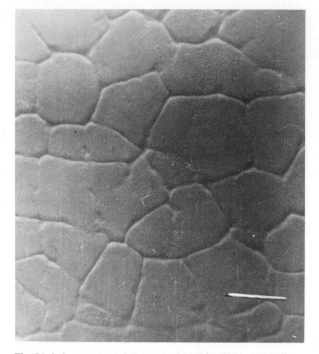

Fig. 75. Subaqueous shrinkage cracks. (After Kuenen 1965)

Such subaqueous shrinkage cracks differ from subaerial desiccation cracks in that they are not so well-developed, the cracks are rather narrow, and they do not possess well-developed V-shapes in transverse sections. In general, subaqueous shrinkage cracks are less regular in form and often incomplete. Sometimes, cracks are developed as open, straight to curved cracks occurring singly or in sets, having a preferred orientation. The cracks are 2–8 cm in length and known as linear-shrinkage cracks. According to Picard and High (1973) linear shrinkage cracks develop when relatively thick, water-saturated thixotropic muds dehydrate, usually under standing water.

Frost and Ice Cracks

Soils and surface sediments in some cold regions are deformed and cracked by frost action, producing V-shaped fissures which can be several meters deep, wide, and long (Lachenbruch 1962; Dylik and Maarleveld 1967). Later, such openings are filled by some foreign material (Fig. 76). In the soils of polar regions, pebbles accumulate along cracks, making polygonal network of pebbles.

However, such features are also known to occur in tropical and subtropical regions (Bremer 1965).

dried-up ponds, coastal and inland sebkha, lakes, and lagoons; abandoned river channels; flood plains; and intertidal zones. In these areas mud cracks are usually associated with other features of subaerial exposure, such as raindrop imprints, hailstone imprints, etc. A combination of such features constitutes one of the best indicators of intermittent subaerial exposure of a sedimentation surface.

Baria (1977) demonstrates that the pattern of desiccation cracks in carbonate muds is strongly controlled by the salinity. The sediments with low salinity show the best-developed desiccation cracks. Hunt and Washburn (1966) discuss the variations and causes of the development of desiccation cracks in Death Valley. It seems that any kind of inhomogenities and flaws in the sediment have a strong control on the pattern of cracking.

Mud cracks (shrinkage cracks) (Fig. 75) can also originate subaqueously as a result of synaeresis (Jüngst 1934). A rapidly flocculated clay layer develops shrinkage cracks due to compaction (White 1961). Similarly, an increase in salinity can also generate shrinkage cracks in the mud layers (Burst 1965). This process can be important in coastal lagoons and inland sebkhas where salinity of water increases markedly during certain periods. Kuenen (1963, 1965) and Dangeard et al. (1964) also produced underwater shrinkage cracks in the laboratory.

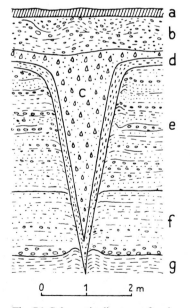

Fig. 76. Schematic diagram of an ice crack. Such V-shaped fissures can be several meters deep. a soil horizon; b upper stratified gravels; c nonstratified gravels filling the crack; d lining layer of the crack; e slightly deformed gravels; f lower well-stratified gravels, sand layer; g underground. (Originally from Peterson 1941.) (After Woldstedt 1961)

Raindrop Imprints

Raindrops falling on a soft sediment surface make small impact craters, which are circular if raindrops fall vertically downward (Fig. 77), and slightly eliptical (Fig. 78) if their path is oblique to the surface (Shrock 1948). The rim of the crater rises slightly above the general surface and is likely to be rough. In this respect they differ from the pits made by bubbles of foam.

Falling waterdrops of spray or splash make impressions similar to those made by raindrops.

If rain falls obliquely on a rippled surface, only one side of the ripples shows imprints. On a plane surface, oblique-falling raindrops sometimes produce imprints where the rim on the backside is broken (Fig. 78). Reineck (1955a) describes them in

Recent als well as ancient sediments (Rotliegendes of Germany). If raindrops fall on a freshly emerged rippled surface, impacts are better made on the relatively drier crests than in the wet troughs (Singh 1969).

Large waterdrops falling from low heights on the same point produce impacts that are deep and possess high rims. Such impacts are reported by Reineck (1956a) where waterdrops fall from a melting iceblock on a muddy surface. However, such impacts are restricted to a narrow zone of a sediment surface.

When there is much rain, raindrop imprints are not so well-developed. Instead of well-developed imprints, the surface is covered with irregular, partly connected depressions (Fig. 79). These depressions are separated by narrow ridges (Wasmund 1930).

Clifton (1977) describes some features named rain-impact ripples. They are produced by oblique impact of wind-driven rain on fine–sand surfaces. They are sinuous, asymmetrical ripple forms spaced about a centimetre apart and 2–3 mm high, and resemble adhesion ripples, but rain-impact ripples possess steeper sides in the down-wind direction.

Raindrop imprints have better chances of preservation in areas receiving only occasional rains. Thus, they have been reported mainly on continental deposits in arid and semi-arid climates.

Hailstones also make similar imprints. As suggested by Lyell (1851), imprints of hailstones are likely to be larger, deeper, more irregular, and have more ragged rims than raindrop imprints.

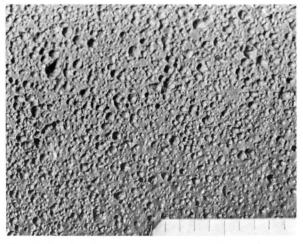

Fig. 77. Raindrop imprints. Imprints are small impact craters, circular in shape, as the rain fell almost vertically

Fig. 78. Raindrop imprints with elliptical impact craters, because of the oblique direction of falling drops. The rim of the impact crater is broken on one side. Laboratory experiment. (After Reineck 1955a)

Fig. 79. Raindrop imprints from heavy rain, leading to distortion of impact craters. Irregular partly connected depressions are separated by narrow ridges. (Photograph by F. Wunderlich)

Foam Impressions

Bubbles of foam leave behind small hemispherical pits with smooth walls without raised rims and with smooth surfaces when they come to rest on a sediment surface. Foam impressions generally occur as clusters of pits, looking like small-pox scars with a wide range of sizes on a small surface (Fig. 80).

Sometimes foam is blown across a sediment surface, leaving behind very shallow, somewhat elongated bubble impressions and tracks (Häntzschel 1935a) (Fig. 81).

Tail-shaped foam impressions (Fig. 82) are produced when a wave pushes foam landward on the beach, and with the backwash of the wave, foam is moved across along the inclined beach surface (Reineck 1955b). Reineck (1954) describes ancient foam impressions in the Rotliegendes of Germany

Fig. 82. Tail-shaped foam impressions made by foam moved by the backwash across the beach. (After Reineck 1955b)

Fig. 80. Foam impressions on a rippled surface. The surface of the impressions is rather smooth. The rims are not well defined

Fig. 83. Foam impressions in ancient sediments. Rotliegendes of Germany. (After Reineck 1954)

Fig. 81. Surface markings produced by foam blown across a sediment surface. Elongated bubble impressions arranged in rows are formed

(Fig. 83). Single bubbles may also leave behind impressions (Figs. 84 and 85) when swept along a sediment surface (Reineck 1954, 1955b).

Bubble impressions are made on sediment surfaces which undergo at least intermittent subaerial exposure. In modern sediments they are most common on sea beaches, lake coasts, and intertidal flats. As these regions suffer repeated reworking, the possibility of preservation of foam impressions is almost negligible.

Fig. 84. A track made by a single moving bubble along the sediment surface. Laboratory experiment. (After Reineck 1954)

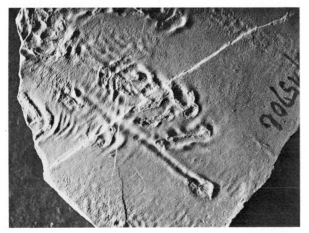

Fig. 85. A track made by a moving single bubble. Rotliegendes of Germany. (After Reineck 1954)

Crystal Imprints and Casts

Under favorable conditions, crystals of various substances such as ice, salt, gypsum, etc., grow on the surface of soft sediments. If these crystals later disappear from processes such as melting, solution, etc., a distinctive system of impressions of crystal forms is left behind (Shrock 1948). When these impressions are filled by sediment, their crystal pseudomorphs are formed. Generally, the chance of preservation of crystal imprints is better in muddy sediments.

Most well known are the imprints of salt crystals and their moulds. The presence of salt crystal impressions is indicative of an environment with increased salinity. On some of the intertidal flats of warm seas they are formed together with gypsum crystals. Illing et al. (1965) reported them from the muddy intertidal flats of the Persian Gulf. They may occur in both marine and nonmarine sediments. Gypsum is formed under both subaerial and subaqueous conditions. In nonmarine evironments gypsum and salt crystals can form in salt lakes–inland sebkhas subject to periodic desiccation. Picard and High (1973) describe gypsum crystals from a depression on point bar of an ephemeral stream.

Ice crystal imprints are formed mainly under subaerial conditions (Fig. 86). In some cases, ice crystals can be formed in subaqueous conditions (unpublished observation). Häntzschel (1935b) describes ice crystal imprints in modern sediments and their ancient analogues. Reineck (1955a) reports them from ancient sediments. Sometimes ice may crystallize in the form of ice flowers, and leave behind ice-flower imprints (Fig. 87) (Reineck 1955c). Snow falling on a muddy surface may also leave impressions of snow crystals (Reineck 1952).

Ice crystal imprints and casts are abundantly formed on the banks of lakes, flood plains of rivers, and intertidal flats in temperate and cold climates.

Fig. 86. Ice crystal imprints. They are developed as elongated needle-like forms, intersecting each other at various angles

Fig. 87. Imprint of ice flowers on a muddy sediment surface. (After Reineck 1955c)

Water Level Marks

At the contact between water level and a sloping sediment surface, a marking is engraved (Fig. 88). With the falling water level, at different levels a series of such markings known as water level marks are formed (Häntzschel 1939). These sloping surfaces may be slopes of channels, longshore bars, or even of ripples.

Water-level marks are good indicators of the sinking water level in an area exposed intermittently subaerially.

Primary Current Lineation

Primary current lineation refers to lineations present on bedding surfaces. Lineation is characterized by ridges alternating with hollows (Fig. 89). Ridges are a few grain diameters in height, a few millimeters to 1 centimeter apart, and 20 to 30 centimeters in length, arranged in an en-echelon fashion (Allen 1968b). In coarser sediments ridges are more widely spaced than in finer sediments. It has

Fig. 89. Primary current lineation. Ridges of a few grains diameter in height alternate with hollows. (After Häntzschel 1939)

Fig. 88. Water-level marks. They are produced as a result of a discontinuously falling water level

been shown that the mean preferred dimensional orientation of the constituent grains is parallel to the elongation of the lineation (Potter and Mast 1963; McBride and Yeakel 1963; and Allen 1964a).

Parting lineation is formed on plane beds, the stoss side of small ripples as well as megaripples. However, parting lineation occurring in combination with plane beds (horizontal bedding) is a good indicator of deposition in the upper flow regime. Primary current lineation is rather common on the beaches, where it is produced by backwash in combination with beach lamination (Häntzschel 1939). Fluvial sands also commonly show primary current lineation in association with horizontally laminated bedding.

Mantz (1978) in a flume experiment with mica flakes found that these flakes are arranged as lineations in a turbulent-smooth flow.

McBride and Yeakel (1963) distinguish two types of parting lineations: (1) parting plane lineation–subparallel linear shallow ridges and grooves, and (2) parting-step lineation–subparallel, step-like ridges where the parting surface cuts across several adjacent laminations (see Pettijohn and Potter 1964).

Wrinkle Marks

Wrinkle marks or Runzelmarken are miniature, irregular ripple-like features. Similar features are also known as Kinneyia ripples in the literature (cf. Martinsson 1965). They are made up of small ridges 0.5 to 1 mm thick and a few millimeters in length. Generally these small ridges run parallel to each other, but occasionally the ridges are curved and make honeycomb-like structures. The sediment surface then looks as though it has been wrinkled (Fig. 90). Wrinkle marks are described by Häntzschel and Reineck (1968).

Reineck (1969) showed experimentally that such wrinkles are produced on sediment surfaces that are partly cohesive. If such a sediment surface is covered by only a very thin film of water (up to 1 cm) and if a strong wind blows over it, the sediment surface develops wrinkles. In general, the thinner the water film present over the sediment surface, the smaller and more closely spaced are the wrinkles made on the sediment surface.

Wrinkle marks can be considered a good indicator of intermittent emergence of a sediment surface.

Structures closely resembling wrinkle marks have been commonly recorded as the sole markings in deep-water flysch sediments. Thus, it seems likely there is some other mode of origin as well active in flysch sediments. Dzulynski and Simpson

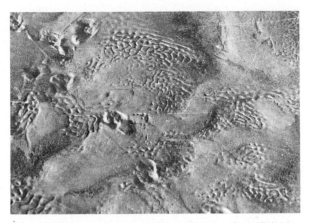

Fig. 90. Wrinkle marks on a muddy sediment surface. Wrinkles are made up of very small ridges (millimeters long) running parallel to each other or irregularly

(1966) discuss the genesis of wrinkle mark-like features in flysch sediments. There, too, these features are produced on a cohesive muddy surface. They result from nonerosive shear exerted on the muddy surface by a turbidity flow.

Millimeter Ripples

Singh and Wunderlich (1978) propose the name millimeter ripples for almost straight-crested ripples with mostly flattened crests, and where troughs and crests are equally broad (2–5 mm) and the height of crests is less than 1 mm. They are only surface forms and do not possess any internal structure. Mostly, ripples are capped off by the last water film during emergence.

Millimeter ripples are known from shallow water environments, e.g., coastal sand, tidal flat, lake coasts etc., and can be used as an indicator of subaerial emergence of sedimentation surface. They differ from wrinkle marks in the regularity of their crests.

Antiripplets (Adhesion Ripples)

Antiripplets are made up of irregular parallel running crests of sand arranged at right angles to the direction of the wind (Fig. 91). The crests are strongly asymmetrical in cross section. The stoss side of antiripplets is steeper than the lee side. Occasionally oversteepened forms are seen. The distance between two adjacent crests varies from 1 mm to about 2 cm or more, while the usual height of the crests is less than 2 mm (van Straaten 1953b;

Fig. 91. Adhesion ripples (antiripplets) on a water-soaked sandy surface. Crests are well-developed and grow upward by capturing wind-blown sand on the water-saturated sandy surface. The stoss side of adhesion ripples is steeper than the lee side. The wind direction is away from observer

Fig. 92. Internal structure of adhesion ripples as seen in a section cut parallel to the crest elongation. The laminae are undulating, stagged one above the other. Knobs correspond to the crest of laminae which are broad convex upward, alternating with very narrow troughs. Laboratory experiment. (After Reineck 1955d)

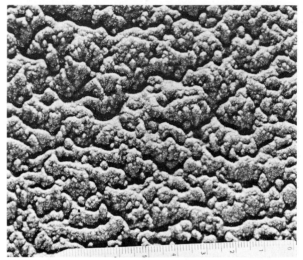

Fig. 93. Adhesion warts are irregular wart-like sand accumulations made by blowing wind on a moist sand surface. Adhesion warts are produced when the wind direction changes rapidly, so that adhesion ripples are not formed. (After Reineck 1955d)

Reineck 1955d). Glennie (1970) calls them adhesion ripples and notes that they may grow 30 to 40 cm in length and a few centimeters in height.

Antiripplets originate when dry sand is blown by wind over a smooth moist surface (van Straaten 1953b; Reineck 1955d). The crests of antiripplets migrate and grow against the direction of wind.

The most important factor in the genesis of antiripplets is the presence of a moist surface. On the moist surface the falling sand grains adhere to the point at which they fall. Because of capillary action of the water new sand grains are also moistened and are capable of receiving new sand grains. Thus, tiny ridges of sand are produced. However, if a water film is present over the sediment surface or if too much sand is transported, antiripplets are not produced.

Under favorable conditions the antiripplets may grow upward rapidly. Internally, antiripplets are composed of undulating laminae stacked one above the other, showing broad, convex upward crests and very narrow troughs (Reineck 1955d) (Fig. 92).

If the direction of the wind changes rapidly, irregular knobby adhesion warts are formed instead of regular antiripplets (Fig. 93). They look like rows of irregular wart-shaped sand accumulations (Reineck 1955d). Hunter (1969) describes an ancient example.

Hunter (1973) discusses that adhesion ripples climbing one above the other are capable of making deposits some few centimetres thick resembling cross-bedding. Such units possess a high potential of preservation.

Bubble Sand Structure and Other Bubble Cavities

It is commonly observed on beaches and steeply sloping sandy intertidal flats that during rapid sedimentation from swash, many air bubbles are entrapped within the sediment. On preservation, these air bubbles impart a spongy texture to the sediment (Fig. 94). The bubbles are a few millimeters in cross section and more or less oval in shape.

Bubble sand is produced on coasts where the surf is not too strong (Emery 1945; Baudoin 1949)

Fig. 94. Bubble sand structure is produced from the entrapment of air bubbles in the beach sand. The sand shows a spongy structure. This is produced when a beach is flooded rather quickly by swash and when wave action is not strong

sible for the deformational processes leading to the genesis of convolute bedding (see p. 88).

Bubble sand structure is a good indicator of deposition on foreshore or similar environments. Birds eye structures (Shinn 1968) in carbonate rocks originate also from the entrapment of air in sediments on intertidal flats.

A feature worth mentioning at this point is gas bubble cavities found in more muddy sediments. Such features have been reported from Rhein delta sediments of the Bodensee (Förstner et al., 1968). Decomposition of organic matter produces gas bubbles that are partly preserved as smaller cavities (Fig. 95). Such gas bubble structures are mainly found in muddy sediments rich in organic matter, in contrast to the bubble sand structure of well-sorted, clean sand in beach environments. Bull (1964) describes bubble cavities and intergranular cavities in the alluvial fan sediments produced as a result of entrapment of air during rapid deposition of muddy sediments.

Deelman (1972) gives a good review of birdseye structures in carbonate and non-carbonate sediments.

and the sediment surface is flooded with water rather rapidly (Reineck 1956b).

De Boer (1979) discusses that entrapment of air in the sand of intertidal zone is related mainly to the fact that the ground water table follows the surface water level fluctuations with a phase lag, which depends upon the permeability and position within the sediment body. During long low water periods, the ground water table sinks, while during rising tide the water level rises at a higher speed and covers the whole sandy surface before the ground water level reaches the sediment surface. Consequently, air is entrapped during flood phase within the upper part of sand bed. Now the water starts to penetrate through the sediment surface downward into the sediment, and the entrapped air is compressed, and the pressurized air expands into bubbles. During falling tide, the cavities are sustained by the friction of the grain fabric and surface tension of the sand–water–air contacts. Investigations of Wunderlich (1979) demonstrate that water moves down from above, the grains with flat surfaces collapse together and air-filled vacuoles are produced. Around the vacuoles, the sand grains are closely packed due to cohesive forces. This causes reduced porosity and entrapment of air. De Boer (1979) also observed this process in laboratory experiment. Sometimes the pressurized air emerges from sediment surfaces as bubbles. On the other hand high content of air cavities in certain layers produces density instability respon-

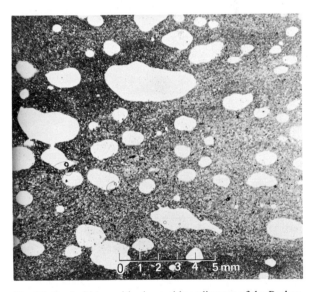

Fig. 95. Gas bubble cavities in muddy sediments of the Bodensee. Gas is produced from the decomposition of organic material and is entrapped within the sediment. Thin section. Rhein delta. Lake Constance. (After Förstner et al., 1968)

Swash Marks

On sandy beaches the lobate fronts of dying waves during backwash leave behind a pattern of tiny imbricating sand ridges. These curved ridges with their convexity landward mark the line of farthest encroachment of the waves. Such ridges of swash marks are never more than 1 or 2 mm in height and are composed mainly of very fine sand grains (Fig. 96). Usually coarser but hydrodynamically lighter material, such as broken shells, wood pieces, seaweeds, etc., accumulates along these curved ridges.

Closely spaced swash marks are formed on steeply inclined beaches, and widely spaced swash marks are seen on gently sloping beaches (Emery and Gale 1951). Generally, fine striations arranged in a centripetal pattern are seen to run from swash mark ridges seaward. Sometimes, instead of centripetal striations, comma-shaped hooks are formed (Fig. 96). These comma-shaped forms of a swash mark always face seaward. (Reineck 1956b).

Swash marks are excellent indicators of the shore line of a water body, which can be a sea, a lake, etc.

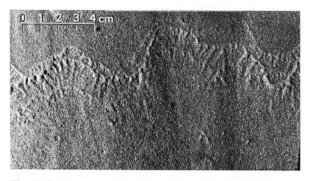

Fig. 96. Swash marks with short comma-shaped striatations on the seaward side of the mark. The sea is toward the lower side. (After Reineck 1956b)

Rill Marks

Rill marks are bifurcating dendritic erosional sculptures made by a system of small rivulets. Rill marks are formed under a flow of a thin layer of water on a sediment surface during the process of sinking water level. They possess a U-shaped cross-section.

Rill marks can acquire diverse forms, which are controlled mainly by the local morphology, slope of the sediment surface, and grain size (cf. Čepek and Reineck 1970). They are formed on slopes, of different angles, but are generally abundant on slopes possessing angles of 2 to 3°.

Several authors have attempted a classification of rill marks (R. Richter 1935; Shrock 1948; Hatai 1960). Recently, Čepek and Reineck (1970) have suggested a classification of rill marks based mainly on their morphology. They distinguish between the following forms:

Tooth-Shaped Rill Marks (Fig. 97). These rill marks develop on sharp, steep edges. Each toothlike projection can be a few millimeters deep and a few millimeters in length.

Comb-Shaped Rill Marks (Fig. 98). Like toothshaped rill marks, comb-shaped rill marks also develop on sharp, steep edges. However, each tooth is several centimeters long, sometimes even up to 45 cm. Occasionally, they show simple bifurcation. The rill between two tongue-like projections can be as broad as 6 cm and up to 2 cm in depth.

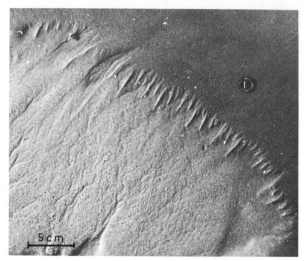

Fig. 97. Tooth-shaped rill marks. A sharp edge showing toothlike projections. The flow is toward the lower left. (After Čepek and Reineck 1970)

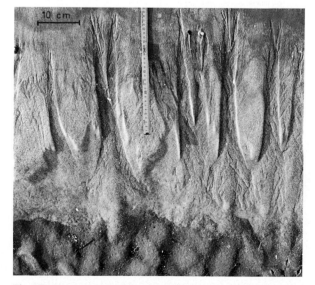

Fig. 98. Comb-shaped rill marks. The flow is from top to bottom. (After Čepek and Reineck 1970)

Fringy Rill Marks (Fig. 99). Fringy rill marks are made up of closely spaced narrow, mostly bifurcating rills, giving a fringy character to the sharp edge. Each rill is rather narrow: 1 to 3 mm broad, up to 4 cm long, and 1 to 6 cm deep.

Conical Rill Marks (Fig. 100). These marks occur in the form of conical depressions whose walls are sculptered by fine rills. Cones can be up to 20 cm across (patterned cones of Boyd and Ore 1963).

Branching Rill Marks (Fig. 101). Branching rill marks are composed of small rill systems which meet together to form a broad main channel. Rills may show extreme bifurcation so as to produce a dendritic pattern. Bifurcation is in the up-current direction.

Meandering Rill Marks (Fig. 102). The rill systems run in a meandering fashion, at the same time showing bifurcation and branching.

Bifurcating Rill Marks (Fig. 103). Rill patterns of these rill marks show strong bifurcation in the down-current direction, and are meandering. The last bifurcations remain open in the down-current direction. They are usually associated with meandering rill marks.

Rill Marks with Accumulation Tongues (Fig. 104). Unlike other rill marks, the sediments eroded here in the process of rill formation accumulate at the end of the rill system in the form of sand tongues which are usually a few millimeters in thickness.

The above-described forms of rill marks are only representative types. Several of them possess transitional forms, and occasionally, different types of rill marks may occur together on the same surface. In general, tooth-shaped, comb-shaped, fringed,

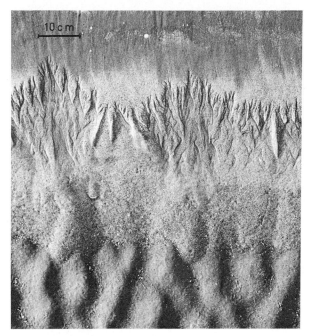

Fig. 99. Fringy rill marks. The edge is fringy in character. The flow is toward the observer. (After Čepek and Reineck 1970)

Fig. 100. Conical rill marks. Walls of conical depressions are sculptered by fine rills. The flow is from top to bottom. (After Čepek and Reineck 1970)

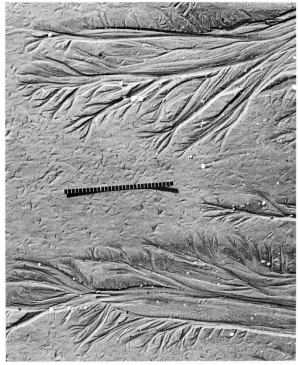

Fig. 101. Branching rill marks. Small rills meet together to form a broad main channel. Rills may show extreme bifurcation. Bifurcation is in an up-current direction. The flow is toward the right. (After Čepek and Reineck 1970)

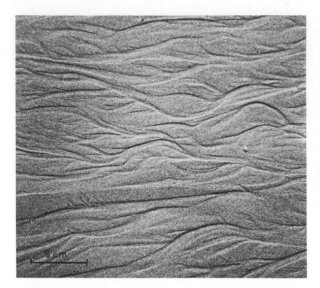

Fig. 102. Meandering rill marks. Rills are meandering and branching. The flow is toward the right. (Čepek and Reineck 1970)

Fig. 104. Rill marks with accumulation tongues. Sediment eroded during rill formation is deposited at the end of the rill system in the form of tongues a few millimeters thick. The flow is away from the observer. (After Čepek and Reineck 1970)

Fig. 103. Bifurcating rill marks. Rills show strong bifurcation in the down-current direction. The flow is toward the observer. (Čepek and Reineck 1970)

and conical rill marks are strongly related genetically and are found on a steep, sharp edge. On the other hand, branching, meandering, bifurcating rill marks, rill marks with accumulation tongues, and partly also conical rill marks are found mainly on gently sloping larger surfaces.

Rill marks originate when a thin film of water flows over a sediment surface. They are generally associated with a change from subaqueous conditions to subaerial conditions. During exposure of the sediment surface by the water's retreat, the water is drained out of the sediment and flows into rills eroded on the sediment surface. Such a change may take place during a falling water level on intertidal flats, swash and backwash of waves on beaches and longshore bars, on river banks and flood plains after a flood period, or from flowing rain water in an otherwise dry terrestrial environment. Sometimes even water expelled from the sediment itself during compaction may be sufficient to produce rill marks. Rill marks are a certain criteria for intermittent subaerial exposure of a surface.

Rill marks are usually generated under a water film only few millimeters thick, but always thinner than 2 cm (Čepek and Reineck 1970). In finer sediments, branching of rills is much better developed than in the coarser sediments.

Bull (1978) regards the dendritic type of rill marks as surge marks, and points out that its morphology depends upon the sediment slope and fine-grained nature of the sediment. They are a product of selective depositional/erosional processes, together with external deformation pressures.

Picard and High (1973) describe various types of rill marks, including dendritic marks from ephemeral stream deposits.

Scour-and-Fill Structures

Scour-and-fill structures resembling small asymmetrical troughs, are generally produced on channel bottoms, with their longer axis running parallel to the current direction.

Under certain conditions water flowing over an unconsolidated sediment surface scours out shallow depressions. These depressions are usually asymmetrical with a steep up-current slope and a gentle down-current slope. In a few cases up-current slopes are more gentle than the down-current slopes. When the current velocity decreases, these depressions are back-filled with somewhat coarser sediments than the substratum (Shrock 1948). Generally, several back-filled scours are made in a row, running parallel to the current direction (Fig. 105). Each scour-and-fill structure is a few centimeters to several meters in dimensions.

During the filling-in process of scours, the sediment generally rolls down along the steep up-current slope, some being deposited from suspension. This produces inclined laminae, which grow flatter in the upper part of the scour fill.

Singh (1977) describes well-developed scour and fill structures from a fluvial sand bar made up of fine sand, and there is no grain-size difference in the scoured-in and filled-in sediments. Picard and

High (1973) report them from ephemeral stream deposits.

Whitaker (1973) proposes the term "Gutter cast " (properly gutter mold) for large-scale scour and fill, grooves, erosional channels etc. where hollows are made on a firm mud bottom and later filled by sandy material. The hollows are generated by water moving along helicoidal paths with horizontal axes.

In other cases, if the current which did the scouring suddenly loses its power and the water becomes rather quiet, the scours are rapidly filled by finer-grained material than the substratum (Shrock 1948). In such a case, laminae are rather ill-defined and more or less conform to the shape of the depressions. Such types of scours and fills are rather common in river environments. Scour-and-fill structures are the most common type of sedimentary structure in coarse-grained alluvial fan deposits and deposits of outwash plains. They are also known as cut-and-fill structures.

Channels

Flowing water over a soft sediment surface may, under certain conditions, erode a channel. These erosional channels may acquire varying forms and magnitudes. Even small channels are several centimeters in cross-section and run for many meters. If channels are somewhat larger in size, they may show meandering characteristics. Lager river channels and larger tidal channels are also erosional channels. However, here we shall consider only the smaller channels, ranging up to few meters in cross-section.

Smaller and larger channels and gullies are important features of an intertidal flat area. They are generally strongly meandering with well-developed point bars. At point bars longitudinal cross bedding is produced (see p. 104). Within the channel itself (in channel fill sediments) a variety of sedimentary structures is found. Small-current ripples are rather common, and the strata may thus show small-ripple bedding. Horizontal lamination may be locally abundant. Parting lineation, groove marks, and other marks may be common. Convolute bedding, slump structures, and load structures are generally found.

The sediment of channel fill is generally of a different nature than the surrounding sediments. Generally, near the base of the channel, coarser material such as mud pebbles, boulders, shells, plant debris, etc., is concentrated as channel lag deposit. Although channel fill sediment is usually sandy in nature, in many cases a channel may be filled up by muddy sediments. In tidal flat environ-

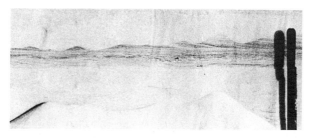

Fig. 105. Scour-and-fill structure on a small scale. Several scours are made in a row and filled very quickly, so that bedding in the filling sand is poorly developed. The direction of flow is from left to right. Point bar deposits of the Gomti River, India

ments some gullies and channels are filled predominantly by muddy sediments.

Smaller channels can be common in fluvial environments or even in deeper water environments (flysch sediments). In fluvial sediments they are common on natural levees as crevasse splays, and in the flood basins. An erosional channel may be symmetrical or asymmetrical in form and may possess U or V profiles.

A detailed study of channels in the carboniferous coal measures of South Wales was made by Bluck and Kelling (1963). McCabe (1977) reports channels up to 40 m deep cutting into underlying sediments from upper Carboniferous of England which are filled by giant cross-beds.

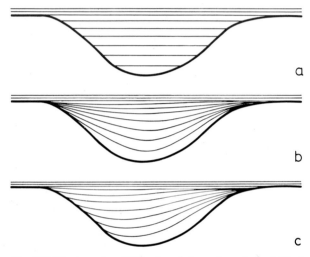

Fig. 106. Three modes of filling up of channels. a channel filled by horizontally layered sediments; b channel filled by layers conforming roughly to the channel shape, and near the top laminae become horizontal; c channel filled asymmetrically by steeply inclined layers. (Based on McKee 1957a)

According to McKee (1957a), channels are produced either by streams in a partly subaerial position or by submerged or submarine currents.

A channel can be filled up in three ways (McKee 1957a):

1. By horizontal layers. This is more common in channels which are not submerged. A stream deposits sediment in a channel bottom because of either increase in sediment load or decrease in stream velocity. The water level is within the channel (Fig. 106). However, we feel that this type of channel filling indicates a rather rapid deposition, leading to quick dumping of sediment load. The channel may or may not be submerged.

2. By layers conforming approximately to the channel shape with concavity upward. Layers may be uniform or equal in thickness in complete cross section of the channel, or they may thin out laterally on the sides. This type of channel filling is found usually in completely submerged channels (Fig. 106).

3. A channel is filled up asymmetrically by steeply inclined layers. It is produced by diagonally passing currents (Fig. 106). In tidal zones when intertidal flats are still under water, and flow of water is controlled more by difference in water levels than by the surface morphology, such diagonal filling of channels is commonly observed.

Generally, the uppermost sediment layers of channel-fill sediments are parallel-bedded and may grade into overlying sediments. In other cases the contact between channel-fill sediments and overlying sediments is an erosional one. An erosional channel may itself show scour and fill structures at its bottom. The difference between a channel and a scour and fill is that in the former, direction of elongation is parallel to the direction of flow, and they are several times longer than the width.

Scour Marks

Scour marks are produced as a result of erosion of a sediment surface by the current flowing over it. The soft but cohesive sediment surface, generally mud, is sculptured and reshaped by the scouring action of current. Erosion and transport is not in the form of individual grains but rather as "chips" or pieces of sediment which are eroded out and transported.

The erosion can be ascribed to the formation of vortices in a turbulent current related to the process of flow separation. The differential erosion of the sediment surface is either by the abrasive action of the larger, sharp grains suspended in the eddies or through the direct action of fluid stresses. Allen (1971) provides a detailed text on the mode of formation of scour marks and their importance.

In the genesis of scour marks there is a transfer of matter from the bed to the flow, and the scours can be arranged either transversely or longitudinally to the flow. Allen (1971) discusses that the scour marks on a cohesive sediment surface can develop either due to fluid stresses on a passive bed, or due to defects in the bed leading to changes in the flow pattern. The character of the scour marks developed on strong muddy beds by corrosion depends on the defects existing in the bed, the duration of the eroding process, and the flow properties, while the character of the structures produced on weakly cohesive mud by fluid stressing depend mainly on the flow properties.

The scour marks are profusely produced on soft muddy surfaces, but are usually better preserved in the form of moulds on the underside of the overlying sandy layers. Because of their common preservation on the lower side of a bed, they are better known in the literature as sole markings. Although in the literature they have been wrongly referred to as casts, we agree with the suggestion of Dzulynski and Walton (1965) to call the sole markings moulds.

It should be noted that, although sole markings are usually found well preserved in flysch deposits, they can be rather common in other environments as well i. e., fluvial, lagoon, tidal flat.

Channels and scour and fill structures are also the result of scouring action, but for practical reasons, they have been described separately (see p. 71).

Recently, Dzulynski and Walton (1965) have given a detailed account of sole markings. They identify scour marks (current scours in terminology of Dzulynski and Walton) as the following different types (we shall consider their obstacle marks under tool marks):

1. Flute marks
2. Transverse scour marks
3. Flute rill marks
4. Longitudinal furrows and ridges
5. Triangular marks tapering down-current
6. Pillow-like scour marks
7. Harrow marks
8. Setulfs

Flute Marks

Flute marks consist of discontinuous, elongated hollows. The upstream ends of the hollows are deep and steep, and they flare out and shoal down-current merging into the sediment surface. They are generally found preserved as flute moulds on the lower surface of a sand layer, where the surface looks as if were covered with discontinuous, bulbous bodies, slightly elongated in form (Fig. 107)

Flute marks show a variety of shapes and sizes. In their deepest part they may range from few millimeters to few centimeters, and in length they are few centimeters to several decimeters.

Based on the shape of the flutes, several classifications have been suggested (e. g., Rücklin 1938; Vassoevic 1953; Ten Haaf 1959). A recent one is by Dzulynski and Walton (1965), who distinguish four major types of flute marks: Linguiform, triangular or conical, elongate-symmetrical, and bulbous. Each of them may be of shallow or deep variety. This is purely a descriptive classification, and all the transitional forms exist.

Flute marks may be symmetrical or asymmetrical in form. Some of them are even terraced. These terraced flute marks are formed where sandy and

Fig. 107. Flute marks preserved as flute moulds on the lower surface of a bed. Flute moulds are arranged parallel to the current. The current is from left to right. (After Dzulynski and Walton 1965)

muddy laminae are differentially eroded (Dzulynski and Walton 1965).

On a surface, flute marks may be isolated or arranged in groups. They are usually arranged in groups in an en-echelon fashion. However, they may also occur arranged in rows–parallel, zigzag, etc. Most workers agree that flute marks, like ripples, are associated at the time of formation with a separation of flow and a vigorously eddying current (Rücklin 1938; Dzulynski and Walton 1965; Allen 1968 b).

Rücklin (1938) produced several types of flute marks experimentally and showed that these structures are a product of erosion on a muddy surface by vortices. He contemplated that horizontal eddies are produced behind some chance obstacle or chance scour of a soft sediment surface. Hopkins (1964) also stresses the importance of nearly horizontal vortices. Fixed eddies are formed on the lee side of the chance scour made on a muddy surface. The fixed eddy would dig out a deep downstream scour. The free eddies would cause some scour near the step, but their effect would gradually taper off downstream. However, as also expressed by Dzulynski and Walton (1965), eddy systems other than horizontal types must also have played some role. Allen (1968 b) discusses in detail flow separation over flute marks (Fig. 108)

Dzulynski and Walton (1965) produced flute marks by experimental turbidity currents flowing over muddy bottoms. It was observed that flutes are produced on a sediment bed by the action of sediment-laden currents (turbidity currents) or sediment-free currents. In all cases the flutes are formed in the strongly turbulent zone near the point of discharge of turbidity currents.

Allen (1971) discusses in detail the development of flute marks from defects in the bed. Allen (1973 b, 1975) demonstrates that even from pairs or trios of defects arranged close together, single flute marks evolve as a result of interaction during growth. There is a reduction in the number density of the flute marks, and with time individual marks grow larger and take up an increasing proportion of the sediment surface. The flute is scoured by a captive eddy rotating about a near-horizontal axis transverse to flow.

Generally, flutes are symmetrical in shape. The oblique or diagonal arrangement of asymmetrical flutes are formed in places where the flow of the current is curved (Dzulynski and Walton 1965). Although found most abundantly in flysch sediments, flute marks are seen abundantly in shallow-water marine as well as in nonmarine environments.

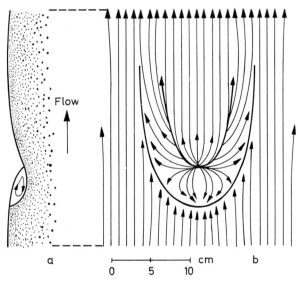

Fig. 108. Flow separation over flute marks, a longitudinal profile of the flute mark and position of the separation stream line; b skin-friction lines on the surface of the mark and on the surrounding flat bed. (Modified after Allen 1968 b)

Transverse Scour Marks

These structures first described by Dzulynski and Sanders (1962) show gradation to normal flute marks. These structures are probably associated with erosion combined with a shearing phenomenon produced by the current moving over a muddy bottom. A series of parallel, rather shallow, and

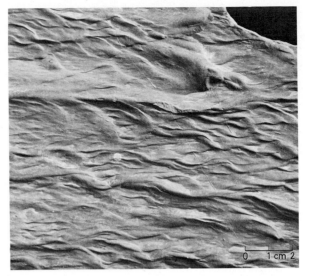

Fig. 109. Moulds of transverse scour marks on the lower surface of a bed. The flow is away from the observer. Superimposed are moulds of various tool marks. (After Dzulynski and Walton 1965)

Fig. 110. Flute rill marks are present as moulds on a lower bedding surface. Rills are sinuous and anastomozing in character. (After Dzulynski and Walton 1965)

asymmetrical scours are formed running normal to the current direction (Fig. 109). In experimental turbidity currents these structures were produced in a zone of slightly slower current flow than that which made "normal" flute marks (Dzulynski and Walton 1965).

Allen (1971) made experimental transverse scour marks (transverse wave-like erosional marks) on weakly cohesive mud surfaces by fluid stressing. Often, superimposed on these transverse scour marks are minor features of small dimensions. These minor features develop due to action on the bed of individual turbulent eddies or structural features in the flow, or failure of bed under stress of the current. These minor features can be equidimensional or elongated, and are formed when fluid stresses are large compared with the bed strength (Allen 1971).

Flute Rill Marks

These marks are called rill marks by Dzulynski and Walton (1965). But in order to differentiate them from ordinary rill marks found in shallow-water environments, we suggest the term flute rill marks. Flute rill marks are made up of continuous narrow, slightly meandering scour structures, elongated parallel with the current (Fig. 110). The trough of the scour is continuous, but is also punctuated at intervals by deeper portions (resembling flute marks) which tend to be broader than the remain-

ing part of the structures (Dzulynski and Walton 1965).

Flute rill marks are rather rare and also differ in form from normal rill marks of shallow-water environments.

Longitudinal Furrows and Ridges

Longitudinal furrows and ridges are structures made up of closely spaced continuous ridges alternating with furrows, arranged parallel to the current (Fig. 111). Regularly spaced ridges are separated by furrows a few millimeters to several centimeters apart. Ridges make the continuous structures. They may continue parallel over long distances, occasionally coalescing. The furrows are rounded in cross-section. They begin upstream with a convex beak and are broken along their length by occasional cuspate bars (Dzulynski and Walton 1965). Locally the ridges may sweep into sinuous curves, or a branching fleur-de-lys arrangement may occur.

Based on their observations in laboratory experiments, Dzulynski and Walton (1965) believe that the formation of parallel ridges is caused by the flow of liquid or suspension in stringers or tube-like bodies. Such stringers are arranged longitudinally in the moving fluid, and in each one fluid moves in the form of two helical spirals possessing an opposite sense of motion. Scouring occurs along the stringers, and the eroded material is piled on the sides in form of longitudinal ridges.

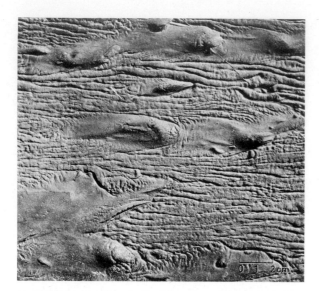

Fig. 111. Moulds of longitudinal furrows and ridges on the lower surface of a bed. They are associated with furrows and ridges and flute moulds. Both structures merge into each other, suggesting contemporaneous formation. (After Dzulynski and Walton 1965)

Dzulynski and Walton (1965) believe that the formation of ridges is to a certain extent–independent of current velocity. But there is a close relationship between the current velocity and the intensity and angle of bifurcation of the ridges. Higher current velocity tends to produce straight parallel ridges, whereas at lower current velocity ridges are more coalescing with rather higher angles of convergence. Furthermore, the flows with low current velocities are easily affected by the floor configuration, and the ridges may curve around minor irregularities.

Triangular Marks Tapering Down-Current

In shape these markings resemble very flat, triangular flute marks, but they are oriented in the opposite direction. Pointed tapering ends point downstream (Fig. 112) (Dzulynski 1965; Dzulynski and Walton 1965).

Pillow-like Scour Marks

Unlike other scour marks, these marks show rather indistinct orientation. At first sight they look more like load structures. Fig. 113 shows pillow-like

scour marks. On the other hand, occurrence of pillow-like scour marks together with flute marks with all possible transition forms, as well as with longitudinal ridges, strongly suggests their current origin due to scouring. Some of these patterns seem to be the result of interference in the current flow (Dzulynski and Walton 1965).

Fig. 112. Triangular markings as moulds on the lower surface of a sandstone. Pointed tapering ends point down-current. (After Dzulynski and Walton 1965)

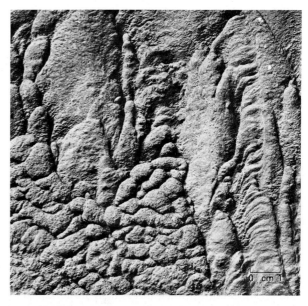

Fig. 113. Pillow-like marks associated with flute marks preserved as moulds. (After Dzulynski and Walton 1965)

Harrow Marks

Harrow marks are current-aligned, fine-grained ridges, and intervening trough-like strips of coarser sediments and commonly develop in ephemeral streams (Karcz 1967, 1972). The patterns are rather regular and continuous and individual markings extend over tens of meters, the ridges are 1–10 cm in height and 3–50 cm in spacing. They are formed by regular systems of longitudinal, helically advancing flow threads with an alternating sense of rotation, related to secondary transversal instability of flow (Karcz 1967).

Setulfs

Friedman and Sanders (1974) proposed this name for positive-relief bedform resembling flutes which are 4–5 cm long, 2–3 cm wide and about 1 cm high; their long axes are parallel to the direction of flow with pointed ends on the upcurrent side. Associated with them are groove-like linear ridges, 5–30 cm long, 5–6 mm wide, and 2–3 mm high running parallel to the direction of flow. These forms are known from the non-cohesive pelletal carbonates of tidal flats, formed by sheet-flow in response to a complex pattern of flow lines in a shallow current.

Tool Marks

Objects lying in the way of a current as obstacles, or being themselves moved by the current, produce markings on the sediment surface. Here, we can also include the markings made by objects lying or coming to rest on a soft sediment surface. All such markings made by objects on a sediment surface are called tool marks. Tool marks can be divided into 3 Types:

1. Stationary tool marks
2. Obstacle marks
3. Moving tool marks.

Stationary Tool Marks

Various objects such as wood, pebbles, mud pebbles, shells, dead animals, etc., coming to rest on a soft sediment surface leave behind their impressions. Such impressions made by temporarily resting passive objects are termed stationary tool marks. Even small ice blocks pushed on the intertidal flats may make impressive markings (Fig. 114) (Reineck 1956a). Larger objects such as large ice blocks may even produce at the same time deformation structures in the underlying layers because of excessive loading (Fig. 115).

In the regions undergoing intermittent subaerial exposure, such as river levees, flood plains, intertidal flats, etc. various floating and moving objects come to rest with the falling water level, so that stationary tool marks are likely to be produced more commonly in environments undergoing intermittent subaerial exposure than in environments always under water.

However, the possibility of preservation of stationary tool marks on surfaces exposed subaerially is very limited, as such surfaces undergo frequent reworking, with the result that surface markings in such areas are more often destroyed. In contrast, the potential of preservation of surface markings in deeper water is rather high. Here, reworking of the sediment surface is much less. In such areas stationary tool marks, once formed, have a good chance of being preserved, as do other surface markings.

Fig. 114. Stationary tool marks of ice blocks on a muddy sediment surface. Ice blocks have melted away, leaving behind impressive depressions. North Sea tidal flats

Obstacle Marks

Obstacles lying in the way of a current tend to deflect the flow lines, which results either in erosion or deposition or both, behind the obstacles.

It is a common observation that small objects such as pebbles, shells, etc., lying in the path of a current tend to accumulate sand on their downcurrent side in the shape of a small ridge which tapers downcurrent. Generally, a semicircular depression also results on the upcurrent side of the obstacle. These features are commonly known as current crescents (Fig. 116). Sengupta (1966) discusses the mode of erosion and deposition around a tool during the formation of current crescents (Fig. 117). Often behind the obstacles, more than one ridge, alternating with grooves tapering downcurrent, is formed.

In other cases, small depressions are eroded out behind the obstacles–longitudinal obstacle scours (Fig. 118) (Rust 1963; Dzulynski and Walton 1965).

Fig. 115. Deformation of sedimentary layers as a result of over-loading by ice blocks. North Sea tidal flats. (Photograph by F. Wunderlich)

Karcz (1968) describes various types of obstacle mark from a fluvial environment which include current crescent and various types of obstacle shadows. They include, (1) poniard-shaped accumulation of sand or silt, aligned parallel to the flow and tapering down-stream, (2) single furrow in the lee of the obstacle, developed only behind triangular obstacles, (3) two furrows of washouts separated by a scour remnant ridge, (4) ridge of sediment in keel-like shape, projecting above the lips of the furrows, similar to "sand tails" made by wind.

The size of obstacle marks is controlled by the size of obstacles, and the geometry of the marks is determined by secondary flow patterns induced within the main flow by the obstacles themselves (Karcz 1968). Werner and Newton (1975) describe

obstacle marks of large dimensions from the Baltic sea, as comet marks (see also p. 378).

On beaches, in deserts, and in other areas where wind is active, sand is deposited behind smaller irregularities making "sand tails". Sand is eroded away between the "sand tails" (Fig. 119).

Moving Tool Marks

Different objects can be moved by the current along the bottom. Various types of markings are produced, depending mainly upon the shape and

Fig. 116. Current crescents–obstacle marks produced in association with shells. The flow is toward the observer. (After Čepek and Reineck 1970)

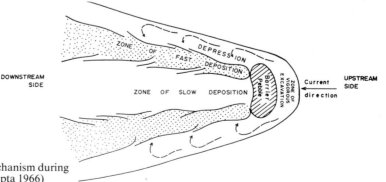

Fig. 117. Pattern of flow and sedimentation mechanism during development of a current crescent. (After Sengupta 1966)

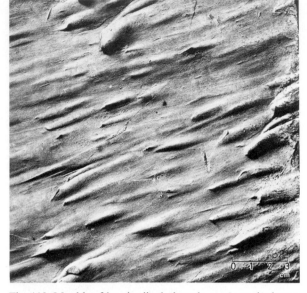

Fig. 118. Moulds of longitudinal obstacle-scour on the lower surface of a bed. Some tools can be seen on the up-current end of a mould. (After Dzulynski and Sanders 1962)

Fig. 119. Sand tails produced by the deposition of wind-blown sand behind small obstacles. In the regions between sand tails deflation is active. (Photograph by K. Gripp)

the size of the moving tools, the mode of transportation, and the nature of the sediment surface. The shape of the markings is generally controlled by the manner in which the tool is moved. i.e., whether by dragging, rolling, or saltation. However, even tools moving in different ways may produce similar markings. Generally, the outline of the tool markings is very sharp. Sometimes the outlines are rilled. Rilling takes place in very water-rich cohesive mud where some flowage

along the margins of tool markings takes place (Dzulynski and Walton 1965).

In ancient sediment, moving tool marks are better known preserved as sole markings on the lower surfaces of sandstones lying over more muddy sediments.

Dzulynski and Sanders (1962), and Dzulynski and Walton (1965) distinguish between continuous and discontinuous moving tool marks. Continuous marks are elongated markings produced by tools being swept continuously along the bottom. They include groove marks and chevron marks. Discontinuous marks are short, distinct marks; single or arranged in sets. They are produced by tools touching the sediment surface at intervals. They include prod marks, bounce marks, brush marks, skip and roll marks. Following are the major types of moving tool marks:

Groove Marks

Groove marks are long, remarkably straight, gutter-like troughs which run for some distance. They range in width from less than 1 mm to tens of centimeters, and can usually be followed for a few centimeters or even a few meters (Dzulynski and Walton 1965).

Grooves are produced by tools carried by the current along a soft bottom (Fig. 120). Both rolling and dragging objects make grooves on a soft sediment surface, but generally it is not possible to distinguish between the two. Sometimes the generating tool is found preserved at the end of the groove. Tools may be shells, pieces of wood, pebbles, mud pebbles, seaweeds, etc. First described by Hall (1843), groove marks have been dealt with in detail by Ten Haaf (1959). Dzulynski and Sanders (1962), Dzulynski and Walton (1965), and others.

Generally, groove marks are found well preserved as sole markings in flysch sediments. On the other hand, they are profusely produced in shallow-water sediments. Especially in areas affected by water-level changes, i.e., intertidal flats, flood plains where even swimming tools have chances to produce groove marks at times, when the water level is so low that a swimming tool just touches the bottom. Thus, groove marks should be equally common in very shallow water regions (especially those effected by water-level changes). The same seems to be true of other moving tool marks.

Sometimes a tool can be moved along a soft bottom by waves. In this case, movement of the tool is punctuated by short time breaks. Each new wave pushes the tool farther, generally with a slight change in direction of movement. In this manner a rather continuous groove with typical knicks is produced. Ice blocks pushed on the tidal flats by

Fig. 120. Scheme showing the development
of a groove mark by a moving tool, a the tool
is preserved at the end of the groove; b the
tool is taken away after the groove is made.
(Modified after Ricci Lucchi 1970)

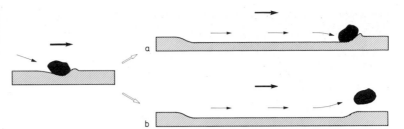

waves are known to produce such grooves (Fig.
121) (Reineck 1956 a).

Dionne (1972 a) describes ribbed grooves and
tracks from the mud tidal flats of Canada, made by
ice blocks scratching the surface when carried by
ebb currents. Harris and Jollymore (1974) report
giant grooves, several kilometers in length, 30 m
wide, and 6 m deep made by iceberg on the continental shelf of Newfoundland (see also p. 416).
Minute grooves are recorded in the ephemeral
stream deposits (Picard and High 1973).

Fig. 121. Groove marks produced by drift ice. The drift ice was
moved by waves, thus producing an irregular, swinging course.
North Sea tidal flats. (After Reineck 1956 c)

Chevron Marks

The typical chevron marks are made up of continuous open V marks arranged to form a straight ridge,
the V forms closing in the down-current direction
(Fig. 122) (Dunbar and Rodgers 1957; Dzulynski
and Walton 1965). They are related to groove
marks.

Dzulynski and Walton (1965) discuss the genesis of chevron marks. They suggest that consistency
of mud plays an important role in the genesis of
chevron marks. If a cohesive film of sediment is
present, this film will be rucked up on the sides of
the grooves to form the ruffled grooves. If the tool
is now lifted from the surface, the eddying effect
behind the moving tool will create a forward suction on the cohesive sediment surface, rucking it into chevrons. They believe that chevron marks are
produced by the tool moving just above the sediment surface, not touching the surface (Fig. 123).
Reversed chevron marks with V forms closing in an
up-current direction are also known.

Prod Marks

Prod marks are asymmetrical, elongated semiconi-
cal to triangular depressions with a shallow pointed upcurrent part and a deeper broad down-
current part (Fig. 124). In the genesis of prod
marks the tool reaches the sediment surface at a
rather high angle and is perhaps halted for a

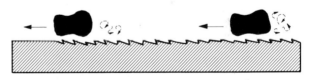

Fig. 123. Scheme showing the genesis of chevron marks. The
tool moved parallel to and just above the sediment surface. The
sediment film was sucked into the chevrons. (Modified after
Dzulynski and Walton 1965)

Fig. 122. Chevron marks preserved as
moulds on the lower surface of a bed. Open
V marks are arranged in the shape of a ridge.
The V closes in the down-current direction.
(After Dzulynski and Walton 1965)

short moment before being lifted upward into the current (Fig. 125) (Dzulynski and Walton 1965).

Dzulynski et al. (1972) report that sometimes the surface of brush and prod marks is sculptured by a delicate pattern of longitudinal dendritic ridges. They are formed in temporarily unstable systems as a result of tool impact.

Fig. 126. Bounce marks preserved as a mould on a lower surface. The mould tapers in both the up- and down-current directions. (After Ricci Lucchi 1970. Photograph by G. Piecentius)

Fig. 124. Prod mark moulds. These marks are asymmetrical, deepening in the down-current direction. (After Dzulynski and Walton 1965)

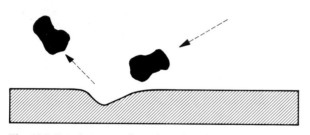

Fig. 125. Development of prod marks. The tool reaches the cohesive sediment surface at a rather high angle, and is then lifted away in the current. (Modified after Dzulynski and Walton 1965 and Ricci Lucchi 1970)

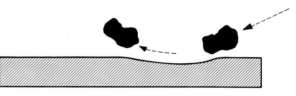

Fig. 127. Scheme showing the genesis of bounce mark. The tool approaches the sediment surface at a rather low angle and immediately bounces back into the current. The groove is more or less a symmetrical depression (Modified after Ricci Lucchi 1970)

Bounce Marks

If a falling tool approaches the sediment surface at a rather low angle and immediately bounces back into the current, the structure produced is a more or less symmetrical depression, tapering and flattening off in both the up-current and down-current directions (Figs. 126 and 127) (Dzulynski and Walton 1965).

Brush Marks

Brush marks are elongated shallow depressions with a small rounded ridge of mud at the down-current end. They are formed when a tool reaches the sediment surface at a very low angle and is again lifted away (the angle of incidence for brush marks is lower than in bounce marks) (Figs. 128 and 129) (Dzulynski and Walton 1965). Dzulynski and Sanders (1962) suggested that in brush marks the axis of the tool may have been inclined up-current.

Fig. 128. Brush mark mould on a lower surface of a bed. On the down-current direction hollow and ridge are present (Dzulynski and Sanders 1962)

Skip and Roll Marks

When a tool makes impact on the sediment surface at regular intervals, a set of markings arranged in a row (skip marks) are formed (Fig. 130). These are

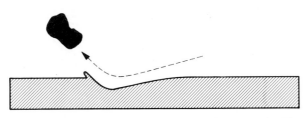

Fig. 129. Scheme showing development of brush marks. The tool approaches the sediment surface at a very low angle, with axis of tool inclined up-current. A rounded ridge of mud is produced at the down-current end of the mark. (Modified after Ricci Lucchi 1970)

produced essentially by tools moving by saltation (Fig. 131).

On the other hand, if a tool, instead of saltating, rolls over a sediment surface, the marking is a continuous track. Such continuous tracks produced by the rolling of a tool are called roll marks (Figs. 132 and 133) (Krejci-Graf 1932; Dzulynski and Walton 1965).

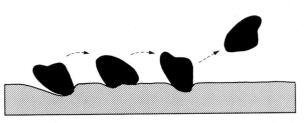

Fig. 131. Scheme showing the genesis of skip marks. The tool moves in saltation, hitting the sediment surface at near-regular intervals. (Modified after Ricci Lucchi 1970)

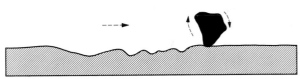

Fig. 132. Scheme showing the genesis of roll marks. The tool is rolling continuously, making a continuous roll mark. (Modified after Ricci Lucchi 1970)

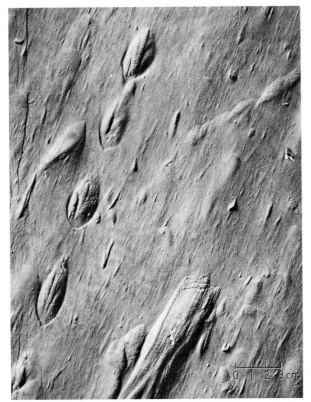

Fig. 130. Skip mark moulds. There are four moulds repeated at regular intervals, besides several smaller prod and bounce moulds. (After Dzulynski and Slaczka 1958)

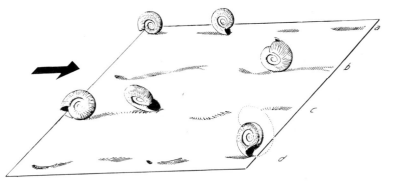

Fig. 133. Scheme showing the roll marks produced by rolling ammonoid shells. (After Seilacher 1963)

Penecontemporaneous Deformation Structures

Penecontemporaneous deformation structures comprise disturbed, distorted, or deformed sedimentary layers produced by inorganic agencies. These features have been formed at the time of or very shortly after depositon of sediment, but in any case, before the consolidation of sediment. Generally, such deformation features are of local character, being primarily confined to a single bed within undeformed beds.

The type of deformation varies from slightly disturbed layers to intricately crumpled, broken, and shifted layers. The nature of deformation structures prduced is mainly controlled by the nature of the sediment. Cohesive sediments can be deformed to a degree so as to show well-developed fracture planes, with little or no bending of layers. In relatively noncohesive sediments deformation is rather continuous. The layers can be intricately bent and reshaped with little or no fracturing (cf. Conybeare and Crook 1968).

Penecontemporaneous deformation results from several processes. An important process is the slumping and sliding downslope under the influence of gravity. Deformation also results from overloading and unequal loading by sediment or even ice. Such bedding can produce both lateral and vertical movements. Even frictional drag, e. g., skin-friction of currents of overflowing fluid or semi-solid material or water escape may produce deformation structures.

Anketell et al. (1970) discuss the importance of liquefaction in a reversed density gradient system in the genesis of deformation structures and produce a variety of deformation structures depending upon non-mobile or horizontally mobile systems. They point out that genetically diverse deformations may produce morphologically similar structures. Depending upon the difference in the kinematic viscosity of different density layers in a reverse density non–mobile system (a, b), a variety of deformation structures can be produced (Fig. 135). In mobile (b, a) system other patterns are produced (Fig. 136).

Penecontemporaneous deformation structures can be grouped into the following types:

1. Load structures
2. Ball-and-pillow structures
3. Convolute bedding
4. Dish structure
5. Slump structures (including contorted bedding)

Lowe (1975) groups several penecontemporaneous deformation structures resulting from the process of liquefaction and fluidization as water-escape structure. Van Loon and Wiggers (1975a, 1976c) use the term metasedimentary structures instead of penecontemporaneous deformation structure.

Load Structures

Load structures are sole markings generally preserved on the lower side of the sand layer overlying the mud layer. On the surface, they appear as swellings, varying in shape from slight bulges, to deep or

Fig. 134. Load structures preserved as moulds on the lower surface of a bed. (After Pettijohn and Potter 1964)

shallow rounded knobby bodies, to highly irregular protuberances (Fig. 134). Generally, bulges of load structures vary in size from a few millimeters to several decimeters (Potter and Pettijohn 1963). They are distinguished from flute marks by their greater irregularity and absence of distinct up- and down-current ends. The underlying mud layer is distorted and bent downward beneath the load protuberances. Some interesting papers on various aspects of load structures are by Kelling and Walton (1957), Plessmann (1961), and Einsele (1963).

Related to load structures are flame structures, which are best recognized in sections cut perpendicular to the bedding surface. Flame structures show curved, pointed tongues of mud projecting upward into an overlying sand layer. Because of unequal loading and liquefaction, the mud layer

has moved up in the form of tongues into the overlying sand layer.

Load structures are the result of deposition of sand over a hydroplastic mud layer. This overloading or unequal loading is adjusted by mainly vertical movements leading to the sinking of the sand layer in the form of lobes, or even pushing upward of mud layer in the form of tongues (Fig. 137). This view has been also expressed by Kuenen and Prentice (1957).

Load structures are also produced as a result of differential deposition, as in the case of ripples. Ripple crests tend to sink down in the underlying soft muddy layer. Piling up of ripples leading to increased sinking produces "load-casted" ripples (Figs. 138 and 139) (Dzulynski and Kotlarczyk 1962). Dzulynski and Walton (1965) believe that

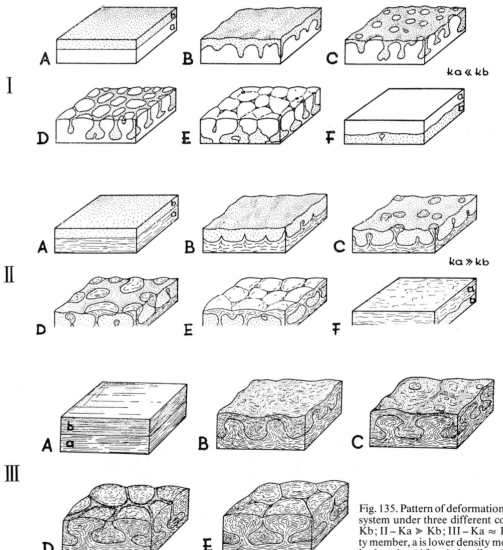

Fig. 135. Pattern of deformation in non-mobile (a, b) system under three different conditions: I – Ka ≪ Kb; II – Ka ≫ Kb; III – Ka ≈ Kb. b is higher density member, a is lower density member, K is kinematic viscosity. (After Anketell et al. 1970)

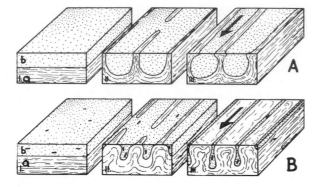

Fig. 136. Pattern of deformation in mobile system (b, a). A – Ka ≫ Kb; B – Ka ≪ Kb. (After Anketell et al. 1970)

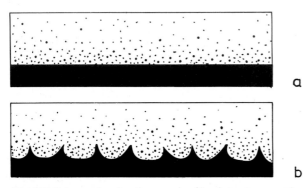

Fig. 137. Scheme showing the genesis of load structures on the lower surface of a sand layer resting on a mud layer (black)

mainly three processes lead to unequal loading producing load structures:

1. Filling of scour marks
2. Differential deposition, as in ripples
3. Formation of a wavy interface by the passage of shock waves.

Sometimes loading results not only in vertical movement, but also in lateral movement as a result of flowing. In this case load structures are modified and sometimes even look like flute marks. Some of the sole markings can be modified as a result of flowage producing new patterns, as in the case of frondescent structures (Dzulynski and Walton 1965). Anketell et al. (1970) provide experimental information on the production of load structures.

Load structures are not restricted to any environment. The only requirement for their genesis is the deposition of a sand layer over a hydroplastic mud layer, leading to unequal loading, resulting in mainly vertical adjustments at the sand-mud interface. They are generally known well preserved in the flysch deposits. However, they are occasionally found in shallow-water environments, especially in areas normally having a rapid mud sedimentation, interrupted by occasional sand deposition. They are known to occur commonly in the channels of muddy intertidal flats. They have been reported from fluvial environments. Coleman (1969) recorded penecontemporaneous deformation structures in the channel bar sediments of the Brahmaputra River, produced from the differential loading. Van Loon and Wiggers (1976c) describe load structures from a Holocene lagoonal deposit.

Ball-and-Pillow Structure

The ball-and-pillow structure is exhibited by sand layers lying above a muddy layer. The sand layer is broken up into several pillow-shaped, more or less ellipsoidal masses (Fig. 140). In size these bodies range from a few centimeters to several meters. These pillows may be slightly connected, or sometimes even completely isolated, floating freely in a

Fig. 138. Load structures produced by piled load-casted ripples. The longitudinal section across the ripples shows that the ripples have been piled up and completely rotated. (After Dzulynski and Kotlarczyk 1962)

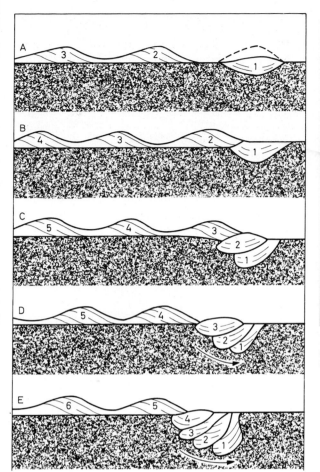

Fig. 139. Scheme showing the development of load-casted ripples in five stages. This structure is probably produced as a result of the tendency of ripple crests to sink down into the underlying soft mud layer. (Modified after Dzulynski and Kotlarczyk 1962).

Fig. 140. Ball-and-pillow structure in a sandstone layer. The sand layer is broken up into isolated pillows. (After Howard and Lohr-Engel 1969)

muddy matrix. Pillows are generally better developed in the lower part of the sand layer, grading upward into a more or less undisturbed sand layer. The underlying mud layer is usually involved in the deformation. It is partly broken and extends upward into the sand layer, between the pillows, in the form of tongues (Potter and Pettijohn 1963).

The pillows themselves may be structureless or show laminated structure. In the latter case, the laminae are generally curved or deformed.

Kuenen (1968) produced the structure experimentally. A layer of sand was deposited on a clay layer. A shock applied to the deposit caused the sand layer to break up into the saucer- and kidney-shaped segments. These semi-isolated sand bodies sank into the clay layer (Fig. 141). Most of the other workers, based on their observations of modern sediments, also agree that in the genesis of ball-and-pillow structure the sand layer is broken and then sinks down into the underlying yielding mud

layer (Emery 1950; Macar 1951; Kaye and Power 1954). Coleman (1969) described similar structures from the modern sediments of the Brahmaputra River. Van Loon and Wiggers (1976c) report them from a Holocene lagoonal deposit.

Most of the workers argree that the vertical displacement is most prominent in the development of the ball-and-pillow structure. However, a few workers attribute the structure to slumping, with vertical displacement combined with lateral displacement.

Ball-and-pillow structures are not confined to any particular environment. They are known both in shallow-water environments and deeper-water turbidites. Rather, they point to the rapid sedimentation of the unit with which they are associated.

Convolute Bedding

Convolute bedding is a structure showing marked crumpling or complicated folding of the laminae of a rather well-defined sedimentation unit (Kuenen 1953a; Potter and Pettijohn 1963). Thus, laminae of a primary bedding parallel laminae, foreset laminae of a ripple bedding, etc., are thrown into convolutions.

Even though laminae may be intensively folded, they are remarkably continuous. Faulting and slip-

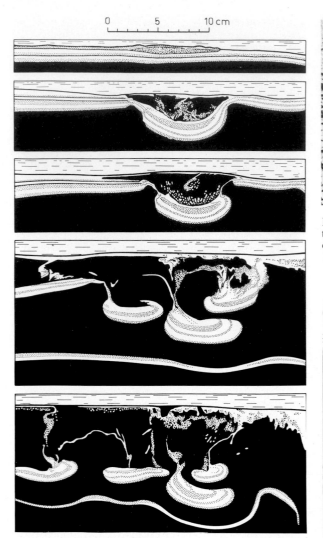

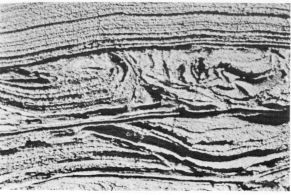

Fig. 142. Convolute bedding as seen in a relief cast. The top of the convolute bedding is capped off before laminated sand is deposited on top of it. North Sea tidal flats. (After Wunderlich 1967a)

Fig. 141. Scheme showing the development of the ball-and-pillow structure in a laboratory experiment. The overlying sand layer on a mud layer sank and broke into saucer-like pillows of sand as a consequence of vibrations from the sand layer and the local burden. (After Kuenen 1965)

Fig. 143. Convolute bedding. The Brahmaputra River. (After Coleman 1969)

ping is normally not associated with convolutions. Generally, convolutions show more or less sharp crests alternating with rather broad troughs (Figs. 142 and 143). Convolute bedding is generally well-developed in fine-grained noncohesive sediment, such as fine sand or silty fine sand.

Several explanations have been proposed for the genesis of convolute bedding. It seems likely that under the action of localized differential forces a hydroplastic sediment layer is thrown into convolutions, producing convolute bedding.

Williams (1960) suggested that convolute bedding is produced by differential liquefaction of a sedimentation unit. Lateral intrastratal flow of these liquified layers produces contortions. Wunderlich (1967a) describes that convolute bedding is

rather a common feature on the steeper slopes of sand bars in tidal environments. He explains that their origin is a result of liquefaction. During subaerial exposure of the sediment surface at low tide, compaction of sediment from the expulsion of water produces local liquefaction of sediment, which results in the development of convolutions.

Liquefaction of a sediment is also achieved by overloading, by seismic waves, or by some other shock, causing a disturbance in the sediment packing. McKee et al. (1962) and McKee and Goldberg (1969) produced convolute bedding in laboratory experiments as a result of differential overloading due to deposition of sand from above. They believe that vertical forces resulting from overloading are important in the generation of convolute bedding.

Some other workers regard convolute bedding as being formed during current activity and sedimentation. Kuenen (1953a) believes that convolute bedding develops from the deformation of ripple marks. A current flowing over ripples causes a vertical suction over the crests and a downward pressure in the troughs. These pressure differences would cause the hydroplastic sediment to yield and thrown into convolutions with continued sedimentation.

According to Sanders (1965), convolute bedding is formed from the shearing action of the current flowing over a sediment layer acting as a cohesive layer (a result of compaction after deposition). The layered sediment is dragged into décollement-like convolutions due to increased shearing by strong currents. Einsele (1963), and Anketell and Dzulynski (1968a,b) also provide some important information on the genesis of convolute bedding. Anketell et al. (1970) emphasize the role of reversed density stratification in the genesis of convolute bedding.

Middleton and Hampton (1973) consider partial liquefaction of very fine to fine sand to be the main cause of genesis of convolute bedding in the turbidites, and liquefaction can take place at several intervals during the deposition of the bed.

Chakrabarti (1977) discusses the genesis of convolute bedding in fluvial sediments, and attempts a mathematical analysis. He relates its origin to an upward flow of sediments related either to the vortex motion present in the turbulent cells on the water surface, or to the effect of vertical expulsion of water or gas caused by sudden impact of loading.

Lowe (1975) regards some of the convolute bedding in Turbidite sequences as water-escape structures, produced during dewatering and consolidation.

Allen (1977) discusses the genesis of convolute bedding in normally graded beds under the driving force of gravity. Convolute bedding is produced after liquefaction, while the deposit is in a liquedized state and involves "passive" fluid. There is disruption of grain fabric by a seismic shock, or increase in pore-water pressure as a result of movement of formation water, or pressure changes due to the passage of internal or surface waves.

Einsele (1963), Wunderlich (1967a), and De Boer (1979) describe convolute bedding from the present–day tidal zone deposits.

De Boer (1979) describes convolute bedding from the intertidal shoals which have been produced due to entrapped air in the bubble sand. Entrapment of air in the sand causes density instability, and a reversed density stratification leading to deformation of the sediments. In such a case deformation process is slow and runs over several tides.

It seems certain that convolute bedding can be produced in more than one way. However, we feel that liquefaction of a sediment is the most important factor in the genesis of convolute bedding.

Convolute bedding was earlier regarded as typical of turbidite sequences. It seems that locally, they can be very abundant in intertidal flats and fluviatile environments. In a fluviatile environment they are commonly found in flood plain deposits and point bar sediments.

McKee (1966a) and Coleman (1969) describe them abundantly from river flood plains. Coleman and Gagliano (1965) found them on natural levees of distributary channels of the Mississippi delta.

Dish Structures

These are distinct sedimentary structures developed in sandy beds parallel to bedding, where they appear as subhorizontal to concave-upward, dark, clay-rich laminae varying in width from a few centimeters to 50 cm (Fig. 150). Each dish becomes sandier upwards. They are commonly separated by vertical streaks of massive sand, named pillars. Within an individual bed dish structures often show an increase in concavity and decrease in width towards the top. Wentworth (1967), Stauffer (1967) reported them from turbidite deposits where they commonly occur in the lower graded horizons of turbidites. Wentworth (1967) related it to antidune under upper flow regime with sporadic water expulsion, while Stauffer (1967) related it to mass flow. Middleton and Hampton (1973) suggested its formation by a continuous process of deposition and formation of faint lamination combined with disruption of lamination by upward loss of water and load casting of the less viscous upper sand layers down into the more viscous, but still partly fluidized layers beneath. Lowe and Lo Piccolo (1974) and Lowe (1975) demonstrate that dish structures are water-escape structures, and represent a penecontemporaneous deformation structure formed during the consolidation and dewatering of quickly deposited sediments which undergo liquefaction and fluidization. Dishes evolving as upward–moving fluidized sediment-water slurries are forced to follow horizontal flow paths beneath semi-permeable laminations to points where continued vertical escape is possible. Fine-grained material is concentrated as laminations, and deformed to make dishes; while, pillars form during forceful, explosive water escape (see also p. 93 and Fig. 150).

Dish structures are reported from proximal continental rise deposits, delta front deposits, and alluvial fan deposits, where periods of rapid deposition alternate with intervals of reduced sedimentation (Lowe and Lo Piccolo 1974). Nilsen et al.

(1977) point out that dish structures are widely distributed also in fluvial, lake, and delta environments.

Slump Structures

Slump structures is a general term which includes all the penecontemporaneous deformation structures resulting from movement and displacement of already deposited sediment layers, mainly under the action of gravity. Different types of irregular contorted bedding also belong to slump structures.

Sometimes even overloading of sediments and dragging by icebergs produces slump structures such as distorted bedding, small-scale faulting, etc. (see p. 79).

Slump structures are generally associated with rapid sedimentation. Such regions may be unstable because of greater slopes, type of sediment deposited, or other reasons. Slumping of a sediment mass may result in the breaking and transportation of sedimentary layers, generally producing a chaotic mixture of different types of sediments, such as a broken mud layer embedded in a sandy mass. In other cases slumping may result in a décollement type of movement, producing distorted or corrugated bedding with little or no displacement (cf. Potter and Pettijohn 1963).

Slump structures are often associated with plastic deformation, brittle fracture, brecciated horizons with rotated and inverted blocks, injection phenomena and small-scale faults. Kennedy and Juignet (1974) discuss the genesis of slump masses in a carbonate bank complex related to shallowing of the area.

Some of the slump structures and slump deposits, especially those associated with flysch deposits, may be very thick and widespread. They may be several meters thick and run for hundreds of meters. Sometimes slump masses may have traveled for kilometers (Kuenen 1949; Ksiakiewicz 1958, and other authors). These masses generally show a variety full of slump structures, such as complicated folding, overturned folds together with thrusting and nappe-like features, broken rolled structures, slump breccias, etc. Kuenen (1965) demonstrated their formation in laboratory experiments (Fig. 144). Large-scale slumps are known from the continental margin of the modern oceans and have been recorded in seismic profiles (for example, Moore et al. 1970; Kelling and Stanley 1970). Large-scale slumps are also described on p. 488.

Intraformational folds can also develop on flat bottom surfaces, in the absence of incipient differ-

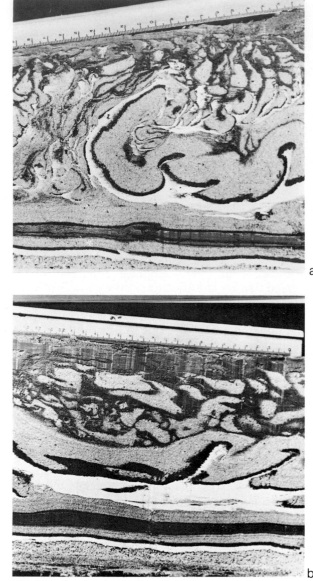

Fig. 144. Slump structures produced in a laboratory experiment. a section parallel to the slope; b section normal to the slope. (After Kuenen 1958)

ential loading, by a unidirectional lateral pressure (Anketell et al. 1970).

Some common slump structures produced under the influence of gravity are found on the oversteepened point bars of gullies of intertidal flats and river channels. Here gravity faults generally develop (Reineck 1958a). Fault planes are rather curved, concave upward. Such gravity faults are better developed in muddy sediments. Figure 145 shows such structures from point bar deposits of a gully on an intertidal flat.

Van Loon and Wiggers (1976c) describe a graben-like block bounded by fault planes and name it gravifossum. Gravifossum is mostly a 50-cm-wide subsided block bordered by two major faults and several small–scale faults. Sometimes within this structure fault breccia is developed due to the

Fig. 145. Slump structures–gravity faults with curved, concave-upward fault planes. Point bar deposits of a gully on intertidal flats of the North Sea. (After Reineck 1958a)

rigid nature of the sediment layers (see Anketell et al. 1970). Some of the layers may show plastic deformation. Small-scale penecontemporaneous deformation structures in the form of synclines and anticlines of a few centimeters to decimeters are developed in the recent sand ridges, and have been ascribed to the action of storm surges (Lindström 1979). They are developed in somewhat coarse-grained sands.

Distorted bedding showing décollement-like folding and small-scale faulting is a widespread feature in glacio-lacustrine and glacio-fluvial sediments (Fig. 146) (Fairbridge 1947; Van Straaten 1949). This may be produced by an overriding mass of ice causing the sediment to slump. More important is the effect of the melting of ice. If ice-masses entrapped in glacial sediments melt the sediment around it slumps. The same effect is achieved if sediment has been deposited on the ice, and from the melting of ice, sediment is transferred to ground, in the process producing deformation structures. Butrym et al. (1964) discuss the genesis of various kinds of "periglacial structures." McDonald and Shilts (1975) discuss the genesis and significance of faults in glaciofluvial sediments. Quite often faults are related to deposition over ice and later melting of the ice mass. Complementary sets of high angle reverse faults that concave upward are related to the ice mass. Some of the faulting is also related to differential compaction and subaqueous slumps. Kowalczyk (1974) describes deformation structures related to ice action (cryoturbation) from Germany.

Allen and Banks (1972) discuss the deformation of cross-bedding, and identify three types (1) simple recumbent folds, (2) a series of folds with or

Fig. 146. Distorted (contorted) bedding showing small-scale folding and faulting in glacial sediments. (After Pettijohn and Potter 1964)

without overturning, and (3) a combination of faulting, folding, and local destruction of bedding (Fig. 147). The first two are common in water-laid sediments while the third is common in aeolian dune sediments.

McKee et al. (1971) provide an account of sub-aerial deformation features produced in sand dunes. They write that avalanching along a lee face takes place, either in the form of a sand flow, in which sand grains move independently of each other, or as slumps, in which a mass of sediment in cohesive form is moved. Observations of avalanches on slip face of modern dunes are made by Gripp (1961) and from Permian dunes reported by Walker and Harms (1972). These processes partly produce contorted primary layers which are various forms of folds, faults, warping, rotation, etc. Structures produced by slumping are rather distinct; those formed by the process of sand flow become progressivly indistinct with the distance of travel. Besides, moisture content plays an important role in the genesis of various types of deformation structures in the aeolian sand.

McKee and Bigarella (1972) report deformational structures in the Brazilian coastal dunes formed due to the wet, cohesive nature of sand. These deformation structures have been further intensified by plant root growths.

Anketell et al. (1970) demonstrate that deformation processes involving a brittle layer and a liquefied layer may lead to the genesis of breccia, similar to those made during sedimentation.

A tepee structure is made up of small or large broken pieces, often saucer-shaped. They occur mostly in the carbonate tidal flat sediments pro-duced during early diagenesis incorporating carbonate precipitation, expansion and dessication, and filling up of the voids by available sediments. Asserto and Kendall (1977) provide a useful summary of tepee structures in Recent and ancient sediments.

Van Loon and Wiggers (1976c) discuss in detail penecontemporaneous deformation structures ("metasedimentary structures") in a Holocene lagoonal deposit, Netherlands. They found that different types of deformation structures occur closely associated. Features produced by plastic deformation, rigid behavior of the sediments and liquefaction occur close together, sometimes within a single layer.

Van Loon and Wiggers (1976c) found that compaction and loading produced depressions, slumping, load structures, and ball and pillow structure. In the extreme case of compaction faults develop due to sudden, short strain. Mostly normal faults develop, but due to lateral forces upthrusts also develop. Fine-grained layers rich in organic matter are more susceptible to plastic deformation.

However, a single layer laterally shows plastic deformation showing folds etc. changing to faults. Similarly, a layer showing plastic deformation changes laterally to show liquefaction and flow structures. According to Van Loon and Wiggers (1976c) this behavior is related to false-body thixotrophy, i. e., a transformation from gel into more fluid gel and back into a gel after only a sluggish movement of the sediments. The silty nature of the sediments and their changing water content are responsible for the development of a wide spectrum of deformation features.

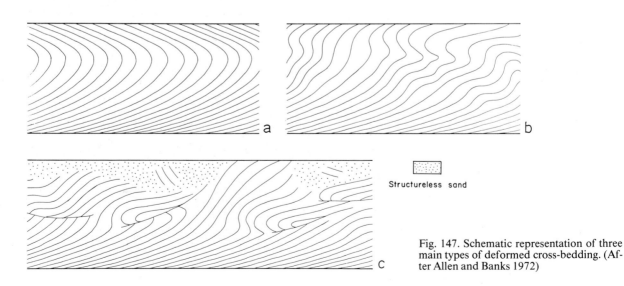

Structureless sand

Fig. 147. Schematic representation of three main types of deformed cross-bedding. (After Allen and Banks 1972)

Water-Escape Structures

Lowe (1975) suggests a general term water-escape structures to include various types of penecontemporaneous deformation structures and slump structures in the sediments. Water-escape structures are post-depositional structures formed in loose, unconsolidated sediments as a result of pore-water escape leading to consolidation of sediment. Water escape causes rearrangement of the grains leading to deformation of the existing laminations or genesis of entirely new structures.

According to Lowe (1975) there are three processes of water escape: (1) seepage – slow upward movement of pore water without disturbing the sediment grains, (2) liquefaction – collapse of a loosely packed framework of sediment grains, the grains becoming suspended in their own fluid and settling. The fluid is displaced upward, and a more tightly packed, grain-supported framework is produced, (3) fluidization – moving pore fluid exerts an upward drag force on the sediment grains, lifting the grains and destroying the grain framework (see Lowe 1976).

Lowe (1975) recognize four general geometric varieties of water–escape structures:

1. Soft Sediment Mixing Bodies – These represent layers of sediment that have been internally rearranged during water escape but have not moved significantly relative to the adjacent sediments. Such bodies can be hydroplastic mixing layers, liquefaction layers and pockets, and fluidization channels and layers. Layers of convolute bedding, layers of deformed cross-bedding, and pillar structures come under this grouping (Fig. 148).

2. Soft Sediment Intrusions – These are formed when hydroplastic, liquefied, or fluidized sediment is mobilized and intruded into adjacent strata. They are clastic sills and dykes. Hydroplastic intrusions are mostly concordant, fluidized intrusions are typically discordant, and liquified intrusions may be concordant or discordant (Fig. 149).

3. Soft Sediment Folds – This includes deformed laminations in the sediments formed in association with differential loading, downslope movement, current drag etc. Convolute bedding, cross–bedding with overturned foresets are included herein.

4. Consolidation – This includes entirely new laminae formed as a result of particle rearrangement by escaping pore water. Dish structures are produced in this manner (Fig. 150).

The genesis of water-escape structures is controlled by grain size, sediment packing, strength and permeability, and hydrodynamic instability. In general, water-escape structures are abundantly

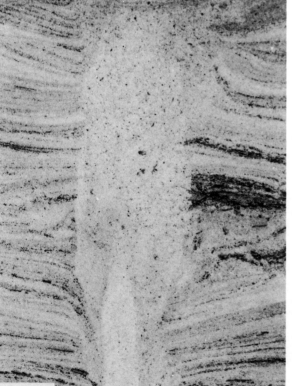

Fig. 148. Flame-like, cylindrical pillar truncating the thinly laminated sand. (After Lowe 1975)

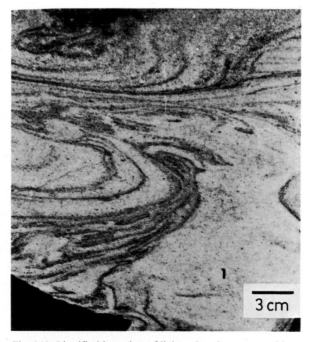

Fig. 149. Liquified intrusion of light-colored coarse sand into deformed laminated sand. (After Lowe 1975).

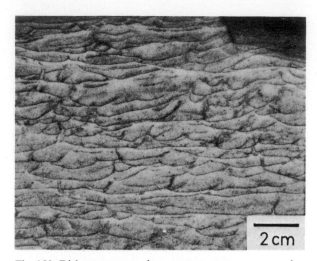

Fig. 150. Dish structures and mer water-escape structures due to complex dewatering history. Horizontal fluidization channels are also seen. (After Lowe 1975)

formed in medium-to-fine sand having high porosity, deposited at quick rates. Areas showing rapid deposition of sand alternating with mud deposition show them abundantly, e. g., deep sea flysch, pro–delta, and some areas of fluvial sedimentation (Lowe 1975). Figure 151 shows the relationship between grain size, water-escape rates, and the resulting structures.

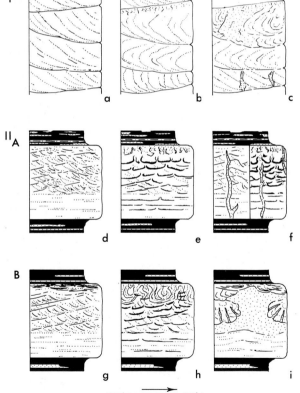

water escape rate

Fig. 151. Common relationships among sediment grain size, strength, primary sedimentary structures, water escape rates, and consolidation structures in natural deposits. *I* Sequence of medium- to coarse-grained cross-stratified sands such as occur in many alluvial and shallow marine deposits: *a* unmodified primary structures; *b* repeated liquefaction coupled with current drag acts to produce oversteepened and recumbent-folded deformed cross-bedding; *c* at higher water discharge rates, local liquefied intrusions and small Type B pillars evolve. *II* Turbidite-type beds showing Bouma B, C, D, and E subdivisions. *A* Bed with noncohesive C and D subdivisions: *d* unmodified primary structures; *e* at discharges just above those characterizing pure seepage, elutriation and redistribution of mobile grains in B subdivision forms flat consolidation laminations and small Type A pillars. Collapse of loosely packed sand in C subdivision results in partial liquefaction and wide development of Type A pillars and strongly curved dishes. Bed surface is fully fluidized forming layer of free surface pillars; *f* at highest discharge rates, structures formed depend on rate of discharge increase. If discharge rises abruptly, large Type B pillars may develop across unmodified primary structures (left); if discharge increases gradually, large Type B pillars can form cross-cutting lower-discharge consolidation structures (right). *B* Bed with cohesive upper C and D subdivisions: *g* unmodified primary structures; *h* similar structures evolve as in *e* at low dewatering rates in non-cohesive layers. Cohesive upper layers resist fluidization and deform hydroplastically into convolute lamination; *i* at highest water escape rates, bed is fully liquefied; large liquefied intrusions and coupled subsidence lobes showing stress pillars develop. (After Lowe 1975)

Bedding

General Information

Bedding is perhaps the most important feature of a sedimentary rock, and is thus the most widely used term in describing a sedimentary sequence. Despite the importance of bedding in sedimentary rocks, no completely satisfactory or agreed-upon classification of bedding yet exists.

A widely accepted definition of a single bed is that by Otto (1938): A single bed is a sedimentation unit which has been deposited under essentially constant physical conditions. Potter and Pettijohn (1963) and Campbell (1967) also agree with this definition. However, in practice it is not always possible to recognize sedimentation units deposited under essentially constant physical conditions. Sometimes beds originating under very different physical conditions may be deceptively alike in appearance. However, generally beds originating under different physical conditions differ considerably in their appearance.

A single bed is separated from adjoining beds by bounding planes – bedding surfaces – commonly known as bedding planes. The bedding surfaces are visible because of some textural or compositional changes from one bed to the other.

McKee and Weir (1953) discuss in detail the problem of bedding and its classification. A very interesting recent paper on this topic is that by Campbell (1967). The following account is based mainly on his paper with certain modifications. As described by Campbell (1967), a bedding plane represents essentially a plane of nondeposition or abrupt change in depositional conditions or an erosional surface, and usually a bedding surface so formed is the depositional surface for the overlying bed. Distinction between beds depends upon recognition of bedding surfaces. Bedding surfaces have no thickness, but have areal extents equivalent to the beds they bound. The geometry of a bed depends upon the relations between bedding surfaces, which may be parallel or nonparallel, and in themselves even, wavy, or curved. Thus, beds may acquire a variety of shapes; most common shapes are tabular, lenticular, wedge-shaped, irregular, or curved-tabular forms. Figure 152 shows different shapes which can be acquired by beds and laminae.

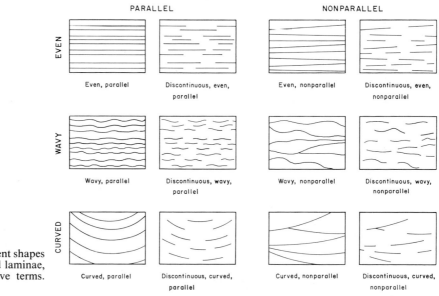

Fig. 152. Diagram showing different shapes that can be acquired by beds and laminae, and the corresponding descriptive terms. (After Campbell 1967)

We agree with Campbell (1967) that beds have no limiting thickness. This contrasts with the definition by McKee and Weir (1953), who define a bed as a stratum thicker than 1 cm. Beds may range in thickness from a few millimeters to tens of meters but they are usually measured in centimeters or multiples of centimeters. Like Campbell (1967), we also prefer the scheme of Ingram (1954) for describing the thickness of beds and laminae. Beds range widely in lateral extent. Usually they can be delineated in only a single outcrop, but sometimes it is possible to follow them for kilometers. A bed terminates laterally by:

1. convergence and intersection of the bedding surface
2. lateral gradation of the material constituting the bed into another in which the bedding surfaces become indistinguishable
3. abutting against an unconformity.

The composition and textures within the bed may be (1) uniform or heterogeneous, (2) rhythmically variable, or (3) systematically gradational (cf. Campbell 1967).

Rhythmic changes occur in beds composed of two or more laminae of different compositions. Systematic gradation is visible in graded bedding. Thus, a bed may be homogeneous internally or may be composed of smaller units–laminae. These laminae within a bed may be parallel or at an angle to the bedding surface. Laminae in repetitive arrangements are similar in texture and composition, as in cross-bedding. Laminae in rhythmic arrangement vary in composition and texture, but form groups of two or more laminae that show some pattern of change. Laminae in gradational arrangement shows gradational variations in composition and texture: Graded rhythmites.

A lamina can be regarded as the smallest megascopic layer in a sedimentary sequence (Campbell 1967). A lamina is bounded above and below by laminar surfaces. They are the same as the bedding surfaces, except for their smaller areal extent and shorter time of formation, because they are contained within beds. Genetically, lamina is a "small bed:" consequently, it is characterized by similar features as a bed, with the following exceptions.

1. A lamina is relatively uniform in composition and texture; however, sometimes some gradational features may be present.
2. A lamina is generally not megascopic internally layered.
3. A lamina has a smaller (or the same) areal extent as enclosing bed.
4. A lamina forms in a shorter period of time than the enclosing bed.

In addition, the thickness of a lamina is usually measured in millimeters, although sometimes they may be several centimeters thick.

We do not agree with Campbell (1967) regarding the use of the term "laminaset." In our terminology a laminaset is essentially a bed. A set of laminae representing a ripple is thus a bed – a "ripple bed" regardless of its scale.

A set is defined by McKee and Weir (1953) as a group of essentially conformable strata or cross-strata (bed and lamina in our terminology) separated from adjoining sedimentary units by surfaces of erosion, nondeposition, or abrupt change in characters–bedding surfaces or laminar surfaces. A *simple bedset (coset)* consists of two or more superimposed beds characterized by similar composition, texture, and internal structure. A bedset is bounded above and below by bedset surfaces and is separated from strata of different natures. A *composite bedset* denotes a group of beds and bedsets differing in composition, texture, or internal structure, but associated genetically, representing a common type of depositional sequence. Depositional sequences on a smaller scale are thus composite bedsets.

An important feature of a bed is its internal structure, which is characterized by conditions of deposition. Based on the internal structure of a bed and its arrangement, several bedding types can be recognized. A bedding type can be recognized and named on the basis of a single bed; or it is only possible when same type of bed is repeated a several times–a bedset; or it is then possible when two or more beds of different natures are repeated in certain sequences. Thus, a bedding type can be made of the same type of beds or of different types of beds (cf.Reineck and Wunderlich 1969).

The term "bedding" in ancient sediments is sometimes misused to describe the way the bed weathers out–parting planes. A bed may be thinly or thickly bedded. Although usually these planes correspond to original bedding planes, sometimes they may be different. Thus, a thinly bedded unit may possess a tendency to develop parting planes only at greater intervals, thus appearing "thickly bedded."

Based on the above discussion we can conclude:

1. A single bed is a sedimentation unit formed under essentially constant physical conditions (in cluding the gliding changes, like waning current required in the formation of graded bedding) and constant delivery of the same material during deposition of a single bed. The thickness of a bed may vary from few millimeters to several meters.

2. A single bed can be internally layered, made up of smaller units-laminae. A lamina is produced as a result of some minor fluctuations in rather constant physical conditions. Generally, a lamina is measured only in millimeters. However, in exceptional cases they can be a few centimeters thick.

3. A bedset consists of two or more superimposed beds of similar or different nature, related

genetically. A bedset is bounded above and below by bedding surfaces which separate them from other bedsets. A bedset can be a simple bedset, consisting of two or more superimposed beds of similar nature; or a composite bedset, consisting of two or more types of beds of differing nature, but related genetically. In a composite bedset different types of beds are repeated in a rhythmic fashion. Some of the bedding types made up of two or more different types of beds is a case of a composite bedset.

We suggest the use of "layer" and "strata" as descriptive terms. The terms "layer" and "layered sediments" can be used when a definite differentiation between bed and lamina is not possible or very difficult. Figure 153 illustrates these terms graphically with examples of various bedding types. In the following, we shall describe and provide genetic insight into the different types of commonly known bedding.

Various types of bedding upon consolidation may show some minor changes in their characteristics. Experimentally determined changes in some bedding structures upon consolidation are shown in Table 6.

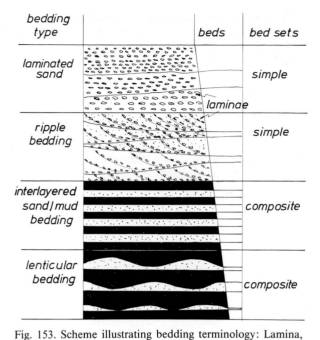

Fig. 153. Scheme illustrating bedding terminology: Lamina, bed, simple bed set, composite bedset, and bedding type

Cross-bedding

Cross-bedding is probably the commonest type of bedding that geologists encounter. Since the last century it has been recognized everywhere and described from sediments of all ages and from every part of the world. Different terms have been used for the same feature.

A cross-bed can be defined as a single layer, or a single sedimentation unit consisting of internal laminae (foreset laminae) inclined to the principal surface of sedimentation. This sedimentation unit is separated from adjacent layers by a surface of erosion, nondeposition, or abrupt change in char-

Table 6. Examples of possible changes in appearance of sedimentary structures induced by laboratory consolidation experiments (Data of Chmelik 1970; Fitzgerald and Bouma 1972)

Before consolidation	After consolidation
1. Parallel lamination	Closely spaced fine laminations. Regular and irregular laminations. Thinning and thickening of irregular laminae
2. Irregular lamination	Thick and thin slightly irregular laminae. Tendency to convolute laminations
3. Vertical laminae and thin beds	Folded, irregular, and often broken laminae with lenticular portions
4. Foreset bedding	Foreset bedding with lower dip of laminae. Undulating lamination
5. Graded bedding in laminae	Thick and thin laminae, often indistinct and grading not visible
6. Convolute lamination	Load casted appearance
7. Slump	Slump with different shape, often flame structures
8. Load casts	Load cast with flame structures. Convolute lenticular appearance
9. Fold	Fold, faults can be introduced
10. Burrows	Burrows, sometimes irregular in shape. Lenses. Strings
11. Cylindrical section	Oval section
12. Mottling	Mottlings. Lenses. Load-casted appearance
13. Mica pockets	Lenses of mica
14. Organic fragments	Lenses of organic matter

acter. This definition is essentially the same as that given by Otto (1938) and later accepted by Potter and Pettijohn (1963). The thickness of a cross-bedded unit may vary from a few millimeters to tens of meters.

We recognize two major types of cross-bedding based on the character of bounding surface of a cross-bedded unit:

1. A cross-bedded unit whose bounding surfaces form more or less planar surfaces is called a planar cross-bedding (Fig. 154). These units are, thus, tabular to wedge-shaped. We feel that normally it is almost impossible to recognize the difference between simple and planar cross-bedding. So for us planar cross-bedding denotes both simple and the planar cross-bedding of McKee and Weir (1953).

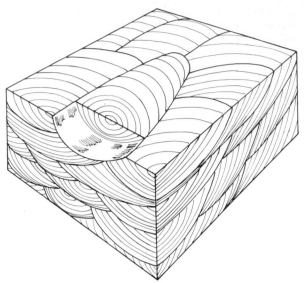

Fig. 155. Block diagram showing trough cross-bedding as seen in horizontal, transverse, and longitudinal sections. Units are festoon-shaped. In transverse section troughs are well developed, with strongly curved bedding surfaces. All transition forms between planar and trough cross-bedding are known. (Modified after Pettijohn 1957)

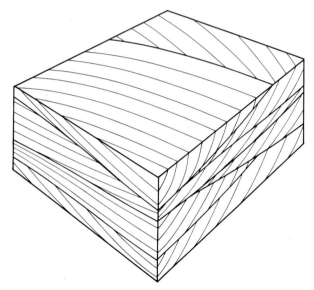

Fig. 154. Block diagram showing planar cross-bedding as seen in horizontal, transverse, and longitudinal sections. Units are tabular to wedge-shaped; bedding surfaces are planar

2. A cross-bedded unit whose bounding surfaces are curved surfaces and the unit are trough-shaped is called trough cross-bedding (Fig. 155).

However, it should be emphasized here that recognition of these two basic forms depends upon the three-dimensional sections available. It should be borne in mind that in certain sections even a trough cross-bedding may look like a planar cross-bedding. Potter and Pettijohn (1963) also recognize these two basic forms.

We use the term "cross-bedding" in a purely descriptive sense as defined by Shrock (1948) to designate structures commonly present in granular sediment and sedimentary rocks, which consist of tabular, irregularly lenticular, or wedge-shaped bodies lying essentially parallel to the general stratification, which themselves show a pro-

nounced laminated structure in which the laminae are inclined to the general bedding.

Today, we know that cross-bedding can originate in several genetically different ways. Recent-investigations have shown that cross-bedding in most of the cases is a result of the migration of small ripples and megaripples. Foreset laminae of a cross-bed are in this case the lee side laminae of the ripples. Micro-deltas also contribute to the formation of cross-bedding, but their role is rather insignificant. In other cases, cross-bedding may be the result of scour and channel-fill features; deposition on the point bars of small meandering channels or on the inclined surfaces of beaches and bars; or lee-side deposition of sand dunes. It is not always possible to recognize the mode of origin of cross-bedding. In Recent sediments or sometimes even in ancient sediments it may be possible to recognize the different types of cross-bedding genetically.

Cross-bedding can be developed on various scales, i. e., individual sets may be only a few millimeters or a few meters thick. Depending upon the thickness of cross-bedded units, one can differentiate.

1. Small-scale Cross-bedding. Individual units are only a few millimeters to a maximum of 4 cm in thickness. Units are usually trough-shaped. They are mostly a result of deposition from migrating small-current and wave ripples. If individual units are only a few millimeters thick, the term "*microscale cross-bedding*" can be used.

2. Large-scale Cross-bedding. Individual units are usually more than 4 cm in thickness, and may be as thick as 1 or 2 m. Units can be planar or trough-shaped. They are of varied origin, e. g., megaripples, sand dunes, longshore bars, etc. Foreset laminae, which constitute a cross-bedded unit, may acquire different shapes from angular through slightly tangential and concave to sigmoidal, depending upon the hydrodynamic factors existing at the time of deposition of units concerned (see Fig. 24). These features and their significance have been discussed in the chapter on ripple marks (see p. 22 ff.).

Two general features are that foreset laminae usually meet the lower basal surface at lower angles and have an upward-facing concavity, and in larger forms lower parts of laminae show considerable enrichment in coarser material. Both of these features may be useful as an up-down criterion (Shrock 1948).

In exceptional cases foreset laminae are seen overturned in their upper parts. This feature is produced by the drag of strong sediment-laden currents across the top of a set of foreset laminae. The moving mass of sand and water drags the top of the foreset laminae into recumbent folds. McKee et al. (1962) produced this structure in flume experiments. They called it "intraformational recumbent folds." This feature has been commonly observed in fluviatile deposits. Frazier and Osanik (1961) report it from point bars of the Mississippi River; Harms et al. (1963) in Red River, Louisiana; McKee (1938) in the flood-plain deposits of the Colorado River, Arizona; Coleman and Gagliano (1965) in the delta front environment of the Mississippi River; Coleman (1969) in the channel bar sediment of the Brahmaputra River.

Allen and Banks (1972) discuss the origin of such recumbent-folded deformed cross-bedding as a result of deformation of a liquefied (or fluidized) sand by current drag following probably an earthquake shock.

Hendry and Stauffer (1975) describe recumbent folds in the cross-bedded Pleistocene sands, where they occur at several levels formed at different times. The direction of overturning is the same as the palaeocurrent direction. They believe that recumbent folds developed due to frictional drag of sediment-laden flow of flood surges, and the process of liquefaction was not important.

Another special form of foreset laminae worth mentioning here is herringbone cross-bedding. This term is applied to cross-bedded units with opposite directions of foreset laminae in adjacent layers (Fig. 156). Sometimes, even if the mean difference in the direction of dip of foreset laminae in adjacent layers is less than 90°, in certain sections they seem to be opposite dipping; festoon bedding seen in diagonal sections may especially look very

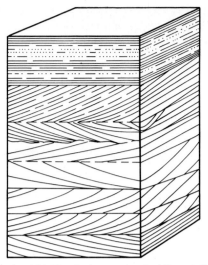

Fig. 156. Herringbone cross-bedding. Adjacent units of this bedding show foreset laminae dipping in opposite directions

much like herringbone cross-bedding. Thus, a true herringbone cross-bedding can be recognized only in 3-dimensional sections. Singh (1969) describes a special case of herringbone cross bedding, where two opposite-dipping cross-bedded units are separated by a thin mud layer. This is typical of tidal environments.

Concave-downward curvatures of the foreset laminae have been recorded in different environme nts (Fig. 157). McKee (1966b) describes them from the parabolic dunes of New Mexico. According to him concavity seems to be best explained by the dune shape: The protuding frontal margin of the dune exposes the slip face and permits occasional cross winds to erode or undermine its base and to oversteepen the lower part, thus producing concavity downward. McKenzie (1964) reports concave downward foresets from the dunes of carbonate sands in Bermuda. He believes that a concave-downward pattern results from early stabilization of the calcareous sand. McKee (1966b) also reports development of concave-downward laminae in laboratory experiments of offshore bars where undercurrents erode the lower parts of foreset laminae. Frasier and Osanik (1961) report concave-downward laminae in point bar deposits. However, cross-bedding with concave-downward laminae are rather rare, and originate under exceptional cases only.

Internally, cross-bedded units often show discontinuities in the pattern and attitude of the foreset laminae commonly termed reactivation surfaces. McCabe and Jones (1977) define it as an inclined surface within a cross-bed set which separates adjacent foresets, with similar orientations, and truncates the lower foreset laminae. Collinson (1970) describes such features in fluvial sediments,

Fig. 157. Cross-bedding with foresets having a concave-downward curvature in a sand dune. White Sands, New Mexico. (After McKee 1966b)

and Boersma (1969), Klein (1970a) discuss them in a tidal environment. Such internal discontinuities in the cross-bedded units are related to random behavior of individual bedforms and its interaction with adjacent flow in rather equilibrium conditions (Allen 1973c). McCabe and Jones (1977) suggest that reactivation surfaces can also be produced in a constant stage where they invariably show concave upward orientation. Thus, reactivation surfaces seem to be related to fluctuations or changes in the flow mechanism or direction and may not be diagnostic of any environment. No doubt they are rather common in tidal flat environments, but are also known from fluvial and aeolian deposits. Figure 158 shows genesis of reactivation surfaces.

Harms (1975) introduced a new term "hummocky cross-stratification" which is characterized by erosional lower bounding surfaces dipping mostly at less than 10°, with laminae almost parallel to the lower bounding surface. The laminae can systematically thicken laterally in a set, and the dip directions of erosional set boundaries and of the overlying laminae are scattered (Fig. 159). Deposition of such units takes place in low hummocks and shallow swales related to increased wave energy. The hummocks are 10 to 50 cm high and spaced one to a few meters apart. Such units are known to occur in the lower part of the ancient shore-line sequences, especially in the fine sandstone of shoreface; Campbell (1966) describes some structures as truncated wave-ripple laminae. His larger forms look rather similar to hummocky cross-stratification.

Hunter (1977b) discusses some aspects of the terminology of cross-bedding, especially when applied to climbing ripple bedforms in general, and

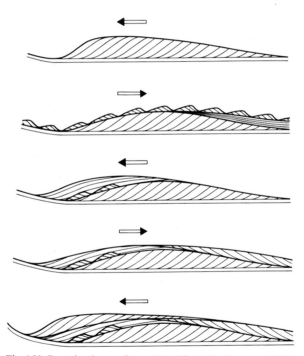

Fig. 158. Reactivation surfaces. (Modified after Boersma 1969; Klein 1970a)

to stratification produced by climbing wind ripples in particular.

He proposes a new name "translatent stratum" for a sedimentation unit, generated by the dominantly translational movement of a depositional surface. Translatent strata can be either laterally translatent or climbing translatent, depending upon the direction of translation. The term climbing translatent stratum is equivalent to the "pseu-

Fig. 159. Hummocky cross-stratification, showing low-angle dipping laminae. (Modified after Harms 1975)

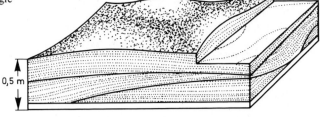

0,5 m

dobed" of the climbing ripple lamination (see Fig. 323).

In the following, the mode of origin of several types of cross bedding will be described and discussed.

planar small-ripple bedding are also not so uncommon (related to straight-crested ripples). In the case of trough-shaped small-ripple bedding, the radius of the trough is usually less than 20 cm (Fig. 162).

Small-ripple Bedding

Cross-bedded units made up of foreset laminae produced from the migration of small-current ripples are called small-ripple bedding (Figs. 160 and 161). Both upper and lower bounding surfaces are planes of erosion. In contrast to climbing-ripple lamination, small-ripple bedding is formed under conditions where relatively less sediment is available and reworking is stronger. Figure 161 demonstrates the generation of small-ripple bedding by migrating small ripples.

In almost any environment with noncohesive sediment where small-current ripple marks are formed, small-ripple bedding is also produced. It is most abundantly developed in sandy intertidal flats, shoals, fluvial sediments–upper point bars, levees–, in deep-sea sediments, where the current is available to produce ripples; and to a lesser degree in lacustrine sediments and fluvio-glacial sediments (see Table 5).

The basal lamina (bottomset) of the ripple is preserved, but the upper parts of the lee-side laminae are normally not preserved. In general, the single cross-bedded unit is less than 4 cm in thickness (Reineck 1963a). It can be as thin as 1 mm. The shape of cross-bedded units depends upon the shape of the ripples producing them. Usually they are festoon-shaped (related to lingoid ripples), but

Fig. 160. Small-ripple bedding. Both planar and trough-shaped units are present. On the top the ripple morphology with foresets is present. Relief cast. North Sea tidal flats. (After Reineck 1967a)

Fig. 161. Showing generation of cross-bedding by migrating small ripples. Each migrating ripple produces a bed with foreset laminae. The following ripple erodes the top layer and deposits a new bed with foreset laminae. The surface morphology of ripples is preserved. The flow is from right to left. Laboratory experiment. (After Reineck 1961)

Megaripple Bedding

Cross-bedded units composed of foreset laminae produced as a result of migration of megacurrent ripples are called megaripple bedding (Figs. 163 and 164). Although, in form, megaripple bedding is similar to small-ripple bedding, it differs from small-ripple bedding in size. The thickness of indi-

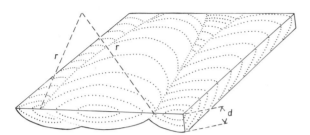

Fig. 162. Diagram illustrating the definition of trough-shaped small-ripple and megaripple bedding. In small-ripple bedding the radius of the trough (r) is less than 20 cm, and thickness (d) is less than 4 cm. For megaripple bedding the radius of the trough (r) is more than 20 cm, and the thickness (d) is more than 4 cm. (After Reineck 1963a)

vidual units is usually more than 4 cm, and may be as much as 1 m or even more.

As megaripple bedding is produced by migrating megacurrent ripples, it can be said that in the sand of the same grain size they are produced at higher energies or velocities than the small-ripple bedding generated by small ripples. However, in certain cases, under the same velocity conditions coarser sand produces megaripple bedding and finer-sand small-ripple bedding (cf. Reineck and Singh 1967). This is because in finer sand higher velocity is required to produce megaripples than in the coarser sand. Both trough-shaped and planar megaripple bedding are known. In case of trough cross-bedding the radius of trough is usually more than 20 cm. Basal parts of the megaripple bedding may show development of regressive small ripples, moving upward along the foreset laminae (backflow); sometimes these regressive small ripples are developed in form of climbing ripples (see Fig. 22) (Lüders 1929; Boersma 1967; Boersma et al. 1968).

All environments favoring formation of megaripples also show megaripple bedding, e.g., beaches, shoals, tidal channels, rivers, etc.

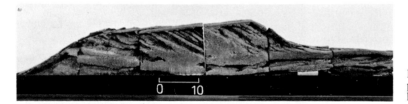

Fig. 163. Megaripple bedding. Surface morphology with foreset laminae is visible. Relief cast

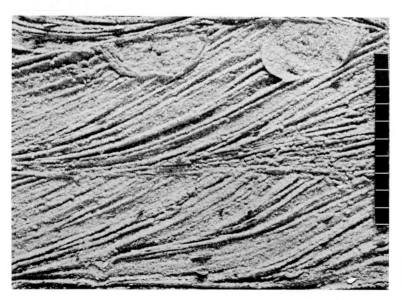

Fig. 164. Megaripple bedding. Inclined foreset laminae of two units of megaripple bedding are clearly visible

Wave-ripple Bedding

Cross-bedding can be produced when wave ripples are produced and migrate on a sediment surface. This type of cross-bedding is called wave-ripple bedding.

In many cases, especially with asymmetrical wave ripples, it may be rather difficult to distinguish them from small-current ripple bedding. An important investigation in this direction has been carried out by Boersma (1970), who summarizes the characteristics of cross-bedding produced by wave ripples and tries to distinguish them from cross-bedding formed by small-current ripples. The following are the main distinguishing features of a wave-ripple cross-bedding (Boersma 1970):

1. The lower bounding surfaces of wave-ripple bedding units are rather irregular or caternary-arcuate. In contrast, small-current ripple bedding possesses regular straight or curved bounding surfaces. This irregular bounding surface is a result of irregular migrational behavior of wave ripples.

2. Ripple foresets of wave ripples are generally composed of cross-laminae arranged in groups. Thus, cross-bedded units show bundle-wise up-building (cf. Fig. 165).

3. Chevron structures are shown by foresets. Although not commonly encountered, the presence of chevron structures in cross-bedded units provides a good clue to the wave-formed cross-bedding (symmetrical wave ripples). Depending upon the controlling hydrodynamic conditions, several variants of chevron structures are produced. In some cases, the chevron structure is also found in downward-curved laminae, formed in ripple troughs.

4. Synchronous sets often possess differing internal structures; some of them may show simple, erosive, or composite forms. Moreover, even the same ripple when cut at different intervals shows differing internal arrangement of laminae.

5. Wave ripple cross-bedded units generally show cross-stratal off-shoots. That is, in case of wave-formed cross-bedding foreset laminae are lofty, wavy in configuration. The inclined foreset laminae generally pass the trough and peak up again on the other flank of the adjacent ripple, sometimes even reaching the top of it.

6. Presence of a discrepancy between ripple shape and internal structure, last stage bundles, and laminae spreading from the top of the ripple are the other characteristic features of wave ripple bedding. However, we feel that these features are also commonly found in small-ripple bedding, especially in environments undergoing continuous reworking, such as tidal zones.

The most characteristic features for wave-ripple bedding are listed in 1, 3, and 5. Other features can also be found in small-current ripple bedding. Figure 165 shows the characteristic features of wave-ripple bedding.

De Raaf et al. (1977) describe a shallow-water sequence built up of wave-formed structures, e. g., wave ripples, and wave ripple bedding. Figure 166 depict various characteristics of wave-built deposits.

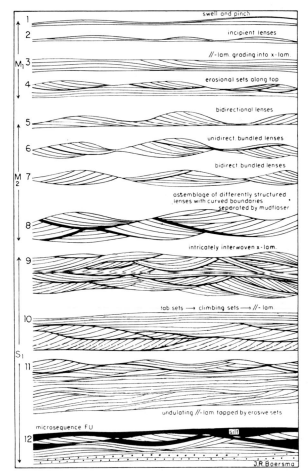

Fig. 166. Different types of bedding features formed as a result of wave action, depending upon the intensity and type of wave activity. (After De Raaf et al. 1977)

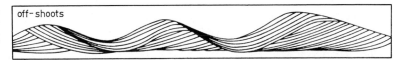

Fig. 165. Wave ripple cross-bedding. Note bundle-wise up-building of foreset laminae in a single unit. The foresets show offshoots passing the troughs and reaching the flank of the adjoining ripples. (After Boersma 1970)

A general term "ripple bedding" is proposed to include all the bedding types produced as a result of the activity of ripples. It includes small-ripple bedding, megaripple bedding, wave-ripple bedding, climbing-ripple lamination, and rippled sand lenses of lenticular and flaser bedding. The term "ripple form-set" denotes that the complete ripple form with its internal structure is preserved as such.

Longitudinal Cross-bedding

The lateral shifting of tidal channels on the inter-tidal flats produces beds, which in contrast to other types of cross-beds, run parallel to the current direction, and is thus called longitudinal cross-bedding (Reineck 1958a).

In meandering channels deposition takes place on convex sides–point bars. Sedimentation takes place laterally (Trusheim 1929b; van Straaten 1954a) in the form of inclined beds on the point bars, and a packet of inclined beds is produced. If this packet of inclined beds is covered by horizontal beds of intertidal flats, it looks like a cross-bedded unit. Sometimes, because of the shifting direction of the meandering channels, several packets of inclined beds dipping in different directions may be deposited one above the other, thus producing a set of crossbeds.

More or less straight channels and gullies can also be shifted in one direction. One side is eroded, generally by wave action, while the sedimentation takes place on the other side. A series of inclined layers is produced during such migration. Beds formed on the deposition side of the straight channels differ from beds of point bars, in that beds of the straight channels are straight and can be followed for greater distances laterally, unlike beds of point bars of meandering channels, which are curved and of short lateral extent.

This type of cross-bedding is very different from the cross-bedding produced as a result of the mi-gration of ripples. In the case of longitudinal cross-bedding, in contrast to ripple bedding, small or mega, each inclined layer is in itself a bed and not a lamina. Each inclined layer may in itself show a different type of bedding, e. g., thinly interlaminated sand/mud bedding, lenticular bedding, or ripple bedding.

Reineck (1958a) discusses the origin and the importance of longitudinal cross-bedding. Important feature always associated with this bedding in intertidal zone are curved slip planes, which are gravity fault planes (Fig. 145).

Longitudinal cross-bedding is very common in intertidal flat environments, especially on mixed flats; and it is the main factor in the inorganic reworking of the sediments on the intertidal flats (Reineck 1958a). Longitudinal cross-bedding is also formed on the point bars of meandering and braided channels of river channels. Wright (1959), discusses the origin of cross-bedding on point bars of rivers. Longitudinal cross-bedding can be important in point bar deposits of smaller river channels. However, in the case of larger rivers inclined layers are difficult to identify. Longitudinal cross-bedding produced by deposition on the point bar related to the lateral migration of a meandering channel is also referred to as epsilon cross-bedding.

Diagnostic features of longitudinal cross-bedding can be summed up as follows:

1. Layers and beds of longitudinal cross-bedding dip perpendicular to the general flow direction of that region (Figs. 167 and 168). If the layers are curved, they show a sweeping strike direction.

2. The inclined layers are not laminae, but are composed of different types of bedding. They can be sometimes thick mud or even sand layers. Curved slip planes are common. Figure 169 shows detailed structures of inclined layers of longitudinal cross-bedding.

3. Current directions obtained from other features, i. e., ripple marks, oriented shells, etc., indi-

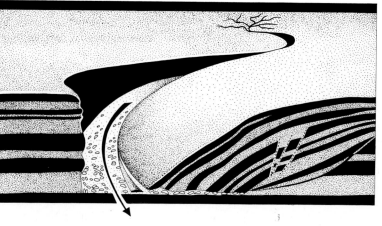

Fig. 167. Schematic diagram showing the generation of longitudinal cross-bedding on the point bar of a tidal creek. The shell bed at the base represents channel lag deposits. Tension faults with curved fault planes are common. (After Reineck 1967a)

Fig. 168. A general view of a tidal channel showing the generation of longitudinal bedding on the point bars. (After Reineck 1958a)

cate a current running parallel to the strike direction of the layers of longitudinal cross-bedding.

4. At the base of a packet of longitudinal cross-bedding is accumulation and enrichment of mud pebbles, shells, and other coarser material. This layer represents the bottom of the channel, which has been covered by lateral sedimentation at the point bar (cf. Jarke 1948; van Straaten 1950a).

Channel-fill Cross-bedding

Cross-beds may be produced in the filling-up of small alluvial or erosional channels. A trough-shaped scoured-out channel is slowly filled up by sets of thin laminae, conforming, in general, to the shape of the trough-shaped floor (cf. McKee 1957a).

In a later phase this trough-shaped channel with its conformable laminae is partially eroded away, and a new younger trough is produced and filled up by thin laminae. However, when viewed in 3-dimensions, it is not festoon-shaped. If this process is repeated several times, a set of cross-strata are produced (Fig. 170). Figure 171 shows the scheme of development of channel-fill cross-bedding.

This cross-bedding from a frontal view looks like large-scale festoon bedding. The layers constituting the cross-bedding are generally thinly laminated mud or sand. There can be enrichment of coarser material such as shells near the bottom of

Fig. 169. Vertical section across the inclined layers of longitudinal cross-bedding. Inclined layers are internally made up of various types of bedding, i. e., small-ripple bedding, lenticular and flaser bedding, tidal bedding (thinly interlayered sand/mud), thick mud layers, etc.

Fig. 170. Channel-fill cross-bedding. Crevasse channel in the natural levee sediments of the Gomti River, India. On the top the channel of the last phase is present, filled with layers conforming to the shape of the channel

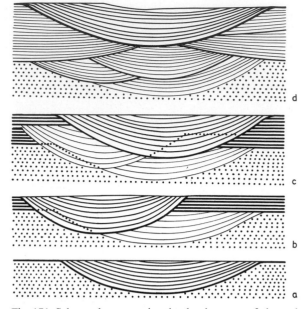

Fig. 171. Scheme demonstrating the development of channel-fill cross-bedding. (After Singh 1972)

each unit. Singh (1972) reports this feature from natural levee deposits of Gomti River, India.

Channel-fill cross-bedding can be locally common in fluvial sediments–especially in the areas of overbank flows and in natural levees of larger rivers.

Antidune Cross-bedding

Migrating antidunes leave behind the laminae arranged in the form of cross-bedded units with characteristics of their own. Usually they are lenticular sandy units, inclined at rather low angles both upstream and/or downstream (Wunderlich 1972). Laminations are usually very faint and dip at low angles (usually less than 10°. Lenticular and wedge-shaped sets are most common. Middleton (1965b) produced antidunes in large flumes and described in detail their internal structure. Reineck (1963a) described antidune cross-bedding from the beaches of the Frisian Islands. Panin and Panin (1967) describe antidune cross-bedding from the Black Sea shore of Roumania. Antidune cross-bedding is shown in Figs. 58 and 59.

The generation of antidunes requires very high stream power. In extremely shallow waters high stream power can be achieved at relatively low current velocities. On the other hand, in nature, in deeper waters conditions of generation of antidunes cannot be achieved (see p. 47). Thus, antidune cross-bedding is generally restricted to some parts of beaches, and natural levees and point bars of rivers.

Microdelta Cross-bedding

As shown by experiments of Jopling (1963, 1965a) and McKee (1957a), migration of a microdelta would produce cross-bedded units.

In surface view, microdelta possesses a characteristic form, and is triangular in shape. The surface is slightly convex upward. The lee side of a microdelta is made up of lobes, rather than a plain surface (Fig. 172). Unlike ripples, microdeltas are always single, solitary bodies, i. e., they never occur in trains. However, characteristics of the foreset laminae of microdeltas are the same as of the foreset laminae of smaller megaripples. Figure 173 shows cross-bedded units produced from the migration of a laboratory microdelta. Microdelta cross-bedding is extremely common in glacio-fluvial and sandur (outwash plain) sediments. Here,

Fig. 172. A small delta built by a tidal creek at its mouth. North Sea tidal flats

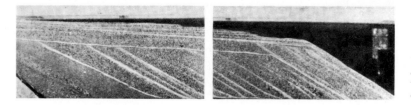

Fig. 173. "Microdelta" cross-bedding produced as a result of migration of a small delta with a well-developed lee face. Laboratory experiment. (After McKee 1957a)

because of the great abundance of sediment, small deltas and lee faces are often established which migrate for certain distances before becoming logged by incoming sediment.

At low-water stages micro-delta develop on the emerging bars of a river (Singh and Kumar 1974), and on the bars of a braided stream in a sandur plain, and may make an important cross-bedded pattern (Bluck 1974). Deposits of micro-delta show a wide range of dip directions.

Beach and Longshore Bar Cross-bedding

It is a common observation that beach surfaces dip gently seaward with long even slopes; on these surfaces evenly laminated sand is produced from wave activity. Small differences in the angles and direction of dips of laminae of adjacent sets gives a cross-bedding appearance. Measurements of McKee (1957b), McKee and Sterrett (1961) show that the angle of dip of laminae ranges from 2 to 5° on the beaches of the Gulf Coast of Texas and 7 to 10° on the coasts of California. Laminae of beach cross-bedding are actually the laminae of evenly laminated sand. Recognition of beach cross-bedding may be possible because of the low angles of dip of the laminae, and the rather long lateral extent of the laminae, which are remarkably even and regular in thickness.

As described by McKee and Sterrett (1961), the longshore bars are mainly made up of layers dipping shoreward at angles of 16 to 20°. On the seaward part the layers dip seaward at gentle angles (4 to 5°). Thus, sets of cross-bedded units are produced, with the constituting layers dipping in different directions and angles, while still retaining the same general strike. Cross-bedded units are of the tabular type, generally wedge-shaped. Trough-shaped units are absent. Thompson (1937) Doeglas (1954), and Werner (1963) have done detailed studies of beach bars and longshore bars and came to very similar conclusions. Figures 174 and 175 show the characteristic features of longshore bar cross-bedding. These features are also termed low and high angle bar bedding (see description of beach bar on p. 364 ff.).

Sand Dune Cross-bedding

Aeolian sand dunes possess a well-developed slip face along which, during the process of its migration, sand avalanches, producing well-developed laminae. These laminae are usually at a steep angle to the horizontal. Several units arranged one above the other produce sand dune cross-bedding (see also p. 350, Fig. 483). McKee (1966b) has described in detail the type of cross-bedding developed in different types of dunes. The most important feature of sand dune cross-bedding is the relatively large-scale cross-bedding, with steeply dipping cross-laminae attaining angles up to 30 or 34°. McKee (1953) in laboratory experiments found

Fig. 174. A beach bar with its steeper slope toward land. In the section cut across the bar steeply dipping laminae with discordances are evident. Laminae are produced as a result of migration of the landward steep slope. Several reactivation surfaces are visible

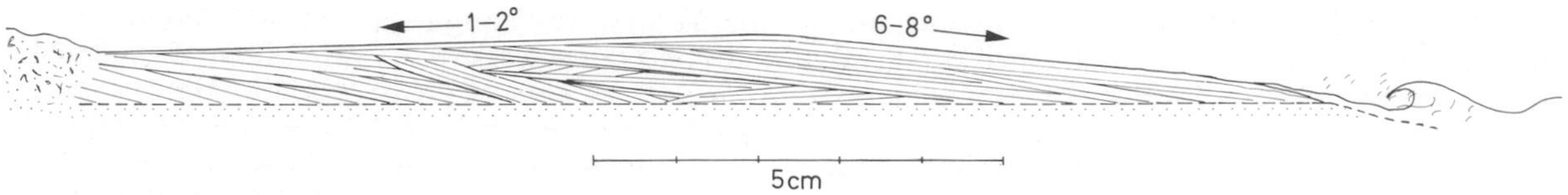

←—1–2°—→   6–8°—→

5cm

Fig. 175. A section across a longshore bar to show the internal structure. The bar is mainly made up of sets of laminated sand gently dipping toward the sea. However, several sets of steeply dipping laminations are also present. (Modified after Werner 1963)

that in subaerial samples, the angle of repose varies between 30 and 40°. Cross-bedded units are generally several decimeters to 1 to 2 m thick; however, at times they can be several meters in thickness. Bedding surfaces separating individual units are more or less horizontal or possess gentle dips. Some of the bedding surfaces may show exceptionally high dips (20 to 28°). Foreset laminae are on a large scale, and are generally straight, with high angles of dip. Where much sand is carried in suspension, the foresets become progressively flatter downwind, thus attaining tangential shapes. In rare cases high-angled foreset laminae may show concave-downward curvature. Individual foreset laminae are rather thick–2 to 5 cm. Figures 176 and 177 show the cross-bedded units of sand dunes.

Goldsmith (1973) studied the internal structures of vegetated coastal dunes and found that the dip distribution in sand dune cross-bedding is bimodal with modes at 10–15° and 25–35°.

The planar type of cross-bedding is very common. Among planar types both tabular and wedge-shaped units are present. Wedge-shaped units

Fig. 177. A closer view of sand dune cross-bedding. The position of successive intrasets marks the stages in the migration of the dune front. The migration is from left to right. The scale is in inches. White Sands, New Mexico. (After McKee 1966b)

Fig. 176. Sand dune cross-bedding. Largescale units show steeply inclined foreset laminae, capped by a thin set of almost horizontal laminae. White Sands, New Mexico. Scale is in inches. (After McKee 1966b)

seem to be rather rare. Trough-shaped cross-bedding is rather uncommon. In places trough-shaped cross-bedding is solitary; in some instances it makes a well-developed festoon pattern (McKee 1966b).

Vertical sections of dunes generally show progressively thinner units of cross-bedding from bottom to top. Foreset laminae of the cross-bedded units tend to be flatter upward, especially on the upwind part of the dune. See also the internal structure of sand dunes, p. 232.

Sand-Drift Cross-bedding

Sometimes cross-bedding of aeolian origin is formed in sand drifts, which are aeolian sand deposits associated with some obstruction. Foreset laminae are low-angled and rather short in dimensions. If formed along the bank of a wadi, foreset laminae may possess large pebbles near the upper part of the foresets (Glennie 1970). See also wadi sediments and sand drift on pp. 217 and 220 respectively.

Scour-and-Fill Cross-bedding

Scour-and-fill structures have been described on p. 71. Occasionally a new scour is made in an already filled scour. Some of the scours possess a steep up-current slope and are filled by laminae dipping 20 to 25°, becoming flatter upward. If a series of scour-and-fills are present one above the other, a set of cross-bedded units is produced. This is rather common in glacial outwash plains, glacio-fluvial sediments-eskers, alluvial fan sediments, etc.

Low-Angle Cross-bedding in Fluvial Sediments

Several decimeter-thick sets of cross-bedded units, where foresets dip at 5–15°, traceable over several meters, are common in some fluvial deposits. The foresets are mostly planar, rarely slightly curved. These sets can be stacked one above the other to make successions of a few meters thick.

This type of cross-bedding is produced due to migration of bars in a fluvial environment. Deposition takes place mainly by low-velocity currents under a shallow cover of water.

Picard and High (1973) suggest that low-angle cross-bedding (angle of dip 5–10°) is very common in the ephemeral stream deposits and can be considered as a characteristic bedding of shallow or ephemeral stream deposits. Singh (1977) reports thick units of low-angle cross-bedding from a sand bar of the Ganga River. Foreset dips are 10–15°, and they are interbedded with scour and fill bedding. Cosets of low-angle cross-bedding can be up to 3 m thick.

Planar Cross-bedding of Fluvial Bars

In fluvial channels there are many types of bar present which possess active slip faces, and are considered unit bars (see fluvial bars on p. 262). They are longitudinal bars, transverse bars (including lingoid), and scroll bars of the point bars. A characteristic feature of these bars is that they possess a steep slipface, which upon avalanching produces planar cross-bedding with foresets dipping around 30°. The units are mostly tabular. The foreset laminae are made up of alternating coarse- and fine-graded sediments, which have been previously sorted by small bedforms on the bar surface (Smith 1972). Reactivation surfaces are quite common in these planar cross-beddings, and may sometimes be made up of thin layers showing small-ripple bedding. Planar cross-bedding in fluvial bars is reported by Smith (1970, 1971a), Collinson (1970), Jackson (1976a), Cant and Walker (1978).

Bluck (1974) distinguishes two types of planar cross-bedding in a braided stream: (1) Planar cross-bedding produced by migrating, steep accretionary bar margins showing coarse sediments. Thick foreset laminae showing alternating coarse and fine laminae. The finer-grained laminae may show ripples. (2) Planar cross-bedding produced by ripple topped bars. They develop when a bar fills up a channel. Units are quite extensive, mud drapes or rippled layers between the foresets are common.

Planar cross-bedding in fluvial channels develops in both gravelly as well as sandy sediments and both in braided and meandering streams. However, they are less common in fine-sand silty sediments. They are most abundant in pebbly-coarse sand sediments of braided streams. In some braided streams, planar cross-bedding is the most dominant bedding structure, e.g., Platte River (Smith 1970).

Backset Bedding

The term backset bedding denotes a set of foreset laminae, dipping in the upcurrent direction The dip of the foresets of backset bedding is relatively low (mostly < 15°). The backset bedding occurs intercalated with horizontal bedding and cross-bedded units showing dip in the down-current direction.

Most of the workers agree that backset bedding is produced in upper flow regime conditions and has been found in the deposits of turbidity currents and other high-energy flows.

The most common origin of backset bedding is migrating antidunes (Jopling and Richardson 1966; Skipper 1971). Picard and High (1973) report backset bedding from ephemeral stream deposits in coarse, pebbly sand, with foresets dipping in upcurrent direction at about 10°. Several centimeter-thick tabular sets are present. They relate it to the migration of antidune and chute and pool bedforms of upper flow regime.

Wunderlich (1973) reports backset bedding from beach sediments, produced by upcurrent migration of rhomboid ripples.

Climbing-ripple Lamination

Climbing-ripple laminations are the internal structure formed in noncohesive material from migration and simultaneous upward growth of ripples produced either by currents or waves. Sorby (1859) was probably the first to discuss the concept of lateral shifting as applied to superimposed ripple laminae. The development of climbing-ripple lamination from ripples requires that abundant sediment is continually available to a current or wave so that the ripples are built upward in an overlapping series rather than merely migrating in a forward direction. This has been recognized by Bucher (1919), Reineck (1961), Allen (1963), Walker (1963), and McKee (1965). Climbing-ripple lamination may be produced by straight-crest, undulatory, or lingoid small-current ripples, or even by wave ripples. Climbing-ripple lamination produced by mega-ripples is extremely rare. Coleman (1969) reports them from the sediments of the Brahmaputra River.

Under favorable conditions ripples are rapidly formed in sand and migrate continuously without building any permanent structure. When much sediment is available, especially in suspension, this extra sand quickly buries and preserves the original rippled layers, either completely or in part, and a series of superimposed ripples that eventually result in climbing-ripple lamination–ripple laminae in-phase and/or ripple laminae in-drift. In the case of climbing-ripple lamination, much sediment is deposited from suspension load, and there seems to be no clear-cut difference between bedload and suspension load.

As suggested by McKee (1965), basically, climbing-ripple lamination can be divided into two types:

1. ripple laminae in-phase
2. ripple laminae in-drift.

All the transitions between these two types are known. Sometimes, within a single outcrop gradual evolution of ripple laminae in-phase to ripple laminae in-drift can be observed. Forms in which crests of ripples have not migrated, grading into ripple laminae where crests have only slightly migrated, to ripple laminae with strongly migrated crests, can be observed. The change from ripple laminae in-phase to those out-of-phase indicates a slight but regular, progressive increase in current velocity with a decrease in depth (a progressive increase in stream power) (McKee 1965).

In ripple laminae in-phase one ripple crest is directly above the other. Usually a slight shift of crest in one direction (the direction of flow) is observed. To obtain a thick sequence of in-phase ripple laminae, various factors such as water depth, wave strength or current velocity, current direction, available sediment, etc., must remain essentially constant despite a growing sediment surface. A very delicate balance of the various factors is necessary. As soon as this balance is changed, a different pattern is formed (McKee 1965). Jopling and Walker (1968) describe a special case of ripple laminae in-phase in which ripple lamination consists of superimposed undulating laminae possessing a symmetrical sinusoidal profile. They call it sinusoidal ripple lamination. In other cases the crest of the constituting ripple laminae can be more peaked and slightly asymmetrical in form.

Characteristic features of ripple laminae in-drift when seen in sections normal to the ripple crests, are the nearly parallel bounding planes which delineate the pseudobeds and which dip in the upstream direction. These planes represent the surface of nondeposition or even slight erosion on the stoss side of the ripples. The angle of inclination of pseudobeds decreases with increasing current strength.

Ripple laminae in-drift, however, when seen in sections other than normal to crests, may be rather difficult to recognize. Laminae in other sections in other directions may run horizontal or may be trough-shaped, depending upon the orientation of the section with respect to ripple crests and troughs, and upon the shapes of ripples which produced them.

Depending upon the preservation of the stoss side of the ripples, ripple laminae in-drift can be separated into two types:

1. Both lee side and stoss side are well preserved (Type C of Jopling and Walker 1968).
2. Stoss side is absent, only lee side is preserved (Type A of Jopling and Walker 1968).

These two types should be considered the end members of a rather continuous series. All the variations exist from cases where the stoss side laminae are completely preserved through a partly preserved stoss side into forms with the stoss side completely absent. Figure 178 shows schematically various types of climbing-ripple lamination.

Jopling and Walker (1968) have very correctly interpreted the change from type 1 into type 2 as a result of a decrease in suspended load / bed load ratio. In other words, if enough sediment is avail-

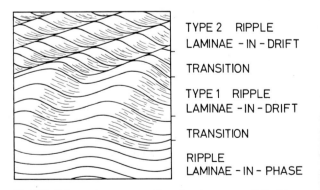

TYPE 2   RIPPLE
LAMINAE – IN – DRIFT

TRANSITION

TYPE 1   RIPPLE
LAMINAE – IN – DRIFT

TRANSITION

RIPPLE
LAMINAE – IN – PHASE

Fig. 178. Schematic representation of various types of climbing-ripple lamination. Ripple laminae in-phase gradually change into ripple drift laminae of type 2. All transitional forms are existent. The variation is mainly in the preservation of stoss-side lamination. (Modified after Jopling and Walker 1968)

able in suspension, almost no erosion of the stoss side takes place, and ripples are completely buried and preserved under a cover of sand, deposited mainly from suspension, which in turn is again rippled, but with a slight drift of crests.

If the ratio of suspended load / bed load decreases, lesser sediment is available in suspension. The net result is that the stoss side of the ripples is eroded before they can be buried and preserved. Thus, climbing ripples whose lee side alone is well preserved are produced.

If the ratio of suspended load/bed load decreases further to a degree that very little sediment is available in the form of suspension, then ripples only migrate and do not grow upwards simultaneously. The net result is the formation of ripple bedding without any climbing-ripple characteristics. The relationship between type of climbing-ripple lamination and suspended load / bed load ratio has been depicted in Fig. 179. Figures 180 and 181 show various types of climbing-ripple lamination.

Stanley (1974) reports that in a silty glacial deposit climbing-ripple lamination is well-developed, and possesses on its stoss side smaller, second-order micro-ripple-drift cross-laminations, which show characteristics similar to the larger climbing-ripple lamination. This indicates migration of very small ripple forms (mostly 1–3 mm in height, and 3–6 cm in length) on the stoss side of the larger straight-crested ripples (2–6 cm in height and 15–60 cm in length) These sediments were deposited by underflows in a shallow glacial lake.

Allen (1971a) proposes a mathematical analysis of the climbing-ripple lamination to determine the instantaneous sediment deposition rates by the measurements of ripple height, angle of climb and grain size, in case climbing-ripple lamination is underlain by parallel lamination of plane bed phase. The experimental studies and mathematical model led Allen (1971b) to calculate rates of sediment deposition from beds showing climbing-ripple lami-

nation. The angle of climb depends upon the net rate of sediment deposition and the velocity of the ripple tangentially over the bed. The tangential velocity is determined by the rate of bed-load trans-

Fig. 180. Climbing-ripple lamination with laminae in-phase. However, ripples are asymmetrical. They are produced under current action. (After Jopling and Walker 1968)

Fig. 181. Climbing-ripple lamination of type ripple laminae in-drift. Type 1 in the lower part of the photograph grades slowly into type 2 ripple laminae in-drift in the upper part of the photograph. (After Jopling and Walker 1968)

PRESERVATION :

Fig. 179. Diagrammatic representation showing dependence of type of climbing-ripple lamination from the suspension/bedload ratio. With an increasing suspension/bedload ratio the stoss side is better preserved and type 2 ripple lamination in-drift changes to type 1, and ultimately to ripple lamination in-phase. (Modified after Jopling and Walker 1968)

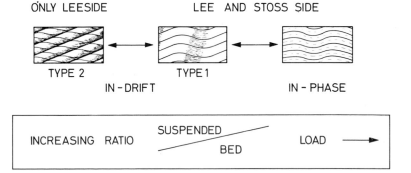

port, and difference of deposition rate between the stoss-side and lee-side faces of the ripples, related to the ripple shape.

Allen (1973a) proposes a classification of climbing-ripple lamination patterns on the basis of vertical stacking of different types of climbing ripples. Three patterns are significant: Pattern I – upward monotonic increase in the angle of climb, accompanied by a parallel decrease in the mean size of the sediment. Kumar and Singh (1978) describe such pattern from the Gomti River deposits (see Fig. 390). Pattern II – maintenance of a more or less uniform angle of climb from the base to the top of layer. Grain size may or may not decrease upwards. Pattern III – upward decrease in the angle of climb, accompanied by an increase in the mean grain size (Fig. 182).

Considering the mode of origin of climbing-ripple lamination, one may conclude that environments of periodic rapid accumulation of sediment are favorable for their development. Environments characterized by the introduction of little new sediment and much reworking are unfavorable for their development. McKee (1938) found that sediments in the upper reaches of the Colorado River delta suffering annual floods show up to 70 % bedding as climbing-ripple lamination.

McKee (1966a) found climbing-ripple lamination to be the common feature in fluviatile sediments. They are particularly abundant in areas of overbank flow and flood plains (McKee 1966a) and natural levees. Coleman and Gagliano (1965) found climbing-ripple lamination in the Mississippi River delta. These laminations are restricted to subaqueous levees of delta front and subaerial levees. In general, climbing-ripple laminations are not found on inter-tidal flats. However, locally, at places of high rates of sedimentation they can be abundant (Wunderlich 1969).

Jopling and Walker (1968) describe them as well-developed features in fluvio-lacustrine deposits – a kame delta formed in a small glacial lake. Walker (1963, 1969) describes climbing ripples as important features in a turbidite environment.

Flaser and Lenticular Bedding

A ripple bedding in which mud streaks are preserved completely in the troughs and partly on the crests is known as flaser bedding. Lenticular bedding shows well-preserved sand lenses embedded

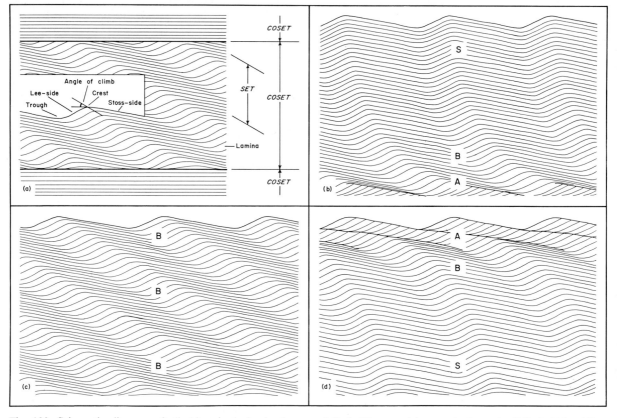

Fig. 182. Schematic diagram of climbing-ripple lamination showing the terminology (a), and three significant patterns: (b) upward increase in the angle of climb, (c) no change in the angle of climb, (d) upward decrease in the angle of climb. (After Allen 1973a)

within the muddy layers. All the transitions from
flaser bedding through wavy bedding to lenticular
bedding exist. A detailed classification of these
features has been given by Reineck and Wunderlich (1968b). Thus, this type of bedding is a case in
which the bedding is made up of not one but two
different types of beds that alternate repeatedly.
The following account of flaser and lenticular bedding is based mainly on the paper by Reineck and
Wunderlich (1968b). Figure 183 shows the scheme
for the classification of flaser and lenticular bedding. The recognition of different types of flaser
and lenticular bedding is possible only when it is
viewed in two sections: One cut perpendicular to
the direction of ripple-crest elongation, the other
cut parallel to the ripple-crest elongation.

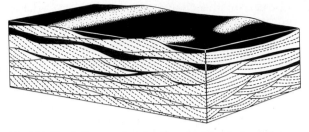

Fig. 184. Block diagram showing flaser bedding in three dimensions. (After Reineck 1967a)

Ripple Bedding with Flasers

This type may be regarded as a transition from
ripple bedding to flaser bedding. Cross-bedded
units show well-developed foreset laminae. Single,
isolated mud flasers are present sporadically in
partly preserved ripple troughs.

Flaser Bedding

Ripple bedding with numerous mud flasers is
identified as flaser bedding (Fig. 184). This struc-

ture implies that both sand and mud are available
and that periods of current activity alternate with
periods of quiescence. During periods of current
activity, the sand is transported and deposited as
ripples, while mud is held in suspension. When the
current pauses, the mud in suspension, mud floccules or faecal pellets segregated by hydraulic difference are deposited mainly in the troughs or
completely cover the ripples. At the start of the next
cycle, ripple crests are eroded away and new sand
is deposited in the form of ripples, burying and preserving ripple beds with mud flasers in the troughs.
Sometimes new sand is deposited above the mud
layer, eroding ripple crests only partly or insignifi-
cantly. Thus, flaser bedding is produced in environments in which conditions for deposition and
preservation of sand are relatively more favorable
than for the mud. Figure 185 shows the scheme for
the genesis and the development of flaser bedding.

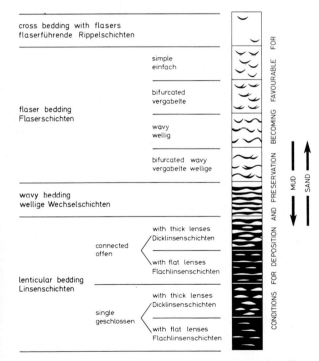

Fig. 183. Scheme of classification of flaser and lenticular bedding. Black mud, white sand. (Modified after Reineck and
Wunderlich 1968b)

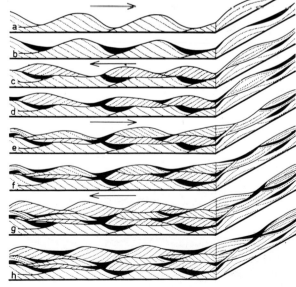

Fig. 185. Scheme showing the genesis of flaser bedding in a tidal
environment. Opposite running flood and ebb currents with
high and low water still-periods. For example, a flood current;
b high water still-stand; c ebb current; d low water stillstand;
and so on (After Reineck 1960a)

Based on the characteristics of flasers, flaser bedding can be divided into the following types:

1. Simple Flaser Bedding. The mud flasers are simple, isolated, showing no contact with each other, with concavity upward. Mud has probably been deposited only in the troughs. Even if mud were also deposited over the ripple crests, it would have been eroded away together with the crests by the next current responsible for sand deposition.

2. Bifurcated Flaser Bedding. The flasers are frequently bifurcated. The bifurcation is the result of the contact of partially exposed flasers of an earlier generation with the later-formed flasers, thus indicating stronger intermittent reworking of sediments than in type 1.

3. Warvy Flaser Bedding. The mud flasers of this bedding are wavy in form. They show concavity upward when occupying ripple troughs and concavity downward when they overlie ripple crests. However, they fail to form continuous beds. Their formation requires conditions in which new sand-depositing currents only partially erode the crests of earlier-formed ripples and the mud layer draping over it.

4. Bifurcated Wavy Flaser Bedding. Flasers of this bedding are wavy in shape and show bifurcations, produced by coalescence with earlier-formed, strongly eroded mud flasers. They indicate conditions similar to those required for the formation of wavy flaser bedding but with stronger intermittent reworking. Various types of flaser bedding are shown in Fig. 186.

Wavy Bedding

In wavy bedding, mud and sand layers alternate and form continuous layers. The mud layers almost completely fill the ripple troughs and make a thin cover over the ripple crests, so that the surface of the mud layer only slightly follows the concavity and convexity of the underlying rippled surface. The thicker the mud layer, the less the form of the underlying rippled surface is traceable on the upper surface of the mud. When only one wavy mud layer is present, it should be recorded as a single occurrence and described as a wavy mud layer. Only when there is a sequence of many wavy mud layers alternating with ripple-bedded sandy layers it is identified as wavy bedding. In contrast to flaser bedding, the ripple-bedded sand layers of wavy bedding are vertically discontinuous and isolated. Various types of wavy bedding are shown in Fig. 187.

In contrast to flaser bedding, the genesis of wavy bedding requires conditions where the deposition and preservation of both sand and mud are possible.

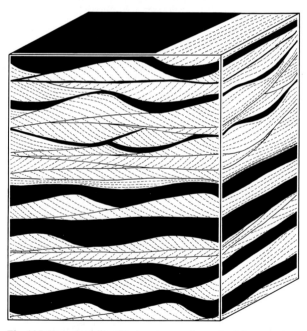

Fig. 187. Wavy bedding. Various types of wavy bedding are produced because of different thicknesses of mud layers. (After Reineck and Wunderlich 1968b)

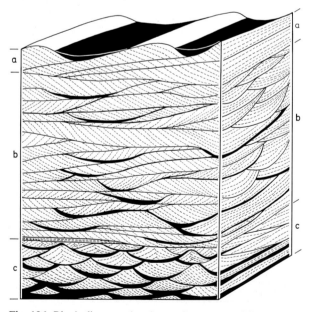

Fig. 186. Block diagram showing various types of flaser bedding. a flaser bedding associated with straight-crested small-current ripples; b flaser bedding formed from small ripples with curved crests; c flaser bedding in association with wave ripples. (After Reineck and Wunderlich 1968b)

Lenticular Bedding

In lenticular bedding the ripples or sand lenses are discontinuous and isolated not only in a vertical but also in a horizontal direction. Thus, ripples are produced in the form of isolated lenticular bodies on a muddy substratum. In other words, lenticular bedding is produced when incomplete sand ripples are formed on muddy substratum and preserved as a result of deposition of the next mud layer. Thus, lenticular bedding is produced under conditions more favorable for the deposition and preservation of mud than for sand. The sand supply is meager, so that only incomplete ripples are produced. The scheme for genesis and development of lenticular bedding is shown in Fig. 188.

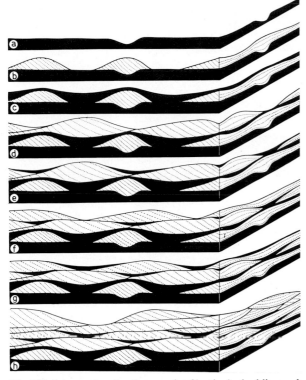

Fig. 188. Scheme showing the genesis of lenticular bedding and flaser bedding (in lower part). a–d shows the genesis of lenticular bedding as a result of formation of incomplete ripples on a muddy substrate, later covered again by a mud layer, followed by deposition of sand in the form of ripples. Periods of current activity alternate with stillstand periods, resulting into deposition of sand and mud, respectively. Only unidirectional current has been considered, from left to right. In tidal environments ripple-bedded units usually show two current directions. (After Reineck 1960a)

Based on the nature of lenses, lenticular bedding can be subdivided into the following types:

1. Lenticular Bedding with Connected Lenses. Up to 75 % of the ripples or sand lenticles are continuous in the horizontal and vertical directions. Based on the morphology of the lenses, they can be distinguished into thick lenses and flat lenses. Thick lenses have a length/height ratio < 20. Flat lenses have a length/height ratio > 20. Considering these characteristics, this bedding can be further distinguished into *lenticular bedding with con-*

nected thick lenses and *lenticular bedding with connected flat lenses.* Figure 189 shows various types of lenticular bedding with connected lenses.

2. Lenticular bedding with Single (Isolated) Lenses. More than 75 % of the sand lenticles are discontinuous. Thus, most of the sand lenticles seem to float in the mud. They are produced under conditions where the sand supply is even less than in type 1 (Fig. 203).

Based on the morphology of sand lenses, we can also distinguish into *lenticular bedding with thick isolated lenses* and *lenticular bedding with flat isolated lenses.* Figure 190 shows various types of lenticular bedding with isolated lenses. Figures 191 to 196 show various types of flaser and lenticular bedding from modern sediments.

As discussed above and by Reineck (1960a, b), the origin of flaser and lenticular bedding requires conditions of current or wave action depositing sand, alternating with slack water conditions when mud is deposited. Thus, sandy layers or lenses are made up of foreset laminae of current ripples or wave ripples. The preferred environments of formation are, therefore, areas where a change takes place between slack water and turbulent water and where the needed sediment exists. Thus, the main environments of its occurrence are subtidal zones (Reineck 1963 a; Reineck et al. 1968) and intertidal zones (Häntzschel 1936a; van Straaten 1954a). In tidal environments the genesis of flaser and lenticular bedding is related to the tidal rhythm,

Fig. 189. Lenticular bedding with thick connected lenses. Ripple-bedded lenses of the upper part are of current origin. Sand lenses of the lower part are deposited as wave ripples, both symmetrical and asymmetrical. (After Reineck and Wunderlich 1968b)

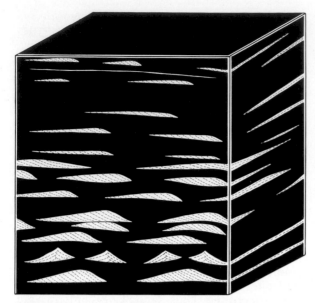

Fig. 190. Lenticular bedding with isolated lenses, the upper part with flat isolated lenses. (After Reineck and Wunderlich 1968b)

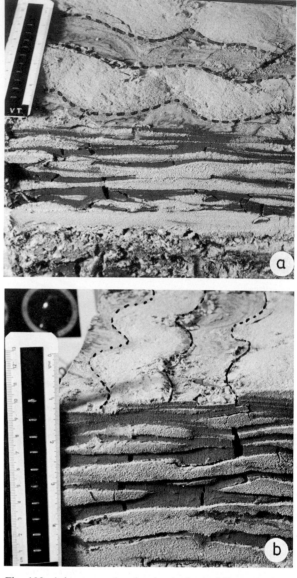

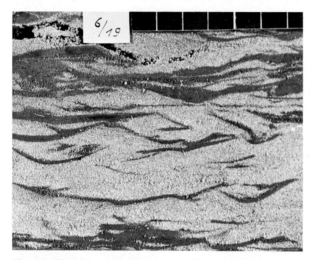

Fig. 191. Flaser bedding. All variants of flaser bedding are visible, i. e., simple flaser bedding, bifurcated flaser bedding, wavy flaser bedding, and bifurcated wavy flaser bedding. North Sea tidal flats. (After Reineck 1967a)

Fig. 192. A box-core showing lenticular bedding caused by small (incomplete) current ripples with undulatory crests. The course of ripples on the horizontal upper surface has been marked by ink. Well-defined sand lenses are seen in a vertical section. a Section parallel to the ripple crests; b section normal to the ripple crests

i. e., to the periods of tidal currents alternating with periods of quiescent or slack water.

McCave (1970) doubted that several millimeter-thick mud layers can be deposited during a single slack-water phase and pointed out that lenticular and flaser bedding may not be related to tidal rhythm. Reineck and Wunderlich (1969), in direct observations on the tidal flats, found that several millimeter-thick mud layer is produced during a single tidal phase. More recently, Wunderlich (1978) demonstrates that in Jade Bay, North Sea, up to 2-cm-thick mud layer can be deposited with-

in 30 min of slack water. Such a quick deposition is because of the high concentration of mud and the presence of clay as grain aggregates. From a 1-m column of Jade Bay water suspended mud settles only within 8 min (Wunderlich 1969) (at a rate of 0.2 cm/sec).

Terwindt and Breusers (1972) conducted experiments on the deposition and consolidation of mud in tidal conditions and the genesis of lenticular and flaser bedding. They conclude that initial consolidation of mud layers is sufficient for their preservation over several tidal cycles, and the flaser bedding is hydraulically different from wavy bedding.

Thus, quick and extensive deposition of mud is facilitated by its high concentration as suspension (Wunderlich 1969; McCave 1971), and the presence of clay particles as aggregates (Wunderlich 1969).

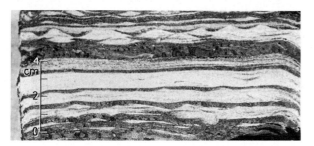

Fig. 193. A box-core from sediments of intertidal flats showing various types of bedding. In the top zone lenticular and flaser bedding are of symmetrical wave ripples origin. North Sea tidal flats

In tidal environment, lenticular and flaser bedding is mostly related to tidal rhythm, though in other environments different rhythms e. g., storms, an alternating supply of material is necessary for their genesis. However, thick cosets of lenticular and flaser bedding are characteristic of tidal sea deposits.

Based on the laboratory experiments and observations in the field, Terwindt (1979) suggests that various types of lenticular- and flaser bedding have definite stability fields in terms of current velocities. According to Terwindt (1979) deposition of mud flocs in the North Sea tidal flats takes place at current velocities 0.2 m/sec. (According to Little-Gadow and Reineck (1974) deposition of mud flocs takes place at current velocity 0.04 m/sec). Erosion of mud drape on a sand layer takes place if the current velocity during the next tide is approximately 0.45 m/sec. Consequently, simple flaser bedding is made if current velocity at $U_{0.5\,max}$ during one tide hardly exceeds 0.45 m/sec. ($U_{0.5\,max}$ is the

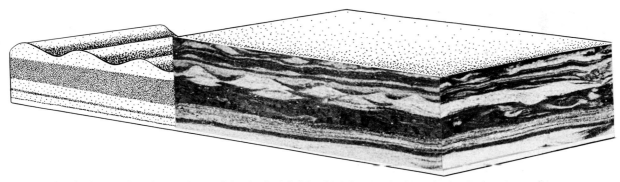

Fig. 194. Lenticular bedding of wave ripple origin. On the left side of the diagram ripples and other features have been reconstructed. Closely spaced dots = mud, widely spaced dots = sand. North Sea tidal flats

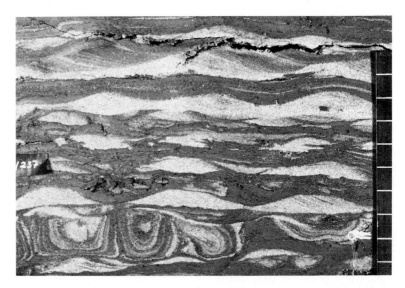

Fig. 195. Lenticular bedding. Sand lenticles are asymmetrical wave ripples. In the lower part convolute bedding is visible. North Sea tidal flats

Fig. 196. Wavy bedding in the sediments of intertidal flats, North Sea

threshold velocity measured at 0.5 m above the bed). The wavy flaser bedding is formed when $U_{0.5\ max}$ value is just below the critical value of 0.45 m/sec, preserving the entire mud drape. Threshold for erosion of mud drape (0.45 m/sec) is well above that for the formation and migration of ripples. So even if mud drape is not eroded, the next rippled sand layer is deposited. Thick mud layers of lenticular bedding mostly represent deposits of several tidal cycles with current velocity below 0.2 m/sec. Occasionally, sand is introduced by fall-out from suspension or migration of ripples. The introduction of sand takes place at velocities just above 0.2 m/sec.

The ripple bedding of flaser and lenticular bedding is mostly of current origin; sometimes foreset laminae may show bipolar directions. However, wave ripples can be locally important in producing ripple bedding of this type (Reineck 1960b).

Lenticular bedding has also been described from the marine delta front (Coleman and Gagliano 1965) and from the lake bottom sediments in front of developing small deltas (Coleman 1966). Mutti (1977) reports lenticular and flaser bedding type features from channel-mouth facies, deposited mainly by traction processes of a fine-grained deep-sea fan deposit.

A descriptive term "rhythmic sand/mud bedding" (or alternating bedding) is proposed to include all the bedding types composed of alternating layers of sand and mud. It includes lenticular and flaser bedding, coarsely interlayered sand mud bedding, and thinly interlayered sand/mud bedding. This term has been used particularly in describing such bedding in tidal environments.

Graded Bedding

Graded beds are sedimentation units characterized by a gradation in grain size, from coarse to fine, upward from the base to the top of the unit. Pettijohn (1957) discusses that theoretically two types of grading are possible. In one type, decrease in the grain size upward is the result of the addition of successive increments of material, each of which is finer than the preceding. In the other, each successive increment is similar to the preceding, except that it contains one less coarse grade. In other words, in the first type there are no fines in the lowest part of the graded bed; in the second type the fines are distibuted throughout (Fig. 197). The first type is probably the result of sedimentation from a current gradually decreasing in velocity and competency. The second type is the product of sedimentation from a suspension in which all sizes are carried, and out of which they settle. Most of the graded beds belong to second type (Pettijohn

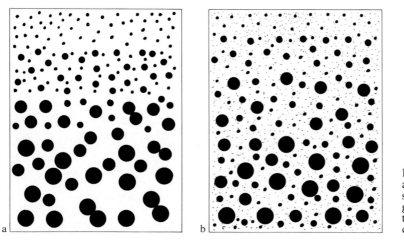

Fig. 197. Two basic types of graded bedding. a There are no fines in the lower part, and sediment sizes decreases gradually. b Fine grains are present as matrix throughout; in the upper part coarser grains gradually decrease. (Modified after Pettijohn 1957).

1957). It seems that all transitions between the two extremes exist.

Middleton (1967) produced these two types of grading in experimental turbidites: (1) distribution grading where whole size distribution shifts to finer sizes progressively from the bottom to the top of the bed. This is formed by layer-by-layer deposition from dilute turbidity currents, (2) coarse-tail grading, where only the coarser grains show decrease in size. This is formed by high concentration flows.

The graded beds always show a sharp contact at the base with the underlying beds. Normally, the lower part of graded beds is made up of relatively coarse sand which grades upward into finer sediments–clay or mud. The individual graded beds of flysch sediments may be as thin as 10 cm, though usually they are 20 to 50 cm thick, and in exceptional cases, over 1 m. However, graded beds of shallow-water environments are only a few millimeters to a few centimeters thick. Normally, graded beds are devoid of any internal lamination other than grading in grain size.

If the gradual grading from sand to clay is complete, the sequence is termed *complete or normal grading* (Dzulynski and Walton 1965; Birkenmaier 1959; Kuenen 1953b; Ksiazkiewicz 1954). In some cases where the clayey band rests directly on sand without intervening silt, the grading is called *interrupted grading* (Kuenen 1953b). When only one gradation is present, the bed is said to show *simple or single grading* (Ksiazkiewicz 1954). On the other hand, many graded subunits may occur in a single-graded bed, and grading is then *multiple or recurrent* (Ksiazkiewicz 1954; Kuenen 1953a,b).

Dzulynski and Walton (1965), based mainly on the description of Kuenen (1953a,b) and Ksiazkiewicz (1954) recognize 8 types of graded bedding.

Sometimes varve-like clayey and silty layers show well-developed grading. They should be considered special cases of graded bedding. Kuenen (1951a,b) suggested that graded varves are produced by deposition from heavy melt-water underflows in glacial lakes, which in fact is a type of turbidity current.

Graded beds commonly occur in thick sequences of flysch type of sediments for which it seems to be the characteristic bedding type. Commonly, the graded beds are separated by parallel-bedded clayey layers, which represent the product of normal sedimentation interrupted by turbidity currents depositing graded sandy beds.

The origin of graded bedding has been attributed mainly to turbidity currents studied in detail by Kuenen and his associates. Kuenen and Migliorini (1950) first expressed the view that turbidity currents are the probable cause of most graded bedding.

Reverse grading seems to be quite characteristic of high concentration flows of turbidity currents and is commonly observed in conglomeratic gravity flows (Fisher 1971; Middleton and Hampton 1973). Small-scale reverse grading is very abundant in the suspended load fluvial deposits (Taylor and Woodyer 1978).

Occasionally, graded bedding is observed in shallow-water sediments. In such cases, however, the graded beds never make thick sequences. They are generally solitary and sporadic. Such sporadic occurrences of graded bedding are usually produced through agencies other than turbidity currents. Graded bedding can be produced by sedimentation of suspension clouds, by deposition in the last phases of a heavy flood, by periodic silting of delta distributaries, by settling of volcanic ash after an eruption, etc. Even on intertidal flats, locally graded bedding showing grading from sand to mud can be observed as a result of deposition by waning current activity. Locally, on intertidal flats and beaches reverse graded bedding (becoming coarser grained from bottom to top of a bed) is also found (Fig. 198). Graded beds of intertidal flats and beaches are very thin–a few millimeters to 2 cm.

Rhoads and Stanley (1965), and Warme (1967) report production of graded bedding due to bioturbation activity of burrowing organisms. A lower-lying sand layer and an overlying mud layer are mixed so as to produce a gradational change from bottom to top, from coarse to fine sediments. Hayes (1967) showed that hurricanes can produce graded bedding on shallow continental shelves. But he believes that graded bedding was produced by turbidity currents generated by hurricanes. Laminated sand deposited from suspension clouds, sometimes, may show well-developed

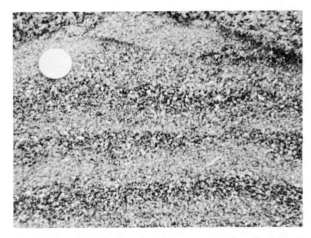

Fig. 198. Beach lamination showing reverse grading. Each lamina is finer-grained near the base (light-colored) and grades upward into a coarse-grained part (dark-colored). Viewed at a low angle to bedding. (After Clifton 1969)

grading (Reineck and Singh 1971). Such graded beds are known from shelf mud (Gadow and Reineck 1969) and lacustrine deltas, e. g., the Rhine delta in the Bodensee (Förstner et al. 1968). These graded beds can sometimes be developed as graded rhythmites (Fig. 199). Here coarser-grained laminae alternate with finer-grained laminae and become thinner and finer grained in their upper part (Gadow and Reineck 1969).

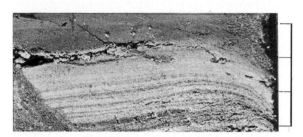

Fig. 199. Graded rhythmites of storm sand layers in a shelf mud. Coarser-grained laminae alternate with finer-grained laminae; the laminae become thinner and finer-grained toward the top of the layer

Graded bedding of shallow-water environments is generally composed of thin units–from a few millimeters to 1 or 2 cm, and seldom makes sequences more than 10 or 20 cm thick. Graded bedding is occasionally interrupted by other types of beds, such as finely interlaminated sand mud bedding, laminated sand, etc.

In a fluvial environment, the pebbly layers often show grading and are related to the decrease in transport velocity (Picard and High 1973). Quite often the pebbly sediments of a fluvial environment show reverse grading.

Channel bank deposits of suspended load rivers, where deposition is by slow-moving currents mainly by suspension load exhibit extensive development of graded bedding. Taylor and Woodyer (1978) discuss that in the upper bank benches of the Barwon River, Australia, graded bedding is the most abundant type of bedding structure, and is present in four configurations:

1. grading up from sand to mud (normal grading), simple
2. grading up from mud to sand (reverse grading), simple
3. grading up from sand to mud to sand (waning-waxing grading) complex
4. grading up from mud to sand to mud (waxing-waning grading) complex.

Further, these four grading patterns may be developed in one of the two following forms:

1. regular transition from coarse to fine or fine to coarse up the bed (continuous grading)
2. a sand (or mud) bed in which there is an increase in the number of fine (or coarse) laminae up through the beds (laminated grading).

Reverse grading is most common, followed by waxing-waning grading in the Barwon River deposits and are developed mostly as laminated grading.

Deposition of these graded beds is from the suspension load, with minor bed load activity reflected by the development of lamination with a graded bed (Taylor and Woodyer 1978). The origin of the graded beds in suspension-load rivers is due to many factors. The development of reverse grading is due to rapid rates of mud deposition in conditions of low bed shear and high concentration of mud before arrival of peak flood. During peak flood, the sand is deposited. Normal grading is produced by deposition of mud after a sand deposition period. The complex grading may be related to a double-peaked flood. Sand is deposited during the rising limb of a flood, followed by mud deposition during the falling limb of the first peak of flood. During the second rise and fall another set of sand-mud is deposited.

However, most of the deep-water graded beds are definitely the result of turbidity currents. Graded beds produced by other agencies are relatively uncommon and possess somewhat different characteristics and are associated with different structures. Some of the deep-water graded bedding is produced by the action of traction currents on the deep ocean floor.

Evenly Laminated Sand and Horizontal Bedding

Evenly laminated sand is composed of parallel and almost horizontal sand layers, 1 or 2 mm thick. Each layer is usually only a few grain diameters thick. Individual layers are usually horizontal or sometimes slightly inclined because of deposition on originally inclined surfaces; they can be followed for a couple of meters. Lamination is generally marked by alternating layers of different grain size, or different heavy mineral content, or both. This bedding is mostly well developed in fine sand and medium sand. Some cases of evenly laminated silt are also known. They are related to laminated sand and originate in a similar fashion. Different sets of evenly laminated sand are usually separated from each other by very low-angled erosional unconformities.

Evenly laminated sand is generally abundantly distributed on beaches or other sandy areas exposed to wave action. Much work has been done on the low-angled beach lamination. Reineck (1963a) ascribed the evenly laminated sand of beaches to swash and backwash action of waves and demonstrated that during one tidal phase many laminae are produced–6 to 16 laminae during one tidal phase. Figure 200 shows evenly laminated sand. Laminae generally show textural as well as mineralogical grading. The wash of a wave brings along a sand layer, which is deposited through backwash in the form of reversely graded laminae (Clifton 1969). Within a single lamina, grain size increases from the bottom to the top of the layer (Sanders 1965). If enough heavy minerals are available, dark colored, heavy, mineral-rich laminae alternate with light-colored, quartz-rich laminae. A mineralogical gradation is apparent within a single lamina. The finer-grained basal part is very rich in heavy minerals, and grades upward into the heavy mineral-poor, coarser-grained part (Clifton 1969). Figure 198 shows well-developed reverse grading in the laminae.

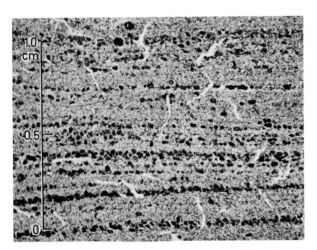

Fig. 200. Evenly laminated sand. Laminae are made up almost entirely of fecal pellets (dark-colored). (After Reineck et al. 1967)

Besides the above-described most common mode of genesis of evenly laminated sand, it can also originate in one of the following ways:

1. Evenly laminated sand can be produced in the plane-bed phase of high flow regime. In the plane-bed phase of high flow regime, some bed load segregation takes place, producing finer-and coarser-grained striations (see parting lineation). Random superimposition of these striations would impart an indistinct lamination (Jopling 1967). If enough sediment is available, a packet of evenly laminated sand can be produced. Evenly laminated sand of high flow regime is invariably associated with parting lineation. The laminae are rather poorly developed and usually of short lateral extent. The reverse grading of beach laminations is not present.

Middleton and Hampton (1973) suggest that some of the parallel lamination in turbidites may be related to migration of long wavelength antidunes, where lamination is diffused.

2. Below the critical velocity of ripple formation, there exists a plain bed phase with sand grain movement (Bogárdi 1965; Guy et al. 1966). If enough sediment is available, a packet of laminated sand can be produced.

3. Another important mode of genesis of evenly laminated sand is through the sedimentation of suspension clouds in current velocities below the genesis of ripples. Reineck and Singh (1972) could produce them experimentally. Laminae produced in such a fashion, if at all, show normal grading–coarser grains near the base grading upward into finer grains. Grading is better developed if enough time is available between the sedimentation of two successive suspension clouds. Suspension clouds are generally produced from shoaling waves. Evenly laminated sand of storm-sand layers in the shelf mud, and some of the laminated sand in the submerged parts of sand bars, shoals, channel slopes, river levees, and point bars are formed as a result of sedimentation of suspension clouds (cf. Reineck 1963; Wunderlich 1969; Reineck and Singh 1972).

4. Evenly laminated sand can also be produced in wind regime, partly by wind ripples. Parts of beaches located just above the high-water line are exposed to wind activity. The bedding below the wind-rippled surface is evenly laminated sand. In favorable conditions alternating heavy mineral-rich and heavy mineral-poor laminae are well-developed. Without additional criteria they may be difficult to differentiate from beach laminations in ancient sediments. Production of horizontal laminae as a result of migration of wind ripples has been also reported by Sharp (1963), Inman et al. (1966), and McKee et al. (1971).

Hunter (1977a) describes that plane bed lamination in wind regime develops at very high wind velocities, too high for the existence of ripples. During a single storm, up to 50-cm-thick units of horizontal bedding are produced. Grainfall deposition, related to the deposition in zones of flow separation on horizontal surfaces, also produces parallel bedding. In grainfall lamination, grain segregation is poor, leading to poor recognition of lamination. The laminations show poor grading.

5. The problem of the genesis of horizontal laminae (evenly laminated sand) in turbidites has been discussed by Kuenen (1966a). He believes that horizontal lamination is not the result of pulsations in the current, as is generally accepted. He concludes that such lamination in sand is a result of de-

position from a perfectly smooth, gradually decelerating suspension current. Development of lamination is a result of the tendency of particles moving along the bottom to segregate into patches of particles of equal weight, density, and shape. Laminae are rather short in lateral extent (a few decimeters), and adjacent laminae differ in median diameter or composition.

6. Horizontal bedding can be produced by low-relief bedforms in very shallow depths. Smith (1971a) observed the generation of horizontal bedding by low-amplitude sand waves in very shallow depths in a river under conditions of lower or transitional flow regime. In such cases laminations are distinct where coarse and fine laminae alternate each other, and lateral pinching could be common.

McBride et al. (1975) produced in a flume parallel, horizontal lamination in sand of 0.2 to 0.8 mm size. Thin lamination developed under shallow flow conditions (< 5 cm water depth) with low-relief bedforms which show prominent lateral segregation of coarser and finer grains. Low-relief ripples produced thin laminae by burial of finer grains of the stoss side by coarser grains of the lee side.

Moss (1972) demonstrated that the plane bed phase (related to horizontal bedding) can develop at all motion intensities from the small ripple stage, the megaripple stage and in the rheologic bed stage (upper flow regime), while due to flow unsteadiness, bedforms sometimes fail to develop. However, grain-size distribution of plane bed of upper flow regime shows all three populations; while the plane beds produced at ripple stages are characterized by dominance of saltation population.

Picard and High (1973) discuss that in ephemeral stream deposits horizontal bedding of low-energy conditions is characteristically parallel and marked by continuous, thin lamination, whereas the horizontal bedding of high-velocity conditions (upper flow regime) occurs in thick laminae and occasional lateral pinching. Horizontal bedding of low-energy conditions shows better differentiation between the adjacent laminae than the horizontal bedding of high-energy conditions (Singh 1972).

Bridge (1978a) discusses the specific formation mechanism of horizontal bedding under turbulent boundary layers. Because of the quasi-cyclic bursting process in turbulent boundary layers, there is temporal and spatial variation in bed shear and lift forces acting on sediment particles in motion over the plane bed leading to vertical variation in the texture of sediment being deposited. This process is especially true in upper flow-regime conditions; though the bursting process can also be active in low-energy conditions.

The above discussion shows that evenly laminated sand can be formed in several different ways, and is found in varied environments: On beaches (Reineck 1963a; Clifton 1969, and others), shoals (Reineck and Singh in Dörjes et al. 1970), shore faces (Reineck and Singh 1972), storm-sand layers of shelf mud (Gadow and Reineck 1969; Reineck and Singh 1972), channel slopes of intertidal flats (Wunderlich 1969). In fluviatile environments they can be locally abundant on levees and point bars (Harms et al. 1963; Coleman 1969). They also occur abundantly in horizontal laminae intervals of flysch sediments (Bouma 1962; Dzulynski and Walton 1965).

The terms "horizontal bedding" or parallel bedding are commonly used instead of laminated sand, especially in case of fluvial sediments. Horizontal bedding is also commonly developed in coarser sediments, e. g., pebbles and gravels. Well-developed horizontal bedding in coarse-grained sediments is generally indicative of a rather high energy of the environment, conditions comparable to upper flow regime.

Coarsely Interlayered Bedding

This bedding is composed of alternating coarser- and finer-grained layers which are several millimeters to several centimeters thick. Coarser layers may be sand or silt. Finer layers may be silt or mud or clay. At times, in the sand layers foreset laminae of ripples are recognizable, but generally the coarser layers are horizontally laminated.

Depending upon the relative thickness of sand and mud layers, three types of coarsely interlayered bedding can be distinguished:

1. Sand and mud layers are almost equally thick
2. Thicker sandy layers are separated by thin clayey or finer-grained layers.
3. Thicker mud layers alternate with relatively thin sand layers.

All the three variations may be encountered within the same sequence, and they seem to be more or less interrelated in their mode of origin. In the storm-sand layers generally all these variations occur together (cf. Reineck and Gadow 1969).

This bedding is common in the sediments of mixed intertidal flats. The mode of origin is not yet clearly understood. Nevertheless, it is quite likely that the sand layers are deposited during current and wave activity. Most mud is deposited during periods of slack water and just before and after the slack water conditions, when there is still a weak current flowing. Similar conditions leading to alternate deposition of thick sand and mud layers can operate in areas other than tidal environments. Figure 201 shows a coarsely interlayered sand/mud bedding.

Fig. 201. Coarsely interlayered sand/mud bedding

Some of the coarsely interlaminated bedding, especially type 2, is certainly also produced under conditions of higher flow regime in the form of horizontal laminae, where enough fine-grained material is available. Fluctuations and pulsations in the current velocity lead to the deposition of extremely thin mud layers. However, in this case finer-grained mud layers separating sand layers are extremely thin and rather indistinct. Sand layers are laminated.

Another mode of origin of coarsely interlayered bedding is when sand is occasionally transported in an environment where mud deposition is normally going on. Such is the case on some parts of the shelf where mud deposition is going on. Mud sedimentation is interrupted by occasional heavy storms, which cause deposition of sand layers. Sand layers are rather thin, usually evenly laminated; sometimes laminae are normal-graded. In Recent environments such examples are known from the North Sea (Gadow and Reineck 1969), and the Gulf of Gaeta (Reineck and Singh 1971).

It is certain that this type of bedding can also originate in environments other than those mentioned above. In ancient sediments such interlayered bedding is widespread in varied environments. In many cases, sand layers several centimeters thick alternate with mud layers several centimeters thick.

It is very likely that in many cases the change of material from coarser to finer grained is not the result of short-term changes, as discussed above; but the result of certain long-term changes in the hydrodynamic conditions in the basin of deposition or in the long-term changes in river systems, etc., leading to changes in the type of material being transported into the basin of deposition. Such longterm changes are then repeated rhythmically to produce a thick sequence.

In several ancient sediments one finds thick carbonate layers alternating with thick or thin mud layers. Their mode of origin is not well understood.

Thinly Interlayered Bedding (Rhythmites)

As the name indicates, this bedding is comprised of all the bedding types composed of alternating thin layers of somewhat different composition, texture, and color. Thickness of individual laminae is usually less than 3 or 4 mm. Because layers of two different kinds of material are alternatingly repeated, they can be named rhythmites. The reasons for such rhythmic repetition are regular changes in the transport or production of material. These regular changes can be of short duration e. g., current fluctuations, variation in flow characteristics, tidal changes, or they can be long-term changes e. g., seasonal changes caused by changes in weather conditions.

Rhythmic bedding produced from tidal changes has been repeatedly described as tidal bedding (Schütte 1929; Johnston 1922; Richter 1929; Lüders 1930). Reineck (1967a) describes it as fine rhythmically laminated bedding. Such a tidal bedding is essentially thinly interlayered sand/mud bedding, made up of sand layers one millimeter thick, alternating with mud layers one millimeter thick. The contact between sand and mud layers is rather sharp (Fig. 202).

According to the definition of bed and lamina (p. 96), sand and mud layers of a tidal bedding are very thin beds. Sometimes sand layers are internally laminated foreset laminae. Nevertheless, because the genesis in ancient sediments is not always clear, the term "layer" is preferred.

124

current and turbulences, when sand is transported both as traction load and suspension (9–20 cm/ sec). There is another phase of current and turbulences, when no sand is in motion, but medium silt and clay are kept in suspension (10–4 cm/sec). At still weaker current and turbulences, clay floccules and fine silt are deposited (< 4 cm/sec).

In other words, in tidal bedding sand and mud layers are separated by a period of non-deposition at intermediate current velocity and turbulence conditions. Sand is deposited at current velocities above 10 cm/sec, no deposition of sediment at current velocities between 4–10 cm/sec, and deposition of fine silt and clay floccules at current velocities less than 4 cm/sec.

The genesis of thinly interlayered sand/mud bedding in a tidal environment has been discussed by Reineck and Wunderlich (1967a, b; 1969). According to them, sand layers are deposited during periods of current activity, both flood and ebb currents. The mud is deposited during stand-still phases of high-water and low-water tides. The rela-

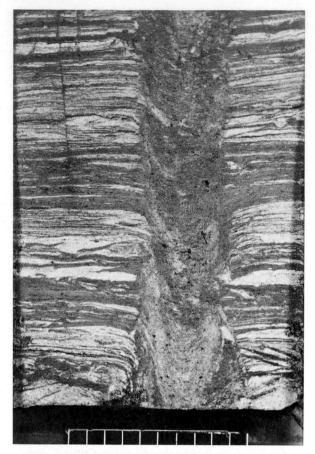

Fig. 202. Box-core from an intertidal flat. A major part of the core is made up of thinly interlayered sand mud bedding (tidal bedding). Some of the layers are only 1 mm or less thick. In the central part a well-defined burrow with spreites is present. North Sea tidal flats. (After Reineck 1972)

The tidal bedding in the North Sea sediments is characteristically ungraded; only rarely reverse or even normal graded bedding is observed in the sand layers. Normally, separation between mud and sand layers is sharp (Fig. 203). Little-Gadow and Reineck (1974) investigated the causes for missing graded bedding in the tidal bedding, although no grain sizes in the range of a long-normal size distribution are missing and sedimentation takes place at a waning current. The main reason for the sharp separation of the sand and mud layer is due to large differences in falling times of silt and sand grains in turbulent water (see Gibbs et al. 1971). Further clay particles are present as floccules and fall together with silt-size particles. In a turbulent flow, sand and coarse silt is deposited, while medium silt and clay is retained in the suspension. Experiments in the flume demonstrated that during a single tidal phase there is a period of

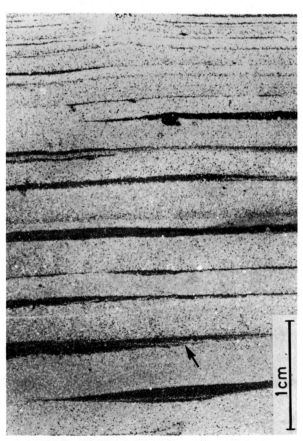

Fig. 203. Thinly interlayered bedding composed of flat lenses embedded in mud. The sand lenses show few foreset laminae (arrow). Black – sand; light – mud. Negative print. North Sea tidal flats.

tive thickness of individual layers can be very different. At certain places and during certain weather conditions sand layers deposited by an ebb current can be thicker than those deposited by a flood current, and vice versa. The same is true for mud layers.

Individual sand layers can be followed laterally for a few decimeters, and in some cases up to several meters. Occasionally, sand layers may show preservation of slightly curved, lower parts of foreset laminae of ripples. In a tidal environment there are at least three more bedding types which resemble tidal bedding:

1. Within the muddy layers a "micro-interlayered bedding" can be observed. This is visible only in thin sections or with the help of a magnifying glass. It is produced as a result of pulsations in the current activity and settling suspension clouds (Reineck and Wunderlich 1969).

2. Finely interlayered sand/mud bedding can also originate when small-current ripples with straight, long crests migrate and enough material is available, and when there is also a simultaneous accumulation of mud in the ripple troughs. The mud accumulating in the troughs is successively covered by foreset laminae of migrating ripples (Wunderlich 1967b). In this way in a single tidal phase several alternating layers of sand and mud are produced. This type of thinly interlayered sand/mud bedding differs from tidal bedding in possessing layers of rather short lateral extent. They are generally only few centimeters to a few decimeters long, whereas sand layers of tidal bedding in favorable conditions can be followed from several meters to several decimeters.

In many parts of intertidal flats (especially on mixed and finer-grained intertidal flats), although the surface is invariably covered with ripple marks, ripple bedding is not always observed. On the other hand, thinly interlayered sand/mud bedding is abundant. McKee (1965) observed the same fact in some of the American intertidal flats. This observation can be best explained by the above-mentioned mode of origin of interlayered sand/mud bedding through extremely flat current ripples.

3. Lenticular bedding with flat lenses also originates as a result of tidal changes and is a special type of tidal bedding. Sometimes in hand specimens or in cores, it may not be possible to differentiate this bedding type from the other two types of interlayered sand/mud bedding (Fig. 203).

The genesis of tidal bedding requires not only tidal rhythm, but also the availability of both sand and mud components. In the tidal environment, where little mud is available, interlayering is less distinct, and individual layers are composed of sand, separated by hardly recognizable, extremely thin mud layers. In this case, they look more like evenly laminated sand.

Thinly interlayered bedding is rather common in intertidal flats and river estuaries. They are rather rare in open-shelf environments.

Thinly interlayered sand/mud bedding has been reported from the foreset and bottomset sediments of the Rhine delta in the Lake Constance (Förstner et al. 1968). Most probably this type of bedding is generated as a result of sorting during sedimentation of suspension clouds, leading to the deposition of sand layers a millimeter thick alternating with mud layers (Fig. 204).

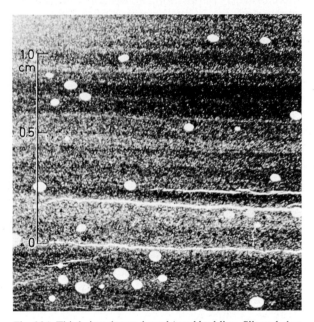

Fig. 204. Thinly interlayered sand/mud bedding. Silt and clay layers alternate with each other. The thickness of layers is variable. Small gas bubble cavities are also visible. Thin section. Rhine delta, Lake Constance

In fine-grained turbidite deposits rhythmites are quite common and silty laminae alternate with mud laminae. Some of them may be the result of deposition from turbid layer flow, each flow depositing one lamina.

Often, in the turbidite deposits rhythmites are developed as graded rhythmites. Piper (1970) describes 10–20 mm thick mud layers from the deep-sea fan deposits showing an upward decrease in the amount of silt and each unit is a single sedimentation unit. Piper and Brisco (1975) describe well-developed sequences of fine-grained rhythmites from the continental margin deposits of Antartica. They are mostly deposits of turbidity currents, while a few may be deposited by contour currents. Piper (1972) discusses the genesis of graded rhythmites (graded laminated bed) of turbidite origin, both from Recent and ancient sediments.

A single layer of graded rhythmite shows alternating silt and mud laminae where there is an overall upward decrease in the abundance, thickness, and mean grain size of silt laminae. The laminae are a fraction of a millimeter to a few millimeters thick, while the unit of graded rhythmite is 1–5 cm thick. The lamination originates in the alternation of cohesive and granular bed conditions. A current, containing both silt and clay, flowing at about 15–30 cm/sec, is able to deposit silt by tractional mechanism on a granular bed and also clay on a cohesive bed. From such a current, firstly, some silt and clay flocs are deposited. The clay flocs cause increased clay deposition and the concentration of suspended clay in the lower part of the flow is reduced. Consequently, silt is deposited to make a silt layer, and the cycle is repeated. In a gradually waning current, the size of silt grains will gradually decrease, and a graded rhythmite is produced.

Gadow and Reineck (1969) describe graded rhythmites produced by deposition of suspension clouds after a storm (see graded bedding).

Stow and Bowen (1978) suggest a new mechanism for the genesis of thinly interlayered mud/silt bedding of fine-grained turbidite deposits. Within a fine-grained turbidity current there are silt grains and clay flocs having equivalent settling velocity, so that the expected deposit from a waning current would be graded, silty mud. However, often such units are distinctly laminated as well as graded. This is related to the depositional sorting by the increased shear at the bottom boundary layer.

The coarser silt settles into the boundary layer as individual grains, and the increasing shear in the boundary layer tends to break up the flocs. The larger silt grains settle down to form a silt lamina. As more sediment is supplied to the top of the boundary layer, the mud concentration increases and reflocculation occurs. At some critical concentration, the clays are able to form aggregates large enough to overcome shear break-up and deposit rapidly through the laminar sublayer as a mud layer over the silt lamina. The process is then continued to deposit another silt layer and then a mud layer. Often, each following silt layer is finer-grained than the previous one and a graded rthythmite is produced. Deposition of these fine-grained sediments can take place in the current velocities of about 15–25 cm/sec. This mechanism can also be operative in other environments where fine-grained laminated sediments are deposited from slowly moving flows.

Rhythmites showing color and textural banding are quite common in the low-energy estuarine and coastal bay deposits. The lamination can be caused by seasonal fluctuations in suspended sediment load or character, or by variations of the amounts and kinds of organic matter supplied by the estuary (Biggs 1978).

Rhythmites produced from rhythmic seasonal changes have been reported repeatedly in Recent sediment. In such seasonal rhythmites a couplet is made up of rather fine-grained material, and individual layers are generally differentiated by light and dark color.

Such rhythmites have been reported from the marine environment by Seibold (1955; 1958) from a constricted bay near Mljet Island, in the Adriatic Sea, where the water depth is 20 m and the environment is rich in H_2S. The rhythmites are made up of alternating dark-and light colored layers (Fig. 205). The light-colored layers are composed mainly of calcite rhombs, with larger grains up to 8 μ in size. They are formed in the summer months, when precipitation of biogenic carbonate is favored as a result of strong assimilation by photoplanktons and increased evaporation. On the other hand, the dark-colored layers are composed mainly of terrigenous material–quartz, iron sulfide, and organic matter. Quartz grains can be as coarse as 10 to 40μ.

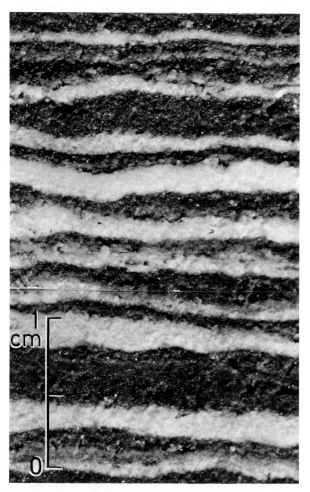

Fig. 205. Seasonal rhythmites. The light-colored layers are made up mainly of calcite, deposited during summer months. Dark-colored layers are made up of terrigenous material, i.e., quartz, iron sulfide, and organic matter, deposited during winter months. Bay near Mljet Island, Adriatic Sea. (After Seibold 1958)

These layers rich in terrigenous material are deposited during periods of higher rains–autumn, winter, and spring.

From Saanich Inlet, British Columbia, a fjord 120 m deep. Gross et al. (1963) and Gross and Gucluer (1964) have reported rhythmites with seasonal couplets (Fig. 206). Here also the bottom water is rich in H_2S. The couplet is made up of an olive green layer rich in diatom skeletons *(Skeletonema costatum)*, alternating with olive black layers with high H_2S content and terrigenous material brought in by rivers. Annual rhythmites have also been described by Strøm (1939) in Drams Fjord, Norway.

Fig. 206. Rhythmites from Saanich Inlet. British Columbia. A couplet is made up of an olive green layer rich in diatoms alternating with an olive black layer with terrigeneous material. (After Gross et al. 1963)

Byrne and Emery (1960) and Calvert (1964) found rhythmites in the Gulf of California from a water depth of 600 m, where there is minimum O_2 content. The rhythmites are made up of light-colored diatom-rich layers alternating with dark-colored clayey layers. In deeper parts of the same basin where O_2 content is higher, sediments are strongly bioturbated, and no visible lamination is found. Emery (1960) and Hülsemann and Emery (1961) also report similar rhythmites from Santa Barbara Basin, California.

In all the cases, such seasonal rhythmites with diatom-rich or calcite-rhombohedra-rich layers are always found associated with stagnation conditions and higher H_2S content. Another important factor is that no benthonic fauna is present which can destroy the bedding because of bioturbation (Calvert 1964).

Another classical example of a stagnant basin is the Black Sea. Here also rhythmites with dark- and light-colored layers have been reported (Arkhangelsky and Strakhov 1938). Müller and Blaschke (1969a, b) demonstrated that light-colored layers are made up mainly of coccoliths ($>$ 50%), diatoms, and calcite grains of 20 to 30 µ in length. The dark-colored layers are made up of detrital material–clay minerals, quartz, feldspars, detrital calcite, and dolomite (Fig. 207).

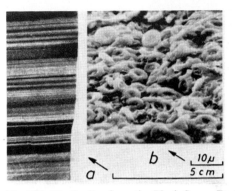

Fig. 207. Rhythmites from the Black Sea. a Dark- and light-colored laminae constitute a single couplet. The light-colored layer is mainly made up of coccoliths, diatoms, and calcite. The dark-colored layer is mainly made up of detrital material. b Coccoliths of the carbonate layer (light-colored layer). (After Müller and Blaschke, 1969a)

Another important type of seasonal rhythmites is the varves of glacial environment (cf. Woldstedt 1961). They are deposited in glacial lakes through the melting water of glaciers. Each couplet is made up of a light-colored relatively coarse-grained layer and a dark-colored relatively fine-grained layer (Fig. 208). The coarser-grained light-colored layer begins with a rather sharp boundary and gradually becomes finer-grained and clayey in texture, grading at last into a dark-colored clayey layer. In other words, the contact between light- to dark-colored layers, is gradational, whereas the contact between dark- to light-colored layers is sharp and abrupt. The light-colored layer is generally coarse to fine silt, and the dark-colored layer is very fine silt and clay (cf. Shrock 1948).

The light-colored layer is known as the summer layer, because it is deposited during the summer months, when much ice suddenly melts, releasing abundant detrital material. This layer gradually becomes finer-grained upward and grades into the dark-colored winter layer, which is deposited during winter months, when no new terrigenous material is available and the suspended finer grained material settles down.

This sequence of events is repeated every year, producing one pair each year. The summer and

Fig. 208. Glacial varves. The light-colored silty layer of summer, together with the dark-colored clay layer of winter constitute a single couplet. The contact between the light to dark layers is gradational, while the light layer shows a sharp lower boundary to the dark layer. (After Pettijohn and Potter 1964)

winter layers can be equally thick, or a thick summer layer may alternate with a thinner winter layer. The relative thickness of individual layers depends upon the variations in weather conditions. Varves can be followed across the whole basin of deposition and generally retain constant thickness.

Harrison (1975) describes rhythmites from glaciolacustrine sediments which look similar to glacial varves. The silt-rich layer grades upwards into a darker clay-rich layer. The lower contact of the silt-rich layer is sharp. Some of the thicker silt-rich layers show foreset laminae, denoting a quick deposition. Thus, these rhythmites represent deposits of turbidity currents in a glacial lake, and not the annual varves.

Ashley (1975) points out that a varve, as a unit, is not a graded bed but consists of two texturally and genetically distinct layers. A couplet is not the result of one sedimentation pulse. She distinguishes three types of rhythmites (varves) in a glacial lake deposit: Group I – clay thickness greater than silt thickness. The silt layer is not graded, though the clay layer shows an upward decrease in grain size.

They are deposited away from the influence zone of melt waters, where normally clay is deposited, interrupted by occasional silt layer deposition. Group II – clay thickness approximately equal to silt thickness. Sometimes it is developed as microvarves. It is also deposited somewhat away from lake deltas. Group III – clay thickness less than silt thickness. The silt layer shows lamination, micrograding of fine laminae, small scale cross-lamination, erosional lower contact. They are formed relatively near delta fronts, associated with high sedimentation rates. Agterberg and Banerjee (1969), and Gustavson (1975a) also provide information on the genesis of glacial varves.

Sturm and Matter (1978) discuss deposition in Lake Brienz, Switzerland. In the central basin, deposition takes place mainly by annual low-density turbidity currents during summer leading to deposition of a graded sand layer, followed by deposition of suspended material. During the winter months a thin winter layer is deposited. Simultaneously on the slopes, out of reach of turbidity currents, varves are deposited, made up of a summer silty layer and a finer-grained winter layer. The turbidity couplets of the basin can be traced laterally into the varve couplets of the slope. They conclude that the varves and turbidites are genetically related and require overflows, interflows, and underflows (turbidity current), and the existence of seasonal thermal stratification is of critical importance.

Sturm (1979) discusses the genesis of rhythmites in general, and varve in particular, in the lakes. According to him, genesis of rhythmical lamination in lakes requires the presence of suspended matter within the water column which can be of varied origin, e. g., river, air, chemical precipitate, upwelling etc. Deposition and genesis of rhythmites depends upon: (1) whether the suspended matter supply is continuous or discontinuous, (2) whether the lake is unstratified, partly stratified (seasonal), or totally stratified. Different combinations of these factors lead to genesis of a number of types of rhythmites (Fig. 209). An important assumption for this scheme is that there is no bottom current activity, and deposition is exclusively from suspension. According to this scheme, deposition of the classical type of varve (showing distinct sand, silt, and clay layers) takes place only if suspended matter is introduced in the lake by discontinuous (seasonal) influx during periods of seasonal stratification. The coarser sediments are immediately deposited as a graded sand-silt layer. The clay is trapped during stratification within the epilimnion, and starts settling after overturn only and causes the deposition of a distinct non-gradational clay layer on the top of the earlier deposited coarser layer.

The varves have been well investigated in the area around Baltic Sea and have been used for age

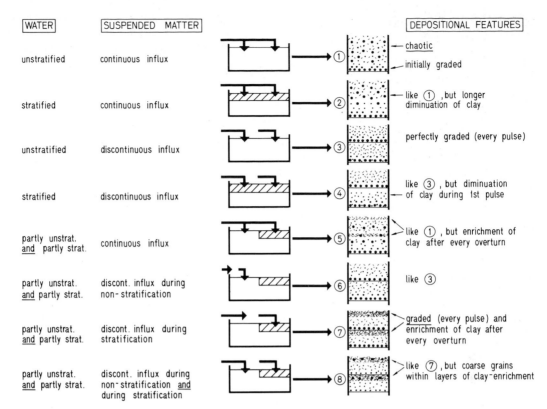

Fig. 209. Schematic representation to show genesis of different types of rhythmites in the lakes as a result of variability in two hydrological parameters (stratification of water and influx of suspended matter). The lake is taken as oligotrophic with clastic deposition. Case 7 represents conditions for deposition of ideal clastic varves. Heavy dots – sand; light dots – silt; light dashes – clay. (After Sturm 1979)

determination (Sauramo 1918, 1923; Troll 1925; De Geer 1940, and others).

Olausson and Olsson (1969) describe varve-like rhythmites in a core from the Gulf of Aden. Dark relative thickness of individual layers depends upon relative content of sand and organic matter. Thickness of the varve is about 1 mm. This bedding appears because of fluctuation in aeolian transport. Aeolian transport is high in summer and very low in winter. Since the outflow from the Red Sea is smaller in summer, sand-rich laminae poor in organic matter are produced in the summer; sand-poor, laminae rich in organic matter are produced in the winter. The absence of benthonic animals from a shortage of O_2 helps in the preservation of these varves. Seasonal rhythmites of dark- and light-colored layers have also been reported from the upper continental slope off Karachi, Pakistan (von Stackelberg 1972). In the same region exists low O_2 content in the bottom water.

Rhythmic bedding is also well known in evaporite sequences. Richter-Bernburg (1955) describes rhythmites made up of alternating layers of dolomite and anhydrite.

The main difference between seasonal rhythmites and tidal rhythmites lies in the grain size and the lateral extent of individual layers. Rhythmites produced in the stagnation basins (marine or non-marine) are composed of very fine-grained material, mainly silt and clay. The layers of a couplet differ from each other mainly in color and type of material. As these layers are deposited in rather calm waters, the layers can be followed for hundreds of meters. Rhythmites of the tidal environment, on the other hand, are made up of alternating sand and mud layers. Thus, layers of a single couplet differ mainly in grain size. As a result of strong tidal current activity, sediment is fractionated; sand is deposited during periods of current activity, while mud is deposited during still-stand phases. The layers of tidal rhythmites can be followed only for a few meters, in some cases for probably only a few decimeters.

A point worth mentioning here is that in ancient sediments the time interval represented in rhythmic patterns might have been very different. It is quite likely that in geological times a couplet of seasonal rhythmites was not produced in a single year (365 days), but during several years. In the same way, a tidal cycle in geological times need not have been 12 hrs, as it is today.

Thinly Laminated Mud

Most of the thick mud layers (more than 1 cm in thickness), which at first glance look massive and to have no internal laminations, are usually internally finely laminated. Internal laminations of such mud layers are best studied in thin sections or with the help of X-rays.

Sometimes, mud layers are interrupted by sporadic layers of fine sand or silt, only one or two grains thick in diameter. Some of the mud layers of the Bodensee, Germany, show well-developed lamination in thin sections. This lamination is apparent because of sporadic, extremely thin coarse silt layers. In some of the mud layers of intertidal flats extremely thin (1 to 2 grains thick in diameter) silty layers impart a laminated character to otherwise rather homogeneous layers (Fig. 592). Sometimes the term "parallel-laminated mud or silt" is used for this type of bedding.

Even pure mud layers generally show some sort of banding. Such banding may be the result of differences in texture, composition, and color. Coleman (1966) demonstrated that massive-looking freshwater clays of the Atchfalaya Basin, when exposed to X-rays, revealed a fine banding. The main factor responsible for such fine-parallel layering, according to him, are alternations of flocculated and nonflocculated layers, and the differing content of colloidal organics.

Homogeneous Bedding

The term "massive bedding" is sometimes used to describe more or less homogeneous-looking sediments. However, most homogeneous-looking sediments show internal laminations if special techniques are applied e.g., exposure to X-rays, (see Hamblin 1965). However, there definitely are cases where lamination or some other type of internal arrangement within the beds is absent or is so faintly developed that it is not possible to decipher it even with the help of X-rays or in thin sections. Homogeneous bedding can be present in fine-grained as well as in coarse-grained sediments. We feel, however, that the term "homogeneous bedding" (massive bedding) should be used in only specific cases, where all methods fail to decipher any internal arrangement.

Homogeneous bedding can be just a diffused mixture of unsorted sediment grains without any well-defined structure. Some of it might be the result of strong bioturbation activity. Animal activity results in the mixing of sediment completely and in the destruction of primary layering, giving it the appearance of a homogeneous mass. Werner (1963) found that micro-organisms living in pore spaces can destroy the primary bedding and produce homogeneous-looking sand. Such homogenization can also result from inorganic processes, e. g., when water is drained out during compaction or when strong generation of gas bubbles moving upward causes destruction of bedding (cf. Förstner et al. 1968). Some of the homogeneous beds may be primary in origin. In the case of very rapid sedimentation, the sediment is dumped as a homogeneous mass. Sediments deposited by grain flow often lack any visible bedding structures.

Thin Sand Layers

In muddy sequences often sand layers are found a few millimeters or a few centimeters thick, which are the result of different processes, depending upon the depositional environment. Internally, such sand layers are made up of different types of bedding features. In general, sand layers deposited in the deeper waters with low turbulences show grading; while the sand layers deposited in turbulent waters show lamination, or even foreset laminations, if enough current is present. Reineck (1974d) compares such sand layers from different environments (Table 7). The sand layers of lake mud mostly show fully developed graded bedding, along with layers showing thin lamination. The sand layers of the transition zone to shelf mud show lamination in the lower part and wave ripples on the top.

Table 7. Comparison of characteristics of thin sand layers embedded in the muds of different modern depositional environments. (Translated by P. E. Potter from Reineck 1974 d)

Depositional site	Grading — Type A*	Type B*	Immature**	Mature**	Graded Rhythmites	Microrhythmites	Cross lamination	Lamination	Erosional discordances within sands	Internal vertical sequence	Contact with underlying mud — Sharp	Graded	Contact with overlying mud — Sharp	Graded	Sand content of Mud	Sequence with Mud	Lateral grain Size Decrease	Thickness	Form, size and occurrence
Deutsche Bucht 15–40m deep	C	–	C	R	C	C	R	C	R	Rᵃ	Cᵇ	–	C	R	C	Cᶜ	C	cm	Parallel to coast in wide 40 km bands
Deutsche Bucht less than 15 m deep	–	–	–	–	–	–	C	C	C	Cᵃ	Cᵇ	–	C	–	C	Cᶜ	R	cm to dm	Bands parallel to coast produced by storm floods
Gulf of Gaeta 30–300 m deep	C	–	C	–	C	C	R	C	–	–	C	–	C	–	–	Cᵈ	C	cm	Parallel to coast in bands 20 km wide
Lake Basin of Bodensee	–	C	–	C	–	–	–	Cᵉ	–	–	Cᶠ	–	–	C	–	–	C	mm to cm	Long tongues of several hundreds of meters
Rhine Delta of Bodensee	Rʰ	–	C	–	–	C	Rⁱ	C	Cⁱ	–	C	–	C	R	R	–	C	cm to dm	Wide tongues of 100 s of meters
Contourite from Continental Slope, Eastern North America	–	R	–	R	–	C	R	C	–	–	C	–	C	–	–	Cʲ	?	mm	Wide bands many kilometers long
Watt, North Sea	–	–	–	–	–	C	C	C	C	–	Cᵇ	–	C	–	C	–	–	mm to cm	Small patches

C = Abundant; R = Rare; – = Absent

 * Pettijohn (1957, Fig. 59); see Fig. 197
** Mature grading goes from coarse to fine without a break, whereas immature has a sharp size gap separating underlying sand from overlying clay

[a] Commonly laminated with small graded zone in upper part
[b] Uneven erosional contacts
[c] Commonly overlain by an undisturbed (?) mudbed
[d] Plant debris common over sand layer
[e] In Lake of Geneva partially laminated, according to Houbolt and Jonker (1968)
[f] Uneven expression
[g] Because of the diverse origins of the sands in the Rhine delta, their characteristics are also very diverse
[h] Graded from fine below to coarser at top
[i] Only found in the channels of the Alpine Rhine up to 14 m depth
[j] Sand layers with globigerina free clay
[k] Lower part of sand layer is laminated; upper part has crosslamination from wave ripples

Sediment Grain Parameters

General Information

The units of sedimentary rocks are mineral and rock grains. The features of individual grains, such as roundness, sphericity, surface markings, etc., are controlled partly by the environment of deposition. In the same way, the maximum grain size, grain-size distribution, fabric, etc., are also controlled by the hydrodynamic conditions existing at the time of deposition. There have been repeated attempts to use such grain parameters as environmental indicators. In the following, a short account of various attempts in this direction and their limitations are given.

Grain Size

The grain size of a clastic sediment is a measure of the energy of the depositing medium and the energy of the basin of deposition. In general, coarser sediments are found in higher-energy environments and finer sediments in low-energy environments. Nevertheless, an important limiting factor is the availability of grains of various sizes.

It has been repeatedly shown that in the direction of transport, the grain size decreases. This feature is well-developed in fluvial sediments, in which the mean grain size decreases in the downstream direction. Pebbles demonstrate this feature very convincingly. Sternberg (1875) demonstrated a decrease in the size of pebbles in the downstream direction of the Rhine River. Plumley (1948) showed the same features in three streams in South Dakota. The decrease in grain size in sand-size sediments is not so dramatic. Over long distances they do show a decrease in grain size, but here the down-current decrease in grain size shows local fluctuations (Pettijohn 1957; Leopold et al. 1964). The fluvial transport over several hundred kilometers of fine sand is incapable of producing decrease in median or other grain size or mineralogical characteristics (Pollack 1961; Kumar and Singh 1978).

The down-current decrease in grain size is ascribed to two processes–abrasion and progressive sorting (Pettijohn 1957). The latter process, however, seems to be more important. The process in which a decrease in grain size is caused by wear and tear of the grains plays a more important role for pebbles than for sands. The other process, which is probably the main factor for the decrease in grain size, especially in sand-size sediments, is sorting during transport. With a decrease in energy and competency of the transporting medium, coarser sediments are deposited, and only finer material is transported farther. In shallow-marine environments a decrease in grain size from sandy beaches to muddy shelf sediments can be ascribed to similar processes. Regional variations in grain size may be helpful in the reconstruction of the basin of sedimentation. However, care must be taken to account for various possibilities: For example tributaries of a main river bringing a different type of sediment (coarser sediment) may cause a downstream increase in grain size and decrease in roundness and sphericity.

Similar difficulties occur in marine shallow-water sediments where new sediment is brought and at the same time some sediment is reworked from the bottom. This produces mixed sediments and residual sediments. In such areas the regional variation in grain size provides little or no help in the reconstruction of the basin of deposition.

Grain-Size Distribution

Various factors related to grain-size distribution, such as the shape of cumulative curves, frequency curves, and histograms, and various other parameters calculated from grain-size distribution, have been tried as environmental indicators.

There are several ways in which grain-size data can be plotted and treated statistically. Most common are the histograms, frequency curves, and cumulative curves. Advantages and disadvantages of various methods have been dealt in detail in various textbooks of sedimentary petrology (Krumbein and Pettijohn 1938; Milner 1962; Köster 1964; Müller 1964). Most commonly cumulative curves are plotted on logarithmic paper. Generally, quartiles Q_1 (25%), Q_2 (50%), and Q_3 (75%) are read from the curve. With the help of these quartiles, the following Trask's parameters are calculated:

Median grain size (Md): Q_2

Sorting coefficient (So): $\sqrt{Q_3/Q_1}$

Skewness (Sk): $\dfrac{Q_1 \cdot Q_3}{Q_2^2}$

Another coefficient which is usually calculated

is kurtosis (k) $= \dfrac{Q_3 - Q_1}{2\,(P_{90} - P_{10})}$ where P_{90} is percentile

(90) and P_{10} is percentile (10).

Currently, statistical parameters of Folk and Ward (1957) are in common use:

Mean size: $M_z = \dfrac{\Phi 16 + \Phi 50 + \Phi 84}{3}$

Inclusive graphic standard deviation:

$$\sigma_I = \frac{\Phi 84 - \Phi 16}{4} + \frac{\Phi 95 - \Phi 5}{6.6}$$

Inclusive graphic skewness:

$$sk_I = \frac{\Phi 16 + \Phi 84 + \Phi 50}{2\,(\Phi 84 - \Phi 16)} + \frac{\Phi 5 + \Phi 95 - 2\,\Phi 50}{2\,(\Phi 95 - \Phi 5)}$$

Graphic kurtosis: $K_G = \dfrac{\Phi 95 - \Phi 5}{2.44\,(\Phi 75 - \Phi 25)}$

Figure 210 shows various types of sediment distribution curves. Some workers prefer the use of moment measures over quartile parameters. A number of authors have attempted to use parameters other than Trask's or even refinements of these parameters (Inman 1952; Folk and Ward 1957, and others). Folk (1966) provides a useful review on grain-size parameters. In many cases, the grain-size distribution curve was found to be log-normal. However, several workers found that the grain-size distribution curve can be analyzed in more than one part, each part showing a log-normal distribution (Doeglas 1946; Tanner 1959; Walger 1962).

Walger (1962) suggested that the polymodality of sediment results because of sampling error, where material from different layers is sampled. Analysis of individual layers showed a log-normal distribution. However, Walger (1962) further noticed that in many cases, even the grain-size distribution of single laminae seems to be made up

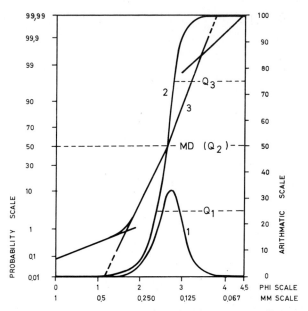

Fig. 210. Diagram showing three common types of sediment distribution curves. *1* Frequency curve; *2* cumulative curve; *3* log-probability curve. Curves 1 and 2 are plotted on an arithmatic scale; curve 3 is plotted on a log-probability scale. Three commonly calculated parameters *Q_1, Q_2 (Md)*, and *Q_3* are marked. (Modified after Visher 1969)

of three log-normal components. He postulated that this can be related to three modes of sediment transport – suspension, saltation, and rolling.

It has been recognized by various authors that the sorting coefficient is strongly dependent upon the median grain size (Krumbein and Aberdeen 1937; Inman 1949; Griffiths 1951; Emery and Stevenson 1950). Inman (1952) found that if the sorting coefficient is plotted against the median diameter, the grain size 0.1 to 0.2 mm exhibits the best sorting. Grain-size analysis of single laminae by Walger (1962) also shows that grain size 0.1 to 0.2 mm possesses the best sorting. However, scattering of the data for single lamina was much less than for total samples, and plots were concentrated more toward lower values (better sorting). The sorting coefficient of a single lamina is called the *elementary sorting* (Walger 1962; Seibold 1963). The elementary sorting represents the optimal sorting which can be attained by sediment of a given grain size. The sorting of average or total sample (average sorting) depends upon the variations in the median diameters of the individual laminae. By relating average sorting to elementary sorting, one obtains the relative sorting coefficient, which is independent of grain size (Seibold 1963). If the relative sorting coefficient is 1, the sample is homogeneous. Generally, because of inhomogeneity of the sample, it is more than 1. In case of residual sediments, the relative sorting coefficient may attain values which are less than 1. Residual sediments with relative sorting coefficient values less than 1 originate when sediment deposited under normal conditions is reworked, winnowing away the finer fractions. By applying these parameters, Seibold (1963) was able to recognize residual sediments in the coastal sands of the Baltic Sea. The relationship between the coefficient of sorting and the median diameter is shown in Fig. 211. The sorting coefficient has often been used as an environmental indicator. However, most of the studies give only relative values. In general, the worst sorting is exhibited by coarse-grained sediments of alluvial fans and moraines. Beach pebbles possess better sorting than river pebbles (Emery 1955; Ruchin 1958). In modern environments it can be said that coastal sands show better sorting than sublittoral marine sands, intertidal flats, and fluvial sands. Dune sand of aeolian origin shows the best sorting.

There have been numerous attempts to relate statistical parameters calculated from grain-size distributions to different environments of deposition (e.g., Folk and Ward 1957; Mason and Folk 1958; Harris 1958; Friedman 1961, 1962, 1967; Sahu 1964; Chappell 1967). Some of these workers plotted different parameters against each other. Doeglas (1968) calculated Wenntworth phi values of Q_1, Md, Q_3, 1st percentile and 99th percentile. These values give 3- and 5-digit indices, which may enable one to distinguish between river, aeolian, and coastal marine deposits. However, all these attempts had only a limited success in environmental interpretation.

Doeglas (1946) tried to relate the shape of the grain-size distribution curve to the specific environment of deposition. Sindowski (1958) found that parameters such as sorting and median cannot be used as environmental indicators, and he, like Doeglas (1946), related the shape of the curves to specific environments. He related the shape of grain-size distribution curves to various shallow-water environments: strand, intertidal flat, shelf, tidal inlet, minor tidal channel, fluvial, relict. His analysis was based on serveral thousand grain-size determinations in modern environments. He also provided a few ancient examples.

Passega (1957, 1964) suggested the use of C/M diagrams for environmental analysis. By plotting M (median) against C (1 percentile, approximate value of maximum grain size), he obtained C/M patterns. The position of points on a C/M diagram depends upon the mode of deposition of sediments, and he tried to distinguish between turbidites, still-water deposits, etc. Position of a sediment in a C/M pattern depends on the mode of transport. Deposits of various environments give characteristic patterns. Thus, C/M diagrams may provide some reference points for the interpretation of environments (Fig. 212). Rizzini (1968) followed a similar approach. Later, Passega and Byramjee (1969) suggested that together with plots of C/M diagrams, other plots F/M, L/M, and A/M can provide additional information for characterizing sediment and its environmental information. C/M characterizes the coarsest fractions. F/M, L/M, and A/M characterize the finer fractions. M is me-

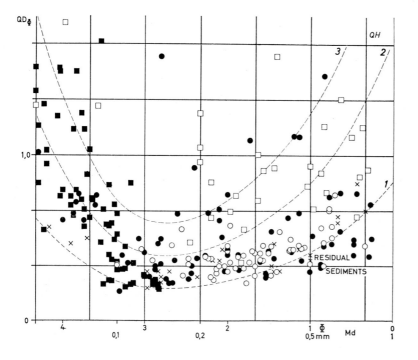

Fig. 211. Diagram showing the coefficient of sorting $(Q\,D\,\Phi)$ as a function of the median diameter (Md). Sediments are from various environments and regions. X data of grain-size analysis of single laminae from various sources. Dashed lines equal QH values (relative coefficient of sorting). Sediments having a QH value < 1 are residual sediments. Originally from Walger (1962). (After Seibold 1963)

Fig. 212. The C/M pattern of sediments and their significance in interpretation of mechanism of transport. Traction current deposits can be differentiated into various parts depending upon mode of transport. N/O rolling sediments; O/P = rolling sediments with some suspension sediments; P/Q = graded suspension with some rolled sediments; Q/R = graded suspension (saltation) deposits; R/S = uniform suspension; T = pelagic suspension. CS is maximum grain size transported as graded suspension; CU is maximum grain size transported as uniform suspension. Segments I, II, III, and IX are denoted by C > 1 mm. Here mainly rolled sediments are incorporated. Suspension is of minor importance. Segments IV, V, VI, and VII are denoted by C < 1 mm. Mainly suspension sediments, rolled sediments < 1 mm may be incorporated. The pattern of turbidity current deposits runs parallel to limit C = M. (Based on Passega 1957, 1964; Passega and Byramjee 1969)

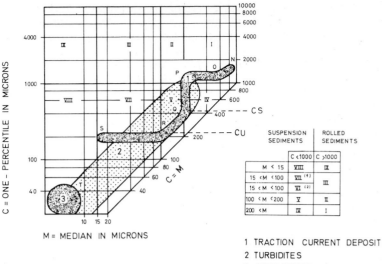

	SUSPENSION SEDIMENTS	ROLLED SEDIMENTS
	C <1000	C >1000
M < 15	VIII	IX
15 <M <100	VII (¹)	III
15 <M <100	VI (²)	
100 <M <200	V	II
200 <M	IV	I

1 TRACTION CURRENT DEPOSIT
2 TURBIDITES
3 QUIET WATER SUSPENSION DEPOSITS

dian; F, L, and A are the percentages by weight in the samples of grains finer than 125, 31, and 4 μ, respectively. According to them, these diagrams form as a group the "grain-size image" of a sediment. These parameters are related to transport and deposition mechanisms. Consequently, the "grain-size image" provides information about the hydrodynamic conditions under which a sediment was deposited.

There are three main modes of transport of sediment: rolling, saltation, and suspension. Despite early recognition of the importance of these processes in sedimentation (e. g., Inman 1949), it was Moss (1962, 1963) who tried to correlate the processes of transport with the shape of the grain-size curve. He found that a single sample may contain all the three subpopulations related to the three modes of transport. Relative abundances of these subpopulations would, thus, reflect the relative importance of the three modes of transport in a given environment. His ideas led to a new approach in environmental analysis with the help of grain-size distribution. Visher (1969) related the shape of grain-size curves to the mode of transport. He published typical curves from various modern as well as ancient environments.

The main idea behind this analysis is the recognition of subpopulations within an individual grain-size distribution. The subpopulations are easily recognized by analyzing log-probability grain-size distribution curves (Fig. 213). Each lognormal subpopulation is related to a mode of transport (suspension, saltation, rolling). In the following we shall give a short account of these three processes of transport and their importance in the grain-size curves.

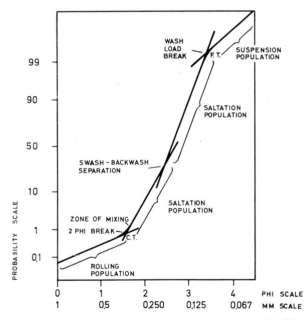

Fig. 213. Grain-size distribution curve plotted on a probability scale. Different log-normal subpopulations become apparent. Each subpopulation can be related to a mode of sediment transport. C. T. is the coarse truncation point; F. T. is the fine truncation point. (Modified after Visher 1969)

Suspension Transport

The maximum grain size of sediment that may be held in suspension depends primarily upon the turbulence energy of the transporting medium. According to Lane (1938), it is usually less than 0.1 mm.

However, this value varies greatly in different environments and depends upon the hydrodynamic conditions at the time of sedimentation. There is always a certain amount of interchange between suspension and bedload. Most of the sediments contain some amount of fine sediments which are deposited from suspension. This can be easily recognized as a separate subpopulation in log-probability curves (Visher 1969).

Saltation Transport

There is very little information regarding the maximum grain size which moves in saltation transport. Data of the U.S. Waterways Experiment Station (1939) shows that grains up to 1.0 mm in diameter have been sampled 60 cm above the bottom. Various hydrodynamic factors, such as current velocity, water depth, and nature of bed, seem to control the maximum size transported by saltation. Plots on log-probability paper show that coarser grains

make a subpopulation (Visher 1965b). In exceptional cases, when opposing flows are active, two distinct saltation subpopulations may develop which differ from each other only slightly in median and sorting. This is the case in beach sediments where swash and backwash are active (Visher 1969). Kolmer (1973) found double saltation population related to swash-backwash processes in a wave tank experiment.

Rolling Transport

Sediment grains moving by rolling transport are the coarsest ones in a sample. Most of the grain-size distributions show a rather coarse-grained subpopulation which is different in median and sorting than the other two subpopulations. Some sediments, e. g., certain fluvial sediments, do not differentiate between rolling and saltation subpopulations.

Table 8. Grain size characteristics of sands of various environments. (After Visher, 1969)

Environment	Saltation population (A)				Suspension population (B)				Rolling population (C)			
	%	Sorting	C.T. Phi.	F.T. Phi.	%	Sorting	Mixing A and B	F.T. Phi.	%	Sorting	C.T. Phi.	Mixing A and C
Fluvial	65 to 98	Fair	−1.5 to −1.0	2.75 to 3.50	2 to 35	Poor	Little	>4.5	Varies	Poor	No limit	Little
Natural levee	0 to 30	Fair	2.0 to 1.0	2.0 to 3.5	60 to 100	Poor	Much	>4.5	0–5			None
Tidal channel	20 to 80	Good	1.5 to 2.0	1.5 to 3.5	0 to 20	Poor Good	Much	3.5 to >4.5	0–70	Fair Good	−0.5 to −1.5	Average
Tidal inlet	30 to 65	Good	1.25 to 1.75	2.0 to 2.5	2 to 5	Fair Good	Average	3.5 to 4.0	30–70	Fair Good	−0.5 to No limit	Average
Beach	50 to 99	2 populations Excellent	0.5 to 2.0	3.0 to 4.25	0 to 10	Fair to good	Little	3.5 to >4.5	0–50	Fair	−1.0 to No limit	Average
Plunge zone	20 to 90	Good	1.5 to 2.5	3.0 to 4.25	0 to 2	Good	Much	3.0 to >4.5	10 90	Fair Poor	No limit	Average
Shoal	30 to 95	Good	2.00 to 2.75	3.5 to >4.5	0 to 2	Poor to fair	Little	3.5 to >4.5	5–70	Fair Poor	0.0 to −2.0	Much
Wave zone	35 to 90	Good to excellent	2.00 to 3.00	3.0 to >4.5	5 to 70	Fair Poor	Much	3.75 to >4.5	0–10	Poor	0.0 to No limit	Little
Dune	97 to 99	Excellent	1.0 to 2.0	3.0 to 4.0	1 to 3	Fair	Average	4.0 to >4.5	0–2	Poor	1.0–0.0	Little
Turbidity current	0 to 70	Fair Poor	1.0 to 2.5	0.0 to 3.5	30 100	Poor	Much	>4.5	0–40	Fair Poor	No limit	Much

C.T. = coarse truncation point, F.T. = fine truncation point, A = saltation population, B = suspension population, C = rolling population

Fuller (1961) suggested that the break between saltation and rolling subpopulations, in many cases, is near 2 phi, or the point of junction between impact and Stoke's law. Krumbein and Pettijohn (1938) give an account of impact and Stoke's law.

A plot of grain-size curve on log-probability paper gives straight-line curve segments, each segment denoting a single subpopulation. Each subpopulation shows a log-normal distribution and differs from other subpopulations in median and standard deviation. The number, amount, size-range, mixing, and sorting of subpopulations of a sample vary systematically in response to provenance, sedimentary process, and sedimentary dynamics (Visher 1969). The analysis of these parameters is the basis for interpretation of the depositional environment of a sediment. Table 8 summarizes the information on grain-size distribution for some environments.

Allen et al. (1972) discuss the use of R-mode factor analysis in the grain-size distribution and find it useful in identifying the subenvironments of the Gironde estuary, France. Three main groups are found, i. e., less than 3.0 φ related to uniform suspension transport, 3.0–0.6 φ related to mixed suspension bedload transport, more than 0.6 φ related to pure bed load transport. Buller and McManus (1972) made plots of Trask's arithmetic measure of quartile deviation QDa against Md mm, and skewness Ska against Md mm on a double log paper and found that plots of each environment vary in a linear fashion. While both the position and slopes of the individual trend curve are different for each environment, this method can be used to distinguish depositional environments. More recently discriminant function analysis has been used for determining the depositional environment. Reed et al. (1975) make multivariate discriminant-function analysis of settling velocity (Psi) distributions to differentiate between sands of different environments. They prefer the settling technique to the sieving in grain-size analysis. Stapor and Tanner (1975) use stepwise linear discriminant analysis to distinguish beach, beach ridge, and coastal dune sand samples from the East coast of America.

Füchtbauer and Müller (1970) compiled the existing information on grain-size parameter (Table 9). Trask's parameters have mainly been considered, as they are most commonly and easily obtained.

In all the three major environments (i. e., fluvial, aeolian, and marine) there are low-energy and high-energy environments. High-energy environments show coarser, better-sorted sediments. Low-energy environments show finer-grained and poorly sorted sediments with much fine-grained fractions (skewness > 1). These data can be helpful in

Table 9. Grain-size parameters of sediments of various environments. (After Füchtbauer and Müller 1970)

1. *Fluvial environment*
   a) River bed and point bars
      Sorting (Q3/Q1) mostly > 1.2; in unregulated rivers mostly > 1.3, skewness < 1, seldom > 1. Typical is the fining upward sequence
   b) Flood plain
      Sorting mostly > 2, skewness always < 1 (fine-grained tail in grain size distribution).

2. *Aeolian environment*
   a) Sand dunes
      Sorting good, skewness mostly < 1. Coarse-grained tail generally absent. Only little variation in grain size in vertical sequence. Median diameter mostly between 0.15 and 0.35 mm.
   b) Loess sediments
      Poorly sorted, skewness mostly < 1 (much fine grained fraction), median diameter generally < 0.1 mm.

3. *Marine environment*
   a) Beach
      Sorting of beach sediments ist the best (mostly 1.1–1.23), skewness mostly > 1, on log probability paper cumulative curve shows two saltation subpopulations.
   b) Shallow marine (tidal flats and shelf) Sorting bad, skewness < 1, in offshore shelf parts sand fraction is almost absent.
   c) Deep sea (continental slope and abyssal plane)
      On continental slope clayey silt, in abyssal plane silty clay; interrupted by coarser-grained deposits of turbidity currents.

providing a preliminary orientation regarding environment.

To conclude, grain-size distribution can be helpful in the identification of depositional environments. However, several limitations should be kept in mind.

The major reason why most of the attempts to use grain-size distribution as environmental indicators have been only partially successful, is that the grain-size distribution is a product of the hydrodynamic factors of an environment. Because the same or similar hydrodynamic factors may be active in a number of different environments, similar grain-size distribution may result in rather different environments.

Another important factor is the availability and mixing of different grain sizes. For example, beach sands are in general well sorted, with little or no fine-grained material. But on the beaches adjacent to river estuaries and deltas, where abundant fine material is brought in, even beach sediments can incorporate much fine-grained sediment.

River sediments generally have much fine material. It may happen that in some rivers fine sediments are not available, so that the deposited sediment lacks fine grades. Thus, provenance and vicinity to source may affect the hydrodynamic picture of an environment of deposition.

Sediments of glacial environments mostly contain an appreciable amount of silt and clay. It is most in the morainic till and decreases in the outwash-plain sediments. Sandy deposits of the till show a wide range in grain-size distribution, though on reworking with water this range is narrowed.

Solohub and Klovan (1970) found that none of the methods of grain-size parameters is helpful in identifying the lacustrine subenvironments.

The applicability of grain-size parameters to ancient sediments is associated with even more hazards. Here together with the above-mentioned factors, post-depositional changes (diagenesis) may affect the grain-size distribution. Moreover, only very scanty information is generally available about the provenance, vicinity to source, and availability of certain grain size grades at the time of deposition. Thus, we suggest the use of grain-size parameters as an environmental indicator only in combination with other parameters, such as sedimentary structures, bioturbation, geological setting, etc. In any case, the grain-size distribution always provides some information on the general hydrodynamic conditions at the site of deposition. One can quite easily distinguish between high-energy and low-energy environments, and determine the relative role of various sedimentation processes–suspension, saltation, rolling. In a fluvial sequence Visher (1965) could successfully distinguish sediments deposited in a lower flow regime from those deposited in a upper flow regime.

Glaister and Nelson (1974) applied grain-size parameters in determining environment of deposition, and suggest that log probability plots give better results than the moment measure cross-plots and factor analysis. They give characteristic curves for braided stream, point bar, beach, tidal flat, dune, stream-mouth bar, distributary channel, interdistributary beach, density and debris flows.

Shape and Roundness of Sediment Grains

The shape and roundness of sediment grains of a sample have been repeatedly used to decipher the environment of deposition. The shape and roundness of sediment grains depend upon the medium of transport and the mode of transport. However, an important controlling factor is the composition and inherent internal structure of sediment grains and the original form of mineral grains. A well-bedded sediment or rock with well-developed fissility and cleavage possesses the tendency to produce tabular, elongated sediment grains, whereas

massive rocks tend to produce spherical grains. The relief of an area, intensity and type of weathering also play important roles in determining the shape of the sediment grains. Some of the rocks are composed of mineral grains which are naturally elongated. This is the case in metamorphic rocks. Thus, quartz grains from metamorphic rocks are more elongated than the quartz grains from igneous rocks. Methods of determination and interpretation of shape and roundness of coarser sediments gravels, etc. differ much from sand-sized sediments. For this practical reason we shall discuss the gravels and sand under different heads.

Shape and Roundness of Gravel

Rayleigh (1943, 1944) gives an account of experimental changes in the shape of pebbles during transport and demonstrates that elongated or irregular grains at the end never attain spherical shape, but remain somewhat flatter in shape. Behrens (1977) discusses the basic concepts of the shape of pebbles, and proposes a new technique for determining their stereometry. Applying this method to the pebbles of Scottish beaches, he found that for each size grade pebbles of quartzites are more spherical than the basalt pebbles, which tend to be slightly flatter and elongated. The shape of these basalt pebbles seems to be controlled by the primary anisotropy in the basalt due to alignment of the felspar grains.

Because of the larger size of gravel-sized grains, they can be easily measured and described with the help of three axes: a–maximum length; b–maximum width; and c–maximum thickness. Based on these three measurements, several indices can be calculated (for details see Köster 1964; Müller 1964). Krumbein (1941) suggested a method of visual estimation of the shape of pebbles, where nine classes are differentiated. This method gives comparable results to those obtained by direct measurements.

Cailleux (1947, 1952) calculates the degree of flatness $(a-b)/2c$ and discusses its environmental significance. The degree of roundness of a pebble is $\frac{2r}{L}$, where L-longest dimension of the pebble, r-the smallest radius of curvature of the least rounded part of the pebble. However, the flatness of pebbles is strongly influenced by the source material. The degree of sphericity of gravels increases with the increasing transport path or the time available for the wear and tear of the gravels. Table 10 gives degree of flatness and roundness of gravels of carbonates from various environments.

Zingg (1935) differentiates between four forms: Tabular, equant, bladed, and prolate. These forms

Table 10. Average values of degree of flatness and roundness of carbonate gravels from various environments. (After Cailleux 1952)

Environment	Degree of flatness	Round-ness
Channel lag gravels	1.2–1.6	290
Ground moraine	1.6–1.8	40– 90
Fluvio-glacial	1.7–2.0	240–300
Seashore	2.3–3.8	170–610
Lake shore (Lake Geneva)	2.3–4.4	300–370
Frosted pieces	2.0–3.1	10– 40
Rivers in warm temperate climate	2.5–3.5	70–200

can be correlated with the sphericity index (see Pettijohn 1957).

The mode of transport and transporting agents can affect the form of gravels. In general, it is said that gravels of the sea coast tend to be flatter than the gravels of rivers (Cailleux 1945; Kuenen 1964a). Lenk-Chevitch (1959) found that beach pebbles are flattened, but possess a subtriangular outline if observed with the short axis oriented vertically. River pebbles are generally elongated or rod-shaped.

On the steeply dipping high-energy coasts spherical gravels predominate. However, on low-energy coasts flat gravels are dominant (Kuenen 1964a). Bluck (1967) discusses distribution of the pebbles on the beaches derived from glacial gravel clays. He found that shape distribution is due to selective sorting and there is some zonation regarding the shape of pebbles.

There are several characteristic gravel forms which orginate in specific environments. To this group belong the ventifacts (Fig. 214). Some of the gravels, generally of hard material, possess one or more facets. Depending upon the number of edges, they are known as one-, two-, or three-edged gravels. Such faceted gravels are typical of warm and cold dry desert climates. Sand grains thrown against gravel by wind produce facets. However,

locally facets can be produced through the action of water (Kuenen 1947a).

Whitney and Dietrich (1973) made a detailed analysis of ventifacts, and simulated development of sculptures on the ventifacts in laboratory. Ventifacts range in shape from subangular to well-rounded with a noticeable percentage of them possessing both well-rounded and angular intersurface junction. Nearly flat, convex and concave surfaces occur. The shape may be prolate, oblate, ellipsoidal, triangular prismatic, or quite irregular. The shape of the ventifacts is controlled mainly by the original shape and is only slightly modified by wind erosion. The surface of the ventifacts may be matte-like and frosted to highly polished, or different surfaces of the same ventifact may show different characteristics, where minute features like scores, pits, and grooves develop. Macroscopic and microscopic features of ventifacts can be formed due to abrasion by suspended dust-size particles (fine silt and clay) (Whitney and Dietrich 1973).

Gravels of moraines are generally tabular in shape. The majority of them show a pentagonal marginal profile (Fig. 215). The broad stoss side generally shows snub scars (Wentworth 1936). Gravels of moraines also show characteristic striations. Figure 216 shows the characteristic shape of pebbles from various environments.

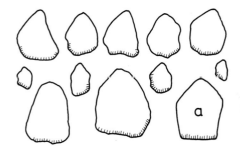

Fig. 215. Diagram showing marginal profile of characteristic glacial gravels. A typical glacial gravel shows a characteristic pentagonal profile (a). The broad stoss side usually shows snub scars. (Modified after Wentworth 1936)

Fig. 214. Ventifacts showing characteristic facets, produced unter wind activity. (After Conybeare and Crook 1968)

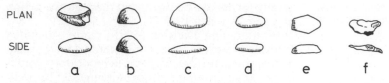

Fig. 216. Shape of characteristic gravels from different environments. Both side and surface views are shown. a Aeolian environment, ventifacts; b high-energy sea coasts, more or less spherical gravels; c low-energy sea coasts, flat gravels; d rivers, rod-shaped gravels; e glacial moraines, tabular with pentagonal outline; f glacial and/or frost areas, unworked frosted pieces. Compiled from various sources

The roundness of gravels is independent of its shape. Several indices are in use for the description of the roundness of gravels. A widely used roundness index is after Cailleux (1947, 1952). The better-rounded gravels indicate a prolonged mechanical wear and tear and with that a long transport. The gravels are well rounded after a transport of even a few kilometers.

The degree of roundness is strongly dependent upon grain size. In rivers the degree of roundness increases directly with increasing grain size. For coastal gravels this relationship is rather complicated. Here, the largest gravels are not always in motion and thus not well rounded. Moreover, sand grains on such coasts are generally moved only in suspension, and possess a poor degree of roundness. There exists a middle grain size in gravels which shows the best rounding (Kuenen 1964a; Füchtbauer and Müller 1970).

King and Buckley (1968), in their study of gravels in arctic regions, found that the gravels of eskers and deltas are better rounded than the gravels of kames. Gravels of moraines show the worst rounding. These differences were assigned to differences in distance of transport.

Boulton (1978) distinguishes three groups of glacial boulders: (1) angular and subangular boulders, derived supraglacially and transported at a high level, (2) subrounded to rounded boulders with several directions of striae which have undergone a phase of traction transport at the base of the glacier, (3) large, rounded and streamlined boulders (0.5–1 m in diameter) with long axes and striae parallel to glacier flow and sharply truncated distal extremities, indicating that they have been deeply embedded in subglacial till. Mills (1978) argues that water is essential for producing rounded and well-rounded pebbles in a glacial environment.

Humbert (1968) made a detailed study of pebbles on gravel beaches. He made use of several indices to study the variations in the shape and roundness of pebbles and he found that the maximum projection sphericity index $(c^2 a^{-1} b^{-1})^{\frac{1}{3}}$ after Sneed and Folk (1958) is most important. He found that the main reason for the change in shape and roundness is selective sorting and that Sternberg's law is applicable for beach pebbles only in a few cases. Richter (1959) was able to differentiate pebbles from different environments in the Pleistocene sediments of North Germany. Figure 217 shows degree of flatness and the rounding index plotted to differentiate various environments.

Dobkins and Folk (1970) made a systematic study of fluvial and beach pebbles on Tahiti-Nui, all of them made up of basalt pebbles. High-energy beach pebbles possess highest roundness, lowest sphericity, and are distinctly oblate. Low energy beaches have intermediate roundness, sphericity ranges from very low to high, small pebbles are oblate, large ones are prolate. Pebbles of an optimal size are disc-shaped, larger and smaller ones are more spherical, depending upon the nature of waves. On beaches prolonged abrasion decreases sphericity, while in a fluvial environment sphericity increases or remains constant. Sphericity values less than 0.66 are typical for beaches. Clifton

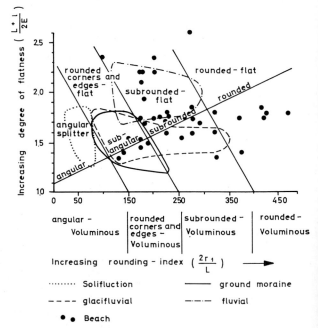

Fig. 217. Distribution of pebbles of different origins, based on the shape parameters. This scheme is based on the measurements made on sandstone gravels. (Modified after Richter 1959)

(1973) reports that wave-worked gravels are better segregated into distinct beds than those in alluvial deposits, and the bedding in wave-worked gravels is laterally more regular than in the stream gravels.

Ostler lenses are small baracanoid mounds of well-sorted, open-work gravel made up of rounded, highly pivotable, small pebbles. They are formed during receding and low flood stages in fluvial channels carrying gravels and pebbles (Martini and Ostler 1973).

Stäblein (1970) discusses the principles of pebble analyses and suggests that genetic differentiation of pebbles is possible only in a single area, and it is difficult to use various parameters on an interregional basis. Brock (1974) discusses the limitations of various methods currently in use to determine the form of pebbles.

Shape and Roundness of Sand

Because of the smaller size of the grains, measurements of shape and roundness in sand are made in only two dimensions.

The most widely used shape parameter for sand grains is the sphericity index after Wadell (1935). Details of sphericity measurements are given by Krumbein and Pettijohn (1938), Milner (1962), and Müller (1964). Another important measure for the form of sand grains is the index of longness (Schneiderhöhn 1954).

With increasing wear and tear during transport, sand grains become more and more spherical. The sphericity of sand grains increases with increasing grain size (Pettijohn 1957). Composition of sand grains also controls the shape of grains.

Roundness of grains is a measure of sharpness of edges and corners. Grains of various forms may possess the same degree of roundness. The most commonly used method for determination of roundness is after Wadell (1932). Russell and Taylor (1937) suggested five groups of roundness, which can also be determined visually. Pettijohn (1957) reproduced these groups in a modified form, which were later accepted by Schneiderhöhn (1954) and Müller (1964). Shepard (1963) distinguishes six groups, as does Powers (1953). These groups are shown in Fig. 218. The sand grains of coastal dunes, in general, show a better rounding as compared with beach sands. However, this feature is well-developed only in prevailing landward winds. With winds parallel to the coast or toward the sea, this difference is rather obscure (Shepard and Young 1961). The roundness of sand grains also shows a strong relationship to grain size. Coarser sand grains possess better roundness than finer ones.

Folk (1978) found that sand grains of the Simpson Desert are subangular to angular and do not

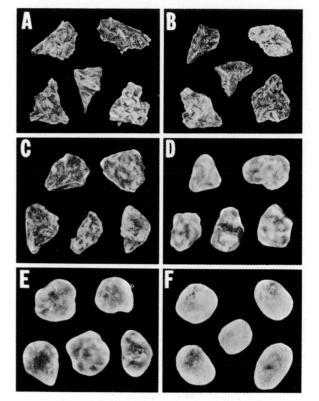

Fig. 218. Six classes are used for roundness determination of sand grains. A = very angular; B = angular; C = subangular; D = subrounded; E = rounded; F = well rounded. (Classes after Powers 1953. After Shepard 1963)

undergo any effective rounding during wind transport, maybe because they are being transported in longitudinal dunes.

Form and roundness parameters do not provide any direct clues to depositional environments. However, they can be helpful in typifying a given sand body in ancient examples. Moreover, better roundness and spherical grains indicate higher maturity of the sediment.

There is also another approach to the study of the shape of sand grains. Several workers have found that geometrical morphometric descriptions of sediment grains are not in accordance with the behavior of sediment during transport and deposition. They developed concepts and measured the functional shape of the sediment grains. McCulloch et al. (1960) introduced the concept of dynamic shape factor, which considers the terminal fall velocity and the fall velocity of a sphere with the same volume and density as the particle.

Shepard and Young (1961) introduced the term "pivotability", which determines the behavior of sediment grains during transport and deposition. Kuenen (1964b) adopted this concept of pivotability and tried to distinguish depositional environments. Pivotability is defined as a tendency of a sediment grain to start rolling on a slope. It is a shape characteristic unrelated to weight, size, or density. More recently, Winkelmolen (1969a) introduced the concept of rollability. He carried out many laboratory experiments and tried to apply it on the coastal sediments of the North Sea. In the following, a short account of the rollability concept taken from Winkelmolen (1969a) will be given.

Rollability deals with the ease with which a sediment rolls after the initial pivot. Rollability has a direct bearing on the dynamic behavior of the grains during transport processes. It shows a positive correlation with settling velocity and a negative correlation with transportability. For the measurements, Winkelmolen used a mechanical device made up of a slightly inclined revolving cylinder through which a sediment is made to roll (Winkelmolen 1969b). For interpretation he transforms absolute rollability values into relative rollability values which are independent of the shape distribution in the source material.

An environment can be characterized by plotting the relative rollability values against the sieve fractions. The form of the graph is called the shape distribution character and has good diagnostic value. It was found, in general, that the receiving deposits are charaterized by low rollabilities and lag deposits possess high rollabilities.

For example, dune sediments show less rollability than beach sediments, which are the sediment source for the dune sand. The beach sands are, in turn, less rollable than the supplying sands in the shallow offshore area. These results are in good agreement with the similar, less-certain results obtained by roundness and pivotability.

Moreover, dune sands show poor rollabilities for the coarser grains and increasingly better rollabilities for the finer grains. Off-shore sands show good rollability for coarser grains and low rollability for the fines. Winkelmolen (1969a) postulated that a combination of well-rollable coarse grains with poorly rollable fine grains point to a lower-current velocity than the combination of poorly rollable coarse grains with better rollable fines.

Surface Texture of Sediment Grains

Minor features developed on sediment grains can sometimes be helpful in deciphering the depositional environment. On pebbles such features can be recognized megascopically, while for sand grains microscopic studies are necessary. More recently, electron microscopy has been applied for the study of the surface texture of sand grains.

Pebbles may show features like polish, striations, frosting, pitting, etc., on the surface. Gravels of deserts often show a glazing surface, known as desert varnish. Such desert varnish is generally chemically composed of iron and maganese oxides. The genesis of such varnish is not clear, but it is a characteristic feature of many desert pebbles, together with their form (ventifacts). Some authors even consider frosting as typical of wind action (Cailleux 1942).

Striated gravels are found in glacial environments. Striations are produced by ice action on the gravels (Fig. 219). Striations are better developed if the gravels are composed of soft material, e. g., limestone. Gravels of hard rocks, e. g., granites and quartzites, generally lack striations (Pettijohn 1957). Boulton (1978) demonstrates that pebbles transported through the tractional zone of a glacier

Fig. 219. Gravel showing striations made by ice. Several sets of striations in different directions are present. (After Conybeare and Crook 1968)

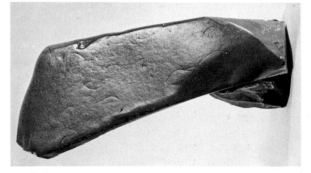

Fig. 220. Crescentic impact marks on a gravel. Such markings are made from high-velocity flows. (After Conybeare and Crook 1968)

tend to be rounded and bear several directions of striation. However, striations can also be produced by tectonic movements.

Crescentic impact scars or percussion marks have been attributed to high-velocity flows (Fig. 220). In high-velocity flows sediment grains blow against each other, making impact scars.

Many workers have tried to relate the surface texture of sand-sized grains, especially the quartz grains, to the environment of deposition. The reasoning behind these investigations is that the agent of deposition makes some characteristic surface features on the quartz grains. The recognition of these surface features may enable one to determine the agent of deposition, and with that the environment of deposition. Cailleux (1942, 1943 and 1952), in a series of papers, discussed the problem of using the surface texture of quartz grains as an environmental indicator.

Cailleux (1942) differentiated between three major types of surface textures:

1. Non-uses: angular grains without any transport; typical for glacial environments.
2. Emousses-luissants: smooth and polished, glazing grains; characterizing water as the agent of transport.
3. Ronds-mats: rounded, frosted grains; indicating wind as the agent of transport.

A fourth type is rond-mats sales–dirty rounded frosted grains, characteristic of ancient wind-worn sand. These surface textures are shown in Fig. 221.

Investigations of Cailleux (1943) and Zimdars (1958) indicated that in a marine environment, the majority of quartz grains show polished surfaces. However, in fluvial environments only a smaller percentage of quartz grains show polished, shining surfaces. The study of quartz-grain surface features has been successfully made by various workers (von Braun 1953; Schneider and Cailleux 1959, and others).

The study of quartz-grain surface features has certain problems and limitations. Many of the quartz grains possess an intermediate texture, and thus cannot be assigned to a specific environment. Sand grains transported mainly by one agent can be deposited without much reworking in another environment. In this case there is a problem of mixing of sand grains of different environments. In ancient sediments diagenesis and cementation may affect the sediment surfaces, so that the original surface features are obliterated.

The application of electron microscopy has provided a possibility of studying the surface textures of quartz grains in much more detail than is possible with ordinary microscopes and binoculars. Biederman (1962) and Porter (1962) provided the first results. Biederman found that quartz grains in an aquatic medium possess triangular etch figures,

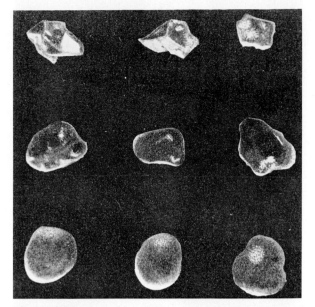

Fig. 221. Diagram showing different kinds of surface texture of quartz grains. Above: non-uses, angular grains, not worn (glacial), middle: emousses–luissants, smooth shining grains (water-worn), below: ronds–mats, rounded frosted grains (aeolian). (After Schneider and Cailleux 1959)

whereas quartz grains in an aeolian medium show irregular grooves and pits.

Krinsley and his coworkers (Krinsley and Takahashi 1962 a, b; Krinsley and Donahue 1968), in a series of publications, provide valuable information on the surface texture of quartz grains from various environments. Bramer (1965), Soutendam (1967), Gees (1969), and many others provide information on the surface texture of sand grains. More recently, attempts have been made to study the quartz-grain surfaces with scanning electron microscopes (Cailleux and Schneider 1968; Busson 1968; Krinsely and Margolis 1969). Figures 222 a–e show typical features of quartz grain surfaces from various environments as observed with the help of scanning electron microscope.

Usually the quartz grains of medium sand size are well-suited for the study of quartz grain surface textures. Schneider (1970) reviews the problem of quartz-grain morphoscopy. Krinsley and Doornkamp (1973) summarize the results of the quartz grain surface features as observed in SEM and their use in the environmental interpretation.

Quartz grains of glacial environment show sharp angular outline, high relief, large breakage blocks, large conchoidal fractures, randomly orientated striations, semi-parallel step-like features, meandering ridges, smooth featureless plains, etc. (see Margolis and Kennett 1971; Krinsley and Doornkamp 1973; Singh 1974). Whalley and Krinsley (1974) provide some detailed data on the quartz grain surface features of various subenvironments of glacial deposit.

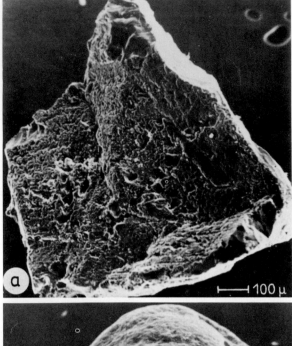

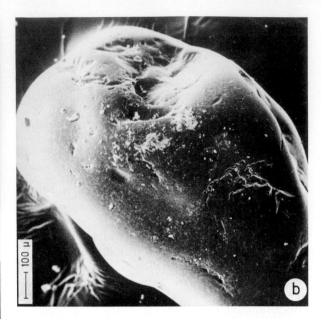

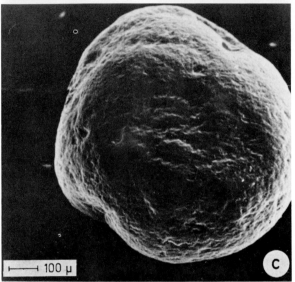

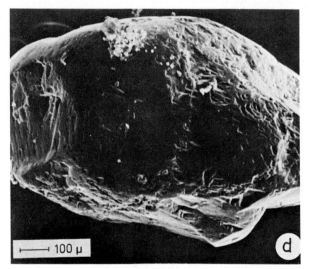

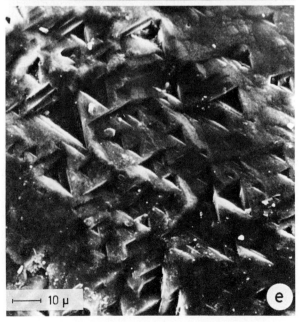

Fig. 222. Scanning electron-microscope photographs of sand grains. (After Schneider 1970) a A sand grain which has not been transported. Granite debris. Fonday, Vogesen, b Naturally polished quartz grain from a modern beach. "Region des lagunes," Ivory Coast. c Rounded, frosted quartz grain from a desert. El Kheneg. Algeria, d Natural etch figures on a quartz grain from a continental terminal, e Enlargement of Fig. 222d

Quartz grains from beach sand show rounded outline, moderate relief, v-shaped pits, straight or slightly curved grooves, blocky conchoidal pattern, and chatter mark.

Aeolian sand grains are subrounded to rounded in outline, moderate relief, frosted oppearance due to slight irregularities on the surface, closely spaced upturned plates, meandering ridges, graded arcs, with minute chemical etching as anastomosing lines. Folk (1978) found that the quartz grains from the Simpson Desert, Australia, show a characteristic greasy luster due to the development of turtle-skin crusts of silica.

Quartz grains of fluvial environment show subangular to subrounded outline, moderate relief, irregular surface, and prominent traces of inherent features in recessed areas (Singh 1974). Kumar and Singh (1978) found that the quartz grains of fine-sand fraction of the Gomti River, India do not show any significant effects of mechanical abrasion during transport. Manker and Ponder (1978) discuss the quartz grain surface features from fluvial environments and conclude that distinctive features are various types of impact marks and solution features. Some of these features are similar to those found in beach and aeolian sands.

Margolis and Krinsley (1974) discuss the genesis of various mechanical and chemical features on the quartz grain surfaces. Nevertheless, there are several discrepancies in the interpretation of various quartz grain surface features, and it may be pointed out that it is the combination of certain features that has a diagnostic value. Care should be taken to differentiate the primary (mechanical) features from the chemical weathering and diagenetic features.

Primary Fabric or Grain Orientation

Every nonspherical object possesses orientation with respect to horizontal and vertical planes. If in a given sample the same special orientation is assumed by a significant number of elements, it is said to show a preferred orientation. This is the primary fabric of a sediment. In ancient sediments metamorphism and deformation may produce a new secondary fabric. The orientation of sediment grains, i. e., the primary fabric, is controlled mainly by the medium of transport, the type of flow, and the direction and velocity of currents. However, the morphology of the surface of deposition may also control the orientation. This is especially true for sand-sized sediment grains (cf. Seibold 1963).

Generally speaking, the sediment grains tend to come to rest in some stable position with reference to flow. Almost any detrital grain is a potential fabric element, though those with maximum dimensional inequalities are the most useful ones.

The orientation of a sediment grain can be described in terms of the direction of orientation of the long axis and inclination with reference to the horizontal. In the case of flat disc-shaped sediment grains the direction is indistinct, but inclination can be easily determined.

Generally, the orientation of pebbles and quartz grains is measured. However, mica grains and organic remains can also be used. Orientation patterns obtained by such measurements can be related to the type of flow or environment. Some patterns seem to be restricted to specific environments.

Orientation of Pebbles

Pebbles can be divided into two main forms: flat equidimensional pebbles, and elongated pebbles. The form of pebbles also partly controls their orientation pattern.

Observation in modern environments has shown that elongated pebbles in fluvial environments can be oriented both parallel and normal to the flow direction. It is generally believed that in rivers with periodic water flows pebbles are usually arranged normal to the flow (cf. Kürsten 1960). Orientation parallel to the flow is common in rivers with regular water flows (Kalterherberg 1956; Schiemenz 1960). However, Sengupta (1966) believes that the orientation of pebbles is controlled by the gradient. If the slope is steep, pebbles are orientated parallel to the current. If the slope is gentle, pebbles are oriented at a right angle to the flow. According to Ruchin (1958), mountain rivers with high gradients show pebbles oriented parallel to the current, whereas flatland rivers with low gradients show pebbles oriented normal to the current direction.

Katzung (1971) observed that flat and rod-shaped pebbles in a small rivulet show orientation of a- and b-axes of the pebbles against the direction of flow, where a-axes are mostly perpendicular to the flow and b-axes mostly parallel to the flow. The a-b plane shows a strong imbrication, varying mostly between 15 and 45°. The angle of imbrication increases with the increase in the slope of the river bed. In a pebbly meandering river Gustavson (1978) found that the pebbles are often imbricated with their a-axes transverse to the flow and b-axes dipping upstream. Rust (1972a) observed that the highest degree of preferred orientation transerve to the current of elongate pebbles is observed from populations of large pebbles isolated on sand beds. Smaller size or increased concentration of pebbles reduces the degree of preferred orientation.

Behrens (1977) reports that basalt pebbles on a Scottish beach show strong imbrication, showing largest axes running parallel to the coast and the largest contour dipping 35° towards the sea. In deltaic environments and in beach bars pebbles are also oriented parallel to the coast line, but they show dips toward both the sea and the land (Ruchin 1958).

In moraines elongated pebbles show a preferred orientation parallel to the direction of ice transport (Richter 1932; Pettijohn 1957). Drake (1974) emphasized that in glacial till orientation of the pebbles is strongly controlled by the particle shape. Rod-shaped pebbles are well-aligned to the ice-flow direction, while blades are less aligned. Sphere- and disc-shaped pebbles tend to be parallel to the ice-flow direction. a-Axes of all the shapes show imbrication. Liboriussen (1975) found that in glaciofluvial deposits particles with axial ratios of less than 0.8 are usable in gravel fabric analysis, and shape does not essentially effect the orientation. Dip direction of a-b plane ist parallel to current direction, especially if the particle dip is greater than 20°.

In talus deposits there is very little preferred orientation of the blocks. However, the variance of long axes of talus blocks decreases downslope with increasing distance (Caine 1972). In turbidites longitudinal orientation – long axis parallel to the direction of flow – seems to be common (Doeglas 1962).

The broader end of elongated pebbles is generally oriented upstream. If in a conglomerate the broad ends of most of the pebbles are oriented in the same direction, it is a case of longitudinal orientation of pebbles (cf. Füchtbauer and Müller 1970).

Flat disc-shaped pebbles are commonly arranged in an overlapping fashion, producing im-

brication patterns with upstream (against the flow) inclinations. Imbrication is well – developed where concentration of pebbles is high. The angle of inclination in imbricated pebbles seems to be environmentally sensitive. In fluvial imbricated pebbles the angle of inclination is generally 15 to 30 (Cailleux 1945; Doeglas 1962; Pettijohn 1957). In glacial and fluvio-glacial sediments the angle of inclination is generally 20 to 25 (Harrison 1957); however, values as high as 40 to 50 are not uncommon. In coastal pebbles the angle of inclination for imbricated pebbles is low, less than 15 (Cailleux 1945; Pettijohn 1957). Fig. 223 shows schematically typical orientation of pebbles in various environments.

Johansson (1963, 1965) made a systematic study of pebble fabric in the laboratory, as well as in the field. He found that in flowing water, orientation of pebbles is mostly transverse to the flow direction, while in the case of glacial ice, it is parallel to the direction of flow. Imbrication gives better coincidence with the flow direction than the long-axis orientation.

Based on theoretical and experimental data Johansson (1976) gives a thorough discussion of fabric pattern of pebbles, granules, and sand grains and their environmental significance. The main points of this study are: particles moving in contact with the bed (rolling, sliding) in a flowing water tend to retain their long axis orientation transverse to flow when the motion ceases. They stop sooner when their long axis is parallel to the flow direction and the drag-generating mantle area is at a minimum. However, with higher tractive force particles stop mostly with their long axis parallel to the flow.

Further, increased shear velocities and smaller particle sizes, and weights tend to strengthen the longitudinal orientation of a-axes. The preferred dip of a-b plane (angle of imbrication) is upcur-

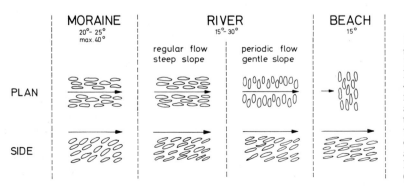

Fig. 223. Schematic representation of orientation of gravels of various environments. Moraine–elongation parallel to the flow, high-angled up-current imbrication (20 to 40); steep slope rivers–elongation parallel to flow, up-current imbrication with moderate angles (15 to 30); gentle slope rivers–elongation normal to flow, up-current imbrication with moderate angles (15 to 30); beach–elongation paralled to the shore line, at right angles to the direction of wave propagation, low-angled imbrication dipping seaward (< 15°). (modified after Ruchin 1958)

rent, tending to increase with higher shear velocities and increased contact frequency, which also causes a stable imbrication and increase in amount of dip. However, in nature, the dispersal in orientation is more than observed in laboratory studies, especially for the small particles.

Several authors have reported inclination of imbricated pebbles in down-current directions (Kalterherberg 1956; Byrne 1963). However, such cases are more the exception than the rule. Such down-current oriented pebbles are usually found in sediments with much fine-grained matrix and deposits of rapid sedimentation.

Nagahama et al. (1966) observed that some siltstones upon weathering break up into elongated grains, known as "dagger blade" and upon deposition their long axis is oriented parallel to the current direction. This dagger blade structure can be used in determining the current direction.

Theisseyre (1975) made a detailed study of pebble fabric in Recent and ancient fluvial deposits. He found that occasionally directions shown by pebble fabric are at an angle to real flow direction, or even that these two directions can be at right angles. The arrangement of pebbles in an alluvial channel is a bank-controlled phenomena. He further suggests that interpretation of pebble fabric needs information about the original bar-and-channel topography, channel symmetry, and the position of a conglomerate investigated with respect to the original bar-and-channel features. If pebble fabric is used with respect to bar-and-channel features, it is possible to reconstruct channel pattern and distinguish between low-sinuosity and high-sinuosity reaches.

Orientation of Sand Grains

Like pebbles, elongated sand grains tend to acquire a preferred orientation parallel to the direction of flow. Experiments by Dapples and Rominger (1945) indicated that the broader ends of quartz grains, like pebbles, point upstream. Moreover, quartz grains with longitudinal orientation often show an inclination in the upstream direction (Schwarzacher 1951). In certain cases sand grains may show a faint imbrication structure (Rusnak 1957). Vollbrecht (1953) discusses various aspects of quartz-grain orientation. Techniques of measuring sand-grain orientation have been dealt with by Dapples and Rominger (1945), Schwarzacher (1951). Zimmerle and Bonham (1962), and Winkelmolen et al. (1968), Starkey (1974) gives a new technique for determining quartz grain orientation with the help of X-ray.

Curray (1956) observed that on beaches, quartz-grain orientation is mainly controlled by backwash and that quartz grains orient themselves parallel to the direction of backwash. Nanz (1955) and Sriramadas (1957) also found that quartz grains orient themselves parallel to direction of wave propagation.

Nachtigall (1962) and Seibold (1963) found that in coastal sands together with a maximum of longitudinal orientation of quartz grains, there is a subordinate maximum normal to the current flow.

In aeolian sands the orientation of elongated sand grains is considered to be less well-developed than in aquatic sands (Dapples and Rominger 1945). However, Curray (1956) suggested that grain elongation in wind-affected sands gives the direction of the prevailing wind. Matalucci et al. (1969) determined grain orientation in Vicksburg loess deposits of the United States. He found a preferred orientation and slight imbrication in the silt-sized sediment grains. Silt particles also orient themselves parallel to the depositing current. His results were in good agreement with the postulated palaeowind directions determined by the thinning of the loess blanket and the decreasing grain size.

In coastal sands the orientation of quartz grains is perpendicular to the trend or elongation of the sand body. In fluvial sands grain orientation is parallel to the elongation of the sand body. The study of sand-grain elongation in sand bodies of known origin may be helpful in tracing the configuration of sand bodies (Curray 1956). Similarly, if the configuration of the sand body and the orientation of its sand grains are known, the genesis of the sand body can be predicted (fluvial versus strand).

Studies of sand-grain orientation in turbidites has shown mainly longitudinal orientation. Bouma (1962) reports cases where grain orientation is normal to the current direction, as determined by sole markings. Several workers have even reported large-scale variations in the direction as determined by grain orientation and by sedimentary structures (Spotts 1964; Colburn 1968).

Sand grains in glacial sediment – moraines and tillites – show very consistent longitudinal orientation parallel to the direction of ice movement. Results of sand grain and pebble orientation of glacial deposits are in good agreement.

Shelton et al. (1974) measured parting lineation, grain orientation and cross-bedding directions in a sand bar of the Cimarron River, and conclude that grain orientation along with parting lineation is the most accurate current indicator in horizontally bedded sediments.

Most of the work on grain orientation has been restricted to quartz grains. The work of Diessel (1966) on mica flakes and Stauffer (1962) on carbonate particles suggest that other sediment particles can also be used in the study of grain orientation.

In general, directions determined from sand-grain orientation show strong variability. The reason may be that even very weak currents and waves, which are not capable of producing any major sedimentary structure, may affect the sand-grain orientation. Seibold (1963) stated that in coastal sands the degree of preferred orientation is strongly affected by the local morphology, wave energy, formation of ripples, etc. Moreover, sand-grain orientation can be easily changed by biogenic activity, penecontemporaneous deformation, and post-depositional effects. These factors make the use of sand-grain orientation in environmental studies much more critical than the use of orientation of pebbles.

To conclude, sand-grain orientation is a much more sensitive parameter than other sedimentary structures. It provides more information on local details and local variations in the direction of flow. The use of sand-grain orientation as a directional parameter on a regional scale must be done with care.

Orientation of Clay Particles

Very little is known of the microtextures and arrangement of clay particles in clays in various environments. Rosenquist (1955) found better parallel oriented clay particles in freshwater clays. In marine clays parallel orientation was not so well developed. Rosenquist explains this as a result of deposition of clays by flocculation in a marine environment.

Von Engelhardt and Gaida (1963) describe "aggregate texture" in clays deposited from solutions rich in electrolytes. Solution with little electrolytes tend to deposit the sediment as single grains rather than as larger aggregates.

Mattiat (1969) experimentally produced honeycomb and card-house structures in coagulation-sedimentation, and parallel structures in single-grain sedimentation (Figs. 224 a-c).

Zabawa (1978) describes microstructure of the agglomerated suspended sediments in Chesapeake Bay and distinguishes between inorganic agglomerates and pellets. The inorganic agglomerates are produced by flocculation and coagulation where composite particles contain angular networks of mineral grains arranged in face-to-face, edge-to-edge modes of grain contacts. The pellet agglomerates are attributable to fecal pellet production by filter-feeding organisms and show a pelleted micro-structure.

Today, we known too little about the environmental significance of these features. Moreover, it seems that the application of data on the orientation of clay particles in ancient sediments would be rather difficult. The reason for this is that clays are strongly affected by compaction and post-depositional diagenetic changes, including the recrystalli-

zation leading to generation of new minerals. The effects of these processes would tend to make the recognition of original texture of clays in ancient sediments almost impossible.

Orientation of Organic Remains

Organic remains are oriented after the same hydrodynamic principles as inorganic remains of similar form. Measurements on the orientation of elongated organic remains may also provide hints to the transport direction.

Elongated organic remains are oriented either parallel to the direction of flow or at right angles to it. Some elongated bodies like elongated shells, conical tubes, fishes, etc., possess a slim and a broad end.

Seilacher (1959) discovered that such polar organic remains, if oriented transversely, show a bipolar orientation with equally developed maxima of their pointed ends. On the other hand, longitudinal orientation (parallel to current) of polar organic remains shows a preferred orientation of their pointed ends (Fig. 225).

A number of workers (e. g. Wasmund 1926; Quenstedt 1927; Schindewolf 1928) found that orientation of polar organic remains is longitudinal and that the pointed ends of conical bodies point up-current. Trusheim (1931) confirmed this observation obtained by the study of fossils, experimentally in a tank. However, conical bodies can also be oriented longitudinally, but with pointed ends oriented down-current. This happens when water is sinking very rapidly and, during the transport itself, the sediment surface emerges. This observation is made on the conical tubes of *Pectinaria koreni* during the backwash of the waves on the beaches (Reineck 1960 c).

In lamellibranchs with well-developed hinge teeth they may function like an anchor, behind which the shell can orient itself (Trusheim 1931; Seilacher and Meischner 1964). Several lamellibranchs show a preferred orientation of the umbo with respect to the current, with this orientation varying from species to species.

Nagle (1967) made a systematic study of the orientation of shells by waves and currents in experiments as well as in rocks. His study demonstrates that the rose diagrams of current-oriented shells show one strong maxima, where most of the shells point into or away from the current depending upon the shell geometry and distribution of shells. In shoaling, non-swash waves, long axes of these shells are aligned parallel to the wave crests, where due to associated current, maxima tend to be skewed. In the swash zone, shell orientations show two maxima of wave patterns, but shells are aligned perpendicular to the ripple crest align-

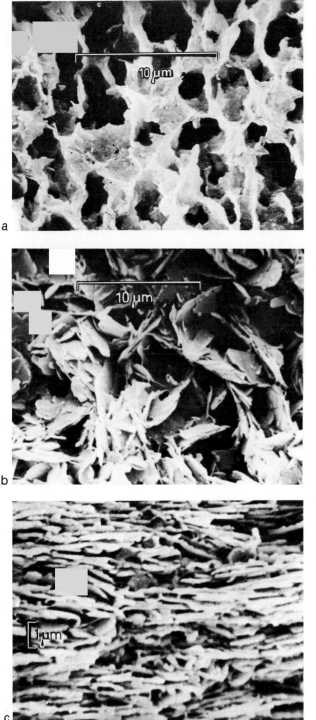

a

b

c

Figs. 224a–c. Scanning electron-microscope photographs showing three main types of arrangement of clay particles, a Honeycomb structure; b card-house structure; c parallel structure, a and b are the result of deposition by coagulation, while c is produced by single grain sedimentation. (Photograph by B. Mattiat)

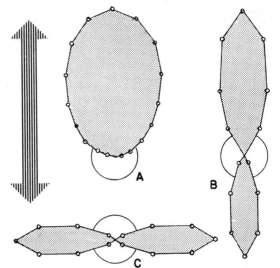

Fig. 225. Diagram illustrating transverse and longitudinal orientation of organic remains with well-defined polarity (with respect to flow direction). A Biogenic orientation of organism with only one broad maxima in one direction. B Longitudinal abiogenic orientation of organic remains with a well-defined maxima in the direction of the current, at the same time a smaller maxima in opposite direction. C Transverse orientation of organic remains with two equally developed maxima at right angles to the flow direction. (After Seilacher 1959)

ment. Various shell forms may develop typical orientation patterns, but commonly show nondiagnostic orientations. Study of *Cardium echinatum* and *Cardium edule* shells demonstrated that azimuth orientation of single valves of these shells is strongly controlled by the position of center of gravity, and even a slight shift in its position in the valve leads to different orientation (Futterer 1974).

Seilacher (1959, 1960) discusses the commonly observed starfish with overturned arms. According to von Koenigswald (1930), the arms of the starfish were overturned from a strong current. However, Seilacher (1959, 1960) demonstrated that in many cases the starfish drifted above the sediment surface, and during drifting the arms overturned backside.

Besides the abiological orientation of organisms, cases are known where animal orient themselves because of some internal response. Seilacher (1960) gives examples from traces of trilobites. Where the animal moved down-current, the distance between the traces is large. If the animal moved at an angle to the current, the traces are slightly laterally displaced. Increasing lateral displacement in the direction of movement points to the turning of the moving animal.

Chemical and Mineralogical Parameters

Chemical factors existing in an environment are important in defining a given environment. Moreover, the chemical environment, especially the salinity and temperature, are the important factors which also control the distribution of fauna, and with that the biogenic sedimentary structures. Thus, biogenic features can sometimes be better understood if detailed information about the chemical environment is available.

Important parameters which help in reconstruction of the chemical environment are mainly the minerals which have been originally precipitated at the time of deposition. They provide information about the chemical equilibrium under which they were precipitated. However, a major problem in such a study is that the chemical equilibrium which existed at the time of deposition changes quickly after deposition. The post-depositional processes establish a new chemical equilibrium, cause diagenetic changes in the minerals, and produce some new minerals. Thus, chemical attributes sensitive to the environment are also susceptible to diagenetic changes. Garrels and Christ (1965), Garrels and McKenzie (1971), Berner (1971, 1976), Riley and Chester (1971), Krauskopf (1967) provide useful information on the principles of sedimentary geochemistry. Pettijohn (1957) states that important chemical factors are Eh (oxidation-reduction potential), pH (acidity-alkalinity), salinity, and temperature. Amstutz and Bernard (1973) is a useful source of information on the mineralogical aspects of the sediments.

Oxidation-Reduction Potential (Eh)

The measure of the oxidizing capacity of an environment is the Eh value. The iron minerals are very sensitive to different oxidation potentials. For example, hematite indicates a fully aerated or oxidizing environment; iron sulfides (pyrite and marcasite) indicate a reducing or oxygen-deficient environment. Krumbein and Garrels (1952) discuss the stability field of various iron minerals in aqueous environments. Berner (1971) gives some additional information on the stability fields of iron minerals, and provides informa-

tion on the basic processes involved in chemical sedimentology. But in ancient rocks the presence of detrital grains of iron minerals and diagenetic changes complicate the problem.

However, features like abundant pyrite associated with the absence of benthonic fauna and traces of benthonic activity are positive evidence of anaerobic conditions of deposition. A high content of preserved organic matter is also good evidence of anaerobic conditions. On the other hand, an abundance of hematite is an indicator of oxidizing conditions.

To a certain extent, oxidizing-reducing environments are also associated with specific hydrodynamic conditions. Thus, anaerobic conditions can be maintained only in the absence of turbulence or current activity, a low rate of deposition, and almost no reworking of sediments.

Acidicity-Alkalinity

The pH of an environment is a controlling factor in the precipitation of certain minerals. For example, calcium carbonate is never precipitated in an acidic environment. For the precipitation of $CaCO_3$, together with an alkaline medium, temperature is also an important factor. Correns (1950) and Krumbein and Garrels (1952) discuss the stability of certain minerals under various pH-values. Garrels and Christ (1965) provide a good discussion of the chemical stability of various minerals.

Salinity

Salinity of water strongly controls the fauna and flora. Thus, detailed biological studies can provide clues to the salinity of the depositional environment. For example, dwarf forms of a marine fauna can indicate a decrease in the salinity of the environment, probably in an estuary or a lagoon. In hypersaline conditions, minerals such as gypsum and anhydrite are precipitated. With still-increasing salinity of the water, precipitation of halite or even potassium salts takes place. Attempts have been made to determine the salinity of an environment by the concentration of trace elements. Determination of the concentration of boron has been partially successful (Goldschmidt and Peters 1932; Bradacs and Ernst 1956; Ernst 1963; Harder 1970; and others). Walker (1972) reviews the problem of paleosalinity determinations.

Temperature

Temperature of an environment has a direct control on the precipitation of certain minerals, and on the distribution and activity of fauna. Carbonate minerals can provide some clues on the temperature of the environment. Evidence of glaciation, and floral and faunal components hint of palaeo-temperatures.

Isotopic composition of oxygen (O^{18} / O^{16}) has been successfully used in the study of palaeo-temperatures (Urey 1951; Epstein et al., 1951; Degens and Epstein 1964).

Index Minerals

Some common minerals in sediments can be used as index minerals, as they provide some information regarding the environment of deposition. For example, glauconite is mainly restricted to marine continental shelf sediments, where there is some turbulence, low rates of sedimentation, and some organic matter (e. g., Valeton 1968; Porrenga 1967; Bell and Goodell 1967). However, in Precambrian rocks glauconite is also abundantly present in the deposits of intertidal zones (Singh 1980b). Distribution of chamosite in modern sediments is restricted mainly to shelf sediments of tropical seas, generally shallower than 60 m but occasionally down to 150 m (Porrenga 1967). In ancient sediments chamosite sometimes constitutes the major mineral in some of the oolitic rocks. However, oolites are generally composed of calcium carbonate. Carbonate oolites in modern sediments are known to occur in shallow, coastal areas with high turbulence, such as the breaker zone. Oolite occurrences on the Great Bahama Banks in turbulent waters are good examples (Purdy and Imbrie 1964).

Primary siderite can be formed only in the lakes with low Eh values and is known from many lacustrine deposits. In marine sediments siderite can develop only by diagenetic processes. Further, in freshwater deposits marcasite (iron-sulphide) is common due to acidic milieu, while in the marine shales pyrite (iron sulphide) is abundant, and can also occur in freshwater sediments.

Phosphorite is also restricted mainly to the continental shelf sediments, in areas of warm water, low rate of detrital sedimentation, and slightly reducing conditions (Bromley 1967). However, in ancient sediments phosphorite is also known to occur in association with stromatolites, which are considered to be deposits of very shallow water with profuse algal growth (Banerjee 1971).

In the Proterozoic of India, as well as in Russia and China, phosphorite is characteristically associated with the stromatolitic carbonates of intertidal zones. Sometimes magnesite and phosphorite make a characteristic association with the stromatolitic carbonates (Valdiya 1972). In these tidal flat deposits phosphate seems to form an integral part of stromatolites. It suggests that precipitation of phosphate took place from phosphate-rich waters, and subsequently during diagenesis phosphate also replaced the carbonates (Chauhan 1979).

Attempts have been made to use clay minerals as environmental indicators, but with very little success. The main reason is that clay minerals are produced on the continents and then transported into basins of deposition, and are very little affected by the environment of deposition. However, in ancient sediments diagenetic changes may complicate the picture still more. On the other hand, clay minerals may provide some information on the climate in the provenance, e. g., kaolinite is produced under warm humid conditions, whereas montmorillonite is produced mainly in arid and temperate climates. Grim (1968) and Millot (1970) provide useful texts on the clay minerals and their various aspects.

In sedimentary sequences where the diagenetic effect on the clay minerals can be neglected, the clay mineral composition may provide useful information on the climate and weathering characteristics in the provenance. Kaolinite develops in soils of extensive leaching and acidic milieu, mostly associated with warm-humid climate Montmorillonite develops in soils where hydrolysis occurs, but drainage is poor and milieu is alkaline associated with poor leaching. Illite is produced by decomposition of feldspars and degradation of mica in initial stages of weathering, in an alkaline milieu and presence of Ca^{2+} ions. Chlorite is formed by hydrolysis of ferromagnesian minerals under alkaline milieu. Presence of palygorskite-attapulgite in the clay fraction is a good indicator of alkaline, hypersaline conditions in the basin of deposition (see Millot 1970 for detailed discussion).

Color

The color of sediments is mainly controlled by the chemical environment under which it has been deposited. Among the minerals the iron minerals are most important in imparting color to a sediment. Sediments having iron in oxidized form (Fe^{3+}) are generally brown to red in color, those having iron in reduced form (Fe^{2+}) are gray to green (cf. Henningsen 1969). Thus, in general, brown-to-red-colored sediments are an indication of deposition in an oxidizing environment, and gray-colored sediments suggest deposition in a reducing environment. Gray- and dark-colored sediments (especially muds) generally also posses high organic matter contents, which also points to deposition in a reducing environment. A green color in sediments is generally due to the presence of minerals such as chlorite and glauconite (Eichhoff and Reineck 1953). McBride (1974) discusses the significance of color in Difunta Group sediments (Upper Cretaceous – Paleocene), where red, green, and purple colors are restricted to delta-plain facies, while dark color is present in all the facies. Green color is due to presence of chlorite and illite; and absence of hematite, organic matter, and sulphides.

The genesis and importance of red-colored sediments have been repeatedly discussed. The red color of these rocks is due to the presence of much hematite and iron pigment as a matrix or coating on detrital grains. Certainly they were formed in a strongly oxidizing environment. However, exact climatic conditions of formation of red-colored sediments is disputable. One group thinks that red-colored sediments originate only in humid tropical conditions (cf. Krynine 1949); others contend that red-colored sediments, originate in desert climates where iron minerals undergo diagenetic changes leading to the formation of hematite and red iron pigment, which impart a red color to the sediments (cf. Walker 1967). Walker (1974) argues that red beds can form diagenetically in a moist, tropical climate by intrastratal alteration processes similar to those operative in a desert climate. Thus, red-colored sediments can originate both in the desert and in humid tropical climates.

However, in general, ancient red-colored sediments are associated with typical desert climate and sediments. But today most deserts do not show the development of red color; red-colored sediments are formed mainly in humid tropical climates. Sands of the Simpson Desert, Australia show red coloration (Folk 1976a). The study of McBride (1974) shows that the red coloration is related to post-depositional reddening of sediments, in parts in soil zones on the delta plain in a sub-humid climate with seasonal droughts. In recent years several authors have summarized and dis-

cussed the problem of red-colored sediments (Krynine 1949; van Houten 1961, 1964a, 1968; Walker 1967; Solle 1966; Hinze and Meischner 1968; Glennie 1970).

Van Houten (1973) reviews the problem of red coloration in sediments. The red coloration is related to the development of hematite and goethite, which are formed in the soils under conditions of weakly acidic to calcareous compositions, good draining in warm subhumid to arid climate. If humus is present goethite is formed. The source of iron is often ferromagnesian minerals, e.g. hornblende and biotite. The process of genesis of iron oxides in the soils is discussed by Schwertmann (1971) and Schwertmann and Fischer (1974).

Many of the fluvial deposits in geological record, e.g. Gondwana sediments, Siwalik rocks, India, are characterized by an alternation of gray-green sandstone units (river channel deposits) and red-colored shale units (flood plain deposits). The paleontological evidences point to a humid climate. These successions usually have little preserved peat or coal. Such successions represent deposits of sub-tropical climate with marked seasonal droughts (see Singh 1976a). In the Precambrian sediments of India red-colored shallow marine deposits are also common (Singh 1978).

To sum up, the red-colored sediments are essentially related to diagenetic processes which can be operative under a wide range of climatic conditions. Detrital red beds are rare. Further, the red color is not related to the content of iron. Aging seems to be another important factor in controlling the red coloration, especially its intensity (Walker 1974; Folk 1976a).

Trace Elements

In recent years the chemical composition of rocks, especially the distribution of trace elements in sediments, has been used for distinguishing depositional environments. Krejci-Graf (1964, 1966), Degens (1968), and Ernst (1970) provide detailed information on the geochemical analysis of depositional environments.

Krejci-Graf (1972) discusses that Ti, Th are characteristic of the sediments and sediment surfaces which have undergone a long peroid of subaerial weathering. They are typically a continent element, and their content decreases with increasing distance from the land. B, Cr, Cu, Ga, Ni, and V are more enriched in marine than in the continental sediments. Cr is concentrated in the sediments of gyttja facies (slightly anaerobic), while V is concentrated in sediments of sapropel facies (highly anaerobic). Concentration of Pb, Ni, and Co always takes place with increasing reducing conditions. Ni, and Cu sometimes become enriched in organic matter in the deposits of reducing environment.

Distribution and concentration of some trace elements, e. g. Mg, Sr, Mn in the carbonate rocks, are often used as environmental indicators. However, distribution of these elements is also strongly influenced by diagenetic processes. Sr has often been used as a facies indicator, and is regarded as increasing from back-reef facies to fore-reef facies. Veizer and Demovic (1974) discuss that Sr is environment-sensitive, although diagenesis and the nature of carbonate minerals present also affect its distribution. Wedepohl (1970), Bathurst (1976) give a useful discussion on the use of trace elements in carbonate rocks as environmental parameter.

In the last few years use of composition of organic matter in geological studies has increased, especially to decipher the paleoenvironment of deposition and correlation. Tissot and Welte (1978) provide a good text on the various aspects of organic geochemistry relevant to a sedimentologist. Preserved organic matter in the sediments can be regarded as geochemical fossils and may convey information on the depositional environment (Tissot and Welte 1978).

Geochemical fossil molecules present in the bitumen extracted from rocks, and such molecules occurring in crude oils are interesting from the point of view of environmental reconstruction. They provide information on the major associations of organisms contributing to the organic matter of sediments. In general three types of organic matter are distinguished (Tissot and Welte 1978):

1. Marine organic matter – Phytoplankton is the major contributor with subordinate zooplankton. n-Alkanes and n-fatty acids of medium molecular weight range, C_{12}–C_{20} with predominance of C_{15} and C_{17} n-Alkanes are abundant. C_{15}–C_{20} isoprenoids, abundant steroids and some carotenoids, ubiquitous triterpenoids particularly of the hopanes series.

2. Continental organic matter – Mainly made up of debris of higher plants incorporated in deltaic and land-derived sediments or peat bogs. Mostly odd n-alkanes of high molecular weight (C_{25}–C_{33}), some tricyclic diterpenes, and ubiquitous hopanes, subordinate cyclic material. The pristance/phytane ratio may reach 4 to 10.

3. Microbial organic matter – is abundant in some lacustrine or paralic environments where plant material is heavily degraded during alternating periods of subaerial microbial activity and water flooding. The organic matter is enriched in lipids. Long chain n-, iso- (2-methyl) and anteiso-(3-methyl) alkanes, sometimes extending up to C_{40} or

C_{50} are present. Isoprenoids and cyclic hydrocarbons are scarce.

However, the continental organic matter can be present in marine sediments if there is considerable continental runoff. Nevertheless, content of continental organic matter increases near the land. Flekken (1978) studied the coastal sediments of the Trias-Lias age in Luxemburg. He developed a "marineness index" ($\sum nC_{15}$–C_{20} % of the total $\sum nC_{15}$–C_{30}) which enables to determine the influence of terrigenous material from the continents during sedimentation and helps in reconstruction of paleogeography.

Stable carbon and hydrogen isotope ratios in the organic matter can also be useful to determine the flux of terrigenous material.

Didyk et al. (1978) also emphasize the role of organic geochemical indicators in paleoenvironmental studies. High wax content in organic matter suggests increased contribution from continents. Oxicity/anoxicity of a sediment can be assessed from the amount and nature of organic matter. Pristane to phytane ratios (Pr/Ph), and porphyrin content are considered to be useful parameters. Sometimes, the nature of specific geochemical fossil may provide a more refined interpretation in terms of plants or animal groups, climate etc.

Generally, it is the relative concentration of certain elements which provides clues to the environments. As pointed out by Krejci-Graf (1964), geochemical diagnosis of the depositional environment can be successful in a given basin of deposition. Distribution of specific elements in the same depositional environment of two different basins may give completely different results and trends. For example, Ernst (1970) states that the same amount of boron in rocks of different ages reflects different environments; in Westphalian A and B the amount of 0.040% B_2O_3 indicates a marine environment, while the same value in the early Stephanian is found in limnic sediments.

Ernst (1970) gives a detailed account of the geochemical analysis of facies. He proposes the term "hydrofacies" for characterizing a geochemical facies or environment. A hydrofacies can be distinguished into three subordinate facies: Salinity, temperature, and oxygen facies. There are various elements whose concentration provides certain information about the environment. The most important are B, Br, Cl, Na, Sr, P, Ni, Co, V, Cr, U, Cu; isotopes of S, C, and O; and content and composition of organic matter. Krejci-Graf (1972) reviews the distribution of trace metals in various sedimentary deposits. A detailed discussion of the various geochemical and mineralogical parameters helpful in environmental reconstruction is out of the scope of this book. Interested readers are referred to the above-mentioned works.

Biological Parameters

General Information

Each modern environment or subenvironment is usually characterized by a given fauna, i. e., a biocoenosis inhabiting the region. Palaeontologists refer to such regions as biofacies. Upon death, burial, and preservation, a biocoenose (cf. Schäfer 1963; Frey 1971; Sarjeant 1975) leaves behind:

1. hard skeletal and biogenic parts, such as shells, carbonate or chitinous skeletal parts, teeth, fish scales, phytoliths etc.
2. sedimentary bioturbate structures
3. excretory matter, such as fecal pellets
4. organic matter
5. bioerosion structures, such as borings, gnawings, and related traces
6. biostratification structures, such as stromatolites, biogenic graded bedding, byssal mats.

Of all the animal communities, benthic communities are most important in characterizing a given environment. Hard skeletal parts of benthic organisms are preserved in ancient sediments in the form of fossils, and provide excellent information regarding the depositional environment, provided no post-mortem transportation of skeletal parts has taken place. However, in several cases, even characteristic species may fail to leave a preserved record, so that a living biocoenosis does not always correspond to a taphocoenosis. Even so, an adequate record of an animal population is generally preserved, so that one can use it as a biofacies characteristic for an environment. Even the absence of a record of a benthic community can be an useful indicator for certain types of environment. For example, in euxinic conditions macrobenthonic animals are absent, and only nektonic and planktonic forms are present. Along with other information, fossils provide an easy and satisfactory method of distinguishing between marine and nonmarine sediments.

In the following pages the sedimentological aspects of the preservable record of organisms are dealt with.

Hard Skeletal Parts

In many cases, coarser skeletal parts are found locally concentrated. Concentration of coarser hard parts of endobionts is achieved in regions where, at the same place or in the adjoining areas, large surfaces are undergoing strong erosion, or where channels are eroding and migrating laterally. These processes may lead to the concentration of hard parts through transportation, or the hard parts may be left behind and concentrated in the form of residual deposits (Richter 1936). However, for the concentration of shells of epibionts, much less hydrodynamic energy is needed (Hessland 1943). Such shell concentrations are commonly encountered on beaches, near mollusc colonies (mussel banks, etc.) and on the bottoms of channels in the form of channel lag deposits. Such shell concentrations have been strongly reworked and are exposed to wave and current action. Thus, broken shells are rather abundant.

In several examples, an echinoderm maxima is present just under the second critical wave depth zone, e. g., in Kattegat (Petersen 1914, 1918), on the Californian coast (Pequegnat 1961 a, b), in the Mississippi delta (Shepard 1956 b; Thorson 1957), in the Baltic sea (Kühlmorgen-Hille 1963; Seibold 1963; Werner 1967). The second critical wave depth zone denotes the water depth where the rhythmic movement of waves is replaced by a comparable strong parallel to coast current (longshore current). The depth of this zone depends upon the intensity and length of waves and the current velocity of the longshore current. Sarnthein (1970) found a distinct relationship between the second critical wave depth zone, an echinoderm maxima, and an echinoderm-skeletal maxima in the sediments of the Persian Gulf.

Van Straaten (1952) describes shell beds intercalated between marsh deposits. Such shell beds are usually of limited thickness and extent. They originate when, during gales, the water rises well above the mean high-tide level and mollusc shells are washed onto the marsh surface (Fig. 226). The percentage of broken shells in such deposits is rather high. Landward such shell beds wedge out over a distance of only a few meters.

Fig. 226. Shell accumulation on a marsh surface. Shells are washed on the marsh by the waves during storms with increased water levels

There is another very interesting occurrence of shell beds on intertidal flats, called *Hydrobia* beds (Fig. 227) by van Straaten (1952). These beds are especially rich in the tests of *Hydrobia ulvae,* but they also contain Foraminifera, peat detritus, and double valves of *Cardium edule* and *Macoma bal-*

Fig. 227. *Hydrobia* bed of the North Sea intertidal flats. The shell bed is made up mainly of tests of gastropod – *Hydrobia.* Some large shells of *Cardium* are also present. This shell layer is produced from the activity of the sand worm *Arenicola.* The *Hydrobia* bed is present 20 to 30 cm below the sediment surface. (After van Straaten 1964a).

thica. These beds are found about 20 to 30 below the surface of the intertidal flat. Genesis of this bed has been assigned to the activity of the sand worm *Arenicola.* This animal swallows sand at about 20 to 30 cm below the surface. After removal of the nutritive constituents, the sand is pushed upward on the surface. The shells are not swallowed, but become concentrated at the lower limit of the U-shaped burrow of *Arenicola* (Van Straaten 1952). Such *Hydrobia* beds only a few centimeters in thickness are common all over the North Sea intertidal flats.

Goethe (1958) describes ornithogenetic shell concentration on sandy surfaces within the intertidal zone, and more often near the dunes. This patchy enrichment of shells is produced by male seagulls, which bring living *Cardium, Macoma,* or *Mytilus* as marriage presents to their brides. Only a few of the shells are eaten. Upon death, they leave behind shell concentrations (Fig. 228).

Seilacher (1970) proposes a classification of fossil enrichments which is based primarily on the mode of genesis and the form of concentration. Table 11 gives salient points of this classification.

It is not in all cases that shells are concentrated in the form of shell beds, where shells of different biocoenoses are mixed leading to a tanatocoenoses. In many cases shells remain at or near the point where the animals were living. Here, shells are autochthonous in origin and represent a taphocoenosis of a biocoenosis. Investigations in the coastal sediments of modern environments have shown that taphocoenoses of areas located below the point of the wave base generally correspond well with living animal communities, i.e., the biocoenoses. However, hard parts of biocoenoses of regions located above the wave base are invariably mixed, and a taphocoenosis here corresponds only partially to the biocoenosis. This problem has been discussed by Hertweck (1971 a) in the coastal sediments of the Gulf of Gaeta, Italy.

If there is little or no transportation, shells may be preserved with both valves joined together. In several cases, even living animals are removed from their living position and are concentrated nearby. Upon death they leave behind shell concentrates with both valves joined together. On in-

Fig. 228. Ornithogenetic shell concentration near the sand dunes. The shells, mainly *Macoma,* have been brought by male seagulls as marriage presents to their brides. North Sea tidal flats. (Photograph by F. Goethe)

Table 11. Classification of "Fossil-Lagerstätten". (After Seilacher 1970)

1. *Konzentrat-Lagerstätten* (concentration deposits).
   Enrichments of disarticulated hard parts of organisms.
   a) *Condensation deposits.* Enrichment of hard parts as a result of absence of or slow rate of deposition of the inorganic detrital matter. Example: condensation horizons.
   b) *Placer deposits.* Concentration of hard parts due to sorting processes of the depositing medium. Example: bone beds, channel lag deposits.
   c) *Concentration traps.* Accumulation due to filling in of cavities. Example: fissure filling, burrow filling.

2. *Konservat-Lagerstätten* (conservation deposits).
   They are marked by preservation of organic matter.
   Complete hard skeletal parts are commonly found.
   a) *Stagnation deposits.* Deposition and concentration of fossils in sapropelic sediments. Water layers just above the sediment surface are marked by anaerobic conditions. Example: black shales, lithographic limestones.
   b) *Obrution deposits.* Due to rapid covering of skeletal parts by sediments deposited in reducing environment – gyttja facies. Example: Hunsrückschiefer.
   c) *Conservation traps.* Rapid embedding and sinking in a conserving medium or in cavities. Example: amber, peat.

Fig. 229. Concentration of *Cardium* shells (partly living) in a channel on the intertidal flat. In such cases both valves are joined together upon preservation. North Sea Tidal flats

tertidal flats, *Cardium* is often found to make such types of concentrations (Fig. 229).

In other cases, after the early decomposition of the hinge ligament, both valves become separated; and if the valves of a shell are of different size or shape, especially in desmodont lamellibranchs, valves are sorted out during transport (Richter 1922). Such sorting of valves of different nature during transport is well exemplified by the mussel *Mya arenaria* in the intertidal flats of the North Sea.

Even in the symmetrical shells, like *Donax vittatus* or *Spisula subtruncata* there is a marked separation of left and right valves: left-right phenomena (Lever 1958). On sandy beaches, there is separation between perforated valves and non-perforated valves of Natica (the hole effect) (Lever et al. 1961). In other experiments even the size effect and diameter effect (diameter of the hole) have been observed (Lever et al. 1964). Seilacher (1973) provides a review of such cases.

The orientation of valves during deposition has led to several discussions and differing points of view. According to Richter (1922), in flowing water 80 to 90 % of the valves are arranged with convexity (the convex side) upward, i. e., in hydrodynamically stable position. He supports this observation with several fossil examples. He further points out that exceptions to this rule are found with valves in the swash zone of sandy beaches. In a swash zone a larger number of valves show convexity downward. Moreover, shells found preserved in the laminae of megaripples also show a lesser percentage of valves with convexity upward, the reason being that during migration of the ripples, the valves roll down the slip face and are preserved without any preferred orientation (Reineck 1963 a).

On the smooth sand surfaces, with low-current velocities molluscan shells are oriented in their most stable position. In such a situation, even the shells which apparently look similar, e. g., *Cardium* echinatum and *Cardium edule,* are oriented with different azimuth, mainly because of the different position of their center of gravity (Futterer 1974). At increased current velocities a small scour is made in which they slide down and are not transported further (Futterer 1976).

On the other hand, Emery (1968 a) reports a preferred orientation of many mollusc valves with convexity downward in the shallow sea environment and the continental shelf. In observations made by underwater photography and direct observation from the research submarine, Alvin, about 90% of the observed valves were found convexity-down oriented (see Table 12). According to Emery (1968a), the reasons for the convexity-downward orientation of shells on the continental shelf is due to the activity of carnivores and scavengers. After the flesh has been eaten, the shells lie there. As the ligament decomposes, the valves are separated and lie beside each other with convexity downward. The saucer-like shape of the valves also causes a bias toward a convexity-downward orientation, when the bottom is randomly stirred by large burrowing animals. This concave-up position of molluscs on the continental shelf is an effective indicator of the presence of minor currents on the continental shelf (Emery 1968a). He gives examples of orientation in wave breaker zones, where up to 80% of the shells show convexity upwards.

A special case of orientation is vertically oriented valves (Fig. 230). According to Mii (1957) and Sanderson and Donovan (1974) they are produced on beaches by swash and backwash, when a small surface irregularity is present to keep the valves in a vertical position. Our own observations show that

Fig. 230. A shell concentration showing the majority of valves oriented vertically on an intertidal flat. North Sea tidal flats

such vertically oriented shells are also produced on intertidal flats where there is no swash and backwash, but instead alternating flood and ebb currents. Also Grinnell (1974) reports vertically orientated oyster shells in the intertidal zone by tidal currents.

On the sandy surfaces located above the high-water line, where mollusc shells are often concentrated, the orientation is normally convexity upward. Wunderlich (1970a), however, observes that sometimes a large percentage of valves are oriented convexity-downward. This reorientation of shells is caused by seagulls that come down in search of food.

Mollusc shells may not get completely preserved but the valves undergo wear and tear producing smaller parts. The valves generally break up along certain growth zones. This post mortem mechanic-

Table 12. Positions of valves on the sediment surface of shelf bottom and beach. (After Emery 1968a)

Species	Area	Measured by	Number of specimens	Percentage	
				Concave up	Concave down
Placopecten magellanicus Gmelin	Continental shelf off Chesapeake Bay	K. O. Emery	118	87	13
Arctica islandica Linné	Continental shelf off Chesapeake Bay	K. O. Emery	1 000	91	9
Spisula solidissima Dillwyn	Continental shelf off Chesapeake Bay	K. O. Emery	1 000	91	9
Spisula solidissima Dillwyn	Beach at Barnstable, Massachusetts	K. O. Emery	119	22	78
Mytilus edulis Linné	Beach at Falmouth, Massachusetts	K. O. Emery	1 000	14	86
Arca ovalis Brugière	Beach at Sapelo Island, Georgia	F. C. Marland	1 000	12	88
Arca campechensis Gmelin	Beach at Galveston, Texas	K. O. Emery	1 000	16	84
Tivela stultorum (Mawe)	Beach at San Quintin, Mexico	D. S. Gorsline	750	20	80

al disintegration of shells should be differentiated from abrasion during life time. Hollmann (1968) carried out such investigations on *Cardiun edule,* which shows a characteristic abrasion on the umbo (Fig. 231). Such abrasion can lead to post-mortem umbonal facets. The shell is later mechanically broken along umbonal facets and along the growth rings, especially the weaker winter rings. Several other shells also show similar disintegration.

Fig. 231. A *Cardium* shell showing characteristic abrasion of the umbo. Upon death the shell disintegrates along umbonal facets and weak winter rings. (After Hollmann 1968)

Besides mechanical disintegration, shells of molluscs can be weakened by boring organisms, especially the alga *Chlorophycae* and *Cyanophycae,* and the polychaete *Polydora* (Schäfer 1938). Moreover, the mollusc shells can also be weakend by predators, e. g., the well-known "Bandschnitte" in gastropods. In this case the outer wall of a gastropod shell near the aperture is broken off in bands so that the spirally arranged columella is exposed. Papp et al. (1947) observed that this destruction is caused by the crab *Pagurus.* Schäfer (1962) recorded that the spiny lobster *(Langust)* can destroy the shells of *Buccinum* in a similar fashion. On the other hand, wave action can also destroy *Buccinum,* exposing the columella in a similar way as that produced by the activity of organisms. Thus, definite recognition of destruction of shells by organic or inorganic agencies is a difficult problem.

Bioturbation Structures

The term "bioturbation structures" (Richter 1936, 1937, 1952) denotes the structures produced by the activity of living animals within the sediment or on the sediment surface. The generation of bioturbation structures causes destruction or re-formation of primary sedimentary structures produced by inorganic agencies (Richter 1936). Unlike hard parts, bioturbation structures are always autochthonous, and cannot be concentrated by reworking, except in very rare cases. Thus, they provide a more reliable record of benthonic communities than the hard parts. Bioturbation structures can be classified along with the other biogenic structure phenomena (Fig. 232). These also include plant trace fossils, of which Sargeant (1975) describes four groups: boring and attachment traces, phytolites, root molds and casts, and stromatolites.

Schäfer (1956) tries to characterize the process of bioturbation by considering the manner of living of animals near the sediment surface. He distinguishes:

1. vagile endobionts – living and free-moving in the sediment
2. endobionts with restricted movement – living within the sediment, but with restricted movement in and around the place of living
3. vagile hemiendobionts – free-moving on the sediment surface; occasionally digging into the sediment
4. vagile epibionts – free-moving on the sediment surface
5. sessile epibionts – sessile animals living on the sediment surface.

He argues that sedimentation processes such as erosion and deposition make the benthonic animals react to the new conditions. Preservation of animals and traces of their activity is accomplished only through the process of deposition. If new sediment is deposited on a sediment surface where benthonic animals are living, vagile endobionts and vagile hemiendobionts are generally the least affected, as they possess specialized organs for digging and thus move to better living positions. Many of the vagile epibionts, although they do not possess specialized organs for this purpose, are also able to move through the newly deposited sediment layer to the sediment surface. Sessile epibionts are generally not able to move to new favorable positions, and thus they die.

Removal of a sediment layer by erosion does not greatly affect the vagile epibionts or vagile hemiendobionts. On the other hand, erosion of sediment causes exposure of endobionts on the sediment surface. They are compelled to dig into the sediment again. However, some of them are unable to dig in downward from the sediment surface. These animals die if erosion is strong and much sediment is removed.

As a result of the activity of benthonic animals, the sediment is disturbed: Layers are deformed or broken and shifted, and new biogenic structures

Fig. 232. Schematic flow diagram showing major relationships among biogenic structures and of these to other phenomena. (After Frey 1971)

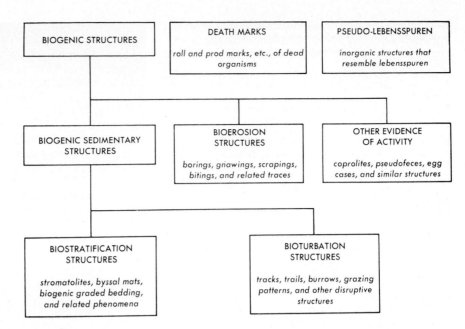

are produced. When such features of animal activity are large enough to be recognized and recorded, they are called lebensspuren. Lithology influences the visibility and preservation of bioturbation structures (Ksiazkiewicz 1977). Most notable are fine-grained, thin-bedded sand layers with visible primary structures and flat soles which alternate with shales of similar thickness.

In Recent sediments, preparation of relief casts of sands usually makes the bioturbation features visible. In muddy sediments, preparation of thin sections generally becomes necessary to make the structures visible. Sometimes X-ray radiographs can be very helpful in locating the bioturbation horizons. It should be emphasized here, that care must be taken to distinguish between disturbed bedding produced by penecontemporaneous deformation processes and bioturbation produced by animal activity.

A bioturbation structure can be recognized by the absence of well-developed bedding. Layers of a stratified unit are disrupted, broken, and partly pulled apart. Layers may be curved downward or upward. Mottled structures are also good indicators of animal activity. In muddy sediments irregular distribution of sandy nests giving a mottled appearance is usually a good sign of bioturbation. Similarly, dark-colored mud flecks in light-colored sand, irregular distribution of shells, nests of broken shells, and sand nests of different grain sizes all provide clues to the activity of organisms.

Broadly, bioturbation structures can be divided into two major groups (Schäfer 1956, 1972):

1. Fossitextura-deformativa – deformative bioturbation structures. These are bioturbation struc-

tures without any definite form. They appear as formless mottled structures or irregular flecks of different grain size, color, etc.

2. Fossitextura figurativa – figurative bioturbation structures. These are bioturbation structures possessing definite recognizable forms, such as burrows. In nature deformative bioturbation structures are more abundant than figurative bioturbation structures.

Along with the distinction between figurative and deformative bioturbation structures, another useful method for studying bioturbation structures is to describe the degree of bioturbation in a given profile. Moore and Scrouten (1957) use a scale where they differentiate between regular layers, irregular layers, distinct mottles, indistinct mottles, and homogeneous deposits. We feel that these classes are too qualitative.

Reineck (1963a) suggests a method in which the percent of the area in a vertical profile in which primary bedding has been destroyed by organisms is recorded. Depending upon the degree to which primary bedding has been destroyed, six grades are distinguished. Table 13 shows this scheme for describing the degree of bioturbation. Werner (1968) differentiates between five grades of bioturbation. He also uses a new factor – bioturbation heaving – which expresses the height of a bioturbation structure in a vertical section cut at right angles to the bedding plane.

As stated by Seilacher (1953a), bioturbation structures can be interpreted from three points of view: Taxonomic, ecologic, and stratonomic. He adds that the ecologic approach should acquire priority; the taxonomic approach often fails be-

Table 13. Classification of bioturbation depending on primary bedding destruction. (Modified after Reineck 1963a)

Grade	Degree of bioturbation in %	Classification of bioturbation
0	0%	No bioturbation
1	1– 5%	Sporadic bioturbation traces
2	5–30%	Weakly bioturbated
3	30–60%	Medium bioturbated
4	60–90%	Strongly bioturbated
5	90–99%	Very strongly bioturbated, but rest of inorganic bedding still recognizable
6(E)	100%	Completely bioturbated

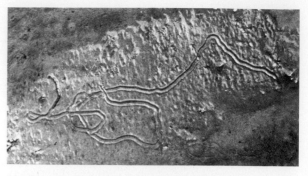

Fig. 234. Crawling traces. The markings are made by moving organisms *(Nereis diversicolor)* on a soft sediment bottom. North Sea tidal flats. (After Häntzschel 1940)

cause it is not always possible to reconstruct the animal species from lebensspuren; the stratonomic approach, although sometimes useful, is not a biological approach (Müller 1951). In the taxonomic study of trace fossils morphological characteristics of lebensspur are described in detail, and individual genera and species are erected. Individual genera and species express only individual morphology. Häntzschel (1962) has compiled the existing trace fossil genera and species. The sedimentological aspect of lebensspuren is rather neglected.

Based on some ecological factors, Seilacher (1953a) suggested a classification which distinguishes five groups of lebensspuren:

1. Resting traces – marks left behind by organisms resting in one place (Fig. 233).
2. Crawling traces – marks made by organisms moving on a soft sediment surface (Fig. 234).
3. Browsing traces – marks made by organisms feeding on a sediment surface (Fig. 235).
4. Feeding structures – burrows and other structures produced by organisms while moving through the sediment for feeding (Fig. 236).
5. Dwelling structures – burrows and other structures made by organisms in which to live (Fig. 237).

Fig. 235. Browsing traces. Markings are made by *Nereis* feeding on a sediment surface. (After Häntzschel 1940)

Fig. 233. Resting traces made by seals coming to rest on sandy bars of tidal flats. North Sea tidal flats

Fig. 236. Feeding structures. *Paraonis* makes well-defined spiral and meandering burrow systems within the sediment while moving through the sediment for feeding. (After Röder 1971).

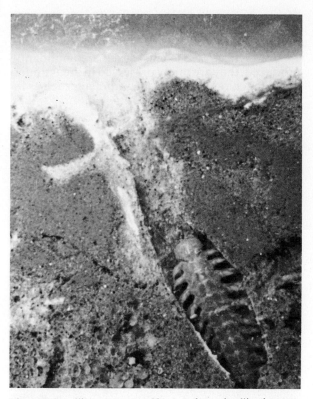

Fig. 237. Dwelling structures. *Nereis* makes a dwelling burrow. The animal is visible in its burrow. Laboratory experiment (After Reineck 1958c).

These five types of lebensspuren even show morphological differences. The only drawback of this excellent genetic classification is that it is not always possible to assign a trace to a definite living activity. Moreover, traces may be multipurposed. This problem is all the more difficult in ancient sediments, where we know very little about the animals producing the lebensspur and their behavior.

The organisms may leave behind traces of their activity on the surface of the sediment or within the sediment. Thus, one may try to distinguish between surface lebensspuren and internal lebensspuren. However, observations in Recent environments strongly suggest that one should not try to make rigid classifications on this basis. Animals moving or resting on the sediment surface and thus normally making surface lebensspuren, may partly dig deeper in the sediment and make internal lebensspuren. On the other hand, animals living normally within the sediment may move or lay part of their bodies on the sediment surface, thus producing surface lebensspuren.

A stratonomic approach to lebensspuren has led to the classification of lebensspuren based on the position of their preservation. Two well-known stratonomic classifications are those by Seilacher (1953 a) and Martinsson (1970). Such an approach

can be useful in the study of ancient sediments. Figure 238 gives the main features of Martinsson's classification.

Based on his observations on intertidal flats, Hertweck (1970) places lebensspuren into three groups:

1. surface lebensspuren
2. internal lebensspuren
3. dwelling structures

This classification has its drawback in that it differentiates between surface and internal lebensspuren. The problem becomes acute as some of the benthonic animals may move within the sediment along the bedding planes. In ancient sediments they are likely to be interpreted as surface lebensspuren, although they were produced not on the sediment surface but within the sediment. Despite this drawback, we shall describe lebensspuren under these three headings.

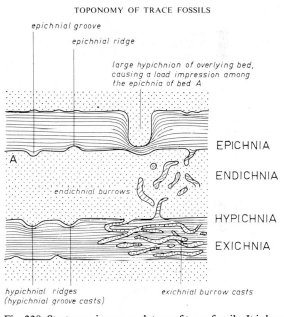

Fig. 238. Stratonomic nomenclature of trace fossils. It is based on the main medium of preservation. Silt and sandstone (stippled) and mudstone (ruled) are interlayered, as seen in the cross-section. The four main terms on the right of the diagram (Epichnia, Endichnia, Hypichnia, and Exichnia) refer to bed A. (After Martinsson 1970)

Surface Lebensspuren

Surface lebensspuren are formed on the upper surface of the sediment as a result of animal activity. Traces of the entire body or of individual organs are left behind. Living activities leading to the formation of lebensspuren are rather varied. They may be resting traces produced by animals lying for a while on a soft sediment surface; crawling traces of animals, leaving behind imprints of organs of locomotion: or browsing traces of animals moving on a soft bottom in search of food. Surface lebensspuren can also be made by flying organisms when they land to rest or are in search of food on a soft sediment surface; they are generally footprints or beak marks of birds. Likewise, swimming animals such as fish also produce traces when they come to rest on the sediment surface or touch the surface during swimming.

To surface lebensspuren also belong the traces which are made on a muddy surface covered by a thin sandy layer, through which an organism ploughs (internal lebensspuren – Seilacher 1953b). These internal lebensspuren are filled in just after their genesis by overlying sand (Fig. 239). On preservation they appear as a well-developed relief on the lower side of the sand layer. Such lebensspuren are generally resting traces made by organisms

coming to rest and digging slightly into the sediment. Such traces should not be confused with the internal crawling traces.

Although surface lebensspuren are abundantly produced, they are rarely preserved in areas undergoing much reworking of the sediment, such as intertidal flat sediments.

Internal Lebensspuren

Internal lebensspuren are traces produced within the sediment, mainly by vagile endobionts. These animals are generally carnivorous forms, but they may also be sediment-eating forms. Examples of internal lebensspuren are "press structures" of the sea urchin *Echinocardium* (Fig. 240) and burrows within the sediment.

Fig. 240. Press structures of *Echinocardium* are made by the moving animal within the sediment. They are typical internal lebensspuren

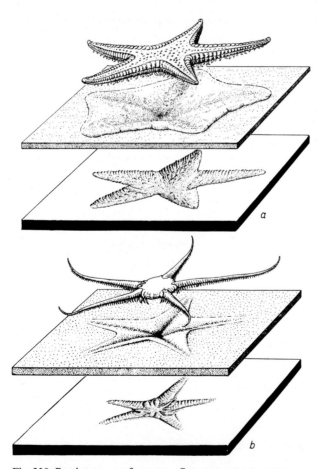

Fig. 239. Resting traces of sea stars. Sea stars come to rest on a sandy surface below which a muddy layer is present. If the animal digs in slightly, it makes a well-developed impression on the mud layer. As soon as the animal leaves the place, the impression is covered by exiting sand and the trace is preserved. (After Seilacher 1953b)

Dwelling Structures

Dwelling structures are the burrows and other similar structures made by organisms in which to live. This is done primarily to protect themselves from falling prey to other animals. However, some of the burrowing patterns are essentially feeding structures made to search for food. Essentially, dwelling structures are distinguished into two groups:

1. Borings. Borings are holes made by organisms in hard surfaces either for protection or in search of food (Fig. 241). They are likely to be formed and preserved in places where a hard rock surface has been exposed to animal activity for a certain length of time before being covered under a blanket of transgressive sediments. The holes are generally smooth and are either vertical or inclined at various angles. Occasionally, the boring organism may be found preserved in the hole itself. Various organisms make borings in rock, shells, word

Fig. 241. Borings filled with sand, made earlier by molluscs in hardened mud

etc. (Warme and Marshall 1969; Warme 1970; Warme and McHuron 1978) Röder (1977) describes the relationship between shell construction and different substrates such as mud, rock, and wood in mechanically boring Lamellibranchiata. Even plant roots can produce borings (Sargeant 1975). Edward and Perkins (1974) found in samples of skeletal substrates from South Carolina to central Florida the distribution of microborings of fungal origin and of green alga and sponges. The distribution pattern of microborings of algal origin within sediments showed a highly bored inner sediment band related to the present-day photic zone and an outer band interpreted as relict photic zone. This relict band marks a "fossil photic zone" associated with a lower sea-level stand, suggesting its potential application as a paleobathymetric indicator.

2. *Burrows.* Burrows are holes made by organisms in soft sediments. They are easy to identify and are commonly reported from various environments. Burrows are made not only by worms, but also by members of other animal groups possessing hard skeletal parts, such as molluscs, echinoderms, etc. A special type of burrow are tubes, which are untransportable dwelling structures made up of sediment grains cemented together by the organisms, e.g., *Lanice.* Sometimes conical pipes (quivers) are transportable, e.g., *Pectinaria.*

Genetically, burrows are holes made by organisms through their bioturbation activity and are often made stable through cementation by slimy secretions. Bioturbation is accomplished in one or several ways, such as scratching, climbing, pushing, blowing, digging, etc. Once a burrow is completed, the movement of the animal is generally restricted to the burrow. Only occasionally – for feeding or excretion – an animal may partly move out of its burrow. Burrows may acquire varied forms; Simple burrows, U-shaped burrows with an opening at one or both ends, branched burrows, meandering burrows, etc.

In the case of burrows, one can distinguish between a *burrow nucleus* – the main burrow made by the animals with the organs of locomotion – and a *burrow halo,* the affected area round the burrow nucleus in which bedding is disturbed (Fig. 242) (Reineck 1958b). In this area, the layers are broken and swept upwards or downwards. The direction in which layers are swept has no direct bearing on the direction of movement of the digging organism.

A clear-cut differentiation between dwelling structures and feeding structures is possible in only some cases. In general, suspension feeders possess simple straight, or U-shaped burrows. Complicated patterns of spreiten are generally feeding structures made by sediment feeders. These patterns are generally planned to use a maximum possible surface area (Seilacher 1954).

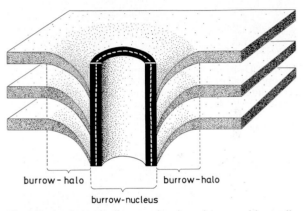

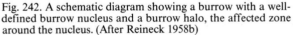

Fig. 242. A schematic diagram showing a burrow with a well-defined burrow nucleus and a burrow halo, the affected zone around the nucleus. (After Reineck 1958b)

Some of the burrowing organisms modify their burrows. They possess the ability to coat their burrows. Reineck (1958 b) suggests that such coating is of two types:

1. A burrow is coated because of some internal instincts to have a constructionally stable burrow (Fig. 243). Such coating is found mainly in feeding burrows, where the animal adds the fecal matter to the burrow. It can show varied patterns.

2. A burrow is coated to keep the burrow clean, and coating is done, because of a reaction of the organism to external factors (Fig. 244). For example, if some sediment falls in the burrow, the animal presses this sediment on the side or to the bottom.

A coating of type 1 is rather regular; type 2 is rather irregular, and added layers are discontinuous, varying in number of layers and thickness. Type 2 coating is common in areas undergoing intermittent reworking, in which sediment falls continually into the burrow.

Not all animals, however, react in the same fashion. Some animals would prefer to leave the burrow and dig a new burrow rather than to clean the

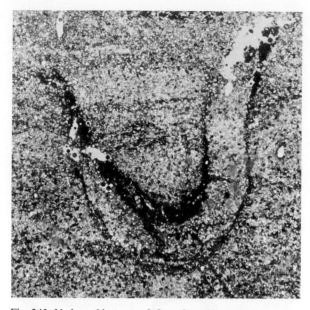

Fig. 243. U-shaped burrow of *Corophium* in sandy sediments. The upper wall of the curved part of the U-shaped burrow has been reinforced by muddy layers, and at this point the burrow wall is thick. Such coating not only cleans the burrow; mainly it keeps the construction of the burrow stable. North Sea tidal flats. (After Reineck 1958c)

Fig. 244. A burrow of *Nereis* showing a multi-layered coating. The coating is made by the organism adding the extra sediment to the wall. This sediment is mainly sediment grains falling into burrows during sedimentation processes. North Sea tidal flats (After Reineck 1958c)

existing burrow. A few animals, like the amphipod *Corophium,* are even capable of blowing out unwanted sediment in order to keep the burrow clean.

If sedimentation is excessive, organisms are compelled to move upward from their living horizons to a new living horizon near the sediment surface, thus producing escape traces (Reineck 1958c). Fig 245 depicts escape traces of *Macoma.* Escape traces are differentiated from normal burrows in that they are more or less straight, unbranched, and almost vertical. They are not reinforced by slimy secretions and never possess multilayered walls. Along the escape traces sediment layers are invariably bent downward, despite the upward movement of the organism (Fig. 245). Escape traces of *Hydrobia* are vertical spiral traces (Hertweck in Reineck et al. 1968). Escape traces are also produced by lamellibranchs, such as *Mya arenaria,* which shows a reaction to the sedimentation (Reineck 1958c). Figure 246 depicts escape traces due to rapid rate of sedimentation.

Fig. 245. Escape traces of *Macoma*. The traces are almost straight, vertical, and are identifiable because of the downward-bent of sediment layers along the traces. At the top end of the trace, the organism is itself present. Relief cast. North Sea tidal flats

Well-defined bioturbation structures are also made by organisms which do not leave any hard parts. In some cases, the lebensspuren of a polychaete can even be a potential characteristic species (Dörjes in Reineck et al., 1968). Later, on preservation, such lebensspuren can become the characteristic fossil of a biofacies (Hertweck 1970).

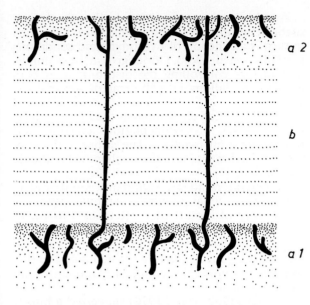

Fig. 246. Schematic representation of escape traces as a result of rapid sedimentation. a_1 Thickly populated living horizon of polychaetes; b rapidly sedimented unit leading to the escape of a few organisms toward the top; a_2 new living horizon of polychaetes populated partly by organisms of the lower horizon (a_1) which were able to escape and partly by new organisms brought in from adjoining areas. Burrows of the living horizon are curved, branched, and high in number (a_1 and a_2), while the escape burrows (traces) are few in number, straight, and unbranched (b). (After Reineck 1958c)

Characterizing of Environments by Lebensspuren

Attempts have been made to characterize a given environment with a spectrum of lebensspuren, and to use them as indicators of bathymetry. The important papers in this direction are by Seilacher. He showed that the spectrum of lebensspuren in molasse sediments is different from that in flysch sediments (Seilacher 1954, 1958, 1964). Frey and Howard (1969) and Basan and Frey (1977) attempted to differentiate various coastal environments near Sapelo Island, Georgia, U.S.A., on the basis of lebensspuren. Seilacher (1967) gave a bathymetry based on the form and patterns of burrows. According to him, there is a general gradation from vertical burrows in the shallow-water depositis to horizontal and increasingly patterned burrows in the deep-water deposits (Fig. 247). This corresponds to a trend from suspension to sediment-feeding habit, which in turn is a response to the availability of food in different depths. While food particles remain suspended in the agitated zone of shallow water, they settle in quieter deeper waters. Hertweck (1975) gives the distribution and zonation of ecologically indicative lebensspuren in a generalized beach-offshore profile (Fig. 248). The environmental significance of trace fossils is given by Rhoads (1975), Frey (1978), and Seilacher (1978). Rhoads and Young (1970), in their in-

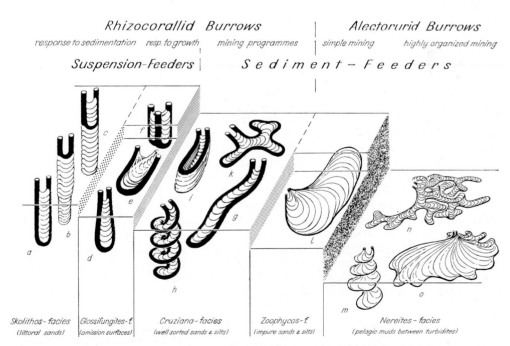

Fig. 247. Bathimetric zonation of trace fossils with spreites. There is a general gradation from vertical burrows in shallow-water deposits to increasingly patterned and horizontal burrows of deep water deposits. (After Seilacher 1967)

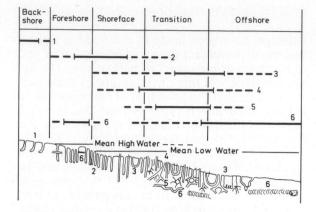

Fig. 248. Schematic representation of distribution and zonation of ecologically indicative lebensspuren in a generalized beach-offshore profile. *1* burrows of terrestrial crustaceans (mostly scavengers); *2* straight vertical burrows of low-level suspension feeders; *3* burrows having two openings (basically U-shaped) of low-level suspension feeders or collectors; *4* tubes of high-level feeders; *5* dwelling burrows of intrasedimentary feeding animals; *6* crawling traces of intrasedimentary feeding animals. (After Hertweck 1975)

vestigations of organisms in Buzzards Bay, Massachusetts, found that sediment-feeding (deposit-feeding) and suspension-feeding benthonic animals show a marked spatial separation. Suspension feeders are largely confined to sandy or firm mud bottoms, while sediment feeders are abundant on soft mud surfaces. They observed that intensive bioturbation of the upper few centimeters of a mud bottom by sediment feeders produces a fluid fecal-rich surface that is easily resuspended by low-velocity currents. Physical instability of this layer tends to: (1) clog the filtering structures of suspension-feeding organisms, (2) bury newly settled larvae or discourage the settling of suspension-feeding larvae, (3) prevent sessile epifauna from attaching to the sediment bottoms. Thus, suspension feeders are unable to successfully inhabit all areas where a suspended food source is available, especially in areas where mud bottoms are intensively reworked by sediment feeders. Such modification of benthonic populations is called trophic group amensalism. Rhoads and Young (1970) point out that this biotic relationship seems to be important in shaping trophic-group distributions in embayments and basins on continental shelves.

The density of animal population and bioturbation depends mainly on the oxygen content of water, the availability of food, and the conditions of sedimentation. Schmidt (1935, 1944, 1958) expresses the view that water circulation, which also controls the oxygen content of water, sediment type, and sedimentary structures, is an important factor. He gives a classification of environments based only on the oxygen content of water.

Another important factor controlling the density of the benthonic population is the grain size, which in turn is a function of hydrodynamic conditions (cf. Dörjes et al. 1969). Dörjes gives a list of important factors controlling the distribution of benthonic populations in well-aerated, subtidal environments.

In well-aerated environments the fauna reflects the processes of sedimentation, erosion, and resedimentation together. The stronger the processes of sedimentation, the smaller the benthonic population.

In general, rapid deposition and rapid erosion reduces the animal population and with that also the density of bioturbation (Reineck 1967b). In Fig. 249 these relationships have been depicted diagramatically. In areas of strong reworking, sediments are relatively coarser grained, and the corresponding inorganic sedimentary structures are well-developed.

Schäfer (1962, 1963, 1972) suggested a binary nomenclature, giving five biofacies. Four groups cover the clastic sediments, whereas the fifth group covers the bioherms. Schäfer uses the term "vital" for animal populated regions, "letal" for regions without animal population. Regions with well-developed erosional unconformities and reworking

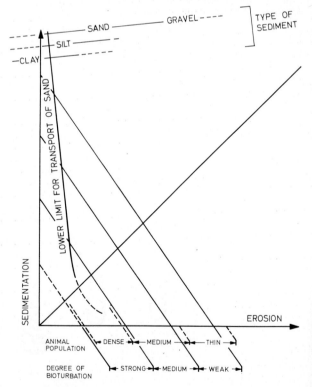

Fig. 249. Diagram illustrating the relationship of grain size and animal population and bioturbation to the processes of erosion and sedimentation. The increasing rate of erosion causes coarsening of sediment, thin animal population and less bioturbation. (Modified after Reineck 1967b)

Fig. 250. Vital-astrat biofacies, characterized by hard bottom with a stabilized biocoenosis of sessile organisms. Agitated clear water without any detrital suspension. From top to bottom: Living corals, living carbonate producing algal mats, coral detritus, and reef carbonate. (After Schäfer 1963)

Fig. 251. Letal-pantostrat biofacies, characterized by taphocoenosis of nektonic and planktonic organisms, well-developed fine bedding, and stagnant water. Macrobenthos are absent. (After Schäfer 1963)

1 Fish carcass having recently sunk to the sea floor, the abdominal cavity is still somewhat bloated by gas and thus not lying flat. *2* Nektonic decapodan crustacean, spread out on bedding plane, abdominal portion lying on its side. *3* Tooth of marine mammal. *4* Skeleton of selachian fish. *5* Scales of teleostean fish. *6* Fragmented coprolite. *7* Vertebra from proximal portion of marine mammal tail, commonly first part of drifting carcass to decay. *8* Shell of *Janthina,* a prosobranch gastropod living as plankton by attaching itself to foam raft. *9 Nautilus* shell (usually drift at the surface for 6 weeks after death, then sink). *10* Bird skull. *11* Teleostean fish skeleton. *12* Skeleton of marine mammal. *13* Fragments of *Lepas* valve

are "lipostrat," and regions without any erosional unconformities, but with continuous sedimentation are "pantostrat." Thus, five biofacies are

1. Vital-astrat-bioherms (Fig. 250)
2. Letal-pantostrat (Fig. 251)
3. Vital-pantostrat (Fig. 252)
4. Vital-lipostrat (Fig. 253)
5. Letal-lipostrat (Fig. 254)

Many features associated with trace-fossils may provide valuable information regarding mode and rate of sedimentation. As discussed by Goldring (1964), the frequent occurrence of abundant sur-

Table 14. Thickness of the living horizons of some marine animals capable of producing completely bioturbated horizons, in which inorganic bedding is completely destroyed. In some cases such bioturbation horizons are produced due to the activity of several species. However, in such cases, only the major bioturbating animal of the living and the bioturbating community has been mentioned. Individual burrows with their maximum depths have not been considered

Area	Species	Number of individuals/m^2	Thickness of the bioturbation horizon in cm	Depth of the bioturbation horizon below the sediment surface
N, M	*Amphiura filiformis* (Echinodermata)	100	5	0
N	*Arenicola marina*[b] (Polychaeta)	50	25	0
A	*Busycon carica*[b] (Gastropoda)	–	10	0
A	*Callianassa n. sp.*[b] (Crustacea)	75	10–20	10
M	*Capitomastus minimus* (Polychaeta)	1 500	10	0
N	*Cardium edule* (Lamellibranchiata)	3 000	3–5	0
N	*Corophium volutator*[b] (Crustacea)	4 400	5–8	0
A	*Donax variabilis* (Lamellibranchiata)	17 000	1	0
N, M, A	*Echinocardium cordatum* (Echinodermata)	25	10	5
N	*Echiurus echiurus*[a] (Echiurida)	1 000 (juvenile)	10	5
A	*Hemipholis elongata* (Echinodermata)	50–100	8–10	0
N, A	*Heteromastus filiformis* and *Nereis diversicolor* (Polychaeta)	2 000–3 000	10–15	0
N	*Lanice conchilega* (Polychaeta)	2 500	30–40	0
N	*Mya arenaria*[b] (Lamellibranchiata)	30	30	20
M	*Onuphis eremita*[b] (Polychaeta)	500	30–40	0
N	*Polydora ciliata*[b] (Polychaeta)	200 000	–	0
N	*Pygospio elegans* (Polychaeta)	10 000	3–5	0
N	*Scoloplos armiger* (Polychaeta)	250	10	2
A	*Scoloplos fragilis* (Polychaeta)	250	10	2
N	*Scrobicularia plana* (Lamellibranchiata)	130	15	10–15
N, M, A	*Spiophanes bombyx* (Polychaeta)	3 000	5	0

[a] Complete bioturbation is achieved only by juvenile animals. Adult animals live at greater distances from each other. For this reason, in spite of their greater depths of burrow (up to 25 cm), adult animals are unable to cause complete destruction of bedding.
[b] Population of these animals produces only partial bioturbation. Remaining primary inorganic structures are recognizable.
A = Atlantic Ocean, Georgia Coast, U.S.A. M = Mediterranean Sea, Gulf of Gaeta, Italy. N = North Sea, German Bay, Germany.

face lebensspuren, or those made very close to the sedimentary surface, indicates fairly continuous sedimentation and the preservation of many sedimentary surfaces. On the other hand, the complete absence of surface lebensspuren, associated with evidences of intermittent erosion, is a testimony of discontinuous sedimentation or much reworking, and the eventual preservation of only a fraction of the actual sediment deposited and original sediment surfaces.

In the case of extremely slow rates of sedimentation, burrowing is very intensive as organisms have sufficient time for bioturbation. Usually in sediments of such areas fecal pellets are also found concentrated. Table 14 gives a list of animals from modern environments, and the depth to which they can cause complete bioturbation of sediments. Single species live deeper than 3 m in the sediment. So the supershrimp *Axius serratus* has a burrow deeper than 2.5 m, as reported from Nova Scotia by Pemberton et al. (1976). Sediment contained in old filled burrows is anomalous in its distribution of grain size and its contents of organic carbon, water, and trace elements. This fact affects the geotechnical properties of sediments on the sea floor.

In regions with rather slow rates of deposition and growth of sequence, every new horizon of bioturbation can be superimposed on the older bioturbation horizon. In this manner a thick zone of bioturbation is produced, although a single horizon of bioturbation is only a few centimeters thick (Fig. 255a). Another possibility is that the growth of a sequence is marked by periodic intermittent erosion. The erosion can remove the populated and bioturbated horizon, so that at the end an almost nonbioturbated sequence is produced (Fig. 255b). A third possibility is that after a new sediment layer is deposited, there is a quieter period. During this time the upper few centimeters of the new sediment layer are thickly populated by benthonic organisms and are bioturbated (see Fig. 255c).

Non-bioturbated layers of such profiles indicate a quick rate of sedimentation of the new sediment. Such cases in the shallow sea and tidal flats are always related to storm activity. With increased wave energy and higher current velocities than normal, large amounts of sediment are transported or redeposited. In the shallow sea (subtidal part) mostly storm sand layers are deposited; sometimes storm sand layers also develop on the tidal flats (Reineck 1977). Such layers in the tidal flats may show a shell layer at the base, indicating erosion; or in areas where the old sedimentation surface is protected by an algal layer, new sediment is deposited on top of an algal layer without any visible erosion.

A population can be initiated by the organisms which were previously living in the lower bioturbated horizon, which is now covered by a new sediment layer. In such a case two horizons of biotur-

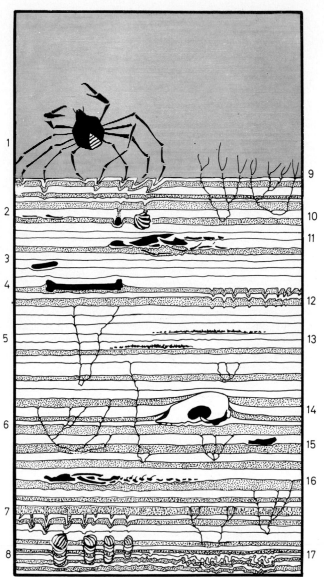

Fig. 252. Vital-pantostrat biofacies, characterized by benthonic biocoenosis and taphocoenosis of nektonic and planktonic organisms. The varying degree of bioturbation destroys primary bedding; however, there are no erosional unconformities but continuous deposition. (After Schäfer 1963)
1. Sea spider, long-legged in adaptation for life in quiet water. *2* Fragments of *Lepas* valves. *3 Nautilus* shell. *4* Limb bone of marine mammal, commonly first object to drop off a carcass drifting on the ocean surface. *5* and *6* Dwelling structures of polychaete worms, prolonged upward with deposition of each new layer of sediment. *7* Crustacean pacing trail, formed as in *1*. *8* Escape trails of bivalves, all simultaneously killed by temporarily anoxic water. *9* Inhabited worm tubes. *10* Bivalves in living position. *11* Brachyuran crustacean. *12* Crustacean pacing trail, formed as in *1*. *13* Brittle star skeleton, below skeleton of starfish. *14* Skull of marine mammal. *15* Coprolite. *16* Skeleton of teleostean fish. *17* Perturbed resting trace

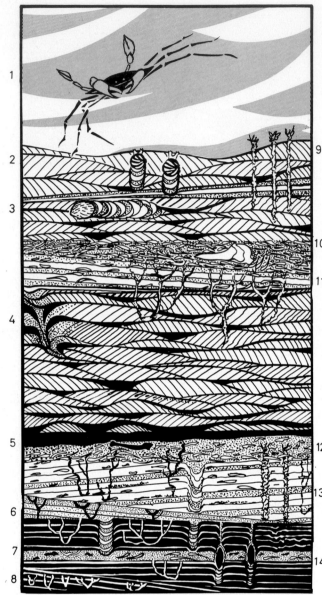

Fig. 253. Vital-lipostrat biofacies, characterized by several benthonic biocoenosis, bioturbation, taphocoenosis, shell layers, and erosional discordances. The deposition is not continuous, being marked by erosional unconformities. The strongly agitated water is ideal for benthonic, nektonic, and planktonic organisms, at the same time causing strong reworking and movement of sediment. (After Schäfer 1963).

1 Swimming curstacean, *2* Bivalves ascending to keep up with sedimentation. *3 Echinocardium,* moving left and leaving press structures: layers of detritus and mud in ripple troughs. *4* Escape trace of *Aphrodite* in cross-bedded sand sediment. *5* Bone of bird limb in layer of coarse sand with shell debris covered by (black) mud layer. *6* Escape traces of bivalves and burrows of worms. *7* Otoliths. *8* Worm burrows. *9* Agglutinated polychaete tubes. *10* Skull of marine mammal in layer of coarse sand. *11* Worm burrows. *12* Agglutinated tubes, damaged in layer of coarse sand and shell debris. *13* Escape trace of benthonic gastropod. *14* Bivalves in living position atop escape traces and surmouted by siphonal shafts

bation are generally joined together through a few straight escape traces. On the other hand, a population can be newly initiated by larvae or by the juvenile organisms that are dispersed into this area. Several months may expire before a new population is initiated through larvae or juveniles. If two strongly bioturbated horizons are found separated by a thin nonbioturbated sediment layer without any escape traces, the upper bioturbation horizon is a result of a new population.

Animals living in their burrows, generally react to rapid erosion and sedimentation by changing the position of their burrows upward or downward, in order to keep the same distance from the sediment surface. Such movements are often recorded in the form of spreiten. These features are good indicators of the process of reworking and can be used to estimate the amount of deposition or erosion during reworking. Two important papers dealing with this problem are those by Reineck (1958a) and Goldring (1964). The geological significance of trace-fossils is discussed by Osgood (1975), Crimes (1975), Howard (1975), and Rhoads (1975).

Reineck (1958c) demonstrated the effect of slow and fast deposition and erosion on the behaviour of some organisms of shallow seas. Some bivalves, e. g., *Mya,* live at a certain depth from the sediment surface. In the case of rapid deposition, *Mya* moves upward and leaves spreiten below its burrow (Reineck 1958a). The net movement of the organism, recognizable in the form of spreiten beneath the lower part of burrow, indicates net growth in sequence. Figure 256 depicts various ways in which *Mya* can react to the process of sedimentation. Figure 257 depicts the reaction of organisms to erosion-deposition as exemplified by *Diplocration yoyo* in the rocks of Devonian age (Goldring 1964). The study of Howard and Frey (1975) suggests that the burrows of *Callianassa major* can be good indicators of erosion and deposition processes (Fig. 258).

Other animals possessing U-shaped burrows, e. g., *Echiurus,* may produce spreiten both above and below the lower curved part of the U. During strong erosion. *Echiurus* is compelled to shift its burrow downward, in the process making spreiten above the curved part of the U. Spreiten below the curved part of the U are produced during rapid sedimentation, when the animal tries to shift its burrow upward.

Some animals produce spreiten when excess sediment falls into their burrows. The animal presses this excess sediment into the lower part of the burrow.

As discussed by Seilacher (1967), not all spreiten are a response to sedimentation. In several cases, especially in the case of sediment feeders, formation of spreiten is a normal process, regardless whether or not sedimentation-erosion is occurring (Fig. 247).

Organisms may in various ways affect the grain-size distribution of the sediments. Some animals, e.g., *Lanice conchilega,* make tubes that are coarser-grained and better-sorted than the surrounding sediments. Such tubes can be easily transported, and accumulation and decay can produce a sand layer in a different place (Wunderlich 1970c). Graded bedding of biogenic origin is also known (see p. 119). Sediment-eating organisms may cause a reduction in grain size during digestion. This is especially true in carbonate sediments.

Furthermore, organisms also produce biochemical changes in the sediments. Frey (1971) provides an useful review on trace fossils.

Fecal Pellets

Excretory matter of organisms can become an important constituent of sediments. The term "coprolite" is generally restricted to designate the excretory matter of vertebrates. Coprolites are large in size, measuring several centimeters in length. A detailed study of coprolites with undigested food may provide some information about the eating habits of a particular animal. Häntzschel et al. (1968) provide an exhaustive bibliography on fossil excrements. They suggest that the term coprolite should be used to cover all sorts of fossil excrements, even the smaller fecal pellets of invertebrates.

In subaquatic environments fecal pellets of invertebrates are much more abundant than the coprolites of vertebrates. Detailed investigations of Recent sediments, especially in thin-section, have shown that fecal pellets are abundant overall, in deep-sea sediments as well as in shallow seas and in intertidal flats (Reineck 1972; Reineck and Singh 1971; Gripp 1956; van Straaten 1954, Schäfer 1953). As fecal pellets are mainly produced by benthonic populations, they are absent in sapropel-sediments, where a benthonic population is also absent (Wetzel 1937).

Based on the study of different shapes and sizes of fecal pellets, attempts have been made in Recent environments to associate them with the producing organisms (Edge 1934; Moore and Kruse 1956; Schäfer 1962, and others). Schäfer (1953) discusses the behavior of organisms at the time of producing excretary matter. For example, gastropods are able to produce fecal matter while still moving. The result is production of threads of fecal matter, which can be compared to fossil examples of *Tomaculum* Groom. Hemi-sessile worms produce excrement in the form of an aggregate of pel-

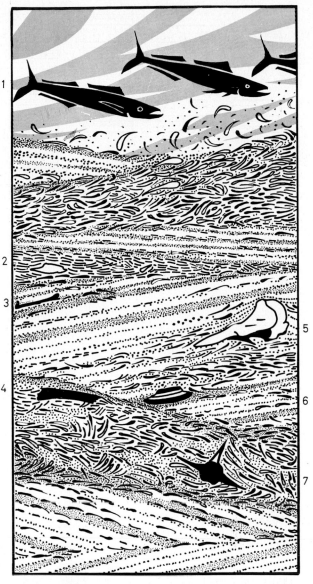

Fig. 254. Letal-lipostrat biofacies, characterized by an enrichment of animal skeletal parts and an abundance of erosional unconformities. Benthonic organisms are absent because of strong rates of reworking and sediment movement. Only very coarse-grained material is deposited. Large shells may be abundant; they are allochthonous, brought in from adjoining areas. (After Schäfer 1963)
1 Swarm of fish (fast swimmers adapted to moving in currents). *2* Clawed limb of crustacean. *3* Limb bone of bird. *4* Rib fragment of a Greenland whale. *5* Dolphin skull. *6* Peat pebble. *7* Vertebra of marine mammal

lets on the sediment surface (Fig. 259). Generally, fecal pellets are moved away from the place of their genesis through bioturbation or inorganic agencies. In such cases, recognition of the fecal pellet-producing animal becomes rather difficult.

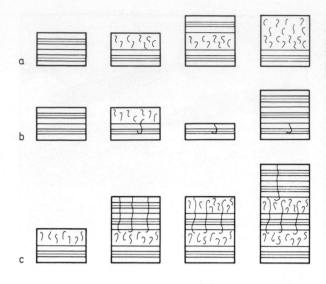

Fig. 255. Scheme demonstrating the possibility of preservation of bioturbation horizons under different conditions of sedimentation and growth of sequences. a The Rate of sedimentation and growth of sequence is slow. Bioturbation is restricted to a thin top layer. New sediment deposited on top of a bioturbated horizon can be completely bioturbated because of slow rates of deposition. Thus, exceptionally thick zones of completely bioturbated sediment are produced, although individual living horizons are much thinner. b The growth of sequences is marked by periodic intermittent erosion. Erosion removes the top bioturbated zone. At last a completely nonbioturbated sequence is produced, although, at times, bioturbation is strong. A few traces of burrowing may be visible. c Sedimentation is rapid, punctuated by long periods of slow deposition or nondeposition. After each phase of deposition (Winter), only the topmost layer of the unit is bioturbated (Summer). Then follows a new phase of deposition. At last, alternating bioturbated and nonbioturbated zones are produced. In the nonbioturbated zone a few escape traces may be present. (Modified after a scheme of van Straaten 1954a)

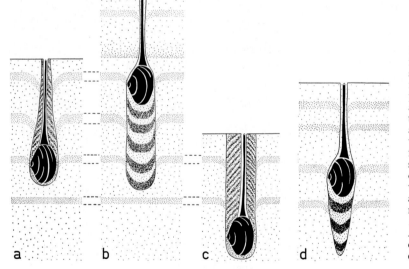

Fig. 256. Diagram illustrating various ways in which *Mya* reacts to erosion and sedimentation processes. a Stationary sediment surface; the growing juvenile pelecypod burrows deeper in the sediment, making a conical-shaped burrow above the animal tapering upward. b Rapid deposition; the animal moves upward keeping pace with sedimentation, leaving below an infilled burrow of the same breadth as the shell. c Rapid erosion leading to degradation of the surface. The animal migrates downward, making a burrow of the same breadth as the shell above the animal. d Very slow rate of sedimentation. A growing juvenile *Mya* keeps the pace of the sediment surface. A concical-shaped infilled burrow tapering downward is present below the animal. (After Reineck 1958c)

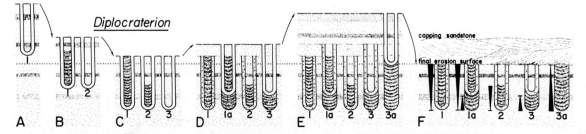

Fig. 257. Reaction of organisms to erosion. Deposition is exemplified by *Diplocration yoyo*. A Development of the burrow. B Because of erosion the burrow is shifted downward, and a new tube (2) is constructed. C Further erosion, a shifting downward, and a new tube (3). D and E Sedimentation causes upward shifting of the tubes, but some tubes are abandoned. F All tubes are abandoned and erosion reduces them to a common base. (After Goldring 1964)

The transported fecal pellets can be deposited in ripple troughs to produce mud flasers or even well-developed layers. Excretory matter of the mussel *Mytilus edulis* can even result in the deposition of thick layers of mud (Fig. 260); however, the form of the fecal pellets is destroyed (van Straaten 1954a). Because of the capture of suspended matter and the production of fecal pellets which falls down between the mussels joined together by byssal threads, the mussels are compelled to grow upward to keep themselves on the surface. Below the *Mytilus* colonies, several decimeter-thick mud layers are produced, composed mainly of fecal pel-

lets. The excretory products of the marine decapod *Callianassa major* Say and the marine annelid *Onuphis microcephala* Hartmann have been studied in shallow marine environments of the Georgia and eastern Gulf of Mexico coasts, U.S.A. (Pryor 1975). These filter-feeding organisms produce depositionally significant quantities of argillaceous fecal pellets that are transported and deposited as granular clay with hydraulically equivalent quartz sand grains.

The fecal pellets are in no way restricted to marine environments. Several occurrences of fecal pellets in freshwater environments are known.

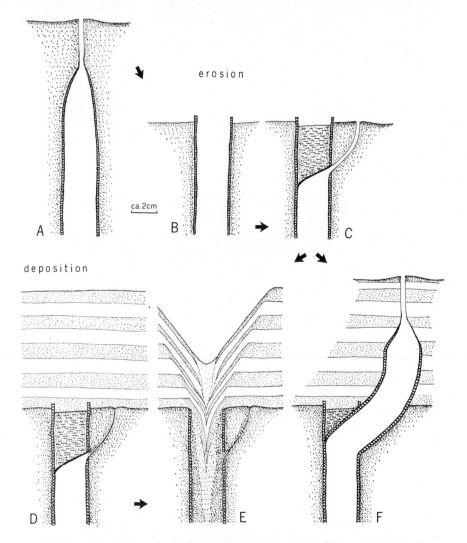

Fig. 258. Burrows of *Callianassa major* as indicators of erosion and deposition. The burrow normally is vertical, with thick durable walls leading up to a more ephemeral, constricted, thinly lined apertural neck (A). With substrate erosion the burrow is truncated, but remnants of the resistant walls commonly project, slightly above the scour horizon (B). The shrimp responds by sealing off the top of the burrow with a clayey plug and forming a new, lateral apertural neck (C). With renewed deposition, the aperture may be abandoned (D). or the shrimp may extend the burrow upward (F). If abandoned (D), the weight of overlying sediments may cause the plug to fail, resulting in a collapse structure and the filling of that part of the burrow (E); the collapse structure strikingly resembles the escape structure produced by certain burrowing actinian anemones. If extended upward (F), the animal prefers to follow the path of the lateral neck rather than merely to extend the old walls directly upward; the sediment laminae are idealized here; in most cases the animal causes bioturbation of adjacent sediments. Reconstructed from various observations. (After Howard and Frey 1975)

Fig. 259. Elongated, elliptical fecal pellets concentrated at one point, producing the polychaete organism *Heteromastus*

These fecal pellets are produced mainly by gastropods. The worm *Tubifex,* which makes rather dense populations in some of the present-day freshwater lakes, seems to transport sediment to the surface in its excrements and produces a rather muddy fecal layer (Solowiew 1929; Zahner 1967; Wagner 1968).

A primary biogenic layer of fecal pellets in marine environments is produced when small amounts of sediment are brought in at long time intervals and are exposed to a dense population of worms. In such a case, several primary biogenic layers of fecal pellets may be present in a profile (Schäfer 1952).

The main difference between such "primary biogenic layers" of fecal pellets and "secondary fecal-pellet layers" (produced by concentration of transported fecal pellets into layers) is that the primary fecal pellet layers overlie a sediment layer with a dense population, the sediment thus being strongly bioturbated. The secondary fecal pellet layers may overlie a completely nonbioturbated horizon.

Fig. 260. *Mytilus* colonies on intertidal flats. Around and in the *Mytilus* colonies mostly muddy sediments are present. They are actually fecal pellets and sediment derived from the destruction of fecal pellets. North Sea tidal flats. (Photograph by F. Wunderlich)

Environmental Reconstructions

In the foregoing chapter we dealt with the various aspects of sediments and their importance in environmental reconstructions.

For environmental reconstruction we need features that were produced at the time of deposition. In ancient sediments many features we observe are a result of processes of compaction, diagenesis, and metamorphism. Thus, it is essential to be able to differentiate between primary features produced during deposition and secondary features produced during the post-depositional history of sediments. Diagenesis is important for the study of rock history, but it tends to obliterate the primary features, which can be helpful in reconstruction of the environment of deposition of a given rock.

For example, mineralogical and chemical composition of a sediment can be changed during diagenesis. Thus, while using mineralogical and chemical parameters for environmental reconstruction, much care is needed. Every attempt should be made to differentiate between minerals which are produced during deposition, those produced later during diagenesis, and those transported into the basin and deposited in the form of detrital grains. Mineralogical and chemical parameters are particularly susceptible to diagenetic changes.

Biological and physical features of a sediment can also be changed during diagenesis. However, they are less susceptible than the mineralogical and chemical characteristics. Fossils may be dissolved and bioturbation structures obliterated. However, physical features are least susceptible to diagenesis and thus make important tools for environmental reconstruction. However, formation of concretion, etc., may make recognition of primary sedimentary structures rather difficult. On the other hand, diagenesis may even help in the preservation of certain physical features which are useful in the study of an environment.

Sediment is an accumulation of solid mineral and rock grains which are transported by a mobile medium, coming to rest at last as sedimentary deposits. Thus, the medium of transport and the medium of deposition are the most important factors. The other important factors are provenance, climate, etc. The ease of preservation and the recognition of features characterizing physical environ-

ment, together with the importance of the physical environment, makes the physical sedimentary environment of prime importance. After establishing the physical sedimentary environment of the deposit, the recognition of geochemical and mineralogical parameters becomes easier. Detailed faunal studies can be done with much certainty if the physical sedimentary environment is well established.

In environmental reconstruction the approach should be pluralistic. Several independent parameters should be studied before any final picture of the depositional environment is made. Environmental reconstruction is more certain if several independent parameters point to the same conclusion.

The environmental reconstructions are based principally on the doctrine of uniformitarianism, which Lyell expressed in the words "the present is a key to the past." Walther (1894) pointed out that progress in palaeontology has been a result of the study of comparative anatomy and ontology. He said further that in the same way, rocks should be studied in a comparative way – comparative lithology. He added that just as a palaeontologist is able to reconstruct the whole animal on the basis of a few vertebrae or other skeletal remains by knowing comparative anatomy, knowledge of comparative lithology would help in the reconstruction of ancient basins and their environments. Based on these thoughts, he gave the law of facies, which led to the study of sediments in vertical profile (e. g., Visher 1965b).

Laporte (1968) called such a comparative approach the rule of analogy. He pointed out further that initial analysis of ancient environments yields data that are of only relative value. For example, one might conclude that some sediments were deposited in relatively shallower water than other sediments. However, by establishing a similarity to a modern equivalent, we are able to interpret in terms of analogy.

However, there are limitations to the uniformitarianism approach in the reconstruction of ancient depositional environments, and it must be applied with caution and restriction. No doubt, geological processes were the same throughout geological time, but their intensities have not always been the same. Relief and distribution of geomorphological features have changed through time. Some of the geological events in the past were unique and we do not have equivalents today.

During certain periods in the geologic past, the distribution of land and sea was much different; in certain periods climatic zones were less differentiated than in the others; and geomorphological relief was probably in more equilibrium than today. Intensity and rhythm of tidal currents might have been rather different. Another very important factor is the intensity and type of vegetation, which strongly controls the rate and type of erosion, and with that the amount of sediment available for deposition. This problem is most acute in sediments of Precambrian and Lower Palaeozoic times, when vegetation was absent or very scarce.

Some of the environments and their sediments are better known in ancient rocks. This is especially true of flysch sediments. Data about flysch type of sediments from modern environments are rather scanty, the main reason being the difficulty and the high cost in obtaining samples from deeper parts of the oceans. However, there are well-exposed ancient examples of deep-water sediments (flysch). In such cases, environmental models must consider the data from ancient rocks. The best way to make an environmental analysis is for the study of past and present to go hand in hand.

However, reconstruction of ancient depositional environments can be achieved only by the detailed study and definition of present-day environments, and various processes active in individual environments. It is quite likely that many environments which existed in the past are not available today. But if we understand the physical processes and their importance in various environments, it would certainly help us in reconstructing even those environments whose analogues today do not exist. Thus, with suitable modifications the results of Recent sedimentation studies can be successfully applied to the interpretation of ancient depositional environments. The study of processes and individual features should be emphasized more than the general comparison of environments.

Most of the geologists working in modern environments agree that geomorphological relief, and distribution and configuration of land and sea today is rather unique in the geological history of the earth and that it is mainly because of the last glaciations. Thus, results obtained from the studies of present-day environments can be applied to ancient environments with restrictions.

In the next part of this book we shall describe and discuss individual modern environments and subenvironments and the sequences.

Part II. Modern Environments

Introduction

Here environmental models based on the study of modern sediments have been compiled. We have been able to use only a few of the models. Several generalizations regarding distribution of bedding types are made. However, the critical reader is referred to Part I, where their distribution and genesis have been described. The utility of such models would be proved by critical study of such models in the light of study of ancient environments.

Moreover, certain models will be modified, as soon as more data become available from modern sediments.

General Information

There is no lack of attempts to classify sedimentary environments, and geological literature is full of such classifications. Almost all the textbooks on sedimentation give one or more classifications of sedimentary environments. Some more important attempts are those by Twenhofel (1950), Krumbein and Sloss (1963), Dunbar and Rodgers (1957). An ideal classification of depositional environments would subdivide the entire sedimentary depositional realm into small useful categories, without any omission or overlap of environments. Moreover, a geologist working in ancient sediments would also demand that such classification be made up of units, that can be easily deciphered and recognized in a sedimentary deposit. It is impossible to meet all these requirements in a single classification of depositional environments. Recently Crosby (1972) gives a good review of various attempts to classify sedimentary environments and discusses the problems in classifying depositional environments.

In this book no attempt has been made to give a sophisticated classification of sedimentary environments. A very simple scheme of classification of sedimentary environments has been adopted. One realizes very soon that necessary investigations of various environments and subenvironments are still lacking. Our knowledge is still incomplete as far as the variations in the deposits of same environment in various climatic zones is concerned. Certainly, deposits of the same environment e. g., lagoon, beach, etc. in arid, humid, or subarctic climate show certain differences.

Another important factor affecting sedimentary environment is the availability of the sediment, i. e., variations caused by little or great availability of sediment. Things become still more complicated, depending upon whether sediment is fine-, medium-, or coarse-grained; and whether energy of the environment is low, medium, or high. If one puts only these factors together, 72 possible variations of each environment can be visualized (Table 15), i. e., 72 possibilities each for each environment e. g., lake, fluvial, delta, coastal sand, tidal flats, lagoon, and shelf. Moreover, there are many more parameters whose influence would further complicate the picture. These might be effects of tectonics and tides, influence of trans-

Table 15. Possible modification in a depositional environment caused by climate, energy of environment, availability of sediment, and type of sediment

Availability of sediment	Type of sediment	Arid			Climate Humid			Subarctic		
		High	Medium	Low	High	Medium	Low	High	Medium	Low
High	Coarse									
	Medium									
	Fine									
	Mixed									
Low	Coarse									
	Medium									
	Fine									
	Mixed									
		Energy								

gressive or regressive tendencies, effect of chemical factors, influence of abnormal weather conditions, etc.

On the other hand it is very important to be able to identify broad environments and their subenvironments, i. e., glacial, desert, lake, fluvial, deltaic, coastal sand, shelf, lagoonal, tidal flat, continental margin, and ocean basin environments.

Thus the subdivision of sedimentary environments based on the present state of our knowledge, used in this book, can only be regarded as pragmatic, not as theoretically complete. At points two broad environments show overlapping region which belong to both environments, e. g., some of the subenvironments of deltaic, and fluvial environments or lake deposits which may occur independently, but are often incorporated in subordinate amounts in fluvial, glacial, or even desert environments.

Importance of Sequence in Environmental Reconstruction

Sequences represent interrelationships of serveral environments. There are two basic approaches in the study of sequences.

The one is the study of aerial variations in an environment. In other words, the relationship of lateral facies is studied. One concentrates on a thin stratigraphic or time unit that covers a wide geographic area, so that several different facies of essentially the same time period are examined. Such an approach can be rather useful in environmental reconstruction. For example, the occurrence of shelf mud facies on one side of a sand body and lagoon facies on the other side would hint that this sand body might be a barrier island. The relationship of lateral facies also helps in defining and delineating various environments of a depositional basin and in finding the configuration of basin and basin analysis, in postulating current directions and dispersal patterns in the basin.

The second approach is the study of vertical profile or vertical facies relationship. Here we study the changes in the environment through time. Different successive stratigraphic units are examined to identify temporal variations in depositional environments.

Walther (1894) was probably the first to emphasize the importance of studying the vertical profile. His law of facies states "daß, primär sich nur solche Facies und Faciesbezirke geologisch überlagern können, die in der Gegenwart nebeneinander zu beobachten sind."

In other words, only those environments that are laterally associated to each other geographically may become associated in a vertical sequence. Thus, in study of vertical profiles we try to study the lateral facies relationship through time.

A broad depositional environment may by subdivided into smaller, essentially uniform subenvironments or sedimentation units, where sediments with their own characteristic features are deposited. For example, shallow marine environment can be subdivided into shelf mud, transition zone, shore face, foreshore, and backshore facies. These units in ancient sediments are lithosomes (Wheeler and Mallory 1956). As sedimentation proceeds, with time facial boundaries migrate laterally under the influence of transgression and regression, and different facies are arranged in an orderly sequence, so that those that are geographically contiguous will become stratigraphically contiguous in a vertical profile. The order in which various facies are arranged depends upon the nature of transgression and regression.

Visher (1965b), using primarily the information in the Walther's law, postulated that it is the fundamental sedimentary process that produces both a specific environmental distribution and a specific vertical profile. He suggested that there are only six fundamental sedimentary processes, resulting in six fundamental models. He emphasized that these idealized models as vertical sequences may be used as a standard for interpreting stratigraphic sections.

Such model approach in environmental reconstructions should be done with care. Rarely are ideal sequences found. Generally, the geological record consits of incomplete sequences. Sometimes, thick sequences result just because of oscillation between two subenvironments. Study of such sections necessitates a thorough understanding of transgression – regression processes. The effect of transgression – regression is maximum in the coastal and near-shore sediments, as they are most easily affected, even by small-scale migration of the shore line.

Curray (1964) gives a thorough discussion on transgression – regression processes and their effect on coastal sediments. His results are based on the study of Holocene sediments of the coastal regions of American continent. In the following some salient points of his discussion are given.

The transgression is a process of migration of the shore line of a water body in a landward direction, and regression is migration of the shore line in a seaward direction.

The products of transgressions and regressions depend upon various factors, such as rate and supply of sediments, intensity of hydrodynamic processes, configuration of basin of sedimentation, and rate and direction of relative sea level changes. All these factors can be grouped into two parameters – rate of deposition, and rate and direction of relative sea level change (Curray 1964). Curray constructed a diagram, that clearly shows that deposition and erosion can control the transgression – regression processes (Fig. 261). In general, rising relative sea level results in transgression; falling relative sea level, in regression. However, this tendency can be modified in the reverse direction by the effect of net deposition and erosion. For example, a slowly rising relative sea level associated with a rather high rate of net deposition can cause shifting of shore line in seaward direction. A good example is building up of the Mississippi delta during Holocene transgression. Despite of increase in sea level, as the rate of net deposition was very high a depositional regression took place, i.e., regression of shore line due to excess deposition.

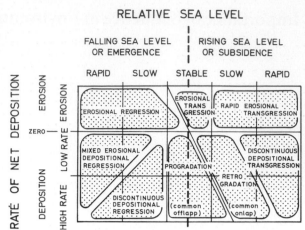

Fig. 262. Classification of transgression and regression. (After Curray 1964)

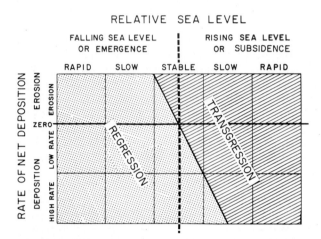

Fig. 261. Diagram of the effects of the rate of change of relative sea level and of the local net rate of deposition on lateral migration of the shore line in a direction perpendicular to the coast. Transgression is defined as a landward shifting of the shore line, and regression as a seaward migration. Rising relative sea level usually results in transgression, but a high rate of deposition can cause progradation of the shore line (regression). Similarly, a falling relative sea level usually results in regression, but excessive erosion over deposition can result in recession of the shore line. (After Curray 1964)

Similarly, a high rate of erosion can offset the tendency of regression caused by a very slowly falling sea level, and the result is a net transgression. Thus it is a combination of factors which decides the relative shifting of sea level. Based on such combinations several types of transgression and regression can be recognized (Fig. 262).

The rate and direction of change of relative sea level can be because of tectonism, eustatic changes, or even sediment compaction. However, it is rarely possible to decipher the cause of change of relative sea level.

Recently, a number of studies have been initiated to determine the cause, extent, and dynamism of the relative sea-level changes during the Holocene (for details see Sindowski and Streif 1974).

The varied effective causes of the sea-level changes in the Holocene of the North Sea area can be put together into two groups (Streif 1978):

Geophysical Driving Factors. Glacial eustatic sea-level fluctuations: caused by climatic changes during the Quaternary, which affect the ice-water proportions on the Earth's surface. Consequently, during glacial periods lowered sea-levels, while during interglacial periods increased sea levels are produced.

Isostatic compensatory movements: Reaction of the Earth's crust to mass changes. The melting of the inland ice leads to reduction in load pressure, causing a regional uplifting of the land surface. On the contrary, the weight of sediment and water cause sinking of the land surface.

Epeirogenic movements: The North Sea basin is an area of prolonged tendency of sinking, which probably also continued during the Holocene and today.

Geoidal effects: Due to vertical and horizontal changes in the rotational ellipsoid, the surface form of the oceanic level also changes in space and time.

Along with these regionally active geophysical driving factors, there are a number of local parameters affecting the sea level, and overlapping with the main factors. These local parameters influence the water body (sea) as well as the position of land surface:

Modulating Factors. Compaction of sediments: as a result of primary differences in water content, grain size, and depositional density, the sediments show a differential tendency to compaction.

Salt dome tectonics: Above the mobile salt domes of the underground, an uplift of surface can take place; while in the adjacent surface depressions may take place.

Suberosion (dissolution): In the rocks soluble in water, solution cavities may be produced, on which the overlying sediments collapse.

Natural and artificial changes in the coastal morphology: Because of the rise in sea level, as well as through construction activity, the coastal configuration of the North Sea, and also the shape of the tidal curve have been changed.

Minor climatic and atmospheric influences: shortlived climatic fluctuations, for example, changes in the main wind direction, or in the frequency of storms.

If a regression is very rapid and if the rate of sediment supply very low, the possibility of a record of regressive deposits is scant, specially if the following transgression is of erosional character. It would remove the thin regressive deposit. Such events are marked in geological record as unconformities. In general, most of the sediments preserved in the geological record is a result of depositional regressions. Transgressions are generally of erosional character, although this is not a rule. Several good examples of transgressive sequences are also known.

Curray (1964) further demonstrated that on continental shelves transgression and regression effects are strongly controlled by the configuration of the shelf, especially by the width of the continental shelf. A careful study of vertical and horizontal sequences together with proper understanding of transgression – regression effects is very important for environmental interpretation.

Two useful terms in the study of vertical sequences are *prograding sequence* and *retrograding sequence*.

In a prograding sequence deposits of shallow water are deposited on the sediments located seaward, which are also lower level sediments. For example, coastal sand is deposited above the shelf mud. If progradation takes place during a rising sea level (depositional regression in Curray's terminology), then sediments deposited near the land, but in relatively lower levels, may grow over the seaward sediments of higher levels e. g., lagoon deposits on the coastal sands. Progradational deposits are also known as off-lap.

In a retrograding sequence the succession is just the reverse. In such sequences, deeper water and seaward deposited sediments lie over the sediments of shallow water. Such sequences are produced during a rising sea level (depositional transgression in Curray's terminology). For example, shelf mud is deposited on the coastal sand. They are also known as on-lap.

However, most important are the prograding sequences, produced due to high rate of sedimentation. Moreover, they possess a high potential of preservation. Because of this, we have tried to construct hypothetical prograding sequences from the study of modern environments which are associated laterally. In the case of retrograding sequences the succession would be just the reverse.

Some of the subenvironments (smaller sedimentation units) of a broad environment can be recognized only because of their vertical and lateral relationships to other subenvironments. Moreover, knowledge of vertical and horizontal relationships helps in delineating the shape or geometry of a sedimentary body. There are various ways in which sedimentary bodies are classified and described. It can be useful to find out whether a sedimentary body is equidimentional of long extent (blanket or sheet type), or an elongated body, with long dimensions several times the width (shoestring or ribbon).

Furthermore, information regarding the relationship between the strike of sedimentary body and strike of basin of deposition can be rather useful. Some sedimentary bodies, such as barrier bars, and cheniers are always parallel to the depositional strike of the basin, while others can be perpendicular to the depositional strike of the basin. Krynine (1948), and Pettijohn, Potter, and Siever (1972) give some useful information on the geometry of sedimentary bodies.

Some information about palaeogeography and dispersal pattern within a basin can be obtained by study of variation of mineralogical composition of sediments, content of heavy minerals, feldspars, specific type of quartz grains, or composition of boulders.

Directional sedimentary structures, such as cross-bedding, ripple marks, various sole markings, and sediment grains, are most commonly used in the reconstruction of palaeocurrent patterns. Potter and Pettijohn (1963) describe in detail the methods and application of paleocurrent analysis. Recognition of the environment of deposition combined with palaeocurrent studies provides a more complete study of a stratigraphic unit. However, palaeocurrents alone give very little information on the environment of deposition, and cannot be used in palaeogeographic reconstruction – land/sea distribution. This is especially true of shallow water marine sediments.

On the other hand, palaeocurrent patterns in relation to the shape of a sand body can give some clues regarding the environment or mode of genesis. For example, in fluvial sand crossbedding direction is parallel to the elongation of the sand body, with a well-marked unipolar direction.

In a coastal environment strand sediments show major crossbedding direction at right angles to the elongation trend of the sand body. Moreover, due to development of ripples in runnels (running parallel to the coast), a second maximum can develop parallel to the elongation of sand body.

In ancient rocks some of the sediment profiles show cyclic composition. Sequence of certain facies-combinations is repeated several times. They are known as cyclothems. Duff et al. (1967) give a detailed description of cyclothems, their importance and genesis. Cyclothems are generally traced over large geographical areas. Generally, large cyclothems are related to sedimentary tectonics.

However, small-scale cyclicity and regular alternation of certain sedimentation units are more often the result of inherent characteristics of a given environment. They are related to changing sedimentation processes with time, in a given environment. For example, fining-upward sequences of fluvial environment or interlayered sand (channel deposit) and mud (flood-basin deposit) cycles are related to the pattern of fluvial sedimentation, and not to sedimentary tectonics. Some of the cyclicity in the sedimentary rocks may also be related to factors like seasonal changes in the climate, fluctuations in the supply of sediment, or periodic occurrence of some catastrophic processes, e. g. storms.

It can be added here that not all thick sequences are produced as a result of repeated transgressions and regressions. Several examples exist where thick sequences in almost the same environment are formed without any major transgression – regression.

The following explains the importance of sedimentary tectonics, especially the relationship between the rate of supply of material and the rate of subsidence. If both of them are maintained, then a thick sequence of sediment is produced maintaining the same environment. Wunderlich (1970b) demonstrated that a 65-m thick profile in Devonian of Germany was formed under almost identical conditions. For the whole sequence water depth remained few meters, and environment was comparable to coastal environment – intertidal flats, and sand bars. He did not find any sequence in the sense of Visher (1965b). Varying lithological and sedimentological units were formed because of changing lateral morphology. This is especially true in shallow tidal environments, where areas of mud and sand deposition, and low- and high-energy areas occur side by side. A slight change in local morphology may change an area of low energy into an area of high energy.

To conclude, the study of vertical profiles in ancient sediments should be based on the detailed study of smaller subunits. The general nature of sedimentary tectonics and paleogeography should always be considered. There are not always bathymetrical changes in a vertical profile. In some cases lateral shifting of morphological features associated with continuous subsidence and sufficient sediment supply may produce a thick sequence of a particular environment.

In spite of these limitations, the importance of study of vertical sequences in the recognition of ancient environments remains unquestionable. As far as it was possible, we have tried to construct vertical prograding sequences, based on the information of the lateral facies of modern environments.

Fisher and Brown (1972) give numerous diagrams and characteristics of different depositional environments, along with significant references. Spearing (1974) provides rather instructive summary sheets of sedimentary deposits.

Dickinson (1975) provides information on the current concepts of depositional systems with application to petroleum geology. Hayes and Kana (1976) made a very useful compilation on terrigenous clastic depositional environments. Friedman and Sanders (1978) discuss different aspects of sedimentology along with useful information on depositional environments. Reading (1978) provides principles of facies analysis and environmental interpretation, mainly from the studies in ancient rocks.

Davis (1978) provides information on coastal environments. Geometry of sandstone reservoir bodies is discussed by Le Blanc (1972), along with information on various environments, i. e., continental, transitional, marine, transgressive marine, submarine fans, and a useful selected list of references.

Lombard, (1972) has provided a very useful account of sedimentary rocks in French and Ricci Lucchi (1978) has published in Italian an excellent book concerning sedimentary environments. Homewood (1979) summarizes sedimentary environments in French, while Medeiros et al. (1971) discuss them in Spanish. The collected papers of Ferm and Horne (1979) on Carboniferous depositional environments in the Appalachian Region is an outstanding example of identifying depositional environments.

Glacial Environment

General Information

Glacial environment is not widespread on the earth's surface today. However, a widespread record of glacial activity during the Pleistocene ice age has been left behind over vast areas, and a detailed study of them has been most helpful in understanding the glacial processes and the characteristics of glacial deposits. This knowledge can be successfully applied in the study of ancient, e.g., pre-Pleistocene glacial deposits. Today, the glacial environment is restricted to areas around the north and south poles, and high mountains in other latitudes covered by snow e.g., the Alps and the Himalayas.

Glacial environments are characterized by the dominance of ice as a geological agent. Accumulation of ice is a result of low temperature combined with high rates of precipitation and extremely low rates of evaporation. Local and occasional presence of ice cannot be regarded as evidence of glacial environment. In temperate climates during winter months activity of ice can be rather important, and it may influence the sedimentation patterns. Ice may cause some transport of material when packed in ice blocks, or when ice blocks strand on a sediment surface they may produce various types of deformation structures in the underlying sediments. In German tidal flats effect of ice on the sedimentation process is sometimes very significant. However, in such regions principal sedimentation is not controlled by the activity of ice. The same is true of the rivers of such climates. In this chapter we shall discuss sedimentation in areas occupied by ice masses and areas around such region: glacio-fluvial.

Permanent ice masses exist above the snow line, which is the level up to which the snow melts in the summer. The occurrence of the snow line in altitude varies with latitude. In higher latitudes the snow line occurs at lower altitudes. Large ice masses are known as glaciers. Such ice masses exist because over long periods the snow and ice have been added faster in a given area than ice is lost by various processes. Glaciers have been defined as masses of ice which, under the influence of gravity,

flow out from the snow fields where they originate (Holmes 1965). A glacier consists of recrystallized and compacted snow, melt water, and some rock debris.

The precipitation above snow line is in the form of snow. The loose snow is transformed into more compacted ice under pressure caused by loading of new snow. The air is squeezed out. Glacial ice is an aggregate of interlocking grains. A thin film of water is present between the ice grains and plays an important role in the movement of glacier.

Geomorphology of Glaciers

Glaciers creep slowly downward and extend as tongue-shaped bodies of ice. Normally, glaciers are fed by the addition of new snow. All processes by which a glacier gains ice are called *accumulation*. In the lower reaches glaciers lose ice by melting, evaporation, and calving of icebergs. The processes are collectively known as *ablation*. The balance between these two processes–accumulation and ablation–characterizes a given glacier. Glaciers advance up to a point where ablation is balanced with accumulation. If conditions change and ablation dominates, the glacier does not reach as far. The glacier is said to be retreating. If accumulation dominates due to heavy snowfall, the glacier advances farther down, and is said to be advancing. Near the lower end of a glacier–the snout–the thickness of ice is considerably less. Figure 263 shows a longitudinal section across a valley glacier to demonstrate the relative importance of ablation and accumulation in various parts of a glacier.

Near the snout, detrital sediment is dumped due to melting ice. If a glacier ends up in a water body (a sea or a lake), sediment is dumped into the water body and is partly reworked by waves and currents. However, if a glacier ends up in a sea deep enough to permit the ice to float, a huge amount of ice is lost by the process of calving. Huge blocks of ice–icebergs–float away from the glacier. Icebergs usually contain much entrapped sediment, and

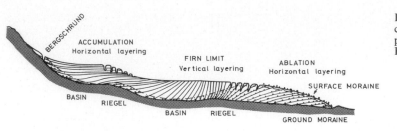

Fig. 263. Schematic longitudinal profile a-cross a valley glacier showing main geomorphological features. (Modified after Streiff-Becker 1952 and Schwarzbach 1964)

may transport glacial sediment over large distances before they are melted. Such glacial-borne sediment becomes incorporated in normal marine sediments. It may be, for example, that much pebble and boulder size sediment is mixed with sandy or muddy sediment of a given region, and sediments of glacial-marine origin are formed.

A characteristic feature of most glaciers is the presence of crevasses and other deformation features in the ice surface. This is mainly due to the fact that the outer crust of a glacier behaves more as an elastic solid than as a rheid. Reasons for the formation of crevasses may be varied; e. g., glacial flow, change in the path of the flow perhaps because of a constriction of the path, widening of the path, or glacier passing around a bend. Depending upon their position, crevasses are named marginal crevasses, transverse crevasses, longitudinal crevasses, etc. Figure 264 and 265 illustrate different types of crevasses.

The classification of glaciers is a difficult problem. Holmes (1965) recognizes following three basic types of glaciers:

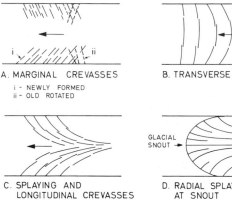

Fig. 264. Scheme showing various types of crevasses in a glacier. (Modified after Sharp 1960)

1. Valley Glaciers. Valley glaciers are ice masses confined within valley walls of a mountain. Thickness of ice can be several hundred meters. Valley glaciers are usually fed by cirques and ice fields located higher up. Examples: Rhône Glacier, Alps; Gangotri Glaciers, Himalayas.

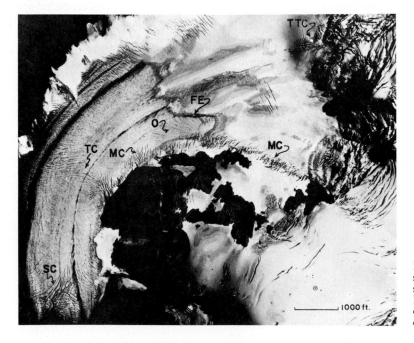

Fig. 265. Photograph of a valley glacier showing various types of crevasses. SC–splaying crevasses, TC–transverse crevasses, MC–marginal crevasses, TTC–tick-tack-toe crevasse pattern, FE–firn edge, O–faint ogives. (After Sharp 1960)

2. Piedmont Glaciers. Piedmont glaciers are sheets of ice formed by the coalescence of several valley glaciers, which spread out beyond their valleys into a lowland area. Example: Malaspina glacier of the coastal plain of southeastern Alaska.

3. Ice Sheets or Ice Caps. Ice sheets are huge masses of ice spreading over large continental or plateau areas. Such masses occur in regions where the snow line is rather low. Smaller ice sheets exist in Spitsbergen, Iceland, from which smaller valley glaciers flow out. Large ice sheets exist in Greenland and Antarctica. These large ice sheets can be several thousand meters thick. They extend to the sea as ice shelves from which icebergs calve out. Figure 266 shows the ice sheet of Greenland. Figure 267 shows various shapes in which glacial ice may occur in nature.

Most of the glaciologists distinguish two types of glaciers (for example, Lagally 1933; Ahlmann 1935; Woldstedt 1961): (1) cold or polar glaciers, (2) warm or temperate glaciers.

In cold glaciers temperature is rather low, well below the pressure melting point. Melt water is almost lacking; there are no well-developed engla-

cial and subglacial streams. The movement of a cold glacier is very slow because lubricating water near the base is not present. Often, there is thick debris-rich ice at the base (several meters thick) present in discontinuous bands, roughly parallel to the glacier base. These are also known as dry-base glaciers.

In warm glaciers temperature is close to the pressure melting point. Melt water is present in sufficient quantities, accumulating in surface lakes and crevasses. A layer of melt water is present at the base of the glacier. This layer of water enables warm glacier to move faster than cold glaciers. Basal zones of warm glaciers have a very thin debris-rich ice layer. They are also known as wet-base glaciers, and at the base there may be a net melting, net freezing, or equilibrium (Boulton 1972a).

However, there are many transitional forms between cold and warm glaciers.

There are several good descriptions of glaciers and glacial environments; Hess (1904), von Klebelsberg (1948), Flint (1957), Sharp (1958, 1960), Holmes (1965), Embleton and King (1968, 1974), West (1968), Tricart (1970), Washburn (1973).

Fig. 266. Profile across central Greenland showing the ice sheet. (Modified after Holmes 1965)

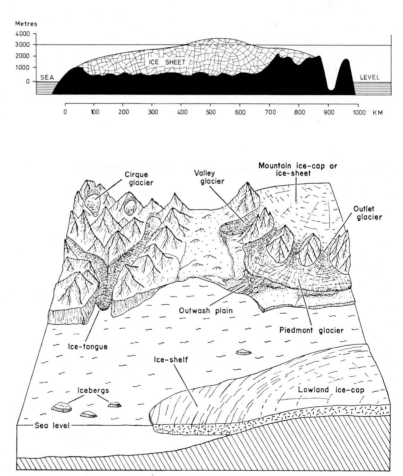

Fig. 267. Scheme showing various forms of occurrence of glacial ice. (After Allen, 1970).

Wright and Moseley (1975) is a good source of information on the glacial sediments. Erosional processes, depositional patterns and landforms of glaciers are described by Boulton (1972b, 1974, 1975). Church and Gilbert (1975) discuss hydraulic and sedimentation processes in proglacial fluvial and lake environments.

The symposium volume *Glaciofluvial and Glaciolacustrine Sedimentation* (Jopling and McDonald 1975) is good source of information on depositional processes and sedimentary sequences in the periglacial areas.

Glacial Flow

As has been pointed out earlier, glaciers move–though slowly. At present it is not possible to explain the motion of glaciers accurately. Sharp (1954), Nye (1952, 1965), Paterson (1969), Tricart (1970), and Allen (1970) discuss various aspects concerning the properties of ice and the movement of glaciers. However, it is generally agreed that pattern of glacial flow is laminar–similar to the flow in liquids and gas. Mechanisms that have been suggested to explain the glacial flow are sliding at base, intergranular adjustments, phase change, internal slip planes, and intergranular gliding, and recrystalization (see Sharp 1960).

Tricart (1970) summarizes two important processes causing movement of ice.

1. Flow by plastic deformation of ice developing due to weight. This develops only if ice is thick enough. It is closely associated to kinematic waves of ice.

2. Flow by gliding, where the glacier slips over its bed. This is expressed in surface tension, which may lead to shearing and the development of open crevasses.

Usually, both of these processes work side by side or in different parts of the same glacier.

The middle part of the glacier moves more quickly than the sides. The major controlling factors for glacial flow are thickness of ice, temperature, slope characteristics, valley shape, and content of sediment debris in the ice. The glacial flow is strongly influenced by presence of a melt water film near the base and the valley walls. The rates of movement are 1 m/day or less; in exceptional cases rates up to 20 m/day may occur.

Glacial Erosion

Glaciers are active erosional agents, and a characteristic landscape is produced under the action of moving ice. Two important processes are active during glacial erosion–plucking and abrasion. The plucking means quarrying out larger rock fragments from the bedrock. The ragged and well-jointed surfaces are more susceptible to plucking than the smooth ones. Plucking is achieved by repeated freezing of melt water in the natural joints of the bedrock, which results in loosening of rock pieces and ultimately carried farther by the moving ice. This process produces mainly larger rock fragments, i.e., gravel, boulders, and blocks.

Rock debris moving along the bottom and valley sides are active tools held firmly in the moving ice. They scrape, scratch, scour, grind, and polish the surfaces. The tools are themselves worn down, becoming flat and striated. The fine rock flour works like an abrasive and polishes the rock surfaces. The process of abrasion produces mainly fine-grained sediment, silt and clay; rarely are there sediments of the size of sand and gravel. Figure 268 demonstrates the plucking and abrasion processes of bedrock.

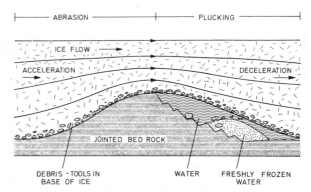

Fig. 268. Schematic representation of the process of plucking and abrasion on the bedrock of a glacier. (Modified after Allen 1970)

Glacial abrasion produces characteristic polished rock surfaces with striations, grooves, and crescentic fractures. The striations run parallel to the path of flow of the ice. Striations are a few millimeters to several centimeters in width, a few to several millimeters in depth, and several meters in length. They run straight, are sometimes curved, dying out continuously or abruptly. Grooves are deeper (1 to 5 cm) and wider forms. Figures 269 and 270 depict various kinds of striations and grooves made on a rock surface. Pressure cracks are also known as crescentic fractures (Tricart

1970). These pressure cracks are crescent-shaped cracks arranged in rows parallel to the direction of moving ice (Figs. 271 and 272). The concavity is oriented either downstream or upstream. However, in every case the steeper side of the crack points in the up-current direction.

Near the snout of a glacier of temperate region subglacial water flowing along the bed suppliments the glacial activity. It causes corrasion of bedrock, producing complex and large grooves and potholes. In some cases melt water is an important erosion agent in a glacial environment. Fi-

Fig. 269. Glacial striations and grooves on a rocky surface. Western Norway

Fig. 270. Rather large-scale striations and grooves of glacial origin. Southern Norway

Fig. 271. Crescentic fractures on a rock surface. Western Norway

Fig. 272. Crescentic fractures of variable shapes. Southern Norway

important sources for rock debris in glaciers: (1) the rock fragments weathering out from the steep slopes above a glacier fall down into the glacial ice and debris added due to headward erosion of cirques, which often nourish the glaciers; (2) the rock debris produced by glacial erosion from the bedrock and valley walls. Because of the high density and internal friction, the glacial ice is able to move sediments of enormous sizes and weight, which water and wind would not be able to transport.

"Moraine" is a general term used by geomorphologists for glacial sediments, both for the sediment in transport and the resulting sediment deposits. Allen (1970) prefers the use of term "sediment load" for the sediment in transport. Sediment load (moraine) of a glacier may occupy different positions within a glacier. Types of moraines are named depending upon their position in a glacier. Figure 274 shows various kinds of moraines.

gure 273 shows schematically melt water streams near the snout of a glacier. Water plays a significant role in the transportation and deposition within a glacier. Röthlisberger (1972) and Shreve (1972) discuss the flow of water within glaciers and suggest that the water tends to become concentrated into larger subglacial conduits (tunnels).

Glacial Transport

Glaciers are very effective transport agents. Competency of glaciers to transport rock debris is rather high. Unlike water and wind transport, there is negligible lateral diffusion and almost no sorting of sediment during glacial transport. There are two

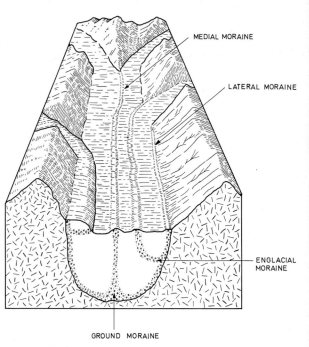

Fig. 274. Diagram illustrating various modes of transport of glacial material and various types of sediment load. (Modified after Sharp 1960)

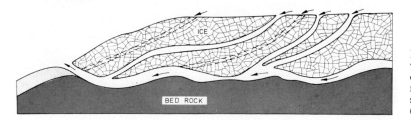

Fig. 273. Schematic diagram showing meltwater streams in the glacial ice near the snout of the glacier. Note that meltwater streams may acquire various positions, i.e., on the surface, within the ice, or at the bottom. (Modified after Woldstedt 1961)

A major part of a sediment load is transported near the surface which is located along the margins. This is the marginal, or lateral moraine. If two glaciers join together, the marginal moraine also occupies a central position–a medial moraine. Thus, surface load is transported as (1) marginal moraines and (2) one or more medial moraines. Figure 275 shows a compound valley glacier with several medial moraines.

Some sediment material is engulfed within the ice as a result of washing into crevasses and tunnels. It is the interior load, or englacial moraine. If certain proportion of debris makes its way down near the bottom, it is called a subglacial moraine. The bottom of a glacier may be heavily charged with sediment debris, obtained by glacial erosion,

i.e., abrasion and quarrying. This is the bottom load, or ground moraine.

Near the snout of a glacier all these types of moraines become mixed together and are dumped down in front of snout in what is known as the terminal moraine. However, the terminal moraine is not a transport feature, but more a depositional feature.

The supraglacially and subglacially derived debris can follow a varied path through the glacier-transporting system. The shape of the clasts and grain-size distribution of a till is dependent upon the transport path of the parent debris (Boulton 1978). Figure 276 depicts the possible flow paths, type of till deposit, and their textural characteristics.

Fig. 275. Photograph of a compound valley glacier fed by many tributary ice streams with moraines. (After Sharp 1960)

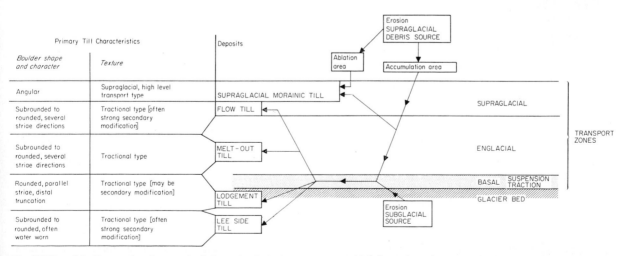

Fig. 276 Possible flow paths of supraglacially and subglacially derived debris through a glacier transporting system. On the left are the clast shape and textural characteristics of the debris types which form the primary source for the various till types. Processes during the depositional phase frequently produce secondary modification of the texture. (After Boulton 1978)

Glacial Deposit

In the lower reaches of a glacier, i. e., the zone of ablation, ice is lost; the glaciers are overloaded with debris, which they begin to drop. The main deposition by glacier takes place during retreat or temporal halt. With the retreating glacier the zone of deposition also retreats, and glacial deposits are left behind. A major part of transported material is dumped in the form of a terminal moraine near the maximum extension of the glacier. If the glacier retreats in stages, a series of terminal moraine ridges are produced.

Boulton (1978) discusses that debris transported in a basal zone tends to be deposited as subglacial till, while debris transported at a relatively high level tends to be exposed supraglacially low in the ablation area, and is finally deposited as flow till or meltout till when the underlying ice melts. Thus basically three types of till can be distinguished (Boulton 1972b):

1. *Flow Till* – released supraglacially, due to high-water content flowing down the glacier surface accumulating at low slopes. Deformation structures are very common as a result of subaerial flow. Equivalent to the ablation till.
2. *Meltout Till* – released either supraglacially from stagnant ice beneath a confining overburden and in which some of the original englacial features are preserved.
3. *Lodgement Till* – deposited beneath actively moving ice and undergoing deformation as a result of shear imposed by this ice.

These till formation processes and incorporation of melt-water within the moraines lead to the development of differentiated sequences such as till – outwash-till, produced during a single glacier retreat, and not related to repeated glacier advance and retreat (Boulton 1972b).

If glacial retreat is very rapid or if the glacier fans out in a vast plain, no well-developed moraine ridges are formed; instead a sheet of glacial sediment is left behind. Usually both terminal moraines and a sheet of glacial sediments are formed.

As the main deposition by glacier takes place during melting, much melt water is available. Thus,

pure glacial deposits have also been partly affected by flowing water. With retreating glaciers lower glacial deposits are left behind to be reworked by flowing water. The deposits which are mainly due to deposition by flowing water are known as glacio-fluvial deposits. Thus, very little sediment is deposited purely under the influence of ice; water action is invariably associated. There is evidence that after deposition of a till, substantial changes in grain-size distribution may occur as a result of winnowing of fines by percolating water, soil-forming processes, and wind activity (Boulton and Dent 1974).

Glacial deposits are commonly known as drift deposits. Figure 277 demonstrates the relationship between the drift sheet, the end moraine, and the ground moraine. Figure 278 shows characteristic features of a glaciated area. In a broad sense glacial deposits can be distinguished into two types (see Flint 1957; Woldstedt 1961; Tricart 1970):

1. Unstratified Deposits. They are often designated as basal deposits (moraines), boulder clay, or till. It is composed of completely unsorted debris. Grain size ranges from boulders or large blocks to silt and clay. No stratification is developed. They are deposits directly laid down by ice at the bottom, and represent essentially the basal moraine.

2. Stratified Deposits. They are often designated as ablation deposits (moraines). Sorting of stratified deposits is better than the associated unstratified deposits, especially if content of silt and clay is

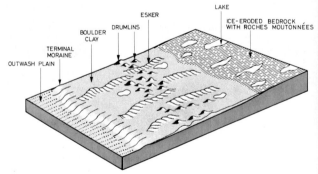

Fig. 278. Diagram showing characteristic features of a glaciated area. (Modified after Holmes 1965)

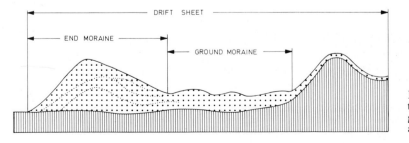

Fig. 277. Idealized profile showing the interrelationship between end moraine, ground moraine, and drift sheet. (Modified after Flint 1957)

rather low. Some stratification is developed. The stratification represents englacial and surface moraines which have been laid down during rapid melting of ice. Some of them are laid down by subglacial streams. Melt water plays an important role in the deposition of stratified deposits. Figure 279 shows deposition of stratified and unstratified deposits.

Flint (1957) distinguishes stratified sediments as:

1. Proglacial deposits. Deposits made beyond the limit of glacier. Outwash plain, glacial-marine deposits.

2. Ice-contact stratified deposits. Water-laid sediments deposited in contact with glacial ice. This includes eskers, kames, etc. They are interbedded or included in nonstratified deposits.

The end result of glacial deposition during glacial retreat is a vast featureless plain of glacial deposit, tens of meters in thickness. Locally the glacial sheet deposits are molded in the form of drumlins, occuring in groups, arranged in an en-echelon fashion. Depending upon the deposition mechanism, deposits such as terminal moraines, eskers, kames, etc., are found associated with glacial sheet deposits.

Downward glacial sheet deposits grade into glacio-fluvial deposits of glacial outwash plains. The melting of ice at the lower end of a glacier produces huge quantities of water, which flow down in more or less temporary, rapidly flowing streams, with characteristics of a braided stream. Action of such streams produces glacial outwash plains, or sandur. Further, downward deposits of glacial outwash plains grade into real braided river deposits.

If a glacier terminates in a water body, delta-like features are produced. And glacial sediments grade into lake and marine sediments.

Glacial-marine sediments can be recognized because of presence of marine organisms in a more or less glacial deposit. Sorting of glacial-marine deposits may be better than glacial deposits. Carey and Ahmad (1961) discuss various aspects of the glacial-marine sediments (see on p. 413 ff.).

Eskers

Eskers, also known as osar or os, are more or less long, curvilinear, wall-like features made up of glacial sediments. Sometimes they show even branching. Sides of the ridge dip 10 to 20°. They can run for several kilometers, they are about 40 to 50 m in height and 500 to 600 m in width; however, they can be of smaller dimensions as well. Eskers are aligned parallel to the flow direction of ice.

Eskers mostly occur in parallel-arranged swarms, and are produced very close to the ice-front. In general, eskers are the product of deposition by a highly organized meltwater flow system within the glacier and common where supply of meltwater is abundant.

Genesis of an esker is attributed to the action of melt water. However, there are many different views as to the actual mechanism of their deposition (Woldstedt 1961).

Bärtling (1905) believes that eskers are built in subglacial tunnels near the bottom of a glacier. The main reason for his thinking this is that eskers are placed on topographically high and low places. However, this can also be explained by considering eskers to be deposited partly on dead ice. After melting of dead ice, parts of eskers sink down.

Korn (1910) believes that some of the eskers are compressional features. Due to differential compression of ice, some bottom sediment is moved up and piled as a ridge.

Philipp (1912) believes that eskers are of englacial origin. They are formed as a result of deposition in tunnels within the ice, and not at the bottom. When ice melts, eskers are transferred to the ground.

Holst (1876) thought eskers to be supraglacial; deposits of streams originating and flowing on the ice, and depositing sediment.

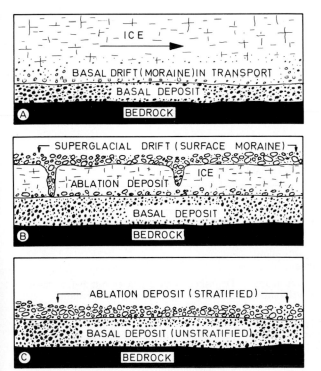

Fig. 279. Schematic representation of deposition of unstratified basal deposit (lodgement till) and stratified deposit (ablation till). (Modified after Flint 1957)

De Geer (1897) based on his study of Swedish eskers, found that eskers are formed as a result of the successive retreat of a glacier, growing backward against the direction of flow of a subglacial stream. Due to melting of ice near the snout of a glacier, "oszentra" (esker-center) deposits of rather coarse sediment are formed. Each esker center represents the deposition of a single phase of melting in summer. By joining the successive esker centers of a receding glacier, a long continuous ridge is produced, which has grown backward. Krause (1911) in his study of eskers, found that at certain distances along an esker, coarse-grained material is concentrated, which corresponds to De Geer's "oszentren."

In Scandinavia most of the glaciers in the last ice age ended in a water body, so that "oszentren" were deposited under water in the form of small well-developed deltas. Investigations of Lewis (1949) on a Norwegian esker, demonstrated well-developed small-delta structures.

Woldstedt (1961) found that in eskers deposited by streams on a land surface there is a progressive decrease in grain size toward the distal end (away from the ice margin). Near the proximal end there is a concentration of large blocks and boulders. Near the distal end the esker may grade into a flood fan leading to a sandur region.

Banerjee and McDonald (1975) provide an excellent review of the sedimentation in esker. They suggest that eskers are common in wet-based or temperate glacier areas and an esker complex includes marginal troughs, kettles, kames, and esker deltas. Further, eskers are time-transgressive features with downstream features being the oldest. Table 16 gives basic morphological types of eskers.

Banerjee and McDonald (1975) consider that the following factors are most important in the genesis of eskers: (1) formation and maintenance of conduit (through which melt water flows), (2) position of esker sediments with respect to the glacier (subglacial, englacial, or supraglacial), (3) nature of the conduit (open channel or tunnel flowing full), and (4) site of deposition (within the conduit or at the glacier terminus). Based on these considerations, three different models of esker sedimentation can be distinguished (Banerjee and McDonald 1975):

1. *Open Channel Model* – deposition in a braided stream, bordered by ice walls. Facies relationship indicates a lateral fining of the sediment. Backset bedding related to antidune is very common. Depositional units are lenticular in transverse section and more tabular in longitudinal section.

2. *Tunnel Model* – the water flows through a subglacial tunnel. The sedimentation units of such esker deposits are tabular and longitudinally persistent. The most common sedimentation units are sheet-like cross-bedded and parallel-bedded gravel and sand layers. Finer-grained sediments (silt and clay) are absent, and there is a low variability of palaeocurrent directions: variance (based on vector mean) is 1,000 to 2,000, and vector strength – 0.80. Saunderson (1977) discusses that poorly sorted, matrix–sorted sand and gravels in an esker deposit represent deposits of the sliding bed stage and may be regarded as a diagnostic feature of transport and deposition in a subglacial tunnel, flowing full of water.

3. *Deltaic Model* – the sedimentation takes place when a glacier reaches a closed-water body. These deposits are characterized by rapid downstream facies change through gradation and interfingering from proximal gravel to lake-bottom rhythmites, composed of silt and clay. Other characteristic features are occurrences of deposits caused by slump, mud flow, and turbidity cur-

Table 16. Basic morphologic types of glaciers. (After Banerjee and McDonald 1975)

Basic morphology	Characteristics of esker complex	Deglaciation environment
1. Continuous single ridge; flanking outwash	Central esker ridge is highest element of complex; flanked progressively outward on each side by elongate marginal kettles, outwash terraces, and scalloped till plain; abrupt topographic discontinuities	No standing water at glacier terminus
2. Continuous single ridge; no flanking outwash	Ridge subdued by subsequent wave action commonly resulting in beach development on esker ridge; flanked by till	Deposited within glacier but below level of standing water body at glacier terminus
3. Broad ridge with multiple crests	May or may not be flanked by outwash; commonly multiple ridges form reticulate pattern	Where broader than 200 to 300 m, probably in part subaerial and some may be part of interlobate moraine complex; narrower varieties flanked by outwash indicate not standing water at glacier terminus
4. Beaded	Pronounced isolated beads are regularly spaced and flanked by till	Deposited where esker stream entered standing water at glacier terminus

rents; presence of typical morphological features such as separate beds and large variability of palaeocurrent directions: variance is greater than 3,000, and vector strength less than 0.65.

As eskers are deposited under the influence of melt water, they are made up of partly worn, graded, and sorted sediments. Boulders of eskers are sometimes well-rounded, suggesting fluvial action. Stratification is moderate to well-developed. Strata of eskers may dip in both directions from the apex, with the dips partly corresponding to the outer slopes. Scour and fill, cross-bedding and horizontal bedding are the major bedding types.

The outer mantle of eskers may be made up of unsorted, nonstratified sediments of the nature of boulder-clay.

Augustinus and Terwindt (personal communication) describe bedding features of an esker in Scotland, and discuss its mode of deposition. Measurements on cross-bedding indicate that sediment was discharged laterally, normal to the esker axis. Scour and fill structures are most abundant and account for the major part of the deposit. Small-scale cross-bedding, sand-loam alternating bedding, loam layers are the other important features. A composite section across this esker is shown in Fig. 280, and four detailed profiles from bottom to top are shown in Fig. 281. Augustinus and Terwindt (personal communication) suggest that deposition of this esker took place in a subglacial tunnel below a thrust plane (Fig. 282). Some important papers giving data on the internal structure of esker deposits are by Szupryczynski (1965), Radlowska (1969), and Shaw (1972).

Esker deposits mostly show a variety of sedimentary facies and sedimentary structures which occur in close juxtaposition. Banerjee and McDo-

nald (1975) distinguish four major types of grainsize facies in esker deposits, each with a specific spectrum of sedimentary structures:

1. Gravel – gravel layers are most abundant, mostly of stable frame-work type, though open frame-work type is also present. In gravelly sediments, cross-bedding is common with up to 7 m thick sets, related to downstream movement of bars. Sometimes backset bedding is observed. Occasionally, gravelly beds are devoid of any stratification. Sometimes units of pebbly mud are present related to mud flow. Other units show dominant or distinctly graded parallel bedding. Some of the gravelly beds represent deposition from a slurry.

2. Sand – coarse-to-medium sand mostly shows cross-bedding related to megaripples or migrating sand bar fronts with 15–30 cm set thickness. The units occur in complex combinations. Superimposed sequences, like bar cross-beds, megaripples, small ripple bedding are sometimes present. A few units also show climbing ripple lamination. Sometimes "massive beds" with faint lamination are present, representing deposition from high suspension. Parallel bedding, related to plane bed phase of lower flow regime is also present.

3. Fine sand and silt – these sediments mostly show small ripple bedding and different types of climbing ripple lamination. Occasionally, these sediments exhibit a variety of penecontemporaneous deformation structures, e.g., contorted lamination, slump structure.

4. Silt and clay – these sediments represent distal facies of esker deposited in glacial lakes. The sediments show well-developed rhythmites and varves.

Banerjee and McDonald (1975) summarize the mode of origin and sequence of facies in two eskers

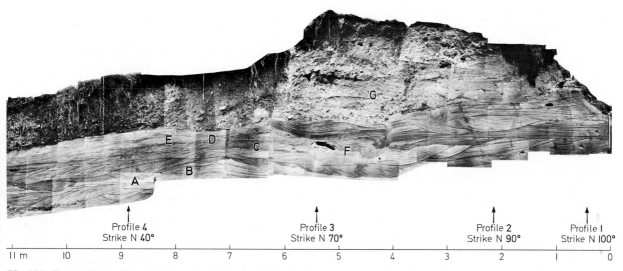

Fig. 280. Composite profile across an esker in Scotland showing internal structure. (Augustinus and Terwindt 1972)

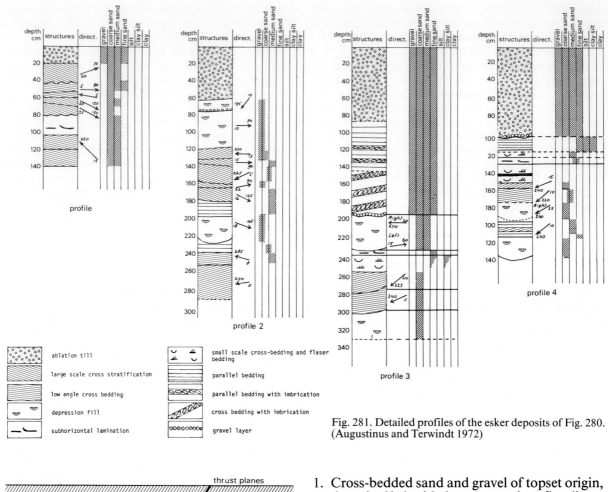

Fig. 281. Detailed profiles of the esker deposits of Fig. 280. (Augustinus and Terwindt 1972)

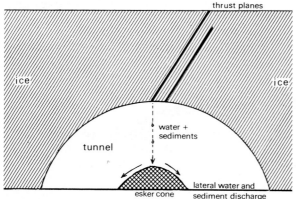

Fig. 282. Model to show the deposition of eskers in subglacial tunnels. (Augustinus and Terwindt 1972)

1. Cross-bedded sand and gravel of topset origin, deposited in braided streams or sheet flooding.
2. Delta-front sands that are poorly sorted and characterized by massive structure, graded bedding, cut-and-fill structures, irregular lamination and parallel lamination of upper flow regime.
3. Cross-laminated cosets of fine sands, mainly climbing-ripple lamination.
4. Prodeltaic rhythmites of sand and silt admixtures, deposited exclusively from suspended load.

This sequence represents esker to varve transition. Deposition of this esker sequence took place in an open channel which prograded through braided streams into a glacial lake.

of Canada. Saunderson (1975) described in detail the sequence of sedimentary structures in Brampton esker, and found that the esker is made up of several time-stratigraphic units, consisting of the following sequence of proximal to distal facies:

Kames

Kames are mound-like features occurring in isolation or in groups. A single kame represents a steep-faced localized cone composed of stratified gravel and sand. The slopes correspond to the angle of repose of the sediment. Kames are closely associated with eskers, and differentiation between the two is not always clear. They are deposited near ice margins under the influence of flowing water. As the ice recedes, a mound of sediment is left behind. Some of the kames represent small delta deposits of superglacial streams deposited within stagnant ice.

Keller (1952) discusses in detail the genesis of various forms of kames. According to Keller (1952), kames possess a concentric peel structure. The stratification of sediments correspond to the outer shape. Often there is a nucleus in the center, produced by pushing up of underlying sediments.

Penecontemporaneous deformation structures are often abundant. Figure 283A and B depict modes of formation of kames. Figure 284 shows the development of various bodies in contact with ice.

Small-Delta Deposits

Building up of small deltas by melt water streams within the region of marginal ice is a common feature. As has already been mentioned, eskers and kames are, at least, partly the result of deposition of small deltas built by melt water. Deltas are also

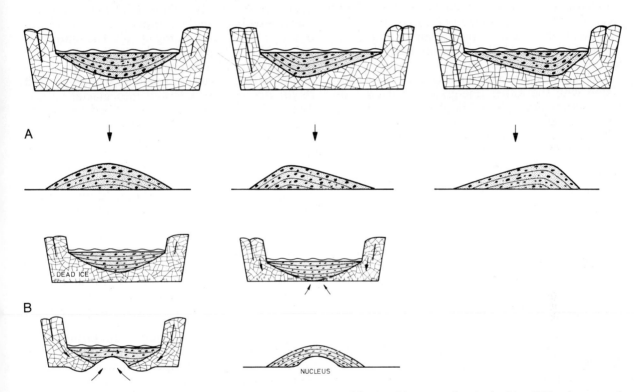

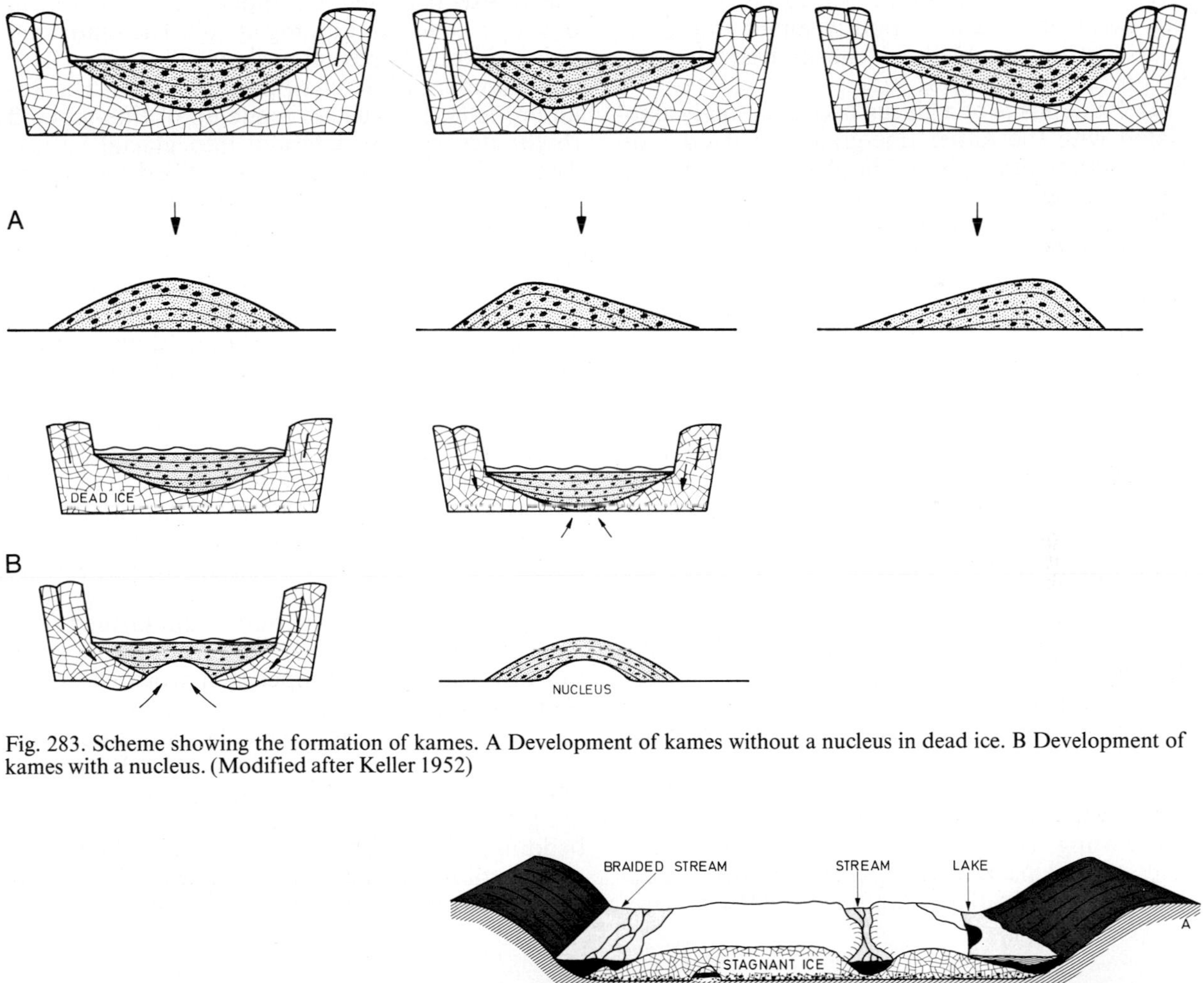

Fig. 283. Scheme showing the formation of kames. A Development of kames without a nucleus in dead ice. B Development of kames with a nucleus. (Modified after Keller 1952)

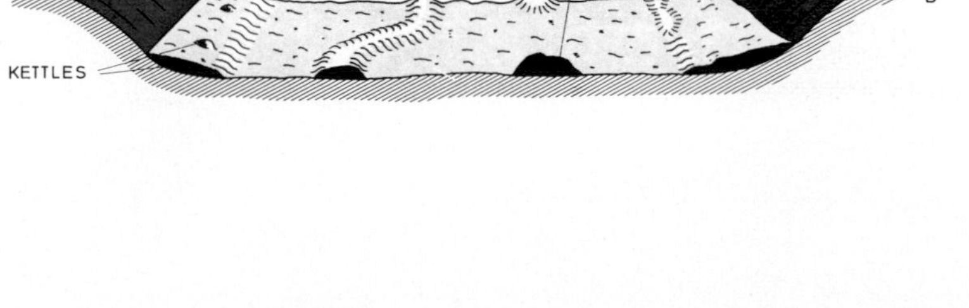

Fig. 284. Genesis of various kinds of bodies made in an ice-contact stratified deposit. (Modified after Flint 1957)

built, if a glacier ends up in a water body. However, such deltas migrate backward during receding ice and are covered by glacio-fluvial deposits on the top. Axelsson (1967) discusses the processes of delta building and geomorphology in a river delta building up in a lake within glacial environment. The results of this study can also be applied to delta building in the sea.

Small deltas are formed easily in a glacial environment because of overloaded streams during ice-melting. Delta deposits with well-developed bottom set, foreset, and topset deposits are found within glacial deposits. The topset is essentially a deposit of a fluvial regime, mostly as gravel bars showing parallel bedding and large-scale cross-bedding. The nature of topset deposits depends upon the characteristics of the river, which may be braided or, rarely, meandering. Sediment is usually poorly sorted and rather coarse-grained (gravelly). Deposition is rather rapid. Foresets often dip at 30°.

Delta deposits of a glacial lake are usually associated with the rather fine-grained, varved sediment of the central part of the glacial lake. The glaciolacustrine deltas are areas of rapid sedimentation. The incoming stream builds a large sediment body in a low-energy area of a lake, where due to lack of wave or tide energy, little lateral redistribution of sediment takes place. The lakeward construction and migration of delta takes place by overlapping lobes of sediment. Such lobe-like deposition is also known in river deltas (see Fig. 457).

Some small deltas are also formed when a melt water stream meets a small glacial lake. Such deposits are fluvio-lacustrine in origin, not occurring in contact of ice, but most often in outwash plains. These melt water streams are extremely shallow, rarely more than few decimeters in depth, and carry abundant bed load and suspended load. If the delta is large, the transition between the fluvial and lacustrine environments is rather gradual; inclination of foreset beds are rather low (5 to 10°). Jopling and Walker (1968) describe sedimentary structures of such a kame delta of fluvio-lacustrine origin (Fig. 285):

Bottomsets are represented by fine sand and silt showing mainly horizontal bedding and some ripple bedding. They grade laterally and upward into foresets, which are made up mainly of ripple-bedded units, mostly of the climbing ripple type. Fore-

sets are overlaid by fluvial topsets made up of coarse grained pebbly sand with well-developed large-scale cross-bedding of fluvial origin.

Jopling (1966) describes a "microdelta" of glacio-fluvial sediments, where a glacial stream entering a depression fills it by forming a delta-like body with well-developed bottomset, foreset, and topset (Fig. 286). Figure 287 shows microdelta bedding. On the top of the delta cross-bedded sediments are present, which are a product of migration of megaripples as soon as the profile of equilibrium was established. Figures 288, 289 and 290 show characteristic bedding features in a large delta-like body of a glacial environment.

Gustavson et al. (1975) discuss sedimentary structures in glaciolacustrine delta, where topset is made up of gravelly sediments of fluvial regime. The foreset beds, in their upper part, are made up of subparallel beds dipping at 30°. The sandy material of foreset beds often shows dips about 15°, and abundant climbing ripple lamination. The distal prodelta deposits are made up of fine-grained rhythmites grading laterally into glacial varves. The sandy deltaic deposits (foreset beds) are made up of characteristic sequences of sedimentary structures, dominated by climbing ripple lamination. A complete sequence starts with a unit of draped lamination. (The term draped lamination denotes parallel laminae of sand, silt, and clay deposited from suspension and draped over an underlying bedform. The thickness of individual laminae remains essentially unchanged). This is followed by thin units of climbing-ripple lamination of type B (2), and type A (1), overlain by a thick unit of climbing–ripple lamination of type B. The B-type climbing ripple lamination shows an increasing angle of climb upwards, overlain by a thick unit of draped lamination. This sequence represents deposition in conditions of increasing flow strength, followed by a period of diminished flow strength. This sequence is produced during summer flow, and is often capped by a winter clay layer (Gustavson et al. 1975).

Shaw (1975) describes glaciolacustrine delta deposits in the Pleistocene of Canada. He distinguishes seven facies within the delta-front deposits: gravel, cross-bedded sand (large-scale cross-bedding), flat bedded sand (parallel bedding), cross-laminated sand (climbing-ripple lamination), alternating beds (thinly interlayered sand

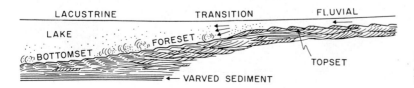

Fig. 285. Schematic representation of a kame delta with a well-defined topset, foreset, and bottomset. (After Jopling and Walker 1968)

Fig. 286. A microdelta in a glacio-fluvial environment. (After Jopling 1966)

Fig. 287. Microdelta bedding made as a result of migration of a microdelta. (After Jopling 1966)

Fig. 288. Rather poorly sorted sediments of a glacial delta, showing cross-bedding and scour-and-fill structures. Southern Norway

Fig. 289. Bedding structures in a glacial delta deposit. Southern Norway

Fig. 290. Details of a part of a delta deposit. Note the high content of clay matrix mixed with boulders. Southern Norway

and silt–clay), parallel lamination (laminated silt and clay), and dimicton.

In simple deltaic sequences, the sequence starts with parallel bedding and shows a fining upward trend, and the palaeocurrents show very consistent trends. This sequence is the result of accumulation by lateral accretion in distributary channels, and is related to continuous channel shifting.

On the other hand, sequences can be more complex, where there are thick intervals of fine-grained sediments (overbank deposits) between lateral accretion sediments. The cyclicity is the result of channel wandering. Sometimes prograding Gil-

bert-type delta units with steep foresets of coarse sediments develop. The deposition in these deltas takes place by lateral-shifting channels, development of distributary mouth bars, and breaching of these bars by channels.

Theakstone (1976) described glacial delta deposits in a lake from Norway where filling-up of the lake took place by delta progradation. The pro-delta silt is covered by cross-bedded fine sand and silt of delta front and near-horizontal sheets of fine sediments laid down between the delta fronts and distal end of the basin. The foresets of the delta front dip at about 20°. No well-defined small sequences of sedimentary structures are developed. At some places ripples related to backflow and climbing ripple lamination are observed. Local scour-and-fill and channels are developed. Minor deformation structures in the sediments developed due to slumping and loading. This example of glaciolacustrine deposit differs from the other examples, in that these deposits do not show pronounced sedimentary structure sequences. This is probably because the deposition took place mainly from suspension, and the bottom currents in the delta front were too weak to produce ripples etc.

Gustavson et al. (1975) discuss that the sedimentary structures in distal outwash deposits, in glaciolacustrine delta, and lake sediments proximal to deltas are similar where the grain size is similar and deposition occurs under similar flow conditions.

Glacial Lake Deposits

Glacial regions are marked by the development of still water bodies–lakes. Most important are the ice-dammed marginal lakes. A glacier occupying a main valley may obstruct the snout of a tributary glacier where the drainage is obstructed and a lake is produced. During receding of glaciers such lakes may come into existence temporarily. Along the margin of such lakes small deltas may be built up. In rare cases if sufficient wave activity is present, lakeshore deposits, i.e., sandy and gravelly beaches are produced. However, in the central part only fine-grained suspended sediment is deposited, usually in form of varved sediments.

Such glacial lakes receive their sediment supply during warmer periods of spring and summer when ice melts and glacial streams bring a large amount of sediment to the lake. The gravel and coarse sand is deposited in the margins. Fine sand, silt, and clay is transported to the central part of the lake in suspension. The coarser fraction fine sand and coarse silt settles down at once. Clay remains in suspension and settles down slowly. During autumn and winter, when there is no sediment supply in frozen lakes, the clay layer is deposited on the top of silt layer. The process is repeated the following year (Fig. 208).

By counting such annual varves, one can calculate the rate of deposition and the time involved in their deposition. The varve deposits of the Pleistocene have been used successfully to establish geochronology in Scandinavia.

In many cases it has been shown that annual varves are themselves finely laminated, which represents fluctuations of shorter duration, e.g., a few days (Vierke 1937; and Schwarzbach 1938, 1940). Schwarzbach (1938, 1940) describes varved sediments from Weistriztal, Eulengebirge, Sudeten. Annual varves are on average 6 to 10 cm thick. Each annual varve is itself composed of 0.2-to-1.0-mm thick laminae, ranging in number from 73 to 158. They indicate fluctuation in the rate of supply of sediment during summer. Some of these varves possess arthropod traces.

Gustavson (1975) studied sedimentation processes and sediments of a proglacial lake, i.e., Malaspina lake. The lake is not thermally stratified, but it is density-stratified with respect to suspended sediment content. Gustavson (1975) discusses that the pattern of meltwater entering into a lake is primarily a function of relative suspended-sediment content of the lake water and the inflowing water. If the suspended-sediment content of the inflowing waters is greater than that of the lake water, meltwater enters the lake as an underflow or continuous turbidity current (density current); if equal to or less than that of the lake surface water, it enters as an overflow. Meltwater enters the lake as an interflow if the suspended sediment of the meltwater lies between the maximum and minimum suspended-sediment content of the lake water. Mostly all three flows are present.

Underflows or turbidity currents flowing along the bottom of the lake deposit the laminated, commonly graded, or rippled sand and silt that characterized the coarse summer layers of the varved sediments. Throughout the lake sediment particles settle from the suspension throughout the year, whenever and wherever water velocity is low enough to allow it. As soon as underflows cease (especially during winter) or shift laterally, fine silt and clay that characterize the winter layer settle to the lake bottom.

In the lake bottom, varves are deposited. The nature of varves is strongly controlled by the relative importance of underflows and overflows. In the case of strong underflows the sand layer of varve is rather thick (up to few centimeters), showing graded (both normal and reverse) thin laminations and small ripple bedding. The clay layer is also laminated, becoming finer-grained towards the top. In the areas where there are no effects of underflows, thin sand layers and clay layers alternate (Gustavson 1975).

Ashley (1975) describes Pleistocene glacial lake deposits of Lake Hitchcock, U.S.A. They are made up of rhythmites (varves), which can be distinguished into three groups depending upon the location of deposition, near the delta front or away from delta front. Thicker silt layers show multiple grading, erosional contact, and foreset laminations, suggesting strong influence of density currents (underflows). This glacial lake was filled by underflows of changing position, causing much overlap and interfingering.

Theakstone (1976) described glacial lake deposits from Norway, where the lake was filled by extensive delta building. The varves are not common in this glacial lake deposit, while rhythmic deposition from underflows rarely occurs. Instead, in the deeper part of the lake parallel, even-laminated silt and clay are present, deposited under stagnant conditions without current activity. The silt and clay are made up of one- or two-grain-thick laminations. The lamination is related to variation in the discharge of suspended sediment which occurred periodically, but not regularly (no annual varves). This succession of laminated silt and clay is followed by a thick sequence of deltaic sedimentation starting with cross-bedded sand and coarse silt of delta front deposit.

Laterally, along the margins glacial lake deposits grade into extensive deltaic deposits.

Sandur Deposits

Glacial and glacio-fluvial deposits grade down-ward into sandur deposits, also known as outwash plain deposits. An outwash plain begins where gla-cial streams become prominent, breaking through morainic ridges. Individual streams fan out in the shape of a trumpet and deposit much of its sedi-ment load in the form of very low flood fans. Sever-al flood fans coalesce together to produce a gentle sloping plain cut through by streams. If the sandur deposits are restricted to valleys, they are known as valley sandur. Outwash plains are made up of stratified glacial sediments, which have been af-fected much by fluvial activity, and deposited only under the influence of flowing water. Outwash plains show a marked decrease in grain size down-stream, away from glacial deposits. At the same time, degree of roundness of pebbles increases considerably, downstream. However, the down-stream grain-size decrease in the sandur deposits is not very uniform. It is less prominent on the higher flood surface areas than in the channel areas, and the channel deposits are always coarser than the adjacent bar deposits. A major part of the coarse sediment is dumped near the upper part. Outwash plains extend over several kilometers and then grade gradually into real fluvial deposits of braided stream. Figure 291 shows an extension of the sandur plain in N. Germany of the Pleistocene age. Figure 292 shows the formation of successive

flood fans leading to the development of a sandur plain. During glacial retreat sandur deposits are layed down successively over glacial sediments.

Sandur sediments are better sorted than glacial sediments, but are more poorly sorted than the river sediments. Especially abundant is fine-grained sediment, together with gravelly sedi-ments. However, sometimes it may not be possible to distinguish sandur deposits from esker or kame deposits.

Hjulström (1952) discusses the processes of de-position in sandur. Krigström (1962), Fahnestock (1963, 1969), and Church (1972) describe the sedi-mentation processes and characteristic deposits of sandur areas. Augustinus and Riezebos (1971) de-scribe sedimentary structures of a sandur deposit of the last ice age from the Netherlands (Figs. 293

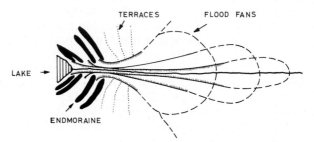

Fig. 292. Diagram showing the formation of successive flood fans of a melt-water stream cutting across a system of end mo-raines, leading to the development of sandur. (Modified after Woldstedt 1961)

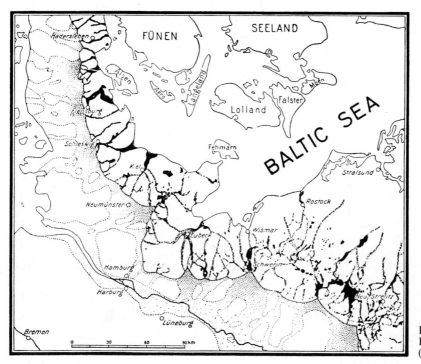

Fig. 291. Extension of the sandur plain of Pleistocene age in northern Germany. (Modified after Woldstedt 1961)

and 294). The most characteristic feature of a sandur deposit is the great variation in grain size of adjoining bedsets. Scour-and-fill structures are most abundant. A horizontal-bedded layer alternates with cross-bedded units (both planar and trough-shaped). Sediments are poorly to moderately sorted, fine sand to gravel. Rapid alternation of moderately, poorly, and very poorly sorted sediment is most characteristic.

In sandur deposits, fining upward sequences of fluvial braided streams are rare or absent. A vertical succession of sedimentary features in sandur deposits is shown in Fig. 295. Occasionally, large ice-rafted boulders are found within moderately sorted finer sediments. Complex channel cut-and-fill structures are abundant, some filled with foresets, others form-conforming.

As already mentioned, the sandur deposits are made up of outwash fans, and within a single fan three main zones can be recognized (Krigström 1962; Boothroyd and Ashley 1975). The active movement of gravels takes place only in the upper part of the outwash fan. Studies of McDonald and Banerjee (1971) and Gustavson (1974) in outwash plains indicate that sedimentation is related to bars and channels, where longitudinal bars are most abundant, showing imbricate, parallel-bedded poorly sorted gravel. The channel is distinguished into pools and riffles, where finer-grained sediments and sometimes fining–upward sequences may develop. The pools are characterized by silty mud showing plane-bed and ripples, and the riffles are characterized by gravel and silt showing transverse ribs as dominant bedform, along with current crescent.

Transverse ribs are linear stripes of pebbles and boulders occurring in groups on the bars and shallow channels, oriented transverse to the current flow, and first described by McDonald and Banerjee (1971). These ribs are spaced at distances of 6 cm to 2 m and are considered to be related to the antidune phase of the upper flow regime (Gustavson 1974; Boothroyd and Ashley 1975; Koster 1978).

Boothroyd and Ashley (1975) describe the processes and sedimentary sequences from outwash fan of Alaska, which is distinguishable into three zones with specific sequences, i. e., upper outwash fan, mid outwash fan, and lower outwash fan.

Upper Outwash Fan – This region is characterized by steep gradients (17 m/km), large clasts (> 10 cm long), well-developed longitudinal bars. The longitudinal bars are made up of well-imbricated, poorly sorted gravel, with clast axes oriented transverse to flow direction. In this area only few important channels are visible. The bars are often covered with transverse ribs. Some bars contain thin layers of parallel–bedded sand, and locally discontinuous, isolated trains of megaripples in sandy parts of the channel develop.

Mid Outwash Fan – This area shows gentler gradients (2–6 m/km), moderate to small clast size (< 10 cm and sand-size sediments), and common

Fig. 293. Scour-and-fill structure in sandur deposits. (After Augustinus and Riezebos 1971)

Fig. 294. Scour-and-fill cross-bedding and horizontal bedding in sandur deposits. (After Augustinus and Riezebos 1971)

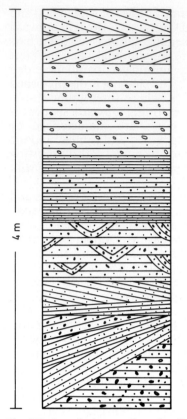

Fig. 295. A vertical succession of sedimentary features in san-dur deposits. (Modified after Augustinus and Riezebos 1971)

The longitudinal bars in this region show planar cross-bedding formed by migration of bar lee-face, overlain by parallel bedding of the bar surface, and climbing-ripple lamination. The lingoid bars show large-scale planar to tangential cross-bedding topped by climbing ripple lamination. The point and lateral bars exhibit planar and trough-shaped large-scale cross-bedding caused by migration of bar surface, bar lee-face and megaripples.

The overbank deposits of silty sand show climbing-ripple lamination.

Clague (1975) discusses characteristics of outwash deposits in British Columbia, Canada, which are characterized by large channels, coarse sediments showing large-scale cross-bedding in a channel-bar complex of high-energy streams. Many channels probably discharged peak volume discharge during Jökullhlaups (catastrophic glacial lake drainages) from galcial lakes in tributary valleys. The deposits are dominated by large-scale cross-bedding (some gravel beds show units up to 5 m thick) produced by migrating avalanche faces of gravel bars and channel fills.

A special case of outwash deposits, formed under subaqueous conditions by glacial water issuing below a standing water is described by Rust and Romanelli (1975). These deposits are made up of proximal boulder gravel and distal sand. The characteristic features are association with flow till, ice-contact deformation, graded units of climbing–ripple lamination, and large-scale channels with apparently massive fill.

Ruegg (1977) describes middle Pleistocene sandur deposits in the Netherlands, and discusses that the sedimentation of sandur is very similar to alluvial fan deposits. Parallel–bedded and low-angle cross-bedded sequences are most abundant and are related to sheet flood deposits. Large-scale channel-fill deposits are less common, and are related to stream channel deposits. Ripple bedded sand and silt are pool deposits. The climbing-ripple lamination sequences are related to glacio-aqueous (glaciolacustrine) deposits formed in accelerated deposition. Rarely, thin till layers are present. They are flow till and can be compared to debris-flow deposits of alluvial fan.

In valley sandur deposits of Iceland, Bluck (1974) observed that directional structures within the bar deposits show more variability than the variability in the direction of channels. The gravelly units are deposited during high flood stage and show less variability in direction, while the sandy units are deposited during falling water stages, and exhibit large variability in directional properties. Cross-bedding is the main bedding structure in these deposits and is produced in four different ways: by small deltas, by megaripples, by ripple-topped bars, and by migrating accretionary bar banks (Fig. 296).

longitudinal bars. The size of gravel decreases and the amount of sand increases. The sand is deposited as parallel-bedded units (upper flow regime) inter-bedded with gravels, or as trough-shaped megaripple bedding in the channels.

Longitudinal bars are most abundant within the shallow channels, showing parallel-bedded, imbricated pebbles. Sometimes sandy slip-faces are poorly to moderately developed at the downstream margins of the bars, showing tangential to planar cross-bedding with reactivation surfaces.

Point bars are also developed, occurring in curved channels at lower stages of flow. Lee-faces are present producing high-angled large-scale cross-bedding.

In the lower part of the mid outwash fan, units of small-scale trough cross-bedding and draped lamination are common. Climbing-ripple lamination of type 1 may be present. Fine sand and silt is deposited as overbank deposits, showing mainly climbing-ripple lamination.

Lower Outwash Fan – This part shows a gradient less than 2 m/km, sand-sized sediments (medium to coarse sand), almost no gravels, longitudinal and lingoid bars in braided reaches and point and lateral bars in meandering reaches.

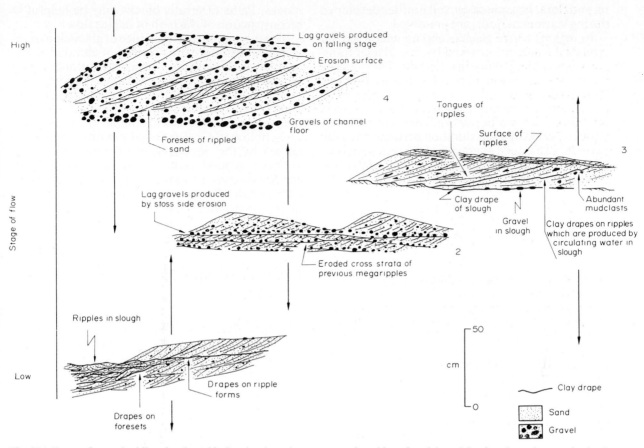

Fig. 296. Types of cross-bedding developed in Sandur deposits and their relationship to flow stage. Arrows indicate range of flow stage in which building of cross-bedding is considered to take place. *1* Section of lateral fill into a slough: topmost units are produced by microdeltas. *2* Section through megaripple. *3* Section through ripple-topped bar. *4* Section through an accreting bar tail margin. (After Bluck 1974)

Pleistocene outwash deposits in southern Ontario, Canada (Eynon and Walker 1974), demonstrate deposition in a large braid bar with adjacent side channel. The main part of the bar shows faint horizontal bedding which in downstream direction shows bar front deposits made up of large gravelly foresets (up to 4 m thick) with reactivation surfaces. The upstream part of the bar shows tabular cross-bedding in coarsening upward sequences. An important point in this deposit is that the side–channel deposits contain no gravels, while the bar deposit is mainly gravelly. Sediments are strongly bimodal, and the sand probably infiltered in the pore spaces of the gravel at a later stage.

Interglacial Deposits

Our knowledge of Pleistocene glaciation has clearly shown that an ice age is not monoglacial, but polyglacial; i. e., during a single ice age several colder glacial [times] episodes alternate with warmer interglacial [times] episodes. Thus glacial deposits possess sediments of other environments interbedded within it. They can be fluvial sediments, wind deposits, etc., with fauna and flora of warmer climates.

During interglacial periods lakes develop extensively in glacial regions. Small deltas are built in marginal parts. In deeper parts of lakes rather fine-grained sediments are deposited; sometimes they are unusually rich in organic matter (mud of gyttja facies). In other cases it may be deposition of diatoms making kieselgur, or abundant carbonate accumulation producing "Seekreide" (lake marl) (Woldstedt 1961; A. Schäfer 1973).

If vegetation becomes better established, dead plants accumulate, and due to insufficiency of oxygen, are preserved. This leads to development of marshes. In ancient sediments marsh deposits are preserved as peat layers, sometimes tens of meters thick.

In some cases interglacial sediments are almost exclusively fluvial sediments, with freshwater fauna and flora. In some cases, soil profiles developed during warmer periods, are preserved.

In regions where glaciers end up near a sea, interglacial stages are marked by interglacial transgressions leading to marine deposits. They may be intertidal flat, lagoon, or beach deposits. The term "interstadial deposits" is used to designate interglacial deposits formed from minor fluctuations in the climate. They represent sediments of relatively a warm period of short duration between two cold periods. The glacial character of the sediment is retained, only some organisms and plants of the cold climate are incorporated.

Evidences of Glacial Activity

If any region undergoes glacial activity, a characteristic landscape is produced as a result of erosional and depositional activity of moving ice. Large-scale landforms point to their glacial origin: U-shaped valleys, hanging valleys, cirques, fjords, glacial lakes. However, they have little chances of being preserved and recognized in the ancient rock record.

However, there are also moderate to small-scale erosional features testifying to glacial activity. They are striated and grooved rock surfaces, sometimes even well polished, concentric marks on rocks, and ice-moulded hummocks, known as roche moutonées. These features have been reported from geological records of various ages.

Sometimes the glacial sheet is moulded into small mounds occurring in a group, arranged in en-echelon fashion; these are drumlins (Fig. 297). They are elongated mounds parallel to the direction of ice movement, with the blunt side toward the upflow direction of ice. The hollows between drumlins may develop in ponds, marshes, etc.

Another characteristic of glaciated regions in the occurrence of scattered boulders and blocks that are foreign to the place, better known as erratic

Fig. 297. A typical drumlin landscape. The movement of ice is from right to left. (Modified after Holmes 1965)

blocks. Paths of erratic blocks may be helpful in reconstruction of the path of the ice flow.

The most important evidence of glaciation in ancient sediments are the presence of abraded, striated bed rock surfaces, roches mountonées, and widespread glacial sediments–till, sandur sediments, local varve sediments, and erratic rock fragments. Associated with regions of glacial activity are periglacial regions with annual mean temperatures below 0°C. Periglacial regions are characterized by the development of permafrost soils, showing ground patterns, ice wedges, pingos, etc. Solifluction is an important process active in such regions. Sand and mud sediments show disturbed, contorted bedding–cryoturbation structures.

Grain Size Characteristics of Glacial Deposits

Glacial sediments are commonly known as drift or boulder clay. Till is a petrographic term denoting glacial sediment; the term tillite is restricted to ancient sediments. Harland et al. (1966) discuss at length the nomenclature problem of till and tillites. Elson (1961) provides a usefull account of petrography and grain-size characteristics of tills. Goldthwait (1971) and Legget (1976) give some additional information of till.

In general, we prefer the use of "glacial sediments" to indicate the glacial origin of the sediments. Usually nonstratified sediments are the result of deposition more or less directly from ice; while stratified sediments are produced, at least partly, by the action of flowing water.

Glacial sediments are characterized by the absence of sorting and their grain size varies from a few microns to several meters. The relative abundance of various grain size varies substantially, although even relatively equigranular sediments possess some amount of clay-sized and pebble-sized sediment grains. The main factors controlling the nature of glacial sediments are (Kukal 1970):

1. Character of bed rock on which the glacier moves.
2. Morphological characteristics and velocity of flow of the glacier.
3. Position of the transported material in relation to the glacier.
4. Mode of deposition.
5. Subsequent reworking by melt water.

The unstratified glacial deposits deposited directly from ice are characterized by the constant presence of a certain amount of gravel fraction and by the equilibrium among the sand, silt, and clay fractions (Kukal 1970). Larger sediment grains are scattered irregularly in the fine-grained matrix. An important feature of glacial sediments is the presence of numerous labile minerals, e. g., feldspar, ferromagnesian minerals as unaltered, angular grains even in silt and clay-sized fractions. The sand fraction is characterized by extremely angular sediment grains. The median size of such deposits varies greatly; sediments are poorly sorted and Phi skewness is more or less zero, ranging from slightly positive to slightly negative skewness.

In the glacial till, two basically different types of grain-size distribution are present (Boulton 1978):

1. Supraglacially derived debris transported supraglacially or englacially at a high level. These sediments show large standard deviation in the vicinity of modal peaks. Individual curves may show one or more secondary modes. The modes are mostly between -4Φ and -1Φ.

2. Debris which is transported in the basal zone of traction. They show one or two modes coarser than 2Φ, and one or two finer than 2Φ. The tractional debris is depleted in the coarse fraction and enriched in fine fraction compared with the debris undergoing high level transport.

The difference in these two types of debris is well-marked in the plots of inclusive graphic standard deviation versus graphic mean and inclusive graphic skewness.

This primary difference in the two types of till is often modified by secondary processes e. g., water flows etc. which remove certain grain sizes and concentrate others.

As mentioned earlier, the main deposition takes place near the snout of a glacier, where melt water is invariably present. Thus, almost all glacial sediments undergo some reworking by water. The main effect of reworking by water is removal of silt- and clay-sized sediments, which leads to the enrichment of coarser sediments. Such reworked sediments are found interlayered with unworked glacial sediments. They can be distinguished by the amount of silt- and clay-sized sediments. Sediments reworked by water show a rather low amount of fine-grained sediments, as compared with the associated unworked glacial sediments. However, the degree of reworking by water varies enormously from point to point.

Mills (1977a,b) discusses that there are differences in the grain-size characteristics of various glaciers, due to bed rock differences. It is possible to distinguish five glacial subenvironments, i.e., basal till, recessional-moraine till, lateral moraine till, ablation till, and outwash on the basis of particle size distribution, roundness, striations, fabric, and lithologic composition. The mean particle size and sorting increase in the order of subenvironments given above. Percentage of silt and clay decreases. The basal till lacks dominant modes in coarse fraction, while others possess dominant modes in the coarser fraction (-5Φ and above). Grain size of sandur deposits is strongly polymodal (Ruegg 1977).

Pebbles are rather characteristic of glacial sediments. They usually reflect bedrock composition and local sediments. However, erratic pebbles can travel over large distances. Glacial pebbles are seldom rounded, the edges and corners are rarely rounded off to smooth curves. The majority of them are disc-shaped. Larger pebbles are better rounded than pebbles a few millimeters in size. In the plan view they are pentagonal (Fig. 216) with a well-developed stoss side. The shape is of course also determined by the source and composition. Striations are commonly present on them (Fig. 219).

The boulder shapes in till deposits can be distinguished into three groups (Boulton 1978):

1. Angular and subangular boulders, which have been derived supraglacially and transported at a high level without undergoing a phase of traction. They are characteristic of supraglacial morainic till.

2. Subrounded to rounded boulders with several directions of striae which have undergone a phase of traction. They are found in flow till, meltout till, and lodgement till.

3. Large, rounded and streamlined boulders of the order 0.5–1.0 m in diameter with long axes and striae parallel to glacier flow, and sharply truncated distal extremities, indicating that they have been deeply embedded in subglacial till. These are characteristics of lodgement till.

Glacial pebbles show a preferred orientation, long axes oriented parallel to the direction of flow of ice. Sometimes a second maxima at right angles to the direction of flow may be present. Imbrication may be present, with rather high angles of imbrication. Some authors think that preferred orientation and imbrication in glacial sediment is only the result of reworking by water. However, there is enough evidence suggesting preferred orientation caused by movement of ice alone. Rust (1975) points out that in the outwash sediments, the discoidal pebbles show a 30° dip in upstream direction and this direction agrees with the mean orientation of surface channels and long axes of braid bars.

Sand grains of glacial sediments show characteristic surface features if studied by the electron microscope (Krinsley and Funnell 1965). Such surfaces show abundantly conchoidal fractures, minor striations, imbricate breakage blocks, and small-scale indentations. Figure 298 shows an electron microscope photograph of such a quartz grain. Whalley and Krinsley (1974) provide some additional information on the quartz grain surface features to distinguish various glacial subenvironments.

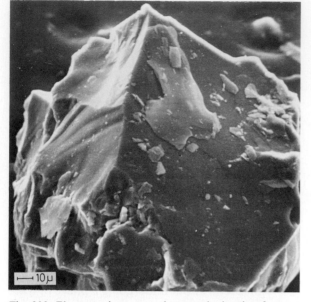

Fig. 298. Electron-microscope photograph showing features characteristic of quartz grains from a glacial region. (Photograph by D. Krinsley)

Wind Activity in Outwash Plains

Outwash plains (sandur) of glacial environments, especially those in high latitudes are exposed to drastic wind activity. Such winds usually flow away from the glacial regions. Outwash plains and adjoining fluvial regions, poorly protected by plant cover, are actively deflated. Coarse gravels are left behind; sand is transported short distances to adjacent regions and is deposited as sand dunes, which actively migrate. The finest sediment (silt and clay) is kept in suspension and transported over long distances before being deposited as loess deposits in the steppes. During the Pleistocene period much silt and clay was transported from glacial regions and deposited in the form of extensive loess deposits. It makes thick deposits in Central Europe, China, and North America. Sometimes wind activity may leave behind evidence in the outwash plains in the form of ventifacts, wind scouring, and sand dunes with quartz grains showing good rounding and even typical frosted surfaces. For a detailed description of loess sediments see p. 237. However, extensive wind deposits in outwash plains are only an exception.

Examples of sand dune deposits associated with outwash plains are known from the Pleistocene age from Central Europe and North America (Flint 1957).

Inactive stream areas in the outwash plains of Skeidararsund show extensive aeolian activity (Nummedal et al. 1974; Boothroyd and Nummed-

al 1978). Over extensive areas dunes develop, elongated parallel to the dominant wind direction. The dunes are 10–20 m long, 5–10 m wide, and 2–5 m high and may or may not be grass-covered. Some of the unvegetated dunes exist as low mounds or as small transverse dunes with or without slipfaces. Internally, dunes show long (3–10m), shallow (50cm–1m) trough cross–bedding and abundant horizontal bedding. Up to 2-m-thick sets of planar cross-bedding are also present.

Ancient Sediments of Glacial Environments

Ancient glacial deposits are not very common, but there is evidence of worldwide glaciation during certain geological times. During such periods widespread evidence of glacial activity, together with glacial deposits are found. There are several glacial deposits known from the Precambrian age ranging in age from 2,500 to 600 m.y.B.P. The last one of Eocambrian age is well developed in Norway (Spjeldnaes 1964; Reading and Walker 1966). Edwards (1975) describes glacial deposits of Late Precambrian age where mixtites are dominant. Glacial loess deposits are also recorded in the Late Precambrian glacial deposits of Norway (Edwards 1979). Perry and Roberts (1968) describe glacial sediments and other glacial evidence, e. g., striated pavements, etc., from the late Precambrian of Australia. Permocarboniferous glaciation is widespread on the continents of Gondwanaland. Hamilton and Krinsley (1967) describe Palaeozoic glacial deposits from Australia and South Africa. Dwayka tillites of South Africa show much evidence of glaciation. Features such as roches mountonées, excavated rock basins, drumlins, and U-shaped valleys have been described together with grooved and ice-faceted boulders and pavements (see Holmes 1965). Frakes et al. (1968) report a fossil esker deposit in the Parana basin, Brazil. Tillites and varved shales are associated with these features. In India evidences of Gondwana glaciation are shown by the Talchir formation (upper Carboniferous). These tillites possess faceted and striated pebbles and boulders, varve-like sediments which are found resting on striated and grooved bedrocks. Banerjee (1966) discusses the deposition of these sediments. Casshyap and Kidwai (1974) describe the sedimentation and sedimentary structures from Late Paleozoic Talchir glacial deposits, where three thick units of polymictite are developed.

Evidence of Pleistocene glaciation are widespread and have been discussed in detail by Flint (1957) and Woldstedt (1961).

Desert Environment

General Information

A desert can be defined as a continental area on which there is a little or no plant cover because of insufficient rainfall, i. e., areas where the potential rate of evaporation exceeds the rate of precipitation, and where wind action is an important geological agent. Several factors may lead to such conditions, like position of the area within wind systems of natural low humidity, the presence of a mountain chain in the path of rain-bringing winds, the absence of sufficient relief to induce rainfall, etc.

Average rainfall in hot deserts is less than 25 cm/year. The potential rate of evaporation in hot desert regions is usually several times the average rainfall.

In arctic regions areas of even lower annual precipitation are present; here plant life is absent because of extreme conditions. Such areas are sometimes called cold deserts. However, here we are concerned mainly with the desert areas of tropical-subtropical regions. Hot desert areas are restricted to subtropical zones in the northern and southern hemisphere. The very existence of a subtropical desert is a result of the wind system of our world.

The major wind systems of the world exist because the air is cold over the polar regions and hot over the equatorial belt. If our earth did not rotate, a simple convection system would exist, i. e., heated air from equator would ascend and blow toward the poles. Chilled air from the poles would return toward the equator.

However, as our earth is rotating a powerful deflecting force is at work, the coriolis force. The rotational velocity of the earth at the equator is very strong, decreasing toward poles, and becoming zero at the poles. Consequently, if something moves relatively freely from south to north in the northern hemisphere, the moving mass tends to move eastward faster than the earth. The rate of deflection increases toward the north. Similarly, if a mass is moving from the north pole toward the equator, the deflection is westward.

The hot air that blows at high altitudes from the equator toward north pole is deflected to the east. At the same time, the returning cold wind from the north pole toward equator flowing near the ground is deflected to the west.

There is another complication in this system. Heated air from the equator flowing toward the north pole passes into higher shorter latitudes. Between latitudes 20 and 35 the crowding of air is sufficient to raise the pressure so that air descends to the surface.

This zone is commonly known as high-pressure subtropical calm belts. Near the surface descending air is divided into (1) trade winds that blow toward the equator, and (2) disorderly westerlies that blow toward the poles. But surface winds are also blowing from the poles toward the equator. Where these two opposing wind systems meet each other, they create variable, disturbed weather. Thus, there are three fundamental cells within a convection cell: A, B, and C. A similar system exists in the southern hemisphere. However, wind systems are further complicated by factors like altitude, etc. Figure 299 schematically depicts wind systems of the earth.

The northeast trade winds blowing from 30° N toward the equator bring no rainfall and where trade winds blow over extensive land masses they become increasingly desiccating. Development of desert characteristics may be further enhanced by other factors, e. g., absence of high areas to cool the occasional humid winds, presence of high mountains in the path of humid winds.

Hot deserts are characterized by violent fluctuations in the wind and temperature, both daily and seasonal. Rains are rather rare and sporadic. Vegetation is extremely scanty or entirely absent. Rains in the desert are usually due to exceptional meteorological conditions. Frequency of rains varies from several times per year to every 10 or 20 years. Such rains can occur very rapidly, pouring out large amounts of water during a short time. Lack of vegetation and soil cover, causes run-off to be very rapid, and often flash floods result. Sporadic water courses in deserts are known as wadis. Wadis sometimes make fanlike features, flood fans (wadi fans). (This term has been preferred instead of the

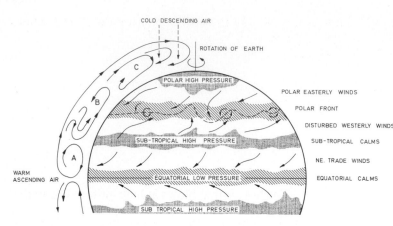

Fig. 299. Scheme showing the distribution of the wind system in the Northern Hemisphere. Note that the basic convection cell is made up of 3 subunits. A similar system exists in the Southern Hemisphere. Trade winds are the main factor for the existence of deserts in tropical and subtropical regions. (Modified after Holmes 1965)

more commonly used term, alluvial fan.) Deserts are areas of inland drainage, i. e., drainage is usually directed toward the center of the basin in low-lying areas, where desert lakes or inland sebkhas often develop.

However, the most important feature of the desert environment is that vast open areas are exposed almost exclusively to wind action. It may be pointed out that wind activity is not restricted only to desert environments. Wind-laid sediments are commonly associated with glacial sediments, coastal sands, and flood-plain deposits. A short account of such windlaid sediments is given in chapters dealing with glacial, coastal, and fluvial sediments.

Under desert environments one should not think exclusively of wind-laid deposits, because locally water-laid deposits are abundantly associated with wind-laid deposits. It is the particular setting and association which is suggestive of a desert environment. Wind-laid sediments may be present, for example, in minor amounts in fluvial sediments, but it would not change the overall environment from fluvial to desert. Figure 300 shows the distribution of various deposits of a desert environment. Note that wind-blown sandy deposits occupy only part of the region.

Some of the inland continental deserts are present in low-lying subsiding basins. Some on them are located even below sea level, and any increase in sea level transgression would not erode them; they would be preserved (cf. Glennie 1970). Walther (1924), Solle (1966), and Glennie (1970) are the important monographs on the desert environment from the geological point of view. Cooke and Warren (1973) give a good account of deserts. Bigarella (1972) gives a good review of aeolian sand deposits, both Recent and ancient.

Erosion and Sedimentation Processes in the Desert

As mentioned earlier, in a hot desert environment wind is an important geological tool. Wind is a very effective transporting and depositing medium, but it is less effective as an erosional tool. However, locally water does play an important role.

In desert, mechanical weathering is a dominant factor, involving processes such as exfoliation, splitting, and crushing of rocks. It is further aided by the abrasive action of wind-blown sand and dust. The presence of some water in the form of occasional rains and early-morning dew activates microchemical processes leading to chemical weathering. Because of an exceptionally high rate of evaporation, dissolved salts are brought to and precipitated near or on the surface. Oxides of iron and manganese are precipitated in the form of "desert varnish" on the upper surface of larger sediment grains. Engel and Sharp (1958) Hunt and Mabey (1966), Hooke et al. (1969), provide some chemical data on the desert varnish.

The development of soils in a desert environment primarily weakens the soil surface, supplying in this way sediments to aeolian and water systems, and then leads to development of a hardened carapace which tends to preserve the desert surfaces (Cooke and Warren 1973).

Sometimes patterned grounds are commonly found in desert environments, e. g., gilgai, which are small-scale surface undulations related to differential expansion and contraction of the soil with periodic wetting and drying. Sediments of playas and wadis exhibit extensive developmen t of desiccation cracks, giant desiccation fissures, and in salt-bearing playas patterned grounds related to salt phenomena are produced (Cooke and Warren 1973).

The transport of sediment by wind is rather similar to the transport of sediment by water. The process of sediment transport in the wind regime has been dealt with in detail by Bagnold (1954a). Some other important papers dealing with this problem are by Moldvay (1957), Horikawa and Shen (1960), and Williams (1964).

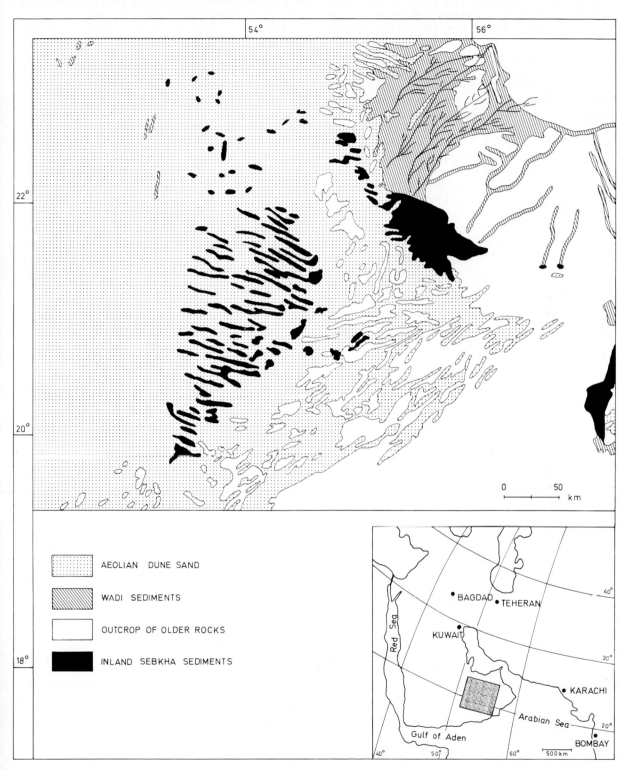

AEOLIAN DUNE SAND

WADI SEDIMENTS

OUTCROP OF OLDER ROCKS

INLAND SEBKHA SEDIMENTS

Fig. 300. Map showing the characteristic distribution of various types of subenvironments in a desert region. Note that the sandy region is only a smaller part of the environment. (Modified and simplified after Glennie 1970)

Sediment in the wind regime is also transported by traction, saltation, and suspension. In the wind regime this separation is much more effective than in acqueous conditions. The result is that three populations (rolling, saltation, and suspension) in wind-laid sediments are more effectively separated.

Wind of sufficient strength moves sediment grains by rolling and saltation. Silt- and clay-sized sediment may be taken into suspension and can be kept in suspension for longer periods and can be transported over long distances. Data of Moldvay (1957) shows that there is a natural break at the grain size 0.05 mm (Fig. 301). Below this grain size sediments once taken into suspension can be transported over long distances because of their small free-fall velocity. This break causes silt- and clay-sized particles to be very quickly winnowed from a sediment by wind action.

Under normal conditions sand-sized sediments usually move by saltation. Generally, grains of 1 mm size move by saltation, and those up to 2 mm by rolling or surface creep. Still larger grains are pushed under the impact of saltating grains. On impact with the grains, saltating grains, by virtue of their kinetic energy can move grains which are six times larger than the saltating grains (Bagnold 1954a). Under exceptionally high wind velocities, even sand-sized sediments can be kept in suspension for a short time.

Thus, if enough time is available for wind action on a sediment-mix, separation into dust (silt and clay), sand, and gravel is very effective.

Although Bagnold (1954a) found that the threshold velocity for sediments smaller than 80 μ in size increases with decreasing size, the finer grains can be easily ejected by bombarding larger saltation grains. And once finer grains are ejected, they are kept in suspension for longer periods.

Thus, wind-laid desert sediments can be differentiated into three deposits:

1. Dust Deposits. Silt- and clay-sized sediments can be transported over long distances from a desert. Usually dust is deposited in adjoining steppes. Some workers have suggested that wind-blown dust is an important source of detrital sediments in the deep sea. Dust deposits within the desert environment itself are of only minor importance.

2. Sand Deposits. Well-sorted sand-sized sediments are common in a desert environment. They are usually piled up in the form of sand dunes, etc., moving mainly by saltation.

3. Deflation Lag Deposits. Gravel-sized sediments are concentrated, making a pavement in areas from where sand and dust has been deflated away.

Deflation is a very important process in the desert. The process of lowering the land surface by removing finer sediment is called deflation. Finer sediments from certain areas are being removed continuously and being deposited as sand dunes in an adjoining area. The area undergoing sediment removal may become a depression. This leads to relief development. However, maximum (base-level of erosion) depth of deflation is limited by the water table.

Such depressions in a desert are often occupied by salt lakes – inland sebkhas, oases, or wadis of braided streams.

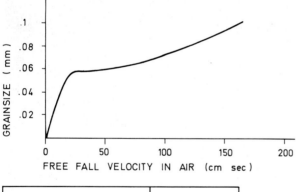

DIAMETER OF SEDIMENTGRAIN m m	FALL VELOCITY c m / s e c
0. 0 1	2. 8
0. 0 2	5. 5
0. 0 5	1 6
0. 0 6	5 0
0. 1	1 6 7
0. 2	2 5 0
2. 0	5 0 0

Fig. 301. Diagram showing the relation between grain size and free-fall velocity in the air, both in tabulated form and in a graph. Note a major break at a grain size of 0.05 mm, which corresponds to the boundary between the suspension load and the saltation load of wind deposits. (Data of Moldvay 1957). (Modified after Kukal 1970)

Deposits of Desert Environments

It should be borne in mind that desert deposits are not only the dune sands. There are varied types of deposits originating in a desert environment; of course, the sand dune deposits are the most impressive and most important geologically. For practical reasons deposits of desert environment can be grouped into following types:

1. Hamada deposits
2. Serir deposits
3. Aeolian sand deposits
4. Wadi deposits
5. Desert lake and inland sebkha deposits
6. Dust or loess deposits

Dust or loess deposits are usually extensively deposited outside the desert environment, in the adjoining regions. Within the desert environment itself dust deposits are of only minor importance. However, they have been described here, since they are an important by-product of a desert environment.

Hamada Deposits

Hamada are the rocky terrains in a desert, characterized by vertical cliffs and flat rock surfaces covered with large boulders, pebbles, etc., and generally associated with inselberg topography. Figure 302 shows a desert area with exposed bedrock. Hamada areas are usually located in deflation basins and in deep valleys of wadis cutting through the rocks.

Sediments of hamada are rather coarse-grained, extremely angular, and unworked; they are usually associated with bizarre bedrock forms (Fig. 303). Boulders are the product of desert disintegration made free *in situ*. Boulders and rock surfaces often show desert varnish. However, desert varnish cannot become preserved in ancient sediments. Ventifacts are common. Locally some sand is deposited behind large blocks in the form of sand shadows.

Hamada are primarily erosional areas and possess very little chance of being preserved, although sometimes under exceptional conditions they may be preserved (e. g., because of a change in the local wind system, hamada may be covered under dune sand).

Thus, hamada deposits may be found at the base of a sequence of desert deposits, with topographical forms. In such cases older beds stand out above the peneplain surface in the form of an inselberg, with thin veneer of coarser-grained debris (hamada deposits) around the inselberg.

Solle (1966) gives examples of possible hamada deposits near the base of Buntsandstein of Central Europe. Walther (1909) describes them from Torridon Sandstone, U.S.A.

Fig. 302. A general view of a desert environment showing bare rocks and wadi systems. Coarse-grained debris usually accumulates near the exposed rocks. Sudan (After Solle 1966)

Fig. 303. Coarse-grained debris of hamada deposits. (Photograph by G. Solle)

Serir Deposits

Serir, also called stony deserts, are more or less horizontal deflation lag surfaces. Serir surfaces are marked by the concentration of coarse-grained sediments, mainly gravel, pebble, and coarse sand (Fig. 304). This is due to removal of finer sediment by wind activity, leaving behind coarse-grained sediments which cannot be moved in suspension or saltation. Such surfaces also serve as a protecting layer for underlying sediments.

The nature of serir deposits depends upon the nature of the source rocks. In every case, pebbles of

Fig. 305. A closer view of a serir surface. (Photograph by G. Solle)

Fig. 304. A typical photograph of a serir surface covered by coarser-grained debris. Egypt. (After Solle 1966)

more resistent material become enriched. Some of the pebbles show impact marks and broken faces; others are well developed in the form of ventifacts. Figure 305 shows a closer view of a serir surface. The lower surface of pebbles may be etched. Serir surfaces are more or less horizontal, and maximum dips may be sometimes 5 or 10°. Serir deposits are usually very thin, a maximum thickness of several centimeters. Serir deposits, especially of low-lying deflation depression, may become preserved, if covered by wind-blown sand. Grain size is coarse sand to pebble-sized. Strata are rather thin parallel beds extending over long distances (Solle 1966).

Sometimes coarse-grained serir sediments are molded into wind granule ripples (see p. 51) (Fig. 306). Largest grains are pushed near the crests. Internally, wind granule ripples are composed of inclined laminae. Serir sediments are poor to moderately sorted.

Serir deposits are abundantly produced in areas undergoing deflation, e.g., interdune areas. Such serir deposits of interdune areas may become incorporated into sand dune deposits. They are pebbly layers few centimeters thick within a sand dune deposit.

McKee and Tibbitts (1964) describe serir deposits from the Libyan Desert. These deposits are poorly sorted, and the grain size is bimodal. Some silt and clay is present. Pebbles are scattered or concentrated into layers of lag deposits. The bedding is horizontal or low-angled dipping laminae (Fig. 307).

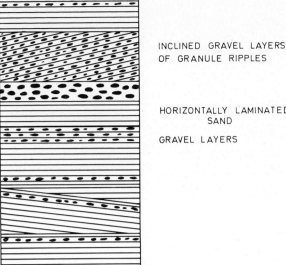

INCLINED GRAVEL LAYERS OF GRANULE RIPPLES

HORIZONTALLY LAMINATED SAND

GRAVEL LAYERS

Fig. 307. Bedding structures in serir deposits shown schematically.

Fig. 306. Granule ripple showing characteristic concentration of coarser grains near the top. Internally, inclined laminae are visible. (After Glennie 1970)

Desert Lake and Inland Sebkha Deposits

It has been pointed our earlier that desert basins are basins of inland drainage, where water flows toward the central parts. Within the desert there are often very shallow depressions of deflation or of tectonic origin. In such low-lying areas water accumulates and very shallow lakes are produced, which are usually dry most of the year. However, some of them may be semi-permanent in nature; these are desert lakes.

If areas are made damp by water and if the water then dries up, leaving behind damp, salt-encrusted sediments, the areas are known as inland sebkhas (Glennie 1970). Terms like takyr, Salztennen, playa, salina have been applied to similar features. Saline lakes of South America are often named salar (Stoertz and Ericksen 1974). Neal (1975) provides a good collection of papers on playas and dried lakes.

Source of water may be inflowing wadis or sepage of ground water. If the water table is higher than the ground surface in deflation hollows a lake results. Figure 308 shows a lake made by inflow of wadi. Figure 309 depicts a sebkha fed by ground water. Development of salt crusts is an important feature of most of the desert lakes. Salt sources are varied.

If a desert lake is associated with a wadi, during floods much detrital sediment is brought into the lake. However, current velocity is almost nil, and main deposition takes place from suspension. Silt and clay is deposited mainly. Individual thin beds often show graded bedding. On the top is a clay layer. During dry periods this clay layer cracks and curls up. If covered by blown sand, such layers may become preserved. Gypsum and halite are often associated with such deposits.

Sometimes sand dunes may be drowned in sepage and inflowing water and become preserved. In such cases, fine-grained sediment is deposited in water-bearing interdune areas.

If lakes are not fed by inflowing wadis, but are a result of seepage of ground water, detrital sediment is rarely deposited. Mainly salt pans are built. Within chemical precipitates some detrital sedi-

Fig. 308. Aerial view of an inland sebkha. Sand dunes are preserved in a temporary lake made as a result of the flow of wadi. (After Glennie 1970)

Fig. 309. Inland sebkha which is fed by ground water. (After Glennie 1970)

ment blown by wind become incorporated as thin layers or as impurities.

Inland sebkhas are characterized by salt crusts formed from the rapid evaporation of water. Gypsum layers are common, and also well-developed gypsum crystals. During long dry periods the salt crust is broken into large polygons. Some of the salt is deflated away by blowing wind. Some salt is dissolved by rain water, inflowing wadi water, or ground water. Part of the salt recrystallizes as salt lenses within the underlying muddy sediments. Gypsiferous mud is the most common sediment in such areas. The bedding is rather irregular and wavy in nature.

Sediments of inland sebkhas are more or less parallel-bedded silty, and clayey sediments, alternating with thin sandy layers, gypsum layers, and gypsiferous clays (Fig. 310). However, bedding is better developed in desert lake sediments than in inland sebkha sediments.

Glennie (1970) points out that there are two important processes bringing detrital sediments to inland sebkhas:

Fig. 310. Bedding structures of an inland sebkha. Sand, clay, and salt layers are interlayered. (After Glennie 1970)

1. Sediment brought by the inflowing wadis and settling down from suspension
2. Wind-blown sediment captured by the formation of adhesion ripples and adhesion warts

Glennie (1970) points out that adhesion ripples are extremely important in the deposition of inland sebkhas (Fig. 311). In favorable conditions adhesion ripples up to 30 or 40 cm in length are formed. These features are preserved because of an accumulation of wind-blown sand and deposition from nonscouring water. Irregular wavy bedding and small diapir-like structures, so commonly found in inland sebkha deposits, are attributed to adhesion ripples (cf. Glennie 1970). Sometimes desert lake and inland sebkha sediments may show well-developed, abundant wave ripples. Current ripples are extremely rare.

Fig. 311. Irregular bedding surfaces in an inland sebkha deposit, formed as a result of adhesion ripples. (After Glennie 1970)

Another important feature in areas of inland sebkhas is the common development of sand dykes (Oomkens 1966; Glennie 1970). In inland sebkhas a salt layer is produced as a result of rapid evaporation, while soft water-saturated muddy sediment is still present below the salt layer. The upper surface is broken into polygons, which can be injected from below by soft mud. These polygons may also be filled by wind-blown sand from above.

Wadi Deposits

Wadis are the streams in the desert environment; they are dry most of the time except just after a rain. As mentioned earlier, the desert environment is characterized by sporadic but heavy rains. Rainfall is somewhat higher near the hills, so that wadis are better developed near the hilly terrains of a desert environment.

Wadis are characterized by sporadic and abrupt fluvial activity and by a rather low water/sediment ratio. The deposition is very rapid because of a sudden loss of velocity and absorption of water underground. Fluvial activity of the desert environment is characteristically flash floods. Williams (1971) describes characteristic features of flash flood deposits of ephemeral streams from central Australia. The major bedforms are small ripples and megaripples, producing cross-bedding. Karcz (1972) describes flash flood deposits from southern Israel. Mainly bars, megaripples, and small ripples are present in superposition to make cross-bedded units. Harrow marks and obstacle marks are quite common. Picard and High (1973) give a detailed account of the deposits of ephemeral streams.

Channels of wadis are not of a permanent nature: These channels may be filled by their own sediment or by wind-blown sediment, and in the next season a new channel system is cut into older

Fig. 312. A system of wadi channels of sub-Recent times. Valley have been partly filled by wind-blown sand. Sudan (After Solle 1966)

sediments. Most of the wadis fan out downslope and deposit most of their sediment in fan-shaped bodies. Figure 312 shows a system of subfossil wadis, which have been filled in partly by wind-blown sand. Figure 313 shows wadi fans. Usually many fanshaped bodies join together.

Such sediment fans are known as wadi fans or flood fans (the term "alluvial fans" should be restricted to a fluvial environment). After a rain these wadi fans are cut by streams of braided nature. Sometimes a wadi is filled by flowing sediment slurry, composed of mud and sand with abundant pebbles and boulders. Such deposits are rather similar to mud-flow deposits of alluvial fans (see p. 302). Wadis originating in hilly terrain are important agents bringing sediment to the desert basin. However, sometimes wadi sediments are just the reworked aeolian sediment.

The study of modern deserts demonstrates that wadi systems were better developed during Pleistocene fluvial periods (Solle 1966; Glennie 1970).

Stream channels of wadis are braided in nature; thus, sediments developed in wadi channels are similar to deposits of braided streams. However, special conditions existing in wadi (for example in a wadi water flows only for few hours or few days) cause several peculiarities in the wadi deposits.

The following description of wadi sediments is based mainly on Glennie (1970).

In wadi channels various bed forms, e. g., small ripples, megaripples, plane beds are produced during various flow conditions. This results in the production of small-ripple bedding, megaripple bedding, and horizontal bedding.

Sometimes during certain phases of the flow, sediment transported through wadi has consistency and characteristics similar to those of mud flows. Most of the sediment of wadi accumulates near the downstream limit of the flow in form of a wadi fan.

Deposits within the wadi channels are mostly rather conglomeratic and fanglomeratic in nature (Fig. 314). However, the nature of the sediment is strongly controlled by source rocks and availability of various grain sizes. Sometimes wadi deposits lack pebbles altogether, and are made up of well-sorted sand, showing ripple bedding and horizontal bedding. Pebbles within conglomerates show imbrication structure. Pebbles are often poorly rounded to angular. There is a tendency of fining upward in deposits of a single phase. An important feature in wadi sediments is the presence of clay or mud layers on the top of a sequence which is extensively mud-cracked and shows features like raindrop imprints. Sometimes a mud layer may be several decimeters thick and have rather deep mud cracks. Mud cracks are usually filled by wind-blown, well-sorted sand from the top. Sometimes

Fig. 314. Conglomerate sediments of a wadi deposit. (After Glennie 1970)

Fig. 313. Wadi fan showing a fan-shaped body cut across by river channels. (After Glennie 1970)

mud layers curl up, are broken into small pieces, and are incorporated in the sediments as clay flakes. Occasionally, even curled mud flakes become covered by wind-blown sand and are preserved *in situ* (Fig. 315).

During long dry periods wadi sediments are exposed to wind activity. Deflation may cause the removal of sand and fine-grained sediment from the top, leaving behind pebble lag deposits. On the other hand, wadi may become covered by wind-blown sand, preserving the whole sequence of wadi deposits. Wind-laid sediments are well-sorted sand showing inclined foresets or horizontal bedding. Sometimes the internal structure in wind-deposited sand is not well developed. It is in regions where high humidity causes plant growth, and bedding is interrupted by plant roots.

During the next flow, water is able to remove only part of the wind-laid sand, and new wadi sediments are deposited on top of the wind-laid sand. While new channels are being cut, scour and fill structures are produced. At places, flowing water deposits just a thin pebble layer on top of a thick wind-laid sand.

Thus, within water-laid wadi sediments wind-laid sand is found sandwiched in thin or thick strata. Water-laid and wind-laid sediments alternate with each other (Fig. 316). Figure 317 is a schematic sequence of wadi deposits showing alternating wind-laid and water-laid sediments.

Glennie (1970) describes a structure commonly developed in wadi sediments from wind activity. Inclined foresets are deposited by blowing wind across the wadi banks in the shade of the steep bank slope. These cross-beds possess large pebbles near the upper part of the foresets (Fig. 318). This is because pebbles occasionally topple down from the wadi sides into the wind-laid sand. In usual water-made cross-bedding, larger pebbles are found in the lower parts of the foresets.

The wadi sediments may undergo rather early cementation. If circulating water is rich in $CaCO_3$, a carbonate cement is formed very shortly due to high rate of evaporation. On the other hand, accompanied wind-blown sand is poorly cemented. It may be pointed out here that early cementation and concretions are also found in natural levee and flood-plain deposits of rivers of semi-arid climates.

Aeolian Sand Deposits

Sand-sized sediments are probably the most important sediments in the desert environment, and constitute the major part of the sedimentary sequences in a desert environment. As mentioned be-

Fig. 315. Curled clay flakes partly covered by wind-blown sand. (After Glennie 1970)

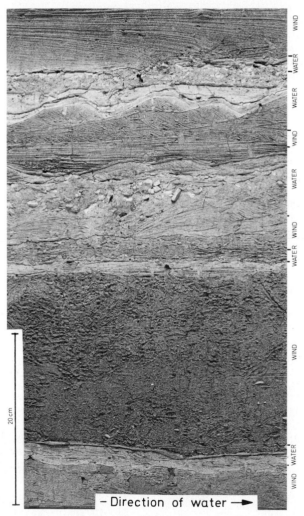

Fig. 316. Interbedded wind-laid and water-deposited sediments of a wadi (After Glennie 1970)

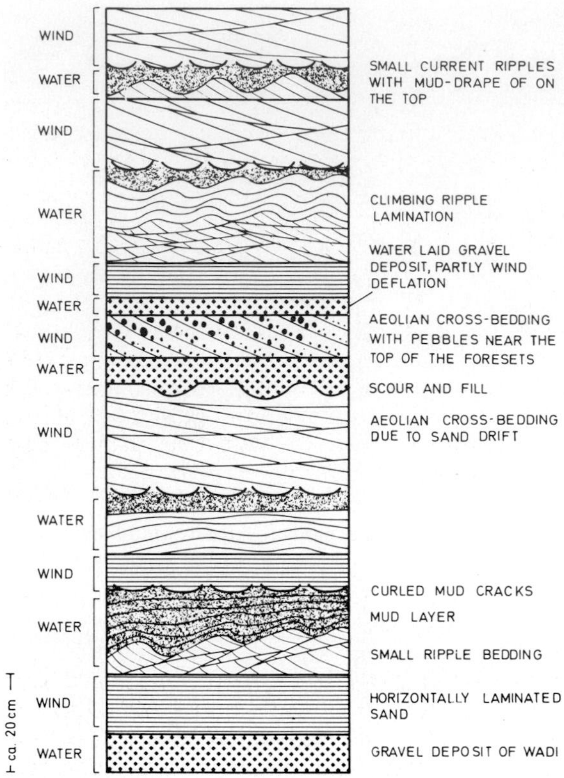

WIND

WATER — SMALL CURRENT RIPPLES WITH MUD-DRAPE OF ON THE TOP

WIND

WATER — CLIMBING RIPPLE LAMINATION

WATER LAID GRAVEL DEPOSIT, PARTLY WIND DEFLATION

WIND

WATER — AEOLIAN CROSS-BEDDING WITH PEBBLES NEAR THE TOP OF THE FORESETS

WIND

WATER — SCOUR AND FILL

AEOLIAN CROSS-BEDDING DUE TO SAND DRIFT

WIND

WATER

WIND — CURLED MUD CRACKS

MUD LAYER

WATER — SMALL RIPPLE BEDDING

WIND — HORIZONTALLY LAMINATED SAND

WATER — GRAVEL DEPOSIT OF WADI

ca. 20 cm

Fig. 317. Schematic diagram showing a sequence of wadi deposits with alternating wind-and water-laid sediments. Based mainly on data of Glennie (1970)

GENERAL
DEFLATION SURFACE →

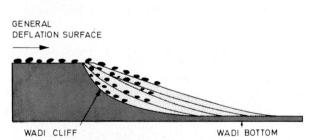

WADI CLIFF   WADI BOTTOM

Fig. 318. Scheme showing the deposition of inclined foresets along a wadi bank in the form of sand drift. Larger pebbles are more common near the upper part of the foresets

fore, wind is a very effective sorting agent. If a variety of grain sizes are available, sand-sized sediment is removed by the process of deflation and is accumulated in adjoining regions in various forms.

There are various factors which control the process and form of sand accumulation, e. g., source and supply of sand, direction and strength of wind, nature of the surface on which sand is drifted. Bagnold (1954a) recognizes the following largescale sand forms produced in the wind regime:

sand shadows and sand drifts, true dunes, whale backs, gozes and sand sheets.

Wilson (1972) made a systematic study of various bedforms in a wind regime which can be arranged into four groups, i. e., draas, dunes, aerodynamic ripples and impact ripples which occur in a hierarchical order. Each of them is subdivided into transverse and longitudinal elements (Table 17). Draas are very large bedforms often covered by sand dunes. Some of the draas are described as star dunes or composite dunes. Impact ripples refer to wind sand ripples and wind granule ripples. Aerodynamic ripples make a distinct group showing long continuous forms with low height/length ratios. Mostly longitudinal patterns dominate, but transverse component may occur. Wilson (1973) gives a good account of ergs – large sandy areas (sand seas) and the distribution of bedforms in them. According to Wilson (1972) aeolian bedforms arise spontaneously by a two-way interaction between surface form and airflow, involving piecemeal material transfer between them. These bedforms reach a dynamic equilibrium, and each of them occurs simultaneously as an independent form. The transverse forms migrate downwind, each type at different speed.

Many organisms are especially adapted for living in the aeolian sand areas and make several different types of burrow. Ahlbrandt et al. (1978) point out that bioturbation structures are very common in aeolian dune fields of different climatic zones. They are especially preserved in somewhat cohesive aeolian sand or reinforced by organisms.

Table 17. Bedforms in the wind regime. (After Wilson 1972)

Order	Wavelength	Height	Orientation	Possible origin	Suggested name
1st	300–5500 m	20–450 m	Longitudinal or transverse	Primary aerodynamic instability	Draas
2nd	3–600 m	0.1–100 m	Longitudinal or transverse	Primary aerodynamic instability	Dunes
3rd	15–250 cm	0.2–5 cm	Longitudinal or transverse	Primary aerodynamic instability	Aerodynamic ripples
4th	0.5–2,000 cm	0.05–100 cm	Transverse	Impact mechanism	Impact ripples
	1–3,000 cm	0.05–100 cm	Longitudinal	Secondary Taylor-Görtler vortices	Secondary ripple sinuosity

We shall describe the large-scale sand forms of the desert environment under following headings:

1. sand drifts and sand shadows
2. gozes
3. sand sheets
4. sand dunes

Sand Drifts and Sand Shadows

Sand drifts are sand accumulations caused by some fixed obstruction in the path of a sand-laden wind. Such obstructions may be small bushes, rocks or boulders, small cliffs, or banks of wadis (Fig. 319). Bagnold (1954a) shows that the sand is deposited as the wind velocity is checked by an obstacle in the path (sand shadow). Usually sand is deposited behind such obstacles. Sand also accumulates in the lee of a gap between two obstacles (sand drift). We prefer to use term "sand drift," meaning any kind of sand deposits associated with obstacles in the way of a sand-laden wind. Most important are the sand drifts formed whenever a sand-bearing wind sweeps over the brink of a sudden fall in the ground, e. g., wadi bank. Sand drifts are tongue-shaped bodies, and show well-developed foreset laminae (Fig. 320).

Gozes

According to Bagnold (1954a), gozes are gentle large-scale undulatory sand surfaces associated with sparse desert vegetation. Thus, gozes are developed in areas having sufficient rainfall to support at least intermittent vegetation.

A sand surface with growing grass serves as a "sediment trap," and oncoming sand is stored between the growing grass. Such areas are areas of continuous deposition, and vast amounts of sand accumulate in form of gentle undulations. However, the presence of vegetation inhibits, the development of dunes with well-developed lee sides. Locally, small lee faces can be formed, producing cross-bedding.

Sand Sheets

Sand sheets are usually very large areas of desert country, more or less flat in nature. Slight undulations or small dune-like features may be present. The surface rarely shows such features as wind sand ripples or wind granule ripples. However, during storms sand strips commonly develop (cf.

Fig. 319. Deposition of sand behind an obstacle, making a sand drift deposit. (After Gripp 1968)

Fig. 320. Scheme showing cross-bedding developed in sand drift deposits

Bagnold 1954a). The surface of sand sheets is sprinkled with coarser sediments–pebbles. Internally a sand sheet is made up of horizontally bedded sand layers, separated by single layers of pebbles. Such evenly laminated sand (horizontally bedded sand) is also a common type of bedding in interdune areas.

Bagnold (1954a) believes that the presence of a pebble layer on sand sheets is of prime importance in the genesis and preservation of laminated sand of sand sheet areas. At the same time sand strips seem to play an important role in the accumulation of sand. A strong sand-laden wind blowing over a uniform rough surface has a transverse instability, so that sand tends to be deposited in longitudinal strips–sand strips (Bagnold 1954a). Narrow longitudinal strips of sand 1 to 3 m wide, and 1 to 2 cm thick are deposited parallel to each other. Sand strips may be several hundred meters in length, and the heads of new sand strips begin in the spaces between the trails of the others (Fig. 321).

There are vast areas of sand sheets known which are devoid of any kind of pebbles. Such sand sheets are made up of well-sorted aeolian sand with well-developed horizontally laminated sand. A combination of rapid sedimentation, high wind velocities, and fairly uniform grain size of the sand cause deposition of sheet sand with an abundantly developed evenly laminated sand bedding (cf. Bagnold 1954a; Glennie 1970).

Sand Dunes

Out of all the aeolian sand deposits, sand dunes are the most impressive and important features of a desert environment. A sand dune may be defined as a hill of sand deposited by wind, which rises to a single summit and possess a slip face. Sand dunes may exist single, but normally they occur in colonies. Sand dunes are found in the flat areas of a desert; such a surface is sometimes covered with coarse-grained sediments. Sand dunes develop wherever a sand-laden wind deposits sand as a random patch. This patch slowly grows in height as a mound, and finally a slip face is formed. Bagnold (1954a) observed that the minimum height of a slip face in a wind regime is about 30 cm. Now, the sand mound migrates forward as a result of the advance of the slip face. Despite their migration, sand dunes are capable of retaining their shape as long as the wind conditions remain the same.

An exhaustive account of the physical conditions and processes in sand dunes is given by Bagnold (1954a). Sand dunes extensively develop also in coastal marine, and river environments (see pp. 311 and 350).

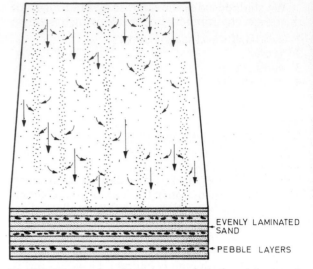

Fig. 321. Scheme of sand strips on a surface of sand sheet, and the type of bedding in the sand-sheet deposition. Hypothetical wind circulation along the sand strips is also shown. (Based on Bagnold 1954a)

Depending upon wind conditions, sand type, and supply of sand, dunes may acquire varied size and shapes. Different authors have attempted to classify sand dunes (see Melton 1940; Bagnold 1954a; Cooper 1958, 1967; Holm 1960; McKee 1966b and many others).

Sand dunes may be oriented perpendicular to the prevalent wind, e. g., barchan dunes, transverse dunes; parallel to the prevalent wind, e. g., seif dunes; or they may acquire complex forms, e. g., dome-shaped dunes, star-shaped dunes.

Following are the more important sand dune types (McKee 1966b):

1. barchan dunes
2. transverse dunes
3. parabolic dunes
4. seif dunes
5. star dunes
6. dome-shaped dunes
7. reversing dunes.

Out of all these forms, barchan dunes and seif dunes are the most abundant and important ones.

Internal Structure of Sand Dunes

Various types of sand dunes have several features in common. Before going into detailed descriptions of various types of sand dunes, an account of features common to all will be given.

The most common characteristic feature of a sand dune is the presence of a steep slip face and a gentle dipping windward side. The height of the dunes depends on the strength of the wind and the grain size. The rate of deposition is zero at the summit of a dune, reaches a maximum on the lee slope, and thereafter falls to zero again at the far end of the lee slope (Bagnold 1954a). This fact results in the faster advance of the upper part of the slip face than the lower part. This causes an oversteepening of the slip face. If the maximum angle of repose (34°) is exceeded, the mass of the sand shears, an avalanche takes place, and a slip face is formed. This process is repeated every time maximum angle of repose is exceeded. On the windward slope sand grains move upward by the processes of saltation and surface creep and are deposited on the upper part of slip face, from where they avalanche. This process also causes forward migration of sand dunes. The avalanche on the slip face of a sand dune affects only a part of the face at one time and a tongue-shaped layer is formed; however, all the parts of the slip face are affected at different times. Superimposition of successive avalanche tongues would produce inclined laminae. Development of avalanche tongues on the lee face is a characteristic feature of aeolian dunes, and individual avalanche tongues possess a concave top at the point of start of an avalanche and rather straight margins, extending along the maximum slope direction. The process of avalanching and its characteristics are discussed by Gripp (1961). However, production of typical aeolian foreset laminae with large sweeping surfaces is probably achieved by reworking of the slip face by oblique or longitudinal winds (Sharp 1966). There is no fixed eddy on the lee side of sand dunes (cf. Sharp 1966).

Bagnold (1954a) recognized that sand dunes are composed of two types of sand beds:
1. accretion deposits
2. enchroachment deposits or avalanche deposits.

On the windward gentle slopes sand grains move by traction and saltation. If more sand grains are retained than ejected by saltating sand grains, a deposit of well-defined laminae is produced. Laminae are rather thin, more or less horizontal, or with gentle dips in the upwind direction (maximum 15°). Sand is rather firmly packed.

Enchroachment deposits are foreset laminae of avalanching sand on the slip face. Laminae are rather thick, and dip 25 to 34°. They constitute the major part of a dune body. The sand is loosely packed.

These two types of deposits can be readily recognized and have been described by various authors (McKee and Tibbitts 1964; Sharp 1966; and Inman et al. 1966). However, we prefer to use descriptive terms, i. e., horizontal bedding (laminated sand) for accretion deposits, and cross-bedding for enchroachment deposits.

Hunter (1977a) studied coastal dunes and found that in dune fields four types of area can be observed, namely surfaces covered by ripples, smooth surfaces except for the grain roughness, slipfaces with avalanches, and floor across which the dune is moving (interdune areas) (Fig. 322).

Hunter (1977a) recognizes six types of bedding structure in small aeolian sand dunes (Table 18). Rippleform laminae, ripple-foreset cross-laminae, and climbing translatent strata can be grouped together as aeolian climbing ripple structure (Fig. 323). In a small aeolian sand dune high-angled cross-bedding of avalanche type may not be the most abundant bedding structure. Avalanche deposits are present only in the main part of the leeface (Figs. 324 and 325). Grainfall deposits are present as bottomset, and make part of the leeface lamination (grainfall deposition is related to flow separation, and may be deposited both on the leeface, intermittent to avalanches and bottomset) (Fig. 326). Topset deposits and the stoss side of the sand dune are made up of mostly aeolian climbing ripple structure. In larger sand dunes avalanching on the leeface seems to be most abundant. Moreover, in ancient sediments differentiation between avalanche laminae and grainfall laminae of the leeface may not be possible. Gripp (1961) also emphasized the role of grainfall deposition and intermittent avalanching in the leeside sedimentation of barchan coastal dunes.

Horizontal Bedding. Horizontal bedding of sand dunes is composed of almost horizontal or low-angled laminae of sand. The angle of dip is usually 3 to 10°. Horizontal bedding is commonly found on the windward slopes, flanks, and near the summit of a sand dune, as well as in the interdune areas. It is also common in deposits of sheet sands (see p. 221).

Individual laminae are rather thin, a fraction of 1 mm to 3 or 4 mm, and composed of well-sorted

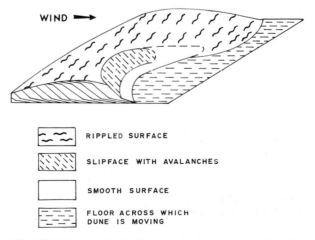

WIND ➡

	RIPPLED SURFACE
	SLIPFACE WITH AVALANCHES
	SMOOTH SURFACE
	FLOOR ACROSS WHICH DUNE IS MOVING

Fig. 322. Schematic block diagram of a dune area showing distribution of different features. (After Hunter 1977a)

fine sand (Fig. 327). Occasionally layers a few centimeters thick of somewhat coarser sand may be found interlayered (Sharp 1966). Lamination is very well developed; sometimes heavy mineral-rich laminae alternate with heavy mineral-poor laminae.

Hunter (1977a) discusses that horizontal bedding in wind regime can develop either as plane bed lamination, related to high wind velocities, or as grainfall lamination, related to the zones of flow separation on horizontal surfaces. In the latter, lamination is poorly developed.

Cross-Bedding. Cross-bedded units of sand dunes are made up of avalanche laminae of the slip face. Foreset laminae are steeply inclined, and the angle of dip usually varies between 25 and 34°. Near the base of slip face foreset laminae tend to flatten out and resemble in nature to horizontal laminae. Individual foreset laminae are rather thick 2 to 5 cm,

and are less well defined than the horizontal laminae (cf. Inman et al. 1966; Sharp 1966). McKee (1966b) gives a detailed account of cross bedding in sand dunes: The bounding surfaces of the cross-bedded units are erosional and planar. Thus, most of the cross-bedded units are of the planar type. Tabular shapes are more common than the wedge-shaped units. Wedge-shaped cross-bedded units are produced if there are changes in wind direction. The bounding surfaces are nearly horizontal or dip in an upwind direction at low angles in the upward part of a dune. Near the slip face bounding surfaces are rather steep (15 to 25°), dipping in a downwind direction.

The Lower Bounding Surfaces. The lower bounding surfaces of sand dune deposits are usually undulating surfaces of older sand dunes. However, sometimes truncation surfaces of successive units are exceptionally smooth and horizontal, especial-

Table 18. Major types of bedding structures developed in aeolian sand dunes. (After Hunter 1977 a)

Depositional process	Character of depositional surface	Type of stratification	Dip angle	Thickness of strata Sharpness of contacts	Segregation of grain types Size grading	Packing	Form of strata
Tractional deposition	Rippled	Subcritically climbing translatent stratification	Stratification: low (typically 0–20°, maximum ~ 30°) Depositional surface: similarly low	Thin (typically 1–10 mm, maximum ~ 5 cm) Sharp, erosional	Distinct Inverse	Close	Tabular, planar
		Supercritically climbing translatent stratification	Stratification: variable (0–90°) Depositional surface: intermed. (10–25°)	Intermediate (typically 5–15 mm) Gradational	Distinct Inverse except in contact zones	Close	Tabular, commonly curved
		Ripple-foreset cross-lamination	Relative to translatent stratification: intermed. (5–20°)	Indiviudal laminae: Thin (typically 1–3 mm)	Individual laminae and sets of laminae:	Close	Tabular, concave-up or sigmoidal
		Rippleform lamination	Generalized: intermediate (typically 10–25°)	Sharp or gradational, non-erosional	Indistinct Normal and inverse,	Close	Very tabular, wavy
	Smooth	Planebed lamination	Low (typically 0–15°, max.?)	Sets of laminae: Intermediate (typically 1–10 cm) Sharp or gradational,	neither greatly predominating	Close	Very tabular, planar
Largely grainfall deposition	Smooth	Grainfall lamination	Intermediate (typically 20–30°, min. 0° max. ~ 40°)	nonerosional		Intermediate	Very tabular, follows pre-existent topography
Grainflow deposition	Marked by avalanches	Sandflow cross-stratification	High (angle of repose) (typically 28–34°)	Thick (typically 2–5 cm) Sharp, erosional or nonerosional	Distinct to indistinct Inverse except near toe	Open	Cone-shaped, tongue-shaped, or roughly tabular

ly in ancient sediments. Stokes (1968) discusses the genesis of such surfaces (Fig. 328). He believes that in areas where groundwater is not too deep, e. g., sand dunes migrating over an inland sebkha, water rises in the sand dunes. The cohesive action of water holds wet sand grains together, and the dry sand above this surface is removed, producing a rather smooth, horizontal surface. If new sand dunes migrate over this surface and the process is repeated several times, several smooth horizontal surfaces are produced. However, not all smooth horizontal surfaces originated in this way. As the excavations by McKee (1966b) in sand dunes of White Sand dunes have shown, horizontal to subhorizontal

bounding surfaces may be present without any groundwater effect.

McKee and Moiola (1975) suggest that the subparallel bounding surfaces are produced by migrating dunes and interdunes.

Brookfield (1977) reviews the origin of bounding surfaces in aeolian sands and suggests that aeolian bounding surfaces of differing inclination and extent can be developed by the migration of aeolian bedforms of differing hierarchical order. He distinguishes three orders of bounding surface. The first-order surfaces are extensive flat-lying bedding produced by the passage of largest aeolian bed forms – draas across the area. Second-order

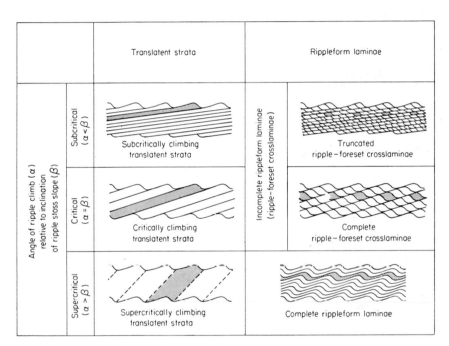

Fig. 323. Types of bedding structures produced by aeolian climbing-ripple structures at various angles of climb. (After Hunter 1977a)

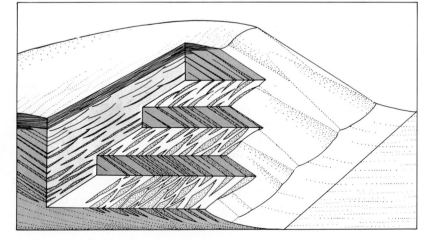

Fig. 324. Schematic block diagram showing distribution bedding in a small dune. Sandflow cross-strata heavily stippled. Grainfall laminae unpatterned (foreset cross-strata) or lightly stippled (bottom set strata). Subcritically climbing translatent strata (topset strata) thin-lined. (After Hunter 1977)

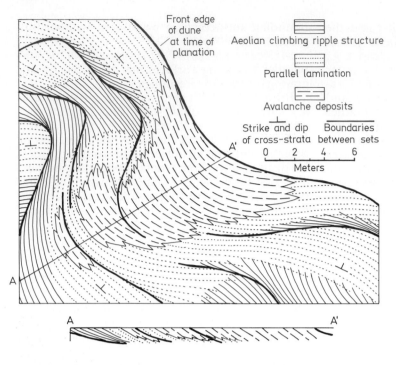

Front edge of dune at time of planation

Aeolian climbing ripple structure

Parallel lamination

Avalanche deposits

Strike and dip of cross-strata

Boundaries between sets

0 2 4 6

Meters

Fig. 325. Schematic diagram showing distribution of various types of bedding in an aeolian dune. Based on exposure on Padre Island, Texas. (After Hunter 1977)

Fig. 326. Horizontal and vertical exposures of a small sand dune showing avalanche deposits (sand flow cross-strata) = light in tone, abuting on and intertonguing with the bottomset and lower foreset deposits of grainfall type (dark in tone). (After Hunter 1977)

surfaces separate two sets of cross-beds and are produced by migrating dunes of various types. The third-order surfaces are bundles of laminae within coset of cross-bed produced due to local fluctuations in the wind velocity and direction and changes in the dune configuration. Hunter (1977a) also reports sets separated by gently dipping erosional bounding surfaces, related to the fluctuations in the wind direction.

Other Bedding Types. Trough-shaped, cross-bedded units are rather uncommon in sand dunes. Trough-shaped, cross-bedding is produced usually in the process of scour and fill, due to change in wind conditions. Brookfield (1977) discusses that trough-shaped cross-bedding in aeolian sand can be produced by migration and deposition in the hollows of sinuous ridges transverse to the wind. These aklé bedforms are common in some deserts (Cooke and Warren 1973).

Fig. 327. Horizontal bedding in dune blow out sand on the foreshore (arrow). The upper-most layer is laminated sand resulting from water. (Wunderlich 1972)

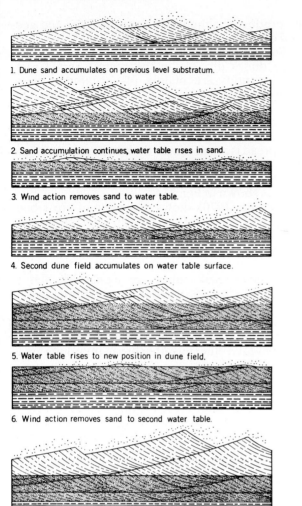

1. Dune sand accumulates on previous level substratum.

2. Sand accumulation continues, water table rises in sand.

3. Wind action removes sand to water table.

4. Second dune field accumulates on water table surface.

5. Water table rises to new position in dune field.

6. Wind action removes sand to second water table.

7. Third dune field accumulates, etc.

Fig. 328. Scheme showing horizontal truncation surfaces in a sand dune deposit. (After Stokes 1968)

Individual cross-bedded units in sand dunes are rather thick. They are never thinner than a few decimeters and may be several meters usually 1 to 2 m in thickness.

McKee (1966b), in his investigations of White Sand dunes, found that in vertical sections thickness of cross-bedded units decreases from bottom to top. Foreset laminae within sets of a vertical sequence tend to flatten upward, especially on the upwind part of the dune.

Another important feature of sand dunes is the presence of penecontemporaneous deformation features. A detailed account of such features is given by Gripp (1961), Bigarella et al. (1969), and McKee et al. (1971). Usually these structures are contorted bedding, various types of rather small-scale folding and faulting, e. g., gentle warping,

overturned folds, stretched laminae, etc. They are mostly associated with slip face avalanching sand. Sometimes even overturned recumbent fold-like features of moderate size are found (Glennie 1970) (Figs. 329, 330 and 331). McKee et al. (1971) suggest that the amount of moisture present in sand strongly controls the type of deformation structure to be produced.

To summarize, sand dunes are mainly composed of planar cross-bedded units usually 1 to 2 m in thickness. Foresets of cross-bedded units are rather steep 25 to 34°. Thickness of foreset lam-

Fig. 329. Deformation structures in a slip face deposit. (After McKee et al. 1971)

Fig. 330. Deformation structures in a sand dune deposit. Note the small-scale folds. (After McKee et al. 1971)

Fig. 331. Deformation structures in coastal sand dunes. (After Bigarella et al. 1969)

be produced, although the general form and general direction of movement is retained.

Foreset laminae of cross-bedded units of barchan dunes dip 34° in the central part. Cross-bedded units are mainly planar-tabular types. However, near the horns of barchans dips are much less than 34°, and run almost 90° to the direction of the prevalent wind. A spread of less than 150° is normally encountered. However, usually the spread is much less. In the barchan dunes of the White Sand dunes the spread of the dip is only 60° (McKee 1966b). A detailed structure of a barchan dune is shown in Fig. 334.

Glennie (1970) shows that plots of cross-bedding on the polar net can be helpful in the identification of the type of dune, especially for barchan dunes and seif dunes (Fig. 335).

Seif Dunes. Seif dunes are elongated, almost straight sand ridges, with long axes oriented parallel to the prevailing wind direction. Sand ridges are continuous but serrated. Several seif dunes occur together in a series of long parallel ridges, separated from each other by rather broad interdune areas (Fig. 336).

There are various views regarding the genesis of seif dunes. Bagnold (1954a) believes that seif dunes are produced when strong winds blow from a quarter other than that of the general drift of sand caused by the more persistent gentle winds. McKee and Tibbitts (1964) and McKee (1966b) suggested that seif dunes are produced in the vector of two converging wind directions blowing

inae ranges from 1 to 5 cm. Associated with cross-bedded units are packets of horizontal bedding. Small-scale penecontemporaneous deformation features are common. Shape, size, and character of sand dunes, climate, and wind conditions control the final characteristics of sand dune deposits.

Sand Dune Types

Barchan Dunes. Barchan dunes are crescent-shaped sand mounds which occur as isolated bodies sporadically, in chains, or in colonies (Fig. 332), where individual barchan dunes tend to coalesce to make complex forms (Fig. 333).

A barchan dune is formed by a unidirectional wind. The barchan dune migrates by avalanching of sand on the slip face. The extremities (horn) of a barchan dune extend forward downwind because the horns migrate more rapidly than the main body.

Simple barchans may be made complex if many sand dunes coalesce. In regions where during certain times of the year wind blows from other directions than the prevailing wind, small slip faces may

Fig. 332. Barchan dunes. (After McKee 1966b)

Fig. 333. Complex barchan dunes, Sudan
(After Solle 1966)

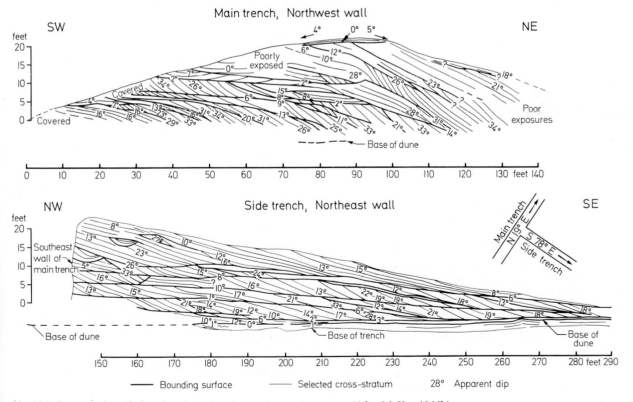

Fig. 334. Cross-section of a barchan dune showing the internal structure. (After McKee 1966b)

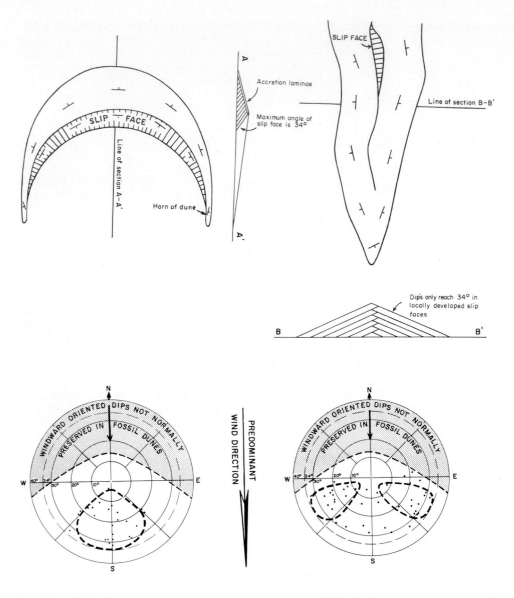

Fig. 335. Cross-bedding directions in a barchan dune and a seif dune in relation to wind direction. (After Glennie 1970)

Barchan dunes

1. "Horns" of dune directed down-wind.

2. Bulk of dune foresets dip down-wind at angles of up to 34°.

3. A small proportion of low-angle dips will be oriented at almost 90° to the predominant wind direction but with a component in the direction of the wind. They represent the bedding on the horns of the dune and are very similar to that seen in a seif dune.

4. Low-angle dips oriented up-wind may be preserved when a barchan climbs onto the windward accretion slope of a more slowly moving dune of similar type.

Seif dunes

1. Axis of dune parallel to the dominant wind direction.

2. The bulk of the dune will have dips of up to 34° oriented almost at right angles to the predominant wind direction but with a component in the direction of the wind.

3. Seif dune bedding may be complicated by the migration of small local barchans over their surface.

4. The frictional effect of the bulk of a seif dune is thought to give the wind a 'corkscrew' rotational component directed towards the dune. Localised 34° laterally dipping slip-faces may result.

5. The slight sinuosity of the dune may result in the preservation of low-angle dips which have an up-wind component.

from two quarters, about 90° apart. Glennie (1970) believes that a more important factor for generation of seif dunes is the existence of a strong wind of uniform direction. He adds further, that in the region of seif dunes, pressure gradients exist between the axes of the interdune areas and the crests of the dunes–caused by the resistance of the wind of the dunes. This results in the formation of wind cells in the interdune areas, and linear sand ridges are made stable.

Glennie (1970) adds that other conditions being equal, barchan dunes develop at lower wind velocities, and seif dunes are found when wind velocities are higher. The higher the wind velocity, the larger the seif dunes and the greater the interdune spacing. He believes that during Pleistocene glaciation, because of stronger winds, seif dunes were produced in abundance. Today, some of the seif dunes are undergoing modification to barchan dunes, etc., because wind velocities are not strong enough to maintain them.

Folk (1971) made a detailed study of the longitudinal dunes in the Simpson desert, Australia, and suggests that the longitudinal dunes are formed by helicoidal windflow.

McKee and Tibbitts (1964) describe in detail the internal structure of a seif dune (Figs. 337 and 338). Sand is deposited alternately on the opposite sides of the sand dunes. Cross-bedding dip normal to the elongation of sand ridge and therefore two maxima of high-angled foresets, almost 180° apart, are found (see also Fig. 335). Locally, some low-angled horizontal bedding is present, especially in the lower part of a dune.

Associated with seif dunes are large-scale features known as whalebacks (Bagnold 1954a). They are platforms of rather coarse-grained sediments left behind by the passage of a series of seif dunes along the same path. Figure 339 shows the schematic internal structure in a whaleback. The platform and the sides are composed of horizontal bedded sediments, below which cross-bedded seif dune sediments are present.

Transverse Dunes. Transverse dunes are elongated, almost-straight sand ridges oriented at right angles to the predominant wind direction. Ridges are regularly spaced, separated by broad interdune areas, which are developed sometimes as inland sebkhas. Bagnold (1954a) considers them to be unstable forms, and suggests that transverse ridges break up into barchan dunes or seif dunes. However, Cooper (1958), McKee (1966b), and Glennie (1970) consider transverse dunes to be stable forms, and they are rather common forms in several deserts. Glennie (1970) believes that transverse dunes originate in areas of inland sebkhas; the damp sebkha surface may inhibit the growth of barchan-like horns. However, later interdune sebkhas disappear and eventually a sand sea with transverse dunes is produced (Fig. 340).

According to McKee (1966b), cross-bedded units are mostly the planar-tabular type. The foreset laminae are relatively long and even in nature, and rather high angled (30 to 34°). The result is that in sections cut at right angles to the wind direction, the traces of foreset beds are nearly horizontal (1 to 6°). In higher transverse dunes horizontal bedding with gentle dips (2 to 5°) in an upwind direction on the windward slopes is well developed (Fig. 341).

The spread of the dip of foreset laminae would be probably the least of all types of sand dunes– one well-developed maxima in the direction of the prevalent wind.

Fig. 336. Aerial view of seif dunes. (After McKee and Tibbitts 1964)

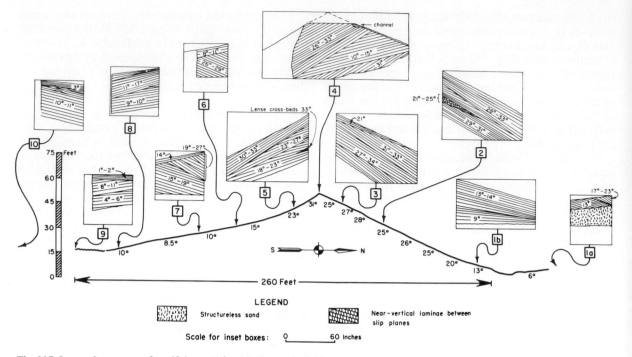

Fig. 337. Internal structure of a seif dune. (After McKee and Tibbitts 1964)

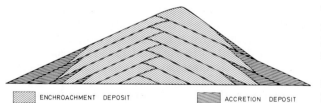

Fig. 338. Schematic internal structure of a seif dune. The central part is made up of opposed dipping high-angled cross-bedded units (encroachment deposit), and the lower parts (i. e., flanks) are made up of low-angled to horizontal-bedded accretion deposits. (Redrawn after Bagnold 1954a)

Parabolic Dunes. Parabolic dunes are U-shaped sand ridges with the concave side toward the wind, associated with blow-up features of sand dunes. The middle part of parabolic dunes moves forward with respect to the sides. McKee (1966b) believes that the arms of parabolic dunes are anchored by vegetation, causing slow movement of the arms; at the same time central parts move faster forward.

The most characteristic feature of parabolic dunes is that the slip face is concave-downward (Fig. 342). Figure 343 shows aerial view of parabolic dunes.

Cross-bedding is the most abundant internal structure (Fig. 344). According to McKee (1966b), foreset laminae of parabolic dunes are low-angled as compared to other dunes. High-angled foreset laminae are only few. Plant growth along bedding planes forms organic accumulations and root tubes in the sand dunes. However, the most conspicuous and characteristic of parabolic dunes is that the foreset laminae are usually concave-downward. This is a result of the concave-downward shape of slip face and the presence of vegetation (McKee 1966b; Bigarella 1969). The spread of the dip of foreset laminae is rather large. For parabolic sand dunes of the White Sand dunes it is about 200° (McKee 1966b).

Ahlbrandt (1975) found that distinctive convex-upward cross-bedded units are common in the parabolic dunes. He suggests that low-angle dipping

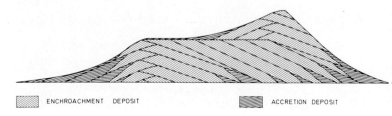

Fig. 339. Hypothetical internal structure of whalebacks associated with seif dunes. Note the arrangement of low-angled accretion deposits and high-angled encroachment deposits. (Redrawn after Bagnold 1954a)

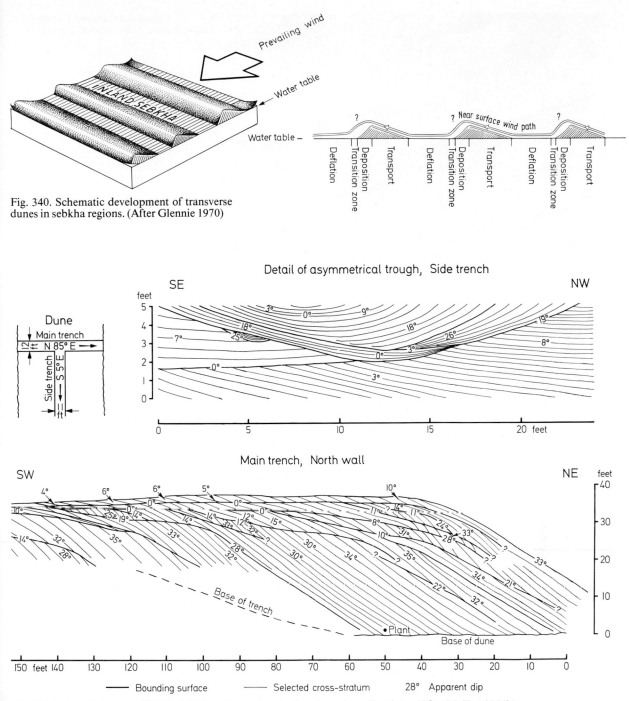

Fig. 340. Schematic development of transverse dunes in sebkha regions. (After Glennie 1970)

Fig. 341. Internal structure of transverse dunes as seen in sections in various directions. (After McKee 1966b)

(< 10° dip) cross-bedding in the wind direction is quite common in these sand-dune deposits.

Dome-shaped Dunes. Dome-shaped dunes are low circular sand ridges. McKee (1966b) believes that dome-shaped dunes develop where dune height is checked by an unobstructed strong wind.

Dome-shaped dunes lack a well-developed downwind steep slip face.

The most characteristic internal structure of dome-shaped dunes is foreset laminae which are low-angled, even in the downwind part of the dune. Dips of 20 to 28° are more common. Cross-bedded units with foreset laminae dipping more

Fig. 342. Concave-downward foresets of a parabolic dune. (After McKee 1966b)

Fig. 343. Aerial view of parabolic dunes. (After McKee 1966b)

than 30° are few. Near the base of the windward side of the dunes, laminae possess a dip of only 8 to 14°. The spread of the dip of the foreset laminae is about 150° (Fig. 345).

Star Dunes. Star dunes are sand ridges with a high central point from which three or more ridges radiate in various directions (Figs. 346 and 347). Holm (1960) and McKee (1966b) believe that star dunes are formed as a result of wind blowing from several directions. Generally every ridge possesses a well-developed slip face which is active at different times.

McKee (1966b) discusses the internal structure of star dunes. Foreset laminae dip steeply and show three or more maxima of steeply dipping foreset laminae. This is a result of changing wind directions causing changing slip faces. The internal structure is rather complicated and is controlled partly by the early history of the sand dune.

Reversing Dunes. Reversing dunes are sand ridges of rather unusual heights but with little migration. Seasonal changes in the direction of the prevailing wind cause it to move in nearly opposite directions alternately. They are usually barchan-shaped, or sometimes even transverse. Merk (1960), McKee (1966b), and Glennie (1970) discuss such dunes.

The overall shape of such sand dunes is controlled by one dominant wind direction, and a well-developed slip face is established. However, as a result of opposite winds of temporary duration, small crestal slip faces are produced just opposite to the main slip face.

The internal structure of reversing dunes is very complex. Wedge-shaped cross-bedded units are more abundant than the planar-tabular types. Two maxima of high-angled foreset laminae are found, although one maxima may be more pronounced than the other. Moreover, low-angled horizontal bedding of the windward side are commonly preserved as a result of reversing wind direction. Penecontemporaneous deformation features are more abundant than in the other types of sand dunes (McKee 1966b).

Ahlbrandt and Andrews (1978) studied sand dunes of a cold climate and conclude that they show few characteristics distinct from the warm-climate sand dunes. Because of freezing of moisture and snow, rate of migration of such dunes is very slow. The sediments show heterogenous composition, poor sorting and negative skewness values (mostly aeolian sand in warm climate shows positive skewness). Further, the sediments show isolated, cohesive sand deformation near the top of the sand dune, dissipation structures immediately overlain by tensional and compressional cohesive sand deformation, and pitted bedding plane surfaces beneath the cohesively deformed layers, all related to freezing, thawing, and melting of the snow. Bioturbation structures are also common.

Dikaka. Glennie and Evamy (1968), and Glennie (1970) use the term dikaka to describe the sand dune deposits with abundant plant-root moulds (Fig. 348). They are deposits of stabilized dunes. Stabilization is achieved by desert plants which possess exceptionally well-developed, deep-penetrating root systems. Plant growth tends to destroy the earlier-formed dune bedding. In most cases,

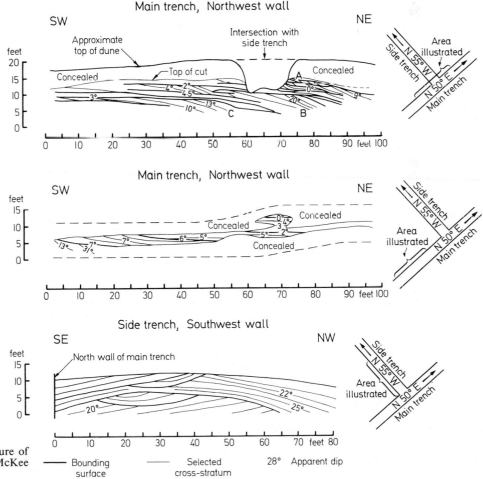

Fig. 344. Internal structure of parabolic dunes. (After McKee 1966b)

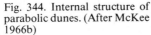

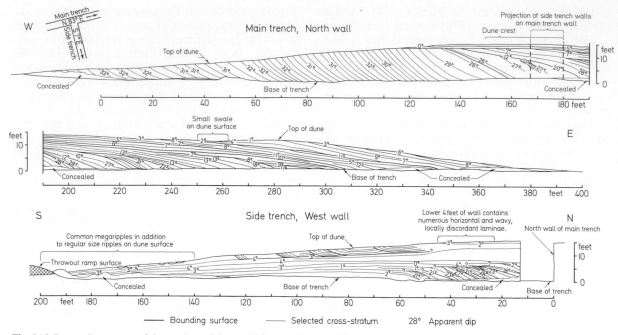

Fig. 345. Internal structure of dome-shaped dunes. (After McKee 1966b)

Fig. 346. Aerial view of star dunes. (After McKee 1966b)

original bedding is still clearly visible, but sometimes roots may be so numerous that original bedding is hard to decipher. Parts occupied by plant roots may show preferential cementation. Dikaka are common near the wadi deposits of a desert environment, or in oases. They indicate the presence of enough nonsaline water to maintain the plant growth.

Interdune Areas

The vast areas of sand seas where sand dunes are abundantly found also have featureless flat areas located between sand dunes–interdune areas. Interdune areas are the areas of active deflation, although some sedimentation does take place. Sometimes interdune areas occupy more surface areas than the associated sand dunes.

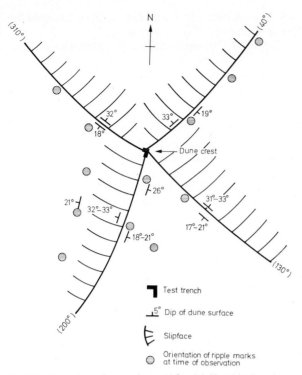

Fig. 347. Plan view of a star dune. (After McKee 1966)

Fig. 348. Dikaka bedding. The bedding is obscured by plant growth. (After Glennie 1970)

Interdune areas are made up mainly of horizontal bedding. Laminae are well-developed and a few millimeters in thickness. Development of lamination is a result of surface creep and saltation movement of sand grains. Deposits of interdune areas may become preserved under the migrating sand dunes. McKee and Tibbitts (1964) found that sorting of interdune sediments is moderate, but poorer than the sorting of the associated sand

dunes. Coarser grains are commonly enriched in interdune sediments.

McKee and Moiola (1975) report that sequence of aeolian sand body, about 8–12-m-thick is present below the present-day White sands dune field, where thick units of clean sand (sand dune deposits) are separated by thin, dark-colored interdune deposits.

Dust or Loess Deposits

As mentioned earlier, silt- and clay-size sediment is winnowed from the desert area, kept in suspension for long periods of time, and blown over long distances. This fine-grained dust is ultimately deposited in adjoining regions, e. g., steppes. In the steppes wind-born dust settles down because of a loss in wind velocity, or it may be washed down from the air by rain and kept in place by the protective grip of grass. Grass keeps on growing higher, and more and more dust is entrapped. Such dust deposits are devoid of any stratification.

Most impressive of all the dust deposits known are the loess deposits deposited mainly during Pleistocene times. Loess is usually nonstratified and commonly unconsolidated sediment composed mostly of silt-size particles, with a minor amount of fine sand and clay particles. Faint lamination may be very infrequently present, interrupted by plant roots. Loess sediments possess high porosity. They may be found interbedded with lacustrine, fluvial, or even glacial sediments. Association with glacial sediments is due to glacial and interglacial stages of an ice age. Loess sediments show a rather uniform size throughout the world. Md ranges between 0.06 and 0.02 mm, and the silt fraction composes 60 to 80 % of the total sediment; the standard deviation ($\sigma\Phi$) is 1 to 3; Phi skewness is mostly slightly positively skewed; maximum grain size is 0.25 mm (Kukal 1970). Grain-size analysis of dust sediments in various areas has confirmed this observation. Data of Swineford and Frye (1945), Laprade (1957), Zeuner (1959) and Game (1964) point to the same result. The maximum grain size of aeolian dust is fine sand. However, its content decreases with increasing distance from the source. The mineralogical composition of loess is extremely varied; however, the main component is always quartz.

Loess deposits occur in sheets covering areas thousands of square kilometers, reaching thicknesses of up to 50 m or more; although sometimes they are only a few decimeters thick. Some terrestial fauna, land molluscs, and mammals may be found. The main source for loess sediments of the Pleistocene age are thought to be extensive outwash plains. Desert basins may also contribute heavily to loess deposits, as well as extensive alluvial regions of semi-arid climates. Grahmann (1932) thinks that loess originating from outwash sediments shows a wider range of grain sizes, i. e., it is poorly sorted, including conspicuous amount of very fine-grained sediment. On the other hand, loess whose source is the desert environment is better sorted and contains a negligible amount of clay sediment.

Extensive loess deposits of the Pleistocene age are found in Central Europe, China, and North America. For a detailed description of Pleistocene loess deposits see Flint (1957) and Woldstedt (1961).

Various aspects of loess are discussed by Bryan (1945), Obruchev (1945), Doeglas (1949), Smalley (1966), Smalley and Vita-Finzi (1968), and Füchtbauer and Müller (1970). Smalley (1975) provides a good collection of papers on the lithology and genesis of loess.

Cegla et al. (1971) and Smalley et al. (1973) give information on the surface textures of the quartz grains in loess. SEM observations reveal that the quartz grains of loess are mostly angular, though a few grains are rounded. Larger grains (20–50 mμ) have adhering material on the surface, fracture features and some abrasion markings are present, carbonate particles grow on the quartz grain surface, and there are signs of particle weathering. Smalley and Leach (1978) provide a review on the loess deposits in the Central Europe.

A large amount of aeolian dust is deposited in oceans. Deep seas have very low rates of sedimentation, and here sometimes aeolian dust may be the most abundant. Radczewski (1937) and Bonatti and Arrhenius (1965) discussed the problem of the contribution of aeolian dust to deep-sea sediment. Windom (1975) further discusses that aeolian dust is transported in huge quantities to the ocean and in some areas it makes a significant part of the oceanic sediments. Sarnthein and Diester-Haass (1977) discuss that aeolian sand from the desert coming to coastal areas is ultimately transported to deeper parts of the ocean as turbidites. Such deposits are usually devoid of any gradation, fine fraction, and mica.

Grain-size Characteristics of Aeolian Sand Deposits

It has been mentioned elsewhere that wind is an effective sorting medium, causing separation into dust, sand, and gravel sediments. Sand-sized aeolian deposits are the most important ones. Mature aeolian sands are marked by their excellent sorting, low content of silt and clay, and sometimes the good roundnes of quartz grains.

Kukal (1970) summarizes the grain-size characteristics of aeolian sands as follows: Median size ranges between 0.15 and 0.25 mm. So (Trask's sorting coefficient)– < 1.25, Phi deviation ($\sigma\Phi$)–0.21 to 0.26. Phi skewness ($\alpha\Phi$)–0.13 to 0.30; skewness is always positive. Ahlbrandt (1975) studied the Killepecker dune field, Wyoming, U.S.A., and found that coarser-grained samples (> 0.31 mm) show positive skewness, while finer-grained samples show negative skewness.

Grain size (Md) is, however, strongly controlled by the provenance and should not be considered a decisive parameter.

Wind is a more effective rounding medium than water. Thus, wind-born sand grains may achieve a high degree of roundness. Wind can produce well-rounded grains of sizes larger than 0.03 mm; for water this minimum boundary is much higher (Kukal 1970).

Folk (1971) discusses that aeolian sand of the Simpson desert is positively skewed, but not well-sorted. The sand of coastal dunes shows better sorting because they derive material from the wave-sorted sand. Folk (1978) describes that quartz grains of the Simpson desert are subangular to angular and do not undergo effective rounding.

Aeolian sands are usually unimodal; however, sometimes they are markedly bimodal. In bimodal aeolian sands two populations of well-sorted sands are present, which differ in size by 2 to 3 Φ. Folk (1968) reports well-sorted bimodal sands from Simpson Desert, Australia. In thin sections such bimodal sands show alternating thin laminae, composed of well-sorted sand differing in median size by 0.1 to 1.0 mm (Glennie 1970).

Folk (1971), Warren (1972) discuss that bimodality is common in desert sediment; however, the sediments in the interdunes and slopes of the dunes mostly show polymodality. Wood (1970) discusses the genesis of polymodality in wind-blown sand from an originally unimodal sand by the process of rectification, meaning stabilization of saltation and surface creep processes during transport. Warren (1972) discusses the genesis of bimodality in aeolian sand due to infiltration of finer particles in the voids during saltation and their protection.

An important characteristic of quartz grains of aeolian origin is well-developed frosting (Glennie 1972). Aeolian-frosted sand grains show a dull surface marked by very small, irregular pits produced by bombarding sand grains during transport (Figs. 221, 222). Sometimes, coarser quartz grains tend to calve. Thus, such grains possess fresh bright fracture surfaces

In a single sand dune Glennie (1970) found that there is fining of sediment upward to the top of a sand dune. At the same time sorting also improves. Near the base of a sand dune poorly sorted coarser sediment of the interdune area is present. This is followed by well-sorted rounded and frosted-sand grains. The sand grains from the top of sand dunes are more angular, and frosting is not so well marked.

Folk (1971) also found that the crests of the dunes are best-sorted, but are coarsest representing the saltation material, while the lower areas are finer-grained and often show polymodality.

Identification of the Desert Environment in Ancient Sediments

Glennie (1970) discusses at length the problem of identification of desert environments in ancient sediments.

The presence of a sequence of red beds in ancient sediments can be taken as a hint of continental deposits, mostly from hot arid or seasonally humid climate. However, not all the ancient desert sediments are red in color, nor did all the red sediments originate in deserts.

Most characteristic deposits of a desert environment are aeolian sand deposits, which possess well-developed cross-bedding with high-angled (30 to 34°) foreset laminae. Locally, horizontal bedding with thin laminae can be common. Bedding of a seif dune is shown in Fig. 349. Sand grains are well rounded and frosted, some with fracture planes. Quartz grains of ancient desert environments are often coated with iron-oxide, giving the sediment a red color. Sorting is extremely good, and silt and clay content is minimal. Grains are usually unimodal; if bimodal, two well-sorted populations are present.

Each dune sequence starts at the same base, with horizontal or low-angled bedded sediments, which are coarser grained, poor- to moderate-sorted, sometimes pebbly. This is followed by well-sorted sand with large-scale cross-bedding showing high-angled foreset laminae, with rather consistent dips. The pattern of palaeo-current may help in the identification of the type of sand dune.

Near the basin margins of a desert sequence hamada and serir deposits with inselberg topography may be present, followed by a sequence of wadi (fluvial) sediments and aeolian sand sediments, alternating in thin or thick units. Wadi sediments and aeolian sand deposits commonly show opposite palaeocurrent directions (Glennie 1970). Higher up mainly dune sand deposits are found, together with local inland sebkha deposits. The contact between conglomeratic beds and sand beds of a desert environment is rather sharp.

Fig. 349. Bedding features of a sand dune as lacquer peel. Dubai, Trucial coast. (After Glennie 1970)

Glennie (1970) gives a series of criterion for wind-deposited and water-deposited sediments of desert environments, and criterion for positive recognition of surface, of subaerial exposure, and of water-laid sediments in the desert.

Wind-deposited Sands

1. Bedding horizontal or usually small- and large-scale cross-bedding, showing constant or multiple orientation.
2. Individual laminae well-sorted, especially in the finer grain sizes; sharp grain size differences between laminae common.
3. Grain size commonly ranges from silt (60 µ) to coarse sand (2 mm). Maximum size for grains transported under the action of wind is in the order of 1 cm, but grains over 5 mm are rare.
4. The larger sand grains (0.5 to 1.0 mm) tend to be well rounded.
5. Clay drapes are very rare.
6. Sands free of clay.

7. Uncemented quartz grains exhibit a frosted surface.
8. Mica generally absent.

Water-deposited Sediments

1. The bedding is that of normal stream-flow sediments, e. g., cross-bedded gravels, small ripples, megaripples, plane-bed, antidunes, etc.
2. Locally, bedding is similar to that of mud flow deposits and flash flood deposits: horizontally laminated and foreset bedding of flood plain deposits with deformation features.
3. Channeling and accretion foresets common in stream flow sediments.
4. Clay laminae or clay drapes are common.
5. Many sands are not well-sorted, being sometimes clayey or pebbly.
6. Sediment commonly cemented by calcite.
7. Sand and silt layers commonly missing from graded conglomerates.
8. Majority of quartz grains (not cemented) exhibit frosted surfaces.
9. Water-laid sediments that are deposited in a desert environment may also possess any of the features of wind-deposited sediments.

Criterion for the Recognition of Subaerial Exposure of Water-laid Sediments

1. Presence of curled clay flakes.
2. Common presence of clay pebbles.
3. Presence of mud cracks with sandy infill.
4. Presence of sand dykes.
5. Aeolian sand interbedded with beds of aquatic origin. The contacts may show evidence of fluvial erosion or wind deflation.

Ancient Deposits of the Desert Environments

Deposits of desert environments are known from different geological times and are distributed all over the world. The most well-known deposits are Navajo Sandstones (Jurassic). Colorado Plateau, U.S.A. (Stokes 1961, 1968). However, recently Freemann and Visher (1975) argue that Navajo Sandstone represent deposits of a tidal sea environment and not aeolian deposits. Coconino Sandstones, U.S.A., is well known for well-preserved wind ripples, vertebrate tracks, etc. (McKee 1945). The Botucatu formation (Mesozoic) of the Parana Basin, Brazil, have been thoroughly studied by Almeida (1953), Sanford and Lange (1960), and Bigarella and Salamuni (1961). Bigarella (1972) provides detailed information on desert deposits, especially the Botucatu, Brazil.

Permian desert deposits of Great Britain have been studied by several workers in detail. Shotton (1956) and Runcorn (1961) think that these deposits were formed under the influence of the northeast trade winds. They suggest that in Permian times Britain lay within 15 to 30° north of the equator. Thompson (1969) describes a dome-shaped sand dune of Triassic age from England. Piper (1970) also studied desert Triassic deposits of England. Brookfield (1977) describes Permian aeolian sandstones from Scotland.

Torridon Sandstone Precambrian (Walther 1909, 1924) and Lower Cambrian Sandstones of Sweden (Thorslund 1960) are other good examples of deposits of desert environments. However, the Torridon quartzites have been also interpreted as fluvial deposits (Selley 1965). Parts of Buntsandstein and Rotliegendes of Central Europe are also deposits of desert environments (Solle 1966).

Glennie (1972) interprets Rotliegendes of NW Europe as deposits of desert environment. Gradzinskii and Jerzyiciewics (1974) describe Cretaceous deposits of Mangolia, where aeolian dune deposits are interbedded with lake and stream deposits.

Walker and Harms (1972) describe aeolian dune deposits from the Lyons Sandstone (Permian), Colorado, where large-scale steep cross-bedding is common. Erosional surfaces representing blowouts, low parallel ripples on the accretion face, raindrop imprints and animal tracks are present.

Lake Environment

General Information

Lakes are standing-water bodies, filled mostly with freshwater. However, many lakes are highly saline in character. Restricted seas and lagoons (also partially standing-water bodies) are discussed on pp. 426 and 501 respectively.

Lakes can be grouped and described from various aspects and points of views. A lake can be described in terms of its shape, expressed as length, breadth, depth, etc., or in terms of its shape in plan view. The shape of a lake may keep on changing with time. Under favorable conditions of sedimentary tectonics, a thick sequence of lake deposits may be produced, provided a long period of time is available. Only such lake deposits have potential of being preserved in the geological record (Table 20).

For the classification of lakes, its mode of origin has been taken as an important parameter. Among those attempting classification of lakes, important authors are Twenhofel (1950), Schwarzbach (1964), Smith (1968), Reeves (1968), Picard and High (1972), Hutchinson (1975, Vol I, Part 1), Lerman (1978), Matter and Tucker (1978). Horie (1974, 1975, 1976, 1977, 1978) gives a detailed account of Lake Biwa in Japan. Reeves (1968) distinguishes the following types of lake basins: Structural basin, mass-movement basin, volcanic basin, meteoric basin, solution basin, glacial basin, fluviatile basin, wind basin, organic basin, animal basin, shore line basin.

However, we are not going to follow such classifications, as they are of little help in identifying the type of lake deposits in the sedimentary record.

A very important parameter is the climate, which strongly controls the characteristics of the lake deposits. Climate controls the amount of precipitation and evaporation, the nature of weathering, and the nature of soil in the catchment area and its vegetation. Further, the amount of clastic sediments coming into the lake basin is dependent upon the seasonal fluctuations in the discharge of the rivers. For practical purposes we can classify lake deposits into three major groups:

1. clastic lake deposits
2. chemical lake deposits
3. biogenic lake deposits.

However, all the transitions from lakes with only detrital sedimentation to lakes with salt precipitations or to lakes with biogenic sedimentation occur. Thus, within the chemical deposits intercalations of detrital sediments e. g., salt clays, are incorporated. Lakes with salt deposits occur in arid to semi-arid climatic zones, where evaporation is in excess of precipitation and water supply by drainage. The lakes with clastic sediments are found in all regions from arid climate to cold climatic zones near glaciers. Lakes with biogenic sediments formed by flora and/or fauna indicate a rich biogenic production and depend on climate and the availability of nutrients.

Clastic Lake Deposits

In the following, a short account of clastic lake deposits is given. Similar deposits are also those of sebkha (p. 215), glacial lakes (p. 201), ox-bow lakes (p. 293), and lagoons (p. 424), which are discussed in other chapters. Twenhofel (1932) gave an idealized picture for the distribution of sediment in lakes: An outer belt of beach pebbles, followed by a zone of sand, an inner zone of sandy marly mud, and mud in the central part (Fig. 350). This zoning corresponds to the zone-like distribution of hydraulic energy, namely the breaker zone, followed by a zone above wave base, and a zone below wave base. In nature there are many variants of this idealized picture. Thus, because of prevailing wind from one direction, the belt of beach pebbles may be developed on only one side of the lake. In the case of a steep slope of the coast, the sandy zone may be completely absent. Many lakes show the sediment distribution proposed by Twenhofel (1932). For example, Lake Constance (Bodensee) shows a belt-like distribution of sediments, although the pattern of distribution is somewhat different (Fig. 351). Many other lakes like Lake Geneva (Jaquet et al. 1975) or Lake Superior (Thomas and Jaquet 1975) also show belt-like distribution. However, no vertical sequence can be constructed from such lateral distribution of sediments, because, in general, a lake does not grow from the margins. A lake is more likely to be filled by the deposition of silt and clay in the central basin as a re-

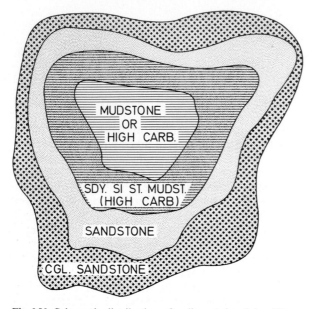

Fig. 350. Schematic distribution of sediments in a lake. (Compare with Fig. 351.) (Modified after Twenhofel 1932)

sult of building up of deltas by one or more rivers reaching the lake (Wagner 1950; Visher 1965a). Figure 352 shows a river delta in a lake. Sandy lake beaches could respond to changes which vary with the year-long conditions of water level changes. The beaches of Lake Michigan, therefore, migrate inland from April through July in an upward-inclined plane with decreasing beach and backshore widths. Although wave action plays an important role as with marine beaches, the annual sea level variation in the Great Lakes is the most important factor controlling the development of lake beach sequences. These variable changes in the vertical sequence of deposition correspond to erosive discordances. Fraser and Hester (1977) indicate a truncated series of beach ridge lineaments in Lake Michigan. At the same time, however, there exists in the younger geological history of Lake Michigan the development of predominantly transgressive and progradational sequences. The transgressive sequences resulted from the water level ascent in the post-Wisconsin period. The progradational sequences have been formed since the end of the transgression. A transgressive sequence of the lake beach–ridge complex, therefore, would consist mainly of fine-grained lower shoreface sediments with abundant redeposited coarse sand and granules from the eroded beach, along with cross-bedded upper shoreface sands deposited mainly in longshore troughs. Sediments deposited in sub-

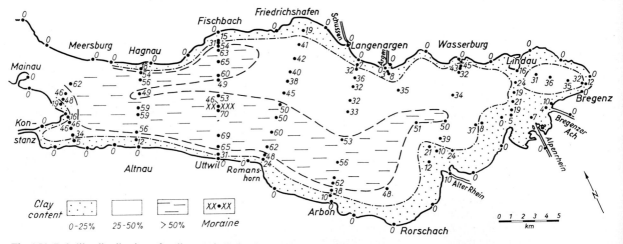

Fig. 351. Belt-like distribution of sediments in Lake Constance. (After Müller 1963)

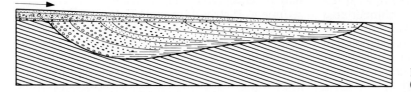

Fig. 352. Filling up of a lake by a delta. (Modified after Wagner 1950)

aerial environments have a poor chance of being preserved during transgression, and would probably occupy only a small portion of the sequence near the base. The progradational sequence of the beach ridge in Illinois is a sand body that has been interrupted only intermittently during its construction by periods of retrogradation.

The progradational sequence of the lake beach–ridge complex would thus consist of lower shoreface sands at the base overlain in turn by upper shoreface sands, and beach sands and gravels. The sequence would be capped by aeolian sands, marsh organic muds, or fluvial sands, depending upon the position of the sand body. Martini (1975) discusses the sedimentology of a lacustrine barrier system at Wasaga Beach, Ontario, Canada.

Along with the deposition in a delta and a regional distribution of silt and clay across the lake, the processes of local sediment slumping and underflow transport by the meeting rivers also play important roles in the sedimentation of lakes. Lake Geneva, Lake Constance, Lake Brienz, and Lake Thun exhibit all these modes of deposition. In the following, depositional processes and characteristics of the sediments of these lakes are discussed.

Lake Geneva

Lake Geneva and Lake Constance are famous since the classical study of Forel (1885, 1888). He found that river channels extend underwater in the lakes, along which water loaded with suspended sediment flows toward the central part of the lake. Houbolt and Jonker (1968) believe that cold river water (thus also heavier) flows along these channels. These sublacustrine channels lead to the central plain of the lake over the delta foreslope. Similar underflows are reported from Lake Thun (Sturm and Matter 1972), Walensee, Switzerland (Lambert et al. 1976; Lambert 1978) and from Lake Superior (Normark and Dickson 1976). A very useful bibliography of Lake Geneva is provided by Vernet et al. (1971).

According to Houbolt and Jonker (1968), the sublacustrine channel of Lake Geneva extend from the mouth of the Rhône River down to a depth of about 250 m. The channel is 200 m broad and 15 m deep with well-developed natural levees, which can be followed up to a water depth of 200 m. In the upper part of the delta a few more channels are present (Fig. 353). These channels most probably stem from the old Rhône channel and its distributaries. During diving investigations in the main channel, Dill found that near the river mouth heavy river water sinks down and flows across the delta in the deeper parts of the lake (Shepard and Dill 1966). The suspended sediment is separated into a coarser fraction (sand), sinking down to the bottom, and a finer fraction (mud), which remains in the suspension. The bottom current is so strong that small current ripples are generated. Such ripples are 2.5 cm in height, 10 cm in length, and can migrate at a rate of 5 cm in 2 min, 25 sec. At places where sublacustrine channels are developed, megaripples 50 cm high are present.

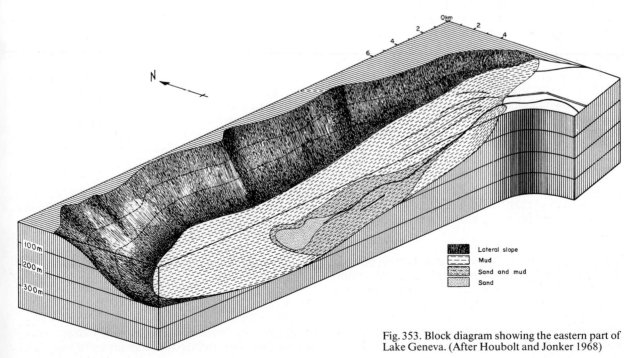

Fig. 353. Block diagram showing the eastern part of Lake Geneva. (After Houbolt and Jonker 1968)

The presence of sublacustrine channels and the sandy layers in the deeper parts of the lake in front of the delta suggest that density currents are actively, transporting material down into central plain to a depth of 309 m.

Houbolt and Jonker (1968) give the following distribution of sediment in Lake Geneva: The bottom of the main sublacustrine channel is covered with fine to medium sand. The natural levees are made up of muddy sediments with intercalations of fine-sand layers.

Below the 200-m water depth, where natural levees disappear, a fan-shaped sandy area is present, extending into the central plain. The fan area is mostly sandy in nature; along the margins sand and mud layers are intercalated.

The delta foreslope outside the channel is made up of muddy sediments; only on the lower–most part of the foreslope are some sandy layers present.

The central plain is covered with thinly bedded muds in which sand sometimes is incorporated. Characteristics of individual subenvironments are as follows (Houbolt and Jonker 1968):

Delta Foreslope Outside the Channel and the Fan. Sediments are mainly silt (Fig. 354). Clay contents is extremely low. The reason seems to be that the Rhône River brings very little clay-size sediment, and it is not quickly flocculated. The silts are mostly well laminated. Some layers of fine sand are also present. In the lower part of the foreslope some graded sequences and small-scale cross-bedding are present. In one core a more or less complete turbidite sequence has been encountered, consisting of an unlaminated interval at the base, followed by an interval of parallel lamination and current ripple bedding. Gas-expansion holes are common.

Central Plain. Deposits of the central plain are more finely grained than delta foreslope deposits. Content of clay-size sediment is higher and some very thin layers of fine sand are present. The central plain deposits may be thickly bedded, finely laminated, and graded. Gas-expansion holes are common.

Lateral Slope. On the lateral slopes fine-grained sediments are deposited. They are fine silts with a high content of clay fraction (Fig. 354). Sediments are thinly bedded and gas-expansion holes are rare.

Channel Bottom. Sediments of the channel bottom are fine to medium sand, mostly horizontally laminated; sometimes they have graded bedding. Small-scale cross-bedding is rather rare.

Central Fan. The central fan is located near the base of the delta foreslope; it is composed mostly

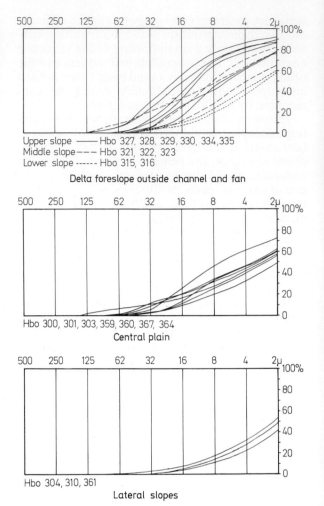

Fig. 354. Diagram showing grain-size distribution curves of the sediments from various parts of Lake Geneva. (After Houbolt and Jonker 1968)

of horizontally bedded sands. Some of the layers show grading from medium to very fine sand. In contrast to the sands of the channel bottom, they are finer-grained and show poorer sorting. Small-scale cross-bedding is commonly encountered. Evenly laminated sand is subordinate. Load structures are common, as well as gas-expansion holes.

Marginal Part of the Fan. Sediments are fine to medium sand, mainly showing horizontal bedding, sometimes also graded bedding. Small-scale cross-bedding is rare. Intercalations of silty and clayey layers are often present, which can be graded in themselves. Gasexpansion holes are very abundant.

Sublacustrine Natural Levees. Levees are made up of silty sediments near the top and fine to very fine sand near the bottom. The silts partly show wavy lamination; interbedded are thin layers of very fine sand. The sand layers are mostly graded. Gas-expansion holes are common.

Seismic investigations show that Holocene deposits represent only a thin blanket of the sediments (Winnock 1965; Serruya 1965; Houbolt and Jonker 1968). The thickness of sediment on the lateral slopes is 12 m and on the fans it is about 50 m. Two older horizons are also visible on seismic profiles. The deeper one probably represents molasse, and the upper one represents the Pleistocene horizon.

If we construct a hypothetical vertical sequence produced after Lake Geneva is filled up, the following succession would be found (from top to bottom):

1. Delta foresets–mainly sandy deposits.

2. Upper delta foreslope deposits–mainly mud, with some minor sand bodies produced as a result of lateral migration of sublacustrine channels.

3. Lower delta foreslope deposits–sand deposits formed as a result of coalescing fans. Intercalated are muddy sediments of lower foreslope deposits. Sand content increases from bottom to top.

4. Central plain deposits–mainly muddy deposits, horizontally bedded, with minor sandy intercalations.

Thus, in this vertical sequence of mainly muddy sediments two sandy parts are present, namely the underflow deposits in the deeper part of the lake and the shallow-water deltaic sands.

Lake Constance

The results of sedimentological investigations in Lake Constance are comparable to those of Lake Geneva. Before the artificial shifting of the mouth of the Rhine River by man in the year 1900 in the Fussacher Bucht, the Rhine River had built up a large delta. Sublacustrine channels are present in the foreslope of this delta. Von Salis (1884) described one such channel, which is 4 km long, 600 m wide, and 70 m deep, and could be followed up to a water depth of 140 m. On more recent charts this channel can be followed to a water depth of 240 m. From a depth of 210 m of this channel, underflow sands have been recovered (Reineck 1974).

The subaquatic delta topsets are also sandy (Förstner et al. 1968). In the deeper parts the lake sediments are clayey silt to silty clay (Müller 1963, 1971). These muddy sediments contain mostly thin (partly graded, partly evenly laminated) sandy layers (Reineck 1974). Cross-bedding is present in only the sandy sediments of the subaquatic delta topsets (for details see the Lake Constance delta on p. 247).

Gas-expansion holes are abundant in delta sediments. At places with abundant food available, mostly in front of river mouths, thick populations of *Tubificides* are present (Wagner 1968; Zahner 1968). The burrows of *Tubificides* are a dense network and can lead to very strong bioturbation (Figs. 355 and 356). Such bioturbation is rare in the deposits of the deeper parts of the lake.

In one of the extensions of Lake Constance, lake chalk has been found. It was produced in a shallower part of the lake and later brought into deeper

Fig. 355. Burrow system of *Tubifex* in the sediments of Lake Constance. The burrow system is filled with resin

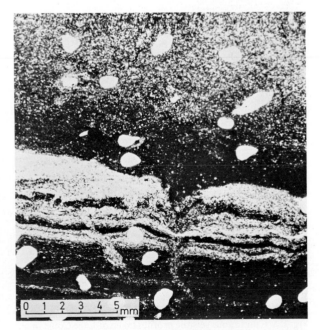

Fig. 356. A thick section of sediment with a *Tubifex* population seen in a vertical plain

parts, increasing the carbonate content of the sediments of the deeper parts. According to A. Schäfer (1972), production of the lake chalk was intensive during the Atlanticum period, and does not take place with that intensity today (Schäfer 1973).

The total thickness of Holocene sediments is a maximum of 90 m. Below it Pleistocene sediments with a maximum thickness of 80 m are present (Müller and Gees 1970) (Figs. 357 and 358).

Construction of a hypothetical vertical sequence of Lake Constance deposits gives a very similar sequences as that of Lake Geneva given by Houbolt and Jonker (1968) (from top to bottom):

1. Fluvial sediments, i. e., river channel, ox-bow lakes, sand bars, natural levees, and flood basin deposits with soil formation and plant-root horizons, or partly with peat deposits.

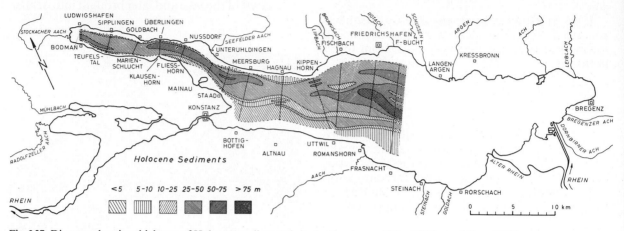

Fig. 357. Diagram showing thickness of Holocene sediments in Lake Constance. (After Müller and Gees 1970)

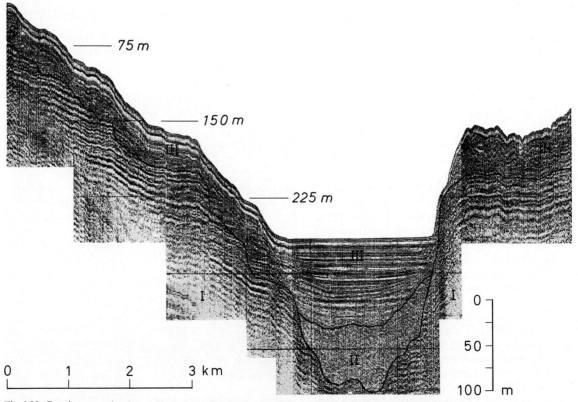

Fig. 358. Continuous seismic profiles taken of Lake Constance. Unit I—molasse sediments of Tertiary age in which the lake basin was eroded. UnitII—Pleistocene sediments, mostly moraines. Unit III—Holocene sediments, well-bedded fine-grained sediments. (After Müller and Gees 1970)

Fig. 359. Distribution of sediment-type in five piston cores taken from the Rhine delta in Lake Constance. (After Förstner et al. 1968)

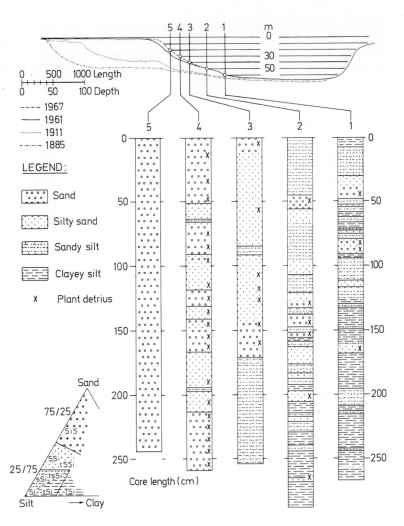

Rhine River Delta in Lake Constance

2. Subaquatic delta topset sands and muds. Mud accumulates in areas between two distributary channels.

3. Mainly sandy and clayey deposits of the fore-set of the prograding delta. In the same water depth, laterally much finer-grained muddy deposits of the lake slopes are present. The foreset deposits in its lower part show a clear reduction in the sand content and an increase in the silt and clay content (Fig. 359). In the lower part of the foreset deposits, i.e., transition to central plain deposits, numerous sandy fan deposits produced by underflow are incorporated.

4. Deposits of the deeper part of the lake—central plain—are made up of very fine sediments with 25 to 75 % clay fraction content. They are thinly bedded and possess only a few very thin graded sand layers.

Because of the artifical shifting of the Rhine River in the Fussacher Bucht in the year 1900, a new delta was rapidly formed in this eastern part of Lake Constance. It was investigated by Förstner et al. (1968). In the following, characteristics of this delta and its deposits are given, as deltas play an important role in the infilling of a lake basin and constitute an important role in the infilling of a lake basin and constitute an important part of the lacustrine deposits (Axelsson 1967).

According to Scruton (1960), important factors controlling the building up of a delta are nature of provenance, energy and intensity of hydrographic factors, and amount of deposition. However, in case of a lake delta the operating hydrographic factors and their intensity are different from the deltas of marine regions, e.g., the delta of a lake hardly suffers reworking by strong wave action as does the delta in a sea. Thus, the characteristic units of a delta, i.e., topset. foreset, and bottomset, are well developed and show a variation in the nature of sediments and type of bedding structures. Samples obtained by box-corer and pistoncorer showed the following types of bedding structures.

Ripple Bedding. Ripple bedding is restricted to sandy sediments and occurs in the form of small ripple cross bedding, mostly of the festoon type. In some cases there is tendency for small ripple bedding to develop like climbing ripple bedding. Palaeocurrent measurement indicates a lakeward direction. The ripple bedding is abundant in topset sediments, especially along the axis of the current direction of the channels.

Graded Bedding. The graded bedding begins with the coarsest grains at the base, gradually becoming finer, and being finest at the top (Fig. 360). Such beds are only a few millimeters to a few centimeters in thickness. For the genesis of graded bedding in the Rhine River delta, it is believed that suspended sediment of different grain size is brought in. In enough water depth the sediment is segregated be-

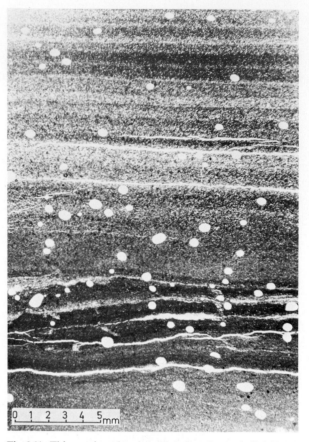

Fig. 361. Thin section showing thinly interlayered silt/clayey silt bedding with gas-expansion holes. The water depth is 34 m. Lake Constance. (After Förstner et al. 1968)

cause of different settling velocities, so that coarsest sediment settles first, followed by the finer grain sizes. Graded bedding is found even at water depths of 2 to 3.4 m; however, these areas are situated away from the main current-axis. Along the axis of the river they begin first at a depth of 10 m and continue into deeper parts.

Thinly Interlayered Bedding. This bedding shows alternating layers of finer-grained and coarser-grained sediments with a sharp contact (Fig. 361). The layers are a few millimeters thick. This suggests that alternating finer and coarser sediment is brought in for deposition.

Possibly they represent deposition of suspension clouds, produced near the upper part of the delta from wave action. While the clay is continuously and slowly deposited, the silty and fine sandy sediments settle down very quickly, producing each time a thin sandy to silty layer.

Laminated Bedding. Laminated bedding denotes a succession of fine laminae of the same material. Here laminated sand and laminated silt are found.

Fig. 360. Thin section showing graded bedding. The water depth is 33 m. Lake Constance. (After Förstner et al. 1968)

The lamination originates because the sediment transport occurs in jerks. Individual suspension clouds are brought at short intervals and deposited.

In the Rhine River delta laminated bedding is especially abundant in silt sediments at depths between 10 and 50 m (Fig. 362). The genesis of suspension clouds of silt is ascribed to wave action in shallower parts. Thus, the genesis of thinly interlayered bedding and silty laminated bedding is very similar. This leads to the fact that all transition between these two bedding types are found, and their differentiation is sometimes very difficult.

Diffuse Bedding. Diffuse bedding denotes sediments without any bedding structures; the sediments are usually poorly sorted (Fig. 363). The mode of origin of such bedding is unclear. It can be due to complete churning of the sediments by bioturbation. However, it can also be produced by gas bubbles destroying primary bedding. As this bedding is especially common in greater water depths, it cannot be a case of rapid deposition of suspended sediment without producing some kind or primary bedding.

Bioturbation Structures. The main bioturbating organism is *Tubifex* sp. The dense population of *Tubifex* in the delta foreslope is present only at depths greater than 30 m. The reason for this is low content of organic matter in the sediments of shallow depths. Another reason may be that the rate of sedimentation above 30 m depth is rather high. This leads to the fact that shallower parts, especially the upper part of the delta, are not preferred by larvae or adult animals for population. The bioturbation structure of *Tubifex* is a network of narrow burrows (Fig. 355). In a transverse section such burrows appear as small holes (Fig. 356). Together with *Tubifex* burrows, there are other bioturbation structures made by some unspecified organisms.

Bubble Cavities (Gas-Expansion Holes) and Plant Remains. Because of its high organic matter content, during its decomposition in sediment, gas which produces bubble structures is generated. The organic debris is mainly plant remains. Part of the plant material is brought in floating on the water surface; the other part is transported along the bottom, together with the sediment. Most of this plant material sinks down into the delta region and is transported into rather calm parts and deposited. Apparently the plant debris, plant leaves, and branches are brought in large quantities during certain seasons of the year–most probably, during autumn and spring. This also explains why plant debris is found concentrated in certain layers within the sediment. In the sediment cores gas-bubble cavities are present even in sediment of 2.5 m.

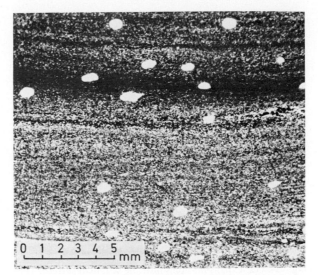

Fig. 362. Evenly laminated bedding in silty sediments with some burrows of *Tubifex* sp. The water depth is 33 m. Lake Constance. (After Förstner et al. 1968)

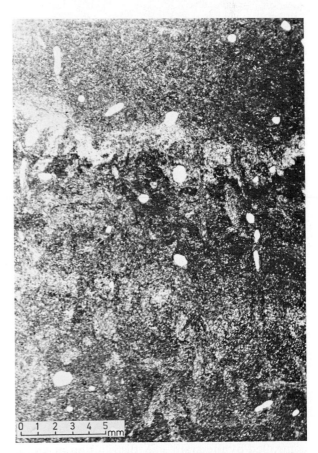

Fig. 363. Diffuse bedding with inumerable gas-bubble tracks. Some individual gas bubbles and burrows of *Tubifex* sp. are also visible. The water depth is 68 m. Lake Constance. (After Förstner et al. 1968)

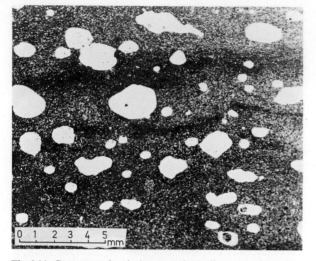

Fig. 364. Gas-expansion holes near the sediment surface. The sample was taken at a water depth of 60 m. The holes have expanded afterward with the removal of water pressure. Lake Constance. (After Förstner et al. 1968)

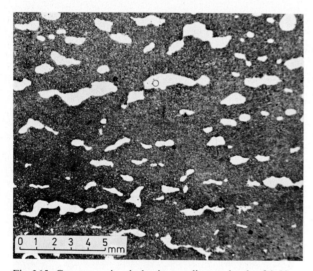

Fig. 365. Gas-expansion holes in a sediment depth of 2.50 m. The water depth is 60 m. The holes have been compressed as a result of pressure created by overlying sediments. Lake Constance. (After Förstner et al. 1968)

However, at such depths in the cores gas-bubble cavities are clearly more pressed together (Figs. 364 and 365).

Distribution of Sedimentary Structures in the Delta Body

As some of the sedimentary structures are directly controlled by the grain size, distribution of grain size also controls the distribution of sedimentary structures. Moreover, the physical factors are very important in the genesis of sedimentary structures.

While physical factors, e. g., current and wave energy, change with depth, the distribution of bedding structures also changes with depth.

Delta Topset Deposits. The topset beds of the Rhine delta extend down to a depth of about 10 m. Along the axis of the Rhine River topsets are made up of ripple bedding and laminated sand. Organic debris is almost absent. Thus, gas-bubble structures and bioturbation structures are extremely rare.

On the sides of the axis of the Rhine River finer-grained sediments are deposited, as the delta is exposed to very little wave activity. In these parts we find sedimentary structures which are normally found at greater depths.

The samples from water depths of 10 to 14 m already show transitional characteristics to the foreset deposits.

Delta Foreset Deposit. The foresets are made up of dark-colored clayey silts with intercalations of light-colored sandy silts.

Ripple bedding is found only sporadically. Main bedding types present are graded bedding (Fig. 360), thinly interlayered silt/sand bedding (Fig. 361), and laminated silt (Fig. 362). Bioturbation structures become common (Figs. 355, 356, and 363). At the same time the content of organic debris (plant debris) also increases. Gas-bubble cavities are commonly present.

Delta Bottomset Deposit. The bottomset sediments show rather similar characteristics as foresets. However, the population of *Tubifex* is much denser in bottomsets than in the foresets.

Various types of structures are present which have been assigned to the gas-bubbles (Figs. 364 and 365). Especially abundant is the diffuse bedding (Fig. 363), together with thinly interlayered bedding, graded bedding, and laminated clayey silt. Ripple bedding is absent, even in the sandy silts.

Central Basin Plain and Lateral Slope of Lake Constance

The silty, clayey deposits of the deeper parts of the lake basin are thinly laminated, made up of alternating layers of dark and light colors of a few millimeters thickness (Fig. 366). The difference between the dark- and light-colored layers is that the light-colored layers contain more calcite than the dark-colored. The calcite is in the form of needles and aggregates. The bedding is a case of seasonal rhythmites (annual varves). In the deeper part of

Fig. 366. Thinly interlayered bedding showing alternating dark and light coloured layers. X-ray radiograph (positive). Core no. 10, water depth 95 m, sediment depth 1.25–1.33 m. 1 div. of scale — 1 cm. After Reineck 1974a)

Lake Thun, Lake Brienz, and Walensee

Lake Thun and Lake Brienz together were a deep longitudinal glacial valley lake, now separated from each other by an alluvial plain. The Aare River still connects both lakes, first reaching Lake Thun, then Lake Brienz. Additional affluents of some importance are the Kander River into Lake Thun and the Lütschine River into Lake Brienz. All affluents contribute a high sediment input forming lake deltas, and represent different source areas from within the Swiss Alps.

Lake Thun, with a depth of about 215 m, has a total sediment layer of more than 300 m (Matter et al. 1971). Lake Brienz on the other hand is at a maximum 260 m deep and shows a sediment thickness of 300 and 550 m in its two separate lake basins (Matter et al. 1973); in comparison the quarternary sediments of Lake Constance are 200 m thick (Müller and Gees 1976); those of Lake Zürich comprise 120 m (Finck and Kelts 1976).

Studies on the sedimentation within the reach of these deltas and on the basin part of these lakes were done by Sturm and Matter (1972) for Lake Thun, and by Sturm (1975, 1976) and by Sturm and Matter (1978) for Lake Brienz.

The high sediment input into the lakes provides excellent conditions for the investigation of the depositional processes. Sturm and Matter (1978) summarized the results of studies concerning Lake Brienz. The sediments brought in by the Aare River build up a delta, and then sediments are taken into the deeper lake bottom by turbidity currents. Some sediments are also supplied from the Lütschine delta. The sediments become finer-grained as one proceeds away from the Aare River delta into distal parts.

Four types of sediment and bedding can be distinguished (Fig. 367) (1) delta sand and silt, (2) massiv or lenticular, laminated mud, (3) homogeneous mud on slopes and shore terraces, and (4) laminated mud interbedded with graded sands and silts on the basin plain. There are two important depositional processes operating in these lakes (Fig. 368) (1) low- and high-density turbidity currents (underflows) moving along the lake bottom and (2) interflows and surface currents (overflows) in and on the lake water. If the river water and the suspended load are less dense than the cold bottom lake water, the river water will flow at or above the thermocline (interflow or overflow) and spread over the entire basin. Coarser particles fall out continuously during the summer months and form the dark grey basal laminae of varve couplets. The light-colored fine-grained fraction that becomes trapped by the density gradient during summer settles out after the thermocline is destroyed in the autumn. High-density turbidity currents,

the lake Müller and Gees (1970) calculated the rate of sedimentation to be about 6 mm per year.

Varve layering is also found in many other lakes e. g., Lake Thun and Lake Brienz (p. 252) also in lakes of New Zealand (Irwin 1972, 1974) and lakes of Japan (Horie 1978). The fact that the varve layers are thinly striped (Fig. 366) points to the changing density of the suspension clouds in the water. Müller (1966) indicated another form of light/dark coloring caused by iron sulfide in strongly reduced environments. This coloring by FeS · n H_2O (hydrotroilite) includes at times more primary sediment layers. Wagner (1971) found a wide range of FeS concretions in Lake Constance sediments. Reineck (1974a) describes them as rounded, grape-like concretions up to 5 mm in diameter. They were found in water depths of 115 to 175 m and in sediment depths of over 4 m.

Sandy layers are also incorporated in the silty-clayey, varved strata and can be attributed to bottom turbidity currents.

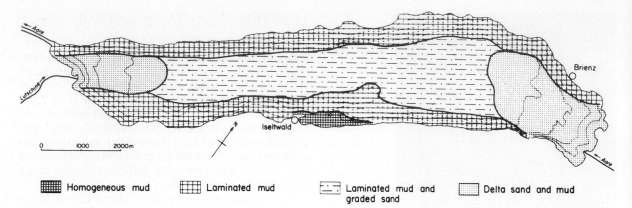

▦ Homogeneous mud   ▦ Laminated mud   ⊡ Laminated mud and graded sand   ▢ Delta sand and mud

Fig. 367. Map of Lake Brienz showing distribution pattern of sediment and bedding types. Contour lines indicate 50 m intervals. (After Sturm and Matter 1978)

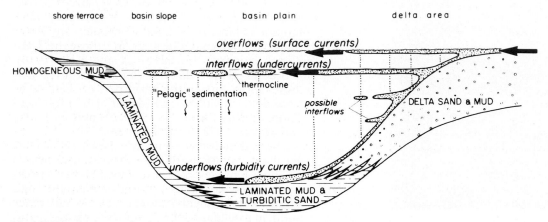

Fig. 368. Distribution mechanisms and resulting sediment types proposed for clastic sedimentation in oligotrophic lakes with annual thermal stratification. Note that hypothetical shore terrace is situated higher than depth of thermocline. Width of basin and sediment thickness are not to scale. (After Sturm and Matter 1978)

which deposit up to 150 cm-thick graded sand layers, occur only once or twice per century after catastrophic flooding. The low-density turbidity currents occur annually related to periods of high river discharge and deposit centimeter-thick faintly graded sand and silt layers, together with dark summer lamination of varve couplets.

The laminated mud deposits are the result of seasonal inputs of sediment-laden river water and the annual sedimentation cycle related to thermal stratification in the lake water (Sturm 1978).

A comparable sediment transport pattern is discussed by Lambert et al. (1976) and Lambert (1978) from Walensee, another Swiss alpine lake. However, Lambert and Hsü (1975) observed that in Walensee, the laminated sediments do not represent annual varves, because both the sedimentological and hydrological records show that graded laminae of silt and clay are not necessarily deposits of annual cycles. Counting of the rhyth-

mic layers showed that in 165 years (during 1811 to 1976), 300 to 360 laminae are deposited. Direct measurement exhibited that upto 5 graded laminae can be deposited during a single year.

These rhythmic laminations are believed to be deposits of continuous-fed turbidity currents, which are produced by hyperpycnal inflow during river-flood stages. The direct current measurements shows that these underflows can occur sporadically throughout the year, but are especially common during the ice-melting and heavy rainfalls. Gustavson (1975) made similar observations in the Malaspina Glacier area, Alaska. The varves of Lake Zürich represent chemical sedimentation prevailing during summers (light-colored laminae), and darker laminae of detrital sedimentation during winter (Kelts 1978). Pharo and Carmack (1979) describe similar sedimentary features from an intermontane lake – Kamloops Lake, British Columbia.

The Life in the Lakes

In the recognition of lake deposits in ancient sediments great help is obtained from palaeontological parameters. In comparison to the marine environment, there are very few species which inhabit lake environments. It is especially so in the case of endobenthonic forms. There are several groups of organisms which never lived in the lakes; among them are cephalopods, corals, brachiopods, echinoderms, pteropods, and polychaetes.

From the middle Palaeozoic times gastropods and lamellibranchs are known to have lived in the lakes. There are several species of fishes living in lake environments; however, they are of little help in the identification of lake deposits.

Most of the macrobenthonic fauna of lakes lives in water depths of less than 10 m. Only a few macro- and micro-organisms live at greater depths in the lakes, and only a few species of bacteria live abundantly at greater water depths. The sediments of the central basin are mostly free of any bioturbation structures. This is exemplified in Lake Thun (Sturm and Matter 1972), Lake Constance (Wagner 1972; Reineck 1974a), Lake Brienz (Sturm 1976), Lake Michigan (Gross et al. 1970), Lakes Ontario, Erie, and Huron (Kemp et al. 1974; Byers 1977).

The plant life is abundant in the shallow parts of the lake bottom and on the water surface. Thus, during the sequence of filling up of a lake, plants become important during the last stages and produce various kinds of plant deposits.

Several plants living in the lakes, e. g., *Potamogeton, Elodea,* and other submerged plants, produce carbonate. During plant decay they produce lake marl (Schäfer 1973; Terlecky 1974). The production of lake marl was significant in the post-glacial Atlanticum period, about 6,000 years before present. The bicarbonate content of the lake water was derived from weathering and leaching of glacial till. Lamellibranchs and gastropods also contribute to the carbonate content of lake sediments. Today carbonate-rich lake environments result from the low precipitation/evaporation ratio. This, for instance, is the case with Lake Balaton in Hungary, investigated by Felföldy et al. (1969), Ronai (1969), Müller (1969), and Müller and Wagner (1978). The carbonates produced are not stoichiometric calcites as is usually the case in lake environments. They are aragonites and high-Mg calcites depending upon the progress of carbonate precipitation which is governed by the Mg/Ca ratio of the lake water (see also Müller et al. 1972; Kelts and Hsü 1978). Lastly, dolomite might be formed in a freshwater lake environment, as is reported by Schroll and Wieden (1971) and Blohm (1974) from Lake Neusiedel, Austria.

Lacustrine blue green algae, together with their carbonates, are delicate witnesses of life in lakes (Krumbein and Cohen 1974). They strongly depend on their narrow-spaced biotope with a shallow, low-energetic, illuminated as well as nutrient- and bicarbonate-rich water. They are also known from the fossil record of lakes, where they are useful facies indicators (Schäfer and Stapf 1978).

In euthrophic lakes (high production of nutrients and low oxygen content) the decay of plants leads to the formation of humus-gel, which may ultimately produce humus-coal, e. g., cannel coal. In the deeper parts of some lakes black sapropel sediments are deposited which are extremely rich in organic matter (Barnes and Barnes 1978). These sediments could develop into oilshale as exemplified in the Eocene Lake Messel (Upper Rhinegraben, SW Germany) (Albrecht 1970; Arpino 1973; Franzen 1976, 1977, 1979; Irion 1977) or in the Eocene Green River Formation of Utah, Colorado and Wyoming, U.S.A. (Bradley 1970; Desborough 1978; Tissot et al. 1978; Surdam and Stanley 1979). An interpretation of diverse fluvial and lacustrine blackshale environments bearing tetrapods is given by Boy (1977).

The shallower a lake is, the broader is the plant-inhabited zone near the shore. After the death and burial of plant population, deposits such as peat are produced. In Central Europe a typical sequence of vegetation develops in an upward-growing lake sequence, i. e., phragmites peat, sedge peat, and wood peat. Thus, clastic lake deposits may be strongly associated with lake marl, and the sequence may be closed by peat or coal deposits.

Chemical Precipitates in Lakes

Chemical precipitates are extremely important in the lake deposits and make workable deposits of various salts. Although it is out of the scope of this book to deal with the chemical deposits, a short discussion is given here because of its importance in lake deposits. Reeves (1968), Picard and High (1972), Hutchinson (1975, Vol. I, Part 2), Jones and Bowser (1978), Eugster and Hardie (1978), Imbowden and Lerman (1978), and Hardie et al. (1978) give in detail an account of lake deposits, including the chemical aspects. The formation and diagenesis of inorganic Ca-Mg carbonates in the lacustrine environment is reported by Müller et al. (1972). Some detailed investigations on the mineralogy and other environmental aspects of various mid-European lakes have recently been done by Abdul-Razak (1974), on Turkish lakes by Irion (1973), on Lake Constance by Förstner and Müller (1974), on Lake Geneva by Vernet et al. (1972),

Vernet and Thomas (1972), on Lake Michigan by Baker-Blocker et al. (1975). Groth (1971), and Förstner (1976) provide a general review on the chemistry of lakes.

Many lakes may have enrichments with respect to iron and manganese (seldom of any industrial value). These deposits are of great use for the discussion of chemical balances in lakes and the relat-

Table 19. The lacustrine evaporites. (After Reeves, 1968; Irion 1973)

Mineral	Composition
Carbonates	
Limestone	$CaCO_3$
Aragonite	$CaCO_3$
Dolomite	$CaMg(CO_3)_2$
Huntite	$Mg_3Ca(CO_3)_4$
Magnesite	$MgCO_3$
Mg-Calcite	4–45 mol % $MgCO_3$ in Calcite
Natron[a]	$Na_2CO_3 \cdot 10H_2O$
Trona[a]	$Na_2CO_3 \cdot NaHCO_3 \cdot 2H_2O$
Sulfates	
Anhydrite	$CaSO_4$
Gypsum	$CaSO_4 \cdot 2H_2O$
Glauberite	$CaSO_4 \cdot Na_2SO_4$
Epsomite	$MgSO_4 \cdot 7H_2O$
Bloedite	$MgSO_4 \cdot Na_2SO_4 \cdot 4H_2O$
Thenardite	Na_2SO_4
Polyhalite	$Ca_2MgK_2(SO_2)_4 \cdot 2H_2O$
Chlorides	
Halite	$NaCl$
Sylvite[a]	KCl
Borates	
Borax	$Na_2B_4O_7 \cdot 10H_2O$
Searlesite[a]	$Na_2O \cdot B_2O_3 \cdot 4SiO_2 \cdot 2H_2O$
Colemanite[a]	$Ca_2B_6O_{11} \cdot 5H_2O$
Nitrates	
Soda niter[a]	$NaNO_3$
Niter	KNO_3

[a] Most uncommon lacustrine evaporites.

ed metal transfer between profundal evenly stratified lake sediments and supernatant lake bottom water. This is discussed in a long series of interesting papers by Einsele (1937), Vasari et al. (1972), Tessenow (1975), Tessenow and Baynes (1975), Cronan and Thomas (1972), Horne and Woernle (1972), Dean and Ghosh (1978), Schöttle and Friedman (1971).

The chemical composition of lake water depends mainly on the dissolved and suspended material brought into the lake by rivers and/or brines. The most comprehensive information on salt lakes is given by Hardie et al. (1978) and Eugster and Hardie (1978), explaining the sedimentology and the chemical balances of salt lake deposits on Recent and fossil examples. Neal (1975) gives a useful collection of papers on chemical lakes. A number of salts are precipitated as a result of chemical and biological processes. Table 19 lists important salts of the lake evaporites.

Examples of Ancient Lake Deposits

Well-developed lake deposits are of only minor importance in ancient stratigraphic records (cf. Picard and High 1972). The reason may be that lake deposits are not easily identifiable. Moreover, the rate of deposition in lakes is not very high. To produce a thick extensive sequence of lake sediments, the lake must have existed over a long period, i. e., over several million years (Table 20). However, lake deposits are known from all geological times.

Primary sedimentary structures are the result of the depositional process; thus, they can be used with success in the reconstruction of the palaeo-environments (Table 21). Lake deposits can be easily

Table 20. Areas and approximate durations of typical lakes. (From Picard and High 1972)

Rock unit, lake	Location	Age	Area (in mi²)	Duration (in m.y.)
Lake Bonneville	Utah, Nevada, Idaho	Pleistocene	19750 (max.)	1.0
Searles Lake	California	Pleistocene	385	0.17 (?)
Lake San Augustin	New Mexico	Pleistocene	225	0.65
Lake Uinta (Green River, lower Uinta fms.)	Utah	Eocene	7700	13.3
Gosiute Lake (Green River Fm.)	Wyoming, Colorado	Eocene	17000 (max.)	4.0
Flagstaff Limestone	Utah, New Mexico, Colorado, Arizona	Paleocene-Eocene	7000 (max.)	2.75
Todilto Limestone		Late Jurassic	34600	0.02
Popo Agie Formation	Wyoming, Utah	Late Triassic	50000 (+)	3.0 (?)
Lockatong Formation	New Jersey, Pennsylvania	Late Triassic	2250	5.1

Table 21. Distribution of bedding type and other sedimentary structures in various lake deposits. (After Picard and High 1972)

Bedding types, Sedimentary structures	Klein (1962)			van Houten (1964b) Lackotong Fm.	Visher (1965a) General lacustrine	Greiner (1962) Albert Fm.	Bradley (1931) Green River Fm.	Recent sediments[a]
	Keuper Marl		General lacustrine					
	Below wave base	Above wave base						
Thin bedding, rhythmic bedding, varves	×	×	×	×	×	×	×	×
Even horizontal bedding	×	×	×	×	×	×	×	×
Cross-stratification		×	×	×	×	r	r	r
trough		×	×	×				r
planar		×						r
Ripple stratification	×	×	×	×			r	r
Disturbed stratification	×		×	×	×	×	×	×
Graded bedding	×		×	×	×			×
Asymmetrical ripple mark		×			r	r	r	r
Symmetrical ripple mark	×	×	×		r	×	r	×
Shrinkage cracks		×		×		×	×	r
Parting lineation		×						r
Rib-and-furrow		×						r
Groove cast		×						r
Load cast	×				×			×
Pull-apart structure	×			×			×	r
Raindrop impression		×						r
Burrowed structure, worm trail	×	×	×	×				×

[a] Based on the observations of Houbolt and Jonker (1968) in Lake Geneva and Reineck (unpublished information) in Lake Constance

× = characteristic; r = rare

differentiated from fluvial deposits. However, sometimes it is very difficult to differentiate lake deposits from shallow marine deposits, especially those of nontidal seas. In such cases, help can be obtained from palaeontological evidences.

Some well-studied examples of ancient lake deposits are: Lockatong formations, Triassic of New Jersey and Pennsylvania (van Houfen 1965), Triassic lake deposits of Wyoming (High and Picard 1965). Eocene lake deposits of Utah, Wyoming, and Colorado (Bradley 1964; Picard and High 1968), and Triassic lake deposits of Canada (Klein 1962). Negendank (1972), Schäfer and Rast (1976), Boy (1977), Rast and Schäfer (1978), and Schäfer and Stapf (1978) describe the sedimentological characteristics of the Lower Permian (Rotliegend) beds in the Saar-Nahe Basin, SW Germany, with lacustrine delta, fluvial, and alluvial deposits. Lützner end Rentsch (1975), and Haubold and Katzung (1978) present sedimentological details about the Lower Permian (Rotliegend) of the Saale Basin, E. Germany, with examples of lake deposits. A source for further information about fluvio-lacustrine beds of Permian age throughout Europe is Falke (1976).

Visher (1965) and Picard and High (1972) give reviews and detailed bibliographies of the ancient lake deposits. Feth (1964) provides a review of an-

cient lake deposits from the western United States. Reeves (1968) gives an account of Pleistocene lake deposits in various parts of the world.

Summary

The sediment grain-size distribution of the lakes is concentric, with grain sizes increasing outward. Lakes with clastic deposition have a graded sediment pattern with the fine-grained, basin plain sediment being overlaid by coarser delta sediments. At the surface, fluvial gravel and flood plain muds will appear. When a lake expands due to organic production, the clastic sediments will be covered by organic sediments (peat).

In chemical lakes a great variety of salts could be present as precipitates or as oolites. In reduced environments, iron nodules can also develop.

In organic lakes, there can be either a strong concentration of organic substance (fossilization to oilshale), or development of stromatolites (also in marine environment), or swamp and peat-bog development.

There are three major facies within a lake environment: lake facies, delta facies, and river facies.

Within these major facies exist a series of subfacies, e. g., central plain, margin, fan, canyon, shoreface, beach, delta, delta-topset, -foreset, -bottomset, channels, shoals, and flood plain facies.

The Central Plain contains mud layers with intercalated turbidites. The mud layers are made up of clay/silt interlayering (varves) of millimeter size.

The beds of the lake margin are made up of massive or laminated clay layers. As a lake is silted up, a vertical structure sequence is produced next to the vertical texture sequence. With the progression of a delta, the sediment becomes coarser, and the structures conform with facies changes. When a lake is silted up with organic substances, the vertical sequence will have a clay facies, overlain by organic facies (swamp, peatbog).

The deep lake bottom (Central Plain) in stagnation basins is mainly free of bioturbation. In the aerobic zone bioturbate structures, for instance inter-twined turbiflex burrows, snail traces, fish traces, and shell traces have a better chance to develop.

In temperate areas, a discontinuity layer (thermocline) works as a suspension trap during the summer months. Bottom currents appear as a result of turbidity flows. In eutrophic lakes in the anaerobic zone often reducing environment is developed (sapropel).

Fluvial Environment

General Information

Rivers are the main agents that transport the sediments from land to the coastal regions of seas and lakes, where these sediments are deposited in thick sequences or transported further to continental shelves and deep-sea basins and produce deep-water sediments. Thus, rivers are the major transporting agents, i.e., they are the major agents to transport the sediment from continents undergoing weathering to water bodies–oceans or lakes.

On the other hand, not all the sediment made available by the weathering processes on the land is ultimately carried to seas and lakes. A part of it is also deposited on the land under the influence of fluvial processes. With suitable sedimentary tectonics sequences of fluvial deposits several thousand meters thick can be formed. This is especially true for the lower reaches of rivers, where rivers build up vast flood plains and alluvial plains and where a large amount of fluvial sedimentation takes place. In some cases, extensive and thick deposits of alluvial fans are built along the valley sides and mountain fronts. In other words, rivers are not only erosional and transporting agents, but also the depositing agents.

Deltas and deltaic deposits are formed from the interaction of fluvial and coastal processes. Thus, in a deltaic environment fluvial processes generally play a very important role.

Before dealing with the various depositional environments of a river we shall discuss in brief general characteristics of rivers and fluvial processes. Leopold et al. (1964) and Allen (1965c) provide useful texts of about fluvial processes. Gregory (1977) and Schumm (1977) give further information on the fluvial processes and the nature of fluvial deposits. Miall (1978a) provides a good collection of papers on various aspects of fluvial sedimentation.

Every stream possesses a drainage basin, which provides the water and the sediment to the stream. In this drainage basin small rivulets and streams join together and eventually meet the main stream as tributaries. Adjoining drainage basins are separated by the watershed or divide.

In classical concept a river system can be distinguished into three stages–young, mature, and old. The young stage of a river occurs in mountainous regions. This is the beginning of the river system, where it tends to grow by meeting of various tributaries. Streams of a young stage are mainly erosional agents.

The mature stage of a river system is characterized by the formation of a flood plain and lateral accretion deposits–point bars.

The old stage of a river system occurs in the coastal region. Here, several flood plains of different river systems generally meet together; divides are obliterated. Usually in the old stage a stream produces a distributive network (just opposite of a contributive network of the young stage). Channels become increasingly smaller through repeated division and dividing of discharge. Ultimately these distributaries meet the sea. Figure 369 shows diagramatically these stages. Geologically, i.e., from the point of a depositional environment, mature and especially the old stages of a river system are more important than the young stage, as major fluvial deposition takes place in mature and old stages.

Schumm (1977) designated youth stage as zone 1 (production and drainage basin), mature stage as zone 2 (transfer), and old stage as zone 3 (deposition). Although the processes of production, transfer, and deposition operate in all the three zones, within each zone usually one process is dominant.

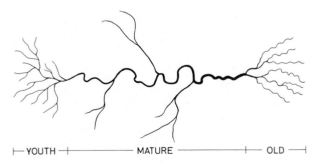

Fig. 369. Schematic representation of young, mature, and old stages of a river

Channel Pattern

By channel pattern is meant the configuration of a river in a plan view. Rivers may acquire varying forms during the flow. The river pattern is a reflection of channel adjustment to channel gradient and cross-section, and seems to be strongly controlled by the amount of sediment load and its characteristics, and the amount and nature of discharge. Most of the workers recognize three channel patterns: Straight, braided, and meandering.

There is a continuous gradation between one type of channel pattern and another. Even the same channel may show changing channel patterns along its length; or same channel reach can be meandering at bankful stages, and floods and appear braided at low stages (cf. Russell 1954). For

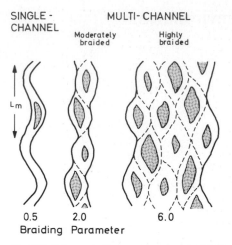

Fig. 371. Schematic diagram showing single-channel and multi-channel (moderately and highly braided) streams. (After Rust 1978a)

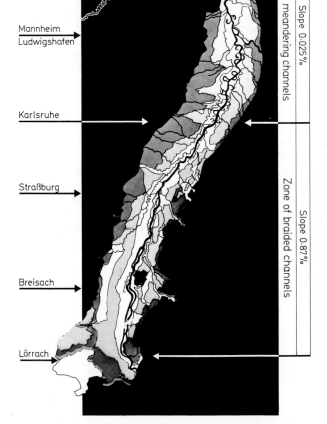

Fig. 370. Schematic diagram showing change in the type of channel pattern of the Rhine River with changing slope. (Modified after W. Schäfer 1973)

example, the Rhine River (Schäfer 1973) shows a variation in its channel pattern along its length, with changing slope (Fig. 370). Because of the continuum of channel patterns Schumm (1963) proposed a classification of channel patterns based on the sinusosity of the channel (ratio of channel length to valley length). He proposed five classes–straight, transitional, regular, irregular, and tortuous channel patterns.

Leopold et al. (1964) distinguished meandering from straight and braided rivers on the basis of sinuosity, defined as a ratio of channel length to down-valley distance. Rivers having a sinuosity of 1.5 or greater are meandering; those below 1.5 are straight and braided.

A new system of classification of alluvial channel is proposed by Rust (1978a), using braiding parameter: the number of braids per mean meander wavelength, the braids being defined by the midline of the channels surrounding each braid bar. Single channel and multi-channel (moderately and highly braided) systems are distinguished (Fig. 371). The channel pattern is further distinguished into low- and high-sinuosity types, at the boundary of sinuosity value of 1.5. Thus, four types of channel patterns are distinguished (Table 22). The meandering and braided types are more common, while straight and anastomosing types are less frequent.

Out of all the parameters which could be used for classifying channel pattern, the characteristic of sediment load is most significant and can be readily recognized in the field and in the geological record. Combining the predominant mode of sediment transport with channel stability, nine subclasses of channel pattern can be recognized (Schumm 1977), as shown in Table 23.

In the following, a short discussion on the three commonly accepted channel patterns is given (Fig. 372).

Straight Channels

Straight channels possess a negligible sinuosity over a distance many times the channel width. As mentioned above straight channels are rare. Thalweg of straight channels is sinuous, and shows deeper parts (pools) alternating with shallower parts (riffles) (Fig. 372). Flow and depositional patterns are similar to those in meandering channels. Straight channels can shift their position by lateral accretion. Erosion takes place along pools and deposition on sediment bars.

Straight channels are rather rare and exist only over short distances. Leopold et al. (1964) suggested that straight reaches never exceeded 10 times the channel width.

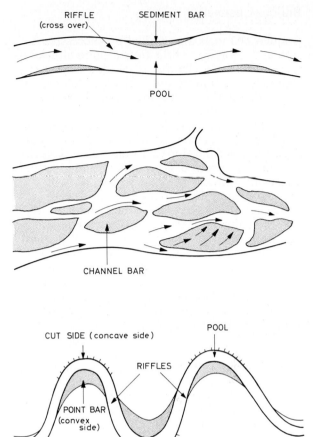

Fig. 372. Straight, braided, and meandering channel patterns

Table 22. Channel classification. (After Rust 1978 a)

	Single-channel (Braiding parameter < 1)	Multi-channel (Braiding parameter > 1)
Low-sinuosity (< 1.5)	Straight	Braided
High-sinuosity (> 1.5)	Meandering	Anastomosing

Table 23. Classification of alluvial channel patterns incorporating sediment type and channel stability. (After Schumm 1977)

Mode of sediment transport and type of channel	Channel sediment (M) (%)	Bedload (percentage of total load)	Channel stability		
			Stable (graded stream)	Depositing (excess load)	Eroding (deficiency of load)
Suspended load	> 20	< 3	Stable suspended-load channel. Width/depth ratio < 10; sinuosity usually > 2.0; gradient, relatively gentle	Depositing suspended load channel. Major deposition on banks cause narrowing of channel; initial streambed deposition minor	Eroding suspended-load channel. Streambed erosion predominant; initial channel widening minor
Mixed load	5–20	3–11	Stable mixed-load channel. Width/depth ratio > 10, < 40; sinuosity usually < 2.0, > 1.3; gradient, moderate	Depositing mixed-load channel. Initial major deposition on banks followed by streambed deposition	Eroding mixed-load channel. Initial streambed erosion followed by channel widening
Bed load	< 5	> 11	Stable bed-load channel. Width/depth ratio > 40; sinuosity usually < 1.3; gradient, relatively steep	Depositing bed-load channel. Streambed deposition and island formation	Eroding bed-load channel. Little streambed erosion; channel widening predominant

Braided Channels

Braided channels are marked by successive division and rejoinings of the flow around alluvial islands. The main channel is divided into several channels which meet and redivide. Channel bars which divide the stream into several channels at low flow are often submerged during high flow. One or more alluvial islands or channel bars may be present in a given channel cross-section. Braiding is most well developed in mountainous reaches of rivers, in streams of alluvial fans, and in streams of glacial outwash plains. Channel bars of such braided streams are commonly composed of gravelly material. Bars tend to be built up by the addition of sediment at the down-stream end and on the lateral parts. The upstream end is partly eroded. The channel bars are composed of coarser-grained lag deposits of the stream which could not be carried by the flow. Once such a channel bar is formed, it may become stabilized by deposition of fine-grained sediment on the top during high flows and may become covered by vegetation.

Williams and Rust (1969) point out that the braid bars are sometimes themselves dissected into smaller bars and shallower channels, and a channel and bar order can be distinguished (Fig. 373).

Braided rivers are characterized by wide channels, and rapid and continuous shifting of the sediment and the position of channels. Fahnestock (1963) noted that during a period of 8 days the stream course in a given reach shifted a lateral distance of more than 100 m. An extreme example of lateral shifting of a braided stream is the Kosi River, a tributary of the Ganges River. During the

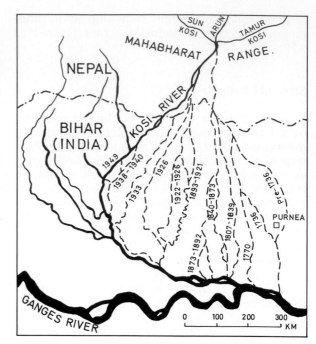

Fig. 374. Sketch map showing lateral shifting of the Kosi River across the alluvial plain. (Modified after Holmes 1965)

last two centuries the Kosi River has shifted its position by about 170 km westward (Fig. 374). The shifting is sporadic; however, in a single year the river may shift 30 km laterally.

Leopold and Wolman (1957) demonstrated that braiding or meandering of river channels depends mainly on the relationship of the channel slope to discharge. In the case of two rivers of the same discharge, braided channels develop on steeper slopes and meandering channels develop more gentle slopes. Steeper slopes cause larger sediment transport and bank erosion, and are often associated with coarser heterogeneous sediments. These factors contribute to braiding. High sediment transport and low threshold of bank erosion are essential conditions for braiding. If discharge is rather high and banks are weak, braiding is common, even in rivers with fine-grained sediments.

Meandering Channels

Leopold and Wolman (1957) called the river reaches meandering if sinuosity of the river is more than 1.5. Figure 375 shows a meandering channel. There seems to be a certain fundamental relationship between the width of a channel and the meander length, and the channel width and the radius of curvature. Meandering channels possess well-defined pools and sediment bars joined by riffles. Sediment bars of meandering channels are better

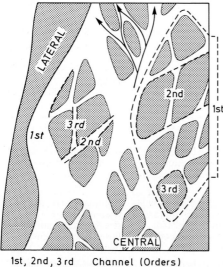

1st, 2nd, 3rd   Channel (Orders)
1st, 2nd, 3rd   Bar   (Orders)

Fig. 373. Schematic diagram showing many of channels and bar orders. (Modified after Williams and Rust 1969)

known as point bars, and constitute the major depositional feature produced as a result of channel action. Meandering of a stream is favored by relatively low slopes, by high suspended load/bed load ratio, cohesive bank material, and relatively steady discharge. The meander wavelength is related to channel width, showing a linear relationship. Bagnold (1960), Leopold and Wolman (1960), and Leopold et al. (1964) discuss the flow mechanism in meandering channels. Although, actual controlling mechanisms of meandering are not well understood, most of the workers consider helical circulation to be the most dominant factor in sedimentation processes in meander.

In short, we can describe the flow mechanism in meanders as follows: Maximum flow velocities are found near the steep concave bank just downstream from the axis of the bend. In the curve there is both a downstream velocity component and a weaker side-ways component, i.e., toward the ou-

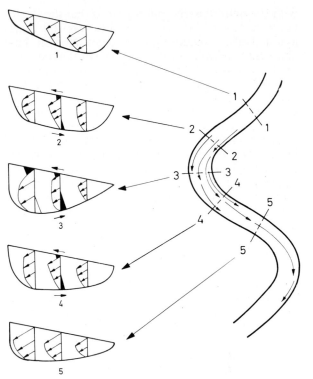

Fig. 376. Scheme showing the pattern of flow in meanders

ter bank (concave side) on the water surface and toward the convex side near the stream bottom. The lateral velocity component is 10 to 20 % of the downstream velocity. Material slumping into the channel by bank caving is caught in the transverse component and carried toward the middle of the channel. However, material eroded from the concave side tends to be deposited on the point bar of the next downstream meander, and not on the point bar opposite concave side. However, vigorous cross-currents in bends may transport some sediment across the channel toward the convex side. The helicoidal flow pattern results from a slight elevation of the water surface against the concave bank (Fig. 376).

Leeder and Bridges (1975) suggest that in the meandering streams flow is not strictly helicoidal, and emphasize the importance of flow separation in the meanders. The process of flow separation causes strong reduction in the effective width of downstream flow, causing increased velocity and erosion on the concave side, with rapid deposition on the convex side with downstream and even local upstream transport.

Meandering channels show slower rates of lateral shifting than the braided channels. However, sometimes even meandering channels may migrate at increased rates. Braiding and meandering is rather interrelated. Coleman (1969) demonstrated

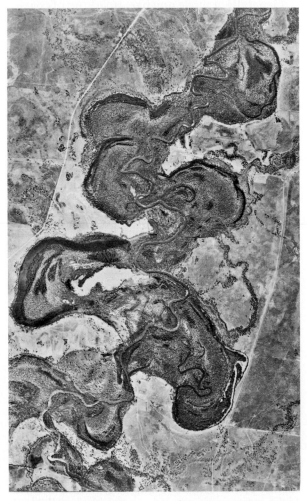

Fig. 375. Photograph of a meandering channel with ox-bow lakes. (After Schumm 1968)

with the help of aerial photographs that the Brahmaputra River in its lower reaches was 100 years ago a typical meandering stream. Now it has changed to a braided river because of the increase in discharge and heavy silt load it is bringing. Rivers in their lower reaches are usually meandering. But if they are heavily charged with sediment and high discharge they show braiding, despite the fine-grained bed material.

Schumm (1968) suggested the term "anastomosing channels" for the channels of alluvial plains that branch and rejoin. They are low-gradient, suspended-load channels carrying only flood water which transports mainly a suspended load and a little bed load. Rust (1978) also recommends the use of the term anastomosing channels for multichannel high-sinuosity patterns.

A river channel may show drastic alterations in the river morphology due to changes in discharge and sediment load, and the river channel pattern would change. Changes in sea level and climate produce still more drastic changes in the channel pattern and sedimentation behavior in an alluvial plain. Schumm (1977) cites several examples for such changes in the channel pattern.

Fluvial Bars

Sediment accumulations to form a topographic high in fluvial channels are termed bars. Meandering rivers are characterized by point bars, developed on the concave side of each bend and related to the flow pattern in a curved channel.

On the other hand, braided streams are characterized by the presence of non-periodic, ephemeral bars occurring at shallow depth, and are usually associated with the pattern of the channel. These bars are mostly tabular bodies which possess one continuous, or several dissected slipfaces near the downstream part of the bar and migrate actively forward or laterally by slip-face avalanching. The hydraulic origin of these bars is not yet clear, and there is also much nomenclature confusion. Further, bars can be depositional forms, erosional forms, or a complex of erosion-deposition processes. The bars in a braided stream may be active bars, which are being modified by sedimentation processes; or inactive bars, which are mostly stationary within a channel. The inactive bars are mostly vegetated islands within a braided stream and often represent erosional remnantes of flood plain deposits or bars which have grown higher due to accretion. The active bars of a braided stream undergoing deposition are called unit bars (Smith 1974), and acquire various shapes and sizes due to local flow conditions, and channel geometry variables.

The active bars are mostly covered by a variety of bedforms, e. g., megaripples or small ripples, depending upon the flow conditions, grain size etc. The exposed bar surfaces of a braided stream exhibit strongly dissected morphology, and lead to complex braided character. This feature is more prominent in the fine-grained braided rivers, e. g., Platte River (Smith 1971b). The dissection of a bar surface is related to the decreasing discharge and low-flow conditions which cause dissection of a single bar into smaller units due to the development of channels.

Krigström (1962), Church (1972), Rust (1972b), Collinson (1970), Smith (1974), Cant and Walker (1978), Bluck (1974) discuss various aspects of bar genesis and nomenclature in the braided streams.

Sometimes, there is a tendency to group bar features and bedforms together, leading to much confusion. The bar features are generally larger forms related to channel width and form; while the bedforms are smaller and are controlled by flow conditions at a given part of the channel.

Generally, a single unit bar is made up of coarse-grained material near the upstream part, becoming finer-grained in the downstream direction. The grain size of the upstream part of a bar possesses the same grain size as the channel, while the downcurrent part of the bars shows finer-grained material than the channel (Smith 1974).

The bars of a braided stream can be given different names according to their morphological form, size, and their relation to the banks of the channel. At present, there is no agreed-upon terminology of the bars, and even the same name is used for morphologically different bar types. Further, there are other inherent problems in naming the bars, as one bar form may evolve into the other (Lewin 1976; Hein and Walker 1977). Smith (1974) distinguishes four basic types of unit bar, i. e., longitudinal, transverse, point, and diagonal (Fig. 377). In the following a short account of some basic bar types in braided streams is given:

Longitudinal bars – These are elongated bodies parallel to the local flow direction, and may acquire different shapes. They often possess a slightly convex surface. They are quite common in the distal parts of gravelly braided streams (Smith 1970; Rust 1972b) and are formed in shallow reaches by unidirectional, bar-parallel flow. Coarse-grained, poorly sorted sediment seems to favor the formation of longitudinal bars. Downstream margin is depositional and shows riffles in gravelly material, and a well-developed slipface in sandy material. Once formed, the bar may grow forward or sideward by slipface avalanching or riffle migration. Internally, the bar is made up of mainly planar cross-bedding of gravels and sands. If the grain size is very coarse, faintly parallel bedded gravels are deposited.

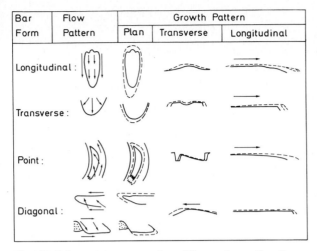

Bar Form	Flow Pattern	Growth Pattern		
		Plan	Transverse	Longitudinal
Longitudinal :				
Transverse :				
Point :				
Diagonal :				

Fig. 377. Schematic representation of flow and growth patterns for typical unit bar forms. In each case, a small amount of vertical accretion is assumed, although each bar may grow laterally without vertical deposition. Dashed lines indicate deposition. (After Smith 1974)

Transverse bars – These show straight, lobate, or sinuous downstream depositional margins (slip faces) and broad upstream surfaces that are either flat or possess axial depressions. Height of bar is usually a few dcm, but sometimes they may be up to 2 m in height. Transverse bars are formed by sediment aggradation to equilibrium, and then grow by down-current extension of slipfaces (Smith 1971b), producing planar cross-bedding (Smith 1972).

Transverse bars are mostly solitary features, but sometimes lobate forms occur in en-echelon fashion and may be termed lingoid bars (Collinson 1970). Lobate or sinuous forms occur in wide or straight channels, while straight-crested forms occur in curved channels. Flow over lobate form is distributed radially, and margins are either high-angle foresets or low-sloping riffles.

Transverse bars are produced in areas of flow-line divergence due to channel widening or abrupt depth increase. In sandy braided streams transverse bars are more common than in the gravelly streams, and are abundantly developed in the distal part of a braided stream.

Martini (1977) discusses that in tight bends of a stream helicoidal flows are important in shaping transverse bars, which are molded into a variety of chute and bar patterns.

Jackson (1976b) describes transverse bars from the meandering Wabash River, U.S.A., developing in shallow water depths, producing up to 2 m-thick planar cross-bedding. Pryor (1973) describes these transverse bars as scale-like bedforms, and points out that their cross-bedding shows thick foreset laminae, which are concave upward.

Boothroyd and Nummedal (1978) describe ex-tensive development of transverse bars from the sandur rivers of Iceland. They are developed as rhombic-shaped forms and are termed "high-stage lingoid bars", which upon migration produce planar cross-bedding. Upon emergence, the bar is dissected into numerous lobate slipfaces producing complicated forms, called "low-stage lingoid bars". In addition to the development of planar cross-bedding, these forms during the last stages undergo some vertical accretion to produce a complex arrangement of bedding structures, capped by small-ripple bedding.

The sandy, braided South Saskatchewan River, Canada, also shows development of transverse bars (bar or cross-channel bars) related to the flow expansion, and internally made up of tabular-planar cross-bedding (Cant and Walker 1978).

Diagonal bars – These possess long axes oriented obliquely to the main flow (Church 1972). Diagonal bars are roughly triangular in cross-section with the downcurrent margin consisting of either slip faces or riffles. They are produced when channel cross-sectional flow is asymmetrical, mostly due to curvature in the main channel. If movement is by migrating slipface planar cross-bedding is produced.

Marginal bars – This term includes lateral bars (Bluck 1974) and point bar (Smith 1974) and denotes bars formed in the curved braided streams almost attached to the stream bank. They may be related to some type of transverse bar. The point bars of a braided stream are separated from the inner convex bank by a smaller channel. They show low slopes dipping towards the outer side, and may have foreset margins directed towards the inner bank. However, the term point bar in the braided streams should be avoided.

The lateral bars are attached or about to be attached to a bank. Bluck (1974) gives a detailed account of the development and bedding structures in lateral bars (Fig. 378). The lateral bar shows a well-developed steep face of gravel (riffle face). Upstream to this riffle face is a shallow channel leading upstream to the pool. The water through this channel shoots over a steep lee-face. The main body of the bar is the bar-platform, constructed by sediments brought downstream and eroded from the cut side. The bar platform is built as the channel migrates laterally towards the cut side. Sand is transported diagonally across the bar to build sand sheets, wedges, and small deltas into the abandoned channels. At the downstream end of the bar is an area of still water – slough. The slough represents an abandoned pool reach formed by lateral channel migration, and filled by muddy sediments.

The alternate bars of straight channels can also be considered a variant of lateral bars and they also possess active slipfaces, which upon migration produce planar cross-bedding.

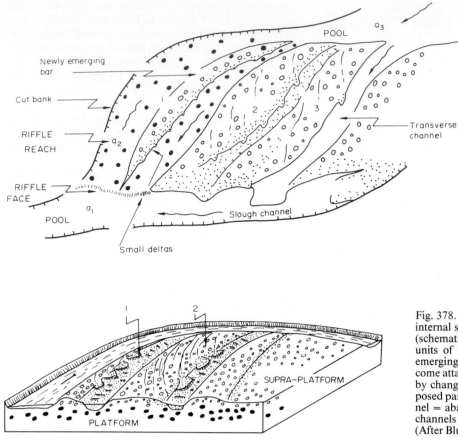

Fig. 378. Morphology, terminology and internal structure of a typical lateral bar (schematic). *1, 2,* and *3* are individual units of the bar where *1* is the newly emerging bar which will eventually become attached to *2,* so that the bar grows by changing phase a[1], etc. Riffle = exposed part of bar platform; slough channel = abandoned pool reach; transverse channels are often former riffle reaches. (After Bluck 1974)

Compound Bar-Like Features (Channel Bars)

In braided streams, especially the sandy type shows large bar-like features, which are built up by a complex depositional-erosional processes. Such areas are comparable to the large point bars of the meandering streams. Most of these large braid bars do not themselves migrate; however, sometimes they do possess active slipfaces and undergo considerable migration themselves. Important types of compound braid bar are:

1. Side bars – These are huge areas attached to the banks of a river, and are exposed during low flow conditions. Internally they are made up of complex assemblage of bar-migration and megaripple bedding. They are described by Collinson (1970).

2. Mid-channel bars – These are huge areas of deposition in the middle of the channel with a rather complex depositional history. Colemann (1969) describes huge mid-channel bars from the Brahmaputra River.

3. Sand flats – Cant and Walker (1978) describe sandy areas in the braided streams, ranging from 50 m to 2 km in length and 30 to 450 m in width, which are complex areas made up of smaller features such as sand flats. They are positive areas of compound depositional and erosional histories within a braided stream. They are initiated on the top of cross-channel bars and grow by accretion along the margin and on the top surface. The sand flats are not unit bars. Such sand flats may be mid-channel or marginal attached to the bank.

Form of a Channel and Fluvial Processes

The shape of the cross-section of a river channel at a given point is a function of flow, the quantity and character of the sediment in movement through the section, and the character of the materials making up the river bank and bed of the channel (Leopold et al. 1964).

Current velocity in a channel depends upon several factors, the most important being the energy gradient (generally approximated by the watersurface slope), water depth, and bed roughness. Current velocity varies from one part of a given cross-section to another and from one cross-section to the other. Most rivers experience a wide range of flows.

Flow in open channels and sediment transport has been intensively studied. Leliavsky (1955), Sundborg (1956), and Chow (1959) provide important summaries on various ideas of flow in channels.

A river bed is composed of noncohesive material and is easily modified into various bedforms depending upon the energy of the flow. However, tranquil flow is generally active, so that small and megacurrent ripples are abundantly produced. In larger rivers like the Mississippi, Brahmaputra, etc., even giant ripples are abundantly produced. Dawdy (1961) demonstrated that rapid flow (upper flow regime) is rather common in streams. Probably conditions of rapid flow are more commonly achieved in fluvial environments than in other environments. One of the highest velocities measured in river channels is of the order of 7 to 8 m/sec (Leopold et al. 1964).

Megaripples are the most abundant bedforms in the fluvial channels, and have been reported from various rivers, e. g., Tana River (Collinson 1970), Ganga, Yamuna, and Son Rivers (Singh and Kumar 1974), Wabash River (Jackson 1976b), Sasketchewan River (Cant 1978). Often, larger ripple forms, i. e., sand waves and giant ripples are also recorded. Sand waves mostly develop in greater water depths and coarser sediments (Jackson 1976b). Sometimes, the transverse bars (including lingoid type) are included as bedforms, and not as bar features (e. g., Jackson 1976b).

Rivers transport large amounts of sediments. The transported sediment can be grouped into bedload–material moving along the river bed by process of saltation and rolling and suspension load–material moving as suspension. During floods the amount of material transported by rivers increases to several times that transported during normal conditions. There are various methods for the measuring sediment load in rivers; even theoretical computations for calculating bedload and suspension load have been made (see Sundborg 1956; Leopold et al. 1964; and Allen 1965). Equations of Einstein (1950), Colby (1964) and Colby and Hembree (1955) are especially useful in calculating bedload and total sediment load.

The suspension load is rather important in the deposition on natural levees and flood basins; whereas bed load is deposited as channel lag deposits in the lower part of point bars. Sundborg (1956) found that the finest material transportable as bed load is about 0.15 to 0.20 mm. However, this limit is rather diffuse and should not be considered as absolute.

Coleman (1969) made a detailed study of bedforms in the Brahmaputra River. He observed small-current ripples, megacurrent ripples, and giant-current ripples, and also the plane bed phase of the upper flow regime marked by large-scale lineations running parallel to the current. These lineations are made up of horizontal laminae. At any locality several bedforms may exist simultaneously because of changing flow conditions, water depth, sediment availability, and the stability of various bedforms. For example, small ripples can be actively superimposed on megaripples, megaripples on giant ripples. This may result in complex bedding structures.

Coleman (1969) found that a sediment layer up to 1 m thick can be deposited during 24 hours. Migrating giant ripples even produce sediment 5 to 6 m thick during a single day. These large bedforms are obviously the major agents for the transport of the bed load. The comparison of echo-sounding profiles over a complete flood cycle showed clearly that there is net accumulation of sediment only during certain periods of flood; during remaining periods the bed is stable or even undergoing erosion.

Fluvial Environments and Their Deposits

What we are concerned with here are the sediments deposited in various river regimen and their characteristics.

Sediments deposited by rivers have been classified in several different manners. Most of the geomorphologists distinguish between vertical accretion and lateral accretion deposits (Mackin 1937; Fisk 1947; Leopold and Wolman 1957). Lateral accretion results from lateral migration of the channel, resulting in redistribution of the available sediment. This process is active on point bars. Vertical accretion denotes vertical deposition and upward growth of the flood plain by deposition of sediment from suspension. This classification is too simple; moreover, it has a drawback that even lateral accretion results in vertical growth, and it is not always possible to distinguish the two.

Leopold et al. (1964) distinguished 8 subenvironments and sedimentary features in flood plains. Happ et al. (1940) distinguished six types of alluvial sediments–channel-fill, vertical accretions, flood plain splays, colluvial, lateral accretion, and channel lag.

For practical purposes fluvial deposits can be grouped into three major groups:

1. Channel deposits. They are sediment deposits formed mainly from the activity of river channels. They include channel lag deposits, point bar deposit, channel bar deposit, and channel fill deposit.

2. Bank deposits. They are sediment deposits formed on the river banks and are produced during flood periods. They include levee deposits and crevasse splay deposits.

3. Flood basin deposits. They are essentially fine-grained sediment deposits formed during heavy floods when river water flows over the levees into the flood basin. They include flood basin deposits and marsh deposits.

Figure 379 shows the scheme of three types of deposits. Figure 380 is map of the Rufiji River, showing the distribution of various subenvironments.

In some rivers, differentiation between bank deposits and flood basin deposits do not exist. This is the case in rivers which migrate actively laterally, so that the flood basin is not well developed. In such cases, fluvial deposits can be differentiated into two groups:

1. channel deposits
2. flood plain deposits.

In such cases, flood plain deposits are rather similar in character to levee deposits.

Allen (1965) proposed a very similar classification of alluvial deposits. In the alluvial plains, in the lower reaches of rivers sediment grain size is rather uniform, and the distinction between various subenvironments is not prominent.

In the following, we shall describe the characteristic features of various deposits.

Channel Lag Deposits

Sediments transported by a river cover a wide range of particle sizes. Silt and clay is carried away much faster in suspension and is deposited in flood basins. In the channel itself the coarser sediment—gravels and pebbles lag behind, while the sand moves as bed load. It represents residual concentration and accumulates as discontinuous lenticular patches in the deeper parts of the channel—channel lag deposit. Such coarse-grained lenticular accumulation quickly becomes covered by finer-grained sediments and is preserved. A similar accumulation of coarse material at the channel bottom also takes place in channels and gullies of intertidal flats.

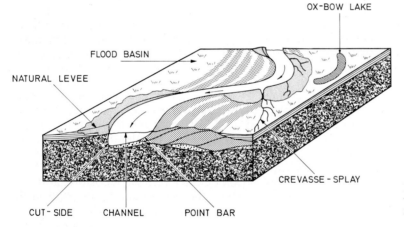

Fig. 379. Diagrammatic representation of various kinds of fluvial deposits. (After Singh 1972)

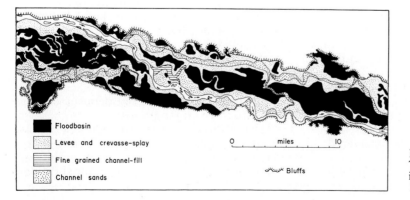

Fig. 380. Map of part of the Rufiji River, Tanganyika, showing the distribution of various fluvial subenvironments. (After Allen 1965c; originally from Anderson 1961)

Fig. 381. A point bar of the Klarälvan River showing bar ridges of scrolls. (After Allen 1965c)

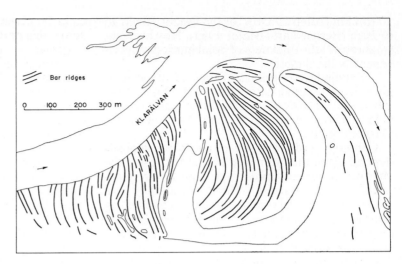

In the lower reaches of rivers where coarser-grained sediment is not available, drifted wood, large sediment blocks of unconsolidated sediment, mud pebbles, and dead organisms are concentrated in channel lag deposits. The slumped blocks from the cut side may become preserved as large, angular blocks, or may be reworked into smaller, sub-rounded mud pebbles. Observations in the point bars of the Ganga River have shown that large mud pebbles are present abundantly in the lower to middle point bar deposits, and locally they are concentrated in pockets and layers.

In areas where adjoining terraces contain abundant kankar (calcrete concretions), they are one of the important constituents in the channel lag deposits. Kankar lenticles are also distributed throughout the point bar deposits. The Chambal and Yamuna Rivers in the western part of Uttar Pradesh, India, contain abundant kankar layers and pockets.

Channel lag deposits never make thick layers, and they are invariably discontinuous. They occupy the lowest part of a channel or point bar sequence. If present, they indicate the base of the channel. Mackin (1937) suggested that the maximum depth of a coarse-grained channel deposit, if there is no deposition on point bars, etc., is equal to maximum depth of effective flood scours. Fisk (1944), Arnborg (1958), Lattman (1960), and others give descriptions of channel lag deposits in modern rivers.

Point Bar Deposits

Point bars are the most conspicuous geomorphic features of a meandering stream; at the same time deposition on point bars is the major process of

sedimentation in meandering river channels. In ancient records point bar deposits may constitute the major part of the fluvial sequence.

Shape and size of point bars vary with the size of the river. In smaller streams point bars are simple depositional features on the convex sides of the meanders dipping gently toward the channel. In large rivers point bars are composed of scroll-shaped ridges (scroll bars) alternating with depressions (swales). Scroll bars can be several meters higher than the adjoining swales. Swales are filled with fine-grained muddy sediments, or marshes may develop in them. Each scroll bar of a point bar represents the results of channel migration during a flood. Repeated migration during floods produces scroll bars. Fisk (1947) describes such features from the Mississippi River; Sundborg (1956) describes them from the Klarälven River (Fig. 381).

In the Wabash River, the scroll bars are developed mainly near the downcurrent part of a point bar and migrate towards the inner bank by avalanching on slipface (Jackson 1976b). The scroll bar is active only over a narrow range of conditions, as it becomes subdued at very high floods, and is exposed at lower flows. The scroll bar shows cross-bedding, oriented 50–90° away from the channel strike and away from the channel. Thicker sets of cross-bedding are separated by low-angle reactivation units made up of erratically oriented small ripple bedding.

A sequence of developing scroll bars may produce a succession of lenticular-shaped sandy units embedded within the muddy sediments of swales (see upper part of Fig. 388).

Deposition on point bars results from lateral migration of a meandering river during flooding. A point bar sequence may be as thick as the depth of the river. For the Mississippi River it can be as thick as 20 to 25 m (Fisk 1944, 1947), for the Niger River,

10 to 15 m (Allen 1965b), for the Brazos River, 15 to 20 m (Bernard and Major 1963). However, for smaller river the thickness of point bar sediments is generally 1 to 3 meters.

A detailed description of point bar deposits is given by Adler and Lattmann (1961), Fisk (1944, 1947), Folk and Ward (1957), Frazier and Osanik (1961), Harms et al. (1963), and McGowen and Garner (1970).

Lithology and grain size of point bar deposits depend upon the grain size available. Beside the channel lag deposit point bar deposits are composed of the coarsest sediment available in a stream. On the very-coarse grained channel lag sediments, coarse-grained point bar sediments are deposited. If a wide range of grain size is available, grain size decreases upward in a point bar sequence. Generally, a muddy or very fine-grained layer is present on the top as a thin veneer. This mud drape, if present, marks the top of a genetically related sequence. If rivers are carrying gravel-sand material, the change is from gravel, coarse sand to fine sand, and silt on the top. In rivers carrying fine-grained material the change is from fine sand layers near the bottom to muddy and clayey sediments near the top.

In the preserved record the uppermost fine-grained part of a sequence may be eroded before the deposition of the next sequence. Thus, only incomplete sequences are preserved. As pointed out by Allen (1965), sorting of individual units of point bar deposits is rather good.

Sometimes in the lower part of the point bar sequence several channel lag layers may become incorporated. Cross-bedding of current ripple origin are the major bedding types in point bar deposits. A major part of a point bar sequence is deposited during the floods, especially when water starts subsiding. The rate of deposition is very high.

Leopold et al. (1964) studied deposition and erosion processes on a single point bar and cut side over several years at Watts Branch near Rockville, Maryland. They found during the period of their study that the channel moved laterally over a distance more than the channel width; however, constant deposition at the point bar kept the channel width constant. The net volume of deposition was equal to the net volume of erosion. Overbank deposition on flood plains was insignificant. The sequence within this bar showed that contacts between sediments of a different nature were more or less horizontal, with a tendency to be parallel to the older surface profiles. Gravelly material normally present in the channel bed had been locally deposited well above the level of channel bed. The nature of sediment of point bar was variable, grading, however, into finer sediments near the top of the section. The bulk of sediment was silty in nature. The individual units, were discontinuous and lenti-

cular in shape. Figure 382 shows a cross-section across this point bar at Watts Branch, Maryland.

In an ideal sequence of point bar deposits large-scale cross-bedding (megaripple bedding) seems to be the most abundant in the lower part. Thickness of units decreases upward. In the lowermost part units as thick as 1 m or more are present, becoming only a few centimeters thick upward. In this zone some scour-and-fill structures may be present.

Above the zone of megaripple bedding there follows a zone of small-ripple cross-bedding and climbing-ripple laminations. In some rivers thick units of climbing-ripple laminations are produced.

This small-ripple bedded zone is followed by units of horizontal bedding. Sometimes horizontal bedded units are present between megaripple and small-ripple bedded units. In many cases, small-ripple bedding is interbedded with horizontal bedding. This horizontal bedding has been interpreted as originating in the upper flow regime (e.g., Harms and Fahnestock 1965). But we feel that much of this horizontal bedding is produced by deposition of suspension clouds due to decrease in turbulence or fluctuations in current velocity. A major part is deposited during the last phase of a receding flood, when velocity is reduced and with that also the competency of the flow. This results in the deposition of suspended sand in the form of horizontal lamination. Locally, antidune cross-bedding may be found.

On the top of the point bar sequence silty and clayey layers are present. However, this ideal profile is not always well developed. Sometimes even in lower parts of the sequence climbing-ripple lamination may be present together with large-scale cross-bedding. As we know, within a flood phase there are fluctuations–decrease and increase of the energy of the environment. Under such conditions, the sequence is made complicated.

Point bar deposits may show accumulation of drifted plant material, freshwater molluscs, mud pebbles, etc.

During receding floods the larger point bars undergo significant modification. Singh and Kumar (1974) describe the development of delta-like lobes in the lower point bar of the Ganga River which migrate laterally, producing a set of cross-bedding oriented at right angles to the direction of flow.

In general sedimentary units of point bars are discontinuous and lenticular in nature. Longitudinal cross-bedding, as commonly observed in the bars of channels of intertidal flats, are not well developed. It may be because of the very rapid rate of sedimentation on fluvial point bars, so that the surface of the point bar is not preserved to be identified. It may also be due to the sandy nature of the point bar sediments of a river. However, in smaller tributaries, especially in coastal plains and deltas where much fine material is available, longitudinal cross-bedding (epsilon cross-bedding) may develop and is recognizable.

Swales within a point bar are filled with fine-grained suspended sediment. In larger point bars, e.g., the Mississippi River, swales may be a few hundred meters in width, several kilometers in length, and 4 to 5 m thick. Swale deposits are generally well-bedded silt and clayey sediments. In ancient sediments they would occur as elongated, rather narrow muddy beds within a point bar sequence. The bottom of such muddy layers is concave-upward in shape. Such clayey sediments are extremely rich in organic debris. Swale-fill deposits have been described by Fisk (1944, 1947).

Fine-Grained Point Bar Deposits

Steinmetz (1967) made a detailed study of a single point bar of the Arkansas River at Wekiwa, Oklahoma. He attempted to correlate the various hydrographic processes to the sedimentary structures observed in the excavation of the point bar. In the following we shall summarize his major results:

Wekiwa point bar was deposited rapidly, during two floods–May 19 to 22, 1957, and October 3 to 6, 1959. During this time roughly 459.000 m³ of sand

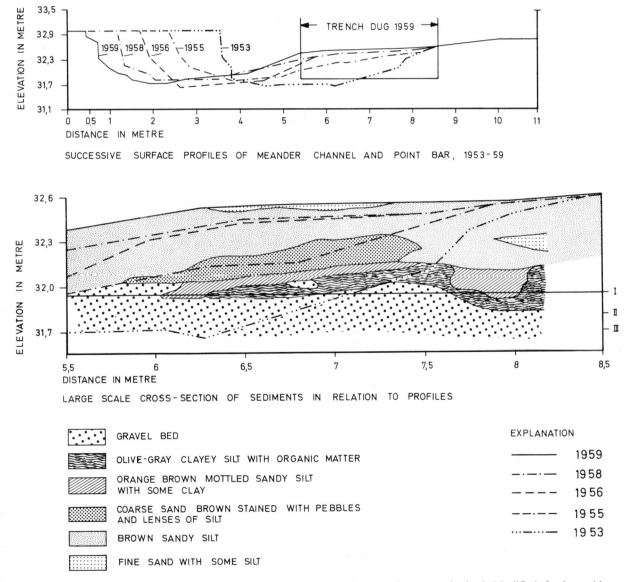

SUCCESSIVE SURFACE PROFILES OF MEANDER CHANNEL AND POINT BAR, 1953-59

LARGE SCALE CROSS-SECTION OF SEDIMENTS IN RELATION TO PROFILES

EXPLANATION

GRAVEL BED	
OLIVE-GRAY CLAYEY SILT WITH ORGANIC MATTER	———— 1959
ORANGE BROWN MOTTLED SANDY SILT WITH SOME CLAY	—·—·— 1958
COARSE SAND BROWN STAINED WITH PEBBLES AND LENSES OF SILT	— — — 1956
BROWN SANDY SILT	—··—··— 1955
FINE SAND WITH SOME SILT	···—···— 1953

Fig. 382. Scheme showing successive cross-sections measured over several years on a point bar at Watts Branch near Rockville, Maryland. Successive profile and distribution of sediments in the cutting are emphasized. (Modified after Leopold et al. 1964)

was deposited, with a maximum thickness of 13 m in a total of 156 hrs. The main bedding type is large-scale cross-bedding. An upward decrease in the thickness of crossbed sets in a single flood deposit is apparent. On the top of the sequence laminated silty layers deposited from slack water are present. Within the sequence scours and pockets with an accumulation of organic debris are found. On the top, wind action has produced a rather thick zone–up to 1 m–of sand dunes.

The grain size show an overall upward decrease in grain size. Sorting in general improves upward, but in some cases decrease in sorting is apparent. In general, sand is moderately to well sorted. Cross-bed data show a large scatter and do not conform with the direction of flow of the river, and two closely spaced profiles do not show a similar direction. Figures 383–387 show various characteristics of point bar deposits of the Arkansas River.

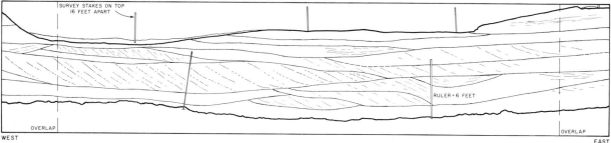

Fig. 383. Sedimentary structures in point bar sediments. The sequence of the point bar sediments is shown. (After Steinmetz 1967)

Fig. 384. Photograph showing details of bedding structures. The main bedding is large-scale cross-bedding. Cross-bedded units become smaller upward in the sequence. The upper quarter of the photograph is a sequence of a second flood. (After Steinmetz 1967)

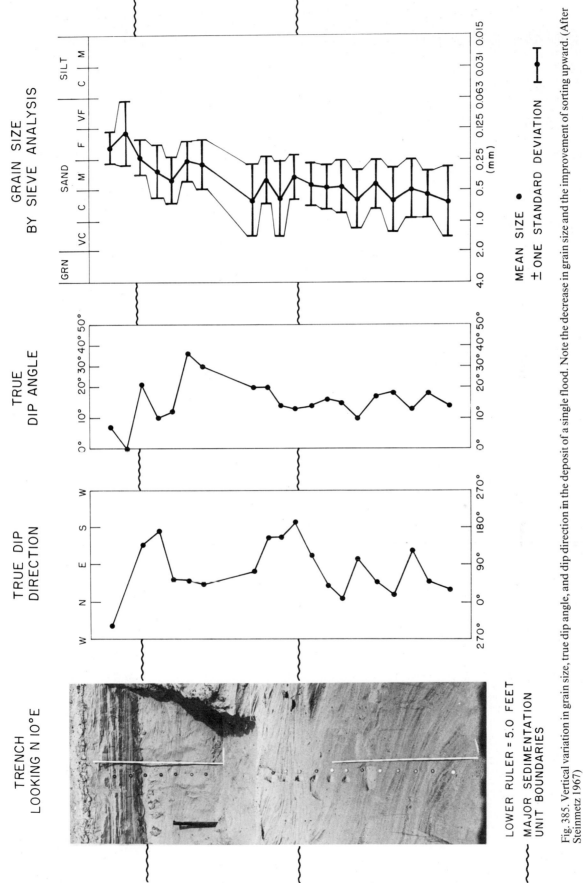

Fig. 385. Vertical variation in grain size, true dip angle, and dip direction in the deposit of a single flood. Note the decrease in grain size and the improvement of sorting upward. (After Steinmetz 1967)

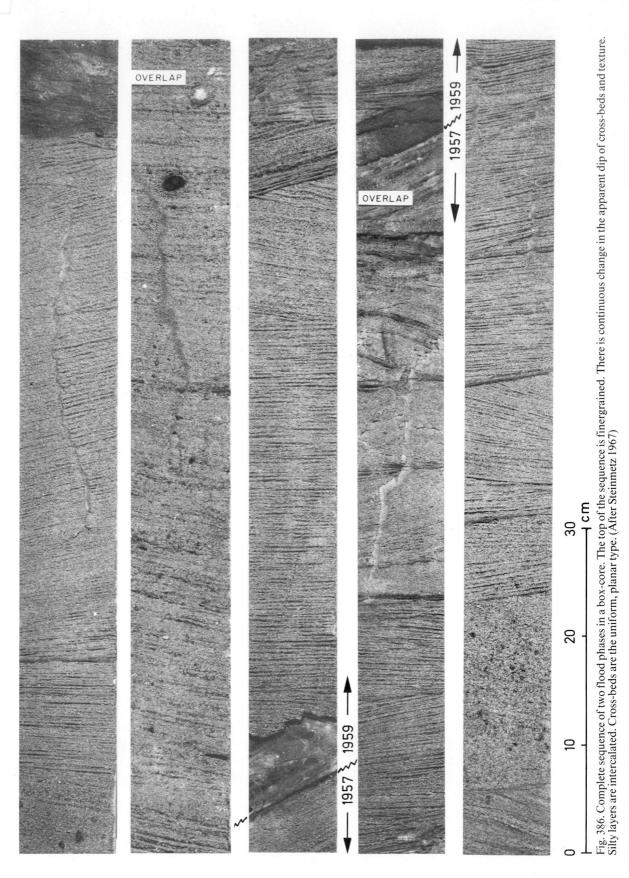

Fig. 386. Complete sequence of two flood phases in a box-core. The top of the sequence is finergrained. There is continuous change in the apparent dip of cross-beds and texture. Silty layers are intercalated. Cross-beds are the uniform, planar type. (After Steinmetz 1967)

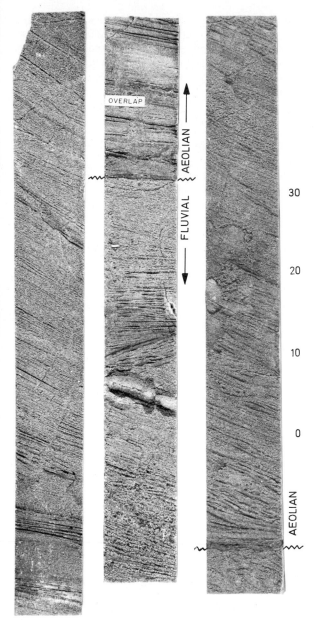

Fig. 387. Point bar sequence showing aeolian sediments within fluvial deposits. Aeolian sediments differ only in their better sorting. (After Steinmetz 1967)

Stray pebbles are present in an otherwise uniformly sized material. Silty layers and climbing-ripple lamination are found, but they are not so abundant.

Deposition at the point bar is very rapid. Single cross-bed units are produced within a few hours, and laminae within minutes. Thus, a characteristic of point bar deposition is periodic, extremely rapid deposition.

Sundborg (1956) made a detailed study of fluvial sediments and processes of the Klarälven River, Sweden. He studied the process of development of

a point bar (Fig. 388). He found that growth of a point bar is a slow process, and starts with the formation of a longitudinal bar (scroll bar) on the convex side of a meandering river with a steeper side toward the river bank. Gradually a steeper slope toward the river bank develops. Slowly it migrates bankward and merges into the natural levee.

As it attains a higher position, only suspended material is deposited which drapes all over the sand bar. Soon the next embryonic bar starts building near the channel and migrates toward river bank. In a preserved record there are elongated curved lenticular sand bodies embedded in a rather fine-grained suspension deposit (Fig. 388).

The Gomti River, India, is a small meandering river carrying only fine-grained sediments, and shows some characteristic point bar sequences (Kumar and Singh 1978). The maximum discharge of the river is 7 m³/sec, and river slope in the middle reaches is 11.5 cm/km. The median of the sediments is about 0.1 mm, and even the coarsest grains can be carried in suspension. Deposition of the sediments takes place by mass-settling from graded suspension. Due to uniformity of grain size of the sediments, there is little variation in the primary sedimentary structures of point bar and natural levee deposits. The deposits mainly show horizontal bedding, small ripple bedding and climbing ripple lamination. In the point bar sequences, due to absence of coarser sediments, there is no fining-upward tendency, although the top of the sequence is always capped by a mud layer (Fig. 389). The sequence of sedimentary structures in natural levee deposits is similar to that of point bar deposits.

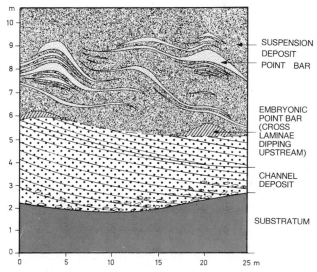

Fig. 388. Point bar sequence of the Klarälvan River. (Modified after Sundborg 1956)

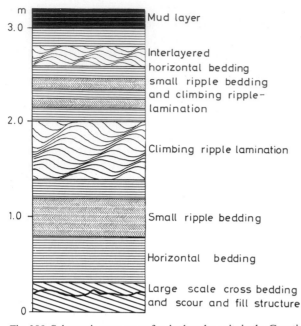

Fig. 389. Schematic sequence of point bar deposits in the Gomti River sediments, India. (After Kumar and Singh 1978)

produced, where sequence of sedimentary structures can be correlated to the flood stages. However, the distribution of various sedimentary structures shows high variability. Several overbank deposits can be stacked one above the other, which are not related to lateral accretion.

Coarse-Grained Point Bar Deposits

Mostly, the coarse-grained rivers are present in the upper reaches of alluvial plains and show extensive braided character. However, rivers bringing coarse-grained sediments may also show meandering patterns, making extensive point bars. Such coarse-grained point bars show a more complex facies differentiation than the finer-grained point bar deposits.

McGowen and Garner (1970) investigated sedimentary structures in point bars of the Amite River, Louisiana, and the Colorado River, Texas, composed of coarse-grained sediment (Figs. 391 and 392). These streams are of low sinuosity (1.4 to 1.7) and rather high gradient. They can be characterized as bed load streams and they transport mainly coarse sand, pebble- to cobble-sized gravels. The rivers are meandering because river banks are made up of erosion-resistant muddy sediments and are vegetated.

McGowen and Garner (1970) say that point bars of these rivers can be distinguished into a lower point bar and an upper point bar. The upper point bar shows development of chute and chute bar sediments. Chutes are characterized by relatively steep sides, flat bottoms, and sinuous trends. They are deepest where they meet the main channel and decrease in depth steadily downcurrent, where scours give way to deposition in the form of chute bars. They can be 2 to 5 m deep, 5 to 7 m wide, and extend for hundreds of meters. However, not all the coarse-grained point bars show well-developed chute and chute bars. Figures 391 and 392 show various aspects of coarse-grained point bar.

The sequence of sedimentary structures in a lower point bar is similar to sedimentary structures of the cross-over (straight channel reaches between two adjacent meander loops).

The vertical sequence of a coarse-grained point bar shows no gradual fining upward characteristics. However, at the top of the sequence a mud drape rich in organic matter is present. Major units are as follows (see Fig. 393) (bottom to top):
1. Trough-fill cross-bedding or homogeneous sediment (scour pool deposit)
2. Small foreset cross-bedding and small trough-fill cross-bedding (lower point bar/cross-over)
3. Large foreset cross-bedding (chute bar)

However, in natural levee deposits units are thinner (20–50 cm) and the capping mud layer thick, while the point bar sequences are 50 cm–2 m thick. Sometimes up to 50 cm-thick units of climbing-ripple lamination are present, representing deposition from a single deccelerating flow, where angle of climb and content of clay increase upwards (Fig. 390).

Ray (1976) studied the deposits of the upper part of a point bar in the Mississippi River, and observed that at least in the upper part of point bars overbank deposition (vertical accretion) plays a significant role. During each major flood a characteristic vertical sequence of overbank deposit is

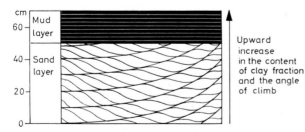

Fig. 390. Sequence of sedimentary structures in a natural levee deposit. The unit is made up of climbing ripple lamination with a top mud layer. There is a systematic increase in the angle of climb of the pseudobeds from bottom towards top of the sequence, in response to gradual increase in the suspension fallout. Gomti River, India. (After Kumar and Singh 1978)

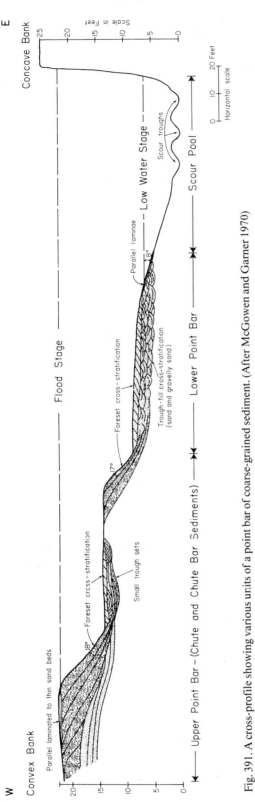

Fig. 391. A cross-profile showing various units of a point bar of coarse-grained sediment. (After McGowen and Garner 1970)

4. Parallel laminae, small foreset cross-bedding, and scour-and-fill (flood plain).

A complete idealized fining upward sequence with changing sedimentary structures is shown in Fig. 394.

The study of the River Endrick, Scotland (Bluck 1971) illustrates very well the variability and complexity of the sequences developed within a coarse-grained meandering river. The river carries gravels and finer sediments and shows a prominent down-current decrease in the grain size along the length of the river. The channel and flood plain are well developed. Within the channel pool and riffle morphology is prominent, along with well-developed point bars. The riffles and point bars are also the main areas of deposition, where also active particle size segregation and changes in facies development take place. A riffle is made up of gravelly sediments near the upstream end (riffle head) with a sandier downstream part (riffle tail). They are made up of cross-bedded gravels and sandy units, often showing ripple bedding.

The point bar is a complex area of deposition and is differentiated into a lower part of uniform material (bar platform), and an upper part with segregated belts of sand and gravels (supra-bar platform). On the supra-bar platform bar-lee and inner accretionary banks may develop. The upstream part of the bar (bar head) is essentially made up of gravels; while the downstream part (bar tail) is mostly sandy in nature. The upper part of the point bar is active only during high-water stages.

The bar head sediments are made up of inclined gravel layers interstratified with sheets of rippled or horizontally bedded sand. The margins of the bar head show steeply dipping gravelly beds with sandy layers. The bar tail sediments are made up of lenses or sheet-like units of cross-bedded sand and pebbly sand with thin layers of pebbles and organic matter. The bar-lee deposits are made up of foresets of silty clay bands or leaf beds with thicker units of rippled sand. The inner accretionary bank deposits are also made up of steeply dipping alternations of rippled sand and silty vegetation horizons. They grade upwards into silts and clays of flood plain deposits. The lower bar platform is made up of a series of thick, low-angled crossbeds.

The cross-bedding in these deposits shows high variability in direction, and is mainly controlled by the local flow conditions. The directions determined in sandy and gravelly units differ significantly from each other. Bluck (1971) could demonstrate that bar sediments grow in the downstream direction into the pool. Due to lateral sedimentation on the point bar and a downstream movement of the meander loop, a complex interlocking litho-

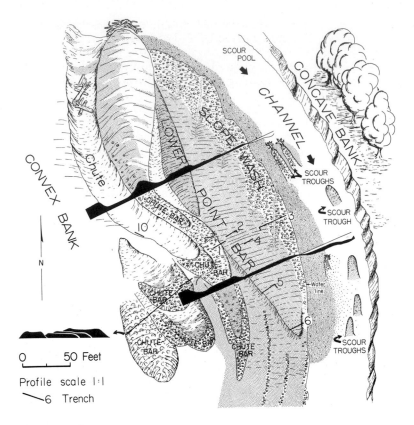

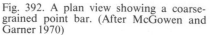

Fig. 392. A plan view showing a coarse-grained point bar. (After McGowen and Garner 1970)

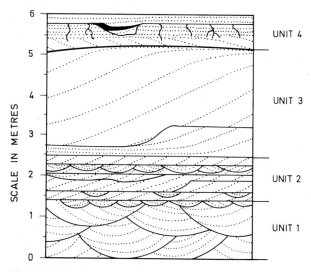

Fig. 393. Vertical sequence of coarse-grained point bar deposits (gravels), showing various bedding features. Four units can be distinguished. Unit 1–scour pool deposit; unit 2–lower point bar deposit; unit 3–chute bar deposit; unit 4–flood plain deposit. (Modified after McGowen and Garner, 1970)

facies pattern is produced. The vertical sequence would vary depending upon the position of section in the point bar, thus producing more variability within a single bar than those between bars. The preservation of these sequences would further depend upon mode of river migration, i. e., by avulsion, by cut-off, or by moving phase of meander.

A coarse-grained point bar (mixed sand and gravel) from the upper Congaree River, U.S.A. is studied by Levey (1978). The gradient is 0.5 m/km, sinuosity index 1.75, and mean discharge 264 m³/sec. The point bars of the upper Congaree River show considerable variation in size, topography, morphologic features, facies and grain size pattern, and four characteristic facies are recognized (Levey 1978):

1. *Bar-apex facies* – The proximal part of a bar made up of either pure gravel or a mixture of coarse sand and gravel showing bimodal distribution. Stratification is poorly developed. These represent lag accumulation of coarse sediments, where clasts show mainly transverse orientation to the flow, and imbrication is poor.

2. *Mid-bar facies* – Middle part of the bar is made up of coarse-to-medium sand, where megaripples are dominant bedforms. Internally, crossbedded units of 10–50 cm set thickness are present, laminae show high angle of dip (32°), reactivation surfaces are very abundant, indicating interruption of steady, continuous bedform movement. Horizontal bedding of upper flow regime is also present.

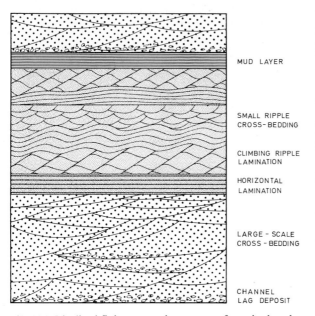

MUD LAYER

SMALL RIPPLE
CROSS-BEDDING

CLIMBING RIPPLE
LAMINATION

HORIZONTAL
LAMINATION

LARGE - SCALE
CROSS - BEDDING

CHANNEL
LAG DEPOSIT

Fig. 394. Idealized fining upward sequence of a point bar deposit. The grain size becomes finer upward. The topmost layer is covered by a mud drape

Transverse bars are also present and upon migration produce large-scale tabular cross-beds with set thickness of mostly > 50 cm. Contorted lamination within the cross-beds are present, related to liquefaction during deposition.

3. Distal bar facies – The downstream part of the bar is characterized by chute channels and chute bars, occurring mainly in the higher part of the point bar. Chute channels are straight-sided channels up to 1 m deep and several meters wide with a chute bar at the terminus which is narrow and long with a lobate downstream end. Chute- and chute bars are active only during high flood conditions. Deposition mainly takes place by migrating megaripples. Scour- and fill structures are common. Direction of cross-bedding is variable. Units showing climbing ripple lamination are present. On the top of sequence mud drapes are present, which separate units deposited during one depositional event.

4. Scour-pool and cross-over facies – Transverse bars, sand-waves and megaripples characterize the cross-over of meander bends. Within the scour-pool, uprooted trees, slump blocks, and gravel accumulation are present.

This study also demonstrates that in a larger coarse-grained point bar there are lateral facies variations (Fig. 395).

In a hypothetical vertical sequence, a coarse-grained channel lag at the base is followed by a unit of coarse-to-medium sand showing trough-shaped cross-beds of megaripple origin (sets 10–50 cm in thickness) and tabular sets related to transverse bars (sets even > 50 cm in thickness). Chute channel and chute bar facies is present in in the highest portion of the point bar showing large, tabular and trough cross-bedding, and climbing-ripple lamination. Units of single depositional event are separated by mud drapes. There is no fining-upward tendency in the sequence.

Jackson (1975, 1976a,b) made a careful study of hydrodynamic regime, bedforms, and facies development in coarse-grained point bars of the Lower Wabash River, Illinois. The Wabash River shows a channel sinuosity of about 2.3, large parts of point bars are exposed at seasonal low flow. On the basis of the pattern of current velocity, bedform and sediment size Jackson (1975) distinguished that the upstream part of a bend (point bar) is a zone of transition from the reversed hydraulic and sedimentologic conditions of the preceding bend. Downstream of this transition zone, a fully developed zone is present, with an intermediate zone in between. Consequently, three major zones are recognized:

1. Transition zone – This is located at the upstream end of a bend. In this zone current velocities are strongest and dunes and sand waves most prominently developed at near-bankfull and higher flows. This zone is transitional for current velocity and grain size.

2. Fully developed zone – Here velocity magnitudes are greatest and dune-like bedform most prominently developed during flows at or below near bankfull with low stream discharges. In this zone each parameter increases from the inner bank to the outer bank, as implied in the ideal model of point bar. This zone shows fully developed current velocities and grain size.

3. Intermediate zone – This zone is located between transition zone and fully developed zone and shows a fully developed velocity zone but a transitional grain-size zone.

The extent of these zones in a bend is determined by channel curvature and water stage and can be expressed by dimensionless ratio of center line radius of curvature (rc) to bankfull channel width along the water surface B (rc/B) (Fig. 396). In both sharply curved and gently curved bends, fully developed zones for bed-material size and megaripple height do not exist, and the entire length of the bend represents transition zone.

Study of sedimentary structures of the Wabash River revealed that three distinct depositional facies corresponding to the three zones in meander bend can be distinguished (Jackson 1976a):

1. Fully developed depositional facies
2. Transitional depositional facies
3. Intermediate depositional facies

In each facies five subfacies are distinguished, whose importance in each type varies.

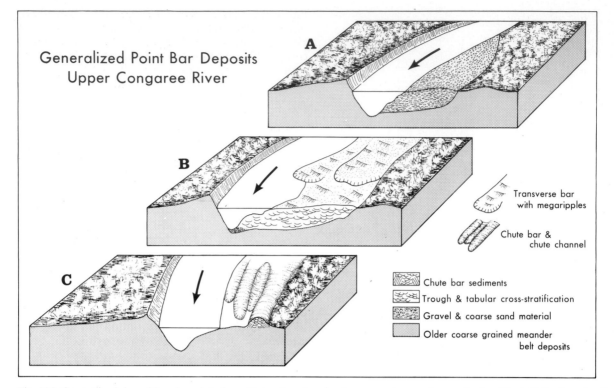

Fig. 395. Generalized depositional model of a well-developed coarse-grained point bar within the upper Congaree River. Block A represents the bar-apex facies of gravel and coarse sand. Block B shows the mid-bar facies characterized by migrating transverse bars and megaripples. Block C displays the distal-bar facies with chute bars and chute channels. (After Levey 1978)

1. Subfacies 1 – represent lag deposit containing pebble gravel with –2 Φ size pebbles. This unit is present in similar development in all the three facies.

2. Subfacies 2 – Trough cross-bedded unit, made mostly by migrating megaripples. In fully developed facies it contains gravels in the lower part. In transitional facies it contains pebbles in the upper part and there seems to be an upward coarsening from the channel base, while in the intermediate facies they are dispersed throughout. In fully developed facies, there can be a systematic upward decrease in the thickness of cross-bedded units.

3. Subfacies 3 – In fully developed facies, this zone may be separated from the underlying trough cross-bedded sequence by a mud drape, and this unit is made up of planar crossbeds due to migrating scroll bars. The cross-beds of subfacies 2 and 3 show current directions at right angles or opposite. However, the planar cross-bedding of this unit may be replaced by trough cross-bedding of megaripples.

In transitional and intermediate facies, this unit shows trough cross-bedded gravelly sand, becoming sandy towards the top. This subfacies corresponds to the deposits of the upper part of point bar exposed at seasonal low.

4. Subfacies 4 – This represents deposits of levee-like features showing sand and mud deposits. There is a faint upward decrease in grain size within this unit. This subfacies is thicker in the transitional facies than in the other two.

5. Subfacies 5 – This unit represents overbank mud deposits. The unit is thinner in transitional facies than in the other two. In may show a fining upward in grain size.

As already pointed out, arrangement of the three depositional facies around meander bends would depend upon the bend curvature. Further, upon migration of meander bends different volumes of these different facies depend upon the migration history of the meander bend.

The most significant point arising out of the study by Jackson (1976a) is that, at least in the coarser-grained point bars of a meandering river, no fining-upward sequences are produced, by point bar migration and the pattern of sequences in a migrating meandering river, where discharge etc. show fluctuations, a rather complex pattern of facies may be produced, a fact also emphasized by Bluck (1971). However, the levee and overbank deposits may show fining-upward tendencies.

Further, in the fully developed facies of the Wabash River, where scroll bars are present, palaeo-

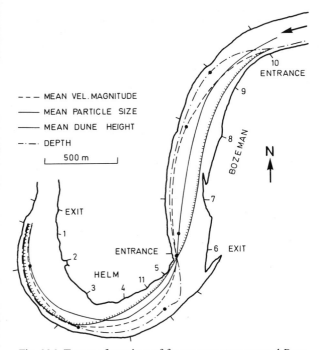

Fig. 396. Traces of maxima of four parameters around Bozeman and Helm bends at near-bankful Mt. Carmel stages. Velocity and size traces determined from surveys along traverses *1* and *2* on 29 July 1973, along traverses *3* through *9* and *11* on 11–14 June 1973, and along traverse *10* on 20 April 1973. The trace of maximum dune height is based on these surveys and on three streamwise fathometer profiles on 12–13 June 1973 (not shown). Thalweg trace determined from survey of 2 January 1973 and from the above surveys along traverses *1* through *11*. Solid circles are the estimated positions along each bend where the parameters have attained their fully developed transversal distributions. (After Jackson 1975)

current is bimodal. Small- and large-scale cross-bedding (related to mega- and small-current ripples) show direction parallel to the channel flow direction, while planar crossbeds related to the migration of the scroll bar towards the inner bank show direction at an angle to the channel direction. Unlike other examples (McGowen and Garner 1970; Levey 1978), the Wabash River point bars do not show development of bar and chute channels.

Another example of coarse-grained point bar deposits is gravel meander lobes of the Nueces River, Texas, U.S.A. (Gustavson 1978), which has a sinuosity index 1·3, average stream surface slope of 1.8 m/km, and discharge varying from almost no flow to 17,000 m³/sec. The sediment is very coarse-grained, predominantly gravelly with mean intermediate diameter 1.2–3.4 cm. Deposition takes place on the broad meander lobes. Transverse gravel bars (giant ripples) are the main bedforms occurring in the channel and lower point bar areas. The transverse bars are up to 2 m high, and wavelength more than 100 m. These giant ripples

taper out towards the upper part of the point bar. Gravels are well-imbricated. Rarely, transverse ribs also develop in the channel. The upper part of the point bar shows gravel sheets – a few clast thick sheets becoming fine-grained downstream, gravel ridges – low, somewhat rounded, elongate accumulation of gravel with long axes parallel to flow and formed in the shadow of an obstacle, usually a plant growth, and current-shadows-minor deposition behind a small obstacle. Less commonly, in the upper point bar, chute bars are developed near the downstream end (Fig. 397).

In a vertical sequence (up to 10 m thick) coarse channel lag is followed by lower point bar deposits, made up of graded planar cross-bedded and horizontal bedded units. The units may be graded (both normal or reversed) or non-graded. The gravels are often open framework type with little or no matrix. The upper point bar deposits are made up of moderately sorted, horizontally bedded, graded or non-graded, clast-supported gravel with some matrix of sand and mud, alternating with units of moderately thick layers of poorly sorted

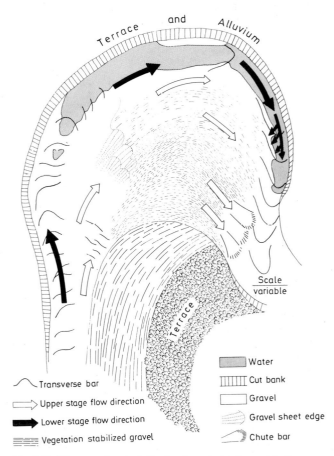

Fig. 397. Diagram showing distribution of physiographic features and flow directions for a composite of Nueces River gravel meander lobes. (After Gustavson 1978)

sandy mud. The sequence is capped by mottled silty and sand mud deposits, which represent overbank or flood plain deposits with root growth, animal burrows, and mud cracks. The schematic distribution of various features in the Nueces River meander and the vertical sequence is shown in Fig. 398. A fining-upward tendency in the sequence is absent.

The above-mentioned examples clearly show that point bar deposits can develop over a wide range of sediment types (coarse gravels, sand, to fine sand-silt) and consequently show a wide range of lithofacies and variability. Further, the large, coarse-grained point bars show variability within a single point bar, e. g., the proximal parts are coarser-grained than the distal parts. Depending upon the flood pattern of a river, a point bar may develop stepped topography if there is more than one flood period of variable intensity. Moreover, preservation of various depositional units within a meandering stream depends upon the pattern of channel migration and sedimentary tectonics. Consequently, the point bar sequences can develop in large variability.

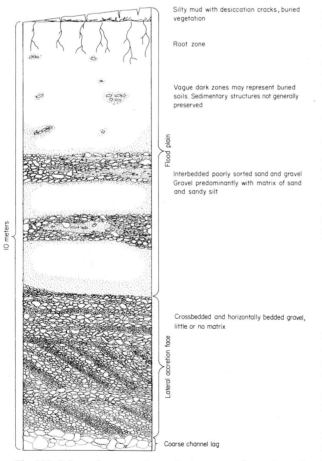

Silty mud with desiccation cracks, buried vegetation

Root zone

Vague dark zones may represent buried soils. Sedimentary structures not generally preserved

Interbedded poorly sorted sand and gravel Gravel predominantly with matrix of sand and sandy silt

Crossbedded and horizontally bedded gravel, little or no matrix

Coarse channel lag

Fig. 398. Schematic composite vertical sequence from channel to floodplain surface for a Nueces River meander lobe. (After Gustavson 1978)

Channel Bar and Braided River Deposits

Braid bars and channel bars (also known as channel islands) are characteristic features of braided rivers. Generally, rivers with steep slopes and coarser bed material show braiding, but sometimes in their lower reaches large rivers with high-sediment discharge may show braiding, even in silty sediments (e. g., the Brahmaputra River; Coleman 1969). Detailed study of channel bars are rarer than those of point bars. Sand flat sequences of the braided Saskatchewan River (Cant and Walker 1978) may be regarded as channel bar sequences.

Deposition on channel bars in braided streams is controlled mainly by processes of lateral and vertical deposition, together with channel cutting and abandonment. The slope of a channel bar in the upstream direction is steep and generally has a pool in front of it. The downstream slope is rather gentle. The channel bar migrates downstream by depositing sediment in front, like a delta, producing foresets. Generally, channel bars also migrate laterally–they possess a steeper concave side and a gentle convex side. Lateral migration also tends to produce foreset laminae.

However, channel bars are invariably covered with various bedforms–giant ripples, megaripples, etc., and its migration is mainly due to migration of these bedforms. The main deposition in channel bars takes place during high-water stages of rivers.

Channel (braid) bars can be distinguished into two types:
1. Braid bars of coarse material, pebbles, etc. Well developed in mountainous streams. Example: Durance and Ardéche Rivers, France (Doeglas 1962).
2. Braid bars of fine-grained material developed in rivers with large seasonal discharge and sediment load in the lower reaches of the river, near deltaic plains. Example: Brahmaputra River, Bangla Desh (Coleman 1969).

Gravelly Braid Bar Deposits

Doeglas (1962) carried out detailed investigations in the Durance and Ardéche rivers, France, especially concerning the distribution of sedimentary structures and vertical sequence. Figure 399 shows the shifting of river channel over several years. Figure 400 shows the cross-section across a dry channel.

The lower part of the sequence of a channel bar is composed of large-scale cross-bedding formed as a result of the lateral and downstream advance of the channel bar into an adjacent anabranch channel. These sediments are coarsest, composed of cobbles, pebbles arranged in inclined layers,

ACTIVE CHANNELS

INDICATIONS OF DRY CHANNELS

COBBLES

SILT DEPOSITS BETWEEN VEGETATION

SILT DEPOSITS IN CHANNELS

EXCAVATIONS

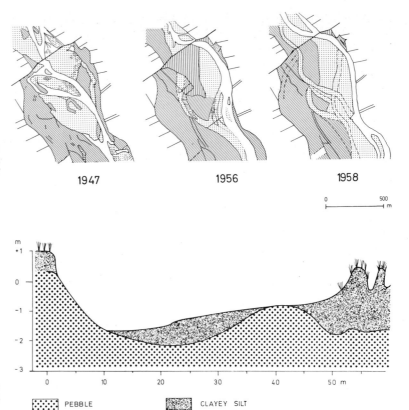

Fig. 399. Diagram showing changes in the pattern of channels of the River Durance, France, over several years. (Modified after Doeglas 1962)

1947   1956   1958

0   500 m

Fig. 400. Section across a dry channel of the Durance River. (Modified after Doeglas 1962)

PEBBLE   CLAYEY SILT

grading into finer-grained sediments deposited in low-water conditions, which are again followed by pebbly sediments.

The boundary between coarse-grained sediments and the underlying finer-grained sediments of an older cycle is very irregular and marked with scour and fill structures. Vertically these pebbly sediments are overlain by more sandy and muddy sediments. Just above the pebbly sediments follows a zone of medium sand with ripple structures–megaripple bedding–followed by fine sand with small-ripple bedding. Small-scale ripple bedding is partly developed in the form of climbing-ripple lamination. At last, a fine sand and mud is deposited in horizontal layers. In the mud layers features like mud cracks, rill marks, and raindrop imprints may develop. Moreover, mud layers occasionally show scour marks and erosional marks. The topmost layers of muddy sediments show convolute bedding, produced by the process of solifluction resulting from the expulsion of water. Sometimes the topmost mud layer shows disturbed bedding due to plant growth and organisms. Figure 401 shows an idealized sequence of such deposits.

Because of the migration of channel bars some of the channels are cut off downstream. In such channels only muddy sediment is deposited, with

occasional sand, until they are completely abandoned.

In channel bar sequences of mountainous rivers, fine-grained sediments are less important, and there is a big range of grain size involved in a single sequence, i. e., pebbles to clay. Flood basins are not well developed.

Doeglas (1962) also measured the orientation of pebbles and sand grains in these rivers. The long axes of the pebbles and the sand grains is oriented perpendicular to the current direction, and imbricated flat pebbles show an inclination upstream.

Williams and Rust (1969) provide another detailed study of a braided stream in a mountainous region. They also found fining upward sequences. Based upon grain size, sedimentary structures, and flora, seven facies are recognized. Individual units are lenticular in nature and cut into each other. Mostly coarse-grained units begin with an erosional contact, followed by fine-grained units with gradational contacts. Such fining upward sequences are most well developed in channel fill sediments. The grain size gradually decreases upward, and the degree of sorting improves upward. Sometimes coarse- and fine-grained units are complexly interbedded.

Main bedding types are large-scale cross-bedding in coarse sand and pebbles, and small-scale

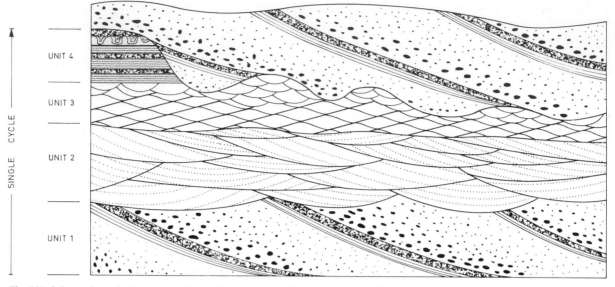

Fig. 401. Schematic vertical sequence of a braided river deposit. Unit 1–large-scale cross-bedding with pebbles; unit 2–megaripple bedding, medium sand; unit 3–small ripple bedding, fine sand; unit 4–fine sand and mud, horizontal bedding, occasional convolute bedding. (Based on data of Doeglas 1962)

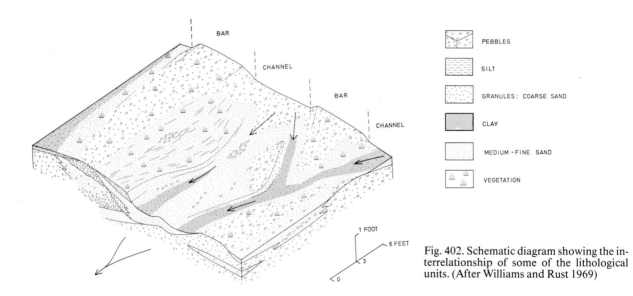

Fig. 402. Schematic diagram showing the interrelationship of some of the lithological units. (After Williams and Rust 1969)

cross-bedding in medium to fine sand and silt, and well-developed lamination in silty and clayey sediments. At some places sediment volcanoes are present in plastic mud. Figure 402 shows the interrelationship of some of the sedimentary units. Figure 403 shows vertical profile of a channel-fill sequence. Figure 404 shows the composite sedimentation model of a braided river deposit.

Cross-bedding data is unidirectional and in good agreement with channel orientation data.

General Characteristics of Gravelly Braided Stream Deposits

Much information in Recent years has gathered on the deposits of modern braided streams, which carry gravels as the main bed load (see also sandur deposits on p. 202). Various lithologies in these rivers show characteristics which are common in most of them (for example, McDonald and Banerjee 1971; Rust 1972b; Smith 1974). Smith (1974) distinguishes five main types of deposit (facies) in the braided river:

1. Laminated silt deposits – They represent thin (< 1 cm) horizontally laminated silt deposited from suspension as surface cover in protected areas, e. g., abandoned channels. Laminae show grading.

2. Rippled sand and silt deposits – The sediment is mostly fine sand, silt with some medium sand and clay contents. This occurs in areas of weak current, e. g., in the small channels and depressions of minor channels on a bar surface, on the margins of a bar during falling stages. The sediments show small ripple bedding (mostly trough-type) and climbing-ripple lamination. Orientation of foreset laminae in these deposits is varied. Sometimes this sand is developed as quick sand and may easily undergo penecontemporaneous deformation structures.

3. Coarse sand deposits – These deposits occur as cutoff channel fills, wedges on the lee side of gravel bars, veneer on bar surface, and occasionally as megaripples or transverse bars. They are mostly cross-bedded. Tabular-planar cross-bedding is most common, though in other areas trough-shaped megaripple bedding can be abundant. Horizontal bedding is less common.

4. Gravel deposits – Gravels are most abundant and comprise main deposits of channels and bars. The bedding is often massive or ill-defined stratification, and horizontal bedding or 3–5° upstream dipping layers. Low-angle cross-bedding is fairly common, and is produced by migration of bars with low channelward dipping slopes or riffle margins. Locally, high-angle dipping planar cross-bedding is common, related to migration of high bars with steep slipfaces. However, high-angle

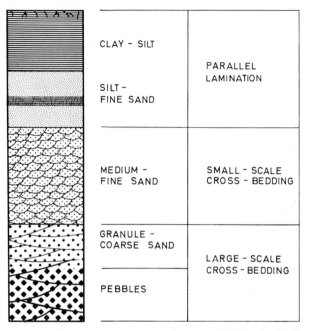

Fig. 403. Fining upward sequence of a channel-fill in a braided river. (Modified after Williams and Rust 1969)

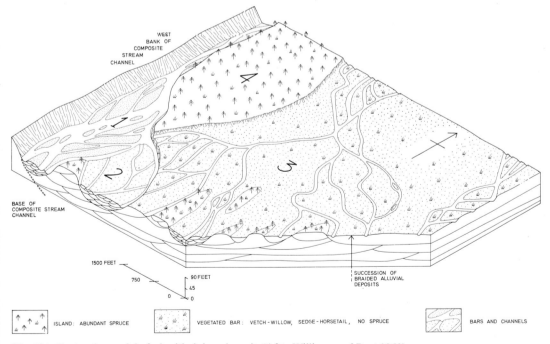

Fig. 404. Composite model of a braided river deposit. (After Williams and Rust 1969)

planar cross-bedding is common in finer gravels. Horizontally bedded gravels are mostly coarse-grained and exhibit good imbrication. The gravel deposits are mostly grain-supported and infilled with fine sand and silt matrix. However, in some deposits layers of open framework gravels alternate with the matrix-filled gravels. The open framework gravels are a product of deposition during high flow when most of the finest are carried away in suspension, as the flow diminishes finer gravels are deposited and finally matrix-sized sediment is deposited with the gravels, filling pore spaces near the surface, but unable to reach the earlier deposited gravels beneath. In the next high flow again first an open framework gravel layer is deposited (Smith 1974).

5. *Floodplain silt and sand deposit* – In the grass-covered overbank areas fine-grained sediments up to 60 cm-thick units are present. Bedding is disturbed by root growth. Thin graded beds are commonly developed. Otherwise finely laminated mud and small-ripple bedded sand are commonly present.

Development of Sequence

The deposits of gravelly braided streams upon preservation show about 80–90 % of the sequence made up of gravels. Because of rapidly shifting channels and bars, the units in such deposits are highly variable both vertically and laterally. The coarsest gravel units usually do not show cross-bedding, but they show horizontal bedding or gently upstream-dipping layers, with intercalated sand layers. Erosional surfaces are very common. The channels associated with bars contain coarsest sediments and they grade into coarse-grained sediments of the upstream part of the bar, which are followed by still finer sediments of the downstream part of the bar. Thus, if preserved, a channel-bar sequence may show a fining-upward tendency. However, it may not be possible to differentiate between the channel deposit and the upstream part of the corresponding bar deposit. Some of the abandoned channels may be filled by silty, muddy sediments.

In braided streams, there is sometimes a channel migration in one preferred direction, causing development of bars to be attached to the other bank, and giving a preferential palaeocurrent direction (Bluck 1974).

Fine-Grained Braid Bar Deposits

Channel bars of the Brahmaputra River represent a case of a braided river in a deltaic plain. Coleman (1969) describes in detail processes and various features of this river. The following account is based on his investigations:

The Brahmaputra River is a braided stream and it actively migrates laterally. Because of a high amount of sediment coming during the rainy season, the river builds up channel bars which migrate actively during floods, mainly due to migration of various bedforms. Some of the channel bars and channels are abandoned during the falling stage of the river, because the channel changes course. Sediment is rather fine-grained, the median varying from 0.340 to 0.028 mm.

The lower part of a channel bar sequence begins with large-scale cross-bedding, with individual sets more than 1 m thick, and whole zone as thick as 7 m and more. These units are produced mainly as a result of the migration of giant ripples, which are important bedforms during the flood. Associated are some irregular scours and scour fill with irregular lower bounding surfaces. Within this zone are also found, locally, small-ripple bedding, small-scale scour-and-fill structures, and deposition of mud along the foreset laminae and bedding planes. Often much organic matter is trapped within this zone, especially at the base of a cross-bed unit and along the foreset laminae. They represent the bottomset layers made up of hydrodynamically light material. Foreset laminae may show distortion features. Deposition of a large-scale cross-bedded unit 10 to 15 m thick takes place very quickly during a single episode of migrating bedforms during one flood cycle (zone A).

Overlying the unit with large-scale cross-bedding is a zone of more silty sediments (zone B). Individual units of this zone are lenticular in shape. Occasionally lenticular bodies with large-scale cross-bedding are present. Sandy and silty units are intercalated. The most common bedding type in this zone is climbing-ripple lamination and small-ripple cross-bedding. This zone was deposited when larger bedforms ceased migrating, and deposition took place mainly in the troughs of the larger bedforms. More silty layers show horizontal lamination. This unit can be as thick as 1 m.

Above this unit is a zone of more silty clays and silt (zone C). It shows horizontal bedding. Layers differ in texture, color, and composition, with occasional thin units showing climbing-ripple lamination. This unit is deposited during slack water, where mainly suspended sediment is deposited. The thickness of this unit varies from 1 to 2 m.

On the top of the sequence are clayey, silty sediments. This is a highly disturbed soil zone (zone D), burrowed and with roots. It forms the final capping of the sequence and is generally several decimeters thick. The idealized sequence of a channel bar deposit is shown in Fig. 405.

Penecontemporaneous deformation structures are rather abundant in channel bar sediments of

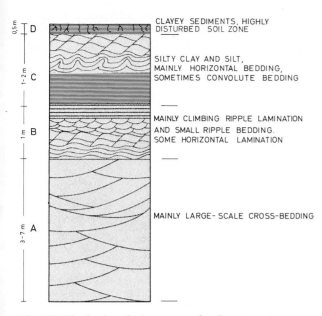

Fig. 405. Idealized vertical sequence of sedimentary structures of a channel bar deposit of the Brahmaputra River. (Based on data of Coleman 1969)

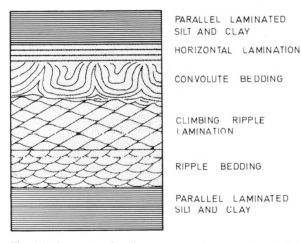

Fig. 406. Sequence of sedimentary structures associated with convolute bedding in the upper part of a channel bar sequence. (Based on data of Coleman 1969)

the Brahmaputra River. They include distortion and overturning of foreset laminae, flow rolls, load structures, and convolute bedding. The rapid rise and fall of the river level, currents heavily laden with fine-grained sediments, and intensive scouring and slope steeping are the main reasons for their abundance. Overturned foresets and contorted foresets are present in abundance in the lowermost unit with large-scale cross-bedding.

However, deformation features were found to be most abundant in zone C, and next in zone B; sometimes they occurred in a definite sequence:

horizontal bedding
contorted bedding
climbing ripples
parallel lamination

A similar sequence is found in point bar deposits of the Mississippi River (Coleman and Gagliano 1965). Figure 406 shows such an ideal sequence of the upper part of a channel bar deposit.

Occasionally, this zone of contorted bedding is represented by convolute bedding. Coleman (1969) believes that this convolute bedding may be produced from increased shear stress due to an increase in current velocity as a result of a sudden rise in turbulence. This zone may represent a transition from deposits of the lower flow regime into overlying upper flow regime beds (horizontal lamination). Another important process for the production of convolute bedding in these deposits is the stresses produced by the emergence of a depositing surface, leading to liquefaction and flow in the underlying sediments, producing convolute bedding.

Coleman (1969) also carried out detailed measurements of cross-bedding for palaeocurrent analysis (Fig. 407). He concluded that in a braided river, the current flow, body trend, and cross-bedding direction agree quite closely if a sufficient number of measurements are taken. At a single station variation in cross-bed vector mean is rather low. However, great variations are present because of riffle pool sequences, on the edges of the bars and on the steep parts of the channel. Thus, the effect of local channel geometry is very significant. Coleman's data are based on the measurement of small-scale cross bedding near the surface. It is quite likely that there is some difference if large-scale cross-bedding from the lower part of the sequence is measured.

The South Sasketchewan River, Canada, is a good example of small-sized sandy braided stream, and has been studied in detail by Cant and Walker (1978). This river is a sandy braided river, made up of moderately well-sorted sand of mean diameter 0.3 mm. The average slope is 0.0003 and the river is 600 m wide with an average discharge of 275 m³/sec, with considerable variation in mean and maximum discharge. The major geomorphological elements of the river are channel, slip-face bounded bars, sand flats, and vegetated islands and flood plain, each of which is characterized by a specific sedimentary structure pattern (Fig. 408). The sedimentary sequence would largely depend on whether mainly sand flat deposits or channel deposits are preserved. Cant and Walker (1978) give three main types of vertical sequence that may develop in this river (Figs. 409 and 410). The channel deposits are characterized by lag deposits and cross-bedded sand, related to the migrating megaripples, while the sand flat deposits are made up of mainly planar cross-bedded units in the lower part

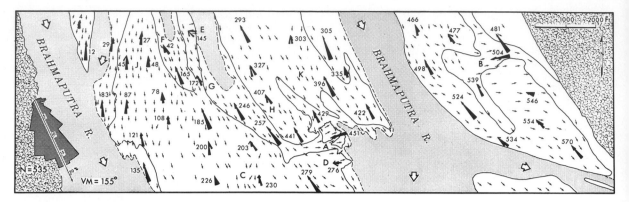

Fig. 407. Cross-bedding measurements in a broad reach of the Brahmaputra River. (After Coleman 1969)

Fig. 408. Major geomorphological features of South Sasketchewan River, Canada. Flow towards North. *1* channels; *2* cross-channel bars; *3* sand flats; *4* vegetated islands and flood plains. (After Cant and Walker 1978)

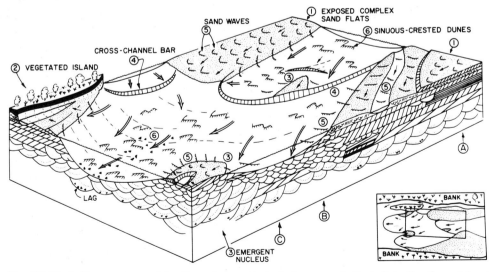

Fig. 409. Block diagram summarizing the major morphological elements, and their associated bedforms and stratification. The hypothetical reach is outlined by a rectangle in the inset, lower right. Stippled areas are emergent. Single-shafted arrows indicate directions of bedform movement, and double-shafted arrows indicate flow directions. A locates stratigraphic sequence dominated by sand flat development, B has mixed sand flat and channel influence, and C is dominated by channel aggradation. (After Cant and Walker 1978)

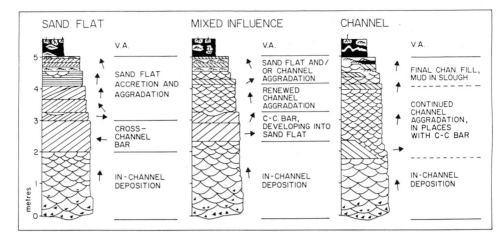

Fig. 410. Schematic lithofacies sequences characterizing areas dominated by sand flat development, areas of mixed sand flat and channel influence, and areas of channel. Arrows indicate in a general way the palaeocurrent variability. (After Cant and Walker 1978)

and a complex associated of large-scale cross-bedding (both planar and trough-shaped), and small-ripple bedding in the upper part. On the top of the sequence flood plain deposits, representing vertical accretion, are present showing silt and clays with few ripple-bedded fine-sand units, and some wind-blown sand layers.

The downstream sandy part of the braided Platte river is characterized by the rather shallow nature of the river, with shallow channels and transverse bars (Smith 1970). The deposits of this sandy braided river are made up of dominantly planar cross-bedding, related to migrating transverse bars, with a few units showing trough cross-bedding and small-ripple bedding.

Singh (1977) describes sequence of sedimentary structures in a channel bar (a braid or side bar) of the Ganga River, made up of fine-grained sediments ($M_z = 2.59 \, \Phi$). The bar possesses an active lee-face and migrates laterally, thereby producing foreset laminae, which are extremely long and inclined at a comparatively low-angle ($10-15°$). The units are planar and show low-angle discordances. The sequence is sandwiched with units of scour-fill cross-bedding (Fig. 411). These units stacked upon each other may produce several meter-thick sequences, underlain and overlain by muddy sediments.

Miall (1977) provides a detailed review of braided stream deposits, and also vertical profile models, recognizing six types which range in composition from gravely deposits to sandy-silty deposits (Fig. 412) (Miall 1978b). Rust (1978b) discusses that braided stream deposits occur in a wide range of grain sizes and can be grouped into three types:

1. Gravel-dominant – Characterized by framework-supported gravel, strong imbrication and horizontal bedding. The distal part of gravel-dominant braided stream deposits show well-developed fining-upward sequences.

2. Sand-dominant – Characterized by mainly horizontal bedding, low-angle planar cross-bedding ($< 10°$), and trough cross-bedding. Proximal types usually contain no or few mud layers, but in the distal type mud content is significant and fining-upward sequences are developed.

3. Silt-dominant – These are dominated by suspension-load deposits. Small-ripple bedding, climbing-ripple lamination, and horizontal bedding are most abundant.

There seems to be no well-defined difference in the sequence of point bar deposits and channel bar deposits. This is especially true of the rivers with very fine-grained sediments. In fact, several rivers possess both point bars and channel bars at differ-

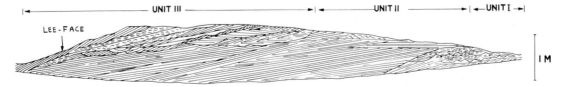

Fig. 411. Sketch showing the bedding structures of the sand bar in Ganga River, Allahabad. The main body is made up of low-angle cross-bedding, along with units of scour and fill cross-bedding. Unit I – mainly scour and fill cross-bedding. Unit II – mainly low-angle cross-bedding. Unit III – mainly scour and fill cross-bedding. (After Singh 1977)

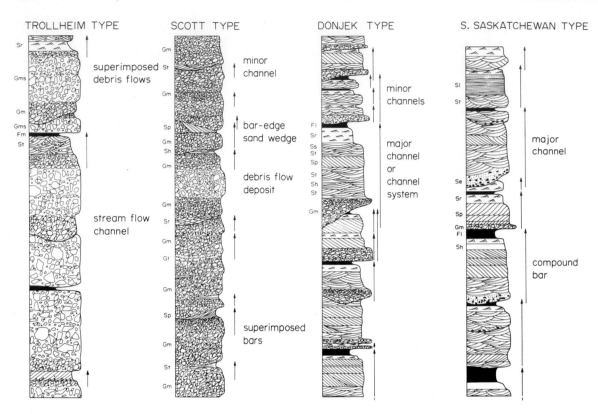

Fig. 412. Vertical profile models for braided stream deposits. Arrows show small-scale cyclic sequences. Sr – very fine to coarse sand showing ripple marks; Gms – massive, matrix-supported gravel; Gm – crudely, horizontal bedding gravel; Fm – mud, massive with dessication cracks; St –medium to very coarse sand showing cross-bedding; Sp – medium to very coarse sand with cross-bedding; Sh – very fine to very coarse sand with horizontal bedding; Gt – gravel showing trough cross-bedding; Fl – sand to mud, fine lamination; Ss – fine to coarse sand with scour and fill structure; Sl – fine sand with low-angle cross-bedding; Se – erosional scours with interclasts. (After Miall 1978)

ent times. Sometimes a point bar may even grow, become detached as a channel bar, and develop further (Sarkar and Basumallick 1968).

Deposits of Suspended Load Rivers

Taylor and Woodyer (1978) describe the pattern of deposition and sedimentary sequence in the Barwon River, Western Australia, a suspension-load river. The river shows a sinuosity of 2.3 with a mean width/depth ratio of 8. Floods are very slow-moving ($< 0.5\,m/sec$) and sediment concentration is low, mostly in clay-size. Rate of sedimentation is low, about $20\,mm/year$. The main deposition in such a suspension-load river takes places within the channel as bank benches (depositional platform within the channel), which occur at three levels. The river channel is more or less stable, and does not move laterally by meander.

The low benches are located in the point bars near the channel and represent lensoid bodies of mainly sandy sediments, thining both bankwards and towards the channel. The sands are mainly fine sand and some medium sand. The sequence is made up of about 10–20 cm-thick sand units alternating with about 1 cm-thick mud layers. Upwards in the sequence, sand layers become thinner, while mud layers are more common and become thicker. The mud layers represent drapes over the sand layers without any intermittent erosion. The mud layers show thin, parallel to wavy lamination. The sand layers mostly show cross-bedding with sets up to 14 cm thick. Parallel bedding is common in fine-sand units. Mud pebbles are abundant mainly in the lower part of each sand layer. The cross-bedding is produced by megaripples in the near-channel part of the low bench.

In the upper part of low bench parallel bedding and small ripples are present. The deposition in low bench is by bed-load activity at times of high flow.

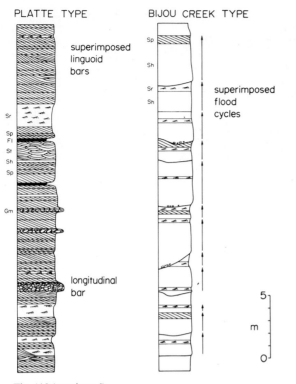

Fig. 412 (continued)

activity. The sand and mud layers are relatively continuous, both laterally and up-dip draping the entire area covered during their deposition. They wedge out up-dip at heights corresponding to the height of the flood from which they were deposited, and wedge out or grade into sandy deposits of low bench. Graded bedding is the most common bedding type and is developed in four different patterns: normal grading, reverse grading, waning-waxing grading, waxing-waning grading (see also p. 120), where reverse grading and waxing-waning grading are more common, and is mostly developed as laminated grading. Parallel lamination is quite common. Small-scale cross-bedding (small-ripple bedding) is sometime developed in the fine-sand layers and shows flow direction towards the channel. Planar cross-bedding (up to 8 cm thick) may develop at the slope between the bank and the bench, showing flow directions towards the channel. Both these cross-beddings suggest deposition by current developed by hydraulic gradient during falling stages of a flood. Erosional surfaces are almost absent. Deposition in middle and high benches is by suspension load with local current activity.

Due to lack of erosion, the benches within the channel accrete both laterally and vertically, leading to avulsion of the channel and gradual development of a new channel. The relict channel is ultimately filled by suspended clay producing a clay plug. The sequence shows an upward fining, and is divisible into three units (Fig. 415).

Natural Levee Deposits

Natural levees are wedge-shaped ridges of sediment bordering stream channels. The maximum elevation of a levee is at or near the edge of the

The middle and high benches are characterized by interbedded sand-mud deposits, which are lensoid-shaped with dimensions of about 1,200 m in length, 30 m in width, and 6 m in thickness. The deposits are made up of fine sand and mud, occurring in general as 5–10 cm thick layers which are internally finely laminated and graded. The layers conform to the inclination of depositional surface (Figs. 413 and 414). These sediments lack any erosional features, and rarely show features of current

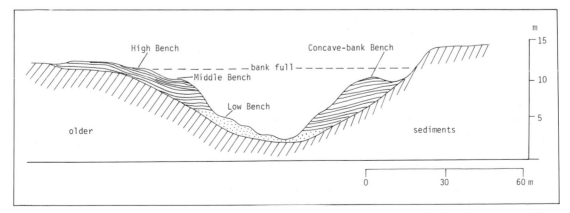

Fig. 413. An idealized channel cross-section showing the positions of low, middle, high, and concave bank benches. Barwon River, Australia. (After Taylor and Woodyer 1978)

Fig. 414. Photomosaics of section through Middle Bench Deposits from hairpin bend. The beds wedge out laterally and dip towards the river. (After Taylor and Woodyer 1978)

channel, where channels commonly form steep high river banks. The levees gently slope from the river bank into flood basins away from the channel.

Figure 415 shows natural levees of a river channel. Levees are better developed on the concave sides of the river channel; on convex sides levees grade into upper point bars. Especially in smaller rivers levees and upper point bars grade into each other and are not easily distinguishable. Not all the rivers possess well-developed levees. Levees of the Mississippi River are especially well developed; they can be up to 1.5 km wide and project 5 to 6 m above the flood basin.

Natural levees are formed by deposition of sediment when flood waters of a stream overtop its banks. The velocity is reduced, causing deposition of much of the suspended sediment near the channel. Coarsest sediment is deposited near the channel, and grain size decreases away from the channel. The rate of deposition also decreases rapidly away from the channel. The maximum height of natural levees indicates the water level reached during the heighest floods.

Fisk (1944, 1961), Shepard (1956b), Lattman (1960), and Allen (1965b) describe the structures of natural levee deposits. In geometry, levee sediments are ribbon-like sinuous bodies, triangular in cross-sectional shape. However, in strongly meandering channels the geometry of levee deposits is rather complicated.

The levee sediments are made up of somewhat finer-grained material than their corresponding point bar sediments. However, the composition of the upper part of point bar sediments is rather similar to the composition of levees. Sedimentary structures of levee deposits are also very similar to the sedimentary structure of the upper point bar. They include small-ripple crossbedding, climbing-ripple lamination, horizontal bedding, and parallel

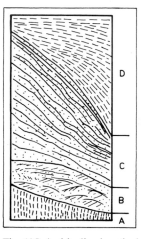

Fig. 415. An idealized vertical profile of channel sedimentation for suspended-load streams. A older sediments, B deposits of the Low Bench, cross-bedded sands with thin mud laminae, C mud and sand interbeds deposited on the Middle and High Benches, D dark clay plug of the abandoned channel fill. (After Taylor and Woodyer 1978)

laminated mud layers. Sandy units are overlaid by muddy units. Generally a sandy layer, a few decimeters thick is overlaid by a muddy layer, a few centimeters thick. Sometimes sand and mud layers can be equally thick, or mud layer can be even thicker than the sand layer. Thus, in a vertical sequence sandy and muddy layers alternate each other. Surface features, e.g., dessication cracks or raindrop imprints, are commonly present on the muddy surface of levee deposits and are likely to be preserved. Soil-forming processes that lead to mottling of the uppermost part of a mud layer, are rather active. Larger levees support much plant growth. Thus, much plant debris and organic matter is incorporated into levee sediments. In general, one can say that in levee deposits muddy layers are much thicker in relation to sandy layers than in the point bar sediments.

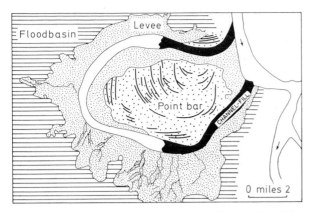

Fig. 416. Natural levee and related features of a meander of the Mississippi River. (After Allen 1965c, originally from Fisk 1947)

Natural levee sediments of the Gomti River, India, show sand layers overlaid by thick mud layers. Sand layers show mainly small-ripple cross-bedding and sometimes horizontal bedding. Mud layers are usually finely laminated (Fig. 417). Within the natural levee deposits of the Gomti River, channels filled with silty muddy sediments are present (Singh 1972), where there is also concentration of plant debris, molluscs, etc. Locally, in small depressions within the natural levee, wave ripples, mud cracks, and rain drop imprints are developed (Kumar and Singh 1978).

Coleman (1969) describes in detail natural levees of the Brahmaputra River which are rather broad and show an irregular extension into the flood basin. The flood water of the Brahmaputra River is mainly funneled via distinct channels cut-

ting across the levees (crevasse splays) rather than by sheet type overflows. As a result levee deposits are made up of numerous interfingering and overlapping lenses of sandy material capped by muddy sediments. Most of the sand on levees is a result of deposition in crevasse splays. Sand predominates in sections near the channel. These sandy layers pinch out away from the channel where more and more muddy sediments predominate; finally levee deposits grade in fine-grained muddy sediments of the flood basin.

A characteristic sequence is found on the levees of the Brahmaputra River, which may vary in thickness from few decimeters to 2 to 3 m (Fig. 418). Coleman (1969) noted that brecciation, small folds and faults (penecontemporaneous deformation structures), are rather common in natural levee deposits. In certain zone disturbance due to liquefaction and flow seem to be a common feature.

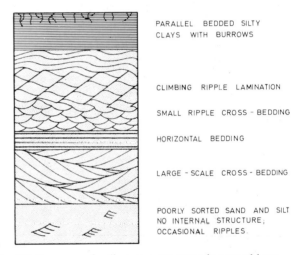

Fig. 418. Sequence of sedimentary structures in natural levee deposits of the Brahmaputra River. (Based on information by Coleman 1969)

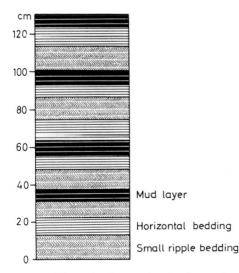

Fig. 417. Schematic diagram showing four cycles of sedimentation in a natural levee deposit of the Gomti River, India. (After Kumar and Singh 1978)

Crevasse-Splay Deposits

During high floods large quantities of flood water and sediment is diverted into an adjacent flood basin. Most overtopping occurs on concave banks. This overtopping of the river bank may occur as flash floods, or water may follow distinct channels cut across the bank through the natural levee deposits.

The water cuts channels called crevasses. The excess water leaves the main channel through such crevasses. These crevasses develop their own channel pattern and system. Sometimes crevasse channels may be several hundred meters broad, but generally they are tens of centimeters to few meters broad. Crevasses extend across the levees into the flood basins. Occasionally, crevasse channels may divert the main river discharge, causing a change in the river course.

Crevasse splay deposits are narrow to broad tongues of sediment, tapering in one direction (in the direction of flood basin). They are tens of centimeters to few meters thick. Crevasse splay deposits are somewhat coarser grained than the associated natural levee deposits in which they are embedded. Crevasse deposits extend in sandy tongues across the natural levee deposits, well into the flood basin. The tongues become narrower and rarer near the flood basin deposits.

Happ et al. (1940), Russel (1954), Kruit (1955), and Coleman (1969) provide useful information on crevasse splay deposits in modern sediments. The Brahmaputra River shows well-developed crevasse splay deposits (Coleman 1969). Figures 419 and 420 show crevasse splay deposits of the Brahmaputra River. Actually, here the major part of natural levees is built up from crevasse splay deposits. On the other hand, in the Mississippi River crevasse splays are not well developed.

Small-scale cross-bedding, climbing-ripple lamination, and some horizontal bedding are the main sedimentary structures. This sandy sequence is generally capped by muddy sediments. Scour-and-fill structures are commonly found. Locally, large-scale cross-bedding due to channel cut and fill may be found (channel-fill cross-bedding). Drifted plant material and fossil remains may be concentrated into crevasse splay deposits.

In ancient sediments crevasse splays may be recognized as sandy channels in muddy natural levee sediments. Orientation of such channels would vary very much from the orientation of the main channel-sand deposit. Some of them may continue as thin sandy layers into flood basin facies.

Channel-fill Deposits (Cut-off Channels and Lakes)

Channel-fill deposits represent sedimentation in stream channels that have been abandoned by a stream because of cut-off processes or avulsions, i.e., by the sudden abandonment of a part or the whole of a channel course, or by filling up due to extreme increase in the rate of sedimentation and reduction in depth (Fig. 421).

The cut-off process is associated with the meandering streams and occurs whenever the stream can shorten its course and thus locally increase its

Fig. 419. Natural levee with crevasse splay of the Brahmaputra River. (After Coleman 1969)

Fig. 420. Natural levee with crevasse splay of the Brahmaputra River. (After Coleman 1969)

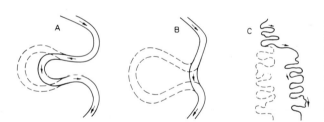

Fig. 421. Diagram showing three modes of abandoning river channels. A Chute cut-off; B neck cut-off; C avulsion. (Modified after Allen 1965c)

slope. Two types of stream cut-off are known (Fisk 1944, 1947):

1. Chute cut-off. If a stream in a meander loop shortens its course by cutting a new channel along a swale of a point bar.
2. Neck cut-off. If a stream cuts a new channel through the narrow neck between two meander loops.

The abandoned channels are slowly alluviated and sealed at both ends, isolating the old channel loop in the form of a cut-off lake, or ox-bow lake. In the beginning, sedimentation is rather rapid at and near the ends of the cut-off lakes. However, later the rate of sedimentation in the cut-off lakes is very slow. Some suspended material is introduced during overbank flows. Mainly clayey sediments and organic matter are deposited until filling is complete. This process produces a sequence of thick clayey sediments in the form of clay plugs, whose thickness is controlled by the depth of the abandoned channel. Clay plugs up to 40 m thick are known in the Mississippi River flood basin. Figure 422 shows various kinds of channel-fill deposits. These features may be up to 35 or 40 km long, sometimes strongly curved, with a thickness of 20 to 30 m, or even more. At the end of the channel-fill deposit sandy plugs are present. Clay and silt is most abundant. Sand is present in only minor amounts. Sandy units are generally cross-bedded. Mud layers are finely laminated. Most of the cut-off channels become lakes with abundant fauna and flora. Pockets of freshwater molluscs and plant remains may become incorporated within the sediment. Otherwise, muddy sediments are very thinly laminated. Lamination may be the result of difference in color, composition, or even grain size. Channel-fill deposits of the Mississippi River have been described by Fisk (1944, 1947). Figure 423 shows bedding structures in channel-fill deposits.

Channel-fill deposits produced by two different types of cut-off show some identifying characteristics. Channel-fill deposits from chute cut-offs are

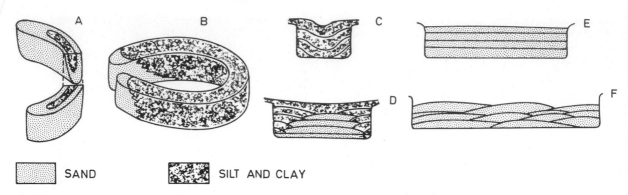

SAND   SILT AND CLAY

Fig. 422. Various types of channel-fill deposits. A Chute cut-off; B neck cut-off; C-F various modes of filling of ephemeral streams. (Modified after Allen 1965c)

relatively short and are less curved. The absolute size depends upon the width of the river and the size of the meanders. Here bed load continues to be deposited for longer periods. The phase of strictly clay sedimentation is short, so that clay plugs of short dimensions are produced. Lithology is mainly silty sand, sand, and clays. In the lower part sand and clay units alternate.

Channel-fill deposits from neck cut-offs are strongly curved and larger. Here the cut-off is complete at an early stage; after that no bed load is available. The filling is done by overbank flows and a fine-grained clay sequence is deposited. In cross section the complete form of the channel is preserved. On the convex side contact with a point bar deposit is gradational, whereas on the concave side contact is erosive, with older alluvium deposits, e. g., flood basin, older channel fills, point bars, levees, etc.

Glenn and Dahl (1959) found that sediments of channel fill are rather similar to flood basin deposits. In the Brazos River channel-fill deposits are made up of laminated and thinly bedded clays and silts, deposits are up to 15 m thick (Bernard et al. 1962; Bernard and Major 1963).

Ephemeral streams in arid climates also make channel-fill deposits. Schumm (1960, 1961) gives a detailed account of these features. The channels are filled up after a flood phase. The mode of filling of ephemeral streams is controlled by the type of sediment available, and the width-depth ratio of the stream. In general, the complete shape of the channel is preserved. Filling may or may not conform to the channel form. Usually smaller channels may be filled by units conforming to the shape of the channels; larger channels are filled by smaller lenticular units not conforming to the shape of the channel.

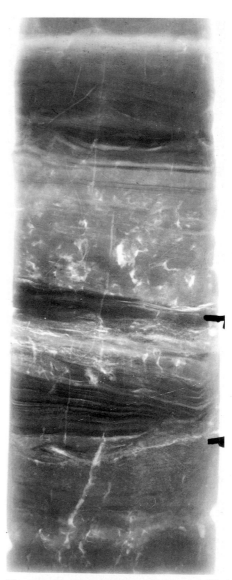

Fig. 423. Bedding structures in channel-fill deposits of the Rhine River as revealed by X-ray

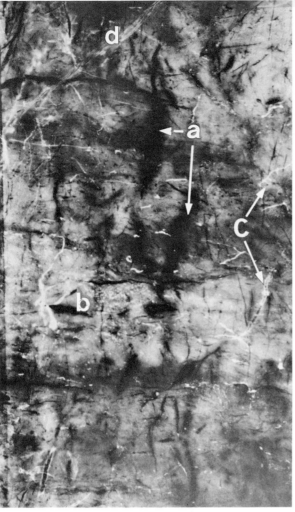

Fig. 424. X-ray radiograph of a poorly drained swamp deposit of a flood basin. *a* Iron oxide or hydrous iron oxide rim around organic rootlet, *b* vivianite nodules, *c* rootlet replaced by pyrite, *d* carbonate coated fracture plane. (After Coleman 1966)

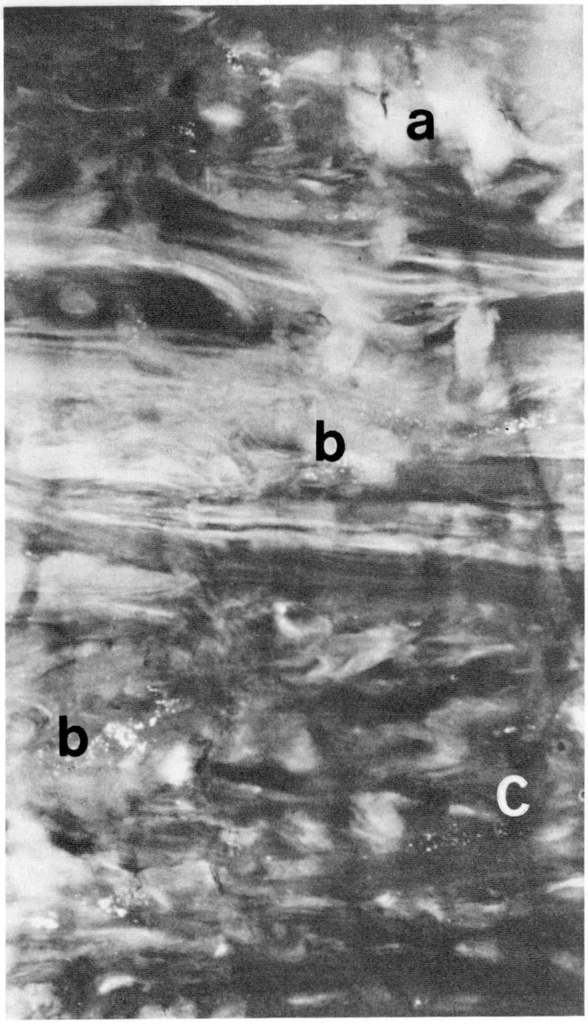

Fig. 425. X-ray radiograph of lacustrine facies of a flood basin. Burrowing activity has decreased as a result of an increase in sedimentation rate. *a* Incipient $CaCO_3$ nodule, *b* vivianite masses, *c* micro-mollusks remains. (After Coleman 1966)

Flood Basin Deposits

Flood basins are the lowest-lying part of a river flood plain. They are poorly drained, flat, featureless areas of little or no relief located adjacent to active or abandoned stream channels. Flood basins act as settling basins, in which suspended fine-grained sediment settles down from overbank flows after the coarser sediments have been deposited on levees and crevasse splays. Thus, flood basin deposits represent the long-continued accumulation of fine-grained suspended sediment. The rate of sedimentation is generally very slow; usually a silt-clay layer, 1 or 2 cm thick is deposited during one flood period.

The size and shape of a flood basin depends primarily on the history of the flood basin. Usually it is elongated, with low-lying areas running parallel to the channel. The other side is defined by a valley wall, alluvial ridges, e.g., abandoned levees, or other channels. In the lower reaches of a river, the area occupied by flood basins increases tremendously in relation to the area occupied by river channels and levees.

The extent and development of flood basin deposits is mainly controlled by the channel form and pattern. Braided streams with their rapid rates of lateral migration inhibit the development of thick flood basin deposits. The same is true of actively shifting meandering channels. In such cases, rather thin flood basin deposits are found sandwiched between channel deposits. However, thick flood basin deposits are produced if streams become more or less fixed in their position, so that longer periods are available for deposition in flood basins.

In a humid climate flood basins are low, wet, and thickly vegetated. In such areas they commonly develop as backswamps. Thick vegetation results in incorporation of much organic matter in flood basin deposits. Sometimes this organic matter may accumulate in peat layers several meters thick, associated with silty clayey sediments deposited during overbank flows. Such low-lying basins may occupy areas of few hundred square kilometers. Thick deposits are produced if a suitable rate of subsidence accompanies the deposition.

In coastal plain areas where major rivers join together, large flood basins develop between the natural levees closed at their downstream end and they are particularly swampy and waterlogged. Sometimes even large shallow lakes develop in such swamps (Russell 1967). Russell noted that the Atchafalaya Basin, Louisiana, is a large backswamp closed by river levees on both sides, and along lower end by a now-abandoned natural levee of the Mississippi River. A large lake exists toward the lower end of the basin, with an outflow through levees.

In such areas thick deposition of clay sediments takes place, associated with accumulation of organic matter, especially plant debris. Coleman (1966) tried to distinguish various subenviron-

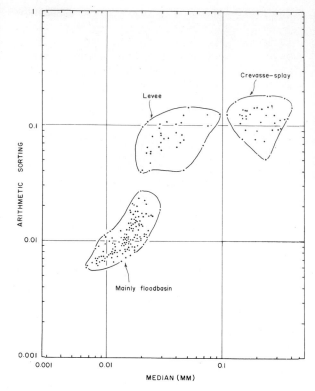

Fig. 426. Grain-size characteristics of various bank and flood plain deposits of Tobitubby and Hurricane creeks, Mississippi. (After Allen 1965c. Data of Happ et al. 1940)

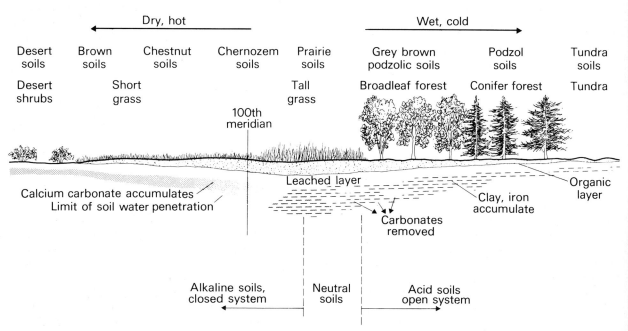

Fig. 427. Two basic types of soil genesis: one operative in wet, cold climate, the other operative in dry, hot climate. (After Collinson 1978)

ments of the Atchafalaya Basin, i. e., poorly drained swamp, well-drained swamp, freshwater lacustrine, lacustrine delta-fill. The study of X-ray radiographs revealed a varied, characteristic assemblage of primary, secondary, and post-depositional sedimentary structures, as well as epigenetic and syngenetic inclusions. Figures 424 and 425 show the characteristics bedding structure of flood basin deposits.

Only in extremely hot climates do flood basins fail to develop as backswamps, and little or no organic matter is incorporated into flood basin deposits. In such hot climates, if flow in the flood basin is sluggish, saline lakes may develop. If the rate of evaporation is high, various salts are produced and become incorporated with fine-grained flood basin deposits.

Sediments of a flood basin contain the finest grains of all the alluvial sediments. Flood basin sediments are finer than the corresponding natural levee, crevasse splay, and point bar deposits, and generally they are fine silt and clay (Fig. 426). There may be a slight upward fining in each flood basin sequence of silt-clayey sediments. Flood basin deposits seldom show horizontal bedding with textural differences. Mostly uniform finely laminated mud is present, interrupted by some sandy or silty intercalations. Sometimes small channels with sandy-silty sediments may be present. Soil horizons with disturbed bedding are abundant. Organic debris and mottled structures are other important features. Locally, pockets of freshwater mollusc shells and bones of vertebrates may be present. Because of repeated exposure, flood basin deposits are subject to dessication. Mud cracks and other surface features are widespread. Locally, wind blown sand deposits may be present.

The sediments of flood basins undergo soil formation and, depending upon the climate, different types of soil profile are developed. Distribution and genesis of various types of soils is discussed by Bridges (1970), Hunt (1972), Birkeland (1974), and Scheffer and Schachtschabel (1976).

Basically, two types of soil genesis can be distinguished: one operative in a wet, cold climate, and the other in a dry, hot climate (Fig. 427). Soils in the humid climate are characterized by the presence of an organic matter-rich zone on the top, making slightly acidic conditions, and leaching and migration of various ions taking place downward. In such soils much organic matter is preserved. In some of the reducing soils pyrite and siderite nodules are formed.

In the soils of a semi-arid climate, contents of organic matter is low, causing increasing pH-value (alkaline conditions), and rate of evaporation is high. This causes precipitation of $CaCO_3$ and Na, K-bicarbonates. Flood plain deposits of a semi-arid climate are characterized by negligible organic

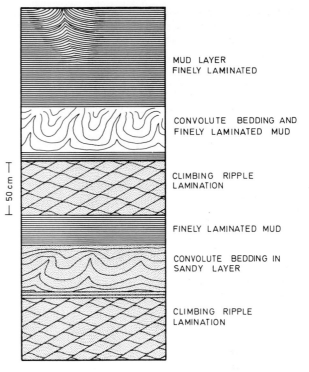

Fig. 428. Sedimentary features of flood plain deposits of the Indus River. (Based on McKee 1966a)

matter content, somewhat reddish in color due to oxidation conditions, and the presence of $CaCO_3$ concretions, known as calcrete (caliche or kankar). Sometimes silica layers may also be present. Goudie (1973) gives a good review of calcrete and other related deposits. Gile et al. (1966), Revees (1970), and Gardner (1972) discuss the genesis of calcrete deposits. There are many views regarding the genesis of calcrete.

In general, it can be stated that calcrete is formed near-surface in stable geomorphic areas, where for a certain period of time there is negligible sediment deposition due to terrace development or distant avulsion of the river. $CaCO_3$ accumulation is due to a pedogenesis process, related to repeated wetting and drying of the upper surface, involving evaporative precipitation of $CaCO_3$. The main source of Ca^{2+} ions is dissolved Ca^{2+} in rain water, along with aeolian dust with some carbonate content, and Ca^{2+} available within the sediment. Gile et al. (1966) distinguish four stages in the development of calcrete layers.

Based on the information on the rate of calcrete formation in Holocene, Leeder (1975) suggested that incipient calcrete development requires 1,000 years, while the complete calcrete layer formation needs a minimum of 10,000 years. He suggested a model for estimating the rate of deposition in the

flood basin deposits of semi-arid climate by observing the nature of calcrete development (using
the four types proposed by Gile et al. 1966):

1. Few calcareous filaments, coatings,
   1.000–4.500 years.
2. Few to common cylindroids and nodules,
   3.500–7.000 years.
3. Many nodules and internodular filling,
   6.000–10.000 years.
4. Massive bed of coalesced nodules, with laminar
   upper horizon and chert lenses, minimum
   10.000 years.

Flood basin sediments produce elongated, tabular, prismatic bodies of rectangular cross-section.
Lateral contact with sandy channel sediments may
be abrupt; or gradational, if levees with crevasse
splays are well developed. Some of the sandy channels of crevasse splay origin continue as thin stringers into the flood basin deposits.

Flood basin deposits of the Mississippi River are
rather extensive. They may form units tens of kilometers long, a few kilometers wide, and 8 to 50 m
thick. However, thickness is generally only few meters (Fisk 1944, 1947). Bernard et al. (1962) report
flood basin deposits up to 7 m thick from the Brazos River flood basin.

Flood Plain Deposits

As mentioned earlier, rivers that move and shift
their position rapidly do not have well-developed
basins. For such rivers a general term–flood plain
deposits–is proposed to denote the sediment deposited during overbank flow. Allen (1965c) calls
such deposits undivided topstratum deposits.

In rivers with rapidly changing channels velocities of overbank flows are rather rapid; very commonly current velocities up to 40 or 50 cm/sec are
achieved (Leopold and Wolman 1957). In such
cases overbank flooding deposits not only muddy
sediments, but also an abundant amount of sand.
In such cases flood plain deposits are similar in nature to natural levee and crevasse splay deposits,
and cannot be differentiated easily. Usually, natural levees are not well developed.

Observations in the flood plain deposits of the
Ganga River show that these deposits are mostly
made up of horizontally bedded fine sand alternating with laminated mud layers. In more sandy
areas, sequences of climbing-ripple bedding
capped by mud layers are present. Sometimes thin
mud and sand layers are intercalated and often
show penecontemporaneous deformation structures.

Generally, fine sand, silt, and clay layers are deposited, and the sequence resembles very much the
sequence in natural levee deposits. Sedimentation
begins with a sand layer, becoming silty upward.
Climbing-ripple lamination is abundantly developed. Small-scale cross bedding and horizontal
bedding are other important bedding types. Sand
and silt units grade upward into finely laminated
clayey sediments. Units may be a few centimeters
to several decimeters thick. Sometimes more than
one coarser and finer units are deposited during a
single flood, which indicates fluctuations during
the flood. Plant debris is abundant. The clay layer
generally shows mud cracks, etc. Near the surface
carbonate concretions and iron concretions may
be formed in areas with high rates of evaporation.
Soil horizons are well developed.

Jahns (1947), McKee (1939, 1965), Happ
(1940), Schumm and Lichty (1963), Doeglas
(1962), and McKee et al. (1967) provide accounts
of undifferentiated flood plain deposits.

In flood plain deposits of the Colorado delta,
Arizona, according to McKee (1939, 1965), up to
80 % of the beddings is represented by climbing-
ripple lamination. The same is true of the Indus
River (McKee 1966a). Figure 428 shows an idealized sequence through flood plain deposits of the
Indus River. However, in the flood plain deposits
of Bigen Creek, Colorado, the major bedding (90
to 95 %) is horizontal-bedded sand, with minor
and local development of climbing-ripple lamination and some large-scale cross-bedding formed
from delta foresets (McKee et al. 1967). Figures
429 and 430 show two sequences in such flood
plain deposits. Convolute bedding is very commonly associated with flood plain deposits and is
generally present in the upper parts of the sandy
units and in the muddy top layer (McKee 1966a).

The flood plain deposits are similar to the sheet
flood deposits of ephemeral streams. The flow is
shallow, thus conditions of upper flow regime are
easily achieved.

Alluvial Fan Deposits

An alluvial fan is a fan-shaped body of rather
coarse detrital sediments, poorly sorted, built up
by a mountain stream at the base of a mountain
front where a steeper slope passes abruptly into a
more gentle slope. Generally, several alluvial fans
occur adjacent to each other, and by their lateral
coalescence a broad sloping plain–alluvial piedmont slope (compound alluvial fan)–is formed
(Blissenbach 1954). Piedmont deposits, fanglomerates, and alluvial cone deposits are all included
under alluvial fan deposits.

Favorable conditions for deposition and preservation of alluvial fan deposits exist where several

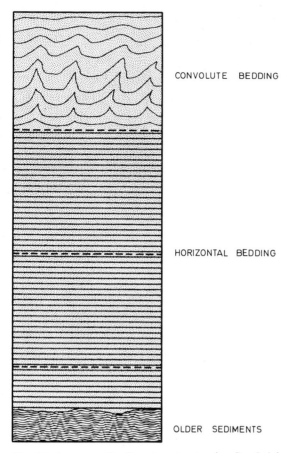

CONVOLUTE BEDDING

HORIZONTAL BEDDING

OLDER SEDIMENTS

Fig. 429. Sequence of sedimentary structure in a flood plain deposit, sedimented during a single flood. Ripple bedding is completely absent. (Based on McKee et al. 1967)

alluvial fans exist beside each other–coalescing–in a tectonically active area, i. e., actively sinking basins. Often fault plains are developed along such mountain chains; the mountains are being elevated and the alluvial fans are sinking. In such cases, thick deposits of alluvial fans are produced and preserved as marginal facies of the basin of deposition.

Alluvial fan deposits may be found with various environmental associations, depending upon the topographical and climatic conditions. The most common association is the fluvial environment, where the alluvial fan deposits are associated with braided river deposits of mountainous regions. Alluvial fan deposits also occur in desert climates, associated with sand dune and playa sediments; in glacial environments they are associated with glacial and fluvio-glacial deposits. Sometimes alluvial fan deposits are associated directly with coastal sediments, e. g., the California coast (Shepard and Dill 1966).

An alluvial fan is of the shape of a segment of a cone that radiates downslope from the point where

the stream channel emerges from a mountainous area. In the lower part the alluvial fan flattens out, becoming more fan-shaped and ultimately coalescing with adjoining alluvial fans in the lower part. The angle of dip of alluvial fans rarely exceeds 10°; generally, it is between 3 and 6°. The radius of alluvial fans varies from a few hundred meters to 100 km or more. The point where the stream emerges from mountains, and the highest point of elevation on the alluvial fan is known as the apex. Bull (1964a) found that the overall radial profiles of alluvial fans are gently concave upward, but that the slopes do not decrease at uniform rates downslope from the apexes. Instead, the radial profile of most of the fans usually consists of three well-defined straight-line segments (sometimes there may be more). The surface represented by each of these segments makes a band of approximately uniform slope and runs concentric about the fan apex. Blissenbach (1954) distinguishes three zones on the alluvial fans:

1. *Fanhead (Upper fan segment):* the area on the alluvial fan close to the apex.
2. *Midfan (middle fan segment):* the area between the fanhead and the other lower margins of the fan.

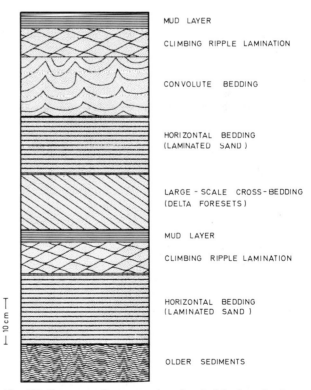

MUD LAYER

CLIMBING RIPPLE LAMINATION

CONVOLUTE BEDDING

HORIZONTAL BEDDING
(LAMINATED SAND)

LARGE - SCALE CROSS - BEDDING
(DELTA FORESETS)

MUD LAYER

CLIMBING RIPPLE LAMINATION

HORIZONTAL BEDDING
(LAMINATED SAND)

OLDER SEDIMENTS

Fig. 430. Sequence of structures in a flood plain deposit of a single flood. (Based on McKee et al. 1967)

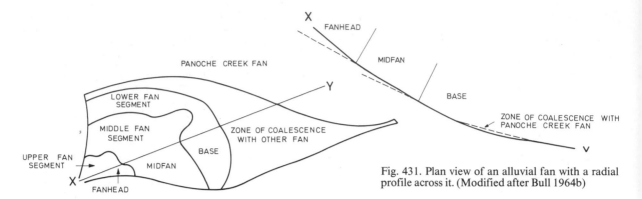

Fig. 431. Plan view of an alluvial fan with a radial profile across it. (Modified after Bull 1964b)

3. *Base (lower fan segment):* the outermost area, or lower zone of the fan. This zone usually grades into the zone of coalescence with other fans.

Figure 431 shows diagrammetically these zones of an alluvial fan and a radial profile through it.

The evolution and geometry of alluvial fans is strongly controlled by the climate, lithology, and tectonic environment.

According to Bull (1964a), the lithology of the source rock is the major controlling factor for shape and size of the alluvial fans. If source rocks are mudstones and shales, alluvial fans are much steeper and twice as large as fans whose source rocks are sandstones. Climate is another important factor. If precipitation is high, the slopes of alluvial fans are gentle, as the flow of water shifts the sediments and decreases the gradient. In arid climates, the slopes of alluvial fans are steeper.

The main conditions for the formation of alluvial fans is a sudden change in the slope, leading to deposition and intermittent stream action. Intermittent stream action may be a result of heavy seasonal rainstorms in an arid climate, or the melting of snow in higher mountains with the onset of summer. Regular heavy rains seem to inhibit the formation of alluvial fans. Thus, they are best developed in arid to semi-arid and subarctic regions. In a humid climate, too, alluvial fans develop, but show different sedimentological characteristics (Schumm 1977).

The main cause for the deposition of a large amount of debris in alluvial fan regions is the decrease in depth and velocity of flow as a result of the increase in width due to the spreading of the flow from the apex of the fan (Bull 1964a). Deposition is further supported by the infiltration of water into loose debris, causing a decrease in the volume of flow, and a pronounced decrease in the slope of a stream channel leaving the mountains.

Alluvial fan deposits are poorly sorted, immature, coarse-grained sediments. Usually gravel, cobblestones, and boulders predominate, with subordinate amounts of sand, silt, and some clay. The composition and nature of alluvial fan sediments is controlled mainly by the local source of debris rock and weathering processes. As there is little transport and sorting, alluvial fan sediment essentially represents a conglomerate of local provenance. The coarsest and thickest deposits occur near the fanhead area. Maximum grain size and thickness of sediments decrease rapidly toward the base of an alluvial fan deposit. However, sometimes a large channel cuts across the fan head, and brings coarser material in the lower part of the fan. The roundness of coarse grains also increases with increasing distance from the apex. Sphericity shows no particular change within an alluvial fan (Blissenbach 1954).

A few important papers dealing with alluvial fan deposits are Twenhofel (1950), Blissenbach (1952, 1954), Bull (1960, 1962, 1964a,b, 1972), Bluck (1964), Melton (1965), Hooke (1967), Beaumont (1972), Wasson (1977), Leggett et al. (1966), and Schumm (1977).

Sediments of alluvial fans are laid down in beds more or less parallel to the surface of the fan, in this way repeating the surface angle of fans in the resulting deposits. Stratification is moderately developed; boulder and pebble beds alternate with sandy, silty, and muddy beds.

Blissenbach (1954) suggested that there are three modes of deposition on alluvial fans:

1. Flash floods. Flash flood deposits accumulate when a large amount of water charged with detrital sediments emerges from the mountains. It tends to spread out in sheets covering part of alluvial fan deposits. Flash floods run for only a small distance and are of short duration. This is analogous to debris flow in the current usage.

2. Streams. Streams are formed if a large enough water supply is maintained. Streams may cut deeply and show braiding. Their action causes deposition of sediments.

3. Stream flood. Streams of alluvial fans are often flooded, making extensive flood plain deposits.

Table 24. Characteristics of three types of deposits in alluvial fan deposits. (After Bull 1960)

Water-laid sediments (stream and flood deposits)	Intermediate deposits	Mudflow depsosits
No discernible margins; usually clean sand or silt; crossbedded, laminated, or massive. Clay content = 6%, sorting index (So) = 1.8, quartile deviation (QDΦ) = 0.8, standard deviation (σ_Φ) = 1.4.	No sharply defined margins; clay films around sand grains and lining voids; graded bedding and oriented fragments. Clay content = 17%, So = 4.0,QDΦ = 2.0, σ_Φ = 3.9.	Abrupt, well-defined margins, lobate tongues; clay may partly fill intergranular voids. May not have graded bedding or particle orientation. Clay content = 31%, So = 9.7, QDΦ = 3.1, σ_Φ = 4.7.

Hooke (1967) studied alluvial fan processes in both field and laboratory. He found that the main channel is generally incised at the fanhead, and appears again on the surface somewhere near the midfan point–the intersection point. Above this point, the mode of deposition is mainly by debris flow, and below it mainly by water-laid deposition–stream, stream flood, and flash floods.

Bull (1960, 1964a) suggested that alluvial fan deposits can be differentiated into three groups (Table 24). This classification is based mainly on the grain-size characteristics, and can be easily identified. Bull (1964a) gives a detailed account of primary structures and textures of alluvial fan deposits. He found that alluvial fan deposits range from well-sorted sands and gravels to very poorly sorted admixtures of clay, sand, and boulders. The C/M plot of alluvial fan deposits gives characteristic patterns (Bull 1962, 1964b). Two zones can be differentiated: A tractive current deposit pattern, and a mud flow deposit pattern. The C/M pattern of mud flow deposits is rather similar to the pattern of turbidity current deposits (Fig. 432). Stratification in alluvial fan deposits is only moderately de-

veloped. Water-laid sediments show well-developed bedding, but bedding in debrisflow deposits is poorly developed. Sediments deposited by the action of a stream show abundant scour-and-fill structures, sometimes developed in form of large-scale cross-bedding. Crudely developed horizontal bedding is rather common in pebble-sized sediments. In finer-grained muddy sediment thin, horizontal lamination is common. Current ripples may be locally present in sandy layers. On muddy surfaces surface features, e.g., mud cracks, rill marks, and raindrop imprints, may develop in abundance, indicating intermittent dry seasons leading to subaerial exposure. In zones where stream activity is common pebble-sized sediments are found as channel lag deposits, and fining upward sequences resulting from migrating channels during floods are well developed. Pebbles usually show horizontal orientation and imbrication structures.

In general, basically two types of deposit can be distinguished in the alluvial fan deposits (see Bull 1972):

a) Debris-flow deposits.
b) Water laid deposits.

Debris-Flow Deposits

Debris-flow deposits are poorly sorted, and may show graded bedding. Pebbles are irregularly arranged and are without preferred orientation; sometimes platy grains are oriented vertically. In more fluid debris-flow deposits, graded bedding is common and flat pebbles are more or less horizontally oriented; but in more viscous debris-flow deposits graded bedding is not present and flat pebbles are randomly oriented–even in a vertical position.

Sharp and Nobles (1953) give an account of mudflows in nature, where flow surges were produced by melting of snow, and the mud flow moved at a rate of about 3 m/sec on a slope of 14 m/km. The individual flows moved in pulsations and deposited thick units. The mudflow deposits are an unbedded mixture of poorly sorted, fine to coarse debris, containing numerous large,

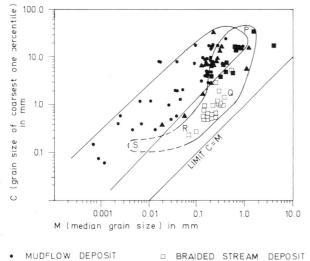

• MUDFLOW DEPOSIT   □ BRAIDED STREAM DEPOSIT
■ STREAM CHANNEL DEPOSIT   ▲ INTERMEDIATE DEPOSIT

Fig. 432. C/M pattern of alluvial fan sediments of Fresno County, California. (Modified after Bull 1964b)

angular fragments. The fine matrix shows pseudo-vesicular structure due to entrapment of air. Process of debris flow is discussed by Carter (1975), Hampton (1975), and Rodine and Johnson (1976). The debris flow is a gravity-induced movement of sediment-water mixture, in which large sediment blocks embedded in a fine-grained matrix can be transported on gentle slopes. It is due to the high density of the debris, poor sorting of the granular material, and the ability of strength and density of the clay-water fluid phase to suspend fine-grained particles. The debris mass flows as a frictionless mass on low slopes (Rodine and Johnson 1976). The term mud flow is used if the proportion of coarser clasts is low, and debris flow is used if the content of larger clasts is high.

The sediments of debris flow are deposited as thin, wide sheets parallel to each other and parallel to the fan surface. Thickness of individual debris flow units is usually a few dcm. The sediments are mostly bimodal with a high degree of variability, poorly sorted showing matrix-supported clasts. Sometimes clasts may show orientation parallel to bedding, and inverse grading in the deposits is often present (Fisher 1971). The debris flows do not erode the underlying surface; they may smoothen out the basal irregular surface by deposition. Thickness of individual debris flows decreases in the downslope direction. Sometimes on the margins of a flow, debris-flow levees may develop.

Water-Laid Deposits

These represent deposits of flowing water activity and can be distinguished into three types:

1. Stream channel deposits–These are deposits in the channels entrenched into a fan. The channels are of varying shape and often contain a channel-lag gravel at the bottom. Mostly channels are truncated and it is not possible to identify the original shape. The deposits are lenticular-shaped, poorly sorted gravels and sands. Scour- and fill-structure is common. Gravels may show some imbrication, and the sands show cross-bedding. They are most important in the upper part and midfan area, because below the intersection point floods are unconfined and change into sheet floods.

2. Sheet flood deposits–These are deposits made by flood flows spreading out from the channels. Deposition is due to widening of the flow into a shallow braided area. A pattern of small braided channels and bars develops. The flow is shallow and mostly upper flow regime conditions prevail. The deposits are sheet-like, lenticular bands of gravel and sand. They are well-sorted, showing cross-bedding and parallel bedding. Imbrication of pebbles is well developed. Thin units of backset bedding related to migrating antidunes are pres-

ent. Fine sand units may show small ripple bedding. They are more abundant below the intersection point, where discontinuous channels pass into sheet flood areas. The units are sheet-like with length/width ratios of roughly 5 to 20.

3. Sieve deposits–If a flood water is poor in suspended material, the highly permeable older sediment causes flow to diminish rapidly as infiltration of water occurs. As a result clast-supported gravel lobes are deposited and named sieve deposits (Hooke 1967). They are well-sorted poorly imbricated gravel and are deposited just below the intersection point. Later, the interstices of these gravels may be filled with finer infiltrating sediment giving a bimodal character, and making their recognition difficult.

Nevertheless, in practice identification of various types of deposit in an alluvial fan may be difficult, though distinction between debris-flow and water-laid deposits is mostly possible. In an interesting study of Pleistocene alluvial fan deposits Wasson (1977) recognizes two types of water-laid sediment:

Group 1 sediments showing poorly sorted, subhorizontal beds often 20–30 cm thick, lying parallel to fan surface. Beds are often lenticular in shape, orientation of clasts is rather variable, content of fine-grained matrix is high. Clast-supported fabric dominates, though matrix-supported fabric is also present. It is believed that these sediments represent deposition in a stream with a high concentration of fine sediment moving as slurry, but not as debris flow, or the fines have been later added by infiltrating water in an openwork gravel.

Group 2 sediments showing well-sorted, well-bedded gravels and sands showing good separation of sizes, lenticular beds, imbrication well-developed, showing upstream dip of clasts. Low-angle planar cross-bedding, horizontal bedding are common. Fining-upward units are also present. They are bedload sediments deposited mainly in upper flow regime conditions.

Group 1 and 2 sediments are intimately related and within a single channel deposits of both of them occur interbedded. In C/M pattern debris-flow deposits and Group 1 water-laid sediments cannot be distinguished. This example shows that transitions occur even between debris-flow deposits and water-laid deposits.

Sometimes boulders and armored mud balls are present in stream deposits of alluvial fans. Their size decreases downslope. Another characteristic feature for alluvial fan deposits seems to be the presence of voids and cavities. They include intergranular openings between grains held in place by dry clay, bubble cavities formed by air entrapment at the time of deposition, interlaminar openings in thinly laminated sediments, buried but unfilled mud cracks, and voids left by the decay of entrapped vegetation (Bull 1964b). Figure 433 shows some characteristic bedding structures of alluvial fan deposits.

Sediments of alluvial fans are deposited in extremely oxidizing conditions, and organic matter and fossil remains are rather rare. Locally, plant remains may be present, especially in more humid climates, but animal remains are extremely rare.

Individual strata of alluvial fan deposits can be as thick as 3 or 4 meters or as thin as a few centimeters, but generally they are several decimeters thick. Water-laid sediments and mudflow sediments are inter-layered in a complicated manner. Along the radial section individual units may be traced for considerable distances; however, in cross-section beds are of limited extent, generally lenticular- to irregular-shaped units.

Locally, slumping may be common in alluvial fan sediments, giving rise to thick horizons with deformation structures. The process of slumping is facilitated by the unconsolidated nature of sediments on a slope and the content of clay. Figure 434 shows a detailed map of an alluvial fan with various kinds of deposits and grain sizes.

Alluvial fan deposits should not be confused with talus cones formed by gravitational sliding of rock debris along the mountain slopes. Talus cones are much steeper than alluvial fan deposits, and their sediments possess other characteristics. Talus sediments move only under the influence of gravity. Sediment is thickest near the base. Fine-grained clayey sediments are almost absent. Sediment particles are extremely angular. Talus deposits are unimportant in the geological record and possess very little potential of preservation.

Distribution of Facies in a Fan

In an idealized sequence of a fan from a semi-arid climate the apical part of the fan (fan head) consists of mostly coarse-grained conglomerates, where debris-flow (matrix-supported) conglomerates are most abundant. Clast-supported conglomerates may be present, but occur in channels.

The mid-fan area consists of thinner debris-flow units and extensive water-laid sediments. Channeled deposits of conglomerates are quite common. Thick and coarse-grained conglomerates showing both normal and reverse grading are present. A few sandy layers showing parallel bedding and planar cross-bedding are present. The sequence is characterized by sheet-like small units of sand and gravels. Thin bands showing backset bedding related to antidunes may be present. Thin occasional units of conglomerates are present.

The distal fan usually shows sand and silt sediments of sheet-flood deposits. Channeling is normally absent; but may be common if reworked later by fluvial processes. Sediments mostly show parallel bedding and low-angle planar cross-bedding. Few units of ripple bedding may be present. Thin occasional units of conglomerates are present.

The distal fan facies shows sharp interfingering with flood plain or lake deposits. The flood plain sediments are usually silty clays, mottled and showing caliche development. Units of flood plain sediments and distal fan deposits are interbedded. If the fan ends up in a lake, fine-grained rhythmic laminated clays and ripple-bedded sands may be present. A few conglomerate horizons may extend into the lake. Figure 435 shows a schematic distribution of various lithofacies in an alluvial fan.

This ideal sequence may show strong variations, depending upon the proportion of water-laid and debris-flow deposits. If in the source area conditions are not conducive for production of debris flow (e. g., non-availability of fine-grained material, higher humidity and stream flow), the fan may contain only water-laid sediments. In the case of somewhat humid conditions the distal fan facies interfingers and overlaps with the braided stream deposits, and it may be difficult to differentiate the two. On the other hand, if supply of fine-grained

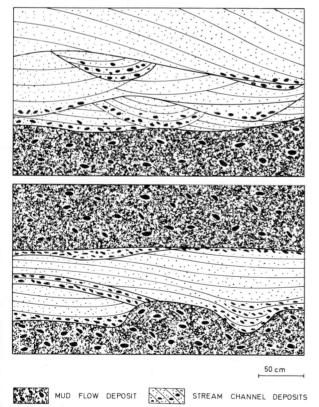

50 cm

▨ MUD FLOW DEPOSIT   ▨ STREAM CHANNEL DEPOSITS

Fig. 433. Bedding structures of the alluvial fan deposits. Mud flow deposits and stream channel deposits are interbedded. (Modified after Blissenbach 1954)

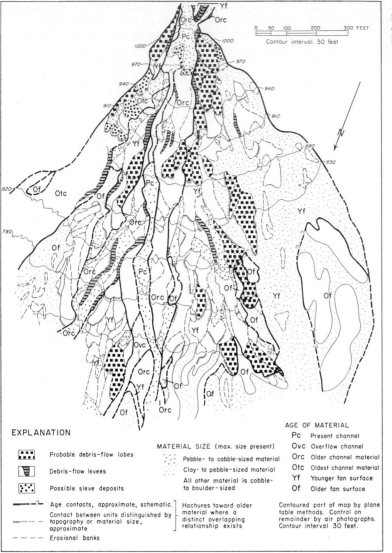

Fig. 434. Detailed map of Trollheim alluvial fan showing the distribution of various types of deposits and sediment types. (After Hooke 1967)

sediment is high and erosion is rapid, debris flow would dominate. In such a case, even in the mid fan area, debris flows would be abundant, and extending even in lower fan areas. The proportion of water-laid and debris-flow deposit varies from fan to fan within an area, and may change during the depositional history of a single fan.

The fan head deposits may be associated with fault breccia, if located near a fault zone. Heim (1882) and Hsu (1975) describe how large rockfalls near steeps slopes commonly generate fast-moving streams of debris (Sturzstrom), capable of moving on gentle slopes, similar to mudflows. These deposits look like breccia, and may become associated with the fanhead deposits.

Humid Alluvial Fans

In a humid climate, development of alluvial fans takes place, but their characteristics differ greatly. Schumm (1977) emphasizes the importance of wet alluvial fans and discusses their characteristics both in the field (modern and ancient) and in laboratory experiments. Many of the ancient humid alluvial fans contain placer deposits of gold, uranium etc.

A humid alluvial fan is formed by a perennial stream and possesses a very short fan head area, while the rest of the fan is characterized by strongly shifting braided streams. One braided channel runs from apex to toe and sweeps systematically over a fan area.

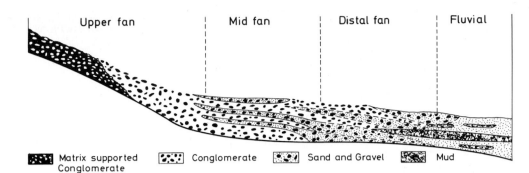

Matrix supported Conglomerate | Conglomerate | Sand and Gravel | Mud

Fig. 435. Schematic diagram showing distribution of lithofacies in an alluvial fan deposit. Upper fan is characterized by gravelly facies where depending upon the climatic conditions, debris flow (matrix-supported) deposits or clast-supported deposits dominate. Mid-fan area shows sheet like gravelly and sandy deposits. Distal fan usually show sand and silt deposits with thin gravel bands. In the lower part distal fan deposits interringers with sand-mud sequences of fluvial environment, or with the lake sediments. (Modified after McGowen and Groat 1971)

The Kosi River fan is considered a good present-day example of a humid alluvial fan, which Gole and Chitale (1966) describe as an inland delta. The Kosi River, draining from the Himalayas, brings huge amounts of sediments, constructs a huge alluvial area, and drains into the Ganga River. At the head of the fan the channel consists of boulders and gravels. These coarse sediments are transported only about 15 km downstream where coarse sediment disappears. From here, the river widens and acquires braiding characteristics and is about 6 km wide. The gradient of about 0.9 m/km near the fan head decreases to about 0.19 m/km. Further downstream, river divides into several channels and widens to about 15 km.

Within the last two hundred years, the Kosi River has moved progressively towards the west (Fig. 374), and has reworked sediments in an area of about 9,000 km².

In the fan head area of humid fans there may be a downstream increase in the grain size (Schumm 1977). The stratification in humid fan deposits is well-defined and there is less variability in the grain size.

Humid alluvial fans may be regarded as deposits of essentially braided streams which develop without any lateral confinement.

Boothroyd and Nummedal (1978) discuss that deposits of proglacial outwash fans may be considered a model for humid alluvial-fan deposits. In this model, the surface gradient decreases from 50 m/km in the proximal area to 2 m/km in the distal margin, accompanied by a downstream decrease in maximum grain size of 45 cm to sand-size sediments over a distance of 21 km. The downstream change in gradient, clast size and other parameters occurs over generally similar distances in a down-fan direction. Thus, the smaller fans show proximal characteristic over their entire length.

The upper fan usually has 1–3 channels, incised in gravels, gradient 6–17 m/km, average clast size

> 10 cm, and longitudinal bars making horizontal bedding and planar cross-bedding.

The midfan shows a large number of braided channels and prominent longitudinal bars of gravel and sand. Planar cross-bedded gravel and sand layers are common, average clast size is 2–10 cm. Locally megaripples may develop.

The distal fan is mainly a sandy area showing extensive bifurcation around lingoid bars (fine-grained). Mainly planar cross-bedding is produced, but units deposited during emergence contain small-ripple bedding etc.

Laterally, the distal fan facies may be associated with swamp and marsh deposits. Thus, in this model too, deposits of humid alluvial fan are essentially deposits of laterally shifting braided rivers.

Fluvial Associations

River deposits occur in certain associations. This led Happ et al. (1940) to suggest four associations in fluvial environments. However, we feel that there are only three specific fluvial associations:

1. Alluvial fan or piedmont association
2. Flood plain association
3. Coastal plain–delta association.

Figure 436 shows schematically the three associations.

Alluvial Fan Association

This association is found along the mountains, where the slope abruptly becomes gentler. Deposits are coarse-grained and poorly sorted. The border of the basin are preserved. Deposits represent almost exclusively bed load deposits. Downslope alluvial fan deposits grade into coarse-grained

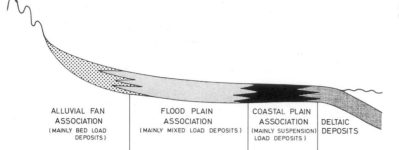

ALLUVIAL FAN
ASSOCIATION
(MAINLY BED LOAD
DEPOSITS)

FLOOD PLAIN
ASSOCIATION
(MAINLY MIXED LOAD DEPOSITS)

COASTAL PLAIN
ASSOCIATION
(MAINLY SUSPENSION)
LOAD DEPOSITS)

DELTAIC
DEPOSITS

braided river deposits. Most abundant are channel deposits composed of lenticular units, with thin units of finer-grained sediments deposited during overbank flows. Fining-upward sequences are well developed. The main structures are cross bedding and scour-and-fill structures.

In some cases alluvial fan deposits are associated with desert sediment–playa deposits, or sand dune deposits, or in rare cases alluvial fans may occur along steep coasts associated with lagoon and beach deposits.

Flood Plain Association

This is the fluvial association developed in mature reaches of rivers. They can be regarded as mixed load deposits, where both bed load and suspension load deposits are well developed. Various subenvironments are well developed.

Point bar and channel bar sequences constitute the major part of the deposits, together with natural levee, crevasse splay, and flood basin deposits. Sometimes differentiation is not well developed and one can recognize only channel deposits–point bar deposits–and flood plain deposits.

In vertical, fining-upward sequences various units are repeated. In suitable subsiding tectonic situations they can make thick sequences. Sandy units of channels alternate with muddy units of flood basins.

Coastal Plain–Delta Association

They generally make the most extensive fluvial deposits, occupying large areas in the lower reaches of rivers and producing thick deposits of fluvial sediments. They gradually grade into upper deltaic plains, coastal sands, and tidal flat sediments.

Sediments of coastal plains may attain thicknesses of several thousand meters. Deposits are rather fine-grained in nature. They can be regarded as mainly suspended load deposits. Sand is deposited only in the bottom of river channels. The main sediment is silt and clay, deposited on the upper parts of point bars and levees. Flood basins in which muddy sediments are deposited are well developed. In humid climates swamps may develop, and peat is commonly associated with flood basin deposits. Lenticular and sheet-like sand bodies are embedded in thick deposits of silty and clayey sediments.

Brahmaputra-Ganges, Mississippi, Hwang Ho, and Indus–all these big rivers make extensive coastal plains in their lower reaches.

Some General Remarks

A fluvial sequence is mostly made up of alternating units of sandy facies and muddy facies. The sandy facies represents deposits of channel activity, while the mud facies represents deposits of flood plain (basin) deposits. The sandy facies may show fining upward sequences or the change from sand to mud facies may be rather abrupt. The mud facies also show much variability e.g. containing thin sandstone units, carbonaceous shale with coal, reddish shale with calcrete layers etc., depending upon the climate and channel characteristics of the fluvial system. The relative importance of sand and mud facies in a vertical succession is highly variable, and depends upon the behaviour of river channels and sedimentary tectonics. Preservation and thickness of muddy facies (flood plain deposits) and shape of sandy bodies are dependent upon the rate of migration of fluvial channel.

Lateral migration of stream channels normally produces a sheet of channel sand that in ancient sediments is known as sheet sand deposits. Thick sheet sand deposits are produced by continuously migrating channels.

Coleman (1969) describes that in 200 years the Brahmaputra River has carved a valley about 200 km long with an average width of 12 km. The average depth of sand laid down by the migrating river is on the order of 20 m, in localized areas up to

40 m. In a rather short time a blanket sand deposit has been formed by the migrating braided river. Recognition and distribution of smaller sand facies is extremely complex, but the overall channel sand body is easily recognizable in borings. Figure 437 depicts the lateral migration of the Brahmaputra River.

Schumm (1968) also suggested that sheet or blanket sand deposits can be produced by any type of migrating river channels (meandering or braided). He adds further that pods and ribbon deposits of Potter (1963) may be discontinuous sand bodies deposited by suspended load rivers, whereas dendroids and belt deposits are deposited by mixed- and bedload channels.

Schumm (1968) summarized that river morphology strongly controls the type and geometry of sediments.

Schumm (1968) provides a useful summary regarding the application of climatic, vegetation, and hydrological data of rivers to the study of ancient fluvial sediments. He concluded that before the appearance of significant plant cover (before the Devonian age), denudation and run-off rates were high and floods were large. Under these conditions mostly bedload channels and channel deposits were produced. With increased plant cover alluvial deposits were stabilized, but periodic large floods helped in marking cyclic deposits. The appearance of hardier plants in Mesozoic and Cenozoic times, and the later appearance of grasses in the late Cenozoic, further influenced the fluvial processes.

Four geologic times can be differentiated (Russell 1956):
1. prevegetation time (pre-Devonian)
2. time of primitive vegetation (Devonian to Cretaceous)
3. time of modern types of flowering plants (Cretaceous to Miocene)
4. time following the appearance of grass (Miocene to present).

It is generally believed that evolution of land vegetation caused a change from the braided to meandering pattern and also helped in the development of fining-upward sequences. The fluvial deposits in Pre-Devonian times are considered to be braided-type, while in Post-Devonian times both meandering-type and braided-type are present (see Cotter 1978). However, we feel that the absence of vegetation in Pre-Devonian times caused more effective lateral shifting of the channels leading to development of dominatly sheet-type units.

In the following, a short discussion on some of the significant features of fluvial deposits is given:

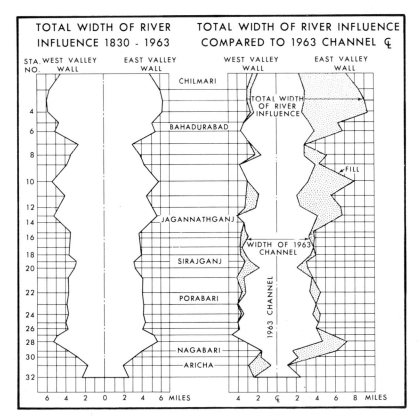

Fig. 437. Diagram showing deposition of sand in the Brahmaputra River valley as a result of lateral migration. (After Coleman 1969)

Palaeocurrent Patterns. Palaeocurrent patterns in the fluvial deposits are much more variable than is generally considered. Directional dispersion of the palaeocurrent data is usually more than the disper-sion of the channel. It also depends upon what features are measured. For example, pebble orien-tation shows lower dispersion than the large-scale cross-bedding, while the dispersion of data of small-ripple bedding is very high. High and Picard (1974) found that in fluvial sediments trough cross-bedding gives more reliable palaeocurrent results (direction of channel) than the planar cross-bedding. There is a tendency in the study of ancient fluvial deposits to measure only the trough cross-bedding. No doubt, such measurement give better information on palaeoslope, but planar cross-bed-ding in fluvial deposits (mostly related to migrating bars) may provide additional information on the pattern of sedimentation.

The variability in palaeocurrent direction is controlled by discharge variation, local relief, and whether the deposits were made by low discharge, high discharge, or falling discharge (Schwartz 1978). Collinson (1971), Steinmetz (1972), Bluck (1974), and Miall (1976) give important informa-tion on palaeocurrent analyses in the fluvial sedi-ments. Thus, the general notion that wide variance in palaeocurrent is produced by meandering streams, and lower dispersion by braided streams should be used with great caution. Within a fluvial sequence thin units may be interbedded which give direction at right angles to the main channel direc-tion towards the channel (related to the delta-like lobes, Singh and Kumar, 1974), or in a direction at right angles to the channel direction away from the main channel (migrating scroll bars, Jackson 1976a), or show upstream directions (related to flow separation in point bars, Davies 1966; Taylor et al. 1971).

Palaeohydraulic Estimates. There have been sever-al attempts to quantify the fluvial sediments in the rock record, and several mathematical models for determining the depositional pattern and palaeo-hydraulics have been proposed. Allen (1970b) pro-posed a quantitative model for calculating charac-teristics of a meandering channel. Bridge (1975, 1978b) proposes a computer model for sedimenta-tion in meandering streams and suggests that com-plete sequence of point bar capped by overbank se-diments would rarely be preserved in a shifting me-ander. Further, due to diversity in the lateral depo-sits, single fining-upward coarse members cannot necessarily be taken as representative of complete point bar or a meandering channel reach.

A mathematical model for calculating various palaeohydrological parameters, e.g., discharge, gradient, meander wavelength, etc. has been pro-posed by Schumm (1972) and successfully used in many examples (see in Ethridge and Schumm 1978). Basic data needed for such palaeohydraulic calculations are: bankfull stream width, bankfull stream depth, percent silt-clay in channel bed, per-cent silt-clay in channel bank.

Estimation of width and depth of palaeochan-nels in rock successions is often a big problem. Longitudinal cross-bedding (epsilon cross-bed-ding) can be used in width calculation of palaeo-channels, as they are considered to present $\frac{2}{3}$ of the width of the former channel bankfull width (Allen 1965e). Width is obtained by determining average width of epsilon cross-bedding multiplied by 1.5. Depth is represented by the thickness of channel sand (point bar deposit). However, various wor-kers made some corrections while calculating depth. Leeder (1973) gives an equation to calculate bankfull width of meandering palaeochannels in fining-upward cycles (bankfull depth).

Using these data in equations given by Schumm (1972), various hydrological informations are ob-tained. However, there are many inherent prob-lems with such calculations, e.g., role of compac-tion and amount of clay formed in diagenesis. As already pointed out by Schumm (1972), if these estimates are at variance with other geological evi-dence, then the geological evidence should be con-sidered to be of a higher order of significance.

Epsilon Cross-Stratification (Longitudinal Cross-Bedding). Longitudinal cross-bedding in fluvial deposits is commonly referred to as epsilon cross-stratification (*ECS*), and shows direction of dip at right angles to the direction derived from the asso-ciated small-ripple bedding. The units are made up of inclined layers, dipping mostly at 15–20°, though sometimes higher dips (> 20°) may be present. The units are generally 1–2 m thick, some-times up to 5 m thick. The internal layers are defin-ed by the grain size changes. Allen (1965e, 1970) emphasized that epsilon cross-bedding is formed by lateral accretion of point bar and is a characteri-stic feature of meandering stream deposits. Many epsilon cross-beddings are made up of alternating silt and clay layers and look like the filling-up of a migrating channel-bar system. In such cases, due to high mud contents, angle of dip of the layers is also high.

It is suggested that epsilon cross-bedding cannot be taken as evidence of a meandering stream, as it can be produced by a migrating bar-channel sy-stem of a braided stream. Large-scale cross-bed-ding in the Kreuznacher Sandstein (Permian), Ger-many, represents, in parts, longitudinal cross-bed-ding produced by the migrating channel-bar sy-stem of braided streams, where sometimes at the end of the cross-bedded unit the filled-up channel is also preserved.

Fining-Upward Sequences. Fining-upward sequences are quite commonly developed in fluvial deposits, and Allen (1970a,b) emphasized that fining-upward sequences are directly related to the lateral migration of point bars in a meandering channel, and most workers considered fining-upward sequences to be the most characteristic feature of meandering streams. Absence of fining-upward sequences in fluvial deposits is sometimes taken as a criteria for braided stream deposits.

A fining-upward sequence develops where an asymmetrical flow pattern is developed, and is related to the radius of curvature of a meander. Theoretical and field data indicate that generation of sequences by migrating meander belts is rather complex and not simple fining-upward (Bridge 1978; Jackson 1976a). Further in larger, coarse-grained point bars, the fining-upward tendency is mostly missing (e.g., McGowen and Garner 1970); similarly in fine-grained point bar deposits, fining-upward tendency is also not apparent.

On the other hand, braid bar sequences of braided streams show the development of fining-upward sequences (e.g., Coleman 1969). Rust (1978b) points out that the distal gravelly braided streams are characterized by fining-upward sequences: trough cross-bedded framework-supported gravel, trough cross-bedded sand, and laminated mud. This sequence is related to coarse sediment deposited by active bars and a decrease in the grain size as the area aggrades and becomes inactive. Further, fining-upward sequences of some sheet-like sandy units (1–2 m thick) are related to deposition by sheet flood in a decelerating flow on the flood plain.

To sum up, fining-upward sequences are abundant in fluvial deposits, though not necessarily produced by migrating point bars.

Minor Sandstone Bodies. Some of the fluvial deposits are characterized by thick successions of dominantly muddy sediments, in which minor sand bodies are incorporated. Sometimes these sand bodies are 1–2 m-thick fine-sand units made up of horizontal bedding, small-ripple bedding and climbing-ripple lamination and show a fining-upward tendency. Such bodies represent deposits of a sheet flood on a flood plain. Small, convex sand bodies pinching out laterally are related to the channels of crevasse-splay. These sand bodies possess a strongly convex erosive base. Some of the laterally pinching sand bodies possess a rather smooth lower surface and strongly convex upper surface, and are internally made up of unidirectionally dipping low-angle planar foresets, with a few trough cross-beds. They represent deposits of a laterally migrating bar in a suspended-load-rich braided stream system. Some of the lenticular sand bodies which become muddy towards the top are

regarded as filled-up abandoned shallow channels. Some of the thin sandy units may represent aeolian sand deposits within the flood plain deposits.

Meandering Versus Braided Stream Deposit. Many of the investigators of the ancient fluvial deposits in the last decade have been concerned mainly with the differentiation and identification of meandering and braided stream deposits. This aspect has often been overemphasized and detailed information on fluvial deposits was sometimes neglected. Moody-Stuart (1966) tried to distinguish between low-sinuosity (braided rivers) and high-sinuosity (meandering rivers) stream deposits and gave a number of useful criteria.

According to him, high-sinuosity stream deposits possess the following characteristics:

1. A meandering belt with coarser material as point bar deposits represents channel deposits. In cross-section it appears as a tabular sand body, with edges defined by the shape of the channel on the outer side of the meander loop. The sandy body is surrounded by flood plain deposits.

2. Sand bodies are often associated with fine-grained channel-fill deposits. The original channel form is present.

3. Longitudinal cross-bedding produced from the migration of point bars is well developed.

4. Thin natural levee deposits may be present on the top of point bar deposits.

5. Paleocurrent data show wide variation.

Characteristic features of low-sinuosity stream deposits are:

a) Channel deposits show an approximately horizontal top, and the lower surface is erosional and trough-shaped. Deposits are rather shallow.

b) Since channels are abandoned by avulsion, the fine-grained sediment is spread across the whole upper surface of the coarser-grained channel deposits.

c) Little or no longitudinal cross-bedding. Large-scale cross-bedding is well developed.

d) Natural levee deposits are rather rare and are preserved only on the sides of the channel deposits.

e) Palaeocurrent data show only minor deviations in the flow pattern.

Smith (1970) considered criteria such as abundance of planar cross-bedding and mud pebbles to be typical of braided stream deposits. However, this is not always true. Rust (1978b) points out that in gravelly braided stream deposits, gravels are dominantly open-framework type, where some sand is infiltered. The gravel layers are thick and continuous. In gravelly meandering stream deposits gravels are scattered throughout the sequence, the sediments are sandy gravel with a few pockets of open-framework-type gravels.

However, as more information from the study of many variants of braided and meandering streams in the present-day environment has been gathered, it has become clear that differentiation between meandering and braided stream deposits is rather a complex problem.

The channel pattern is not always controlled by the sediment types, as pointed out by Schumm (1968). Rust (1978a) discusses that three types of braided alluvium sequences are developed, i. e., gravel-dominant, sand-dominant, and silt-dominant.

Similarly, Jackson (1978) points out that five lithofacies classes of meandering streams can be identified: muddy fine-grained streams, sand-bed streams without mud, gravelliferous sand-bed stream, and streams with coarse gravel and little sand. Each of these five lithofacies classes shows its own characteristic features; many of these features are generally considered typical of braided stream deposits, e. g., negligible mud, large facies changes laterally, no vertical sediment variation, abundance of gravel etc.

Jackson (1978) expresses the opinion that the existing geomorphological and sedimentological criteria for differentiating braided vs. meandering streams have little validity. The high proportion of mud facies (flood plain deposits) in a fluvial sequence is often taken as a criterion for meandering stream deposits. The ratio of sandstone to mud facies in a fluvial sequence is more controlled by the frequency of migration of a channel laterally than by the channel characteristics. No doubt, high mud facies are common in meandering stream deposits, but in suitable conditions they can be found in the braided stream deposits.

Shelton and Noble (1974) describe depositional sequences of the Cimarron River (river gradient 30 cm/km, sinuosity 1.5, fine to medium sand), which show transitional characteristics between meandering and braided streams. The deposits also show transitional characteristics, and a fining-upward tendency in the sequences is present.

Cotter (1978) recognizes four basic types of fluvial styles in ancient fluvial deposits:

1. *Fine-grained meander belt*–Thick shale units alternate with thinner units of fine-grained small-ripple bedded sandstone.
2. *Coarse-grained meander belt*–Thick fining-upward sequences starting with conglomerate-coarse sandstone-fine sand-siltstone and shale. Sometimes multi-storied sandy units are present.
3. *Channeled-braided*–Medium-grained sandstone with minor shale, planar cross-bedding dominant, channel-shaped depositional units, no ordered pattern in vertical sequence.
4. *Sheet-braided*–Sand and conglomerates or only sand units laterally extensive, muddy units

are few and often lenticular-type, sandy units show planar cross-bedding with some trough cross-bedding and horizontal bedding, no vertical variation in grain size.

However, Cotter's (1978) type (1) may be the product of deposition in the flood plain by occasional sheet floods. Type (2) is known both in meandering and braided stream deposits. Types (3) and (4) signify mainly quickly shifting shallow channels, which may be meandering or braiding.

Our experience has shown that in many fluvial deposits a 10–15 m-thick sandy facies alternating with a 3–5 m-thick muddy facies is well developed. The units are laterally very extensive (sheet-like). The sandy facies is made up of horizontal bedding, trough cross-bedding and a little small-ripple bedding. The muddy facies shows thin sandy layers, exhibiting climbing-ripple and horizontal bedding. Such sequences represent deposits of moderately deep channels with flood plain which have been quickly shifting laterally.

It can be said that each river pattern does not produce a single characteristic sequence, but a number of sequences of highly variable character. Similar-looking sequences may be produced by rivers of different patterns. Further, the probability of preservation of a certain type of sequence is strongly controlled by the pattern of migration of the channel.

Ephemeral Stream Deposits

Ephemeral streams occur mostly in an arid to semi-arid climate, and experience short periods of flow, following intense rainfall, alternating with long periods of dryness (see also wadi deposits on p. 217). The ephemeral streams and the adjoining flood plain areas are characterized by sheet flows. The channels of ephemeral streams are mostly shallow with a variety of bar features, on the top of which various bedforms are produced during floods. Sediments are mostly gravelly and sandy. Picard and High (1973) describe in detail sedimentary structures, sequences of the ephemeral stream deposits, and distinguish three types of sedimentary sequence:

Point bar sequence–Point bar sequences of ephemeral stream are mostly less than 1 m thick, showing a fining-upward tendency. The sequence starts with low-angle cross-bedding, followed by small-ripple bedding and climbing-ripple lamination, horizontal bedding, capped by a thin mud layer. The sequence represents deposition during waning stages of a flood.

Channel bar sequence–These are deposits of longitudinal and transverse bars. The sequence starts with a high-angle cross-bedding (large-scale), followed by a thin unit of faintly laminated sand, capped sometimes by small ripple-bedding. The sequence is mostly about 30 cm thick.

Channel-fill sequence– This is similar to point bar sequence. The sequence starts with a basal erosional contact with lag deposits, inclined or horizontal discontinuous laminated sand, followed by sediments showing small-ripple and horizontal bedding. Sometimes, gravels are absent and the sequence is made up of horizontal bedded sand and small-ripple bedding.

These three types of sequence can be interlayered, and some of the sequences may show on the top development of flood plain (muddy sediments with thin sand stringers). Ephemeral stream deposits are common in desert environments, and in some alluvial fan deposits.

Role of Aeolian Sediments in Fluvial Deposits

The active sedimentation in fluvial environment takes place only during high floods, i.e., only a couple of days or weeks in a year. Most of the time fluvial sediments are exposed to the activity of wind. Sellards (1923), Jahns (1947), Kruit (1955), and Schumm and Lichty (1963) have investigated wind activity and features of wind deposits in the fluvial environment.

In humid climates wind activity is hindered by plant growth and a cover of vegetation on exposed fluvial deposits. However, in dry climates or in areas having long dry periods, upper point bar deposits, levee and crevasse splay deposits, and flood basin deposits are subject to wind activity.

The most important features are the development of sand dunes along the stream channels, i.e., on point bars, channel bars, and natural levees, or even in flood basin.

Such sand dunes are generally elongated forms running parallel to the channel, sometimes several meters in height. Generally, sand is removed from point bar deposits during low-water stages and accumulated along the river. Some of these dunes rest on more muddy sediments of the flood basin. Sand dunes are made up of well-sorted sediment showing sand dune cross-bedding. They are partially vegetated and show abundant roots. These dunes or parts of them have a fair chances of being preserved under the cover of the next overbank flows. Figure 438 shows the development of sand dunes along the Red River, U.S.A.

Sand a few decimeters thick in upper point bar deposits and natural levees is actively reworked by wind action, partly reworked into low dune-like

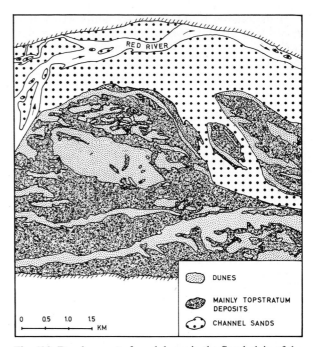

Fig. 438. Development of sand dunes in the flood plain of the Red River. (Modified after Allen 1965. Originally from Sellards 1923)

features, or just moved by wind activity. Wind activity causes winnowing of silt and clay-sized grains, leading to better sorting of the sediment. Roundness as well as the relative concentration of heavy minerals is increased. Wind activity results in the destruction of unstable minerals, especially mica. This wind activity causes better sorting of these sediments than the underlying water-laid sediments. They are evenly laminated or show cross-bedding. They are very likely to be preserved in point bar sequence, and may be difficult to identify, except for their better sorting. Steinmetz (1967) found them preserved in his point bar sequence of Wekiwa point bar, Arkansas River.

Much sand and silt is blown to the flood basins, and surface features become covered by a thin veneer of well-sorted sand. This helps in the preservation of various surface markings in flood plain deposits. Care should be taken while analyzing such water-laid and wind-laid deposits.

Grain-size Characteristics of Fluvial Sediments

Several attempts have been made to distinguish fluvial deposits and various subfacies of fluvial deposits on the basis of grain-size parameters.

In general terms, channel sediments are sand to gravel size, sorting is moderate to good, and clay content is low. Bank deposits are fine sand to silt, moderately sorted. Flood basin deposits are silt to clay, poorly sorted, and have a high content of clay fraction.

Sometimes plots on C/M diagrams have been successful in identifying fluvial environments and subenvironments (e. g., Royse 1970).

Visher (1965a) shows that various subenvironments of a fluvial sequence can be characterized by detailed grain-size studies. Factor analysis of characteristics of grain size distributions provided four well-defined classes of curves, each related to a different sedimentary structure: Large-scale trough cross-bedding, horizontal bedding, smallscale cross-bedding, and climbing-ripple lamination (Fig. 439). If grain-size analysis is plotted on a probability scale, fluvial sediments show a distinctive pattern (Visher 1969). It shows a well-developed suspension population (up to 20 %); truncation between suspension and saltation is between 2.75 and 3.5 phi. There is no distinction between surface creep population and saltation population; if surface creep population is present, it is coarser than 1.0 phi. Such patterns are recognizable in many ancient fluvial sediments.

Several other plots of grain-size parameters may also be useful in characterizing and identifying fluvial sediments. Most important ones are: phi median (Md Φ) versus Folk's skewness, Md Φ versus $\Phi\sigma$ (Phi deviation measure), Md Φ versus $\alpha\Phi$ (Phi skewness measure) (see Kukal 1970; Royse 1970). Fluvial sediments are commonly bimodal and positively skewed, especially the channel sediments. In a vertical sequence there is a tendency of finingupward of grain size.

In rivers made up of sandy material the transportation is mainly in suspension, though there is a frequent interchange with the bed load sediments, and often 80–90 % of the sediment transport in rivers is by suspension (Sundborg 1956). Middleton (1976) analyses the hydraulic significance of the break in the grain-size distribution curves of the fluvial sediments, separating the two coarse subpopulations: traction, and intermittent suspension populations. For the average bed material the settling velocity of the sediment grains at the "break" is approximately equal to the shear velocity of the dominant flows moving the bed material.

Ancient Fluvial Deposits

Fluvial sediments have been reported from all geologic ages, from Precambrian to Quaternary. However, there are only few detailed studies on such deposits.

A thick succession of fluvial sediments of Devonian age are known from western Norway, Great Britain, and the eastern United States. Allen (1964b) describes fluvial sediments of the Devonian in England. More information about the build-up and role of soil processes of these sequences is given by Allen (1970, 1974). Friend (1965) and Moody-Stuart (1966) provide detailed accounts of fluviatile sediments of Devonian age from Spitsbergen. On the western coast of Norway isolated faulted basins with fluvial sediments of Devonian age are present. They provide excellent examples of alluvial fan and braided stream sediments. Nilsen (1969) gives an account of one of such basin from western Norway. The Hornellen basin (Devonian), western Norway, is a well-defined basin where sediments of 25 km thickness are present (Steel and Aasheim 1978). Margins of this basin are flanked by fanglomerates, breccia, conglomerates, etc., with well-documented alluvial fan deposits (Larsen and Steel 1978). There are

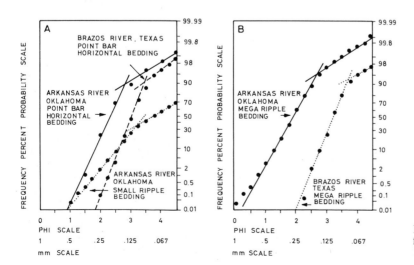

Fig. 439. Characteristic grain-size curves of fluvial deposits. (Modified after Visher 1969)

150 m-thick upward-coarsening cyclothems related to tectonic influence during sedimentation (Steel et al. 1977). The central part of the basin shows longitudinal filling of the basin as a result of alluvial plain or sandy fan delta dominated deposition, dominated by ephemeral, low-sinuosity stream prograding into the flood basin areas (Steel and Aasheim 1978). Bryhni (1978) also discusses the sedimentation pattern in the Hornelen basin and recognizes cyclic succession, where *megarhythms*– related to progradation of fan and valley floor alluvial wedges, and *minorrhythms*–related to flood deposition in uniform or decreasing flow regime are recognized. Friedman and Johnson (1969) provide a recent detailed account of Devonian sediments of the Catskill Mountains, U.S.A., part of which is a fluvial sequence.

Potter (1963) provides a detailed account of Palaeozoic fluvial sediments of the United States, with a detailed palaeocurrent analysis. Carboniferous fluvial sediments associated with coal deposits are well developed in Britain and Germany. Gondwana formations (Carboniferous to Triassic) of India are mainly thick successions of fluvial sediments, deposited in isolated faulted basins. Various aspects of Gondwana sedimentation have been discussed by Niyogi (1966), Sengupta (1970), Cassyap (1970, 1973), and many others.

Siwalik sediments (Molasse of Himalaya) represent extensive fluvial deposits, where lower and middle parts are made up of fine-sand sediments, while the upper parts consist mainly of conglomerates. The Siwalik sediments show well-developed fining-upward sequences and marked channel and flood plain facies (Johnson and Vondra 1972; Halstead and Nanda 1973). The nature of sediments and the development of sequences in the Siwalik is comparable to the present-day Gangetic alluvium, and the Siwalik sediments represent deposits of quickly shifting rivers (Singh 1975). Tandon (1976) discusses petrological aspects of Siwalik sediments, while Prakash et al. (1975) gives information on the palaeocurrents.

Royse (1970) gives a detailed description and analysis of fluvial deposits of Paleocene age from North Dakota.

Belt (1965, 1968) gives an account of ancient alluvial fan deposits from Carboniferous sediments of Canada. Williams (1969) describes Precambrian alluvial fan deposits from Scotland. Steel (1974) describes fluvial deposits from New Red Sandstone, Scotland, and recognizes sediments of alluvial fan, playa, and flood plain environments.

McGowen and Groat (1971) give an excellent account of an alluvial fan deposit of a humid climate, where proximal, mid-fan, and distal fan facies are recognized. The conglomerates are made up almost exclusively of stratified conglomerates and characterize the proximal facies. The mid fan facies show less gravelly beds, more scouring and parallel-bedded sandstone. The distal fan facies shows only a few thin bands of gravels within a cross-bedded sandstone facies. Heward (1978) describes an alluvial fan deposit from Spain, associated with lake deposits. In this alluvial fan deposit, mostly mid fan and distal fan facies are prominent, where deposits of debris flow and turbulent streams dominate. Schumm (1977) reviews the characteristics of several alluvial fan deposits of both wet and dry type.

Selley (1965, 1970) describes fluvial sediments, of the Torridon group (Precambrian) of northwestern Scotland. Allen (1965c) provides a useful list of fluvial deposits from the geologic record.

Puigdefabregas (1973) describes a well-documented sequence of fluvial deposits, made by small streams. Puigdefabregas and Van Vliet (1978) describe meandering stream deposits from the Tertiary of the southern Pyrenees, Spain, where in many cases lateral accretion bedding (longitudinal cross-bedding) is well documented and indicates substantial discontinuity during point-bar growth. Some of these units are embedded as single sand bodies within a muddy flood basin facies.

In another example from the fluvial Eocene sediments of the south Pyrenean basin, Spain, Nijman and Puigdefabregas (1978) discuss a well-preserved sequence of large, coarse-grained point bar, where different geomorphological parts, e. g., scour pool and chute-bar features are recognizable.

Normally, lateral accretionary bedding (longitudinal cross-bedding) is not preserved in the fluvial deposits; although the sedimentation takes place dominantly by lateral accretion. However, in several cases they are well-preserved and seem to be related to fluctuating and discontinuous deposition on the point bars, for example, Karoo basin, South Africa (Hobday 1978), Jurassic of Yorkshire, England (Nami and Leeder 1978), Tertiary of Spain (Puigdefabregas and Van Vliet 1978). These features are also described as epsilon cross-bedding and make units of several meters thick.

McCabe (1977) describes an unusual fluvial sequence from the Namurian of England which is characterized by large-scale channels and bars. Large tabular sets up to 40 m thick, running over a kilometer, are present, and are interpreted as alternate bars of a river.

Cant and Walker (1976) discuss the deposition of Devonian fluvial sediments, and interpret them to be deposits of a low-sinuosity river, where units made by migrating megaripples and laterally shifting transverse bars are distinguished. A sheet-like sand made up of composite units of fluvial channel deposits, about 100 km wide and 160 km long and 60 m thick is described from the Jurassic of New Mexico (Campbell 1976). He gives a model showing development of these fluvial sands (Fig. 440).

Miall (1976) attempts to make palaeohydrological analysis in braided stream deposits of the Cretaceous, Canada. Cotter (1971) describes the fluvial sandstones of Cretaceous age Utah, U.S.A., and attempts to calculate palaeohydraulic parameters, e.g., channel sinuosity, meander length, discharge, etc.

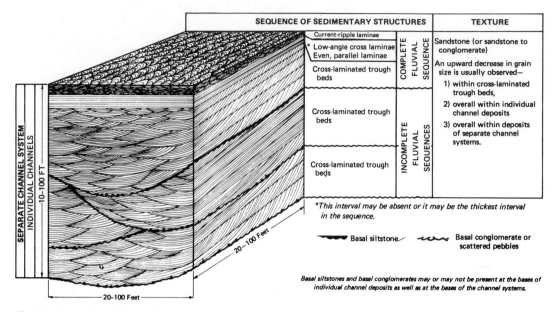

Fig. 440. A complete fluvial sequence of sedimentary structures and textures in the Westwater Canyon Member, Jurassic. Basal siltstones and basal conglomerates may or may not be present at the bases of individual channel deposits as well as at the bases of the channel systems. (After Campbell 1976)

Estuarine Environment

General Information

Geomorphologically, an estuary is a funnel-shaped opening of a river in the sea. Hydrographically, an estuary is characterized by tidal movements, and a highly differentiated development of water stratifications and mixing between river and sea waters (Berthois 1978).

The present-day estuaries came into existence after the Flandrian transgression, and represent mostly drowned river valleys, which have been reshaped by tidal current movements.

Pritchard (1967) defines an estuary as a semi-enclosed coastal body of water which has free connection with the open sea and within which sea water is measurably diluted with fresh water derived from land drainage.

Classification and Hydrography of Estuaries

In accordance with the definition of an estuary by Pritchard (1967), the following types of estuaries can be differentiated:

1. drowned river valleys
2. fjord-type estuaries
3. bar-built estuaries
4. estuaries produced by tectonic processes.

There are many transitional forms between the above types of estuary. For example, according to Reinson (1977) many rivers in southern New South Wales and eastern Victoria, Australia, are drowned river valleys; but their mouths are partly closed by extensive barrier beach and dune ridge complexes.

Allen (1972) points out that there are two most important factors which control estuarine circulation: (a) the volume of water brought in by a tide and (b) the fresh-water volume and the form of estuary (see also Ketchum 1953; Pritchard 1955, 1967; Bowden 1967; Allen 1972; Dyer 1972; Hayes 1976).

In general, four types of estuarine circulation pattern are recognized.

1. River flow dominant – a salt wedge estuary is developed.
2. Equal flow of river discharge and tidal flow on the water surface – a seaward current is developed, while near the sediment bottom a current of salt water is present, directed to upstream of the river. On the boundary of these two water wedges, mixing of freshwater and salt water takes place. The stronger the mixing of waters, the stronger is the seawater current on the sediment bottom in the upstream direction of the river. This process also causes sediment transport on the bottom in the upstream direction of the river (Meade 1969, 1979; Buller 1974, 1975; Buller and McManus 1974, 1975).
3. Low fluvial discharge and high tidal flow – a partly mixed estuary is produced. The vertical gradient of salinity is strongly reduced (for example, the Parker estuary, Daboll 1969). However, it is also possible that a more saline tidal channel may develop on the right-hand side of the estuary, and a relatively fresh-water riverine channel develop on the left-hand side of the estuary (in the northern hemisphere), e.g., the Gironde estuary (Allen 1972), the Merrimack river (Hartwell 1970).
4. Very little fluvial discharge and high tidal flow – a homogenous estuary is produced. Jade river, North Sea, is such an example. In such estuaries, there is no vertical salinity gradient. However, at ebb-tide, individual water bodies are visible; but during flood tide, at current velocities of about 1 m/sec, these water bodies are completely mixed and homogenized by the turbulences.

Sedimentation and Bioturbation

Suspended material brought in by rivers in an estuary is readily flocculated. Illite and montmorillonite are flocculated at a salinity of $> 3\,\%_0$ (Kranck 1973, 1975); while kaolinite can be flocculated at much lower salinity values (Whitehouse et al.

1960; Edzwald and O'Melia 1975). Absorbed organic molecules also help in the flocculation process (Chiang and Anderson 1968; Neihof and Loeb 1975). The floccules, once formed can be, up to a certain extent, stabilized through the mucus of fungus, bacteria, and algae (Pearl 1973, 1974, 1975).

This turbidity maximum, described by Glangeaud (1938) in Garonne as "bouchon vaseux" is observed in many estuaries, e.g., in the Elbe by Postma and Kalle (1955), in the Loire by Berthois (1956), in the Ems by Kühl and Mann (1954). The suspended material is transported almost in a closed circle. With the ebb-current it moves towards the sea near the water surface; and with the flood-current it moves back, in upstream direction of the river, along the bottom waters (van Straaten 1960a).

During periods of high river discharge, for example in the Gironde estuary, France, large amounts of suspended material are transported through the near-surface seaward directed movement by ebb-currents in the shelf areas (Allen and Castaing 1973).

The floccules possess a larger grain-fall velocity than the individual particles. The flocculated material is concentrated in a fluid mud. This fluid mud can be deposited during the still-stand phase of the tide, and can be completely or partly resuspended with the current activity of the tides.

The sediment of the estuaries is brought in both by the rivers and from the open sea. The transport of sediment from the sea into the river estuary depends upon the ratio of tidewater to river water. The Rhine estuary shows a characteristic pattern of sediment distribution, and according to van Veen (1936) the following zones can be differentiated: (1) a fluviatile zone, (2) a zone without sand, and (3) a marine sand zone. The zone without sand and the marine sand zone extend over a distance of about 20 km.

According to Friedman and Sanders (1978), those rivers making estuaries show mostly a lower concentration of suspended sediment (< 160 mg/l) than the rivers building up deltas.

Allen (1971) tried to apply grain-size parameters to the estuarine sediments of the Gironde estuary, France. He distinguishes three groups of sediment

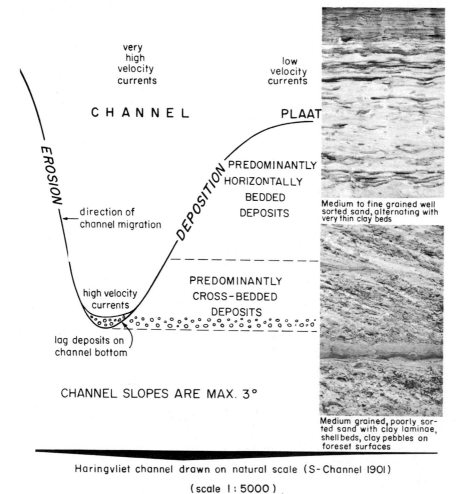

Medium to fine grained well sorted sand, alternating with very thin clay beds

Medium grained, poorly sorted sand with clay laminae, shell beds, clay pebbles on foreset surfaces

Haringvliet channel drawn on natural scale (S-Channel 1901)

(scale 1 : 5000)

Fig. 441. Schematic cross-section of a migrating estuarine channel and their sedimentary structures. (After Oomkens and Terwindt 1960)

with characteristic transport processes. The group finer than 3 Φ is characterized by suspension transport; 3.0 Φ to 0.6 Φ by mixed suspension bed load; and coarser than 0.6 Φ by bed-load transport. The group 3.0 Φ to 0.6 Φ can be further subdivided into 3.0 Φ to 2.0 Φ, characterized by graded suspension; and 2.0 Φ to 0.6 Φ characterized by saltation. Graded suspension denotes increased gradient of suspended particles near the sediment bottom.

Near the mouth of the Haringvliet, a lower estuarine part of the Rhine and Maas rivers, a well-developed vertical sequence of estuarine sediments is present (Oomkens and Terwindt 1960; Terwindt 1971a). From top of bottom, the following sequence is described:

1. thinly interlayered sand/mud bedding or wavy bedding. The mud layers are rarely more than 2 cm thick. The sand shows either current ripple bedding or evenly laminated sand. Partly, flaser bedding is also present (Fig. 441), especially in areas where current velocities are low. For example, on submerged shoal surfaces and on the upper part of the channel slopes. These bedding structures showing alternating sand/mud layering in different dimensions are related to the tidal phases showing alternation of current and slack-water activity (see p. 124).

2. Megaripple bedding in the channel bottom sediments, showing poorly sorted coarse-grained sand, intercalations of mud layers and many mud pebbles. Molluscan shells are quite common, and peat pebbles, rolled wood-pieces, and spines of sea-urchins are also found. Bioturbate structures are absent. The palaeocurrent direction of the cross-bedded units is mostly related to flood-current direction.

During lateral migration, extensive surfaces of channel-bottom sediments are formed, and additionally the units of sand bars are also extensive. In nature, usually there is repeated lateral shifting of channels, leading to complex intertonguing of various litho-units.

Van Beek and Koster (1972) describe a vertical sequence from an excavation in the embanked flood plain of the Oude Mass near Barendrecht, The Netherlands. This sequence is located in a river tract and can be regarded as transitional between fluvial and estuarine environments. They subdivide this sequence in three units (Fig. 442):

1. The upper unit, characterized by flaser-, wavy-, and lenticular bedding, and small-scale trough cross-bedding.

2. The middle unit, made up of alternating layers of small-scale and large-scale cross-bedding, shows bimodal palaeocurrent directions, indicating the influence of tidal currents. An upward decrease in dimension of structures is very apparent, denoting a decrease in the current velocity.

3. The lower unit, made up of large-scale trough cross-bedding. Individual sets can be up to 1 m thick. The foreset laminae are made up of fine-to-very coarse sand. The bottomsets are rather thick, and are made up of well-sorted fine sand. The palaeocurrent direction is uni-directional, in the downstream direction of the river.

In general, this sequence represents an estuarine environment of low intensity.

Besides the above-mentioned more typical examples of estuarine environment, there are estuaries which can be termed immature estuaries, e. g., the St. Lawrence Middle Estuary (D'Anglejan and Brisebois 1978). In this case, a rather massive muddy sequence is present in most parts of the estuary, related to the previous times when a high content of suspension was supplied and deposited. The present-day sedimentation is very low. Present-day sediments represent a thin blanket of sand, which is continuously being reworked, and has been derived from the reworking of moraines.

The present-day supply of suspended sediment of the river is partly deposited on the intertidal flats of the southern side, near the river mouth; and partly it is brought out to the lower estuary and to the gulf of St. Lawrence.

Estuaries of the Georgia Coast, U.S.A., have been studied from the point of view of geology, sedimentology, and biology and the results are published in a series of papers in Senckenbergiana Maritima 1975. Howard (1975a) distinguishes three types of estuaries on the Georgia Coast: (1) estuaries with mainly bioturbated muddy sand, (2) estuaries with dominantly inorganic primary sedimentary structures, e. g., flaser and lenticular bedding, and (3) estuaries with mainly coarse-grained sand.

Within the estuaries, there is an increase in the diversity and abundance of bioturbation structures, in the direction of open sea. In muddy sediments, mostly burrows and dwelling tubes are common, while in sandy sediments, there are a number of additional bioturbation structures (Dörjes and Howard 1975; Howard and Frey 1975; Howard et al. 1975a; Majou and Howard 1975).

Based on these studies of the estuaries of the Georgia coast, Greer (1975) constructed a vertical sequence of an estuarine coast. This shows a progradational estuarine fining-upward sequence, which begins with an erosional lower contact on the top of a progradational shelf sequence. The shelf sequence is coarsening upward.

In an area with marked tidal ranges, the estuarine sediments are laterally associated with intertidal flats or salt marshes.

In regions where the estuarine mouth is partly closed by a barrier-beach and dune-ridge complex, lagoons may develop. In such a case, between the

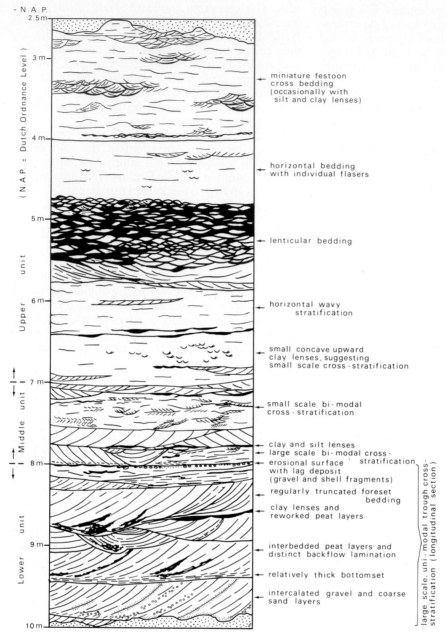

Fig. 442. Schematic representation of the sedimentary structures and their sequences in an estuarine deposit of Oude Mass, Netherlands. The succession can be recognized in three sequential units. The lowermost unit represent deposits of unidirectonal fluvial flow. (After van Beek and Koster 1972)

marine sand of the estuary mouth, and the riverine coarser sand in the deeper parts, lagoonal black shale may be present (Fig. 443).

Estuaries may be preserved in transgressive sequences in the form of drowned river valleys, as shown by Weil (1977) in a transgressive sequence of Delaware Bay, U.S.A. On the top of a zone of Pleistocene oxidized sands and gravels (Fig. 444) follows a basal peat layer and sediments of fringing tidal marshes with mud, peat layers, organic remains, burrows, along with the sediments of shallow lagoon and tidal creeks. These deposits can be

unconformably overlain by coarse sands, which are deposits of the estuarine washover barrier complex of the coast (Kraft 1971). After an erosional discordance, the thinly bedded subtidal sediments follow, becoming sandier upwards. Then follows a unit of cross-bedded fine sands of tidal current ridges. The sequence on the top is closed by a unit of coarse sand which is brought in from the continental shelf, and is still exposed in the bay mouth shoal complex and in the eastern part of the lower bay.

In the Eocene of southern England Goldring et al. (1978) describe estuarine sediments with channel-fill sand, interlayered sand and mud layers, and mud clasts. The facies changes rapidly both in lateral and vertical directions. Trace fossils of these deposits include *Ophiomorpha nodosa* and *Arenicolites* sp.

Fjords

Fjords are drowned cliff valleys of glacial origin, showing U-shaped cross-section and often several hundred meters deep. Fjords often show a rock sill on the seaward side, and a freshwater source (river) on the landward side. Distribution of sediments mostly shows sandy cover on the rock sill and near the river mouth, and muddy sediments in the deeper parts (Lüneburg 1972).

Because of the rock sill on the seaward side, in some fjords anaerobic conditions may occur (Strøm 1936).

The fjords of south Sweden are characterized by the presence of a marked water stratification. The upper 15–20 m water belongs to the Baltic sea with 20–25‰ salinity; while the water in the deeper parts is from Skagerak with 34‰ salinity. The interstitial water shows 1–3‰ higher salinity. The rate of sedimentation is about 30 cm/1,000 years (Lüneburg 1972). Holtedahl (1975) gives an average rate of sedimentation in a Norwegian fjord as 1 cm/year for last 10,000 years. If one excludes the contribution of turbidites and slumps, only 0.5 mm/year is left as rate of sedimentation.

Holtedahl (1975) discusses in detail the geology and sedimentation of the Hardanger Fjord, Western Norway. These sediments contain many good examples of turbidite deposits.

Ancient examples of fjords are some of the side-valleys of the Rhine rift-valley (Geib 1954; Geib and Atzbach 1969; Kuster-Wendenburg 1974).

Summary

The characteristic feature of estuarine deposits is its position in the transition from fluvial to marine deposits in a transgressive sequence, both laterally and in a vertical sequence. As soon as the supply of sediment from river becomes dominant, the estuary changes into a delta and a progradational deltaic sequence is produced, on the top of a transgressive sequence. As already shown in the examples discussed, there are variabilities in this model. Terwindt et al. (1963) differentiate the fluvial and estuarine sediments on the basis that estuarine sediments are finer-grained, contain rounded clay pebbles, and shell beds in ripple areas, and the bottom sediments show ripples with bipolar directions or with upcurrent direction (of the river). Fur-

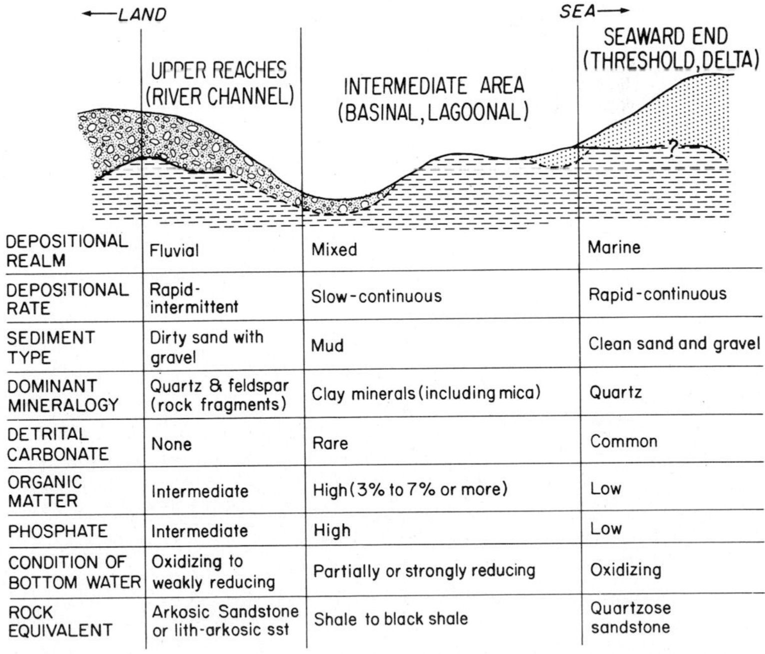

	UPPER REACHES (RIVER CHANNEL)	INTERMEDIATE AREA (BASINAL, LAGOONAL)	SEAWARD END (THRESHOLD, DELTA)
DEPOSITIONAL REALM	Fluvial	Mixed	Marine
DEPOSITIONAL RATE	Rapid-intermittent	Slow-continuous	Rapid-continuous
SEDIMENT TYPE	Dirty sand with gravel	Mud	Clean sand and gravel
DOMINANT MINERALOGY	Quartz & feldspar (rock fragments)	Clay minerals (including mica)	Quartz
DETRITAL CARBONATE	None	Rare	Common
ORGANIC MATTER	Intermediate	High (3% to 7% or more)	Low
PHOSPHATE	Intermediate	High	Low
CONDITION OF BOTTOM WATER	Oxidizing to weakly reducing	Partially or strongly reducing	Oxidizing
ROCK EQUIVALENT	Arkosic Sandstone or lith-arkosic sst	Shale to black shale	Quartzose sandstone

Fig. 443. Sedimentation model for typical lagoonal estuary southeastern Australia. (After Reinson 1977)

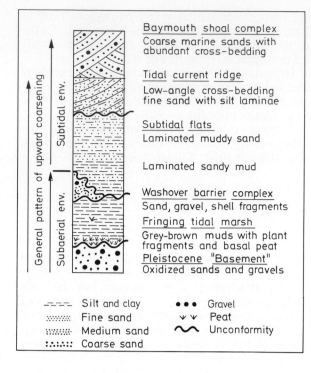

Silt and clay
Fine sand
Medium sand
Coarse sand

Gravel
Peat
Unconformity

ther, the estuarine sediments contain thick layers of different types of bedding, characteristically found in tidal environments, i. e., lenticular and flaser bedding, wavy bedding, thinly interlayered sand/mud bedding. The source of sediments within an estuary is from the river and from the sea.

For example, the profile in Fig. 441 show a few fining-upward sequences, which are not made by progradation, but due to lateral migration of estuarine channels.

A very useful compilation of papers on estuaries is given by Lauff (1967).

Fig. 444. A conceptual model of an ideal vertical sequence of transgressive sedimentary sediments in Delaware Bay. (After Weil 1977)

Deltaic Environment

General Information

At the confluence of rivers with large water bodies large sediment-cones, commonly known as deltas, are deposited, Moore and Asquith (1971) define a delta: "The subaerial and submerged contiguous sediment mass deposited in a body of water (ocean or lake) primarily by the action of a river." In this chapter only deltas formed in an ocean are discussed. Deltas formed in a lake are considered in the chapter on lake environment.

The present-day deltas are relatively young, as they have been formed only after the increased sea level during post-Pleistocene times. However, their size is rather impressive, because of increased rate of deposition. Table 25 gives some data on the size of some of the present-day deltas.

The most important factors for the development of a delta are a large supply of sediments by streams and subsidence in the area of deposition. The ideal form of a delta is considered to be a cone; this ideal shape is seldom achieved. The configuration of a delta is controlled by several factors:

1. coastal morphology, configuration of coast line, angle of slope of the continental shelf
2. direction and intensity of the waves coming from the open sea
3. the degree of coastal transport of sediment in relation to the sediment transport by distributaries
4. tidal range.

Points 2 and 4 are important in determining the sediment transport. Fisher et al. (1969) distinguish between "high-constructive deltas" and "high-destructive deltas". The high-constructive deltas are characterized by the dominance of fluviatile processes over marine processes. In such a situation mostly birdfoot delta, e.g., modern Mississippi type or lobate form delta, e.g., La Fourche delta are made. High-destructive deltas are produced if the tidal range or strong wave action plays an important role in the delta-forming processes. In the case of high tidal ranges (macrotidal), linear, subaqueous sand ridges are made, arranged perpendicular to the coast line. In the case of wave-dominated deltas, sand bodies running parallel to the coast line are formed. In this case, sands are well-sorted, and the offshore slope is steep and concave. Galloway (1975) distinguishes three types of deltas, i.e., fluvial-dominated, wave-dominated, and tide-dominated (Fig. 446). Coleman and Wright (1975) and Coleman (1976) give a comprehensive account of the variability of processes in deltas and the characteristics of sand bodies in modern river deltas.

Morgan (1970) discusses the factors that control the depositional processes and the deposits of a delta. He groups the controlling factors into four groups: river regime, coastal processes, structural behavior, and climate factors. Some of these factors may be more influential in one delta and other in other deltas. Thus, many variable types of deltaic deposits can be produced.

Table 25. Dimension, annual water, sediment discharge, and annual rate of growth of representative deltas. (Modified after Smith 1966)

Delta	Subaerial area km² · 10³	Average water discharge m³/sec · 10³	Annual sediment discharge t · 10⁶	Annual rate of growth, m
Chao Phraya	25	1	5	–
Danube	4	6	91	–
Ganges-Brahmaputra	91	39	635	–
Hwang Ho and N. Yellow Plain	127	4	2	268
Irrawaddy	31	28	272	46–61
Lena	28	9	–	–
Mekong	52	11	–	61
Mississippi	29	17	469	91
Niger	19	6	23	–
Nile	16	3	54	–
Orinoco	57	17	–	–
Po	14	1	61	26–61
Red	8	2	118	–
Rhine	22	2	–	–
Rhône	3	2	41	–
Rio Grande	8	0.1	17	–
Ural	9	–	–	–
Volga	11	–	–	–
Yangtze and S. Yellow Plain	124	22	544	23

The literature on deltas and deltaic deposits is very extensive. The more important compilations on this topic are Scruton (1960), Coleman et al. (1964), Coleman and Gagliano (1965), Broussard (1975), Coleman (1976), and Morgan (ed.) (1970). More than sixty papers on modern and ancient deltas are summarized by Le Blanc (1975).

The Structure of a Delta

According to classical concept, a delta is made up of topset, foreset and bottomset deposits (Fig. 445). These three major subdivisions of a delta are themselves composed of smaller units which are deposited in rather varied environmental conditions. Besides, there are marginal deposits and other environmental units bordering deltas.

Topset Deposits. Topset deposits of a delta are mainly made up of marsh deposits and deltafront silts and sands. Also present are river channel deposits and natural levee deposits, together with crevasse splay deposits. Muddy sediments with shell layers are deposited in the bays between the distributaries and the tidal channels (Scruton 1960). These various sediments are associated with each other in a very complicated way. Their lateral and vertical boundaries are both gradational and sharp. Thus, topset deposits are characterized by their heterogeneity and complex relationship of sedimentary units.

Foreset Deposits. Foreset deposits are made up of pro-delta silty clays and rather coarse sand, silt, and clay deposits formed off the major deltaic distributaries. In the delta foresets (also known as delta front slopes) are also incorporated delta front gullies. Such gullies are 2 to 8 m deep and located about 1 km from each other in the case of the Mis-

sissippi delta (Shepard 1956b). In the Rhône delta these gullies are 1 to 1.5 m deep and are located some 200 m from each other (van Straaten 1959a). Such gullies are also found in the delta of the Fraser River (Mathews and Shepard 1962).

Bottomset Deposits. Bottomset deposits are made up of offshore clays under the influence of active deltas. In the Mississippi delta such offshore clays are up to 10 m thick.

Marginal Deposits. These are transitional deposits between bottomset deposits and the deposits of the subsurface, which were deposited before the building up of the delta. In many present-day examples the underlying deposits of a delta are Pleistocene in age. The marginal deposits of the Mississippi delta are strongly bioturbated, and made up of sandy and muddy sediments.

Shepard (1956b), based on the study of the Mississippi delta, discusses the subenvironments of the submerged area bordering a delta (Fig. 447). There is a platform situated directly adjacent to the forward-building passes. On the inner part of the platform, submarine continuation of the river distributaries can be followed. These channels are as deep as 7.5 m and become shallower away from coast. The channels are bordered by submerged natural levees. At the outer end of the channels the submerged natural levees are connected to an arc-shaped shoal or lunate bar.

If the delta lobes have been built seaward in such a way as to enclose narrow embayments in between, a special environment exists in such bays which is quite similar to the environment of a platform. Such bays are commonly termed *interdistributary bays*. They are devoid of any river current or tidal current activity, and are also largely protected from the wave action of the open sea. Thus, they become settling basins for the sediments washed over the natural levees and sediment brought by small-channel branches of the distributaries.

CROSS SECTION OF A DELTA

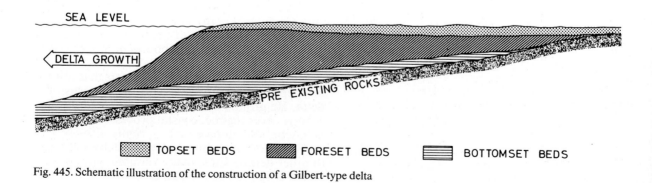

Fig. 445. Schematic illustration of the construction of a Gilbert-type delta

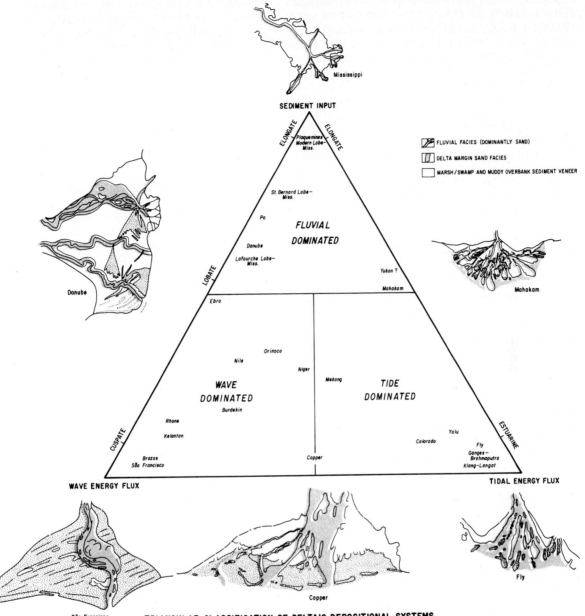

Fig. 446. Schematic diagram depicting the threefold division of deltas into fluvial-dominated, wave-dominated, and tide-dominated types. The relative importance of sediment input, wave energy flux, and tidal energy flux determine the morphology and internal stratigraphy of the delta. (After Galloway 1975)

Beyond the platform is a slope known as the pro-delta slope, or delta front slope. Parts of this shelf which have been influenced by Recent sediment deposition are referred to as open shelf with Recent delta influence. Actually, they can be termed bottomset deposits. This part of the open shelf shows some lithological and faunal changes with depth. Thus, parts less deep than 20 m are referred to as the shallow part of the shelf, and the rest is referred to as the deeper part of the shelf.

Subenvironments of a Delta

As mentioned earlier a delta environment is made up of many distinct subenvironments. In the following, a short account of various subenvironments is given. This account is based mainly on the results obtained by the study of the Mississippi delta. The nature of the deposits of some of the subenvironments of deltas of other climatic zones may be much different. Detailed studies of the Mississippi delta include Fisk (1955, 1960, 1961), Fisk et al. (1954, 1955), Shepard (1956b), Shepard and Lankford (1959), Lankford and Shepard (1960), Coleman and Gagliano (1965), and Kolb and van Lopik (1966).

Many of the other present-day deltas have been also investigated: Fraser River delta, British Columbia (Johnston 1921, 1922; Mathews and Shepard 1962), Orinoco delta (van Andel and Curray 1960; van Andel 1967), Nile delta (Einsele and Werner 1968); Rhône delta (van Straaten 1959a,b, 1960); Colorado River delta, Texas (Kanes 1970), Guadalupe River delta (Donaldson et al. 1970), Niger River delta (Allen 1965b, 1970); (Oomkens 1974), Gum Hollow Fan delta (McGowen 1970), Ebro delta (Maldonado 1972).

Topset deposits can be differentiated into two parts–subaerial and subaqueous. Subaerial topset deposits are marsh and river channel deposits. Subaqueous topset deposits are delta front deposits, accumulating almost entirely under the influ-

ence of fluvial processes. They include channel deposits, natural levee deposits, and flood basin deposits including lake and swamp deposits. A detailed account of such deposits has been given in the chapter on fluvial environment.

The following account is based mainly on the papers of Coleman et al. (1964, 1965), Coleman (1966), and Kolb and van Lopik (1966), which deal with the study of sedimentary structures in the Mississippi delta region, including that of Atchafalaya River basin.

Subenvironments of Subaerial Topset Deposits

The topmost part of a deltaic sequence is mostly made up of swamp deposits, represented by organic clays and peat deposits. Fisk (1960) discussed the deposition of peat in deltaic plains, i. e., topset deposits. Clays rich in organic matter are deposited in areas where clastic sediment only occasionally is brought in. Such sediments generally lack

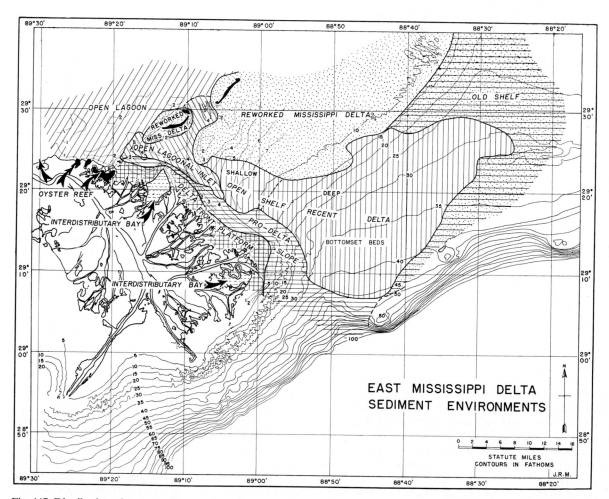

Fig. 447. Distribution of various sedimentary environments in the eastern part of the Mississippi delta. (After Shepard 1956b)

well-defined bedding (Fig. 454). The sediments are extensively burrowed by plant roots and sediment-dwelling organisms. Mostly these deposits are found as a homogeneous mixture of clay and silt with plant remains. In arid climates extensive plant cover is missing. Here the topset deposits show extensive mud cracks and algal mats.

Coleman (1966) studied the swamp deposits of the Atchafalaya basin in detail and was able to recognize 5 environmental units–poorly drained swamp, well-drained swamp, freshwater lacustrine, lacustrine delta fill and channel fill.

Poorly drained swamp deposits are made up of black mud with a high content of organic matter. Occasionally thinly laminated silt is present, brought in during floods (see Fig. 424). There is a high degree of bioturbation. Plant debris are well-preserved because of the reducing milieu. Abundant plant life causes complete churning of the sediments. Iron concretions are present. Pyrite content is relatively high (0.5 to 3.5 %), coupled with a high content of organic matter. Pyrite may occur as small cubes, isolated globular masses and replacing the small root. Vivianite is also commonly found (up to 2 %). Content of calcium carbonate is rather low.

Well-drained swamp deposits are similar to those of the poorly drained swamp. However, the content of organic matter is very low. Deposits are mainly clay with isolated silt lenses. Sometimes insect carapace and remains of charophytes are found. Pyrite and vivianite content is very low. Nodules of $CaCO_3$ are common (Fig. 448a). $CaCO_3$ is also concentrated along bedding and laminar planes. Iron-oxide nodules are abundant.

Freshwater lacustrine deposits are sedimented in very shallow (3 to 4 m deep) but extensive lakes. Such lakes have only very weak wave and current activity. In a sequence, on the top of the lacustrine deposits follow the lacustrine delta fill deposits. Thus, as mentioned in the chapter on lake environment (pp. 242 and 247), filling in such a lake is primarily by a lake delta.

The freshwater lacustrine deposits are made up of dark gray to black clays with silty lenses. In X-ray radiographs clay layers show extremely fine laminations. Coleman (1966) concluded that this lamination is a result of alternating flocculated clay and nonflocculated clay layers (Fig. 448b). This may be a result of rapidly changing pH, the change in nature of cations, and sediment concentration. In flocculated clay layers synaeresis cracks are found. The sediments are commonly bioturbated (see Fig. 425). Sometimes lamellibranch shells in a living position are encountered. Pyrite and vivianite are commonly found; however, iron nodules are absent.

Lacustrine delta fill deposits are created if a branch of a stream enters a lake. The incoming stream branch brings coarser sediments, and beds that are produced become thicker. However, the change from lacustrine deposits into lacustrine delta fill deposits is usually not sharp. The bedding is imparted because of the change in color of clay layers, the flocculated and nonflocculated nature of clay, and the change from silt to clay. Other sedimentary structures found in these sediments are scour-and-fill structure, load structure, distorted bedding, lenticular bedding; even small-ripple bedding is not rare. However, fauna and evidences of life are not so common.

Channel fill deposits are made up of coarser and poorly sorted sediments than the adjoining deposits. Deposition in the channel takes place by the migration of point bars and braid bars. Various kinds of cross-bedding are the most common sedimentary structure. Clay balls and wood particles are often found. Less common structures are scour- and-fill structures, parallel bedding, and lenticular bedding. Associated with channel fill deposits are subaerial natural levee deposits which are deposited during high-water stages of the river.

Subaerial levee deposits show various kinds of current ripples and ripple bedding. Irregular lamination is very common, which results from interference by grass-roots in sedimentation. Locally, intensive burrowing by plants and animals completely obscures all the primary inorganic sedimentary structures (Figs. 454a,b). Iron nodules and carbonate nodules are abundant. (For a detailed account of characteristics of channel and natural levee deposits see the chapter on fluvial environment.)

Subaqueous Topset Deposits and Their Environments

The subaqueous part of the topset deposits are represented by the delta front environment. It is a complex of subenvironments, each with individual characteristics. In an actively prograding delta the delta front environment is the focus of most active deposition (cf. Coleman and Gagliano 1965). This region can be differentiated into the following subenvironments–distributary channels, subaqueous levee, distributary mouth bar, and distal bar (Fig. 449). Each of these subenvironments is dominated by a different sedimentation process: thus, the spectrum of sedimentary structures in each of these sedimentation units is distinct.

Distributary Channel. A distributary channel is a natural stream which leads a part of the sediment and water discharge of a major stream into the sea. This is actually an extension of the river channel into the sea. The distributary channel broadens, becomes shallower, branches off and at last looses its

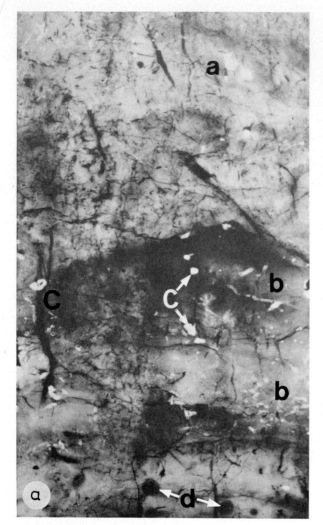

Fig. 448a. X-ray radiograph of the sediments of the well-drained swamp deposits. Atchafalaya Basin, Louisiana. The width of
the core is 12.5 cm. Dark- and light-colored zonations are from
carbonate (white) and iron-oxide (dark) concentration. a calcium carbonate nodule; b rim of calcium carbonate around
rootlet; c calcium carbonate filling an insect burrow; d iron-oxide rim around rootlet. (After Coleman 1966)

Fig. 448b. X-ray radiograph of freshwater lacustrine clays and
silts. Atchafalaya Basin, Louisiana. The width of the core is
12.5 cm. Clays are finely laminated. The high-absorption laminations (white) are layers of flocculated clays and a concentration of carbonate cement. The low-absorption laminations
(gray) are nonflocculated clays. Very low absorption, extremely
thin dark lamination are layers with a high concentration of colloidal organic matter. White dots in the lower, more or less massive part, are pyrite crystals; dark dots are small shells. However, in most cases primary bedding is usually almost completely destroyed due to a high degree of bioturbation. (After Coleman 1966)

identity as it moves further into sea (within the upper part of the delta front environment).

In the active Mississippi delta, such channels
vary in size from 1,200 m broad and 24 m deep to
1.5 m wide and 0.6 m deep. The depth decreases
toward the sea.

Near the upstream part of the channels current
direction is persistent downstream. However, as
the channel flares out downstream, current direction becomes variable and current velocity is reduced; thus, rate of deposition of sediments is increased.

The most common sedimentary structures in the
distributary channel deposits are cross-bedding,

current ripple bedding, scour-and-fill structures
and erosional surfaces. Some of the clay layers deposited during the low-river stage may escape erosion and be well preserved. The upper surface of
such clay layers usually shows scouring features.
Clay fragments are incorporated within the sediment. Intraformational deformation structures are
commonly found. They include slump structures
near the channel walls. Another feature is cross-
bedding with overturned foresets (also termed intraformational recumbent folds).

Subaqueous Levees. These are the submarine
ridges bordering the distributary channel, formed
in response to broadening and shoaling of the
channel. Tides modify and control their morphology. Some parts of the subaqueous levee are ex-

Fig. 449. Diagrammatic representation of the various sedimentation zones in a delta front environment. Based on the study of mouth of Southwest Pass, Mississippi delta. (Base map after Talcott 1839. After Coleman and Gagliano 1965)

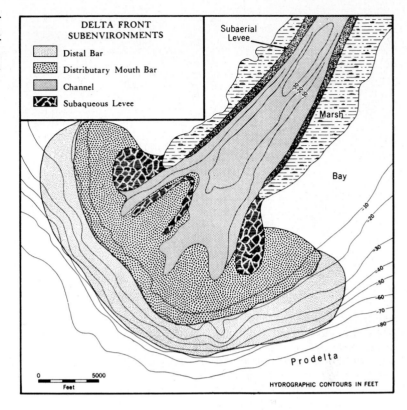

posed as sand flats at low tides. Subaqueous levee deposits are made up of very fine sand and silt, with clay and plant debris intercalations. Sedimentary structures produced by current action are dominant. Locally, combined current and wave action produce complex types of cross bedding. Other characteristic features are wavy bedding, scour-and-fill structures, burrows, and clay balls. Among the deformation structures convolute bedding is most abundant.

Distributary Mouth Bar. A distributary mouth bar is a sandy shoal formed near the seaward limit of the distributary channel. Formation of the shoal is the direct result of the decrease in current velocity and in carrying capacity of the stream as it leaves the channel. The rate of sedimentation is exceptionally high–probably higher than in any other subenvironment of the delta. The sediments are subjected to continuous reworking. i. e., by stream currents and sea waves. The deposits are, consequently, made up of sand and silt. Thin laminations of plant debris are often present. Wood fragments exhibit pronounced rounding effects.

The most common sediment structure found is trough cross-bedding. Both wave and current ripple bedding are present. Occasionally, well-preserved wave and current ripples are also found.

Out of the deformational structures the gas-heave structure is characteristic of distributary

mouth bar deposits. As the delta progrades seaward, the bar deposits are underlayed by bay or gulf deposits, rich in organic matter. As the organic matter decays, gas is released which passes upward through the overlying sandy distributary mouth bar deposits and produces the characteristic gas-heave structures. Other deformational processes like slumping, diapirism, radial grabens, and tension faulting are described by Coleman (1976).

Distal Bar. Seaward of the distributary mouth bar is a zone of predominantly laminated silts and clays with high rates of sedimentation. This is the seaward-sloping margin of the delta front environment. Common sedimentary structures are cross bedding, scour-and-fill structures, erosional surfaces, and ripple marks. Sedimentary structures are repeated in a definite sequence, suggesting seasonal layering.

This zone is extremely favorable for a dense benthonic population. Small and large burrows, completely bioturbated layers, and shell deposits are present throughout.

Interdistributary Bay. Interdistributary bay deposits are closely associated with delta topset deposits. Such areas are open water bodies surrounded by levees or marshes, but open or connected to the open sea by tidal channels, etc. Deposition in such bays is mainly by two processes: The fine-grained

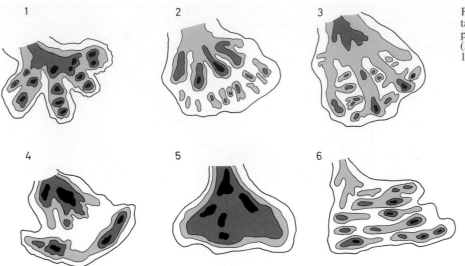

Fig. 450. Schematic representation of net sand distribution patterns in modern deltas. (After Coleman and Wright 1975)

sediment deposited from suspension due to absence of wave activity, and the coarse sediment brought in by crevasse channels. Locally, as a result of the presence of wave activity, lenticular bedding is the most abundant type of bedding. Other bedding types encountered are parallel bedding, showing textural as well as color differences in adjacent layers. Locally, current ripples and scour structures are common (Fig. 454a). They are produced by tidal currents or currents produced by overflow during high-water stages. Shell deposits and bioturbation structures are also present (Fig. 454b). Elliott (1974) describes a number of sedimentary sequences from an interdistributary bay.

Deltaic Sand Bodies

It is known from both Recent and ancient sediments that progradation of a delta produces extensive sand bodies. They are also known to form important oil traps. These sand bodies are essentially associated with delta topset deposits. Coleman and Wright (1975) found the following climatic control in the delta deposits. In a tropical climate, thick peat layers are produced; in a temperate climate, thin peat layers are found interfingering with fine-grained clastics; in an arid climate, evaporites are present.

If the supply of fine-grained suspended sediments is high in a delta, penecontemporaneous deformation features are quite common. High and occasional wave energy in the delta shoreline leads to the development of pure quartz sand bodies with mainly marine character. Strong littoral currents produce delta sand bodies, oriented parallel to the depositional strike. Such sand bodies are in strong contrast to the other types of delta sand bodies running perpendicular to the shoreline. Dominant perpendicular to shoreline sand bodies are produced in areas where strong littoral current is absent. High tidal ranges cause genesis of sand-filled distributaries, while in the case of low tidal ranges, distributaries are mostly mud-filled.

A systematic subdivision of the deltas and the shape of sand bodies is given by Coleman and Wright (1975), and examples of six deltas are given to illustrate the relationship between the process setting and the resulting sedimentary sequence (see Table 26 and Fig. 450):

1. Mississippi delta, U.S.A. – Low wave-energy, low tide range, low offshore slope, low littoral currents, high fine-grained suspended-sediment load, unstable receiving basin, temperate climate.
2. Klang delta, Malaysia – Low wave energy, high tide range, high littoral currents, narrow seaway receiving basin, tropical climate.
3. Ord delta, Australia – Low wave energy, extreme tide range, low littoral currents, restricted receiving basin, arid climate.
4. Burdekin delta, Australia – Intermediate wave energy, high tide range, low littoral drift, stable receiving basin, semi-arid climate.
5. Sao Francisco, Brazil – High, persistent wave energy, intermediate tide range, low littoral currents, steep offshore slope, dry tropical climate.
6. Senegal delta, Senegal (Africa) – Extreme wave energy, intermediate tide range, high littoral currents, steep offshore slopes, arid tropical climate.

Two types i. e., delta front sheet sand and bar finger sand, are the more important types of sand bodies produced in a delta deposition.

Delta Front Sheet Sand. Fisk (1955) studied in detail the delta front sheet sand of the Mississippi delta. By studying extensive borings he showed that in the eastern part of the La Fourche delta such sheet sand covers a total area of 675 km² and has a thickness varying from 7 to 30 m. The sand body becomes thinner toward sea, and it is exposed as sand bars and spits near the outer mouth of La Fourche and Moreau (Fig. 451). Toward land, the sheet sand is overlayed by approximately 12-m-thick marsh deposits. The faunal evidence is sparse, as is also the case with beach deposits. Van Straaten (1959a) and Oomkens (1967) describe similar sheet sands from the Rhône delta. Here extensively cross-bedded fluvial channel sands are incorporated into the sheet sand. Delta front sheet sands have

Table 26. Significant process parameters of the delta examples. (After Coleman and Wright 1975)

Delta	Climate	Discharge (m³/sec)	Sediment	Wave power (ergs/sec × 10⁷)	Tide (m)	Alongshore currents	Shelf slope (%)	Tectonics
Mississippi	Humid subtropic	15,631	High suspended	0.034	0.43	Low	7.0	Rapid Subsidence
Klang	Humid tropic	1,100	High suspended	0.218	4.2	Extremely high	4.1	Low Subsidence
Ord	Dry subtropic	166	Bed load and suspended load equal	1.06	5.8	Low	3.9	Stable
Burdekin	Dry tropic	476	High bed load	6.41	2.2	Low	9.2	Stable
São Francisco	Dry tropic	3,420	High suspended	30.42	1.9	Low	11.2	Stable
Senegal	Arid tropic	867	High suspended	112.42	1.2	Extremely high	17.0	Low Subsidence

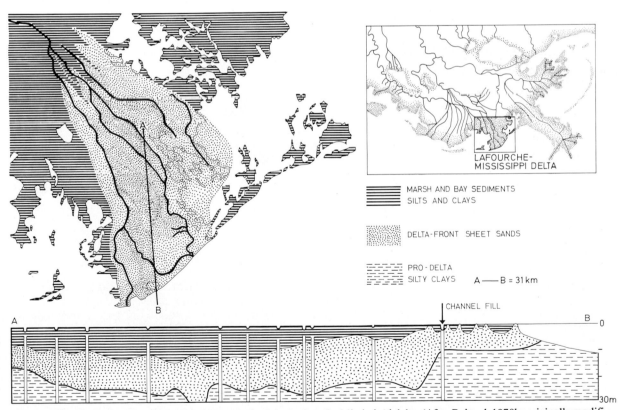

Fig. 451. Schematic diagram of delta front sheet sands of the La fourche Mississippi delta. (After Reineck 1970b; originally modified after Fisk 1955)

also been reported from the Volga River delta (Ruchin 1958).

Delta front sheet sand is made up of a distributary mouth bar, and delta front silt, and sand. These sediments are drifted laterally by waves and accumulate along the coast line (Oomkens 1967).

In the Rhône delta clean sands extend up to water depth of approximately 10 m, where they grade into slightly muddy sand (Kruit 1955; van Straaten 1959a). Van Straaten (1959a) interpreted the boundary between clean sand and muddy sand as the wave base. In completely open areas sand is found up to a water depth of 20 m; in protected bays this boundary between clean sand and muddy sand is at much shallower depth.

Delta front sheet sands are also known from ancient sediments. Nanz (1954) described them from Oligocene sands in southwestern Texas. These sands increase in thickness toward distributaries and extend over a large area. Sandstones of the Horsetooth sequence, lower Cretaceous, Colorado, have also been interpreted as delta front sheet sand (McKenzie 1965). Potter (1963) made a comparison of Palaeozoic sediments of the Illinois basin with the alluvial sediments of the Mississippi delta. He found sheet sands and channel sands, but no bar finger sands in the Palaeozoic sequence.

Bar Finger Sand. Fisk et al. (1954) and Fisk (1961) discussed the genesis of elongated sand bodies–bar finger sands–of the bird foot delta of the Mississippi River. Continued development of sand bars at the mouth of the distributaries as the delta advances seaward produces elongated sand bodies. Such sand bodies tend to sink into the muddy prodelta sediments, and sand bodies thicken (Figs. 452, 453a,b and 454a,b). They can be > 90 m thick, 7.5 km wide, and extend over a distance more than 50 km. In cross section bar finger sand exhibits a typical biconvex form. Because of compaction of the prodelta muds, they also intrude into sand body as mud diapirs; and are sometimes even exposed at the surface. In the upper part of the sand body channel and natural levee sediments are incorporated, which grade laterally into the deltaic plain and marsh deposits.

Shepard and Lankford (1959) studied several borings in the Mississippi delta and think that bar finger sands may be doubtful. However, Coleman and Wright (1973, 1974) found fingerlike protusions of distributary sand. Friedman and Sanders (1978) prefer to use the term "bed-load sand in greatly overthickened pods" instead of bar finger sand.

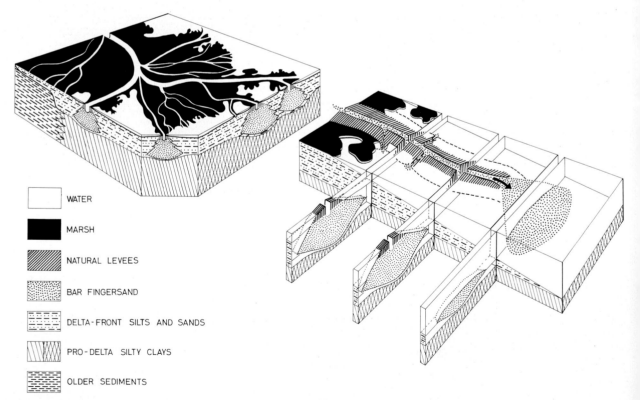

WATER

MARSH

NATURAL LEVEES

BAR FINGERSAND

DELTA-FRONT SILTS AND SANDS

PRO-DELTA SILTY CLAYS

OLDER SEDIMENTS

Fig. 452. Schematic diagram of bar-finger sands in the Mississippi delta. (After Reineck 1970b, originally modified after Fisk et al. 1954)

Fig. 453a. Characteristic features of bar-finger sands and associated facies. (After Fisk 1961; modified by Gould 1970)

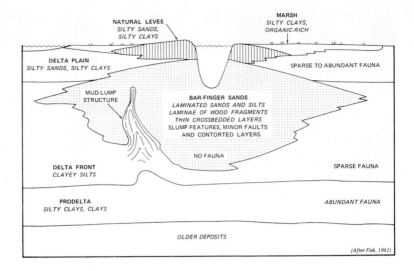

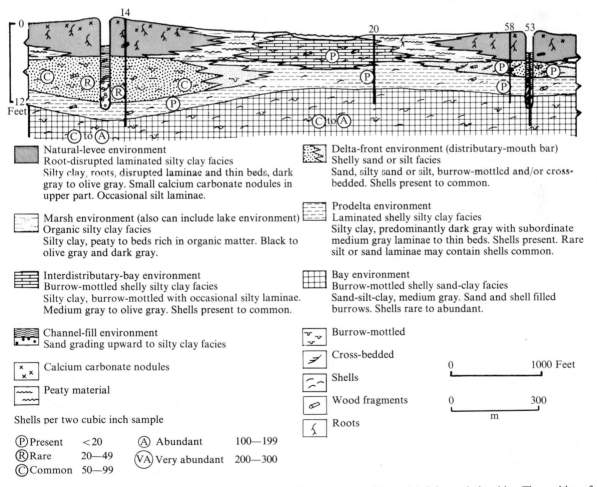

Natural-levee environment
Root-disrupted laminated silty clay facies
Silty clay, roots, disrupted laminae and thin beds, dark gray to olive gray. Small calcium carbonate nodules in upper part. Occasional silt laminae.

Marsh environment (also can include lake environment)
Organic silty clay facies
Silty clay, peaty to beds rich in organic matter. Black to olive gray and dark gray.

Interdistributary-bay environment
Burrow-mottled shelly silty clay facies
Silty clay, burrow-mottled with occasional silty laminae. Medium gray to olive gray. Shells present to common.

Channel-fill environment
Sand grading upward to silty clay facies

Calcium carbonate nodules

Peaty material

Shells per two cubic inch sample

(P) Present   < 20   (A) Abundant   100—199
(R) Rare   20—49   (VA) Very abundant 200—300
(C) Common   50—99

Delta-front environment (distributary-mouth bar)
Shelly sand or silt facies
Sand, silty sand or silt, burrow-mottled and/or cross-bedded. Shells present to common.

Prodelta environment
Laminated shelly silty clay facies
Silty clay, predominantly dark gray with subordinate medium gray laminae to thin beds. Shells present. Rare silt or sand laminae may contain shells common.

Bay environment
Burrow-mottled shelly sand-clay facies
Sand-silt-clay, medium gray. Sand and shell filled burrows. Shells rare to abundant.

Burrow-mottled

Cross-bedded

Shells

Wood fragments

Roots

0   1000 Feet

0   300
        m

Fig. 453b. Schematic cross-section showing various deposits of the Guadalupe delta and their interrelationships. The position of selected cores is marked (see also Fig. 454a,b). (After Donaldson et al. 1970)

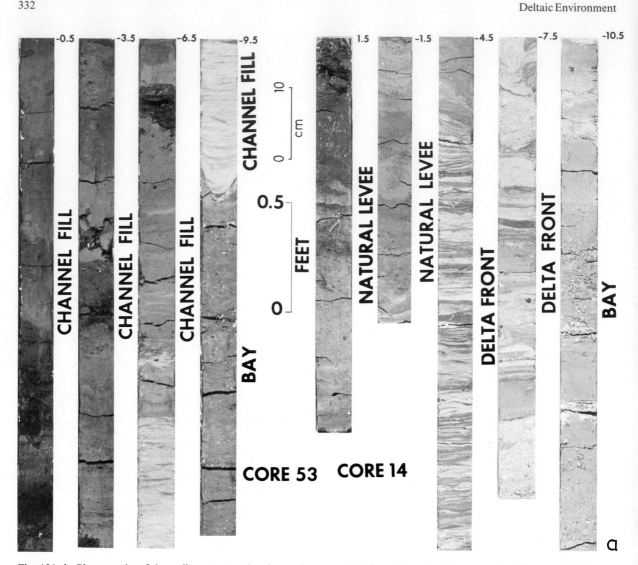

Fig. 454a,b. Photographs of the sediment cores showing sedimentary structures in the deposits of the Guadalupe delta and bay. (After Donaldson et al. 1970) a Core no. 53 from the middle of Flat Bayou, Guadalupe·delta. Elevation 15 cm. Cross-bedded sand of lower channel-fill sediments directly overlie the bay deposits, suggesting that the channel eroded through delta at this point. Cross-bedded sand gradually grades upward into organic-rich clays with abundant wood fragments, indicating a transition from active to abandoned distributary channel. Core

no. 14 is from the natural levee south of Sommerville Bayou. Elevation 45 cm. The core penetrates through a sequence of natural levee, delta front, and bay deposits. The upper part of the delta front deposits are mainly cross-bedded, whereas the lower part is extensively bioturbated. Recognition of prodelta sediments is not possible because of a high degree of bioturbation. The location of the cores is shown in Fig. 453b. The Datum line is sea level. Core no. 58 was obtained from the marsh in the Sommerville Bayou system

The predominant sedimentary structures in the bar finger sand are ripple bedding, wave and current ripples, gas-heave structures, and minor amounts of evenly laminated sand (Coleman et al. 1964). Sand/mud alternating bedding is almost absent in the upper part. Bar finger sands are silty sand near the base, becoming progressively cleaner sand in the upper parts. Load structures and other deformational features are present. Sediments of the channel deposits are coarse sediments with intercalations of muddy sediments. Here small-ripple bedding is most abundant, in parts de-

veloped as climbing-ripple lamination, which is still more abundant in natural levee sediments.

Bar finger sand bodies, though of much smaller dimensions, are present in the Rhône delta deposits (Kruit 1955; van Straaten 1959a). Donaldson et al. (1970) report bar finger sands from the Guadalupe delta, Texas.

Booch sand, upper Carboniferous, Oklahoma, is considered to be bar finger sand deposits (Busch 1951). Wurster (1964) describes bar finger sands from the delta system of the Keuper sandstones of West Germany.

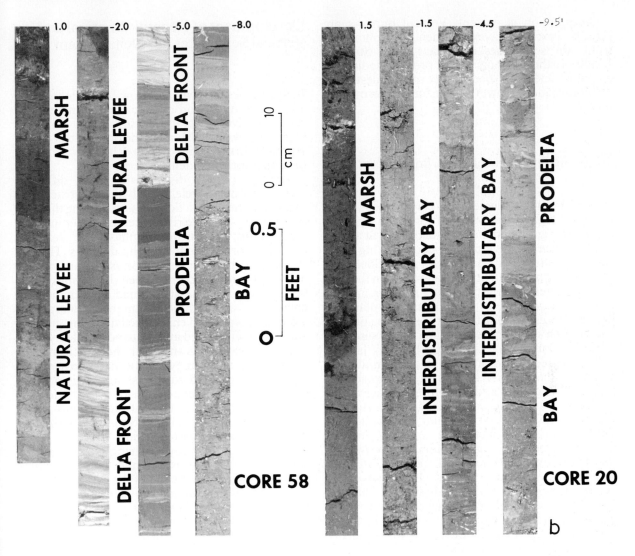

Prodelta Environment

The region seaward of the delta front environment is the prodelta environment, which is closely associated with the prograding delta system. It is also known as the delta front slope. Prodelta deposits are characteristically fine-grained muddy sediments, i. e., clay and silty clay. The prodelta deposits are transitional into the shelf-mud deposits. Sediments show layering due to differences in both color and grain size. Thinly laminated silts and clays are the most noticeable feature. Near the delta front environment sediments are more silty, and parallel and lenticular laminations are common. Occasionally, ripple bedding, current ripples, and small-scale graded bedding are common in more silty layers.

In the sediments away from delta front clays predominate and textural stratification is less frequent. Mostly color layering predominates in the clays. Shell remains are common throughout; wood fragments are also present. Bioturbation is present, usually restricted to certain zones. Well-developed burrows are also found.

The prodelta environment may be difficult to distinguish from the shelf-mud environment; it is distinguishable only when vertical and horizontal sequences are well established.

Van Straaten (1959a) also reports similar features from the Rhône delta. In addition, he reports sandy horizons from the upper part of the prodelta which can be traced over a distance of 2 km up to a water depth of 70 m. He assigns them to the river floods of high current velocities, which are capable of transporting large quantities of even coarser sediment (sand) over large distances into the sea. Current ripples are present in these sands.

Bioturbation is present. Vertical burrows are rare; however, in clayey sediment horizontal burrows are sometimes abundant (van Straaten 1959b). In the lower part of the prodelta sediments of the Rhône delta, bedding ist mostly destroyed by bioturbation. However, horizontal burrows are present.

Shelf-mud Environment (Delta Bottomset)

This environment is essentially clay deposits in the offshore region (seaward of prodelta region), where fine-grained material is deposited at a slow rate. Sediments are homogeneous-looking clay, silty clay, shelly clay, and shell layers. Shells are common throughout. Stratification within the clay is mainly a result of the difference in color, and the presence or absence of shell fragments, plant debris, etc. Occasionally, thin silty layers are present which often show graded bedding; otherwise they are thinly laminated. Lenticular bedding is rare.

The degree of bioturbation in shelf-mud sediments is rather high. Small distinct burrows stuffed with shell fragments and other foreign material are common. Otherwise, sediments show intense bioturbation leading to mottling. Thus, delta bottomset deposits are the product of the slow deposition of suspended sediment. Bioturbation is an active process in these areas. Moore and Scruton (1957) describe similar structures in the shelf deposits of the central Texas coast and the Mississippi delta region. Shepard (1956b) notes that shelf mud shows extensive mottling and homogeneous bedding. Shelf mud of other parts of the world, especially that with delta influence, also shows extensive bioturbation; e.g., Northern Paria-Trinidad shelf (Koldewijn 1958), Western Guinea shelf (Nota 1958) open shelf of the Rhône delta (van Straaten 1959b), the Gulf of Gaeta (Reineck and Singh 1971). In the shelf mud of the Rhône delta and the Gulf of Gaeta swarms of minute, dark-colored fecal pellets are common, which are assigned to the gastropod *Turritella comunis* (see p. 386).

All the environments and subenvironments of a delta system discussed above are present in most of the deltas. However, the relative importance of one subenvironment over another within a specific delta system varies considerably from one delta system to the other. The Holocene delta of the Guadalupe River, Texas, is a rather small delta; however, it shows a good development of most of the subenvironments of a delta system. Donaldson et al. (1970) made a detailed study of the Holocene Guadalupe delta. Figures 454a and b illustrate core profiles of this delta to depict a summarizing sequence produced within a prograding delta.

Development of Repetitive Lateral and Vertical Sequences in a Delta System

In most cases the seaward growth of a delta stops after a certain period of time. This is mainly because the river shifts its course laterally and constructs a new delta. Many of the present-day deltas are made up of a series of such delta lobes, referred to as an imbricated delta system.

Scruton (1960) distinguishes two different phases in a delta cycle, namely the constructional phase and the destructional phase. During the constructional phase the river extends farther into the sea; during destructional phase the sea enchroaches on the built-up delta deposits. The destructional phase begins as soon as forward growth of a delta is retarded or stopped due to shifting of the river course. The abandoned delta undergoes compaction and at the same time sediments are reworked by wave and current activity. Fine-grained material is winnowed away into adjacent bays and other protected areas and a thin sheet of clean sand is produced, which may be modified into a system of beach ridges and barrier islands. This process preserves the delta sequence below. The vertical sequence of a delta is a result of progradation (Fig. 455). In the Mississippi delta Scruton (1960) illustrates such a sequence (Fig. 456).

Later, if the river course changes again, a new constructional phase is initiated, and deposition starts laterally or on the top of the earlier delta deposits, which have already sunken down and undergone reworking. The Mississippi delta offers a good example of such an imbricated delta system with different stages of delta growth (Fig. 457).

A delta sequence may grow, even during a rising sea level. In such a case a retrogradational sequence is produced, i. e., deltaic plain deposits are overlayed by coastal sand deposits, followed by prodelta and then shelf-mud deposits on the top of the sequence.

Such a sequence is present in the lower part of the present-day delta of the Rhône River (Fig. 458).

Sometimes several deltaic sequences may be repeated in a vertical profile, even during a rising sea level. Several units are superimposed on each other in an onlap manner, each unit corresponding to a relatively short time period when the delta was prograding at that locality. The coarsest sediments are near the top of a unit. Such periods of delta progradation alternate with periods when the river channel has migrated somewhere else laterally. During these periods the coast line shifts quickly toward the land; there is no or little deposition or even erosion (Figs. 459 and 460). This results in a sort of cyclic sedimentation during a continuously

Fig. 455. Seaward progradation of a depositional environment. During the growth of a delta, different environment migrate toward the sea. (After Scruton 1960)

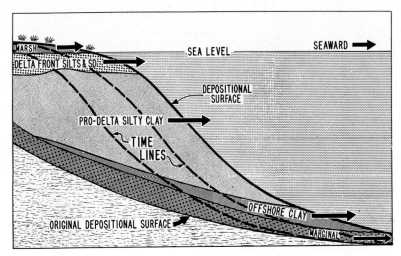

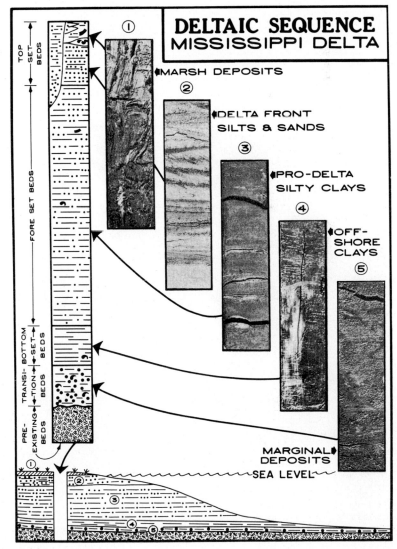

Fig. 456. The vertical sequence in a delta is the same as the seaward sequence from surface water to deeper water. (After Scruton 1960)

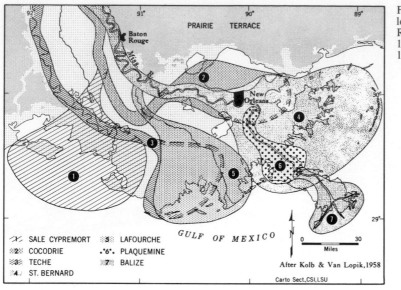

Fig. 457. An imbricating delta. Seven deltaic lobes have been built by the Mississippi River during the 5,000 years. (After Coleman 1968; originally after Kolb and van Lopik 1958)

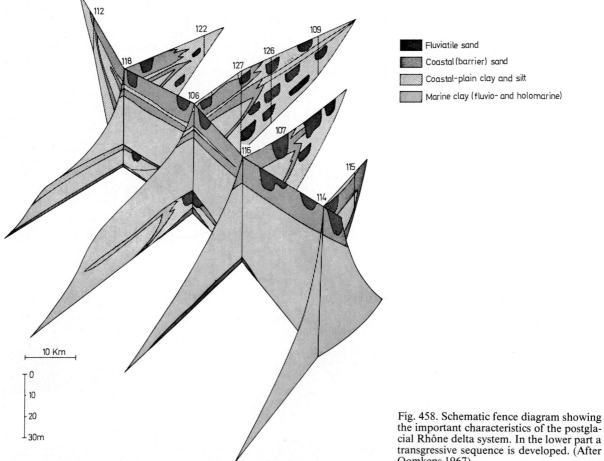

Fig. 458. Schematic fence diagram showing the important characteristics of the postglacial Rhône delta system. In the lower part a transgressive sequence is developed. (After Oomkens 1967)

Fig. 459. Hypothetical cross-section parallel to the coast, showing submarine deposits of the Rhône River delta-type. (After van Straaten 1960b)

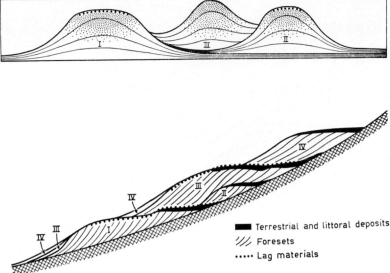

Fig. 460. Hypothetical section depicting deltaic deposits formed during a rapidly rising sea-level (After van Straaten 1960b)

■ Terrestrial and littoral deposits
/// Foresets
····· Lag materials

rising sea level, without any oscillations on land or sea.

Sometimes a retrograde sequence developing during a transgression can grade into a prograde phase of deposition as soon as rate of rise in sea level is slowed down.

Such repetitive progradational and retrogradational sequences are very common in modern and ancient delta deposits (Fig. 461). From the late Quaternary Niger delta complex, progradational sequences are described by Oomkens (1974).

Examples of Ancient Deposits of the Delta Environment

The building up of deltas has taken place throughout geological history as long as sea and rivers have existed. Thus, a substantial amount of coastal sediments incorporate deltaic deposits. However, the recognition of delta deposits in ancient sediments may be very difficult, because for definite recognition of a delta a regional variation in depositional environments has also to be established.

Economically, delta deposits are most important, since oil and gas are often associated with them (Rainwater 1975). One of the well-studied deltaic complexes is the Devonian Catskill deltaic complex, U.S.A. (Barrel 1914; Friedman and Johnson 1966; Allen and Friend 1968).

Other examples include Booch sand, Oklahoma, (Busch 1959), and Keuper delta sandstones, West Germany (Wurster 1964). Wanless et al.

(1970), Weimer (1970), and Curtis (1970) discuss various delta deposits of the U.S.A. Other well-studied examples are the Wealden delta of southeast England and France (Allen 1959) and the Carboniferous delta deposits of northern England (see Selley 1970).

Rainwater (1966) gives a series of criteria that can be helpful in the definite recognition of a delta deposit. Besides the study of lateral and vertical sequences, geometry, and fauna, he stresses that diapiric structures are usually found in association with delta sediments because of rapid rates of sedimentation.

Peterson and Osmond (eds.) (1961), Fisher et al. (1969), Morgan (ed.) (1970), Selley (1970) and Broussard (ed.) (1975) give useful information on deltaic sedimentation, their characteristic and ancient examples.

Summary

Deltas make the largest resting places for the clastic sediments (Dickinson 1975a). The thickness of modern delta deposits varies from a few meters to several tens of meters. The Holocene Mississippi delta is about 50 m thick. Ancient deltas mostly attain thicknesses of several hundred meters (Smith 1975). The areal extension of modern deltas ranges from 388 km^2 (Ebro delta) to 124,000 km^2 (Hwang Ho and N. Yellow Plain). The shape is mostly a lobe-like extension of coast with several divisions. The main axes of the delta run normal to the regional depositional strike.

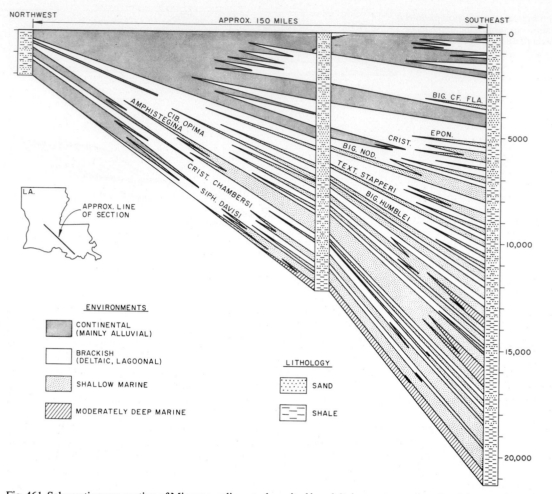

Fig. 461. Schematic cross-section of Miocene sediments deposited in a deltaic system, southern Louisiana. (After Rainwater 1966)

The construction of the delta is always progradational. The vertical sequence of marine deltas is marine-fluvial-lacustrine-subaerial. In lake deltas deposition begins with lake-prodelta deposits where turbidites are also incorporated (see lake environment). A large number of different facies are involved in the construction of a delta. In vertical sequences a coarsening-upward pattern is developed. In the channel deposits, fining-upward sequences are present, which become coarser in areas of destructive processes and where marine processes rework the fluvial deposits. The channel deposits are further distinguished into channel, meander belt, flood basin, interdistributary bay, and marsh and swamp sediments. Marsh and swamps are the areas of development of peat, lignite, and coal.

The high rates of sedimentation in delta areas results in low rates of bioturbation, even in marine parts, where also the wood pieces are often present.

The Coast

Introduction

The hydrographic conditions on the shore and in the shelf are very different; so are the sediments and their characteristics. However, they are, somewhat interrelated; the sediments of the shore laterally change into shelf sediments and in a prograding sequence coastal sediments follow on top of shelf sediments in a vertical succession.

In a sense the differentiation between coastal sand and shelf mud takes place in the rivers that bring sediment to the sea. Sand is transported mainly by rolling and saltation up to the shore line. Silt and clay is transported in suspension, by-passing the coastal sand and being transported into deeper parts, i.e., the shelf.

In the coastal region, besides the coastal sand deposition, sedimentation also takes place in lagoons and tidal flats, which are protected areas where mainly fine-grained material is deposited. Lagoons and tidal flats are also in a certain way part of the coastal sediments, since they are laterally associated and may become incorporated in a vertical sequence of coastal sediments.

As a result of repeated transgressions and regressions of epicontinental seas in geologic times, an extensive record of ancient coastal and shelf sediments is present on the present-day continents. Together with deltaic sediments they probably compose a major part of sedimentary rocks.

Coastal sediments are very important in oil geology. According to Curtis et al. (1961), of all the oil-traps in U.S.A. 10 % are stratigraphically controlled, 34 % are a combination of stratigraphically and structurally controlled, and 56 % are structurally controlled. 61 % of the stratigraphic traps are in one way or another related to coastal sand bodies.

The Coast: Definition and Classification

The coast separates continents from seas and may develop in various geomorphic and sedimentological variants. There are many excellent works dealing with the classification and types of coast. Widely accepted classifications are those of Valentin (1952) and Shepard (1963, 1976).

Normally, in ancient sediments only the coastal sediments which are in equilibrium are preserved in prograding or retrograding sequences. The cliff coasts are rarely preserved, e.g., the Mainz basin (Weinkauff 1859; Geib 1950).

Coastal sediments are usually made up of sand, and sometimes of gravelly sediments. The coastal sediments are gravelly only where source rock is nearby. It may be cliffs exposed on the sea coast which is broken by wave action, e.g., cliff coasts of England, the Pacific coast of the U.S.A. It may be where rivers mainly bring gravels to the sea, e.g., many Italian rivers meeting the Adriatic Sea. In some cases, reworking of the gravelly deposits by a transgressive sea also produces coasts with gravels, e.g., the southern Baltic Sea, where reworked moraines on the coast have provided pebbles. However, such gravelly coasts are not important in producing thick sequences. Important for the formation and preservation of thick sequences are the sandy coasts.

In general, one can recognize following types of coasts:

1. Mainland coast
   a) Coasts with spit development
   b) Coasts with development of chenier
2. Barrier island coast. Given enough time a mainland coast may develop into a barrier island coast (Hoyt 1967, 1969).
3. Coasts with cliffs
4. Coasts with sunken morphology (fjords, etc.)
5. Coasts with bioherms of warm seas, e.g., coral reefs–coastal reef and atoll. There is long bibliography on sea coasts and coastal sediments. Following are some of the selected references dealing with various aspects of coasts:

Andree (1920), Bagnold (1963), Bird (1969), Dickinson et al. (1972), Dolan and McCloy (1964), Emery (1960a), Gorsline ed. (proceedings) (1962), Guilcher (1958), Hunt and Groves (glossary) (1965), Inman and Bagnold (1963), Johnson (1919), King (1972), Kuenen (1950a), Russell (glossary) (1968), Shepard, Phleger, and van Andel (1960), Shepard (1963, 1967), van Straaten (1959, 1964b, 1965), Trask (ed.) (1939), Wiegel (glossary) (1953), Zenkovitch (1967), and the fifth Congrés International (1959).

A series of reviews and regional studies are compiled in the Field Trip Guide Book (South-Central Texas Coast, 1964). Along with deltaic sandstones these deposits represent potential petroleum occurrences (Weimer 1975). Kraft (1972) describes the sandy coastal areas of Greece, while Kent (1976) studies modern coastal sedimentary environments of Alabama and Northwest Florida. Field Trip Guide Book on Coastal Environments of Northeast Massachusetts and New Hampshire (1969), compiled by coastal Research Group of the University of Massachusetts provides useful information on the coastal sediments.

Coastal Sand

The sand body of coastal sand is usually tens of meters to hundreds of meters broad, hundreds of kilometers long, and 10 to 20 m thick. Along its long axis coastal sand is cut across by river deltas, estuaries, bays, lagoonal outlets and tidal inlets. The lower boundary of a coastal sand body is usually uneven. This is because tidal inlets change their position by migrating laterally or because new tidal inlets are made by breaking through the coastal sand. Thus, below the coastal sand, broad filled-up tidal channels may be present. Schwartz (1973) gives a useful collection of 40 papers on the problem of barrier islands.

A much-discussed question is the source of sand, both for mainland beaches as well as barrier beaches. Pierce (1969) calculated that 50 % of the sand of barrier islands on the eastern coast of the U.S.A. is derived by reworking basal transgressive relict sands. Curray et al. (1969) found that up to 50 % of the old Holocene beach bars of the coast of Nayarit, Mexico, is derived from the coast-nearer part of the shelf sediments. Later, when shelf sand was covered by muddy sediments, longshore transport of sand contributed to the further development and deposition of the beach. Zenkovitch (1969) also believes that a major part of the coastal sand is derived from the shoreward parts of the shelf. Reineck (1963a) could demonstrate the transport of sand from the shelf toward the coast by measuring the cross-bedding in the samples obtained by a box-sampler (Fig. 462). The observations are true for only the development of the modern coastal sands which have been deposited in the last 7,000 to 3,000 years.

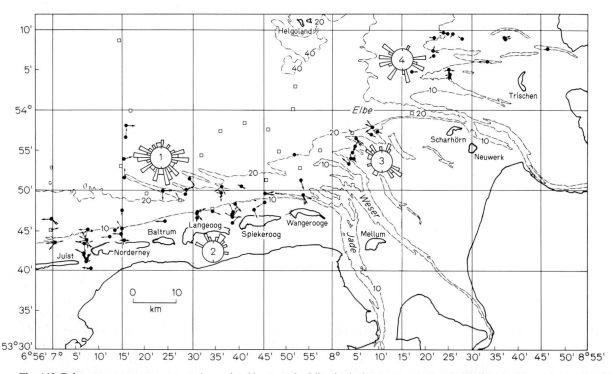

Fig. 462. Palaeocurrent measurements determined by cross-bedding in the box-core samples. Rose diagram 1 is representative for the sandy area in front of the East Frisian Islands. Rose diagram 2 is representative for the tidal inlet between the islands of Norderney and Juist, and for the tidal inlet between the islands of Langeoog and Spiekeroog. (After Reineck 1963a)

However, for the prograding coasts, which are so common in the geological record, the major part of the sand has to be brought in by the rivers. The transport of sand from the shelf toward the coast takes place only at the end of a marine transgression. During this period the irregularities of the coast line extending into the sea are eroded and straightened. This provides another minor source of sediment. The development of a mainland beach can be easily explained as the result of sorting action of wave and current activity of the sea. However, the genesis of barrier islands is rather complex. Hoyt (1967a) observed that there is no mainland beach present behind the barrier islands on the landward side of a tidal flat or lagoon. He concludes that barrier islands can not have originated as offshore bars after transgression. To begin with, barrier islands were mainland beaches, and during the last stages of marine transgression, they became separated from the mainland. In the last phase of marine transgression, the area behind the barrier island was converted into a lagoon or tidal flat, while the barrier island grew upward with the rising sea level (Fig. 463). The question of whether this process is responsible for the genesis of all the barrier islands is still open. For example, Colquhoun (1969) and Pierce and Colquhoun (1970) demonstrated that some of the barrier islands on the coast of South Carolina originated from longshore bars (Fig. 464). On the other hand, Hails and Hoyt (1968) showed that in the coastal plain of Georgia, chains of barrier islands associated with marshes can be ascribed to various marine transgressions during the Pleistocene period (Fig. 465). Thom et al. (1978) are able to differentiate three different types of bay-barrier in the New South Wales, Australia, which can be applied to the different possibilities in the genesis of barrier islands. The three

types of barriers are: prograded, stationary, receded (= retrograded).

Papers by Otvos (1970a,b), Field and Duane (1976), Otvos (1977), and Field and Duane (1977) clearly suggest that there are several modes of origin of barrier islands:

1. Shoal and longshore bar aggregation (van Straaten 1965; Colquhoun 1969a,b; Pierce and Colquhoun 1970; Otvos 1970a).
2. Spit segmentation (Otvos 1977).
3. Mainland ridge engulfment (Hoyt 1967a)
4. Welding or veneering of Holocene dune, beach, and foreshore sands into and over pre-Holocene topographic highs (Oertel 1975; Otvos 1977).

Field and Duane (1976, 1977) are of the opinion that during a transgression, the coastal sands are laterally shifted in the form of barrier islands with the transgression. This may be true for the sand material, but not necessarily for a specific barrier island.

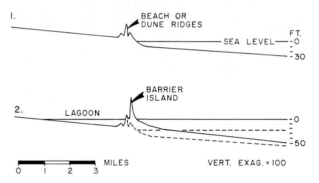

Fig. 463. Development of barrier islands by the process of submergence. *1* Beach or dune ridge formed near the shore line. *2* During submergence, an area landward of the ridge is flooded by water, and a barrier island with a lagoon behind it is formed. (After Hoyt 1967a)

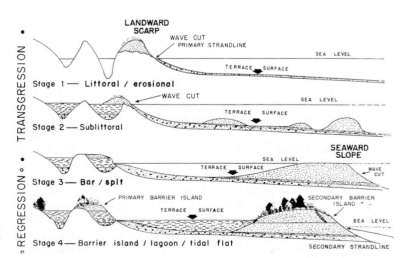

Fig. 464. Primary and secondary barrier islands formed during a transgression-regression cycle. (After Colquhoun 1969b)

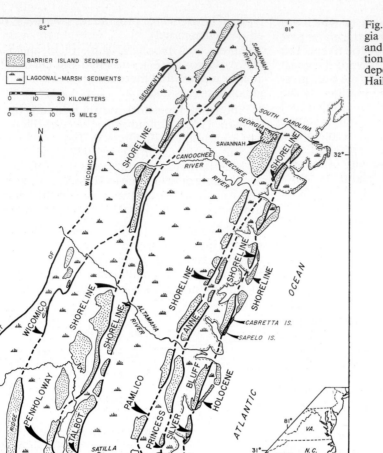

Fig. 465. Map of the coastal region of Georgia showing the position of the Holocene and six Pleistocene shore lines. The distribution of barrier islands and lagoonal-marsh deposits is shown for each shore line. (After Hails and Hoyt 1968)

Göhren (1975a,b) studied the aggradation of longshore bars in front of North Sea tidal flats with the help of aerial photographs. At first, "sandriff" (sand bar) is exposed to wave action, and migrates, in the beginning, at a rather high velocity (ca. 150 m/year) towards the coast. With the increasing distance from the low-water line, the rate of migration of sand bar is reduced. With the result, the succeeding sand bar approaches the earlier one, and is integrated with it.

A new classification of coastal morphology according to tidal range started with Davis (1964, 1980) and Hayes (1975, 1979). Davis (1964) differentiated three categories of coast: microtidal (range < 1.8 m), mesotidal (1.8–3.65 m) and macrotidal (> 3.65 m). Hayes (1975, 1979) applied this terminology to three types of shoreline morphology.

Macrotidal shorelines are those where tidal effects predominate over effects of waves. On macrotidal shorelines barriers are not developed due to the predominance of strong tidal currents in a shore normal direction. Linear sandbodies oriented parallel to tidal currents may be present in the mouth of funnel-shaped embayments and estuaries. Examples are the Bay of Fundy, Bristol Bay, estuaries of the Jade Bay, Weser River and Elbe River in the German Bight (Dijkema 1980).

On mesotidal barrier coasts barriers are short and have a drumstick configuration. Inlets are closely spaced with well-developed ebb tidal deltas. Examples are the East Friesian Islands and barriers before the Georgia coast (Nummedal and Fischer 1978).

Microtidal coasts are characterized by long, rather straight barriers with widely spaced inlets, pro-

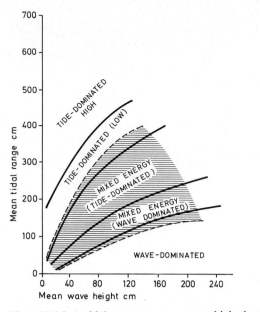

Fig. 466. Mean tidal range vs. mean wave high showing five fields: wave-dominated, mixed energy (wave-dominated), mixed energy (tide-dominated), tide dominated (low), tide dominated (high) (after Hayes 1979). *Dotted area:* approximate occurrence of barriers (B. W. Flemming, pers. communication 1986)

er a spectrum of tidal ranges and wave heights as the relative effect of these processes are important, rather than the absolute values. This means that the dominance is strictly relative. It is not based on any wave or tidal range parameters.

The lateral environmental association of coastal sands is the transition zone and the shelf mud toward the sea, and marshes and alluvium toward the land. In the case of barrier islands, coastal sand toward the land is followed by tidal flats or lagoons, marshes, and alluvium.

On a prograding coast of a barrier island coast, the following vertical sequence is produced (progradational) (Figs. 576 and 577):

Alluvium
Marshes (peat and coal)
Tidal flat and lagoonal deposits
Coastal sand
Transition zone
Shelf mud

Most of the modern barrier islands show only the beginning of such a progradational sequence. Most of these barrier islands have grown upward and partly forward (prograding) only after the end of last transgression. Complete vertical sequences are not yet developed, although the lateral sequence is well developed.

The barrier island Sapelo Island, on the coast of Georgia, U.S.A., shows its growth on Pleistocene sediments and its growth backward (toward land) with increasing water level during transgression (onlap). The coastal sand body is placed on the top of lagoonal and salt marsh sediments (Fig. 467). However, at present progradation is very meager.

The barrier island Padre Island, Gulf of Mexico, U.S.A., shows an upward growth with the last increase in water level during transgression (Fisk 1959). The deposition of coastal sand started some 6,500 years B.P. Since the still-stand of the water level, Padre Island shows progradation (Fig. 468).

Galveston Island, Gulf of Mexico, came into existence 3,500 years ago and has been prograding

minent washovers and well-developed flood tidal deltas. Examples are the coast of Florida and the Outer Banks, North Carolina. These shorelines are wave-dominated, but the given generalization is restricted of coasts with moderate wave energy.

Heward (1981) gave a comprehensive review of wave-dominated coast lines and Davies and Hayes (1984) stated that wave energy and tidal prism must be included in characterizing coasts. "Wave-dominated shorelines are those where wave action causes significant sediment transport and predominates over the effects of tides" (Heward 1981, p 223). Based on Hayes' diagrams (1979, Figs. 15 and 18), five fields represent the range from tide-dominated to wave-dominated (Fig. 466). All fields cov-

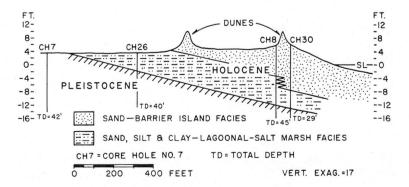

Fig. 467. Idealized section across a narrow Holocene barrier and salt marsh, Sapelo Island, Georgia, U.S.A. (After Hoyt 1967b)

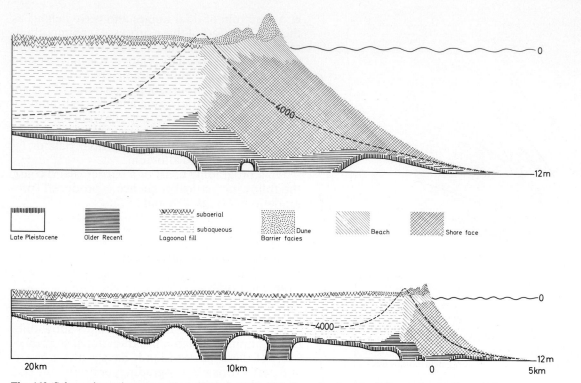

Fig. 468. Schematic section across Padre Island, Gulf of Mexico. During the last phase of rise in sea level, the barrier island was built upward. After the sea level still-stand, the barrier is-

land was built forward toward sea. (After Reineck 1970b; originally modified after Fisk 1959)

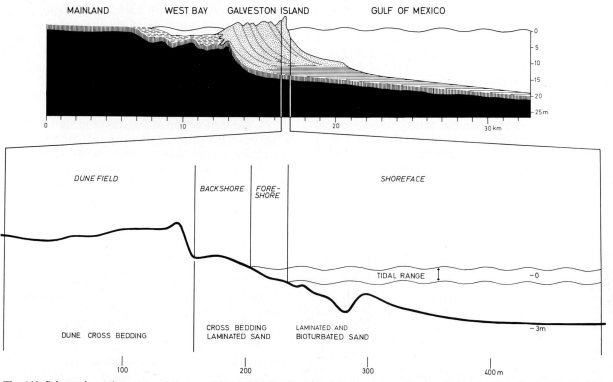

Fig. 469. Schematic section across Galveston Island. Distribution of bedding structures is very schematic. (Based on Bernard et al. 1962. After Reineck 1970b)

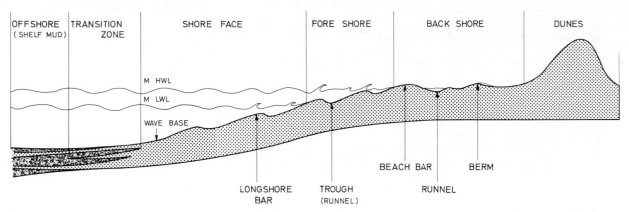

Fig. 470. Schematic representation of the terminology of the various geomorphic units of a beach profile. Various geomorphic features of a beach as well as transition to shelf mud are shown. The terminology is mainly based on Emery (1960a)

ever since (Bernard et al. 1962). In the process of progradation a sequence shelf mud, transition zone and coastal sand has been produced (Fig. 469).

Before 2,800 years, the sandy Dutch coast was situated further outside the present-day coast. Since this time coastal sand migrated back toward land (van Straaten 1965; Hagemann 1969).

In a progradational sequence of the mainland beach, tidal flat and lagoon deposits are missing. The following units are found:

Alluvium
Marshes, peat and coal
Coastal sand
Transition zone
Shelf mud

During transgression, in retrograde sequence the arrangement is just the reverse (Oomkens 1967). Depending upon the rate of transgression, one or more units of the ideal sequence may be ill-developed or completely missing (Fischer 1961). In such a "contraction sequence", for example, shelf mud may be present on the alluvium. In such a case, transgression was too rapid, so that marshes or coastal sand could not develop and be deposited.

Sometimes a rather condensed sequence develops where during retreating coastal sand a transgressive sand layer is deposited. Swift et al. (1971) report that during the retreat of a coast of low sand supply, sand of the upper shoreface and beach is moved landward by washovers, etc., while the sand from the lower shoreface is transported seaward to produce a transgressive sand sheet.

The various depositional units of the coastal sand, e.g., foreshore, shoreface, etc., show similar development and characteristics on mainland beaches and barrier islands. Thus, the sequence of coastal sand on both types of coasts can be de-

scribed collectively. It may be pointed out here that in some areas, especially in tidal seas, coastal sand may develop in various other forms as well, e.g., sand bars (shoals) and channels, sand tongues and channels, etc., making a characteristic sequence of their own.

Geomorphology of Coastal Sand (Beach)

A typical beach (strand) on a sea coast can be divided into several units–sand dunes, backshore, foreshore, and shoreface (Fig. 470).

The boundary between backshore and sand dunes is located near the lower limit of the sand dunes. Lüders (1956) considers the boundary be-

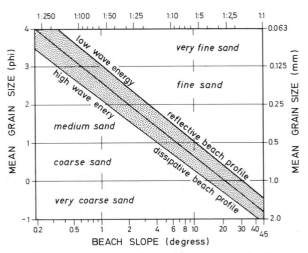

Fig. 471. The relationship between beach slope, grain size and wave energy. (Modified after Flemming and Fricke 1983, recalculated and rearranged after Wiegel 1964)

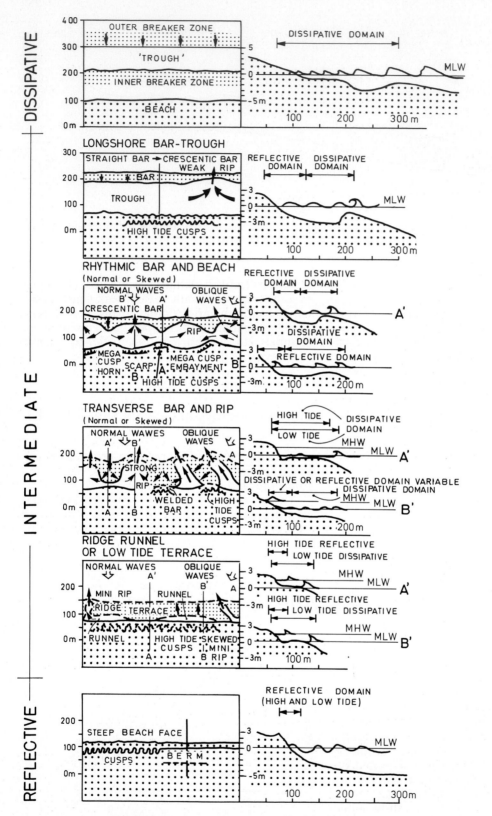

Fig. 472. Plan and profile configuration of the three beach types (dissipative, intermediate and reflective) and six beach stages. (Modified after Wright and Short 1983)

tween the two to be at the lower storm flood level. Quite often the level of high water exceeds the mean tidal high-water level; however, this is not the storm flood water. On the East Frisian Islands of the North Sea the lower limit of the storm flood water is located 1 m above the mean tidal high-water level. In other regions this limit may be located at some other level. Toward the sea the backshore extends up to the mean high-water line. Thus, the backshore represents the supratidal zone. In the backshore features such as the beach bar and berm may develop. In tidal seas the region located between the mean high-water line and the spring tide mean low-water line is called the foreshore. It is the intertidal zone. The foreshore is followed by the shoreface, which extends to the transition zone. The lower limit of the shoreface corresponds to the average maximum wave base. The boundary between the shoreface and the transition zone is commonly associated with a sharp knick in the angle of slope, i.e., the steep slope of the foreshore changes into a rather gentle slope downward. However, in some beach profiles such a knick is altogether absent. In regions where mud is available together with sand the change from shoreface to transition zone is documented by a change in grain size, i.e., the shoreface is made up of clean sand, whereas, below the wave base, the transition zone is composed of silty sediments (sandy silt to silty sand). If longshore bars are present, they are situated above the wave base in the shoreface and foreshore regions. In nontidal seas the foreshore is not developed. The backshore grades directly into the shoreface.

A beach shows regular variations in energy conditions from sand dunes down to the shoreface (Kent 1976). Visher (1969) analyzed samples across Forest Beach, South Carolina, but his grain-size distribution curves are ambiguous.

Beach Dynamics and Sediment Transport

Wave energy is the major factor controlling the development and changes in the beach (Johnson 1919; Johnsen 1961; Zenkovitch 1967; King 1972). A systematic relationship was further observed between grain size, beach slope and the exposure to wave action (Bascom 1951; Wiegel 1964). For a given grain size a beach slope will increase with decreasing wave action. Alternatively, under similar wave conditions a coarser sand will always form a steeper slope than a finer sand (Fig. 471). Beach profiles thus change periodically with changes in the temporal wave climate and in accordance with the above grain size/slope relationship.

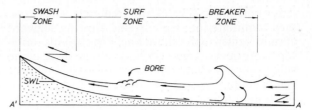

Fig. 473. Diagram showing terminology of the near-shore current system as seen in profile. (After Ingle 1966; originally modified after Shepard and Inman 1950)

The large variety of possible beach states has been reduced to a schematic morphodynamic beach classification (cf. Wright and Short 1983) with three beach types and six beach stages (Fig. 472). The beach types are dissipative, intermediate and reflective (Guza and Inman 1975). A simple empirical derivation of this morphodynamic beach stage (B_s) is $B_s = H_b/W_s T$, where $H_b =$ breaker hight, $T =$ wave period, $W_s =$ settling velocity of the mean grain size in m/s. If $B_s \leq 1$ the beach stage is reflective, if $B_s > 1$ and < 6 the beach stages are intermediate and if $B_s \geq 6$ the beach stage is dissipative.

The incoming waves at the beach start becoming steeper after they touch the ground (above the wave base). Ultimately, they are oversteepened, and break up in the breaker zone (Fig. 473). The breaker zone is followed by the surf zone, which at the shore line is followed by the swash zone.

Coastal currents may be present in the offshore region (beyond the breaker zone). The *longshore currents* and the rip currents develop within the surf zone by wave action only. Longshore currents run almost parallel to the shore line, and at a few

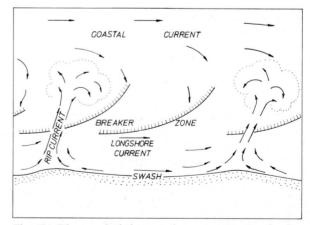

Fig. 474. Diagram depicting nearshore current system in plan view (see also Fig. 472). The incoming breaker generate longshore current, which produces pulsatory rip current moving towards sea. (After Ingle 1966)

tens of meters to several hundred meters distance, they terminate into rip currents, flowing seaward at right angles to the shore line. Rip currents represent a product of piled-up water on the shore in the ridge-runnel system by waves and wind. As this water cannot flow back along the same way, it flows parallel to the shore in the runnels and troughs (longshore currents); and at the end of a runnel it turns around at right angles into the sea (Fig. 474) in the form of rip currents (Shepard et al. 1941). This process produces channels at right angles to the shore line and may be present even in water depths of 5 m (Ingle 1966).

Rip currents are grouped into accretionary, erosional and mega rip systems (Figs. 475 and 476). Erosional rips develop during rising seas at the transverse bar stage and continue into the higher beach stages, being usually accompanied by beach erosion. Rip spacing is 300–500 m. Mega rips are large-scale systems (1 km) often observed at fully dissipative states when strong erosion is common. According to Short (1985), accretionary rips prevail during falling wave conditions. They are more closely spaced (<250 m) and are associated with beach accretion (Short 1979, 1985).

According to Dette (1974), Dette and Führböter (1975), and Führböter et al. (1979) longshore currents can be taken as the most important sediment transport agent, as they are subjected to highly turbulent fluctuations. Current measurements have shown that velocity fluctuations of 100 %, and up to nine periods within one wave period may be present. This phenomena is related to the fact that in shallow water areas, steep waves are made up of two or more waves (solitons) with different height and periodicity. It is reasonable to believe that such highly turbulent currents possess a much higher capacity of sediment transport than the corresponding average current velocities.

Both longshore currents and rip currents are pulsating in nature, and their maximum velocity may exceed values of 1 m/sec (Reineck 1963a). This velocity is sufficient to produce megaripples, and in many cases troughs and runnels of a sandy beach are covered with megaripples of varied shape and size, following the direction of longshore currents and rip currents. Gorsline (1966) measured velocities of longshore currents on the beaches of the Gulf of Mexico, a low-energy coast. The average current velocity is 40 cm/sec, with values ranging from 10 cm/sec to 150 cm/sec. The current velocity of longshore currents is strongly dependent upon the height of the breakers.

The wave action is not limited to the fact that they produce longshore currents and rip currents which are capable of transporting sediment. But the waves themselves cause active sediment-transport when they are shoaling (waves are touching the sediment bottom with sufficient energy). In rel-

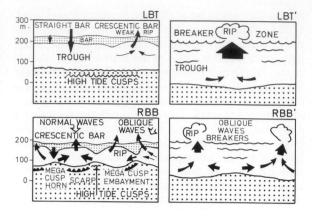

Fig. 475. Configuration of the four intermediate beach stages and accretion rips. *LTB* Longshore bar – trough; *RBB* Rhythmic bar and beach; *TBR* Transverse bar and rip; *LTT* Low tide terrace, see also Fig. 472. (Modified after Short 1985)

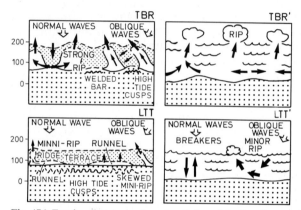

Fig. 476. Erosional states and erosional rips of the four intermediate beach stages, see also Figs. 472 and 475. (Modified after Short 1985)

atively deeper parts of the beach, symmetrical wave ripples are formed with straight, long crests running parallel to the shore line. As the longshore currents cause lateral displacement of water along the coast, the sand grains not only move back and forth by waves, but also suffer a lateral movement from longshore currents (Reineck and Singh 1971). In the Gulf of Gaeta, Italy, symmetrical wave ripples are present in water depths up to 4 m. On the coast of California, wave ripples are known up to a water depth of 30 m (Inman 1957). If the beach slope is very gentle, the shoaling of the waves results in asymmetrical waves thus producing asymmetrical wave ripples (see p. 33). Clifton (1976) describes in a conceptual model a sequence of wave-formed structures. In this case four parameters are most significant: maximum bottom orbital velocity, velocity asymmetry, medium grain size and wave period.

In the surf zone the undertow of backwash transports coarser sediment toward the breaker zone. At the same time sediment is brought from the open sea into the breaker zone (Fig. 477). Thus, in the breaker zone the coarsest sediment is more abundant (Fig. 478). If waves reach the beach at rather low angles, stronger longshore currents are produced. This results in an active sediment transport parallel to the coast (Fig. 479). This leads to sediment sorting, so that only the sediment grains which are in equilibrium in the surf zone and with longshore current are transported along the coast, while moving back and forth. While the other grain sizes are transported to the breaker zone or swash zone (Fig. 480). Pebbles with attached algae (Fig. 481) are transported towards land by the oscillatory water movement (Kudras 1974).

In the swash zone water movement is also back and forth, and parallel to the shore. Besides, during washover of longshore bars (just before emergence or just after submergence of the bars), and in runnels with rather shallow water, conditions of rapid flows are achieved with beforms of the upper flow regime. Antidunes are commonly observed under these conditions, and sediment transport is enormous.

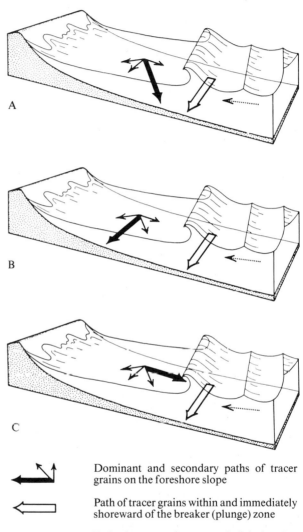

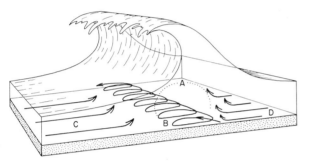

Fig. 477. Schematic representation of the motion of sediment grains beneath a breaking wave. Largest grains present move by saltation along the bottom at position B. Majority of sediment grains move along various horizontal to flat-elliptical parallel-to-coast paths during collapse of each wave. Finest sediment grains move in suspension at position A. Sediment grains shoreward C and seaward D of the breaker zone move toward the plunging wave essentially along the water particle motion. The size of the sediment grains at any point is essentially a function of available wave energy. (After Ingle 1966)

Dominant and secondary paths of tracer grains on the foreshore slope

Path of tracer grains within and immediately shoreward of the breaker (plunge) zone

Path of tracer grains seaward of the breaker zone

Fig. 479. Block diagrams depicting the major paths or vectors of tracer grain movement on planar foreshore-shoreface slopes under three different surf conditions. A Sediment grain movement under surf conditions where the longshore current and the wave motion exert an equal influence. B Sediment grain motion under conditions of a high-velocity longshore current (velocity > 60 cm/sec). C Sediment grain motion when the longshore current velocity is < 30 cm/sec, and the onshore-offshore motion of waves controls the sediment grain transport. The three diagrams represent the qualitative average of tests performed on planar slopes. (After Ingle 1966)

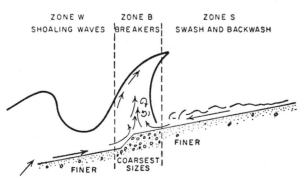

Fig. 478. Mechanism for the sediment pattern in the breaker zone (zone B). Arrows indicate net sediment movements at the moment the wave breaks. (After Miller and Ziegler 1958)

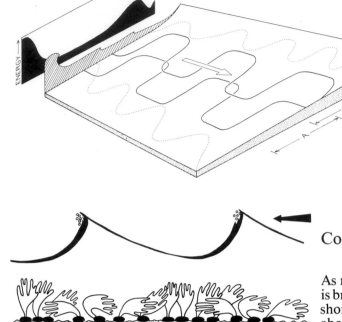

Fig. 480. Schematic diagram showing the sorting action beneath the surf zone. Solid lines denote hypothetical paths of two sand grains in momentary equilibrium beneath the surf zone. The position of a sediment grain on the beach slope is determined by the average onshore-offshore bottom-current velocity, and to some extent by the angle of slope. Net paths of the two grains in equilibrium overlap (A and B). Along-shore vectors of grain transport are a function of longshore current velocity. Grains midway between the breaker and collision zones possess the strongest longshore vectors. Grains not in equilibrium on the slope follow oscillating paths (dotted lines) either seaward or shoreward, depending upon their specific characteristics and point of origin. (After Ingle 1966)

Coastal Sand Dunes

Fig. 481. Diagram showing movement of pebbles with attached algae in shoaling waves. (After Kudras 1974)

To conclude, the mechanism and direction of transport of sand in the region of the beach are strongly controlled by the movement of the water, which is controlled mainly by waves and currents, produced by wave action. The transport is in the form of suspension, surface creep and saltation. Brenningmeyer (1976) observed sand fountains in the surf zone. However, Komar (1978) points out that the suspended sand in the surf zone is only 25 % of the sand in transport; while 75 % of sand is transported as bed load. The whirling up of the sand in the breaker zone is mainly related to the development of an air-water mixture, so that for a few seconds there is a three-phase system, i. e., water-air-sediment (Führböter 1970).

Meyer (1972) gives a collection of 12 papers on waves on beaches and resulting sediment transport. Davis and Ethington (1976) give a number of interesting papers on beach sedimentation; and Hails and Carr (1975) provide papers on nearshore sedimentation processes.

As mentioned elsewhere, along the seashore sand is brought from shoreface into foreshore and backshore regions by wave action. The sand located above sea level is exposed to wind activity, and sand is reworked into low dunes. Coastal dunes develop where there is a sufficient supply of sand, and where a dominant strong wind is present in the onshore direction. According to Gripp (1968), coastal dunes develop on stable, and especially on prograding, shores. The coastal dunes may rather occupy a broad zone along the coast (Cooper 1967). For example, on the Pacific coast of Oregon and Washington, where dunes are associated with large trees, they are a maximum 4.5 km wide. On the other hand, on the southern Pacific coast, California, dunes occupy zones as broad as 9 km and are predominantly the transverse and parabolic types (for a discussion on various types of sand dunes, see desert environment, p. 222). Cooper (1967) provided a detailed study of coastal sand dunes. He distinguished two types of sand dunes in nonvegetated regions:

1. Transverse ridges. They develop where wind is blowing constantly from the same direction, and are normal to the wind direction. They migrate slowly.
2. Oblique ridges. They develop where wind blows from two quarters at different times. They are more or less stationary and show no migration. Sand dunes of the Gulf Coast, Texas, are of this type.

In vegetated regions two more types are found:

1. Parabolic dunes. They develop in sparsely vegetated regions. They show blow-out features and possess a concave-downward lee-face. They are common on the Pacific Coast of the U.S.A.

2. Precipitation dunes. They develop where sand is blown against densely vegetated areas with trees. Sand is made to be deposited by trees in the way of sand-laden winds. They show movement against the forest (Fig. 482).

In vegetated regions conditions controlling growth and deflation of sand are rather complex. Vegetation tends to stabilize sand. Changing wind directions and the effect of heavy storms may strongly modify dune shapes, and a confused assemblage of sand hills and hollows may result. In general sand accumulation in low dunes begins usually around some small plants and other obstacles (Gripp 1968). However, Cooper (1967) contended that development of sand dunes does not require any plants etc.

The bedding of coastal dunes is mainly sand dune cross-bedding with abundant erosional unconformities (Fig. 483). The erosional surfaces are often curved. The foreset laminae of cross-bedding are rather steep–up to 30 or 40°. Low-angled or horizontally laminated units are present, but not common. Coastal dunes often show cross-bedded units made up of convex-upward foreset laminae. Mackenzie (1964) and Ball (1967) show this feature in the coastal dunes of Bermuda and the Bahamas. Bigarella et al. (1969) report this feature from coastal dunes of Brazil, especially in the deposition in parabolic dunes. Small-scale deformation fea-

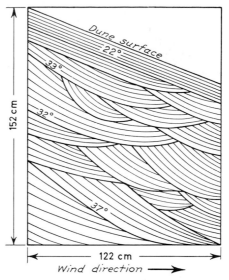

Fig. 483. Internal structure of coastal dunes, Mustang Island, Texas. (Modified after McKee 1957b)

tures in the lee-face sediments are rather common. Bigarella et al. (1969) in their detailed description of coastal dunes of Brazil give several examples of small-scale penecontemporaneous deformation structures in the sand dune sediments. In the interdune parts and hollows enough water is usually present to support vegetation, which upon decay leaves behind peat layers and root horizons (Fig. 484). A study of Shuyskiy (1975) reveals the quantitative parameters of aeolian transport of sand.

Dune sand is usually very well sorted; sand is mostly fine- to medium-size. Dune sediments are essentially beach sediments which are piled up by wind. The wind activity on beach sediment produces the following effects: Median size is increased, sorting is improved, skewness changes from negative or zero to positive values, relative enrichment of heavy minerals and improvement of roundness of quartz grains. All these changes are only relative, and they may not be always well-marked.

Bigarella et al. (1969) investigated textural characteristics of coastal dunes, sand ridges, and beach sediments and conclude that in areas where beach sand is composed of fine-grained sand, there is no significant difference between the degree of sorting and mean grain size of the dune and beach deposits.

According to Moiola and Weiser (1968) a combination of mean diameter vs. standard deviation is most effective in differentiating between beach and coastal dune sands. There are numerous studies on the textural characteristics of beach-dune sediments; however, results of such studies are applicable to specific regions only. Shideler (1973)

Fig. 482. Precipitation dunes. Near a forest, wind looses its sediment-carrying power and sand is deposited to build precipitation dunes. The foremost row of trees is slowly covered by windblown sand. Coastal region of Oregon

Fig. 484. Root horizon of Palametto in the sand dunes, Gulf of Mexico

water-laid beach sediments and wind-laid dune sediments in an interlayered dune beach sequence, for example, Beal and Shepard (1956) (Fig. 486). Eisma (1965) states, that at low wind velocities (5–9 m/sec at 1 m height) actually dune sands are formed that are more angular than the nearby beach sands. But storm winds are leading to the formation of dune sands that are rounder than the corresponding beach sands.

Coastal dunes are not always composed of quartz sand. In warm climate, coastal dunes may often be composed almost exclusively of carbonate grains (also oolites), and contain tests of typical marine organisms. In more evaporatic environment coastal dunes may be made up of gypsum sand, derived from adjoining sebkha and lagoonal areas. However, bedding, grain size characteristics of carbonate and gypsum dunes are similar to those of quartz sand dunes.

concludes that size characteristics of the source material are the most important factors influencing the grain-size differentiation of the beach-dune sands. This factor is far more significant than the effect of aeolian and hydraulic processes.

The criteria of better roundness of quartz grains of dune sand has been repeatedly discussed (e. g., Russel 1939; Pettijohn 1957). A detailed study by Beal and Shepard (1956), Shepard and Young (1961), Kuenen (1964b) confirm this statement. Sand grains of coastal dunes show a better roundness than the sand grains from the beach (Fig. 485). This difference is well-marked if prevailing winds flow onshore at right angles to the coast (Shepard and Young 1961). Moreover, dune sand is more enriched in silt-size material than the associated beach sediments, however, in dune sand broken shells, and tests of smaller organisms, e. g. foraminifera may be abundant. On the basis of these indices it may be possible to differentiate between

Beach Ridges and Cheniers

A beach ridge is a continuous linear mound of rather coarser sediment near the high-water line. The sediments have been heaped up by waves during high waters beyond the mean high-water line, and storm tides. This beach ridge may involve the development of a berm. In front of a beach ridge a sandy beach is always present. It may occur as a single ridge or as a series of ridges running roughly parallel. Cheniers, on the other hand, are sandy linear ridges well above the high-water line, separated from the shore as a result of deposition of fine-grained sediment on its seaward side (Fig. 487). A fine-grained marshy sediment is always present in front of a chenier (see Russell 1968).

Beach ridges are made up of sand, gravel, and shell debris. They develop mainly during storms and exceptionally high waters (Köster 1955; Gierloff-Emden 1961; Psuty 1966). Gierloff-Emden (1959) reports beach ridges in El Salvador which

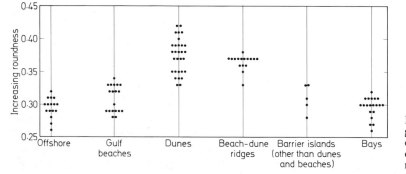

Fig. 485. Roundness values of the sand grains from the surface samples of the sandy coastal region of Texas. Out of all the geomorphic units, sand from dunes shows the best roundness. (After Beal and Shepard 1956)

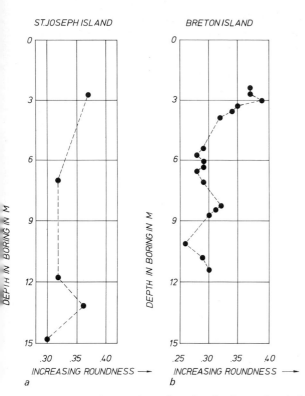

of a beach ridge, mainly horizontally laminated sand is present; it is separated from underlying sediment by an erosional unconformity. Then cross-bedded units follow, with laminae dipping at angles of 7 to 28° toward land.

Occasionally, one can recognize several phases of development in a beach ridge, which are usually separated by erosional discordances. In general, a beach ridge migrates toward land. During migration the beach ridge may cover the shallow runnel located behind the beach ridge toward the land. In such runnels during high water enough current is present to produce small-current ripples. These ripples are covered by the migrating beach ridge (Fig. 488). Both Werner (1963) and Thompson (1937) stress that the low-angled laminated sand

Fig. 486. Roundness values of sand grains in two boreholes. The values more than. 33 suggest dune sand. a St. Joseph Island, Central Texas Coast; b Breton Island, East Mississippi delta area. (After Beal and Shepard 1956)

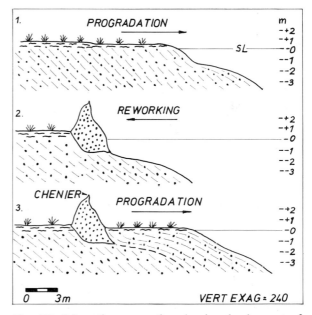

Fig. 487. Schematic cross-section showing development of chenier. *1* Mud flat progradation; *2* erosion and reworking of mud flat deposits and development of a ridge parallel to the shore line; *3* the next stage of mud flat progradation, where the ridge becomes chenier. (Modified after Hoyt 1969)

have originated by continuous slow deposition during spring-tide high waters. Beach ridges, which are produced mainly during storms, are composed of the coarsest material available on the beach. It may be pebbles, e.g., in the Baltic Sea (Köster 1955) or it may be shells, e.g., in Essex, Great Britain (Greensmith and Tucker 1968) and the Gulf of California, U.S.A. (Thompson 1968). The internal structure of sandy beach ridges has been described in detail by Werner (1963) and Psuty (1966). According to Psuty (1966), near the base

Fig. 488. A beach ridge (with bedding dipping toward the right) migrating over a beach runnel. The rippled bottom of the runnel is still visible. Near the base are gentle seaward-dipping sets of lamination. Sapelo Island, Georgia, U.S.A.

shows a bipolar dip direction and a consistent strike direction (dipping toward land and sea). Moreover, the high-angled cross-bedded units dip consistently toward land (Werner 1963). On a progradational shore several beach ridges may be formed one after the other (Figs. 489 and 490). This produces a characteristic beach ridge morphology (Köster 1955; Gierloff-Emden 1959; Psuty 1966). Beach ridges may be several meters in height, tens of meters in width, and hundreds of meters to kilometers in length.

Beach ridges can also develop behind the prograded midtidal mangrove flats. Jennings and Coventry (1973) describe internal structure of such beach ridges, developed through several phases of deposition. The beach ridges are made up of three types of bedding units, i.e., units made up of landward-dipping laminations, almost horizontal topslope laminations, and gentle, seaward-dipping laminations which occur in a ratio of 4:8:1. Obviously, the mangrove vegetation does not protect the beach ridges from wave activity of storm periods with raised water levels, and even the mangrove vegetation itself does not receive any severe damage.

The word "chenier" means covered by oak; it originates from the French word le chêne (the oak). Price (1955) used this term for the coastal plain of southwestern Louisiana. Cheniers are comparable to beach ridges in dimensions: 150 to 200 m broad, 3 m high, and up to 50 km long. However, cheniers are located in marshy regions. They owe their origin to a variable sediment supply: When the nearby rivers bringing fine-grained sediments reduce their sediment supply, waves rework the existing sediment and a beach with a sandy ridge is formed. As soon as the supply of fine-grained sediment in-

creases, a muddy marshy zone is build up in front of the sandy ridge. As this process is repeated, a series of sandy ridges with marshy sediments in front are produced. Later, sandy cheniers usually sink down partly into the muddy sediments, thus producing biconvex forms in cross-section. Another characteristic of the chenier is low-angled

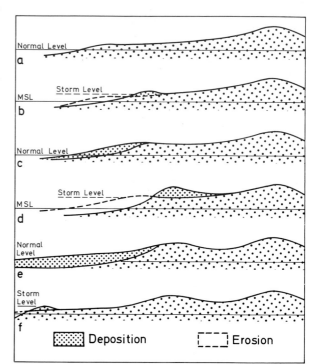

Fig. 489. Intermittent progradation and retrogradation causing development and abandonment of beach ridge. (Modified after Psuty 1966)

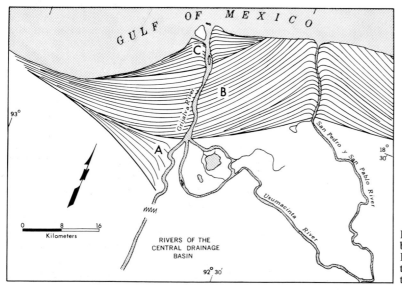

Fig. 490. Diagrammatic illustration of the beach-ridge system along the lower Grijalva River channel. Ridge lineations indicate the trends and not the specific ridges. (After Psuty 1966)

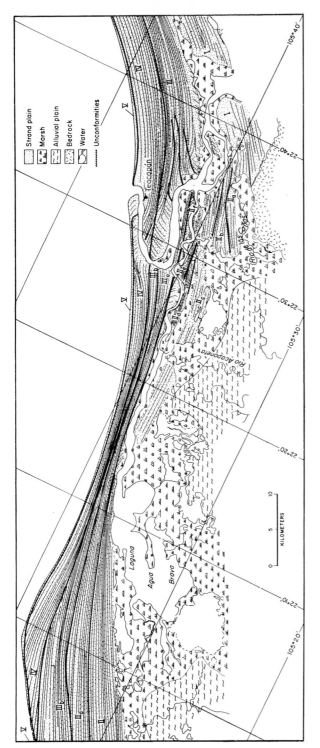

Fig. 491. Diagrammatic representation of the physiographic features of the strand plain with beach ridges, Nayarit. (After Curray et al. 1969)

branching (cf. Fisk 1955). The seaward side of a chenier is regular, straight, and steep; the landward side is rather irregular. Locally, washover fans are present across the marsh and chenier. Cheniers are made up of sand and shell debris (see p. 359).

Beach Ridges of the Coast of Nayarit, Mexico

On the western coast of Mexico, north of Rio Grande de Santiago, is a progradational coast about 225 km long. In the coastal plain some 250 almost parallel running beach ridges are present (Fig. 491). Curray et al. (1969) have investigated these coastal sediments of Nayarit. On an average the zone of beach ridges is 5 km wide, and in the main part–130 km long–, the coast has prograded some 10 km. The maximum distance of progradation is 15 km. Individual beach ridges are spaced 30 to 200 m from each other. The height of beach ridges varies from 1 to 2 m. In contrast to the observations of Psuty (1965, 1966) on the beach ridges of the Tabasco coast, Gulf of Mexico, Curray et al. (1969) believe that the beach ridges on the Nayarit coast started building up on shoreface by moderate wave action. These longshore bars slowly migrated upland from the shoreface, ultimately reaching the shore line. As more and more sand is made available from the shelf and due to progradation, new longshore bars develop on the shoreface and migrate toward the shore line. In the later phases of development of beach ridges wind plays an important role, so that Curray et al. (1969) call them beach-dune ridges.

Beal and Shepard (1956) also suggested contribution of wind activity in the development of beach ridges, as they found that sand grains of beach ridges are as equally well rounded as the dune sand.

The development of the Nayarit coast started with the post-Pleistocene transgression. The rise in sea level was very rapid until 7,000 years B.P. Then the rate of transgression slowed down and the sandy shore started developing about 4,750 to 3,600 years B.P. Although even after this period there was a rise in the sea level, the coast had already started prograding. Besides the lateral sequence of facies, there is also a beginning in development of vertical sequences which can be compared with ancient sediments (Fig. 492).

The beach-ridges-coastal plain of Nayarit shows the following lateral facies sequence (Fig. 492): Older sediments (pre-transgression) are represented by fluvial coastal plain deposits, on top of which are deposited marsh and lagoonal sediments. This zone is followed laterally by a broad zone of beach ridges. Lagoonal and marshy sediments develop within the troughs of beach ridges. The sandy zone continues into the shoreface region, up to a water depth of 7 m. Here coastal sand grades into the transition zone. The zone of shelf mud starts somewhere between 10 and 13 m water depth. This lateral sequence corresponds well with the coastal sand–shelf mud sequence of other regions (e. g., see p. 386). The following description of individual units is based on Curray et al. (1969).

The *river alluvium* represents flood plain deposits of a number of small rivers. The sediment is mainly clayey silt, with lesser amounts, of silty sand. Locally gravels, and coarser sand are present, which represent the channel-lag deposits. It includes fluvial deposits formed both before and after the transgression.

The *lagoon and marsh* deposits are characterized by alternating layers of sand and mud, together with peat layers. The present-day lagoons are bordered by mangroves, and show a paucity of sandy sediments. The lagoons formed during the transgressive phase were perhaps more restrictive, and their deposits include many interbedded sandy layers, which are deposits of washover fans or partly blow-over sand. A typical mollusc of these lagoonal deposits is *Crassostrea columiensis,* whose shells are found locally concentrated into layers and pockets.

The *coastal sand* deposits are made up of well-sorted, fine- to medium-grained sand. Content of shells is variable. Locally, thick layers of shells, broken shell debris, together with wood, peat and pumice are present. The typical molluscs of the coastal sand deposits are *Tivela* sp. *Donax carinatus, D. contusus, D. punctatostriatus,* and *Mulinia pallida.*

The *transition zone* deposits are silty sand. Curray et al. (1969) consider this to be shelf mud re-

worked during progradation of the coastal sand. He shows this zone only in one section (Fig. 492).

The *shelf mud* deposits are made up of clayey silt and silty clays. Typical fauna of the shelf mud include *Chione guidia, Andora obesa, Laevicardium elenense,* and *Nassarius* cf. *N. versicolor.* In its outer parts shelf mud grades into sandy shelf sediments.

A vertical sequence produced as a result of transgression and later progradation, is shown in Fig. 492. The older alluvium is followed by a sheet sand layer several-meter-thick, on the top of which is Recent coastal sand. The Recent part of the coastal sand is prograding on the shelf mud. On the landward side coastal sand is covered by lagoonal and fluvial deposits. This complete sequence presents a well-developed model, which can be compared with comparable ancient deposits.

Another good example of beach-ridge landscape is described by Thom et al. (1978) from New South Wales, Australia.

Chenier Plain of Southwestern Louisiana

A well-developed chenier plain is located west of the Mississippi delta in southwestern Louisiana, Gulf of Mexico (Fig. 493). It represents a progradational marshy region with numerous cheniers–the sandy beach bars (see p. 353). The cheniers are up to 3 m in height; but as they are partly sunken into marshy sediments, they may be up to 4.5 m thick (Fig. 487). They are about 180 m in width, and up to 50 km long.

The chenier plain of southwestern Louisiana represents a sediment wedge made up of various sediment deposits built up during the Holocene period (Russell and Howe 1935; Fisk 1955; Byrne et

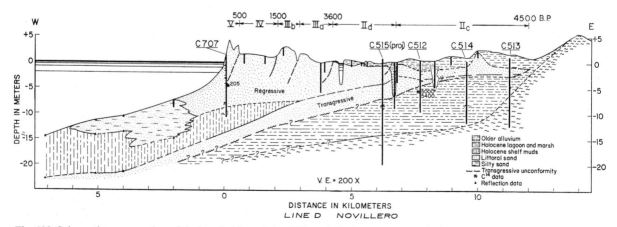

Fig. 492. Schematic cross-section of the beach ridges plain of Nayarit. It shows a retrogradational and a progradational sequence. (After Curray et al. 1969)

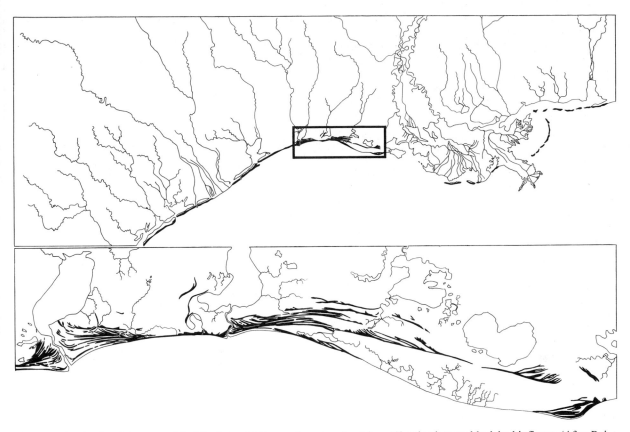

Fig. 493. The chenier plain, west of the Mississippi delta, southwestern Louisiana. The cheniers are black in this figure. (After Reineck 1970b; originally modified after Gould and McFarlan 1959)

al. 1959; and Gould and McFarlan 1959). At the base of this sediment wedge are transgressive, brackish to marine deposits on the top of Pleistocene sediments. These transgressive deposits were laid down 5,600 to 3,000 years B.P. Since then, large amounts of sediments have been deposited. This sediment has been derived mainly from the Mississippi delta by coastal currents parallel to the coast. As a result of high rates of deposition, the shore has prograded about 15 km toward the sea (Fig. 494). During progradation the shelf mud has been overlaid by coastal muddy and marsh deposits, deposited during periods of abundant sediment supply. During periods of limited sediment supply under wave activity, beach bars–chenier– are deposited. The age of cheniers ranges from 2,800 to 300 years B.P. During periods when reduced sediment supply caused the formation of cheniers, transition zone deposits (between chenier and shelf mud) were formed in deeper water. Byrne et al. (1959) call them gulf bottom sand and silty clay deposits. Such a transition zone is developed in front of modern sandy beaches. However, during periods of active progradation the zone of shelf mud extended up to the shoreline, followed

by densely vegetated marsh deposits. In the development of chenier plains two more sedimentation units are associated. They are bay-bottom and mud-flat clayey silt, and bay-mouth silt. The former represents fine-grained deposits laid down in embayments and bays, produced as the result of growth of progradational cheniers at the end of the transgression. Chenier served as barriers behind which embayments were produced. They are brackish water deposits. The bay-mouth silt represents a gradational facies between open-shelf mud deposits and protected embayment deposits.

Deposits of various sedimentation units show a characteristic assemblage of grain size, sedimentary structures, macrofauna, and foraminifera (Byrne et al. 1959). In the following a short account of these units is given.

Basal transgressive deposits rest on top of Prairie formations of Pleistocene age. They are mainly clays and silty clays with high organic matter content–mainly plant debris. Locally, peat layers and thin sand layers are present. The bioturbation is intense, so that mottled structures are common. The mineralogical composition indicates that they have been derived mainly from the reworked un-

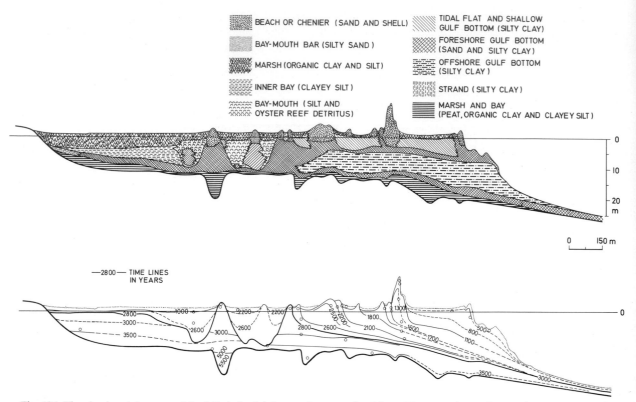

BEACH OR CHENIER (SAND AND SHELL)
TIDAL FLAT AND SHALLOW GULF BOTTOM (SILTY CLAY)
BAY-MOUTH BAR (SILTY SAND)
FORESHORE GULF BOTTOM (SAND AND SILTY CLAY)
MARSH (ORGANIC CLAY AND SILT)
OFFSHORE GULF BOTTOM (SILTY CLAY)
INNER BAY (CLAYEY SILT)
STRAND (SILTY CLAY)
BAY-MOUTH (SILT AND OYSTER REEF DETRITUS)
MARSH AND BAY (PEAT, ORGANIC CLAY AND CLAYEY SILT)

Fig. 494. The chenier plain, west of the Mississippi delta, southwestern Louisiana. Transgressive and regressive sequence. (After Reineck 1970b; originally modified after Byrne et al. 1959; and Gould and McFarlan 1959)

derlying Prairie formation. The fauna of basal transgressive deposits suggest a changing salinity condition, resulting from alternate flooding by marine, brackish, and fresh water and increased evaporation to raise the salinity above the normal marine range. Parts were laid down in shallow ponds and lakes and restricted bays, which developed at the top of Prairie formations as the sea began to transgress into the area. Near the base of the basal transgressive deposits is a discontinuous sand layer about 30 cm thick. It is interpreted as a beach sand deposit of a transgressive sea, one of the few examples of transgressive sand layers from the sub-Recent (see p. 420). The sand is well sorted and fine. The composition of silty clay is 48.5 % clay, 40.5 % silt, 8.3 % sand, 1.3 % wood and 1 to 4 % shell.

Transition zone deposits (gulf-bottom sand and silty clay) show interbedding of sand and silty clay, abundant stuffed burrows, and high shell content. The sand layers are made up of well-sorted very fine sand. Average composition of the sediment is 55.5 % silt and clay, 40.8 % sand, 1.5 % shell, 0.6 % forams, 0.9 % wood, 0.2 % limonite and 0.5 % mica. The transition zone deposits are present as transgressive deposits on top of the basal transgressive deposits in the lower part of the section, while

in the upper part of the section they are present in the progradational sequence. In the progradational sequence they underlie and extend seaward of each of the major cheniers.

Shelf mud deposits (gulf-bottom silty clay) is made up of 70 % clay, 26 % silt and only traces of sand. The median diameter of the sediments is less than 2 μ. Shell content and organic matter content are low. The bedding is massive to very finely laminated. At places, where shelf mud grades directly into marsh deposits, a transitional unit is developed. This unit is characterized by abundant dark organic flecks, plant roots and irregular patches of green montmorillonite. It represents tidal mud flat deposits.

Bay bottom and *mud-flat clayey silt deposits* are rather similar in structure to the tidal mud-flat zone of shelf mud deposits. Root fibers are abundant, together with irregular patches of green montmorillonite. However, the grain size is somewhat coarser than the tidal mud-flat deposits. Average composition of sediment is 37.3 % clay, 53.8 % silt, 4.9 % sand, 2.6 % wood, 0.7 % limonite and 0.7 % mica. In contrast to shelf mud, authigenic pyrite is abundant. The sediment is almost barren of any fauna; only a few foraminifera, and some shells are present. Basal beds of this unit were deposited as tidal mud-flat deposits, whereas upper beds accumulated in restricted embayments under brakish water and partly stagnent conditions.

Bay-mouth silt deposits represent deposits transitional with the bay- and gulf-bottom deposits, both laterally and vertically. Average composition is 75 % silt, 10 % clay, 9 % sand, 3 % wood, and 3 % mica. They are characterized by excellent sorting, fine grain size and local abundance of oyster shell debris.

Marsh deposits are represented by organic clay and silt sediments. They cover most of the surface of the chenier plain and are accumulating today in low areas between cheniers. They are characterized by a brackish water assemblage of foraminifera. The deposits are made up of dark-gray-colored highly organic clay and silt, capped by a several-centimeter-thick layer of plant debris – vegetal muck. Roots are abundant; authigenic pyrite is common, suggesting deposition under reducing conditions. Average composition of the sediment is 48.3 % clay, 42.2 % silt, 6.7 % wood, and 2.9 % sand.

Chenier deposits consist mainly of sand and shells, with only minor amounts of silt and clay. Average composition of the sediment is 71.1 % sand, 22.7 % shell, 5.2 % silt and clay, 0.9 % foraminifera, and 0.1 % wood. The shells occur as distinct layers several centimeters thick and also as fragments disseminated in the sand. On the seaward side of a typical chenier near the base, sediments are fine-grained, grading upward into coarser sediments. This coarsening is mainly because of increasing content of shells in the front slope. The sediments of the landward side of a chenier are finer-grained and poor in shell content. Shells are deposited mainly on the seaward side and at the crest of a chenier. Where waves rework the sediments shells are concentrated as lag deposits. Where waves wash over the crests only finer sediments are carried and deposited on the backside of the chenier.

Chenier deposits are characteristically associated with transition zone deposits and rest on top of them; they are overlapped by marsh deposits on the landward side. On the seaward side chenier deposits abut sharply against shelf mud and marsh deposits laid down during the later stages of coastal progradation.

Washover Fans

During heavy storms much sediment can be eroded from the coastal sand and distributed further inland in form of washover fans. For the development of washover fans or genetically related tidal inlets, a number of parameters are to be satisfied (Pierce 1970). If a barrier island is attacked from the seaward side with increased water level and heavy sea, washover fans are made, in the case of extensive tidal flats being present behind the barrier. Small inlets may develop if no tidal flats are present behind the barrier. Storm surges from the lagoonal side, channeled along tidal creeks can easily cut inlets through a barrier.

The most important washover fans today are generated by hurricanes, which are tropical storms with cyclone-like wind circulation. Wind velocities of 12 Bft (> 118 km/hr) and more are commonly achieved. Intensive hurricanes occur every 2.5 years, on average; thus they are important geologically.

During hurricanes, sea level can be elevated up to 4 m for a period of 2.5 to 5 hrs as a result of wind pressure and barometric low pressures. There is a heavy sea. This phenomena often takes place on the Atlantic coast of the southern United States, and especially, on the Gulf coast of Texas. During such a hurricane rather coarse-grained material, e. g., macro-invertebrates, coral fragments, and rock fragments are transported from water depths of 25 m or more on to the beach and beyond the chain of coastal sand dunes. At places, coastal dunes are broken through and a washover channel is formed (Fig. 495). The sand eroded from the beach and the coastal dunes are transported and deposited in the form of a washover fan on the backside (landward side) of the barrier island (Fig. 496) on the salt marsh and/or shallow lagoons (see Price 1947; Hayes 1964; Hoyt et al. 1966; Pierce 1970; Andrews 1970).

The washover channel branches off in several distributaries arranged in the shape of a fan. In plan view a washover fan looks like a part of a cir-

Fig. 495. Washover channel with megaripples breaking through coastal sand dunes. Cape Henlopea, Delaware, U.S.A.

Fig. 496. Washover fan on top of a salt marsh with a washover channel. (Photograph by J. D. Howard)

cle with strongly segmented borders. In a cross-section it is wedge-shaped.

Andrews (1970) measured the thickness of the sediments across a washover fan on St. Joseph Island, on the central Texas coast. This washover fan extends over an area of 7 km²; the thickness in the central part of the fan is 125 cm, and near the margins, about 75 cm. The washover fan is made up of several superimposed sandy blankets of successive washover fans, each of which can be ascribed to a hurricane. Each sandy blanket begins with a shell-rich layer, which possesses an erosive contact to the lower sediments. In the shelly horizon shells of macro-invertebrates from different biocoenoses are mixed together. Then a sandy layer follows with well-developed evenly laminated sand.

The bedded units are made up of individual sedimentation units, which are partly graded. A throat core is made up of inversely graded units (Leatherman et al. 1977). The sedimentation units begin with a fine-grained heavy-mineral enriched zone, related to erosional concentration, and also characterized by negative skewness. Cores recovered from the fan area are mostly made up of normally graded beds, related to the overwash surges and denote primarily depositional nature of the sediment.

In the washover channel and its distributaries, as well as on some parts of the fan surface, packets showing extensive small-ripple bedding and antidune crossbedding are common. In the depressions (small ponds) within the washover fan muddy layers may be deposited. Bubble sand structure is locally common. Bioturbate structures including root mottles and layers of pelletoidal sand can be locally common (Frey and Mayou 1971). Among the lebensspuren small cylindrical insect-burrows and isolated large inclined Y- or U-shaped burrows of *Ocypode quadrata* and *Uca pugilator* are found, e.g., in the washover fans of the Georgia coast, U.S.A. (Frey and Howard 1969). Hence Deery and Howard (1977) distinguish an active phase, when waves break through the dunes, and sediment and water flow landwards into the salt marshes. This is followed by a passive phase, when aeolian and biogenic processes set in. It is quite likely that in a sequence of prograding coast, within the sediments of salt marsh and lagoons, sandy blankets of washover fans are incorporated. Such sandy blankets are overlaid by salt marsh and lagoonal deposits during further progradation of the coast. Davies at al. (1971) describe ancient washovers from a lagoonal fill sequence of Lower Cretaceous Muddy Sandstone.

Backshore

The backshore represents the upper part of a beach which remains normally dry, except under unusually high water conditions, when it can be flooded and acted upon by waves and weak currents. However, most of the time it is exposed to wind activity. The dominant bedding structures of the backshore are almost horizontally laminated sand (cf. Russell and McIntire 1965; Andrew and van der Lingen 1969). The monotony of laminated sand is broken by the presence of small-scale cross-bedding produced by small-current ripples. Shallow morphological depressions, when filled up, may give rise to low-angled cross bedded units. Deflation surfaces showing a concentration of shells are common. Wind removes the finer sediments, leaving behind larger shells as deflation deposits. The shells are oriented convex side up. During stronger winds a blanket of blown sand or small sand dunes or sand drifts may accumulate parallel to the coast. Wind ripples are commonly produced, though they have a very low potential of preservation. Thus upward growth of the backshore takes place during unusually high waters and storms. The backshore is flooded by rather shallow water in which surges are active, which causes destruction of most of the wind-formed features. Along the margins of small pools and puddles, created from water seepage, blown sand produces antiripplets (van Straaten 1953; Reineck 1955d). Strong wind may pucker the water of puddles and move it in the form of a sandwater slurry, which is extended in form of streaks (Fig. 497) (Gripp 1963). The horizontal lamination in the sand is produced as a result of the action of surges; however, some are horizontally laminated wind-blown sand blankets.

In shallow depressions a thin veneer of mud and algal mat may be deposited. Sudden and quick flooding of the backshore leads to the development of bubble sand (Fig. 94). Locally, the escape of air produces spring pits, sand domes, etc. At places, rill marks may develop. After the flooding, salt crusts develop on the surface as a result of evaporation. The wind action may erode loose sand below the crust, producing cavities and "sand pebbles". This also happens if water-saturated sand is present on top of a dry-sediment layer. Insect-burrows and the burrowing of small rodents are common, but not usually preserved (Reineck 1968a). Amphipods and isopods produce abundant burrows. A sandy horizon burrowed by amphipods shows a characteristic bioturbation structure (Frey and Howard 1969).

The above-mentioned characteristics of the backshore are present mostly on the broad beaches. On the contrary, narrow backshores

Fig. 497. Wind-blown sand saturated with water moved by wind in the form of a sand-water slurry. The rough surface shows adhesion ripples. (After Gripp 1963)

made up of fine sand mostly show strong development of sedimentary structures related to the ridge-runnel system with many erosional discordances (Figs. 498 and 499). In the backshore sediments the bedding structures related to moving water (Fig. 508) dominate over the aeolian structures, because during high-water stages the backshore is flooded by water, planed-off and bedding structures related to flowing water are produced; and later only the upper surface of the backshore is reworked by wind action.

The relative sorting by water during storms produces heavy mineral placers in the backshore region (Trusheim 1935b; Inman and Filloux 1960; Hayes and Boothroyd 1969). In many regions such placer deposits are quarried (Andrée 1920; Mero

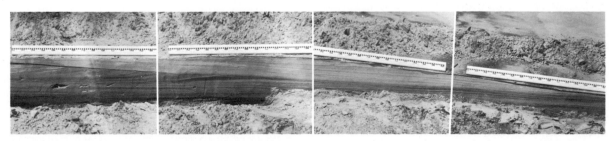

Fig. 498. Low-angle bar bedding with erosional contact to high-angle bar bedding on the left side. Sea towards the right side. Backshore, Sapelo Island, Georgia, U.S.A. (Archiv no. 2328–31)

Fig. 499. A runnel in the backshore region filled up by wind and water action showing various types of structure. (Archiv no. 2325)

Towards sea ⟶

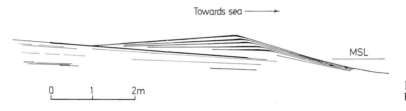

MSL

0   1   2m

Fig. 500. Internal structure of the lower berm. (After Panin 1967)

1965; Gadow and Reineck 1968; Weyl 1969). The primary enrichment of heavy minerals takes place near the upper limit of the backshore toward land, where only storm waves are active. This is usually the case near the base of the sand dunes. Wind action plays also an important role in heavy mineral concentration (Stapor 1973; Woolsey et al. 1975).

Heavy storms usually modify the morphology of the beach. Both backshore and foreshore are planed off and slightly concave profiles develop (Hayes and Boothroyd 1969). During the storm or the receding phase of the storm, when the water level is yet rather high, a beach ridge is generated, which possesses a steeper landward slope made up mainly of steeply dipping laminae. During the later development of the beach ridge, rather low-angled laminae are laid down, dipping toward both land and sea. Behind the beach ridge a shallow runnel is usually present in which water flows during high water conditions, generating small-current ripples. During the migration of the beach bar toward land, ripples of the runnel may be covered and preserved. In a strong progradational coast a series of beach ridges may develop (see the chapter beach ridges and cheniers).

Panin (1967) discusses in detail the effect of destructive and constructive waves, and wind activity on the backshore of a beach in the Black Sea. The upper beach ridge is complexly built by the highest-reaching waves. As soon as these waves are

short, erosion begins on the beach, producing a cliff. During slackening of waves and retreat of water level, a lower beach ridge is produced with a concave seaward slope, dipping 5 to 18°. The beds below the crests dip toward land at angles of 1 to 9° (Fig. 500). Sandwiched within the beach layers are aeolian layers and ripple-bedded units (Fig. 501) (Wunderlich 1972).

In the zone of backshore–foreshore transition, and sometimes in the foreshore itself, beach cusps develop (see Johnson 1919; Gellert 1937; Thompson 1937; Kuenen 1948; Guilcher 1949; Mii 1958; Russell and McIntire 1965; Cloud 1966). Beach cusps are cuspate deposits extending as tongues or horns of coarser sediment toward the sea at equal distances, separated by shallow bow-shaped de-

Direction of wind towards sea

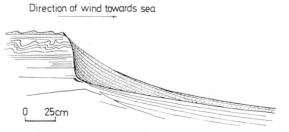

0   25cm

Fig. 501. Wind-blown sand accumulating at the base of a small cliff on the beach. The wind is blowing toward the sea. (After Panin 1967)

pressions of finer-grained material. They develop abundantly on shores of medium wave-energy without a strong longshore current, when waves approach the shore line at almost right angles. Often beach cusps develop abundantly during retreating storms. The spacing of the horns is related to wave energy at the time of their origin, especially to the wave height. Cloud (1966) thinks that a breaking wave approaching a shore line can be considered a cylindrical wave form which is broken into regularly spaced segments which correspond to the spacing of resulting beach cusps. Intersecting wave trains have been mentioned as one possible mechanism for cusp formation (Branner 1900; Dalrymple and Lanan 1976).

Mii (1958) gives an account of various variants of beach cusps (Fig. 502). Their internal structure is not very well known; however, the base of beach cusps shows an erosional contact with underlying sediments (Thompson 1937; Mii 1958). In the depressions of the cusps trough-shaped laminae are laid down between the horns (Fig. 503). Mii (1958) pointed out that the deposits of embayments with their erosional contact to underlying sediments and gravelly horns may become preserved in sediments. He states further that such structures have been described by Fairchild (1901) from Medina Sandstone of New York State. Beach cusps situat-

ed on the backshore have better chances of being preserved than those of the foreshore.

Russell (1968) pointed out that the composition of beach cusp sediment is coarser than the surrounding beach which they overlie. This suggests that beach cusps are depositional features made by the accumulation of sediment brought in by waves from the shore face. Otvos (1964a) studied the internal structure and demonstrates that the horns can be formed by truncation as well as by accumulation. Williams (1973) and Komar (1973) provide some additional information on the beach cusps.

Foreshore

The foreshore of high-energy and of low-energy beaches during gales is more or less flat, without any morphological features. The main bedding type is 1-to-15-cm thick bedsets of evenly laminated sand with low-angled discordances (Fig. 504) (Thompson 1937; McKee 1957b). The individual laminae can be followed up to 30 m parallel to the shore, and up to 10 m normal to the shore line. Sand laminae have been deposited from suspension clouds, brought by incoming waves. In the

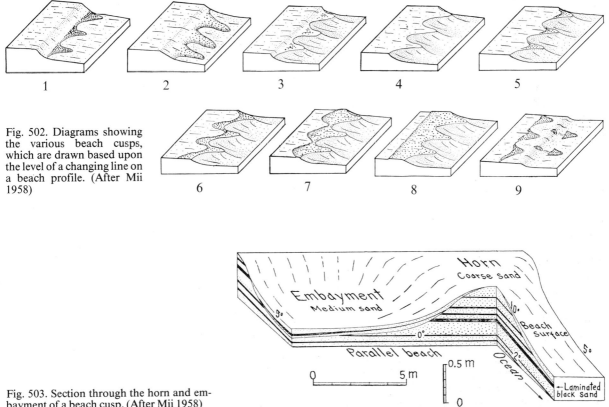

Fig. 502. Diagrams showing the various beach cusps, which are drawn based upon the level of a changing line on a beach profile. (After Mii 1958)

Fig. 503. Section through the horn and embayment of a beach cusp. (After Mii 1958)

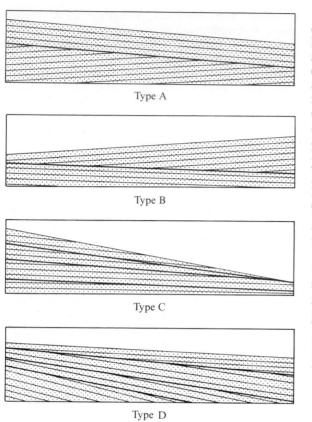

Type A

Type B

Type C

Type D

Fig. 504. Schematic illustration showing four major types of cross-bedding in beach deposits. The sea is toward the right. (Modified after Thompson 1937)

Quite often the foreshore shows development of one or many longshore bars. On beaches where longshore bars have been destroyed by storms they are regenerated within 4 tidal cycles, i.e., 48 hrs (Reineck 1963a). Firstly, after a storm when waves are still strong, rather coarse sediment is brought in. In the following days with decreasing wave energy longshore bars are built up by finer sand and at the same time they migrate toward land. Hayes and Boothroyd (1969) found that regeneration of longshore bars takes from 2 days to 6 weeks. Each time waves get stronger, coarser-grained sediment is deposited. Vollbrecht (1957) describes the morphology, development, and destruction of sand bars in the shoreface of the Baltic Sea. Greenwood and Davidson-Arnott (1979) investigated the sediment movement in wave-formed bars.

In cross-section a longshore bar is asymmetrical in shape. From the apex of the ridge, the seaward slope is gentle. Corresponding to the seaward gentle slope are also the low-angled laminated sand bedsets (4 to 6°) (McKee and Sterrett 1961). The wide seaward gentle-dipping part of longshore bars are made almost exclusively of low-angled laminated sand (Fig. 505). However, sometimes small-ripple bedding and antidune bedding may be present (Fig. 58). The landward slope of a longshore bar is composed mainly of planar cross-

swash zone swash and backwash produces a characteristic sorting in individual laminae (Clifton 1969) (see p. 119). The angle of dip of individual bedsets is strongly controlled by grain size (Russell and McIntire 1965): The coarser the sediment, the higher the angle of slope of the beach; thus also the higher the angle of dip of bedsets of laminated sand.

Fig. 505. A vertical section across a beach ridge showing evenly laminated sand, and intercalated layers of coarser material, Sapelo Island, Georgia, U.S.A.

Fig. 506. Beach runnel with megaripples. In the background is a beach ridge (longshore bar). The view is toward the sea. Norderney Island, North Sea

bedded units (= high angle bar bedding, Fig. 174) with dips of 10 to 30° (Thompson 1937; McKee and Sterrett 1961; Hoyt 1962; Bigarella 1965; Psuty 1966). In the longshore bars, together with planar cross beds, festoon-shaped cross bedding of megaripple origin may also be present (Reineck 1963a). Megaripples on longshore bars are oriented at an angle to the axis of the bar and also at an angle to the shore line. Toward the landward side of a longshore bar, a channel (trough) is present in which megaripples and/or small-current ripples may originate, oriented in the direction of flowing water more or less at right angles to the shore line (Fig. 506). Moreover, symmetrical and asymmetrical wave ripples are abundantly developed on the foreshore (Fig. 507). The orientation of wave ripples may be parallel or at right angles to the axis of runnel. Wave ripples may be preserved if they are covered by migrating longshore bars. The thickness of ripple-bedded (both mega and small) channel sediments is rather small The sand which is brought from the longshore bar into the adjacent runnel may be formed into climbing wave ripples (Fig. 508).

Two common bedforms developed on the emerged surface of the foreshore are antidunes (Fig. 58) and rhomboid ripples. Rhomboid ripples are present on both landward-dipping and seaward-dipping surfaces. On the landward slopes mega rhomboid ripples are common (Gierloff-Emden 1959; Hoyt and Henry 1963; Reineck 1963a; Otvos 1964b).

Other characteristic features of very shallow water, and features associated with falling water level and intermittent subaerial emergence of a sedimentation surface are commonly found: Swash marks, adhesion ripples, various foam marks, rill marks, flat-topped wave ripples, wave ripples with rounded crests and pointed troughs, modified wave ripples and current ripples during the last phase of emergence by flowing water. Crests of megaripples are cut through by small channels. Wave ripples are flattened and extended forward-

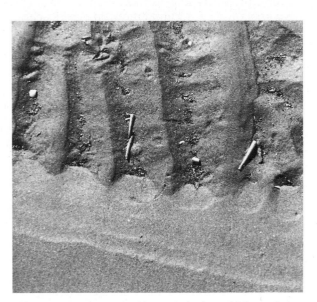

Fig. 507. Beach runnel with wave ripples, which are being covered by a landward-moving ridge. See also Fig. 491. Norderney Island, North Sea

Fig. 508. A unit of climbing wave ripple lamination in a filled-up beach-runnel. Sapelo Island. Sea on the right hand side. (Archiv no. 2353)

into very flat current ripples (lingoid and rhomboid forms). The rise and the fall of tides cause the migration of energy zones, which result in characteristic sedimentary structures (Wunderlich 1972).

Because of the swash activity on the intertidal part of the beaches, large quantities of water move through the beach sand (Riedl 1971), and the redox-system of the sediments is strongly influenced up to a depth of 4 m. Wadell (1976) points out that the water table in the beach sediments fluctuates periodically with the swash action and helps periodic erosion and accumulation on the beaches.

Frey and Howard (1969) and Hill and Hunter (1976) describe characteristic bioturbation structures from various subunits of a beach, including the foreshore (Fig. 509).

According to Thompson (1937), sediments of the lower part of the foreshore are poorly sorted and contain much broken shell, mica, organic debris, etc., as compared to sediments of the upper part of the foreshore. In general, the foreshore is marked by the presence of shells, which become concentrated in form of broad swash marks–the position depending upon the water level. Sometimes during extraordinarily high water conditions shells are also concentrated in the backshore. They are usually oriented convex upward; only in dense accumulations are vertical-oriented shells abundant. Sanderson and Donovan (1974) describe vertically oriented shells. In swash marks of the foreshore various types of dead organisms are concentrated: Heart urchin (Häntzschel 1936b), jelly fishes (Schwarz 1932; Müller 1970), insects (Trusheim 1929a; Franz 1931); *Pectinaria* pipes (Reineck 1960c). Locally, plant debris may also become concentrated into swash marks.

Thompson (1937) suggested that in ancient sediments, if abundant shells are present in a sand body and moreover, if they belong to various biotopes of marine as well as freshwater regions, it

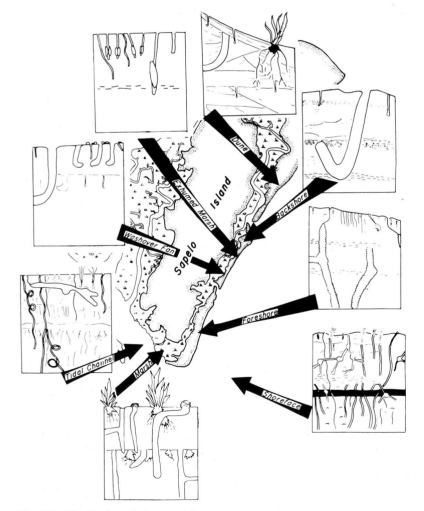

Fig. 509. Distribution of characteristic trace fossils in various geomorphic units of coastal sediments. Sapelo Island, Georgia, U.S.A. (After Frey and Howard 1969)

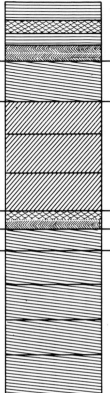

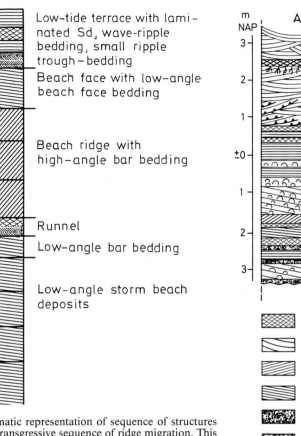

Fig. 510. Schematic representation of sequence of structures produced in a transgressive sequence of ridge migration. This sequence is similar both in tidal and non-tidal beaches. (Modified after Davis et al. 1972)

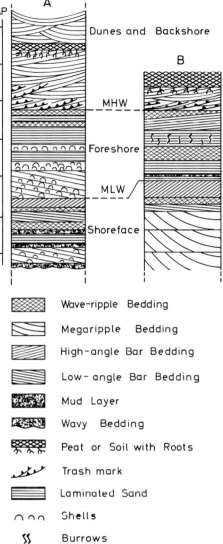

Fig. 511. Vertical succession in two sub-recent deposits of sand-dune-backshore-foreshore-shoreface sequence with inferred MHW and MLW-lines. (Modified after Roep et al. 1975)

suggests that the sand body may be a coastal sand body (see also Hertweck 1971).

Distribution of bedding features on the foreshore with ridge and runnel system is investigated by Davidson-Arnott and Greenwood (1976). According to them, within the seaward-dipping laminated sand of the ridge, landward-dipping ripples are incorporated. In the crestal part of the ridge, along with the laminated sand, cross-bedding of lunate megaripples is also present (Reineck 1963a; Davidson-Arnott and Greenwood 1976). Davis et al. (1972) compare the ridge and runnel systems of the tidal and non-tidal beaches. Except for the dimensions, there is no difference between the two regions. Figure 510 shows a schematic transgressive sequence of both types of beach.

Roep et al. (1975) studied sub-Recent regressive beach sequences in two pits in the Netherlands. The sediments are 2,300 years B.P. and 3,560 years B.P. old. For both the profiles, the following sequence is recognizable (Fig. 511): continental sediments with dunes, soil, and peat. The foreshore exhibits individual strongly bioturbated sediments with wave ripples and low-angle wave-built bars. The foreshore sediments represent planed-off or

almost planed-off winter beach. The mean high water line is recognizable by swash marks, highest burrow-level and lowest level of aeolian activity.

The swash mark is made up of a concentration of light material, e.g., drift-wood pieces, peat, shells, and landward steeply upward-curved ends.

Below the low-water line, high-angle slipface laminations are present. Although the mean low-water line is not very clear, in the expected position of the mean low-water line, concentration of shell is recorded.

Barwis (1976) describes two vertical sequences. One sequence (Fig. 512) is from the backside of Kiawah Island, South Carolina, where washover deposits extend over the marsh sediments. The sec-

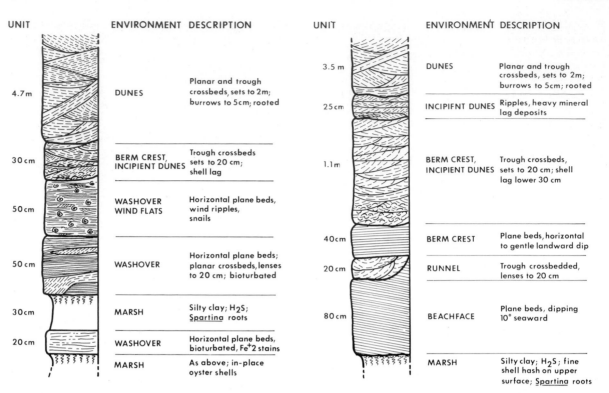

Fig. 512. Lithologic sequence in the landward edge of a Kiawah Island beach ridge. It is a salt marsh-washover-windflat-dune sequence. In this succession, the sequence salt marsh to windflat is transgressive, while the sequence windflat to dune is regressive. (After Barwis 1976)

Fig. 513. Lithologic sequence in the seaward edge of a Kiawah Island beach ridge. The boundary marsh-beachface is transgressive due to backward migration of the island. The rest of the sequence is regressive (= progradational). (After Barwis 1976)

ond profile is from the seaward side of the island (Fig. 513). Here also, the sequence begins with marsh deposits, an indication that the island has been shifted towards land. Thus, the boundary between marsh and shoreface deposits is a transgressive one.

Concentration of pebbles in the foreshore (see also p. 411) is described by Bluck (1967) and Hobday and Banks (1971), and in the swash zone by Novak (1972). Bluck (1967) describes the distribution pattern of flat and spherical pebbles of different dimensions (Fig. 514). Hobday and Banks (1971) show that the pebbles brought by a river show poor sorting, and they are sorted and oriented by marine processes. The fine-grained sediment is accumulated as spits, while the small pebbles are concentrated in the landward-shifting swash bars.

Shoreface

The shoreface is always submerged under water, and the surface is usually made up of submerged longshore bars with channels on their landward side. Commonly two, sometimes even more, longshore bars are present. The weaker the wave energy of a coast, the fewer in number are the longshore bars of the shoreface. On rather low-energy coasts only one longshore bar is developed near the low-water line.

The characteristics of submerged longshore bars and longshore troughs, and their relationships to waves, tides, and currents is discussed by Shepard (1950). The longshore troughs are the result of plunging breakers and the longshore currents, which are feeders to rip currents. During heavy storms at least the upper most longshore bar, together with longshore bars of the foreshore, are made level. As the storm recedes they are rebuilt.

Longshore bars develop in the breaker zone, where sediment from both off-shore and land is brought in (Ingle 1966). Adjacent to the primary breaker zone, a secondary breaker zone may develop toward land. This situation has been shown during high water in Fig. 470. At low water, only one breaker zone is shown in this schematic diagram. During strong waves a breaker zone develops on the top of the most deeply located longshore bar. This longshore bar is activated only by strong waves of a storm.

The channels of the shoreface also show development of small-current ripples, sometimes also megacurrent ripples. The crests are aligned at right angles to the long axis of the channel. In the uppermost channel and also on top of the uppermost longshore bar of the shoreface, during strong waves, wave ripples and undulatory megaripples are produced. The cross-bedding of the upper shoreface shows a maxima toward land; however, another maxima is present parallel to the shore line. There is much dispersion in the values.

Toward the deeper part of the shoreface, cross bedding is very rare. Here laminated sand is mostly the dominant bedding type (see Reineck and Singh 1971). At the same time the degree of bioturbation also increases. Lastly, the foreshore grades into finer-grained sediments of the transition zone. The lower shoreface, where normal waves do not reach and sediment is moved only during storms, is abundantly covered with symmetrical wave rip-

ples. However, these ripples are seldom preserved in the sediment. It seems likely that these ripples are destroyed before deposition of the next sediment layer. Reineck and Singh (1972) give the following explanation. Normally, sand is mainly deposited in the upper shoreface and the foreshore. During heavy storms in an onland direction much sand is eroded on the upper part of the beach, e. g., the foreshore and the upper shoreface, and is taken into suspension by turbulent water. The suspended sand is brought to the lower shoreface region and is deposited as evenly laminated sand. During increased water level caused by onlap storms, only individual turbulences of very strong waves reach the sediment bottom and make the sediment surface level, destroying the ripples. At the same time, and afterward, suspension clouds settle down in the form of laminated sand. Later, with a decreased water level, but still strong waves reaching the sediment, the bottom of the lower shoreface again produces wave ripples at the top of freshly deposited sediment. During the next phase of deposition, surface wave ripples are again destroyed before the next packet of laminated sand is deposited.

Campbell (1966, 1971) considers truncated wave ripple laminae as the characteristic structure of the shoreface. Chowdhury and Reineck (1978) describe this feature from the Recent shoreface of the Wangerooge island, North Sea. Ghibaudo et al. (1974; Fig. 581) report truncated ripple laminae

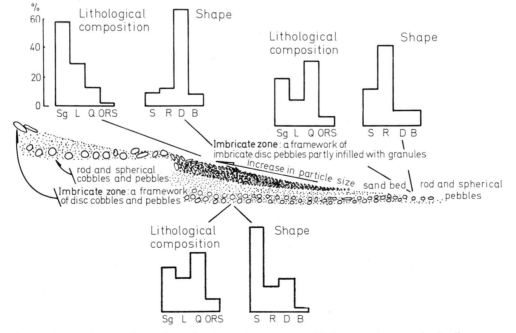

Fig. 514. Diagram showing structure and distribution pattern of gravels on a gravelly beach at Newton. Internal structure has been obtained from trenches cut into the beach. S – spherical, R – rod, D – disc, B – blade, Sg – subgreywacke, L – limestone, Q – quartzite, ORS – old Red Sandstone. (After Bluck 1967)

in fossil shoreface sediments of Upper Cretaceous from the Pyrenees.

Sometimes megaripples and megaripple bedding may also be present in the lower shoreface region. They may be produced as a result of rip current (Ingle 1966; Reimnitz 1971). Further, megaripples may be produced in small and large channels, e. g., in tidal inlets, tidal estuaries, or channels of sand tongue-channel system.

On the high-energy coasts megaripples can be produced in deeper parts (up to 3 m and more), on the seaward side of breaker line. The ripple crests run parallel to the shore line, and the lee face is located on the landward side (Clifton et al. 1971).

Vertical Sequences of Coastal Sand

The vertical sequence of shoreface, foreshore, backshore, and coastal dunes is shown in several schematic diagrams: Fig. 566 (Sapelo Island, U.S.A.), Fig. 534 (Gulf of Gaeta, Italy), Fig. 583 (schematic diagram), Figs. 575 and 581 (ancient sequence). These sequences are also referred to in beach-shelf profiles in the chapter the shelf. The *Field Trip Guide Book (South Central Texas Coast)* (1964) gives some examples of coastal sand sequences. Weimer (1975) considers coastal sands along with deltaic deposits as potential petroleum rocks. Kraft (1972) describes the sandy coast of Greece, Kent (1976) discusses modern coastal environments of Alabama and Northwest Florida. The *Field Trip Guide Book on Coastal Environments of NE Massachusetts and New Hampshire* (1969) gives useful information on the coastal sediments. Even so, the information on coastal sand is still full of lacunae, and in future more work is needed on the coastal sand to be able to record the spectrum of variability in coastal sand deposits.

The Shelf

Hydrodynamic Conditions on the Continental Shelf

The continental shelf is the underwater region from the shore line down, to about 200 m, where there is a marked break in the angle of slope, i.e., the gentle slope of the continental shelf changes into the steep slopes of the continental slope. In this discussion "shelf" denotes the region followed by coastal sand and the transition zone, i.e., from water depths of 10 to 20 m down, to the continental slope. The current systems on the continental shelf vary strongly in their nature and intensity. As discussed elsewhere, in the shore zone littoral current systems are active; however, away from the coast, on the shelf other current systems with very different current direction and intensity are present. A good account of the shelf sediment transport is given by Swift et al. (1972). Weggel (1972) gives the classification of currents on the shelf (Fig. 515; see also p. 463). An interdisciplinary synthesis of the sea bed, sediments and seawaters off the NW European continental shelf is compiled by Banner et al. (1979).

In many shelf regions current velocity is so low that it does not produce any substantial effects on the sediment bottom. Emery (1968a) found that on many shelves pelecypod shells show a concave-up preferred orientation. In the presence of strong currents this orientation is hydrodynamically unstable. Moreover, many shelves show the presence of uncovered older sediments and well-preserved geomorphological features, such as submerged beach ridges and terraces. These features plus the direct measurements of current velocity led Emery (1968a) to conclude that current velocities, at least, on such shelves are extremely weak. In contrast, very strong currents develop in the shelf of narrow seas and straits, e.g., the English Channel, the Irish Sea, the Strait of Gibralter. In such regions tidal currents, density currents, or other meteorological

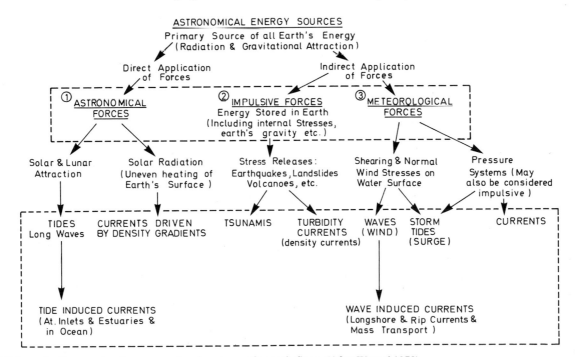

Fig. 515. Schematic diagram showing causes of various types of oceanic flows. (After Weggel 1972)

currents may reach values up to 150 cm/sec or more. In the North Sea between Holland and England current velocities of 75 to 100 cm/sec are recorded. In the Geman Bay, southeast of Helgoland, velocities of 50 to 75 cm/sec are recorded (Fig. 516). However, these measurements are from the middle of the water column and do not give any direct velocities near the sediment bottom. But areas with such high values of current velocities show megaripples and giant ripples which actively migrate. Furthermore, tidal currents supported by strong waves during heavy storms may actively influence the sea bottoms, taking much sediment, even in suspension, and moving it to adjoining places. Hadley (1964) calculated the velocity of currents produced by oscillatory waves for the Celtic Sea with southwest winds of Beaufort Number 10 (89–102 km/hr) in water depth of 100 fathoms (1 fathom = 1.83 m). He found that maximum oscillatory velocity at the sediment bottom with 100-fathom water height is 50 cm/sec.

In continuation to the studies of Hadley (1964) and Draper (1967), Ewing (1973) comes to the conclusion that in the margins of shelf regions, exposed to the storm surges, oscillatory currents are present which cause sediment movement on the shelf bottom. In the areas where canyons are present, the currents can be many times stronger, because of the refraction effect.

This demonstrates that the rarer, heavy storms may play an important role in the sedimentation of shelf sediments. The more common but weak storms do not produce any significant effect. With the method of Rankine (see Zenkovich 1967) a rough estimate of the progressive decrease of wave energy with increasing water depth can be made:

Depth of water layer–in fractions of wavelength:

$$\frac{1}{9} \quad \frac{2}{9} \quad \frac{3}{9} \quad \frac{4}{9} \quad \frac{5}{9} \quad \frac{6}{9} \quad \frac{7}{9} \quad \frac{8}{9} \quad \frac{9}{9}$$

Diameter of water particle orbit–in fractions of wave height:

$$\frac{1}{2} \quad \frac{1}{4} \quad \frac{1}{8} \quad \frac{1}{16} \quad \frac{1}{32} \quad \frac{1}{64} \quad \frac{1}{128} \quad \frac{1}{256} \quad \frac{1}{512}$$

This demonstrates clearly that with an arithmetic increase in water depth, the diameter of the water particle orbit decreases in a geometric progression. Thus, very high and long waves may be effective in much deeper water. In the following, two examples are given.

A trochoidal wave at wind velocity of 7 to 8 Bft (50–74 km/hr) possesses a wavelength of 90 m and a wave height of 3.20 m, with an oscillatory period of 8 sec. At a water depth of 20 m, the following conditions exist. 20 m water depth corresponds to $\frac{2}{9}$ of the wavelength. The diameter of the orbit is reduced to $\frac{1}{4}$, i. e., 80 cm. The circumference of an 80-cm diameter orbit is $80 \cdot \pi = 250$ cm. 250 cm are traveled in 8 sec, equal to a velocity of 31 cm/sec.

However, the orbits are deformed near the sediment bottom and the velocity is much reduced (Fig. 517). Thus, in this particular case oscillation velocity shall be insufficient to produce wave ripples.

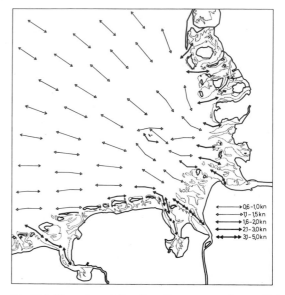

Fig. 516. Strength and direction of average maximum tidal currents in the German Bay, North Sea

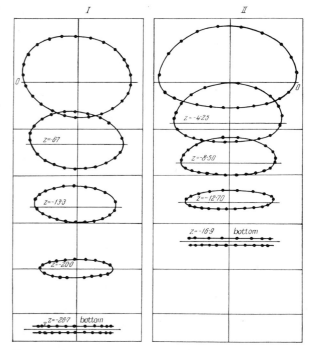

Fig. 517. Diagram showing mean trajectories of particle motion in a shallow-water wave according to Kondrat'ev (1953). *I* Wave at the beginning of distortion; *II* strongly distorted wave. The distance between two points on the trajectories is proportional to the time. 0–0: calm level, *z*: depth in cm. (After Zenkovich 1967)

The second example is based on the measurements at lightship Elbe I (Bruns 1955). The wind velocity is 9 Bft (75–88 km/hr), the wavelength is 120 m, the wave height is 4.5 m, the oscillation period is 9.8 sec, and the water depth is 26.6 m.

26.6 m water depth is $\frac{2}{9}$ of the wavelength, orbital diameter decreases by $\frac{1}{4}$, from 4.5 m to 1.1 m. The circumference of the orbit is 1.1 m $\cdot$ π = 3.5 m traveled in 9.8 sec, gives a velocity of 36 cm/sec. Even if this velocity is reduced to 75 % of its value due to deformation of the orbit near the sediment bottom, it is strong enough to move sediment on the sediment surface. Moreover, in tidal seas one direction of the wave oscillation can be supported by the tidal current, so that even rather weak oscillations may be effective in transportation and erosion of the bottom sediments of the shelf.

This phenomena is observed in the SW side of England, where in 3 % of the time, in 100 m water depth, currents strong enough to move the sediments of grain size Md Φ of 1.40 are produced (Channon and Hamilton 1976). On the Washington continental shelf, Sternberg and Larson (1976) made measurements near the shelf break in a water depth of 167 m, and found that on 5 out of 205 days of measurements, the threshold of grain motion was exceeded. Carter and Heath (1975) in their studies on the New Zealand continental shelf found that waves produced during regular, annual storms, coupled with the tidal currents move sediments up to 20 m water depth on the shelf. During occasional heavy storms, waves are effective up to much deeper water depth.

On the shelf, along with the intensity of tidal currents, the direction of tidal currents must also be given some consideration. Mostly ebb current runs in the opposite direction of the flood current. However, it is known at various points in the North Sea that sometimes the directions of the ebb and flood currents vary considerably from the usual. Furthermore, there are cases where current flows around the clock from variable directions; i.e. there are no well-developed ebb and flood current phases separated by still-stand phases of low and high tides, although there are current maxima development at ebb and flood times.

To conclude the geological knowledge, especially the hydrodynamic conditions on present day shelves, one might say that there are parts of the shelf which show regular strong currents, so that if suitable sediment is available, mega- and giant ripples develop. Besides, there are other parts which show negligible currents and parts which are occasionally affected by waves of heavy storms, especially when supported by tidal currents.

Velocity of tidal currents can also be affected by storm tides. The increase in velocity during storm tides is dependent primarily on the quantity of water and the time period during which large quanti-

ties of water are piled up in the coastal zone and later flow back. The height of a storm tide is controlled by wind set-up, atmospheric pressure, gustiness of wind, inhomogeneity of the wind field, and height of the swells. However, wind set-up contributes about $\frac{2}{3}$ of the force in piling up water (Koopman 1962).

As of now, very little is known about the effect of internal waves on the continental shelf, especially on the outer part. There are many large current systems flowing on the shelves, but their current velocity is rather low. Although such current systems cause large-scale movement of water masses, their effect on bottom sediments is almost nil.

Transition Zone

The transition zone denotes transitional sediments between coastal sand and shelf mud sediments. The sediments of the transition zone are usually clayey silt to silty sand. They are finer-grained than the sediments of coastal sand, but coarser-grained, than the shelf mud sediments. However, a prerequisite for the development of the transition zone is that both sand and mud is available and conditions are suitable for their deposition. On many modern coasts this is not the case. In many cases coastal sand grades directly into off-shore Pleistocene sands and no mud is deposited. In other cases where only mud is deposited, no sand is deposited right up to the shore line. Such conditions existed during deposition of the chenier plain of southwestern Texas.

The actual depth of the transition zone depends upon the energy of the coast. The lower the energy of the coast, the lesser the depth of the transition zone. The upper limit of the transition zone varies between 2 and 20 m. On an average it is between 8 and 10 m. The lower limit of the transition zone is somewhere between 8 and 30 m. In coasts of medium energy the lower limit is between 10 and 15 m water depth.

Biologically, the transition zone is characterized by a maximum of species, as well as individuals (Dörjes 1971). Bioturbation is very high. The upper limit of the transition zone is located just below the normal wave base; thus, mainly silt and clay settles down. Sandy layers are often present and are a result of deposition during heavy storms. Such storm sand layers are thicker and more numerous in the transition zone than in the shelf mud. Shell layers may be present; they are usually autochthonous. However, sometimes shell layers are almost exclusively made up of allochthonous shells. In general, sandy layers and muddy layers are equally represented. If bioturbation is strong, primary bedding may be completely destroyed and a homogeneous mixed sediment is produced, e. g., the Gulf of Gaeta.

Shelf Sediments

According to Emery (1968b), about 70 % of the continental shelves of today are covered with relict sediments. The term "relict sediments" denotes sediments that were deposited by agents or under conditions different from those that characterize their present environment. The main reason for this feature is rapid or hectic processes of transgressions and regressions during the Quaternary period. For example, a worldwide transgression during the last few thousand years caused a rise in sea level by about 125 m. This rapid rise in sea level left no time for the development of an equilibrium between the rate of sediment influx and the rate of sea-level rise. Only some 7,000 years B.P., the rate of increase in sea level slowed down to an extent that sediments of some reasonable thickness could be deposited down on the continental shelf and the coastal regions. Such modern deposits are mainly present in regions where much sediment is made available from the inland regions by rivers making deltas etc. For example, the chenier plain of southwestern Texas, the beach ridges of the Nayarit Coast, Mexico.

Only the sediments which have been laid down during last 7,000 years, i. e., after the slow down of the transgression, can be used for a comparative study of ancient shelf and coast sediments. With this slowed-down rate of transgression, conditions of active deposition are produced; which can be compared to slow transgression and regression, and sinking and exposure of continental shelves.

The modern shelf mud deposits are silty clay to clayey silt. Nearshore shelf mud deposits often contain layers of coarse silt or fine sand, commonly known as the storm sand layer, which originates during heavy storms (Hayes 1967; Reineck et al. 1967; Gadow and Reineck 1969). Such storm sand layers can be found as far as 40 km away from the coast, and can be traced toward the coast (see p. 395). The storm sand layers are graded or developed as graded rhythmites (Reineck and Singh 1971). Figure 547 shows storm sand layers. Storm sand layers are sometimes referred to as tempestites.

The rate of deposition of shelf mud in southeastern Heligoland, North Sea, has been estimated at 20 to 50 cm/100 years (Reineck 1963a; Gadow 1969). ^{210}Pb and ^{137}Cs chronology gives a rate of sedimentation in this area as 170–180 cm/100 years, and includes periods of hiatus, as a result of erosion by storms (Dominik et al. 1978). Moore (1955) gives values of 6 to 135 cm/100 years, with an average value of 30 cm/100 years for the shelf mud sediments of the Gulf of Mexico.

The animal population and the degree of bioturbation of shelf mud can be highly variable. In the Gulf of Gaeta, Italy, the animal population in shelf mud is very sparse; however, the degree of bioturbation is extreme. In the shelf mud of the North Sea (southeast of Heligoland) the animal popula-

Fig. 518. Sea floor with brittle stars (*Ophiura albida*), starfish (*Asteria rubens*) and a snail (*Lunatia catena*) as well as openings of polychaete burrows and tubes and other biogenic traces. North Sea, S.E. Heligoland. 35 m water depth.

Fig. 519. Diagram showing water discharge
and sediment discharge of various rivers into
the Atlantic Ocean, between Cape Cod and
Cape Kennedy. The rivers from the north
with high water discharge bring fewer sediments than the rivers of low water discharge
of the south. (After Meade 1969)

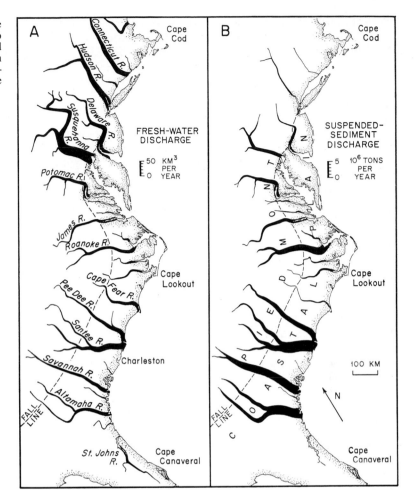

tion is very high (Fig. 518) and also the bioturbation is high. Only the storm sand layers are little
bioturbated.

At the base of a storm sand layer, a thin shell layer sometimes is present, pointing to intermittent
erosion leading to concentration of shell as lag. Although shells of shelf mud are mainly autochthonous, even allochthonous shells and other organic
substance e. g., wood, freshwater diatoms, pollen
and spores can be present in minor amounts. In the
North Sea, even larger shells, of tidal flat environments are transported into the shelf region by
floating ice during winter months. Furthermore,
Dörjes and Hertweck (in Reineck et al. 1968)
found that among the allochthonous shells, juvenile shells are most abundant. Most probably they
are brought in as larvae, develop into juveniles, but
because of a different environment do not grow into adults.

Fecal pellets are very abundant in the shelf mud
sediments. Bioturbation structures are abundant,
most of them deformative bioturbation structures.

Besides, there are several figurative bioturbation
structures, such as burrows of various sizes and
shapes, spreiten, and stuffed burrows.

The major source of shelf mud sediments is the
suspension load of rivers, which bypasses the
coastal region and is deposited on the shelf. Smaller rivers bringing large suspended loads are as important as larger rivers. The distance of the sediment source from the river mouth, the transport capacity of the river, and probably most important
the availability of various grain sizes inland are other controlling factors.

Thus, Meade (1969) pointed out that on the Atlantic coast of North America, rivers located north
of Cape Lookout drain a glacial terrain, hence, despite their higher water discharges they bring very
little sediment into the sea. On the other hand, rivers of lower water discharges located south of
Cape Lookout bring much more sediment into the
sea as they drain the deeply weathered piedmont
area and coastal plain (Fig. 519). Meade (1969)
states, further, that a large amount of sediment is

being deposited to fill the drowned mouths of rivers–in estuaries, and very little sediment is made available to be taken to the shelf. However, under exceptional weather and hydrodynamic conditions some sediment is certainly taken out on the shelf; e. g., the Elbe and Weser rivers have built up estuaries, where much sediment has been deposited and they have contributed enough sediment to build up a blanket of shelf mud roughly 20 m thick in part of the German Bay.

Part of the shelf sediments, especially those in warm waters, are of biogenic origin. Locally, volcanic or aeolian sediments may be important contributors. Phosphorite, glauconite, and chamosite are the important authigenic minerals of the shelf environment. An important source of shelf sediments is the subaqueous weathering of both soft and hard materials. This process is supported by the activity of boring organisms.

Emery (1952, 1968b) distinguished the following types of sediments on the modern continental shelves: *Detrital* (laid down by water, wind, and ice), *biogenic* (mainly carbonate shells and tests), *volcanic* (volcanic debris near volcanoes), *authigenic* (mainly phosphorite, glauconite), *residual* (product of *in situ* weathering of bedrock). In general terms, two groups of sediments can be distinguished. (I) *Modern sediments*, which are in equilibrium with the existing conditions of sedimentation unit; they may have been primarily produced *in situ* or brought in as a result of operating depositional agencies. (II) The other group is *relict sediments*, which are not in equilibrium with the conditions of today. They represent older sediments deposited under very different conditions than those of today. Figure 520 shows the distribution of relict sediments of modern shelves, occupying some 70 % of the total area. Sediments of the continental shelf also show a definite relationship to climate and current systems (Fig. 521).

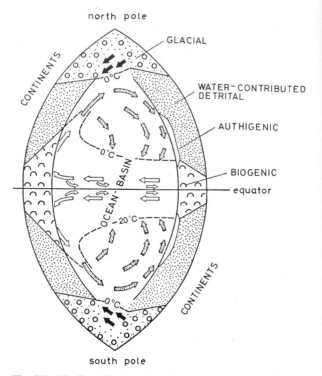

Fig. 521. Idealized distribution of various classes of sediment on continental shelves, where sediments are in equilibrium with their environment. White arrows–warm water; dotted arrows–upwelling water; black arrows–cold water. (After Reineck 1968b; originally modified after Emery 1968b)

The classification by Emery (1968) has not taken account of relict sediments of certain parts of the shelves that have been intensively reworked by hydrodynamic and biological conditions of today. For such reworked relict sediments of the continental shelf, the term *palimpsest sediments* has been suggested (Swift et al. 1971).

McManus (1975) gives a more complete classification of these sediments, where he also incorporates supplier and distributor processes:

1. Neoteric sediment (Recent in origin) is a modern deposit that consists of particles supplied to the depository at present. Example–sediments of a modern delta.

2. Proteric sediment is a modern deposit that consists of particles that were supplied to the depository in the past. Example–sand waves in the Well Bank area of the North Sea (Houbold 1968).

3. Amphoteric sediment is a modern deposit that consists of some particles that are supplied at present and some that were supplied in the past. Example–SW of Heligoland, North Sea, the deposit is a mixture of sand grains from Pleistocene sediments and Recent mud, mixed by bottom-dwelling organisms. A similar mixed sediment is reported by Kulm et al. (1975) on the mid and outer continental shelf of Oregon.

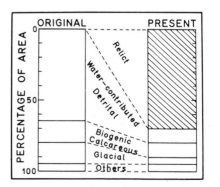

Fig. 520. Diagram showing the distribution of the percentage area of the world's continental shelves that are covered by various types of sediments, as estimated for the time of original deposition (left) and for the present time (right). (After Emery 1968b)

4. *Palimpsest sediment,* according to McManus (1975) is a relict deposit that contains some particles now being supplied to the depository. Example – A relict deposit into which the reworking by benthic organisms has added foraminiferal tests.

5. *Relict sediment* or relict deposit is solely composed of particles supplied before the present. Example – beaches drowned by rising sea-level.

The distribution of sediments on the present-day continental shelf has been extensively studied, and clearly demonstrates that many huge areas of the shelf are free of any influence of Recent sedimentation. However, there are many exceptions, where there is a continuous sedimentation of terrigenous-calcareous nature. For example, the shoals of Djerba, Mediterranean Sea are one example of carbonate sedimentation, which have been diluted by the aeolian sediments (Fabricius and Schmidt-Thomé 1972). Sarnthein and Walger (1974) report that there is a very heavy transport of aeolian sediments from the west Sahara on to the shelf of the Atlantic coast. Schneidermann et al. (1976) discuss that on the Puerto Rico insular shelf there is a patchy distribution of mud (fluvial-derived terrigenous clastic) and calcareous skeletal material, depending upon the hydraulic energy. Milliman et al. (1972) report that on the shelf and continental margin off the eastern United States, due to the low sedimentation rate, a primary source is calcareous skeletal material. Birch (1977) discusses that, besides the nearshore zone of modern sediments, i.e., mud on the West coast and continental margin of South Africa, and a broad zone of erosional surface on the mid shelf, parts of the outer shelf and slopes are covered by Recent foraminiferal and coccolithophorid debris.

Sand Deposits on the Continental Shelf

As pointed out elsewhere vast parts of the continental shelves today are covered with relict sediments. These relict sediments in various regions have been and are being reworked under present-day dynamic conditions into various forms of sand bodies, e. g., sand ribbons, giant ripples and megaripples, and tidal current ridges. These are mostly proteric sediments, in the terminology of McManus (1975). In the following, a short account of these features is given. Many of these sand bodies are shown in the picture atlas *Sonographs of the Sea Floor* (Belderson et al. 1972).

Sand Ribbons

In basal conglomerates of transgressive seas sand ribbons may develop, provided enough current velocity is present and that sand-size sediment is available. Sand ribbons are linear bed forms on sandy to gravelly shelf bottoms. They can be as long as 15 km and up to 200 m broad. The height is usually less than 1 m and the length to width ratio is usually more than 40. Sand ribbons have been described by Stride (1963), Belderson and Stride (1969), and Kenyon (1970) from the continental shelf bottoms around the British Islands. The surface current velocity in these areas is more than 1 m/sec. Werner and Newton (1975) describe sand ribbons from the Langeland Belt, the southern part of the main connection between the North Sea and the Baltic Sea.

Kenyon (1970) distinguishes four types of sand ribbons (Fig. 522):

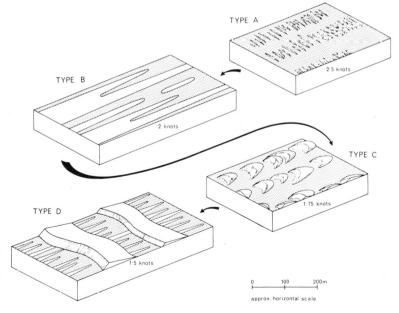

Fig. 522. Diagrammetic representation of the four basic types of sand ribbons. The typical maximum near-surface current velocity in the area of each type is given. (After Kenyon 1970)

Type A. Sand ribbons are ribbon-like, made up of transversely arranged straight-crested short ridges present on a rocky floor. The current velocity near the sea surface is 1.25 m/sec.

Type B. Sand ribbons occur in regions with current velocities somewhat less than type A. They are the most common type. Often they are relatively thin sand layers. Occasionally sand ribbons are covered with asymmetrical ripples a few centimeters high, with wavelengths more than 1 m.

Type C. Sand ribbons are ribbon-like, elongated bodies made up of sinuous-crested waves or ripple-like features. The length of the ripple-like waves is 150 m, like that of giant ripples, but the height is less than 1 m.

Type D. Sand ribbons are found in the troughs of the giant ripples. They are relatively continuous along their length, and only a few centimeters thick. The width of such sand ribbons seems to be related to the length of giant ripples with which they are closely associated.

The sand ribbons are present on reworked, coarse-grained sediments. Such surfaces are exposed between two adjacent sand ribbons and represent gravelly lag deposits. The sediment is of Pleistocene and early Holocene age. Sediments are mainly gravels, blocks, pebbles and shells. Occasionally they are populated with a delicate epibenthonic fauna, e. g., bryozoans, lamellibranchs, etc. When larger animals die, their shells accumulate as shell layers.

Comet Marks and Sand Shadows

Werner and Newton (1975) describe two new large dimensional markings, i. e., comet marks, and sand shadows from the Langeland Belt, Baltic Sea. Comet marks are defined as erosional markings in the current shadows of an obstacle. In the side-scan records, they are several to more than a hundred meters in length. The comet tail is mostly 1.5 to 3 times broader than the obstacle, but do not widens towards the end. In general, sediment of the comet tail is coarser than in the adjacent areas.

Sand shadows are sand bodies which are deposited on the lee-side of an obstacle to the current.

Giant Ripples

On the vast parts of the sandy shelf of the North Sea giant ripples are abundantly developed. They are found in areas of relatively high-current velocities (60 to 120 cm/sec), and enough water depth (> 18 m) so that giant ripples are saved from strong wave activity. One of the largest fields of giant ripples is located beyond the Dutch coast in the southern bight of the North Sea; it occupies an estimated area of 15,000 km² (McCave 1971; Terwindt 1971). They have been described from other parts of the North Sea by Jordan (1962), Harvey (1966), Stride (1970), and Ulrich (1972). The rate of migration of giant ripples is roughly 10 cm/day (Cartwright and Stride 1958).

The giant ripples of the North Sea vary in height from 2.90 to 9.10 m, and in length from 125 to 1,250 m (Terwindt 1971). During heavy storms the height of the giant ripples can be lowered by 2 m (Terwindt 1971). Giant ripples are mostly covered with the actively migrating megaripples, as described by Reineck (1963b) for the giant ripples of the Outer Jade. Giant ripples may range from symmetrical to strongly asymmetrical in form. The asymmetrical giant ripples develop where one of the two opposing tidal currents is stronger than the other (Fig. 537; see also p. 389). Instead of the term giant ripples, very often the terms dunes and sand waves are used (Belderson et al. 1972; Knebel and Folger 1976; Bokunewicz et al. 1977; Hunt et al. 1977). For the description of internal structure of giant ripples see p. 386.

Tidal Current Ridges

Unlike giant ripples, tidal current ridges are arranged parallel to the current direction. They are elongated ridge-like bodies of sand occurring in groups on a tidal-swept shelf floor. In the shelf part of the North Sea several such groups of tidal current ridges are known to occur (van Veen 1936; Veenstra 1964; Houbolt 1968). Such ridges have a maximum length of 65 km, a width of up to 5 km, and maximum heights of up to 40 m (Fig. 523). The following description is based mainly on Houbolt (1968).

Sparker sections of such current ridges in the Well Bank region, North Sea, show a distinct reflector horizon at the base of the ridges. This horizon outcrops in the troughs and represents a clay layer of Pleistocene age. In troughs this horizon is usually covered by a thin sheet of lag deposits. The ridges of the Well Bank region are asymmetrical. The northeast side is much steeper than the southwest slope. The sparker sections also show internal cross-bedding of ridges, laminae dipping to the northeast (Fig. 524). Twelve cores taken in the Well Bank ridges show predominant cross-bedding. Half of the measured cross-bedding dips in the northeast quadrant. The rest of the dips point southeast almost parallel to the direction of elongation of the ridge. The southeast dips are probably the result of migrating megaripples. Along the foreset laminae, mud layers of variable thickness are occasionally present. Such clay drapes are deposits of the still-stand phase of tides. Other ridges also show similar internal structures.

Houbolt (1968) noted that tidal currents transport the sand along the gentle southwest slope of the ridge obliquely toward the crest of the ridge. This transport is mainly in the form of migrating megaripples. From the crest sand is deposited on the steeper northeast slope in the form of foreset laminae. Along the steeper slope sand is transported to the southeast (Fig. 525). Thus, sand actually seems to go around the ridge; and the ridge migrates very slowly to the northeast, normal to the long axis and the tidal currents. However, the rate of migration is rather slow. The ridges are lowered du-

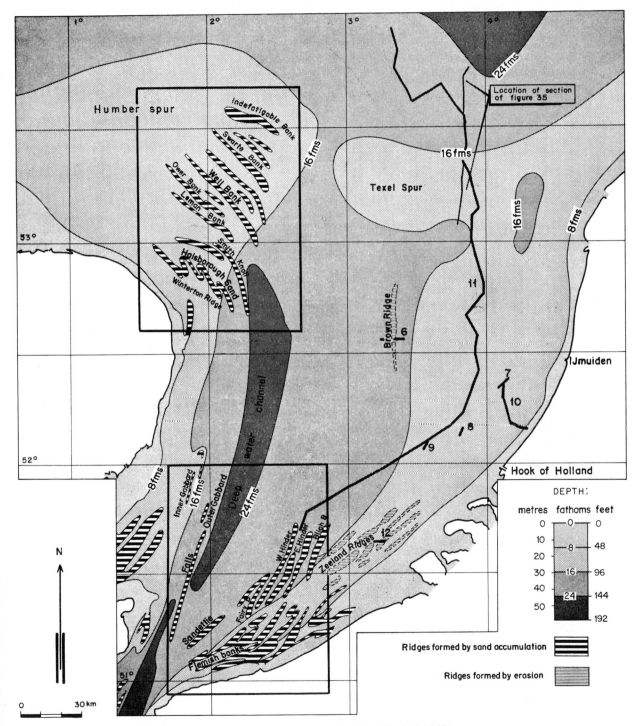

Fig. 523. Distribution of tidal current ridges in the southern North Sea. (After Houbolt 1968)

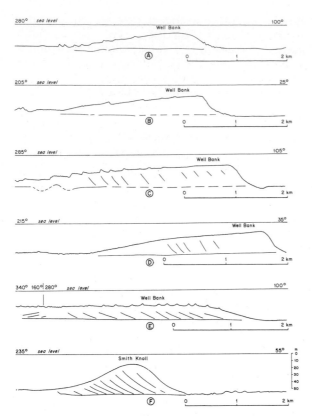

Fig. 524. Interpretation of sparker profiles over tidal current ridges in the North Sea. The vertical scale is exaggerated. The dip of slope averages only 5 degrees. (After Houbolt 1968)

ring storms, but are built up again during calm weather conditions. Houbolt (1968) assumed a system of spiral currents for the development and maintenance of tidal current ridges. The current is stronger over the troughs than over the ridges. This is compensated by the flow of water in two long spirals in such a fashion that the water in the trough bottom is directed outward from the trough toward the crest of the ridges.

Tidal current ridges of some other regions, e. g., Zeeland ridges, Brown ridges, are mainly erosional in nature. They show the presence of older sediments, internally. It seems quite likely that tidal currents under very similar hydrographic conditions are capable of producing ridges by eroding and sculpturing older sediments as well as by accumulating reworked sand on top of older sediments.

The overall pattern of sand transport in the southern North Sea, and its relationship to the main current direction near sediment bottom is given by Caston and Stride (1970) and Caston (1971). The main sand movement in the eastern part of the channels is the northwesterly direction, while in the western part, transport direction is towards the southeast. Convergence of sand streams causes sand accumulation with the result that there is a continuous growth of banks parallel to the direction of tidal currents. The movement of asymmetrical banks takes place by the migration towards the steeper slope.

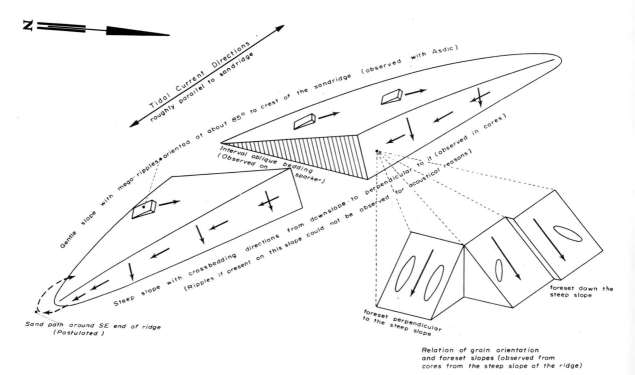

Fig. 525. Schematic illustration of sedimentary features of a tidal current ridge. Southern North Sea. (After Houbolt 1968)

Examples of Ancient Sand Deposits of the Shelf

One may try to interpret ancient transgressive sand blankets in the light of knowledge obtained by the study of present-day sand blankets with sand ribbons and giant ripples. The genetic relation between a marine transgression and the up-building of three ancient tidal ridge complexes in Europe is discussed by Nio (1976).

Banks (1973) compares the tide-dominated offshore sediments of the Lower Cambrian, Norway, with the sandy continental shelf around Great Britain. Pryor (1971) compares the Permian Yellow Sands of NE England with the tidal current ridges of the North Sea. Spearing (1975), and Johnson (1978) give probable examples of ancient offshore marine sands. Hobday and Reading (1972) describe shallow marine sand sequence in the late Pre-Cambrian of Finnmark, North Norway.

Also Pryor and Amaral (1971) compare large-scale cross-stratification in St. Peter Sandstone (Middle Ordovician) with the North Sea tidal current ridges.

Examples of Beach-Shelf Profiles from Modern Environments

Introduction

In the following, some selected modern beach-shelf profiles are described and discussed. Despite numerous papers on the coastal and continental shelf sediments of modern environments, there are only a few systematic detailed descriptions of such regions which can be used to make models for the environmental interpretation of ancient sediments. Particularly rare are papers giving systematic descriptions of sedimentary structures, i. e., primary bedding and bioturbation structures.

A systematic study of the underwater sandy part of coastal regions has been possible for only a decade or so. Box-sampler has played important role in obtaining samples for such studies (Reineck 1958b; 1963a,b). Such sandy samples are studied by making relief casts and X-ray radiographs. Cores from shelf mud are obtained by a piston corer, whose construction, is based on Kullenberg. Lately, special corers have been designed for obtaining longer cores from sandy regions (Bussche and Houbolt 1964); among these is the vibration corer (e. g., Schmidt and Kolp 1965; Kögler 1963).

Can samplers are used for obtaining samples from the shoreface during diving (Reineck and Rosenboom 1969; Howard and Reineck 1972).

Beach Sand–Shelf Mud Profile, Gulf of Gaeta, Mediterranean Sea, Italy

The Gulf of Gaeta is situated north of the Gulf of Naples in the Tyrrhenian Sea, which is not affected by tides. Coastal currents running parallel to the coast develop during storms and other unusual weather conditions. The transport is mainly northward. The littoral currents (longshore currents) also run from south to north. The summer months are calm weather periods with negligible precipitation. In autumn, winter, and spring seasonal storms are rather common and precipitation is high.

On the *backshore* a small beach ridge is developed. In the shoreface region there are two longshore bars. The outer longshore bar is active only during strong storms (Fig. 526).

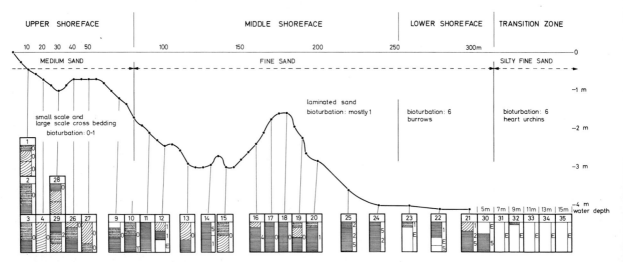

Fig. 526. Profile across the shoreface showing the distribution of the type of bedding, sediment type, and bioturbation. The samples from nos. 30–35 are not drawn to scale with respect to distance from shore line and water depth. Licola, Gulf of Gaeta, Italy. (After Reineck and Singh 1971)

Fig. 527. Sedimentary structures of shore-face sediments in relief casts. In the zone of 0 to 2 m water depth megaripple bedding and small-ripple bedding are most abundant. Bioturbation is insignificant. Gulf of Gaeta, Italy. (After Reineck and Singh 1971)

The backshore part of the beach shows a changing morphology and sedimentary structures during summer and winter periods (Wunderlich 1971). After a period of heavy storms the beach is made level. Backshore and swash zones are made up of almost horizontally laminated sand, dipping 2 to 4° toward the sea. Individual sets of laminated sand may show faintly developed discordances. Locally, heavy minerals–e. g., magnetite have been concentrated into workable placer deposits (Gadow and Reineck 1968).

During the calm period of summer, part of the winter beach is reworked into a summer beach, with local development of small cliffs. Both lamination and bedding of beach ridges dip toward land and sea. The angle of dip ranges from 5 to 30°; the steeper angles dip toward land. As very little new sediment is coming to the coast during the winter, structures of the summer beach are almost completely destroyed and replaced by low-angled laminated sand, dipping toward the sea of the winter beach. During a slow progradation of

the coast, generally only the structures of the winter beach will be preserved (Wunderlich 1971).

The *shore face* shows small ripple bedding, large-scale trough cross bedding, evenly laminated sand and longshore bar cross bedding, composed of planar sets (Reineck and Singh 1971). Among bioturbation structures, burrows, "press structures" of the heart urchin, and deformative bioturbation structures are common.

In 0 to 2 m water depth, the upper shore face, main bedding types are small-ripple bedding and megaripple bedding (Fig. 527), as well as subordinate planar cross bedding. The bioturbation is very low, although this region is densely populated by lamellibranchs (Dörjes 1971). The sediment is medium sandy fine sand.

Below water depth of 2 m, the middle shore face, ripple bedding is present only in the uppermost part of the samples. Ripple bedding is probably destroyed during new phases of sedimentation (see p. 419). Laminated sands is the main bedding type (Fig. 528). With increasing water depth, in the

Fig. 528. Samples from water depth of 2 to 4 m. The degree of bioturbation increases; mainly burrows are found. The less bioturbated upper few centimeters of sediments were deposited during a storm which affected the area some 6 weeks before the samples were taken. Shoreface, Gulf of Gaeta, Italy. (After Reineck and Singh 1971)

lower shoreface, the degree of bioturbation also increases (Fig. 529). The shore face extends up to 6 m water depth. The region between 2 to 6 m water depth is composed of fine sand with very minor amounts of medium sand, i. e., this region is finer-grained than the region 0 to 2 m water depth. In the trough of the outer longshore bar a 20-cm-thick shell layer is present and partly covered by the sediments of landward migrating outer longshore bar. Together with deformative bioturbation structures, well-defined burrows are also present.

Below 6 m water depth the shore face grades into the *transition zone*. Here sediment is silty fine sand to fine-sandy silt. The degree of bioturbation is very strong; the inorganic bedding is seen only as traces in nonbioturbated parts (Fig. 530). The dominant bioturbation structure is the "press structure" of *Echinocardium cordatum*. In opposition to the mixed dead fauna of shells in the coastal sand part, shells of the transition zone are mostly autochthonous (Hertweck 1971a). From the lower part of shore face shells are transported toward the shore line. This leads to mixing of shell fauna of different zones of the shore face.

The boundary between the transition zone and the shelf mud is rather transitional and variable from place to place. In the northern part of the Gulf of Gaeta this boundary is located at 10 m water depth, whereas in southern part this boundary lies at about 20 m water depth (see Gadow 1971). Below this boundary shelf mud is deposited. The sediment is silty clay.

In opposition to northerly transport along the coast, the direction of transport on the shelf seems to be toward the south, as demonstrated by decreasing silt content. The variation in thickness and intensity of silty layers in shelf mud also suggests a transport toward the south (Reineck and Singh 1971). The thickness and number of silty layers/m in shelf mud decreases continuously away from coast, as well as toward the south from the Volturno delta (Fig. 531). Reineck and Singh (1971) believe that silty layers are produced mainly by sediment coming from the Volturno delta, and to a lesser amount by sediment taken into suspension during heavy storms on the coast and transported away from coast. A southerly drift of suspended sediment coming out of the Volturno River has

Fig. 529. Samples from water depth of 4 to 7 m. Sample no. 23 (water depth 4 m) still shows deposition by storms in the upper few centimeters. The degree of bioturbation is high in all the samples. Sample no. 30 shows heart urchin traces in the upper ⅓ of the profile. Shoreface, Gulf of Gaeta, Italy. (After Reineck and Singh 1971)

Fig. 530. Samples from water depth of 9 to 13 m. Sediments are silty in nature. Bioturbation is high. Heart urchin traces are most abundant. Transition zone. Gulf of Gaeta, Italy (After Reineck and Singh 1971)

Fig. 531. Map showing zones of different thickness of silty layers. Zones are based on total thickness of silty layers per meter. Thickness of silty layers/meter decreases toward the open sea and to the south. This suggests the direction of silt transport from land toward the sea and to the south. (After Reineck and Singh 1971)

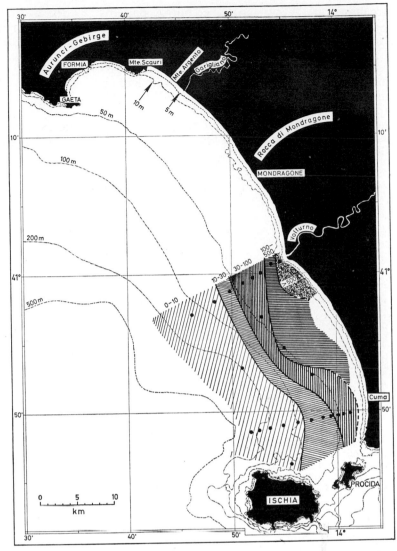

been recorded by Düing (1965). This silty sediment is deposited in shelf mud in the form of silty layers. The time interval by which the two silty layers are separated by the mud layer must be of order of several years, in some cases even tens of years.

Silty layers of shelf mud are evenly laminated; sometimes a weak grading is developed (Fig. 532). Occasionally, low-angled discordances are present. In some cases silty layers are capped by a thin veneer of detrital plant material. In 24 piston cores from shelf mud, 251 silty layers were measured. Silty layers 2 mm thick are most abundant. More than 50 % of the silty layers are less than 3 mm thick. Thick silty layers, i. e., more than 30 mm, are found only in regions near the coast. The shelf mud looks rather monotonous in character. Study of thin sections demonstrated a high degree of bioturbation in shelf mud. Open and stuffed burrows

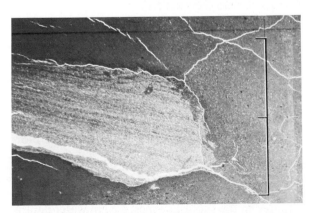

Fig. 532. Thin section showing well-laminated coarse silty layers (storm sand layers). Silty layers are partly disturbed by bioturbation, piston-corer sample. Water depth is 300 m; length of the scale is 1 cm. Gulf of Gaeta, Italy. (After Reineck and Singh 1971)

are rather rare. A few star-shaped burrow patterns were observed which resemble the fucoids of ancient sediments (Fig. 533).

Shell layers were not observed. In thin sections fecal pellets were abundantly recorded, arranged mostly in nests. Most probably they were laid on the sediment surface by the gastropod *Turritella comunis*.

The hypothetical resulting progradational vertical sequence of sediments of the Gulf of Gaeta is shown in Fig. 534.

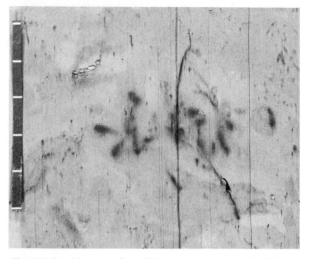

Fig. 533. Freshly cut surface of the piston-corer sample of fango (fine-grained mud). Degree of bioturbation is high, and marked by circular and elongated spots (stuffed burrows). The black star-shaped burrows resemble the fucoids. The water depth is 60 m, the sediment depth is 30 cm. Scale is in centimeters. Gulf of Gaeta, Italy. (After Reineck and Singh 1971)

Channels and Sand Bars (Shoals)–Channels and Sand Tongues, North Sea

In the inner part of the German Bay, in the region of the mouths of the Jade, Weser, and Elbe rivers, north of Elbe estuary, and along the coast, exposed tidal flats are developed, without barrier islands in front of them (Fig. 535). In the region of Nordergründe sand tongues are developed alternating with channels. In the tidal estuary of the Weser and the Outer Jade sand bars (mouth bars) are well developed, together with channels. Both regions are situated in the subtidal zones, i. e., in water depths similar to the shore face and transition zone of the beach shores. The tidal range in this region is about 2.40 m. The energy of the environment is medium to high.

The hydrodynamic conditions, sediments, and structures in the channels are similar to tidal inlets. But at the outer and inner ends of tidal inlets shoals are developed–underwater deltas dissected by channels.

The Channels and Sand Bars (Shoals) of the Outer Jade, North Sea

The shoals of Outer Jade are elongated sand bodies 2.6 to 10.6 km in length. The shallow parts of the sand bars have water depths of around 5 m. The channels are locally up to 15 m deep and can be as broad as 2 km. The average maximum current velocity in such channels is about 1.40 m/sec. On the sand bars the current velocity is much lower. The channels show a tendency to bend out in a northeasterly direction. At the same time the sand bars, situated between two channels, also show a tendency to bend out in a northeasterly direction and to migrate. The southwest-side of the sand bars is eroded, whereas new sediment is added on the northeast side. The average rate of migration of sand bars is about 115 m/year (Reineck 1963a).

The channels are cut in the coarse-grained sediment of Pleistocene age. The sediment is enriched in large lamellibranch shells from intertidal flats, the open sea, and older Pleistocene sediments. Thus, the channel sediment is made up mainly of variable proportions of medium sand, coarse sand, and gravels (Reineck 1963a; Gadow and Dörjes et al. 1970). The sediment of sand bars is finer-grained than that of channels. It is medium sandy fine sand (Fig. 536).

In the channels giant ripples are well-developed, running at right angles to the direction of current flow (Lüders 1929; Hülsemann 1955; Reineck 1963a; Völpel and Samu 1966; Samu 1968).

The internal structure of giant ripples is made up of cross-bedded units of megaripple origin (Fig. 537).

The length of giant ripples in these channels varies from 93 to 248 m. Height varies from 1.7 to 5.5 m. The crests of giant ripples can be followed over the entire width of the channels (Fig. 538), i. e., up to 1.5 km. They are mostly asymmetrical in shape. The steeper slope exists in the direction of the stronger of the two opposing currents, i. e., ebb or flood. The surface of giant ripples, as well as other parts of the channels, are covered by megaripples (Fig. 537). These megaripples are reformed and migrate in opposite directions with every change in tidal current directions (ebb and flood directions).

The sand bars are situated between the channels. The longer axes of sand bars, which are somewhat curved, run parallel to the direction of the current. Whereas in the channels, because of coarser-grained sediment and high-current velocities, megaripple bedding primarily is developed, the sand bars show a rather high percentage of small-ripple bedding and evenly laminated sand, together with megaripple bedding (Fig. 539). The small-ripple bedding can be assigned to lower current velocities, as well as finer-grained sediment (Fig. 536). In the finer sediment of sand bars small ripples can be produced in contrast to medium and coarse sand of the channel (see p. 14).

The evenly laminated sand is produced mainly because of the deposition of suspension clouds.

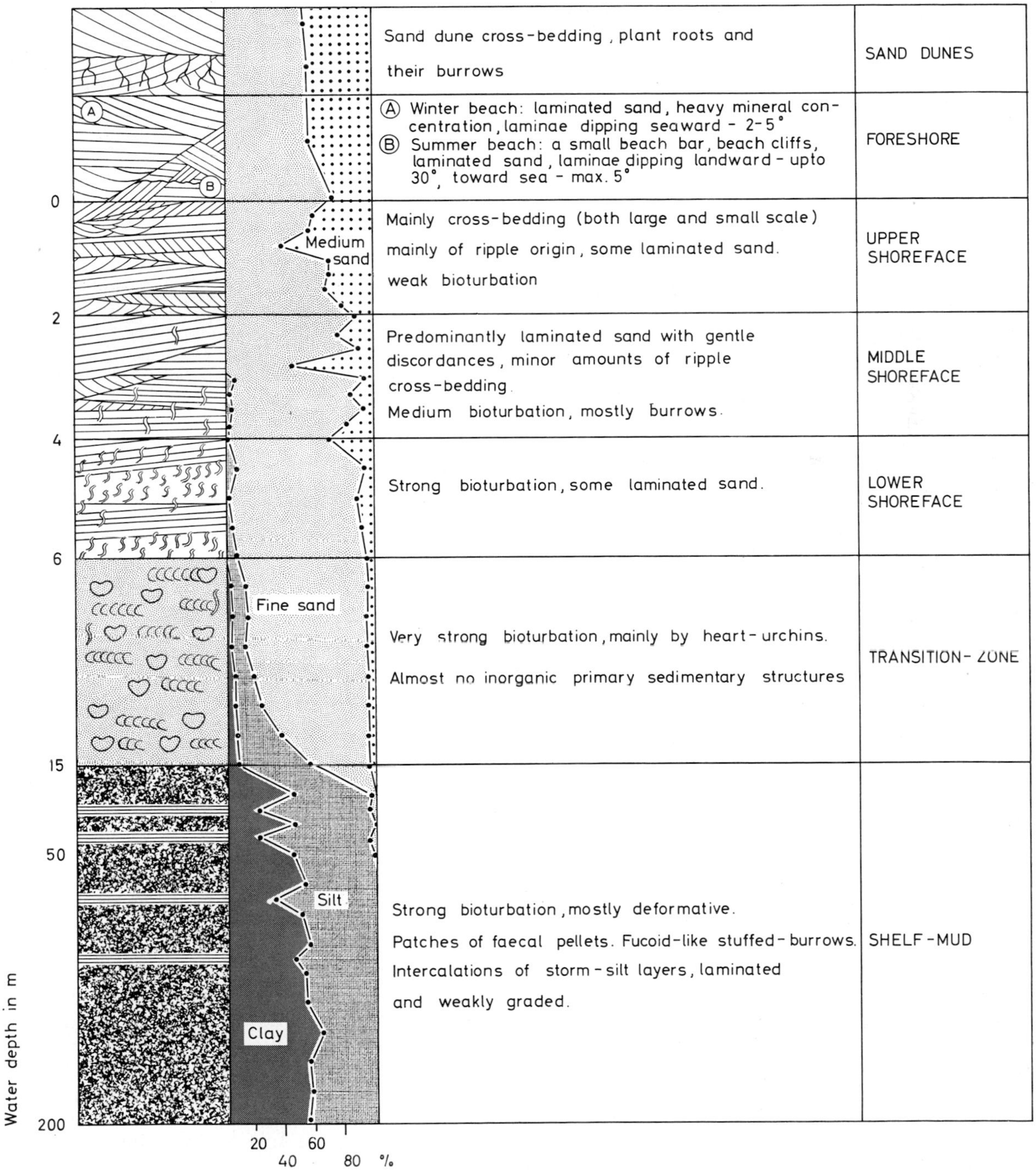

Fig. 534. Compilation of the sedimentary features of the beach-shelf mud profile off Licola, Gulf of Gaeta, Italy. The results are presented in a vertical sequence, which would develop in a prograding coast

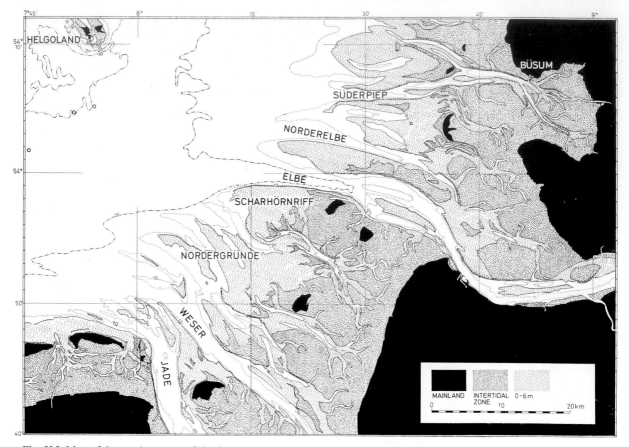

Fig. 535. Map of the southern part of the German Bay, North Sea. Channels and sand bars are in the Outer Jade: channel-sand tongue systems of the area are between the Weser and Elbe rivers (Nordergründe), and the area north of the Elbe (Büsum)

The investigations of Wasser und Schiffahrtsamt show that suspension clouds are produced at the time of maximum tidal currents. Reineck (1963) believes that suspension clouds are produced mainly by wave action when shoaling waves take much sand into suspension.

In the channel sediments lebensspuren are rather few; they are mainly burrows of polychaetes. Lamellibranchs and echinoids do not live in the channels with coarse-grained sediments. On the sand bars with relatively finer-grained sediment, *Echinocardium cordatum* is the main bioturbating animal. The palaeocurrent measurements in channels and sand bars are bipolar. There are slight differences in the direction of palaeocurrents on sand bars and in adjoining channels.

In the Weser estuary as well, sand bars (mouth bars) and channels are developed. The distribution of sediment and sedimentary structures is rather similar to that of the Jade region. If one adds this region to the Jade region of sand bars and channels, an area of approximately 12 × 18 km of shoal-channel system is present. In this region sequences of channel-lag deposit 1 to 3 m thick constitute the lower part of the sequence, as a result of migration. Such channel deposits have been recognized below the sand bars in the borings. If the whole region progrades, the sand bar sediments will be present on top of the channel deposits. Figure 539 shows idealized progradational sequence of shoal-channel systems of such an area.

Sand Tongues (Shoals) and Channels in the Region of Nordergründe, North Sea

The region of the Nordergründe is situated south of the Elbe estuary, and Büsum is located north of Elbe estuary (Fig. 535). In the regions of Nordergründe and Büsum the sand is deposited coming from coast-parallel transport along the East Frisian Islands eastward (Gripp 1944). Toward the sea both regions are followed by shelf mud deposits (see p. 395).

In the subtidal zone channels are present. They are narrow near the intertidal flats and become broader toward the open sea. Near the intertidal flats they are about 1 km broad, and become 3 km broad near the far end of the channel. The depth of channels is rather variable: locally, depths up to 20 m are present. In general, water depth in the channel increases from 6 m near the upper end to 13 m near the far end toward the sea.

Unlike the channels of the Outer Jade, the channels of the Nordergründe migrate rather slowly (see Lang 1970). Homeier (1969) estimated the rate of migration of 8 large channels of Nor-

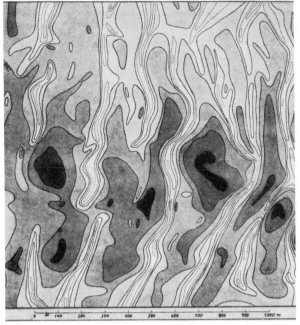

Fig. 538. Giant ripples in the estuary of the Jade. (After Lüders 1929)

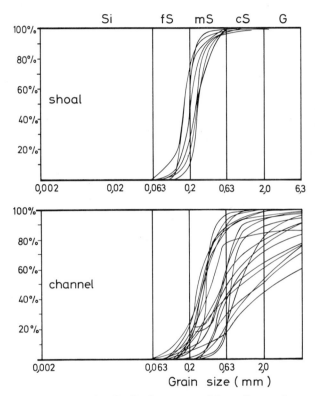

Fig. 536. Grain-size distribution curves of the sediments from the Outer Jade area. (After Gadow in Dörjes et al. 1970)

dergründe by comparing the charts of the last 100 years; he found that the average rate of migration is about 27 m/year, in the northeast direction. This means that also here, below the sequence of sand tongues, channel sediments are present. Sand tongues are seaward extensions of intertidal flats into the subtidal zone, located between two channels. Sand tongues are 16 to 18 km long and on the average of 2 km in width. Channels are funnel-shaped, with the broader end toward the open sea. The widening of channels seawards causes simultaneous narrowing and tapering of sand tongues in the same direction.

In the channels of the tidal inlets often huge quantities of coarse and medium sand, along with

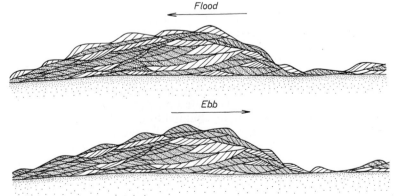

Fig. 537. Internal structure of an ebb-current-oriented giant ripple. Vertical exaggeration is 10 times. The steeper slope is in the ebb-current direction. During flood tide megaripples move across the giant ripple (white ripples); during ebb tide megaripples move in the opposite direction (dotted ripples). In the lee face of the giant ripple megaripples move up the lee face. (After Reineck 1963a)

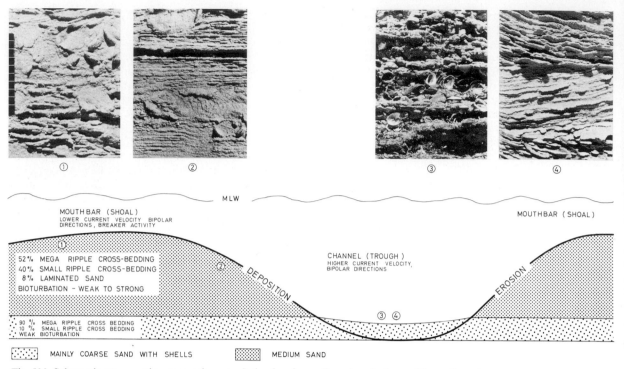

Fig. 539. Schematic cross-section across the outer Jade, showing sedimentary features of the sediments

shells are present, often in the form of giant ripples. The shells of these channels are dug for the purpose of road-building material, animal food making etc. Schubart (1927), Jüngst (1942), Köhler (1963) describe these deposits from the North Sea coast. Such accumulations of the Gulf Coast estuaries are discussed by Bouma (1975).

According to Göhren (1969), the average maximum current velocity in the channels is 60 to 100 cm/sec. On the sand tongues maximum current velocity is 30 to 50 cm/sec, which is much less than in channels. The grain size distribution in this region is different than in the Outer Jade (Gadow in Dörjes et al. 1970). The sand tongues are composed of silty fine sand to medium sandy fine sand, as are the sand bars of the Outer Jade (Fig. 540). The channels are also made up of silty fine sand to medium sandy fine sand (Fig. 540). Unlike the Outer Jade, there is no difference in grain size of the channel and the sand bar (sand tongue) sediments.

The reasons for such sediment distribution are: Channels do not cut into the coarser-grained Pleistocene sediments and the current velocity is not so strong as in the channels of the Outer Jade. Moreover, the wave action is limited only to sand tongues, and not appear in the channel floors. Because of wave action the finer-grained material from sand tongues is winnowed away (Lüneburg 1961).

Distribution of fauna is rather similar in sand tongues and channels. In the Outer Jade channels, fauna is almost absent compared to sand bars. The wave base marks an important boundary in the distribution of fauna, i. e., below wave base (about 6 m water depth); on sand tongues the number of living species is much greater than the number of species found in the upper part of sand tongues, above wave base (Dörjes, in Dörjes et al. 1970).

The bedding types present include megaripple bedding, small-ripple bedding, laminated sand, interlayered sand/mud bedding (alternating bedding), and thick mud layers. There is no clear-cut differentiation in the bedding types found in sand tongues and channels. Only the wave base and the low water line mark the zones showing a weak differentiation in the distribution of bedding types. Above wave base the percentage of megaripple bedding is low, as compared to its amount in sediments below wave base; alternating bedding is absent. At the same time the amount of evenly laminated sand shows a clear increase above wave base. Above low water line the amount of megaripple bedding shows a marked increase in a narrow zone. The palaeocurrent measured from megaripple and small-ripple bedding is bipolar in the channels, running parallel to the elongation direction of channels (Reineck 1963, 1965; Reineck and Singh in Dörjes et al. 1970). On the sand tongues the palaeocurrent direction is also bipolar. In both channels and sand tongues, the ebb current direction is more dominant than the flood current direction. The palaeocurrent direction on sand tongues runs at some acute angle to the elongation axis of sand tongues. A hypothetical progradational sequence of the sand tongue-channel system of the Nordergründe shows the following units (Figs. 541 and 542).

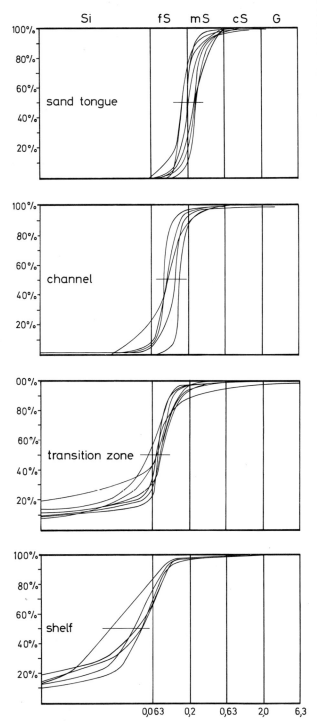

Fig. 540. Grain-size distribution curves (cummulative curves) of some typical sediments of the Nordergründe area. (After Gadow in Dörjes et al. 1970)

1. Shelf mud with storm layers of sand and silt are present in offshore deeper water parts (see p. 395).

2. Shelf mud is followed by the transition zone, with slightly clayey, silty fine sand, or fine sand sediments. The degree of bioturbation is high. No megaripple bedding, but small-ripple bedding, is present. In opposition, small-ripple bedding is nonexistant in the transition zone of the Gulf of Gaeta. Its presence in the Nordergründe region can be assigned to the activity of tidal currents. In addition, alternating bedding and mud layers are abundant. The upper boundary of the transition zone, i.e., the boundary between the transition zone and the channel deposits of coastal sand is located at a water depth of about 20 m.

3. The lowermost part of coastal sand is composed of sediments of channels and sand tongues below the wave base. They are slightly silty fine sand to slightly medium sandy fine sand. Here, megaripple bedding is well represented. As compared to the profile of the Gulf of Gaeta, megaripple bedding in the Nordergründe is present at relatively greater depths (20 m water depth). Right from the basal part of coastal sand megaripple bedding is present and constitutes 20 to 30 % of the bedding. The main bedding type is small-ripple bedding (50 %). Evenly laminated sand constitutes 15 to 20 %, and alternating bedding, 5 to 20 %. Degree of bioturbation in this part is higher than in the sand tongues above wave base; in the channel it is 10 %, in the sand tongue below wave base it is 20 %. In the channel sediments there is characteristic enrichment of mud pebbles and shells (Fig. 543).

4. Above the wave base, sediments become coarser-grained, i.e., medium sandy fine sand. The bioturbation is very weak to absent. Mud layers are absent. No significant difference exists between sediments of channels and the lower part of the sand tongue to the sediments of the upper part of the sand tongue (above the wave base). Only the amount of evenly laminated sand shows a marked increase from 20 % in the lower part to 40 % in the upper part of sand tongue.

5. Above the low water line, sand tongue-channel deposits are capped by sediments of the intertidal zone. First follows the zone of sand tongue which undergoes intermittent subaerial exposure. This zone shows abundant development of megaripples and megaripple bedding. Then follows a sequence of intertidal flats until the sedimentation surface is elevated by deposition. In the larger channels of the tidal flats current velocities are rather high (probably because of the decrease in cross-section), producing abundantly megaripples and giant ripples; cross-bedding of ripple origin constitute about 80 % of the total bedding types. In muddy intertidal flats even channel sediments are of muddy character.

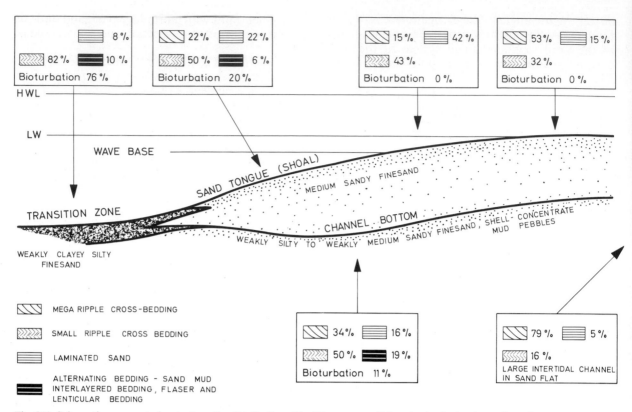

Fig. 541. Schematic representation to show the distribution of bedding type and bioturbation in the channel-sand tongue system of the Nordergründe, North Sea. (After Reineck and Singh in Dörjes et al. 1970)

Tidal Inlets and Tidal Deltas

Tidal inlets are narrow channels between two adjacent barrier islands, and connect the open sea with the lagoon or tidal flat located behind the barriers (Price 1963a). The tidal inlets are mostly oriented at right angles to the coast line. In the southern North Sea, inlets can be up to 30 m in water depth. On the seaward and landward ends of the tidal inlets, deltas (tidal deltas) are developed, which are made up of sand bars (shoals) alternating with channels. Many workers have studied the form and migration of shoals and the tidal deltas. Oertel (1977) demonstrates that the form of the tidal delta depends upon the onshore energy, offshore energy, and longshore currents.

On coasts with strong longshore currents, there is a net sand movement laterally, though not directly from one barrier island to the next. Studies of Winkelmolen and Veenstra (1974) show that sand up to 0.4 mm size is brought into the tidal inlet and then in the tidal flats, where it is fractionated. The sediment is partly brought into the open sea through outer tidal delta. Within the tidal delta itself, sediment is fractionated into several populations. A coarse-grained lag deposit remains on the

frontal part of the tidal delta. Other fractions are brought back into the tidal inlet and partly moved on the next barrier island and build up the beach. In offshore regions, sand movement is restricted to a shallow zone and takes place by wave ripples. In still deeper parts, the sediment is not in movement. Partly, sediments of deeper parts show characteristics of the early Holocene transgressive phase.

Terwindt (1973) studied the tidal area of the SW part of the Netherlands and came to the conclusion that sand movement in the tidal inlet follows a very complicated pattern. On the shoals of the tidal delta, there is very little sand transport by currents; however, waves are important in sand movement on these shoals. Movement of fine-grained sand in the underwater part of the delta is in the same direction as on the shoals.

On the shoals of the tidal delta megaripples and sand waves are abundant, and are oriented mostly towards land (flood current), because besides the flood currents, wave bores are also present which reduce the ebb-current energy (Fitzgerald 1976). In the channels, between the shoals, which may be developed as ebb or flood channels, unidirectional bedforms are commonly present. Mostly, transverse megaripples, or lingoid or rhomboid megaripples are present (van Straaten 1953a). Hubbard

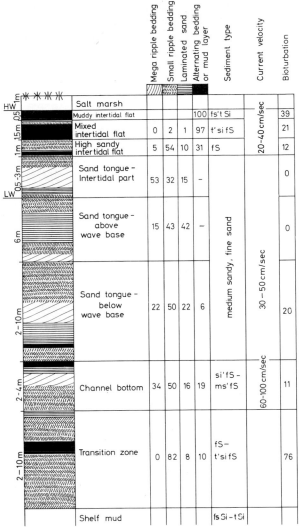

Depositional environment	Mega ripple bedding	Small ripple bedding	Laminated sand	Alternating bedding or mud layer	Sediment type	Current velocity	Bioturbation
Salt marsh							
Muddy intertidal flat				100	fs't Si	20-40 cm/sec	39
Mixed intertidal flat	0	2	1	97	t'si fS		21
High sandy intertidal flat	5	54	10	31	fS		12
Sand tongue – Intertidal part	53	32	15	-			0
Sand tongue – above wave base	15	43	42	-	medium sandy, fine sand	30-50 cm/sec	0
Sand tongue – below wave base	22	50	22	6			20
Channel bottom	34	50	16	19	si'fS - ms'fS	60-100 cm/sec	11
Transition zone	0	82	8	10	fS - t'si fS		76
Shelf mud					fs Si - t Si		

Fig. 542. Hypothetical vertical sequence of the Nordergründe area which would develop in the case of progradation. (After Reineck and Singh in Dörjes et al. 1970)

Table 27. Hypothetical regressive sequence based on a composite of South Carolina tidal inlets. Inlet channel-fill facies and ebbtidal delta facies are intimately related laterally, so no simple vertical succession is possible. (After Hubbard and Barwis 1976)

Depositional environment	Dominant sedimentary structures
Non-marine	?
Marsh	Rooting, burrows
Tidal flat	Burrows, small scale cross beds
Flood delta	Planar cross beds Planar and trough cross beds
Open bay	Varies
Dune and washover	Large scale planar and festoon cross beds (dune)/landward dipping high angle bar bedding and small planar crossbeds
Beach	Shallow seaward dipping plane beds; landward dipping high angle planar bar bedding
Ebb delta	Plane beds, multidirectional trough cross beds
Shallow channel	Large scale planar crossbeds grading upward to bidirectional trough cross beds
Deep channel	
Channel bottom	Coarse sand and shell lag varies

oped: (A) channel floor, characterized by a lag gravel composed of shells, pebbles, and other coarse particles; (B) deep channel (−10.0 to −4.5 m), characterized by lenticular sets of ebb-oriented reactivation surfaces; (C) shallow channel (−4.5 to −3.75 m), characterized by plane, parallel laminae; (D) spit-platform (−3.75 to −0.6 m), characterized by steeply dipping planar cross-bedding and small-scale, flood-oriented laminae near the top and ebb-oriented laminae towards the bottom; (E) spit (−0.6 to +2.0 m),

and Barwis (1976) based on their studies in the South Carolina tidal inlets, reconstruct a progradational and a retrogradational (transgressive) sequences (Tables 27 and 28).

Moreover, there is exchange of sand material in the water depths of 10–20 m between the underwater tidal delta and the relict sand of the North Sea bottom.

On the shoals and in the channels, megaripples and giant ripples (dunes and sandwaves) are very abundant. Thus, large-scale cross-bedding is the most dominant bedding type in these sediments. Kumar and Sanders (1975) studied the vertical sequence of the channel of tidal inlet and the barrier island, as a result of lateral migration of a tidal inlet.

According to Kumar and Sanders (1975) in ascending order the following sequence is devel-

Fig. 543. Mud pebbles and shells on the surface of a channel bottom. Box sample. Nordergründe, North Sea

Table 28. Hypothetical transgressive sequence based on a composite of South Carolina inlet/bay systems. Note that the inlet-filling sands disconformably overlie marsh/flood-tidal delta facies, and are, in turn, overlain by a ravinement surface. This ravinement surface may cut through the entire flood-tidal delta/marsh/inlet sequence, depending on its thickness. Note also that flood-tidal delta sands are completely enclosed in mud. The evolution of sand stringers from sheet sands is a presumption based on observations by Morton and Donaldson (1972). (After Hubbard and Barwis 1976)

Depositional environment	Dominant sedimentary structures
Offshore marine	Varies
Nearshore marine: lower contact disconformable	Variety of small scale cross beds
Tidal inlet: lower contact disconformable	Shallow: small scale cross beds Deep: large scale planar and trough cross beds Bottom: coarse channel lag
Marsh: point bars and tidal deltas in channels	Marsh: rooted bioturbated Sand: bimodal trough cross beds, sets up to 1 m
Flood tidal delta: lenshaped shoot sand	Trough and planar cross beds, sets up to 1 m
Protected lagoon: shallow open bay behind newly formed barrier	Ripples, bioturbated

characterized by steep and gentle seaward-dipping laminae and steep and gentle landward-dipping laminae in its various subenvironments. Size analysis curves can be used to characterize different units, using the method of Visher (1969).

During transgressions, tidal inlets play an important role by the side of washover fans to account for the landward sediment movement. In the transgressive barrier island system, southern Gulf of St. Lawrence, Canada, 90 % of the landward-oriented sediment transport is related to the present-day or earlier tidal inlets (Armon and McLann 1979). Bartberger (1976) for Assateague Island, and Pierce (1969) for the coast west of Cape Hatteras obtain comparable results.

Profile of the Büsum: Coastal Sand – Transition Zone – Shelf Mud with Storm Sand Layers

Coastal Sand

The coastal sand of the Büsum region shows development of sand tongues and channels (Fig. 535) similar to that of the Nordergründe region (Reineck 1963a; Reineck et al. 1967, 1968). However, coastal sand of the Büsum region is much more silty than that of the Nordergründe. Higher silt content in the coastal sand of the Büsum is attributed to fine-grained sediment brought by the Elbe River and transported northward to the Büsum region. Small-ripple bedding is the most predominant sedimentary structure. With increasing water depth, content of flaser bedding progressively increases; at the same time the percentage of evenly laminated sand decreases with water depth. In deeper parts the nature of sedimentary structure and grain size is the same on both sand tongues and broad-end channels. Both sedimentation units show small-ripple bedding, flaser bedding and alternating sand/mud bedding. However, channels show presence of shells and mud pebbles.

Transition Zone

The transition zone of the Büsum area is located at much less water depth than the Nordergründe. The upper limit of the transition zone, i. e., the boundary between coastal sand and transition zone, is located at a water depth of 10 m or even less. (In the Nordergründe this boundary is at 20 m water depth.) Up to a depth of 10 m, even sand tongues can be followed morphologically. The lower limit of the transition zone, i. e., the boundary of shelf mud, is located at 15 m water depth. The bioturbation in the transition zone sediment of the Büsum is much less than in the Nordergründe. Alternating bedding is abundant.

Sediments of the transition zone are rather silty to muddy in nature. Together with small-ripple bedding and flaser bedding, lenticular bedding is also present.

Sand layers with ripple bedding or with rippled surface are abundant (see p. 419). These sand layers are storm sand layers, becoming thicker toward land, ultimately grading into coastal sand. Toward the shelf mud such storm sand layers become thinner and finer-grained (Gadow and Reineck 1969). However, the storm sand layers of shelf mud do not show ripples and ripple bedding. Occasionally, at the base of the storm sand layers of the transition zone a shell layer is present, sometimes made up mainly of tests of the gastropod *Hydrobia*. *Hydrobia* lives in the intertidal zone, from where it has been brought in the transition zone during heavy storms.

Aigner and Reineck (1982) demonstrated systematic changes in storm layers (tempestites) along coastal to offshore transects from the Helgoland Bight (North Sea). The storm sand characteristics (Fig. 544) are controlled by water depth and distance from land. Storm sands are transported from coastal sands and spread across the shelf by offshore flowing gradient currents (Gienapp 1973)

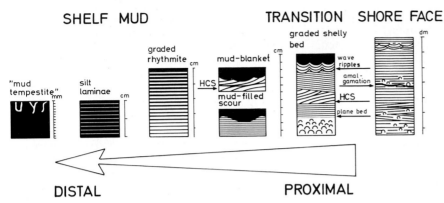

Fig. 544. Schematic diagram showing proximal and distal facies of typical storm sand sequences in relation to water depth and distance from shore as the two important gradients controlling the characteristics of storm deposits (tempestites). Helgoland Bight, North Sea. (After Aigner and Reineck 1982)

which compensate for coastal wind-generated and atmospheric pressure generated water build-ups, and are enhanced by ebb tidal currents (Allen 1982, Aigner and Reineck 1982). Density currents as Walker (1980) explains, are not required.

In an unpublished paper Reineck describes that in the example of the North Sea, tempestites are often interrupted for short distances, as in fossil shelf sediments.

Seilacher (1982) discusses distinctive features of inundites, tempestites and turbidites. Sandy tempestites are accompanied by hummocky cross stratification, the tops are usually sharp, climbing ripples are very rare. Tempestites and the origin of shell beds in the lower Lias of S-Germany are reported and discussed by Bloos (1976, 1982). Aigner (1985), using two case histories of storm sedimentation in modern shallow marine environments, presents a comprehensive analysis of an ancient storm depositional system on a basin wide scale of the Upper Muschelkalk, SW Germany.

Shelf Mud with Storm Sand Layers

The shelf mud sedimentation unit is located somewhat north of the line joining the Elbe estuary and island of Heligoland (Fig. 545). Toward the north and west, the shelf mud wedges out. Toward the western edge, below the shelf mud, medium sandy silty fine sand is present, which occupies a vast surface area west of the shelf mud unit (Reineck et al. 1967). In its central part the shelf mud is about 20 m thick and rests on top of Pleistocene sediments (Fig. 546) (Reineck 1969b).

Shelf mud is mainly fine sandy silt to clayey silt (see p. 391). In nonbioturbated parts more sandy and more muddy layers are present as well-developed strata. Dark-colored muddy layers (clayey silt) alternate with light-colored fine sandy layers (silty fine sand to strongly fine sandy silt). The mud

has been derived as suspension load from the Elbe estuary. Tidal currents moved this suspended load back and forth, with a tidal rest-current causing northeast drift, finally depositing it in the region of shelf mud. The rate of sedimentation of shelf mud is roughly 20 to 50 cm/100 years (Reineck 1963a; Gadow 1969).

Most sandy layers are present at intervals of 5 to 10 cm within the muddy sediments (Fig. 547). These sandy layers are evenly laminated; sometimes with vague discordances. Occasionally, fecal pellets are seen showing well-developed lamination and bedding. In a few cases laminated sand layers are developed as graded rhythmites, i. e., the unit starts with coarser-grained and thicker layers, grading upwards into finer-grained and thinner layers. Quite often such sand layers are capped by a unit of non-bioturbated mud. These units sometimes show characteristic escape traces (Fig. 548) (Hertweck in Reineck et al. 1968). This suggests that the sand layer and the overlying nonbioturbated mud layer are deposited very quickly. The sand layers, i. e., storm sand layers, decrease in thickness away from the coast. In the nearshore part of shelf mud they are on the average > 20 mm in thickness; in offshore parts, some 30 km away from shore, thickness decreases to < 10 mm (Fig. 549). Maximum water depth of their occurrence is 40 m.

The source of sand of storm sand layers is assigned to coastal sand; during heavy storms and storm tides sediment is eroded in the nearshore region and transported to the shelf region by retreating waves and ebb current (Reineck et al. 1967, 1968; Gadow and Reineck 1969). The deposition of sand in shelf mud is mostly preceded by slight erosion, which is documented by the occasional presence of an autochthonous shell layer at the base of a storm sand layer (Fig. 550). With decreasing wave energy, sand clouds settle down in the form of laminated sand (Reineck and Singh 1972). Together with sand, much mud is present as suspension. In the last phase of deposition, on top of a sand layer a mud layer is quickly deposited. This mud layer is usually nonbioturbated. Then, normal shelf mud deposition begins again; and the surface of the shelf mud is populated by numerous endo- and epibionts. However, most of these animals cause bioturbation of a thin mud layer near the surface.

The most important organisms producing deformative bioturbation structures in the top mud layer are ophiurides (Figs. 518 and 551).

However, the characteristic species of I. order of shelf mud region is the polychaete *Echiurus echiurus* (Dörjes in Reineck et al. 1968; Hertweck 1971b). *Echiurus echiurus* produces well-developed U-shaped burrows, 1 to 2 cm in diameter, reaching up to a depth of 20 cm. The lower, curved part of the U is surrounded above and below by spreit-

en, produced as a result of upward and downward shifting of the burrow in response to sedimentation (Figs. 552 and 553).

Other important bioturbation features are thin spiral burrows (3 mm diameter) of *Notomastus latericeus,* vertical regular burrows of *Ceriantus,* trifurcating burrows of *Thalassinides,* stuffed burrows of *Echinocardium,* other simple burrows of various polychaetes, and shell nests produced from activity of *Pectinaria.* Furthermore, there are various deformative bioturbation structures. Besides, there are many other species in the fauna of shelf mud which do not produce any definite bioturbation structures.

A hypothetical progradational vertical sequence of the Büsum region is shown in Fig. 555. Table 29 gives important characteristics of various sedimentation units of the Büsum region.

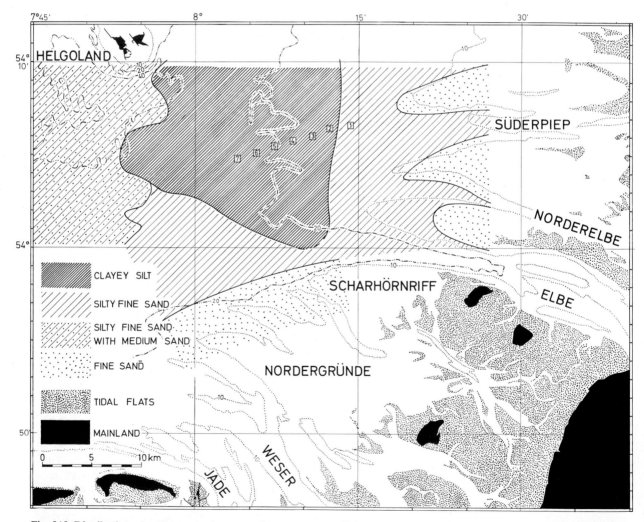

Fig. 545. Distribution of sediments in the eastern inner part of the German Bay, North Sea. A silty fine sand zone forms the transition zone between the coastal sand (medium sandy fine sand) and the deeper water muddy zone (fine sandy to clayey silt in water depth of 15 to 40 m). These three regions are bordered by an area of medium sandy, silty fine sand in the west. (After Gadow and Reineck 1969)

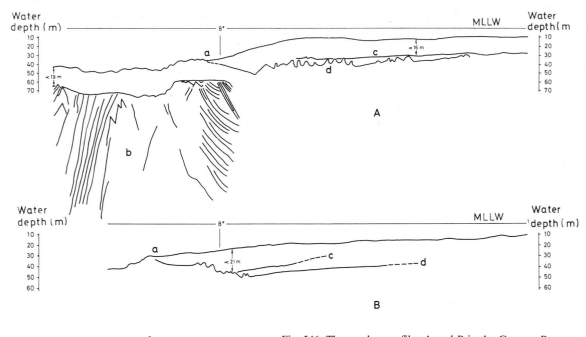

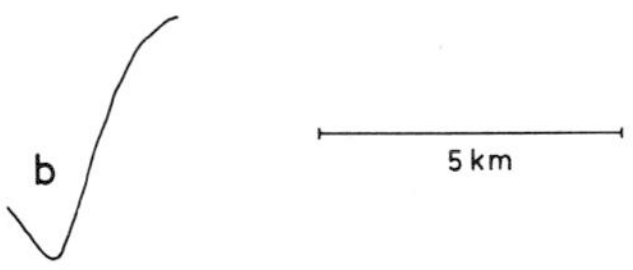

Fig. 546. The sparker profiles A and B in the German Bay. a Pinching out of the muddy deposits in the west; b sedimentary rocks of Heligoland; c a sediment unit deposited in the last interglacial; d moraines of the Saale glaciation. MSp TLW Average spring tide low-water depths. (After Reineck 1969b)

Fig. 547. Storm sand layers. These layers are embedded in the shelf mud and exhibit only evenly laminated sand bedding or graded rhythmite bedding. (After Reineck and Singh 1972)

The Miocene of Denmark shows a remarkable similarily with the deposits of Nordergründe, North Sea, both in primary sedimentary structures and in the trace fossils (Radwanski et al. 1975).

Profiles of the Two East Frisian Barrier Islands, Norderney Island and Wangerooge Island, North Sea

Two profiles across shoreface-offshore areas of a medium to high-energy coast are investigated, taking box core and can-core samples. In front of both the barrier islands offshore shelf mud is absent (Figs. 555 and 556).

In the case of Norderney Island, Recent sediments grade into Pleistocene sands of the offshore region, which were reworked during the last transgression and are also being reworked today. Systematic samples were collected from low-water line to 20 m water depth. Four zones showing an increasing degree of bioturbation, and a decreasing rate of sediment transport with increasing water depths have been differentiated (Reineck 1976b):

Zone I: Located between the low-water line and the 1.2 m water depth includes the uppermost longshore bar and runnel of the shoreface. It is characterized by coarse-grained sediments, high proportion of large-scale cross-bedding of mostly megaripple, and high- and low-angle bar bedding origin is common along with laminated sand. Degree of bioturbation is low. Sediment reworking is very high.

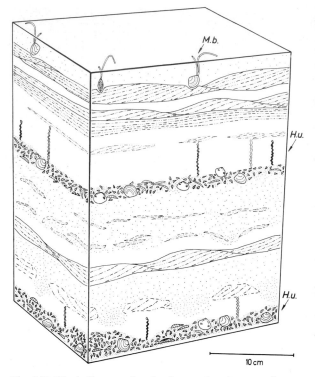

Fig. 548. Block diagram showing inorganic primary sedimentary structures and bioturbation structures in the region of fine sand with mud (transition zone) between Büsum and Heligoland, North Sea. The water depth is 10 to 15 m. The most characteristic bedding of the region are flaser and lenticular bedding, laminated sand, and ripple bedding. During heavy storms living and dead shells of *Hydrobia Ulvae* are brought in from the tidal flats. Living *Hydrobia* attempt to escape through the newly deposited mud cover to the surface, and produce characteristic spiral "escape traces". Escape is only successful partly, and the escape traces are filled with sand. (After Hertweck in Reineck et al. 1968)

ly in the winter months. Dörjes (1976) discusses the benthonic communities inhabiting this area.

These four zones are also recognizable in the profile from Wangerooge Island (Chowdhuri and Reineck 1978). Table 30 gives the comparison of various sedimentological parameters of these four zones in the two profiles. In both cases, coarser sediments in the offshore area start at a water depth of 7.5 m. The shallower zones (zone I and II) contain longshore bars in both cases; in zone III (water depth 3–7.5 m) and zone IV (water depth 7.5–20 m) of the Wangerooge, unlike Norderney, no saw-tooth bars and tongue-like bars are present, respectively. On the contrary, in the case of Wangerooge Island, a deep channel with mega- and giant ripples is present. In this channel sediments, corresponding to zone IV of Norderney Island, due to high-current velocities the degree of bioturbation is very low and megaripple bedding is most abundant. In both cases, mud layers are present only in the upper part of the shoreface.

In the shoreface sediments of both the islands, bedding features similar to "truncated wave-ripple laminae" (Campbell 1966), and "hummocky cross-stratification" (Harms 1975) are recorded (Fig. 557), which are considered to be typical of shoreface deposits (Harms 1975).

Galveston Island (Gulf of Mexico, U.S.A.)

Galveston Island is one of the barrier islands of Gulf of Mexico which have been investigated in much detail to obtain a clear picture of the progradational sequence of coastal sand. Major work has been done by the geologists of oil exploration companies. To the east of Galveston Island is Bolivar Island. Both islands show well-developed beach ridges and sand dunes, running parallel to the shore line. Many washover fans have carried sediment to the marshes located behind the islands. These barrier islands are separated from the mainland by lagoons, i.e., West Bay, Galveston Bay, and East Bay. Galveston Island is separated from Bolivar Island by a well-defined tidal inlet, with well-developed tidal deltas, both toward the open sea and the bay.

The Galveston Island beach dips at 3° toward the sea. Normally, four longshore bars are developed, migrating toward the coast. During stormy weather longshore bars are made level. The tidal range varies from 30 to 75 cm. A summary on the investigations of Galveston Island is given by Bernard et al. (1962). In a series of borings a vertical sequence with sand dunes, foreshore, shore face, transition zone, and shelf mud has been encoun-

Zone II: Located between 1.2 and 3.0 m water depths is finer-grained than zone I. Bedding is mostly low-angle laminated sand and some small-wave and current-ripple bedding. It extends over the second bar and runnel of the shoreface. During calm weather, there is no sediment transport.

Zone III: Located between 3.0 and 7.5 m water depths is still finer-grained than zone II. Degree of bioturbation is higher than in zone II and I. Bedding types are the same as in zone II. In this zone saw-tooth bars, running obliquely to the depositional strike, are developed.

Zone IV: Located between 7.5 and 20.0 m water depths is mainly medium-grained sand, representing reworked Pleistocene sediments. Tongue-like bars are present, cross-bedding of megaripple origin is common. However, the high degree of bioturbation suggests that sediment transport takes place only sporadically during heavy storms, most-

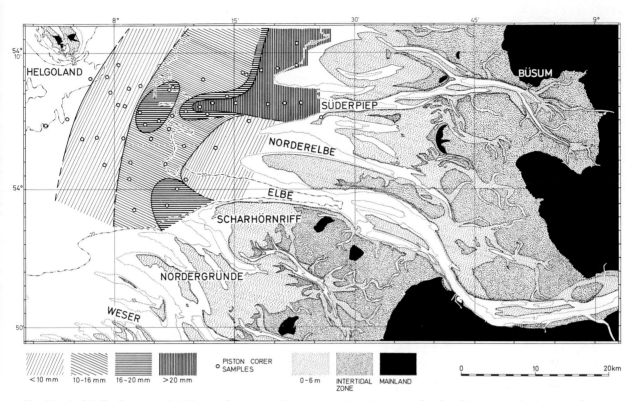

Fig. 549. Aerial distribution and thickness of storm sand layers in the eastern muddy part of the German Bay. The sand is primarily derived from tidal flats west of the Büsum with westerly flowing ebb-currents, and the Scharhörnriff and Nordergründe with northwesterly flowing ebb currents. In the zone of convergence of both transport paths the thickness of individual sandy layers increases to 16 to 20 mm. (After Gadow and Reineck 1969)

tered, which corresponds well with the present-day lateral sequence of these sedimentation units.

On top of the vertical sequence is a 0.60-to-2.40-m thick aeolian sand dune deposit (with some beach, i.e., backshore deposits). Bedding features are mostly obscured because of weathering, soil formation, bioturbation, and movement of vadose water. Plant roots are abundant in this zone. Sediment is well-sorted fine to very fine sand. Content of fine-grained organic matter is rather high. Near the base light- and dark-colored pseudo-laminae are developed, probably because of fluctuating vadose water.

Below the aeolian sand deposit is a 0.9-to-3.0-m thick deposit of foreshore and upper shore face (about 1.5 m below the low-water line). Sediment is fine to very fine sand, well sorted, with some shell content. There are very low-angled unconformities among the sets of laminated sand. Bioturbation is very low. Occasionally small-ripple bedding is present, which developed in troughs, running parallel to the shore line. Then follows a 3-to-9-m thick zone of lower shore face, extending up to a water depth of 9 m below Low-water-line. Bernard et al. (1962) call this zone the middle shore face. Deposits are made up of thinly and evenly laminat-

ed fine sand, with occasional cross-bedded shelly sand and shell layers. Sediment is moderately sorted very fine sand. Bioturbation is very abundant and well-developed burrows are present. Intercalated are some graded beds, starting with coarse shell, grading upward into fine sand and silty clay, with a top layer of very fine macerated plant material resembling peat.

From 9 to 12 m extends the transition zone (Bernard et al. 1962, call it the shore face toe, or lower shore face). Sediments are thinly interlayered sand and silty clay. Bioturbation is very abundant, mainly deformative bioturbation structures with some well-developed burrows. Thin sand layers are sometimes graded.

The shelf mud zone continues from 12 m offshore. Silty clay predominates over a few sandy layers. Sand layers are evenly laminated; some possess a thin shell layer near the base. We think these sand layers as storm sand layers (see p. 395). The bioturbation is patchy; strongly and weakly bioturbated units alternate.

Because Galveston Island has been prograding since its origin, a vertical progradational sequence is present below the island and has been investigated in borings (Fig. 469) (Bernard et al. 1962).

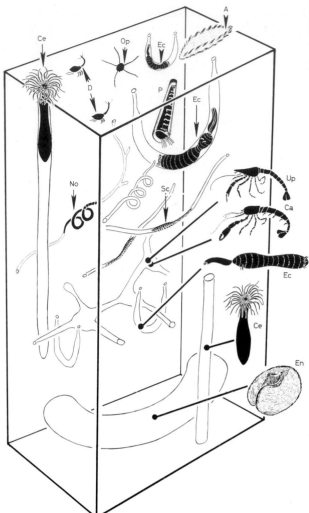

Fig. 550. Schematic representation of the genesis of storm sand layers, and associated features in the muddy shelf region under conditions of continuous mud deposition. a Situation before the heavy storm; dense population and intensive bioturbation. b Beginning of the heavy storm conditions, causing erosion and sedimentation of eroded shells. c Deposition of sand brought in from the coastal region. Mud is deposited quickly on the top as the waves of the storm die out. Molluscs move quickly through the newly deposited sediment and produce escape traces. Organisms living in the burrows extend or shift their burrows upwards. d Additional continuous deposition of mud brought in as mud-suspension by the rivers. There is an increase in animal population and bioturbation activity in the topmost newly deposited sediment layer. (After Reineck et al. 1968)

Fig. 551. Three-dimensional diagrammatic representation of some more important bioturbating animals. German Bay, southern North Sea. In the uppermost part of the diagram animals are shown in living position. Bioturbation patterns of some organisms along with their producer are figured. D = *Diastylis rathkei* (Cumacea), P = *Pectinaria koreni* or *P. auricoma*; En = *Echinocardium cordatum,* Ec = *Echiurus echiurus;* Ca = *Callianassa subterranea* and *C. helgolandica;* Up = *Upogebia deltaura;* and Ce = *Cerianthus lloydii.* Relative sizes of various animals are highly exaggerated, and the living depth is out of scale. (After Reineck et al. 1967)

1. Inner planar zone
2. Inner rough zone
3. Outer planar zone
4. Outer rough zone with lunate megaripples
5. Inner offshore zone with asymmetrical wave ripples.

These five zones are dependent upon the hydrodynamic conditions and change their position with changing tidal conditions and wave energy. This leads to the fact that sedimentary structures of one zone strongly interfinger with those of adjacent zones (Fig. 559); or sometimes some zones are

Beach Profile of a High-energy Coast, Oregon, U.S.A.

Clifton et al. (1971) investigated the high-energy coast (shore face) of southern Oregon, U.S.A., in an area where the shoreface does not show any longshore bars. They divide the shore into five zones (Fig. 558):

ill-developed or absent altogether. In the following, a short account of these five zones is given.

1. Inner Planar Zone. This zone is characterized by the planar surface, and comprises the swash zone. The main bedding developed is beach or swash lamination (evenly laminated sand), well known from many beaches.

2. Inner Rough Zone. The inner planar zone is followed seaward by a zone showing bed irregularity. On steeper beaches this zone consists of a series

of 3 to 10 symmetrical ripples. Such ripples are steep-sided, 15 to 20 cm in height, 30 to 60 cm in length, and composed of coarse sand or fine gravel. The crests of these ripples are rather continuous and can be followed over long distances, running parallel to the shore.

The first ripple seaward of the planar zone is generally made up of coarsest sediment. In the seaward direction the grain size of sediment composing the ripples decreases continuously, together

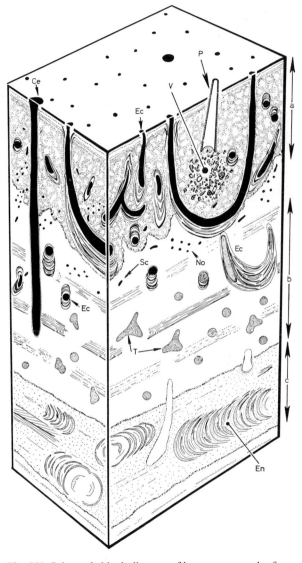

Fig. 552. Three-dimensional diagrammatic representation of burrows found in box-corer samples, German Bay, North Sea. Black burrows are occupied by living animals at the time of sampling. Ec = *Echiurus echiurus;* Sc = *Scalibregma inflatum;* No = *Notomastus latericeus;* Ce = *Cerianthus lloydii;* P = *Pectinaria koreni* and *P. auricoma.* White abandoned burrows are in the lower part of the section. T = *Upogebia deltaura* and/or *Callianassa subterranea* and *C. helgolandica;* No = *Notomastus latericeus; Echinocardium cordatum* (En). (After Reineck et al. 1967)

Fig. 553. Schematic block diagram of box-corer samples from the muddy sediments, south of Heligoland. The topmost layer (a) shows primary bedding completely destroyed by *Echiurus.* Below it another layer (b) shows a lower degree of bioturbation, produced by other benthic organisms (burrows of Thalassinidae). Thick sand layer (c) with bioturbation structures of *Echinocardium cordatum* (En). Multi-walled, U-shaped burrows of *Echiurus* (Ec); shell nests (V) by *Pectinaria* (P); burrows of *Cerianthus lloydii* (Ce) and *Notomastus latericeus.* (After Reineck et al. 1967)

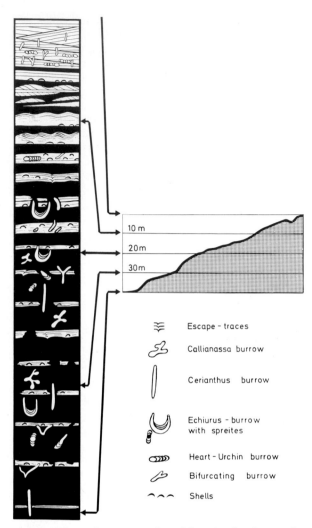

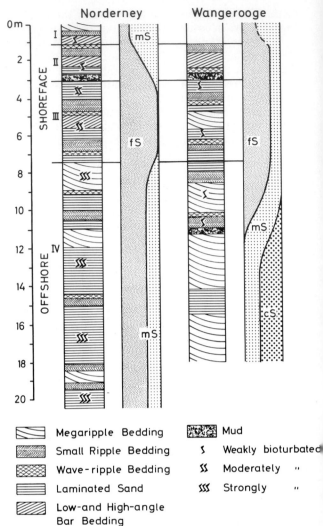

Fig. 554. Schematic representation of deposits of various environments, i.e., shelf mud deposits, transition zone deposits, coastal sand deposits, and their relation to the water depth. (After Reineck et al. 1968)

Fig. 555 and 556. Hypothetical vertical sequence of shoreface-offshore deposits of barrier islands of Norderney and Wangerooge, German Bight, North Sea. The shoreface deposits (zone I–III) are characterized by megaripple bedding, bar cross-bedding (low angle and high angle), laminated sand, and small current and wave ripple bedding. The grain size decreases with increasing water depth, at the same time the degree of bioturbation increases. In the Norderney Island sequence, there is an increase in the laminated sand and degree of bioturbation in the offshore area. In front of Wangerooge Island, a large tidal channel with megaripples and giant ripples is developed. Consequently, in the offshore area, bioturbation is reduced and megaripple bedding is the dominant bedding structure. (Based on Reineck 1976; Chowdhuri and Reineck 1978)

with the ripple height. The internal structure of the ripples is complex; the foresets may dip either landward or seaward.

On gently sloping beaches the inner rough zone shows sets of depressions (troughs) which are 1 to 2 m across and 10 to 15 cm deep. The longer axis of the troughs runs parallel to the shoreline. The troughs are separated by broad flat "ridges". The seaward side of the depressions are steeper and coarser-grained, and troughs slowly migrate seaward. Thus, in such regions major bedding of the inner rough zone is trough cross bedding, with sets varying in thickness from 4 to 100 cm, and dipping seaward. Longshore currents which show maximum velocity over the inner rough zone may modify the shape of the troughs and the direction of their migration. The steeper side shifts toward the

upcurrent side of the trough and cross-bedding shows a pronounced longshore component. Under very strong longshore currents the internal structure may be oriented essentially parallel to the shore. The inner rough zone is 3 to 10 meters wide. On some beaches this zone may be completely absent. Hydrodynamically this zone is controlled by the interaction of surf and swash.

Table 29. Characteristics of various sedimentation units of Büsum region, North Sea. (After Reineck et al. 1968)

	Coastal sand	Transition zone	Shelf mud
Water depth	0–10 m	10–15 m	15–40 m
Grain size	Fine sand–silty fine sand	Silty fine sand	Fine sandy silt–clayey silt
Primary sedimentary structures	Laminated sand, flaser bedding, ripple bedding	Flaser and lenticular bedding, sand layers: evenly laminated or ripple bedded, wavy or plain mud layers	Mud layers, graded rhythmites, laminated sandy silt
Biocoenosis	Depth-variants of *Macoma baltica* – coenose		*Echiurus echiurus* – coenose
Biofacies	Depth-variants of *Macoma baltica* – biofacies Some autochthonous shells, bioturbation structures of *Echinocardium*	Autochthonous shell-layers, and allochtonous *Hydrobia* layers, escape traces of *Hydrobia*, undifferentiated bioturbation structures	*Echiurus echiurus*–biofacies Shells, fecal pellets, burrows and spreites of *Echiurus echiurus,* spiral burrows of *Notomastus,* bioturbation structures of *Echinocardium,* irregularly arranged burrows, changing degree of bioturbation

Table 30. Comparison of the four zones of the shoreface of Norderney and Wangerooge, North Sea, with their characteristic morphological and sedimentological differences. fs = fine sand; mS = medium sand; cS = coarse sand. Megaripple bedding and laminated sand also includes high and low angle bar bedding. (After Chowdhuri and Reineck 1978)

	Zone	Water depth (m)	Morphology	Sediment	Mega ripple bedding	Small ripple bedding	Laminated sand	Mud layer	Mean bio-turbation
Norderney	I	0 – 1.2	1 shoreface bar	fsmS-mS	39%	11%	48%	2%	12%
Wangerooge	I	0 – 1.2	1 shoreface bar	not examined	not examined	not examinded	not examined	not examined	not examined
Norderney	II	1.2– 3.0	2 shoreface bar	msfS	0%	27 %	67%	6%	10 %
Wangerooge	II	1.2– 3.0	2 shoreface bar	msfS	3%	27.6%	59%	10%	3 %
Norderney	III	3.0– 7.5	saw-tooth bars	fS	2%	34 %	64%	0%	30 %
Wangerooge	III	3.0– 7.5		msfS	10%	24 %	66%	0%	16.5%
Norderney	IV	7.5–20	sand tongues	msfS	31%	9 %	60%	0%	43 %
Wangerooge	IV	>7.5	channel	mS-cS	53%	20.5%	22%	4%	11 %

Fig. 557. Hummocky cross stratification (Harms 1975) or "truncated wave ripple laminae" (Campbell 1966). Shoreface of barrier Island Norderney, North Sea. Water depth 3.8 m

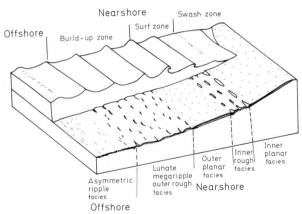

Fig. 558. Diagram showing the distribution of various sedimentation zones on a high-energy beach, in relation to wave type and activity. (After Clifton et al. 1971)

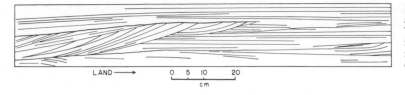

LAND →   0 5 10   20
              cm

Fig. 559. Internal structure of the inner planar zone (swash zone) near its seaward limit. The evenly laminated sand of the swash zone is resting on the top of cross-bedded sediments of the inner rough zone. (After Clifton et al. 1971)

3. Outer Planar Zone. Seaward from the inner rough zone follows the outer planar zone. This occupies the outer part of the surf zone and the inner part of the wave build-up zone, and conditions similar to the upper flow regime are achieved. This zone is 10 to 30 m broad and almost planar. The sand transport is mainly by sheet flow over a planar surface. Bedding in this zone is horizontally laminated sand.

4. Outer Rough Zone with Lunate Megaripples. This zone is generally 30 to 120 m broad. The surface is covered with dominant bedforms, i. e., distinctive landward-facing lunate megaripples (Fig. 560). These lunate megaripples show much variation in size and spacing. Each megaripple possesses a landward-dipping lee side, usually 10 to 30 cm high. The megaripples migrate landward, generally at a rate of about 30 cm/hr. Thus, bedding of this zone is cross bedding with steeply dipping foresets toward land.

5. Inner Offshore Zone. This zone is characteristically occupied by asymmetrical wave ripples, which are 2 to 5 cm in height, 10 to 20 cm in length.

The direction of asymmetry changes in relation to direction and strength of the latest surge. Occasionally, sets of interference ripples are observed. The bedding of this zone is small-scale cross bedding.

Clifton et al. (1971) point to several ancient examples of high-energy shore face sediments. They are Pleistocene terraces on the coast of southern Oregon and central California and marginal marine Branch Canyon Sandstone (Hill et al. 1958) in the southeastern Caliente Range of southern California. If one compares the beach profile across a high-energy coast with that of a low-energy coast, a certain parallelism can be seen. The inner planar zone and the inner rough zone are present in both cases. The outer planar zone of a high-energy beach seems to correspond to the outermost longshore bar of a low-energy beach where laminated sand is sometimes also a dominant feature. However, there is no similarity to the outer rough zone. The inner offshore zone is also present in both cases.

A "High-Energy" Beach-to-Offshore Sequence, California, U.S.A.

A high-energy beach-to-offshore sequence north of Los Angeles was investigated by Howard and Reineck (1981). A total of 226 cores and additional vibrocores were studied by making relief casts and X-ray radiographs.

The sequence can be divided into three zones: (1) coastal sand, up to 9 m water depth, (2) transition zone, between 9–18 m water depth, and (3) shelf mud zone, located below the 18-m water depth. The coastal sand is made up of medium sand, the transition zone is silty sand, and the shelf mud shows silty sand and sandy silt. A hypothetical progradational sequence is plotted in Fig. 561.

The foreshore is made up of laminated sand, showing low-angle dips towards the sea. Near the low-water line small current ripple bedding is common. At the low-water line, layers of pebbles may be present, although otherwise the pebbles occur as distributed particles. In the shoreface sediments laminated sand, along with wave-ripple bedding and small-ripple bedding, are present. In the upper ⅓ part of the shoreface, cross-bedding is more common than wave ripples. The bedsets of lami-

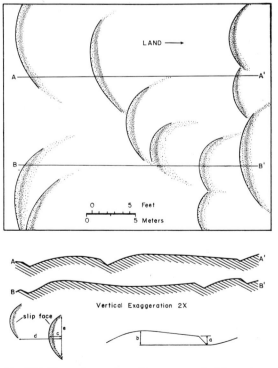

Fig. 560. Plane view and profile across part of the outer rough zone with lunate megaripples. (After Clifton et al. 1971)

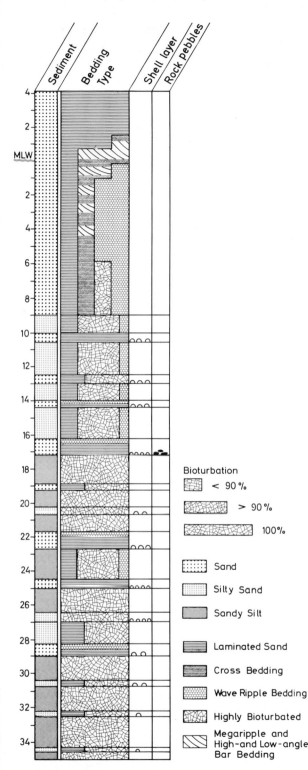

Fig. 561. Vertical sequence model for the Ventura – Port Hueneme portion of the California shelf. This generalized sequence incorporates the results of all the box cores, can cores and vibrocores taken in this area. (After Howard and Reineck 1979)

nated sand are characterized by the presence of low-angle discordances. In the lower $\frac{1}{3}$ part of the shoreface, burrows and bioturbation structures can be recognized and become more abundant with increasing water depth.

The transition zone is bioturbated up to 50 %, a considerable increase in relation to the lower part of the shoreface. Vibrocorer samples exhibit the presence of storm sand layers in the fine-grained transition zone deposits.

The intensity of bioturbation in the sandy silt sediments of shelf mud is greater than in the transition zone deposits. Mostly, they show totally churned sediments – 100 % bioturbated. However, the presence of storm sand layers in the shelf mud sediments can be mostly identified.

In examination of 23 vibrocorer profiles from the transition zone, numerous storm sand layers are recognized. These storm sand layers appear to be correlative in different cores over long distances. The storm sand layers are 10–50 cm thick; mostly about 45 cm thick. In the area located in front of the mouth of the Santa Clara River, the thickness of individual storm sand layers is sometimes up to 190 cm, suggesting a source of sandy material from the Santa Clara River.

A characteristic storm sand layer begins with a unit of shell layer, plant debris, wood pieces, pebbles, and mud pebbles (Fig. 562), suggesting an erosional phase at the beginning of storm sand layer deposition (see also Fig. 550). Significantly, the beginning of a storm sand layer denotes a change from highly churned (bioturbated) sediments below to the sediment showing well-developed primary physical structures above. The major part of a storm sand layer is made up of laminated sand; only near the top of the layer sediments become silty and show wave ripples and wave-ripple bedding (Fig. 562).

Characteristic of this high-energy coastal sand-shelf mud sequence is the large thickness of the coastal sand (about 9 m thick below the low-water line = shoreface + several meter-thick foreshore, backshore), great thickness of transition zone (ca. 9 m thick, from 9–18 m water depth), and thick storm sand layers (about 50 cm thick, maximum up to 2 m).

Beach-Shelf Profile, Sapelo Island, Georgia, U.S.A.

Introduction

Sapelo Island is part of the chain of Pleistocene and Holocene barrier islands on the Atlantic coast of Georgia. They were formed as a result of succes-

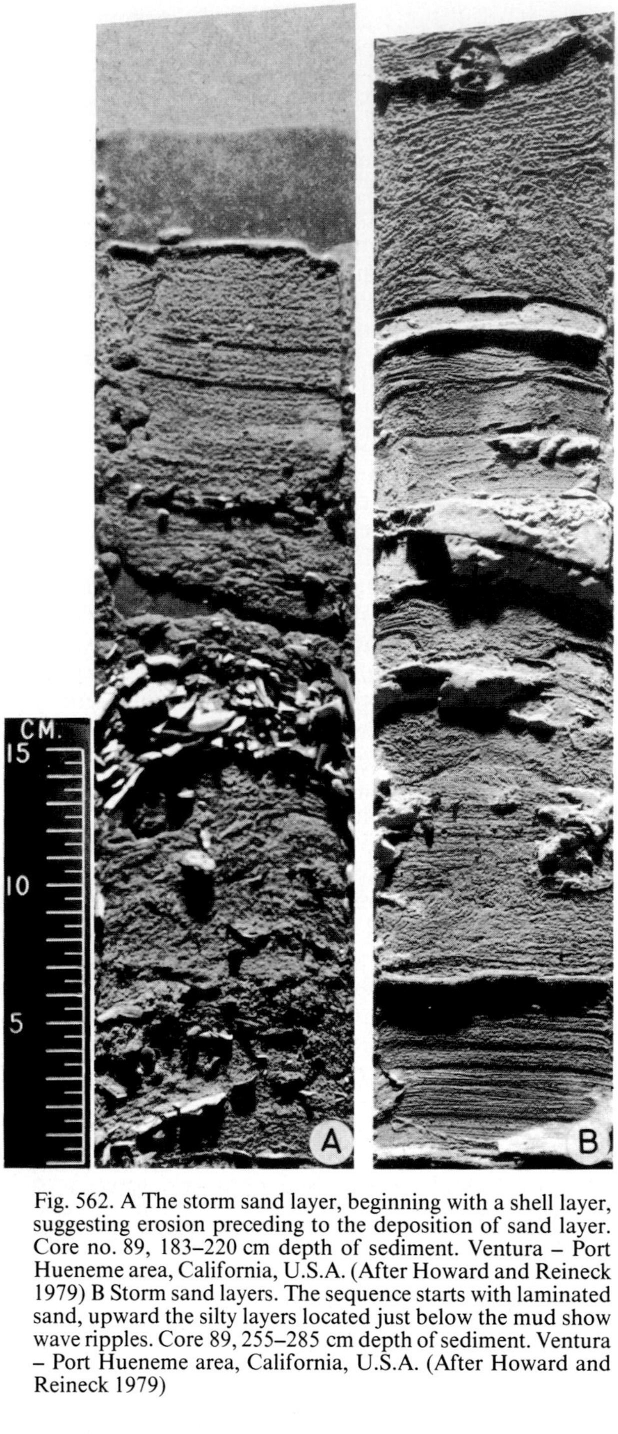

Fig. 562. A The storm sand layer, beginning with a shell layer, suggesting erosion preceding to the deposition of sand layer. Core no. 89, 183–220 cm depth of sediment. Ventura – Port Hueneme area, California, U.S.A. (After Howard and Reineck 1979) B Storm sand layers. The sequence starts with laminated sand, upward the silty layers located just below the mud show wave ripples. Core 89, 255–285 cm depth of sediment. Ventura – Port Hueneme area, California, U.S.A. (After Howard and Reineck 1979)

Sapelo Island is bounded by Sapelo Sound in the north and Duboy Sound in south. These sounds are large tidal channels, which act as passages for water which floods the extensive intertidal marsh and subtidal estuarine and tidal creek complex. On the north and south sides of these entrances complex shoal-channel systems are present (Oertel and Howard 1972).

The shoal system of the south end of Sapelo Island is much larger and more complex than that present at the north end. The system of shoals and channels has a direct effect on the pattern of sedimentation in the Sapelo Island area. These shoals cannot be regarded as offshore bars.

Detailed sampling was done of the region with the help of can cores, box cores, and vibro cores, and inorganic and biogenic sedimentary structures were described in detail (Howard and Reineck 1972a,b; Frey and Howard 1972; Hertweck 1972). Bioturbation structures play an important role in characterizing sedimentation units of this area.

Distribution of Sedimentary Structures

Each sedimentation zone of the area shows a characteristic association of sedimentary structures.

The beach is characterized by typical low-angle, seaward-dipping, wedge-shaped sets of evenly laminated sand (Fig. 564a). In addition, ridge and runnel systems which migrate across the foreshore produce some characteristic structures. Associated with them are antidune cross-bedding (Fig. 58), bioturbation structures, mud layers, current ripples and cross-bedding, wave ripples (Fig. 508), high-angle landward-dipping cross-bedding, wavy bedding, and concentration of shells (Wunderlich 1972). These features are normally not considered characteristic of beach deposits.

Formation and migration of ridges and runnels is a complex process active throughout the tidal cycle. This process is very important in the development of the beach of Sapelo Island. An important factor of ridge and runnel system is their potential for preservation. On Sapelo Island these features are readily destroyed under higher-energy conditions during storms. Thus, their preservation in a beach sequence may suggest deposition associated with normal rather than storm conditions.

Beyond the beach (foreshore) there exists a series of sedimentation units in response to current energy, wave energy, geomorphology, depth, grain size, and nature and abundance of benthonic fauna (cf. Howard and Reineck 1972a).

From the low-water line to approximately 1 m of water depth sedimentary structures are dominantly evenly laminated sand (Figs. 563 and 564a). This zone is the upper shoreface. In the lower shoreface (1 to 2 m water depth) small-ripple bedding is

sive sea level changes during the last regression across the Georgian coastal plain (Hails and Hoyt 1968, 1969). A broad area of salt marshes and a complex tidal drainage system (8 to 11 km wide) separates Sapelo Island from the mainland. Sapelo Island proper is a Pleistocene remnant, being separated from the present-day barrier island by a narrow strip of salt marsh (Howard et al. 1972).

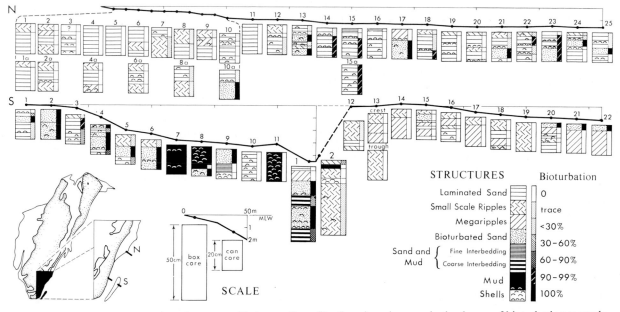

Fig. 563. Geographic representation of can cores. Various sedimentary structures found in each can core profile are shown by symbols for each station. The degree of bioturbation is shown on the right of the column of the can-core profile. In North-profile, there is an increase in the degree of bioturbation towards sea. In South-profile, bioturbation is absent in samples 12–19, because of the developments of a shoal with megaripples. (After Howard and Reineck 1972a)

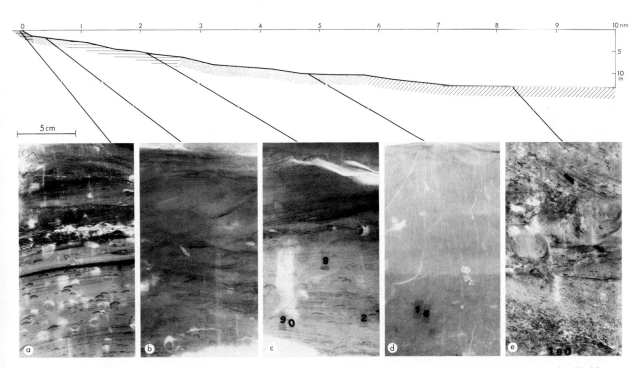

Fig. 564. X-ray radiographic prints showing sedimentary structures in a representative beach offshore profile. a Upper shoreface (0 to 1 m water depth), evenly laminated fine sand. b Lower shoreface (1 to 2 m water depth), small ripple bedding. c Upper offshore (2 to 5 m water depth), evenly laminated fine sand with bioturbation. d Upper offshore (5 to 10 m water depth), bioturbated muddy fine sand. e Lower offshore (> 10 m water depth), clean medium sand with trough cross-bedding and heart urchin bioturbation. (After Howard and Reineck 1972a)

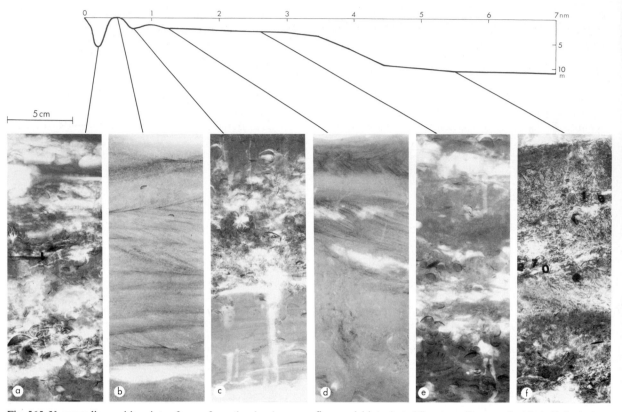

Fig. 565. X-ray radiographic prints of cores from the shoal profile S. The profile illustrates the variation in grain size and sedimentary structures in relation to position in the shoal area and water depth. a Core from the trough between the beach and the shoal. Bioturbated fine sand and shells. Note the mud pebbles at the top of the core. b Trough cross-bedding and small-ripple bedding in the intertidal part of the shoal. c Evenly laminated fine sand, bioturbated fine to medium sand with shells in the intershoal tidal channel. d Trough cross-bedding and bioturbated fine sand from the subtidal part of the shoal. e Bioturbated muddy fine sand in the upper offshore region (5 to 10 m water depth). f Fine and medium sand showing trough cross-bedding and intense bioturbation in the offshore region. (After Howard and Reineck 1972a)

abundant (Figs. 563 and 564b). In the sediments of the shoreface the degree of bioturbation is low. The zone from 2 to 10 m water depth is referred to as the upper offshore. From 2 to 5 m (the upper part of the upper offshore) sediments are dominantly evenly laminated sand and bioturbated sand (Fig. 564c). Layers of evenly laminated (non-bioturbated) sand alternate with layers of bioturbated sand. The evenly laminated sand is produced during storms or greater-than-normal wave energy, and the bioturbated sand is produced during interstorm periods of normal energy, in which animals churn the sediment by burrowing activity. In the lower part of the upper offshore (5 to 10 m water depth) (Fig. 564d) sediments are almost entirely bioturbated. In this part biogenic processes are dominant. The lower offshore (> 10 m water depth) is made up of reworked relict sands, where no new sediment is being deposited. Sediments are clean, medium to coarse sands. Megaripple bedding and heart urchin bioturbation are the only important sedimentary structures of this part (Fig. 564e).

A very significant subenvironment of deposition is represented by shoals built in front of estuaries. Distribution of sediments in a shoal-channel system is variable, but in general it is clean fine to coarse sand. Important sedimentary structures are megaripple bedding and small-ripple bedding (Figs. 563 and 565 b). The degree of bioturbation is low, except in small tidal channels cutting through the shoals. In such regions sediments become more muddy and the degree of bioturbation increases (Figs. 563 and 565a). These shoals are also significant in that their seaward extension is at right angles to the beach alignment and acts as a focus for east and northwest winds. This helps in shifting the beach-offshore sequence seaward. In the shoals on the north end of Sapelo Island biogenic processes and bioturbation structures are more important than the inorganic sedimentary structures.

Distribution of Fauna

Detailed biological sampling was done along three transects from beach-offshore sequences, and macro-invertebrates were studied (Dörjes 1972). Nearly 100,000 individuals belonging to 268 species of 14 taxonomic groups were identified. 87 % of the total number of individuals belong to the 10 most common species. Molluscs, polychaetes, and crustaceans are the three most abundant groups.

From the beach to the offshore area there is a regular, distinct zonation of organisms, which corresponds directly to the sedimentation zones. In the backshore and upper foreshore region there are only a few individuals, which belong to only a few species. All of them are terrestrial and marine migrants. In the lower foreshore individual organisms are abundant; however, they belong to only a few species. This is also true of shoals, which in grain size and hydrodynamic environment resemble the foreshore. Seven of the 10 most abundant species are typical inhabitants of the foreshore and shoal regions.

In the shoreface, by contrast, there are only a few species represented by a few individuals. The upper offshore shows the maximum for both species and individuals. Here polychaetes are most abundant. In the clean coarse sand of the offshore area, only a few species and a few individuals are present. Four animal communities are established, based on the characteristic species. These are the *Ocypode quadrata* community of the backshore; the Haustoriidae community of the foreshore, shoreface, and shoals; the *Hemipholis elongata* community of the upper offshore; and the *Moira atropos* community of the lower offshore.

Distribution of Trace Fossils

Biological parameters may be recorded in two ways: Body fossil and trace fossil. In an environmental reconstruction of the nearshore environment emphasis is laid more on trace fossils. Body fossils are often not preserved in nearshore clastic sediments.

A detailed study of trace fossils was made in the Sapelo Island region (Hertweck 1972). Only a few of the recorded benthonic species are capable of leaving distinct lebensspuren. This may be due to burrowing nature of the organism or because population is so dense that structures are masked by each other. However, specific sedimentation zones can also be defined by dominant lebensspuren. These are the burrows of the ghost crab *Ocypode quadrata* in the backshore, burrows of the ghost shrimp *Callianassa major* in the foreshore and shoals, complete bioturbation of the sediments in the upper offshore by *Callianassa biformis* and other species, and distinct bioturbation structures of the heart urchin *Moira atropos* in the lower offshore. Only shoreface sediments lack any distinct

lebensspuren. This is due to a high rate of reworking by waves of this region.

Summary

The Sapelo Island region, Georgia, is a low-energy tide-dominated coastal region, characterized by specific environmental zones defined by grain-size characteristics, sedimentary structures, macrobenthonic organisms and lebensspuren. These zones are interrelated and interdependent (Fig. 566) (Howard and Reineck 1972b).

Backshore. Sediment is clean fine sand, showing sedimentary structures, e. g., wave-deposited evenly laminated sand and ripple-bedded sand, and wind-deposited laminated sand and wind-sculptured features. Macrobenthos is represented by a few individuals of a few species. Both faunal and trace fossil evidences are characterized by the ghost crab *Ocypode quadrata*. The degree of bioturbation is low.

Foreshore. Sediment is clean fine sand. Two groups of inorganic sedimentary structures are developed. The first is low-angle seaward-dipping sets of evenly laminated sand. The second is the ridge and runnel system, where high-angle landward-dipping laminated sand, antidune cross-bedding, ripple bedding, mud layers, wavy bedding and concentration of shells are commonly developed. Bioturbation is subordinate, and is mainly by amphipods. A few species with a great number of individuals are present. The most characteristic lebensspuren is the burrow of *Callianassa major*.

Shoreface. Fine sand arranged in subparallel sets of evenly laminated sand in the upper part (up to 1 m water depth), and small-ripple bedding in the lower part (1 to 2 m water depth). The number of species and individuals is few, and there are no lebensspuren characteristic of this region.

Upper Offshore. Muddy fine sand shows laminated sand and bioturbated sand. In the upper part (2 to 5 m water depth) both laminated sand and bioturbated sand are present. In the lower part (5 to 10 m water depth) bioturbated sand is dominant; locally, evenly laminated sand may be present in minor amounts. The number of individuals and species is exceptionally high throughout the upper offshore region. A high degree of bioturbation is characteristic of this region; burrows of *Callianassa biformis* are the specific lebensspuren of this zone.

Lower Offshore. Sediment is clean medium to coarse sand showing abundant megaripple bed-

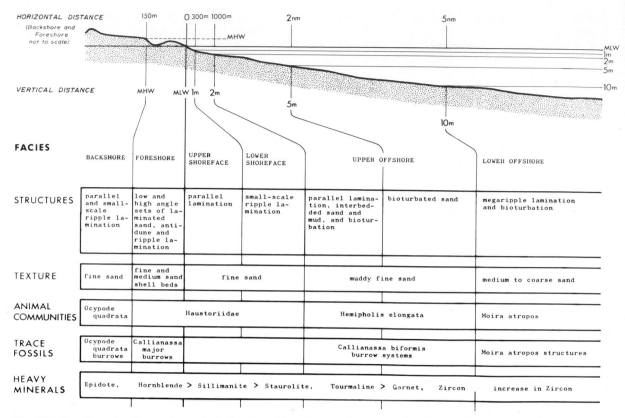

FACIES	BACKSHORE	FORESHORE	UPPER SHOREFACE	LOWER SHOREFACE	UPPER OFFSHORE		LOWER OFFSHORE
STRUCTURES	parallel and small-scale ripple lamination	low and high angle sets of laminated sand, antidune and ripple lamination	parallel lamination	small-scale ripple lamination	parallel lamination, interbedded sand and mud, and bioturbation	bioturbated sand	megaripple lamination and bioturbation
TEXTURE	fine sand	fine and medium sand shell beds	fine sand		muddy fine sand		medium to coarse sand
ANIMAL COMMUNITIES	Ocypode quadrata	Haustoriidae			Hemipholis elongata		Moira atropos
TRACE FOSSILS	Ocypode quadrata burrows	Callianassa major burrows			Callianassa biformis burrow systems		Moira atropos structures
HEAVY MINERALS	Epidote,	Hornblende > Sillimanite > Staurolite,		Tourmaline > Garnet,	Zircon		increase in Zircon

Fig. 566. Zonation of sedimentation units in beach and nearshore environments, Sapelo Island, Georgia. (After Howard and Reineck 1972b)

ding. Only a few individuals of a few species live here. The heart urchin *Moira atropos* is the important animal of this zone, and also makes characteristic lebensspuren.

Shoals. Shoals formed at the mouth of the estuaries make significant interruptions in the beach-offshore sedimentation model. Shoals are composed of clean fine sand showing abundant small- and megaripple bedding. The degree of bioturbation is low. Only a few species with a great number of individuals are present. The important lebensspuren is the burrow of *Callianassa major*.

Beach-to-Offshore Profile on Gravelly Coasts, Costa Brava, Spain

On the Spanish Costa Brava six small bays showing gravelly coasts are studied, especially the sedimentological and biological aspects of the swash zone and shoreface (Reineck and Dörjes 1976). These six bays show different wave energies, ranging from totally exposed to wave action to highly protected. With the exception of the Bay of Sa Tuna, where calcareous phyllites are exposed, all other bays have granitic cliffs, and there is no incoming fluvial sediment or coast-parallel transport. Thus, the granite and granodiorite gravels are derived from the exposed cliffs.

Sphericity and Roundness of the Gravels in Swash Zone. The gravels < 20 mm are significantly more angular than the larger gravels, as the smaller gravels lie protected within the larger gravels.

If one plots the arithmetic mean of the sphericity and roundness of the gravels of individual bays, then the following sequence becomes evident: Sotavento–San Telmo N–San Telmo S–Alger–Atumi–Sa Tuna, where roundness is best at Sotavento and poorest at Sa Tuna (Table 31A,B). Similarly the sphericity is also best in Sotavento and poorest in Sa Tuna (Table 31B). This sequence also corresponds to the extent to which the bays are protected from wave action; i. e., Sotavento is least protected, while Sa Tuna is best protected. This suggests that with the increasing unprotected nature of a bay leading to increasing wave energy, the gravels are more intensively moved, causing improved roundness.

Size Distribution of Sand, Gravels, and Blocks in Foreshore. The sand in these gravelly coasts is

Table 31

A. Indices of roundness of pebbles from the swash zone and the shoreface. The bays are arranged from high to low energy. With increasing sheltering of the bay the roundness of pebbles decreases and the standard deviation increases. (After Reineck and Dörjes 1976)

	Swashzone rounding	Standard deviation	Shoreface rounding	Standard deviation
Sotavento	670	± 155	555	± 147
San Telmo N	574	± 161	534	± 146
San Telmo S	570	± 172	530	± 180
Alger	493	± 165	393	± 129
Atumi	459	± 145	–	–
Sa Tuna	353	± 210	–	–

B. Indices of flatness of pebbles from the swash zone. The bays are arranged from high to low energy. With increasing sheltering of the bay the flatness and standard deviation increases. (After Reineck and Dörjes 1976)

	Flatness	Standard eviation
Sotavento	1.66	± 0.33
San Telmo N	1.68	± 0.37
Alger	1.80	± 0.40
Atumi	1.90	± 0.45
Sa Tuna	2.58	± 0.93

present in greater water depths (Figs. 567 and 568). In unprotected bays, free sand surface is present in water depth > 3 m; while in protected bays it is present at about 2 m water depth. In the swash zone and upper shoreface sand is present only within the spaces of gravels. The grain size of the sand decreases with increasing depth. The sand of the swash zone is 1.4–2.0 mm in size. There is a prominent grain size break between sand-size and gravel-size material, i. e., between 2 and 4 mm size.

The gravels of the bays investigated are classified into size intervals of 6 cm, 6–20, 20–40, 40–100 and > 100 cm (the longest axes of the gravels is considered).

The distribution pattern of gravels depends upon the degree of protectedness of the bay (Figs. 567–570). Most significant is the distribution pattern of gravels in < 6 cm, and 6–20 cm sizes. In the unprotected bays of Sotavento and San Telmo, these gravel sizes are almost completely absent below the swash zone. It seems that the wave action during storms causes strong sorting effects. Essentially, the gravels of 6–20 cm size are moved on the beach, and some of these gravels together with sand are transported into deeper waters. Thus, 20–40-cm-size gravels become concentrated in the breaker zone. A similar sorting effect by waves has been observed by Longinov (1956, 1958, cited in Zenkovich 1967). In the protected bays, such sorting effects are absent, and even in water depths of < 1 m, numerous gravels of 6–20 cm size are present.

Rounding in Shoreface. Except for the Sa Tuna bay, gravels from the shoreface of the other bays are also studied. The results show that in the shore-face area, gravels of the same gravel size class (6–20 cm and 20–100 cm) show poorer sphericity than the gravels in the swash zone (Table 31B). However, the gravels of the shoreface also show a decrease in the degree of roundness, depending upon the protectedness of the bay.

With increasing water depth, the number of large, angular blocks increases. The blocks larger than 300 cm in length are not considered. In exposed bays, the lower limit for angular blocks is even more than 100 cm; although in protected bays this limit is even less than 60 cm. In protected bays, angular blocks are present in much shallower water depths (1 m water depth) than in the exposed bays (Figs. 567–570).

However, despite the sequence of poorly sorted gravels and large, angular blocks increasing with the water depth, in water depths of 2–3 m scattered well-rounded gravels of 6–20 cm size are present. These gravels are a product of reworking in the < 1 m water depth and swash zone, and have been brought into deeper parts by strong wave action during storms.

Distribution of Fauna and Flora, and Palaeontological Aspects. In the zone of active reworking (causing rounding of the gravels) no epifauna and epiflora is encountered (Fig. 567). The lateral extent and the water depth of this zone is very narrow in protected bays. Below this active zone, conditions are suitable for the development of epifauna and epiflora.

The biological zonation is also dependent upon the energy conditions (degree of protectedness) of the bays (Figs. 567–570). In the energy-rich bays, the gravels of the swash zone are free from any or-

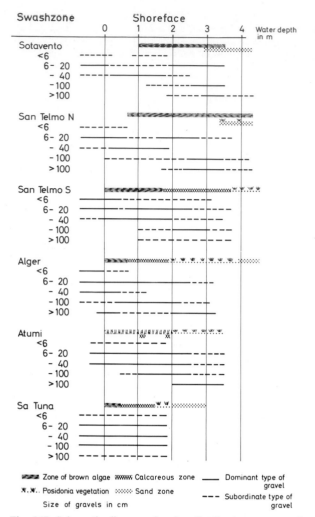

Fig. 567. Schematic diagram showing distribution pattern of gravel sizes, area of braun algae, distribution of Posidonia, and free sandy surfaces in different bays on Costa Brava, Spain. The bays are put in increasing order of protectedness (sheltered). The diagram clearly demonstrates that in the unsheltered bays the sorting is better than in the sheltered bays. In unsheltered bays *Posidonia oceanica* is only sporadically present. Carbonate encrustations are absent in unsheltered bays. With increasing sheltering of the bays, the brown algae and carbonate encrustation are present in decreasing water depths. (After Reineck and Dörjes 1976)

ganic growth, except for the presence of a few blue-green algal filaments. This zone extends to 1 m water depth in Sotavento, and 0.65 m in San Telmo (North).

In the protected bays, brown algal growth is extensive up to the waterline, and extends downwards to different depths in different bays.

The calcareous zone showing carbonate crusts of calcareous algae and calcareous animal tubes is absent in exposed bays of Sotavento and San Telmo (North). In San Telmo (North) bay a calcareous zone is present between a water depth of

1.7–3.7 m. In the bay of Alger and San Tuna this zone begins at a water depth of 0.7 m and 0.5 m, respectively. In the well-protected bay of Atumi, there is no clear separation between brown algae zone and calcareous zone.

Depending upon the hydrodynamic situation, in the deeper parts, the calcareous zone changes into Posidonia grass areas or sandy areas.

As already mentioned, a few well-rounded gravels are also present in deeper parts, brought in by storm action, and in such a situation such well-rounded gravels show epibiont growths (Figs. 571 and 572). Such cases are also known in ancient sedimentary records.

The carbonate content of the sand in the coarser fraction (0.5 mm) is made up of *Balanus,* gastropod shells, and echinoid spines. The fraction 0.5–0.125 mm shows foraminifera, ostracods, skeletal remains of calcareous algae and very small polychaete tubes. In the finest fraction, sponge needles are present. Significantly, only the broken pieces of small polychaete tubes, i. e., *Spirorbis* are present, but not the pieces of *Pomatoceros,* and calcareous algal crusts. In other words, the carbonate content of the sand is mainly due to organisms living in the sand, or those which fell down from the hard ground after death. On the other hand, the reworked hard ground calcareous fragments contribute very little to the sand fraction.

Coastal Upwelling

In the continental shelf areas and the basins near the continental shelves, where there are mostly seaward winds the surface water is transported in the offshore direction, and is replaced by cold, nutrient-rich waters from a water depth of 100–200 m. The nutrient enrichment of these upwelling waters is due to phytoplanktons, which in these water depths are enriched in P, N, F, SiO_2, and some other trace elements. If this upwelling water reaches the sunlit water surface, high productivity of phytoplankton takes place, and become food for zooplankton – fish – birds. However, due to overproduction of dinoflagellates, mass mortality of fishes may take place (Brongersma-Sanders 1957).

Coastal upwelling is quite common in Western United States, Peru, Morocco, South Africa, off the coast of Somaliland, Red Sea, Western Australia, as well as sometimes in SE Asia, east coast of India, Thailand, and Vietnam (Lafond 1966).

The sediments in the coastal upwelling areas are rich in diatoms, and radiolarians, and show low-contents of planktonic and benthonic foraminifera, increased amount of fish debris, and phosphorite grains (Diester-Haas and Schrader 1979). The

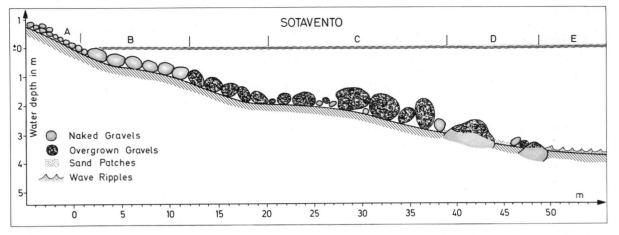

Fig. 568. Swash zone and shoreface morphology of the bay of Sotavento – an example of an unsheltered bay. Costa Brava, Spain, Mediterranean Sea. The sphericity, roundness, and sorting of the gravels are very good. In the swash zone, gravels of 6–40 cm size dominate. In the upper shoreface face (up to water depth of 1.90 m), gravels of 20–40 cm size are most abundant. In the lower shoreface, the sorting becomes poorer, and the gravels are covered by epilithic organisms. At a water depth of 3.5 m, a rippled sand zone is present.

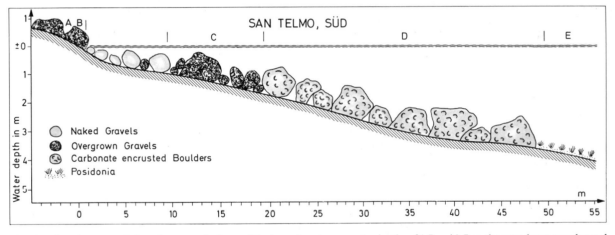

Fig. 569. Swash zone and shoreface morphology of the bay of San Telmo (southern profile) – an example of partly sheltered bay. Costa Brava, Spain, Mediterranean. In the swash zone gravels are grown with brown algae. Thus sorting of the gravels in this profile is poorer than in the Sotavento bay (Fig. 568). Between water depths of 1.7 and 3.7 m, the gravels are angular and are encrusted with carbonate producing organisms. At 3.7 m water depth a sandy bottom begins with eelgrass, *Posidonia oceanica*.

sediments of Walfish bay contain upto 20 % organic carbon, and about 70 % opal, derived from the diatom tests (Seibold 1974; Richert 1976).

Glacial Marine Sediments

In many areas, especially in the higher latitudes, glacial material is intricately mixed with the marine sediments, and can be referred to as glacial marine sediments. Vanney and Dangeard (1976) differentiate the following types of glacial marine sediment:

1. Proximal glacial marine deposits: Sediments which are deposited by the glaciers in a marine area, or sediments which are laid down as under-water moraine below a shelf-ice blanket.
2. "Glaciel" or distal glacial marine deposits: These are deposits where a coarse-grained fraction has been derived by drift ice.
3. Stranded drift-ice deposits: These are deposits produced by stuck-up drift, ice masses on the beaches, and tidal flats.

The term "glaciel" was suggested by Hamelin (1976) at the 1st International Symposium on the geological action of drift ice (Dionne 1976a). On

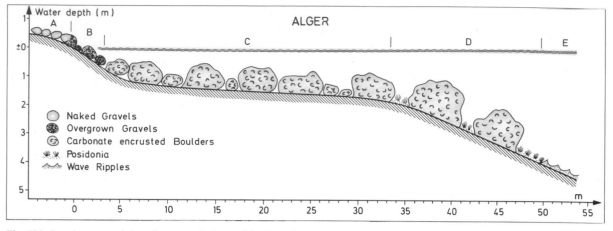

Fig. 570. Swash zone and shoreface morphology of the bay of Alger – an example of a sheltered bay. Costa Brava, Spain, Mediterranean Sea. The carbonate producing organisms are present already at a water depth of 0.7 m water depth. At 1.7 m water depth sandy bottom covered by eelgrass is present. Below 3.9 m water depth, the population density of eelgrass decreases, and the sandy bottom is covered by wave ripples. (After Reineck and Dörjes 1976)

Fig. 571. Well-rounded pebbles of granite with Recent barnacles (*Balanus perforatus*). San Telmo, Costa Brava, Mediterranean Sea. Water depth 1.5 m. (After Reineck and Dörjes 1976)

Fig. 572. Pepples of porphyry with fossil barnacles *(Balanus stellaris)*. Mainz Basin, Eckelheim, Lower Marine Sands (Tertiary, log. K. W. Geiß). (After Reineck and Dörjes 1976)

the present-day continental shelf there are several examples for distal glacial marine deposits ("glaciel"), which are recognizable by the coarse-grained material dispersed in fine-grained sediments. Loring and Nota (1973), and Nota et al. (1970) describe such sediments from the Gulf of St. Lawrence, and Baker and Friedman (1973) from Baffin Bay. Several other examples are discussed in Dionne (1976a,b) and Vanney and Dangeard (1976). Distribution of coarse-grained material in muddy sediments by drift ice is not restricted to the shallow sea only; but it also occurs in the deep-sea and lake sediments (Dionne 1976b). On the basis of UW photographs, Dangeard and Vanney (1975) distinguish several types of glacial-marine scree:

1. Seed bed, a boulder and pebble pavement.
2. Cluster, patches of boulders and pebbles.
3. Dropstones, individual, isolated boulder or pebble grains.

Catchment of sediment in the drift ice takes place by infreezing of the underlying sediment in the ice blocks, wind-blown sand, in-freezing of the suspension, washing over of the sediment on the ice blocks by the waves (Dionne 1970; Reineck 1976), or by the river water flowing over the ice masses (Reimnitz and Bruder 1972).

Another indicator of ice action is ice scouring. Elongated grooves are made by gouging of the sea floor by icebergs, mostly found in water depths of 10–30 m (Reimnitz and Barnes 1974). The grooves are 0.5–1 m deep; rarely up to 5.5 m deep. They are quite common on the arctic shelf (Fig. 573) (Pelletier and Shearer 1972; Kovacs 1972); and also reported from the Labrador Sea (van der Linden 1974), and the Norwegian Coast, North Sea (Belderson and Wilson 1973).

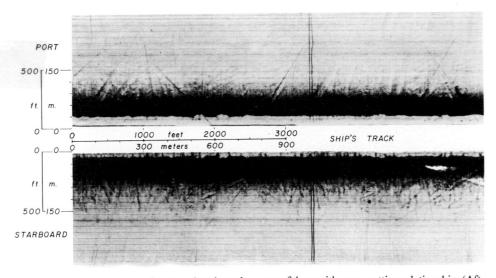

Fig. 573. Side-scan sonar photo reproduction showing numerous plough marks, some of them with cross-cutting relationship. (After Pelletier and Shearer 1972)

Another form of scouring by ice is described by Moign (1973, 1976) from Spitzbergen. Here, drift-ice blocks with their pointed ends, due to melting, produce 3 to 5 m-across and up to 5 m-deep holes in the shelf sediments (Fig. 574). These holes are reportedly smaller in shallower water and larger in deeper water, and are flushed out by the cold water of the melting drift-ice blocks. Another type of scour is described by Reimnitz et al. (1974) as strudel scour. The northerly flowing rivers of Alaska flow over the wide areas of the shelf ice. At places, large quantities of water are drained through the holes in the ice and produce holes in the sediments of the sea bottom, which can be up to 4 m deep.

Several authors have pointed out that during ice formation in the shallow sea, the coast is protected against the action of waves (Knight and Dalrymple 1976; Moign 1976; Owens 1976). Zenkovich (1970) discusses the coastal processes on a permafrost region, and points out that the long duration of the ice cover on the sea breaks the waves, but hardly affects the shoreline. This is why mud is deposited almost on the shoreline itself without any marked beach-ridge development.

Mineral Deposits on the Shelf

Many of the mineral deposits in the present-day shelf sediments are associated with the relict sediments, and their information cannot be utilized in the interpretation of the mineral deposits in the shelf sediments of the geological past.

There are many examples of beach placers in the shelf sediments. Ludwig and Vollbrecht (1975) describe the genesis of placers in the shoreface area; while Woolsey et al. (1975) discuss heavy mineral concentration in the backshore. Lamcke (1940) gives a useful list of older publications on this topic. In the shelf sediments, sometimes some authigenic minerals like phosphorite are concentrated (Bernard 1973). Mero (1965) and Schott (1976) give information on the mineral deposits of the sea. Weeks (1965) considers the shelf as a potential offshore petroleum resource. Rona (1972), and Amstutz and Bernard (1973) provide useful compilations on the mineral deposits (see also Table 32).

Examples of Ancient Coastal Sand and Shelf Sediments

In the last three decades the number of publications identifying ancient coastal and shelf deposits has increased enormously. Examples of ancient coastal sand and shelf sediments have been reviewed in the following publications:

Peterson and Osmond (eds.) (1961), van Straaten (ed.) (1964b), Visher (1965b), Potter (1963), Laporte (1968), Reineck (1970b), Selley (1970), Friedman and Sanders (1978), and Johnson (1978) give a good review of Recent and ancient shelf deposits, and a useful list of references.

In the following, a short account of a few well-known occurrences of coastal sand deposits is given. Identification of coastal sediments is easier in relatively younger sediments, i.e., Tertiary and

Fig. 574. Schematic representation of elongated groove markings made by icebergs, and circular erosional holes made below the foot of the icebergs. (After Moign 1973)

Table 32. Occurrence of continental shelf mineral deposits. (After Rona 1972; modified from Cruickshank 1970)

Unconsolidated		Consolidated		Dissolved
Bottom	Sub-bottom	Bottom	Sub-bottom	Seawater
Shallow beach or offshore placers: Heavy mineral sands Iron sands Silica sands Lime sands Sand and gravel	Buried beach and river placers: Diamonds Gold Platinum Tin	Exposed stratified deposits: Ironstone Limestone	Disseminated massive, vein or tabular deposits: Coal Iron Tin	Metals and salts: Magnesium Sodium Calcium Bromine
Chemical deposits: Manganese nodules (Co, Ni, Cu, Mn) Phosphoite nodules Phosphorite sands Glauconite sands	Heavy minerals: Magnetite Ilmenite Rutile Zircon Leucoxene Monazite Chromite Scheelite Wolframite	Chemical deposits: Manganese oxide Associated Co, Ni, Cu Phosphorite Hydrocarbons: Coal	Gold Sulphur Metallic sulfides Metallic salts	Potassium Sulphur Stronsium Boron Uranium Fresh water
	Hydrocarbons: Oil Gas			

Pleistocene, for the purpose of comparing them with modern environments.

On the coast of the North Sea interglacial marine transgressive phases are covered under the present-day sea level (van Straaten 1957; Gripp 1964; Sindowski 1965; Linke 1970). However, on the east coast of the U.S.A., especially the coast of Georgia, various interglacial transgressive shore lines are present one after the other (Hoyt and Weimer 1965; Weimer and Hoyt 1964; Hoyt et al. 1964, 1966, 1968; Hails and Hoyt 1968).

A total of six interglacial shore lines are recognized on the coast of Georgia (Fig. 465). They are located at various elevations: Wicomico (29 to 31 m), Penholoway (21 to 23 m), Talbot (12 to 14 m), Pamlico (7.5 m), Princess Anne (4 m), and Silver Bluff (1.4 m). Each shore line is marked by a series of barrier islands with lagoon, saltmarsh, and tidal inlet deposits behind it. The age of Silver Bluff has been determined by radioactive carbon dating, and corresponds to Wisconsin period (middle Würm). The position of the shore line has been determined by the height of lagoon sediments behind a barrier island. The highest position of burrows of *Callianassa (Orhiomorpha nodosa)* served as additional criteria. Weimer and Hoyt (1964) found that the burrows of *Callianassa major* are present only up to the mean sea level, and not above that. In some well-exposed outcrops the vertical sequence of a barrier island can be seen.

Colquhoun (1969) summarized the late-Tertiary-Holocene sediments of South Carolina and Georgia, which are present in some well-defined terraces. A large part of these sediments have been positively identified to be barrier islands, and other associated sedimentation facies.

Selley (1966) described a sequence of shore line sediments from central Libya of Miocene age. In the north-south profile various depositional units can be recognized: Coastal sands, lagoons, intertidal flats, estuarine channels, and fluvial sediments. There are no well-developed progradational tendencies to produce vertical sequences of various units.

Carter (1978) differentiates five depositional environments; namely, surf zone, foreshore, backshore-dune, freshwater marsh, and salt-water marsh in Tertiary Cohansey sand, and assigns a situation of barrier complex. Four other depositional environments, i. e., low and intermediate flow strength subtidal channels, intertidal sand flats and abandoned channels are ascribed to a barrier-protected complex.

Roep et al. (1979) reconstruct a progradational coastal sequence of wave-built structures in the Miocene of SE Spain. This reconstruction is compared with Recent and sub-Recent coastal sequences (Fig. 575).

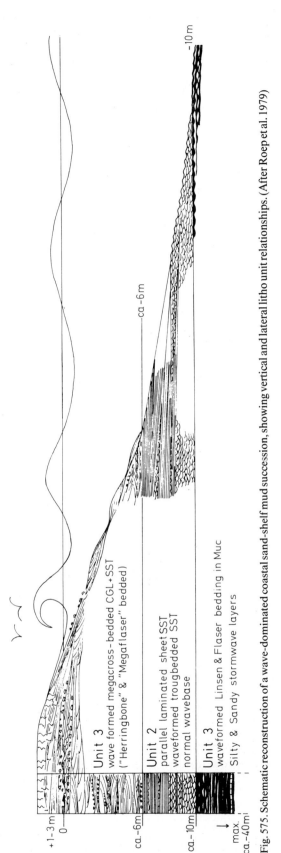

Fig. 575. Schematic reconstruction of a wave-dominated coastal sand-shelf mud succession, showing vertical and lateral litho unit relationships. (After Roep et al. 1979)

Unit 3
wave formed megacross-bedded CGL+SST
("Herringbone" & "Megaflaser" bedded)

Unit 2
parallel laminated sheet SST
waveformed trougbedded SST
normal wavebase

Unit 3
waveformed Linsen & Flaser bedding in Mud
Silty & Sandy stormwave layers

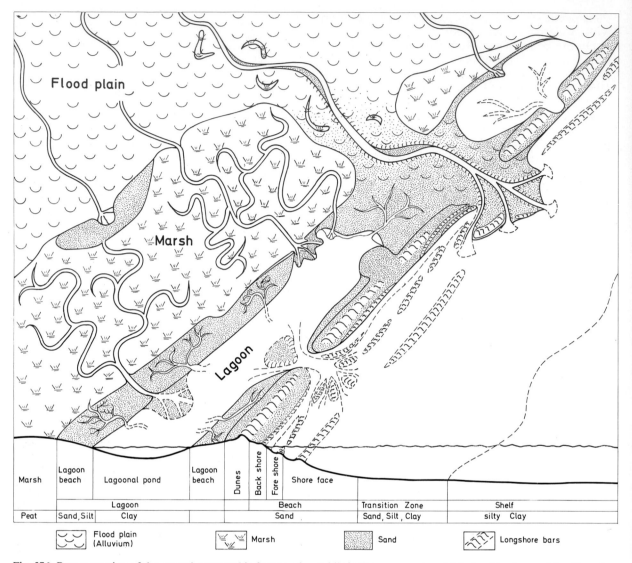

Fig. 576. Reconstruction of the coastal geomorphic feature of Cretaceous times showing various environments of deposition. In the lower part coastal sand is developed as barrier islands, while in the uppermost part it is developed as mainland beach. (After Reineck 1971; originally modified after Masters 1965)

From the Cretaceous period of the United States many coastal sand shelf mud deposits have been described. Hollenshead and Pritchard (1961) pointed out from the study of Cretaceous shore line sediments that a strong progradational tendency also exists during rising sea level (transgression). This leads to the fact that coastal sand bodies can prograde, if enough sediment is available, or they may grow upward if enough sediment supply is coupled with rising sea level; e. g. Padre Island, Texas. Moreover, coastal sand bodies may prograde forward, together with an upward growth, so that an inclined upward growth of sand bodies results (Figs. 576 and 577). Besides, sand bodies may grow backward in a retrogradational sequence in

an inclined upward growth. The type of growth of coastal sand bodies is strongly controlled by the rate of subsidence in relation to the rate of deposition, coupled with the rise or fall in sea level.

In the upper Cretaceous sediments of Utah and Colorado four major regressive and transgressive phases are present which caused a corresponding forward or backward growth of shore line deposits (Weimer 1960). Masters (1965, 1967) was able to recognize various environments and subenvironments in the Mesaverde Group and the Mancos formations. Both mainland beach and barrier island sequences are present (Fig. 578). Well-developed progradational vertical sequences are present, starting with shelf mud at the base (Fig. 579).

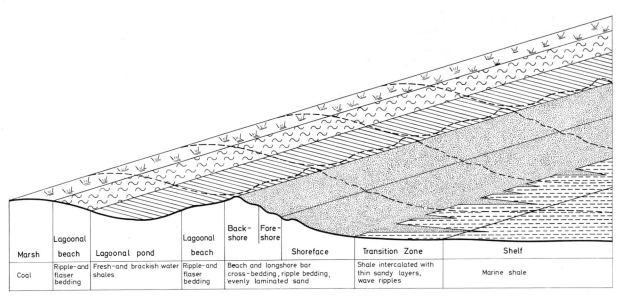

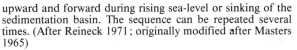

Marsh	Lagoonal beach	Lagoonal pond	Lagoonal beach	Back-shore	Fore-shore	Shoreface	Transition Zone	Shelf
Coal	Ripple- and flaser bedding	Fresh- and brackish water shales	Ripple- and flaser bedding	Beach and longshore bar cross-bedding, ripple bedding, evenly laminated sand			Shale intercalated with thin sandy layers, wave ripples	Marine shale

Fig. 577. Development of a vertical sequence made up of sedimentation (environmental) units which were originally located in lateral association. Development of a vertical sequence of laterally associated environments demands availability of enough sediment so that individual units may build inclined-upward and forward during rising sea-level or sinking of the sedimentation basin. The sequence can be repeated several times. (After Reineck 1971; originally modified after Masters 1965)

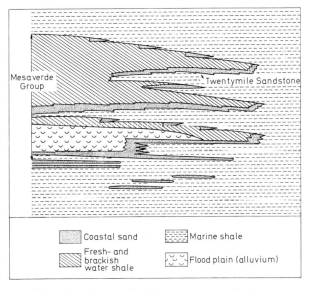

Fig. 578. Inclined forward building-up of coastal sediments of the Mesaverde Group, Upper Cretaceous. Such building-up of a sequence takes place in stages, alternating with periods of transgression. In the lower part of the diagram coastal sands are directly overlaid by alluvial sediments. This is a case in mainland beaches. In the upper part coastal sands are overlaid by lagoon deposits. This is the case with barrier-island beaches. (After Reineck 1971; originally modified after Masters 1967)

On the shelf mud follows the transition zone, with muddy sediments showing thinly interlayered sand/mud bedding, wavy and lenticular bedding. Sand layers are intercalated; mainly evenly laminated. Symmetrical wave ripples are on the top of the sand layer. Sediments of the lower shoreface are less bioturbated than their modern equivalents. In the upper part of beach sand megaripple bedding and landward-dipping, cross-bedding becomes more abundant than the laminated sand, which is the major bedding type in the lower part of beach sediments, i.e., shoreface and lower foreshore. The foreshore sediments show seaward dips, corresponding to the primary slope of the shoreface. In sections normal to the depositional strike original angles of shoreface slopes are visible (Fig. 580). In case of the barrier island sequence on the top of beach sediments lagoon and tidal flat deposits are present. Associated with lagoon sediments are deposits of tidal inlets and tidal deltas. On the top of the sequence swamp and flood plain deposits are present.

Ryer (1977) distinguishes a series of environments, i.e., lower offshore marine, upper offshore marine, offshore-shoreface transition, nearshore marine, coastal swamp and lagoon, coastal plain fluvial in the Cretaceous sediments of the Coalville and Rockport area, Utah, U.S.A.

Of the same age as Mesaverde Group are the Book Cliff sediments of Utah. They show a comparable vertical sequence and are one of the well-known coastal sand sequences (Young 1955, 1967; Howard 1966a). In addition, sandstones of the Book Cliff formation show abundant trace fossils and bioturbation structures (Howard 1966b).

Another example of coastal sequence is described by Ghibaudo et al. (1974) from Aren Sandstone (Upper Cretaceous), Pyrenees, Spain. It is a

Fig. 579. Photograph showing coastal sand deposits (thick sandstone beds) resting on the top of the shelf mud deposits (dark-colored sub-twenty mile shale on the right, below). Between the coastal sand and the shelf-mud is a zone with thin sandstone beds. This is the transition zone. Mesaverde Group, Upper Cretaceous. (After Reineck 1971)

case of progradational sequence with landward imbricated progressively younger depositional units: backshore, foreshore (with truncated wave ripple laminae), shore-face, offshore-beach transition, and offshore siltstone and mudstone. This sequence is shown in Fig. 581. This coastal sequence grades into turbidites, which in the beginning were deposited at a water depth of only 500 m.

Campbell (1979) describes a well-documented beach shoreline sequence from the Gallup Sandstone (Upper Cretaceous) of NW New Mexico. The sequence is made up of 0.1 to 0.6 m-thick layers, arranged in an imbricated fashion, so that younger horizons occur slightly displaced in the seaward direction. The sequence is made up of deposits of coal swamp, backshore, foreshore, shore-face, offshore-beach transition, down to offshore siltstone and mudstone. Campbell (1971) gives typical characteristics of individual environments (Fig. 582).

Aeolian dune deposits are made up of trough-shaped beds, looking like festoon bedding in the front view (Figs. 582A and 483). Contrary to the fluvial sands, aeolian dune deposits do not show any vertical sequence of sedimentary structures and textures.

Backshore deposits are made up of complex bedding structures, made by the filling up of swales (beach runnels). Cross-bedding showing angle of dip up to 25° in both landward and seaward direction are present. The swales run parallel to the shoreline (Fig. 582B).

Foreshore deposits are made up of even, parallel beds (laminated sand) with occasional low-angled cross-bedding showing seaward dips (Fig. 582C). They strike in the direction of shoreline. In Recent beaches, the situation is comparable to winter beach. Backshore and foreshore exhibit *Ophiomorpha*-burrows.

Fig. 580. Coastal sand deposits showing inclined bedsets of laminated sand, dipping toward the right. They suggest the forward-building beach. Mesaverde Group, Upper Cretaceous. (After Reineck 1971)

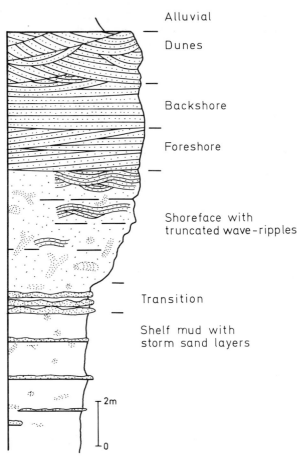

Alluvial

Dunes

Backshore

Foreshore

Shoreface with
truncated wave-ripples

Transition

Shelf mud with
storm sand layers

2m

0

Fig. 581. A progradational sequence of shelf mud – transition zone – shoreface – foreshore – backshore – sanddune in the upper Cretaceous sandstones of south-central Pyrenees. (After Ghibaudo et al. 1974)

Shoreface sediments are also made up of even, parallel beds (laminated sand), which are either strongly bioturbated or show truncated wave-ripple laminae (Campbell 1966). The crests of wave-formed features are placed at 1.5 to 4.5 m distance, and the height of the ripples is about 0.25 m. Besides, small wave ripples, burrows, and convolute bedding are also present. Burrows are mostly *Ophiomorpha*, and *Teichnichnus*.

In offshore-beach transition, the sandy beds gradually grade into siltstone and mudstone. The truncated wave ripple structures show lesser wave length than in the shoreface sediments, mostly 0.3 m in wave length, and less than 7.5 cm in height. Burrows, wave ripples, rarely current ripples, and contorted bedding are present.

The offshore siltstone and mudstone are made up of even and wavy parallel beds. Sometimes they are not parallel-bedded. Thickness of the beds is 2 to 10 cm, and they may show truncated wave-ripple or current-ripple bedding and on the top of a

bed wave ripples are present. These sediments show a rather high degree of bioturbation.

Campbell (1971) describe coastal sand-shelf mud sediments from the Cretaceous Gallup shore line deposits, Ship Rock area, New Mexico.

Davies et al. (1971) describes two examples of shore line sediments: Cretaceous of Bell Creek Oil Field, Montana and Jurassic of South England.

Häntzschel and Reineck (1968) describe sediments of Liassic age from Helmstedt, Germany. There are five sequences, which are repeated–each sequence starting in the shelf mud, passing into the transition zone, and again into the shelf mud. Possibly a decrease in sediment supply discontinued the progradational tendency and a reverse sequence was produced with increasing sea level. Sediments of the transition zone show a wide range of bioturbation structures and trace fossils.

An upper Palaeozoic transgressive sequence is described from the SW County Cork, Ireland (Graham 1975), where a fluvial deposit is overlain by a tidal flat deposit, followed by a barrier beach deposit, and after a minor non-depositional phase offshore subtidal sediments are present on the top.

Gieteling (1973) describes a Cambro-Ordovician prograding coastline sequence from the Cantabrian Mountains, NW Spain; which is made up of four facies groups, each of which shows several subfacies. Facies Group I represents littoral deposits with four subfacies, i.e., barrier beach deposit, beach deposit, tidal deposit, and rip-current channel-fill deposit. Facies Group II represents tidal deposits, consisting of high-energy tidal channel deposits, low-energy tidal channel deposit, non-channel tidal deposit, and highly wave-influenced tidal deposit. The other facies groups are fluvial-influenced deposits, delta deposits, and lagoonal deposits. For each facies group, a vertical sequence has been reconstructed (Gieteling 1973).

Storm sand layers in the form of shell layers are described by Brenner and Davics (1973) from the Upper Jurassic-Red Water Shale Member. Allochthonous graded coquinas in the upper Muschelkalk (Triassic) are interpreted as storm deposits (Aigner 1979). Goldring and Bridges (1973) give an interesting account of ancient storm-sand layers.

Kelling and Mullin (1975) studied storm deposits from the Carboniferous of Morocco, and point out that there are two types of effect by storms on the sediments: (1) storm-stirring and erosion and (2) lateral transfer of storm-mobilized sediment. By this process, coarse-grained sediment (quartz sand and carbonate detrital grains) is brought into slightly deeper environments, in which mud is being deposited by more protected processes.

De Raaf et al. (1977) describe a shallow marine sequence from the Lower Carboniferous of Ireland, which is essentially a wave-generated se-

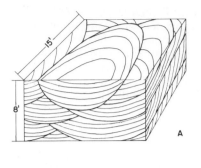
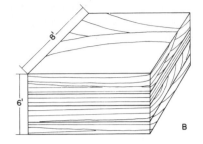
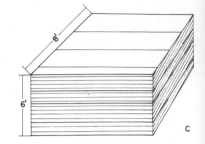
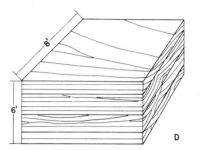
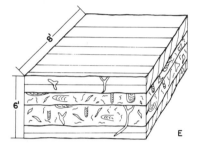

Fig. 582. A Dune deposit. In exposures perpendicular to wind direction, dune sandstone shows a crisscross pattern of laminae that compose the trough-shaped beds. In exposure parallel with wind direction, dune sandstone consists of sets of cross laminae that dip in a single direction and are contained between nearly parallel bedding surfaces. Downwind is toward viewer.
B Backshore deposit. Cross laminae dip in diverse directions between parallel bedding surfaces or are parallel with the bedding surfaces. Seaward is toward the viewer.
C Foreshore deposit. Cross laminae dip uniformly seaward between parallel bedding surfaces. Seaward is toward viewer.
D Shoreface deposit. Large-scale truncated wave-ripple laminae present between bedding surfaces. Churning by organisms sometimes present. Seaward is toward the front of the block.
E Offshore-bar structure. Cross laminae in offshore-bar sandstone dip uniformly in a single direction and at a high angle between wavy-parallel bedding surfaces. Cross laminae may be partly to completely destroyed by the churning of burrowing organisms. Downcurrent is toward viewer; seaward is to right. (After Campbell 1971)

quence. An exact modern counterpart of such sequences is not yet known.

Some of the shoestring sands are considered to be chenier deposits. Dillard et al. (1941) describe well-developed shoestring sands within the Pennsylvanian shales. Fisk (1955) interpreted these shoestring sands as chenier deposits. These shoestring sands show branching, and end abruptly. They are plano-convex in shape, with a flat base and a convex surface in a cross-section. Thickness is irregular. They are made up of well-sorted sand, showing well-developed cross-bedding.

Remarks

The various environments of a coastal complex which are laterally associated, i.e., alluvium, marshes, lagoon, lagoonal beach, tidal flat with tidal channels and creeks, washover, aeolian dunes, backshore, foreshore, shoreface, transition zone, and inner and outer shelf mud may form highly variable vertical sequences in the geological re-

cord. Some of such examples are discussed in the chapters on coast, shelf, coastal lagoon, and tidal flats.

One of the main factors controlling the variability of coastal sequences is whether the sequence is progradational or retrogradational. Most of the examples discussed earlier are progradational sequences, as they are also most common in the geological record. The retrogradational or transgressive sequences are mostly rather thin and poorly developed.

Other important parameters affecting the coastal sequences are the tidal range and the energy (mainly wave energy). The higher the tidal range, the thicker the tidal flat sequences and foreshore deposits. Energy is the most important factor controlling the thickness of the coastal sand deposits. In low-energy coasts, shelf mud is located at much less water depths (p. 405), while in the high-energy coasts, shoreface deposits are very thick, and shelf mud is present at greater water depths (p. 406).

In extreme cases, exceptional situations may develop, e.g., Nair (1976) describes mud banks from the SW coast of India, which are present just below the beach sands. Because of the very high content of suspension load in the waters, wave-action is reduced and mud banks are not eroded.

Synsedimentation behavior of a depositional basin is sometimes very important in controlling the development of sequences. If the depositional basin sinks at a pace of rate of deposition, then hundreds of meter-thick sequences of some environment are deposited, without the development of vertical sequences. Singh (1976) discusses this problem in the Vindhyan basin (Precambrian), India.

Inner shelf-beach sequences are developed in two major variants: (1) Sandy inner shelf-beach sequences and (2) muddy inner shelf-beach sequences.

A schematical vertical sequence of shelf-beach sequences is shown in Fig. 583. The water depths of different zones depend upon the effective depth and frequency of waves and currents causing erosion and deposition. The frequency is empirical and should be regarded as relative. Currents are mostly longshore and tidal currents. The wave base is marked, as above this waves *and* currents cause reworking. The high energy which is present on the foreshore during storms has been disregarded.

In the upper part of the foreshore the deposits of winter beach have much higher chances of preservation.

Further, information on the distribution of grain size in the vertical sequence is also given in case A (Fig. 583), where originally a lateral sand-mud sequence was developed. Curves depicting the distribution of a number of species and a number of individuals are also shown basing on data given by Dörjes (1975, 1976). Preservation of bioturbation structures depends upon a number of factors, where the number of individuals and rare physical reworking of the sediments are the most important factors. Although in the shelf mud the number of individuals and number of species decrease, the degree of bioturbation is high, because of the low rate of deposition and reworking. The number of species is important because the high number of species is associated with animals making larger burrows, and the larger burrows are much more effective for bioturbation than the smaller ones. The agglutinated and high-mucus burrows are better preserved than low-mucus burrows and surface traces.

The vertical sequence B can be made by progradation of a lateral sequence of sandy inner shelf-coastal sand.

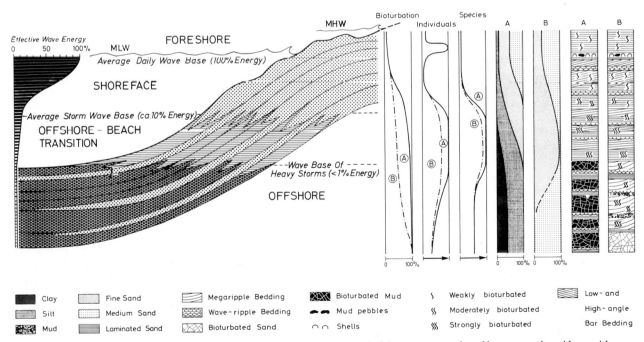

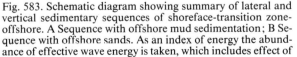

Fig. 583. Schematic diagram showing summary of lateral and vertical sedimentary sequences of shoreface-transition zone-offshore. A Sequence with offshore mud sedimentation; B Sequence with offshore sands. As an index of energy the abundance of effective wave energy is taken, which includes effect of waves and of the currents produced by wave action. Along with the wave energy tidal currents can also be effective agents. Near the sea-level the energy is considered as 100%, that means daily and the abundance decreases with increasing water depth.

Coastal Lagoons

General Information

Coastal lagoons are shallow water bodies, running parallel to the coast, and connected to the open sea with an outlet. They are separated from the open sea by sand bars or barrier islands (Fig. 576).

According to Phleger (1969), important factors for the development of lagoons is that enough sand-size sediment is available (brought by rivers) and enough wave activity is present for the formation of a sand bar or a barrier island. Phleger believes, as does Hoyt (1967a), that barrier islands develop as a result of a slowly rising sea level. Zenkowitch (1969) believes that one of the important sources of sand for the development of barrier islands is the shelf sediment in front of the coast.

The lagoons of the tidal regions are developed more or less like tidal flats. In such regions it may be difficult to differentiate between lagoons and tidal flats of the areas behind the barrier islands. Phleger (1969) includes the protected tidal flats behind barrier islands as lagoons. He even includes the intertidal zone among the lagoons. We suggest that in tidal regions behind the barrier islands, only those shallow depressions which remain water filled even at low tide should be considered lagoons. Moreover, lagoons of tidal regions gradually develop into tidal flat (intertidal zone) environments if enough sediment is available. The lagoons show large lateral extensions. Laguna Madre, in the Gulf of Mexico, is about 200 km long; Santo Domingo Lagune, in the Gulf of California, is more than 100 km long. Coastal lagoons are distributed all over the world. According to Zenkowitch (1969), 13 % of the coast line today is developed as a barrier island coast line, with lagoons behind them. Phleger (1969) states that $\frac{1}{3}$ of the Mexican coast and $\frac{1}{2}$ of the eastern coast of the U.S.A. south of New York is made up of lagoonal coast. However, in both cases no differentiation is made between tidal flats and lagoons.

The bottom of lagoons can be irregular in character. Deep channel-like features running at right angles to the coast line represent sunken river channels during transgression, which are kept open because of slow rates of sedimentation, e. g.,

San Antonio Bay, Texas (Shepard and Moore 1960). In other cases, such channels are maintained because of tidal currents, e. g., Laguna Ojo de Libre, Gulf of California (Phleger 1969).

The connection of lagoons with the open sea is known as inlet. On both extremities these inlets show the development of underwater deltas. Such deltas are made up of radially arranged channels and sand bars. The size and number of inlets of a lagoon depends upon the quantity of water which flows through it during a given time. At the same time, the amount of water is controlled by tidal range, number of tides per day, and water discharge of inflowing rivers. The salinity of lagoons is controlled by the water exchange with sea water, and amount of inflowing fresh water; there can be differences in salinity in various parts of a single lagoon. Evans and Bush (1969) measured salinities of 42 to 66 ‰ at water temperatures of 22 to 36 °C in the lagoons of the Persian Gulf. Where a river brings fresh water to the lagoon, salinity decreases from the inlet to the river mouth. Usually, a saltwater wedge is present in the lagoon below the fresh water. If no fresh water comes into a lagoon, this may lead to hypersalinity, e. g., the lagoons in Persian Gulf and Laguna Ojo de Libre, Gulf of California. The salinity is an important factor controlling the distribution of fauna in a lagoon. Lagoons with normal salinities show fauna similar to those in the open sea. Fauna are strongly impoverished with a change in salinity. In front of river mouths an altogether different fauna than the lagoonal fauna may develop. This has been demonstrated by Phleger (1964) and Parker (1959, 1960, 1964) by the study of foraminifera and macrofauna, respectively.

The lagoon Cienaga Grande covers about 450 km² area and is the largest coastal lagoon of Columbia. The salinity of this lagoon shows strong fluctuations, i. e., 30 ‰ in May and almost freshwater characteristics in December (Wiedeman 1973). Only few species have tolerance for such salinity fluctuations. Crabs, barnacles, encrusting serpulid worms, molluscs, e. g., *Polymesoda aequilaterata, Congeria sallei,* and *Hydrobia* sp. inhabit this lagoon. Near the tidal inlet, where salinity is higher, the number of species shows an increase.

Distribution of Sediment and Sedimentary Structures

Distribution of sediment and sedimentary structures within a lagoon is controlled mainly by hydrographic conditions and availability of the sediment. In the channels of the lagoons where velocity is rather high sandy sediments dominate and current ripples are commonly developed; in channels with lower current velocities sediments are silty to muddy in character. Extensive lagoon bottoms provide ideal conditions for the deposition of silty and muddy sediments, which can be extensively bioturbated. Intercalated within the fine-grained muddy sediments are sandy layers. The sand is mostly brought into the lagoon during storms. It may be aeolian sand blown in the lagoon, suspended sand eroded on tidal flats and lagoonal margins due to increased wave activity, or sand washed into lagoons by washover fans. Such sandy layers are either horizontally laminated, or developed as wave ripples and wave-ripple cross-bedding (Fig. 584).

Sediments deposited in front of river mouths bringing fine-grained sediment into lagoons show increased rates of sedimentation. Such sediments show well-developed inorganic primary structures that are very weakly bioturbated.

The intertidal zone associated with lagoons shows the same features as the intertidal zone of a tidal environment. In warm humid climates intertidal and supratidal zones of lagoonal environments are covered with plants, e.g., *Zostera, Spartina,* and *Salicornia.* In tropical climates, mangrove vegetation is present (Allen 1965; Curray 1969a). In several regions algal mats show a wide

Fig. 585. Microbial mat layers (dark-colored layers in the photograph) with mud cracks. Carbonate mud. Florida Bay

distribution. During dessication surfaces covered with microbial mats develop mud cracks with curled margins. If covered by new incoming sediment, such surfaces may be preserved (Figs. 585 and 617).

In a prograding coastal line, lagoonal sediments are overlain by freshwater marshes with peat formation.

In arid climates salt pans develop toward the landward side of lagoons. In Laguna Madre, during dry seasons, hypersalinity is achieved, leading to the development of evaporites and oolitic structures (Rusnak 1960).

The filling up of a lagoon is controlled by the availability of sediment. If a sediment source is available, e. g., a river entering a lagoon and bringing much sediment, a lagoon is rapidly filled up. Shepard and Moore (1960) calculated by comparison of hydrographic charts, a rate of sedimentation of 38 cm/year for the lagoons of central Texas.

Brown (1969) observed that lagoonal sediments may develop large-scale penecontemporaneous deformation structures, when sand dunes move across the lagoonal muds, as is the case in the Coorong Lagoon, South Australia. The lagoonal muds are modified into a series of subparallel anticlinal ridges, while the dune sand is crossed by a number of gravity faults in the process of settling over the lagoonal mud.

Example of a Modern Coastal Lagoon

In the following, a detailed description of Mugu Lagoon, southern California, is given. The description is based on the publications of Warme (1966, 1967, 1969a,b). A comprehensive account is given in Warme (1971).

Mugu Lagoon is situated on the Pacific coast of southern California. Annual rainfall is 25 to 50 cm.

Fig. 584. Lagoonal sediments showing wave ripples, Almere Deposits, Netherlands

Summers are dry; mean temperature is about 22 °C. Twice a day the region is affected by tides, which are of unequal heights (Fig. 586). Only the upper part of the tidal range affects the inside of the lagoon, which is true of most of the lagoons of Pacific coast. In the tidal inlet, during neap-tide, current velocities up to 1 m/sec are common; during spring tides current velocity reaches values up to 5 m/sec. During spring tides upper marshes and salt pans are flooded. Three types of sediments are present: Coastal sand, lagoonal mud (silt and clay), and salt pan silt. Most sediments of the lagoon are bimodal in nature, being a mixture in varying proportions of beach sand and lagoon mud. In the Mugu Lagoon several depositional units can be distinguished: Barrier beaches, tidal inlets, tidal deltas, tidal channels, ponds (lagoon bottom), tidal gullies, lower intertidal zones (barren zone), marshes and salt pans. All these units are located at a certain height in relation to mean low-water and mean high-water. At the same time they are arranged in a series away from barrier beach and inlet, in the direction of decreasing energy of the tidal currents. In the process of sedimentation these depositional units of the lagoon migrate laterally as the lagoon is filled, leaving behind a characteristic vertical sequence. The plants of salt marshes on the margin of the lagoon show zonation according to tidal level. The upper part of the intertidal zone is developed as a salt marsh. There are only a few species of animals with preservable hard skeletal parts but there is a high number of individuals. The vertical zonation of organisms in the intertidal zone is controlled by the water level of the tides. In the subtidal zone of the lagoonal bottom, vertical zonation is rather indistinct. However, lateral zonation in the lagoonal bottom is controlled by such factors as distance from inlet, current velocity, grain size. There is almost no post-mortem transport of hard

parts from one depositional unit into the other, and distribution of dead fauna corresponds well with the living fauna.

The inorganic primary sedimentary structures, i.e., bedding, are destroyed by bioturbation, which can be locally rather extensive. In the following, depositional units of Mugu Lagoon are briefly described.

Barrier Beach

The barrier beach is composed of medium sand, mainly laminated bedding with low-angled discordances. Individual laminae may differ in grain size and in mineral content. Sometimes small pebbly laminae are present. Development and filling-up of beach cusps produce trough-shaped cross-bedded units with axes normal to the shore line and laminae dipping seaward. In cross-section they resemble festoon-shaped units.

Biogenic structures are made by isopods, amphipods, molluscs *(Tivela stultormus* and *Donax gouldii),* and crabs *(Emerita).* Tracks of insects, birds, and seals are common.

Barrier and Sand Bar

The barrier bar is made up of clean sand, some pebbles, many mollusc shells, and some drift wood. Storm waves coupled with high tides cause extensive erosion of beach by backwash. Backwash is increased by a rise in the beach water table (Grant 1948). Much sand is transported seaward and the seaward profile becomes very steep sometimes almost vertical. Under such conditions and during high tide, washover commonly occurs and water and sediment wash down the back side of the barri-

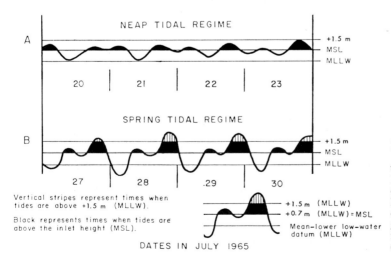

Fig. 586. Tidal curves for several days in July 1965, showing at what tidal heights in relation to the "sill depth" (high of the inlet channel), Mugu Lagoon was flooded. A-Times of flooding and drainage during neap tides. Drainage of the higher parts of the lagoon continues for several days. B-Times of flooding and drainage in the lagoon during spring tides. Also the curves showing times of flooding above the 1.5 m datum line. (After Warme 1971)

er, reaching the lagoon. Sand washed into the lagoon forms steep foreset beds on the backside of the barrier bar. The gentle backslope of the barrier is affected by wind activity during dry periods. Wind ripples, small dunes and deflation surfaces are produced. The next washover produces a smooth surface, and a thin layer of coarse material is deposited. When wave energy is reduced, new sand is deposited on the seaward side (Fig. 587). Arthropods are the important bioturbation animals, affecting the upper few centimeters. Plant roots are common.

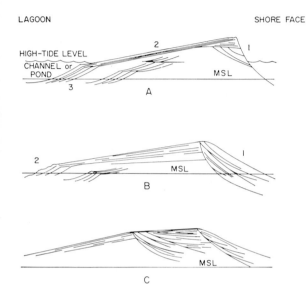

Fig. 587. Development of sedimentary structures in the barrier beach of Mugu Lagoon, California. A Transverse section of the barrier, illustrating erosion at times when high tides coincide with large waves. This is the winter beach profile: (1) Steep-angled beach due to erosion of beach front. (2) Lamination of sand and gravels formed by washover of the waves. (3) Foreset cross-bedding develops as sediment is washed into ponds or channels of the lagoon. B Normal or summer beach profile: (1) Gentle seaward slope develops as sand is sedimented on the beach front; (2) high-tide bench showing cross-bedding, notched on the surface by high-tide levels inside the lagoon. C Multiple beach sequences are preserved. (After Warme 1971)

Tidal Delta and "Algal"-bound Sand Flats

This unit is usually made up of clean sand, although some mud is also incorporated. Sand is horizontally laminated with occasional thin mud layers. The mud is more abundant on the inactive part of tidal delta ("algal"-bound sand flat), and this represents mud suspension trapped by microbial mats. In the inlet channels and partly on the sand flats megaripples and small ripples are formed, mainly in the flood direction. In the higher parts of inlet laminated sand is also present. The activity of organisms is very low.

Tidal Channels

The tidal channels coming from the inlet leading into the lagoon are made up of clean sand somewhat coarser-grained than sand in the sand flats. Some pebbles, gravels and shells are associated with sand. Ripple bedding and laminated sand are the major bedding types. However, bioturbation by *Callianassa* burrows is rather high (up to 160 specimens/m^2). Besides, many benthonic lamellibranchs, tube-dwelling polychaetes, sand dollars, gastropods and flat fishes also cause destruction of primary bedding.

Lagoonal Bottom (Subtidal Pond)

Sediments of the ponds are muddy in nature and black in color. Large areas are covered by eelgrass. The mud near the surface contains up to 90 % water. Mud is finely laminated and thinly bedded. However, the degree of bioturbation is very high; hardly any traces of primary bedding are visible. The main bioturbation is caused by polychaetes, arthropods, rays, fishes and plant roots.

Gullies (Tidal Creeks)

Sediments of tidal creeks show a wide spectrum of grain sizes, ranging from clean sand, through muddy sand to pure muddy sediments. In the sandy fraction abundant forams and ostracods are found. Ostracods are more common in muddy creeks.

The primary bedding of tidal creeks is highly variable, like the sediments. Many erosional discordances are present in complexly cross-bedded units. On the point bars of meandering gullies longitudinal cross bedding is produced. Creeks become progressively muddier as they are elevated by sedimentation, at last becoming just mud-filled troughs. Various benthonic animals inhabit tidal creeks in great numbers, producing highly bioturbated sediments.

Intertidal Zone

The intertidal zone comprises the barren zone, marsh, and salt pans. The barren zone extends from the nonvegetated lowest intertidal level up to the vegetated part of intertidal zone (marsh), which extends up to the highest intertidal level. The marsh is followed by salt pans, which are reached only by spring tides.

Barren Zone

The barren zone is made up of clean sand near to the inlet or barrier bar and of pure muddy sediments near the tidal creeks or mud flats away from the inlet. The lower part of the barren zone becomes sandier toward the tidal channel near inlet, and muddier toward the subtidal ponds. Small-ripple bedding, laminated sand, and thin mud-algal mat layers are the major bedding types. Interlayered sand/mud bedding may be present.

Bioturbation is strong. The main bioturbating organisms are *Callianassa,* Annelid-*Notomastus tenius,* molluscs and fishes. The resulting sediment is completely bioturbated or strongly bioturbated in structure.

Marsh

The sediments of marshes are made up of thinly bedded, sandy and muddy layers, the bedding planes are wavy in nature. Burrows and plant roots are the major features. Mud cracks are abundant.

Salt Pans

The sediments of salt pans are clayey silt, with a negligible amounts of sand-size grains. The source of the silty sediments is rivers. Sediments are well-bedded and thinly laminated; laminae rarely thicker than 5 mm. Mud cracks with upturned edges are abundant. The upper few centimeters of sediment is rather porous, with many cavities. Dried surfaces are eroded by wind and mud flakes are concentrated in the form of "clay dunes" (silt dunes). When wetted by rains or flooding spring tides, such dunes are again dispersed.

Examples of Ancient Lagoonal Sediments

Coastal lagoons are geomorphologically closely associated with tidal flats; thus, in ancient tidal flat deposits parts of the tidal sequences represent lagoonal deposits. However, examples of well-established lagoonal deposits from geological record are rather scarce. Masters (1965) interpreted dark-colored, pyritiferous shales from the Mesaverde Group, Cretaceous, U.S.A., as lagoon sediments. These shales are finely laminated with intercalations of thin silt and sand layers. Some of sand layers are sunken into mud layers, producing ball-and-pillow structures. Beds are strongly bioturbated. The brackish water lamellibranch *Crassostrea subtriangonalis* is present. Above and below this

unit are beds with wavy bedding, lenticular bedding, etc., belonging to deposits of the intertidal zone. Wave ripples are abundant, which suggest more of a lagoon environment.

Van Straaten (1954b) compared Psammites du Condroz (Devonian) of Belgium with the Dutch modern tidal flat deposits. However, despite a major similarity between the two, he points out several important differences. There are fewer features pointing to strong current activity and fewer channels in the Devonian of Belgium than in the tidal flats. Additionally, there are many sandy beds continuous over long distances and an abundance of wave ripples and wave-ripple cross bedding. Van Straaten (1954b) concluded that parts of these deposits are probably lagoonal deposits, sedimented in lagoons comparable to modern lagoons on the coast of Texas and Louisiana.

Van Loon and Wiggers (1975a,b,c, 1976a,b) describe a Holocene lagoonal deposit from Zuidersee, Netherlands, with many illustrations. This lagoon was dominated only by waves, so that wave ripples are predominantly present. Locally, the waves caused irregular bottom topography due to abrasion, and distribution of a large amount of sediments. A few scattered pebbles observed in the deposits have been brought in by drift ice. Erosional features such as peat holes and rocking holes point to wave action and some bottom current. These deposits are characterized by the common presence of penecontemporaneous deformation structures, named "metasedimentary structures" by Van Loon and Wiggers (1976c; see also p. 92). Singh (1973, 1976) interprets several litho-units of the Vindhyan system, India (Proterozoic) as lagoonal deposits. The most extensive and significant is the Bijaigarh shale (Singh 1980), which is a pyritiferous, black shale succession. Sandier parts of this succession (corresponding to the lagoonal margins) are characterized by the dominance of wave ripples, and features indicating intermittent subaerial exposure. Lagoons are often associated with wind tidal flats and salt pans, where salt deposits and stromatolites commonly develop.

Summary

Lagoonal sediments are deposits of low-energy areas with little current activity. Wave ripples dominate over the current ripples. Channels may be present. In the area of tidal inlet, features of current action are present, but they have a very small part in a lagoonal complex. The lagoonal beach often shows ripples. Laminated sand related to swash-and-backwash is rare.

Lagoonal deposits are made up of several subenvironments, e.g., the lagoon proper–lagoonal pond, mostly mud, the inlet with sand bodies, channels, the lagoonal margin, and the tidal flats or wind tidal flats. Besides, deposits of coastal dunes, barrier island, delta, swamp, and alluvial plain are closely associated. Because of hypersalinity or strongly fluctuating salinity, fauna is uncommon. A progradational sequence is depicted in Fig. 577. A retrogradational sequence is shown in Fig. 512.

Tidal Flats

General Information

Tidal flats develop along the gently dipping sea coasts with marked tidal rhythms, where enough sediment is available and strong wave action is not present. This may be the case in estuaries, lagoons, bays, or behind barrier islands or other sand bars.

Tidal flats facing open sea may also develop, especially if wave activity is reduced due to development of long subtidal zones extended toward the sea (e.g., sand tongues in the region of the Nordergründe, German North Sea).

The main part of the tidal flat is located between intertidal zone. Toward land is the supratidal zone and below the low water line is the subtidal zone.

In warm and temperate climatic zones the supratidal zone is vegetated by halophytic plants. In arid climates such regions can be devoid of any vegetation, e.g., the Gulf of California. The intertidal zone is mostly without any significant vegetation. In humid warm climates mangrove vegetation is present. A selective bibliography of mangrove literature is given by Frith (1977). The bibliographic list contains more than 2,600 titles. A subject index is added. However, on the eastern coast of the U.S.A., even in the temperate climate, the intertidal zone is vegetated–spartina salt marsh.

The surface of tidal flats slopes gently from high-water level toward low water level, although the slope is rather irregular. Depending upon the tidal range this difference can be several meters. However, the slopes are rather gentle; e.g., the North Sea intertidal zone is up to 7 km wide and the tidal range is 2.4 to 4 m. Within the intertidal zone channels are present and they are extremely branched toward their landward ends (Fig. 588). The meandering channels are eroded usually by ebb currents. However, the flood current also leave their mark in these channels in the form of tidal wedges (Fig. 589) (van Veen 1950).

We agree with Gripp (1956), not to include such regions among tidal flats which are subaerially exposed only occasionally due to wind effects, etc., because such regions lack characteristics of typical tidal flats.

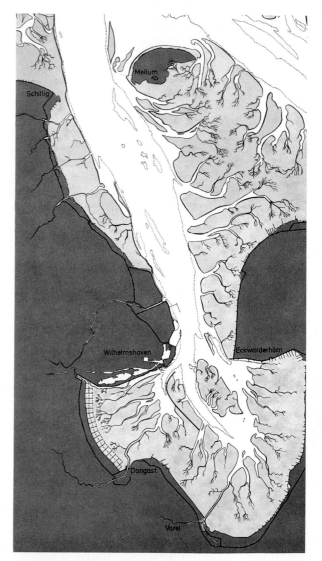

Fig. 588. The Jade, with Jade Bay and Inner Jade. The Inner Jade extends from the line joining Wilhelmshaven – Eckwarderhörne in the south, up to the line joining Schillig and Mellum in the north. North of this begins the Outer Jade. Dark gray: Mainland (areas situated above the mean high-water line). Medium gray: Intertidal flats (areas located between mean high-water and mean spring tide-low water). White: Subtidal zone. Dotted line: Five-meter depth contour. Chart datum of the hydrographic charts refers to the mean low water springs. (After Reineck 1970c)

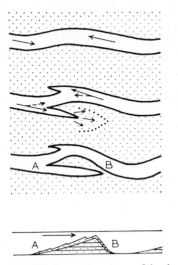

Fig. 589. Successive stages of the development of tidal wedges. A–B: Vertical section through tidal wedge. The flood current is to the right. (After van Straaten 1954a)

The subtidal zone is mostly occupied by channels and subtidal sand bars and shoals. Rarely do lagoons constitute the subtidal part. In the case of tidal flats located near deltas and estuaries and open tidal flats, marginal parts of the intertidal zone can be included in the tidal flat environment. However, deeper parts, i. e., the shelf and the river channels should not be included within the tidal environment.

Physiography and Morphology

Sediments in the intertidal zone are located between the high and low water lines over a vertical range of usually 2 or 3 m, and up to 10 or 15 m, depending upon the tidal range. Fluctuation in the water level due to tides causes tidal currents, which produce numerous gullies and channels (Fig. 588). The currents of a high tidal range erode deeper channels than the currents of low tidal range.

The current velocity of tidal currents on the flats usually possesses values of 30 to 50 cm/sec. Therefore, on sandy tidal flats small-current ripples are abundantly produced. The current velocity in gullies and channels may acquire values up to 1.5 m/sec or even more, so that in gullies and channels often megaripples and sometimes even giant ripples, are produced (Lüders 1929; Häntzschel 1938; van Straaten 1950b; Hülsemann 1955; Reineck 1963a). Although wave action is never too strong, it is nevertheless an important factor in controlling the distribution of sediments on intertidal flats.

In the gullies and channels bipolar tidal currents are important. This is manifested by the abundance of herringbone cross-bedding in channel sediments. On the tidal flat surface the current pattern is rather complex. Here, both current and waves are highly variable in their direction and intensity. The direction of current is controlled, as long as no area is subaerially exposed, by water-level differ-

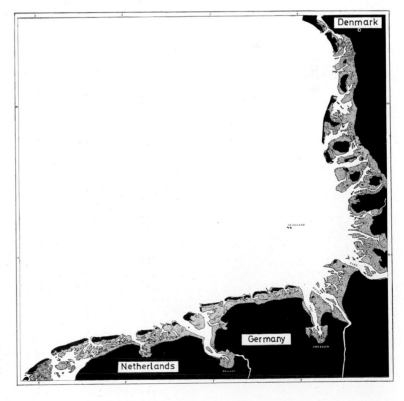

Fig. 590. General map showing tidal flats along the Dutch, German, and Danish coasts. The tidal flats extend over a distance of approximately 450 km and have an average width of 7 to 10 km. (After Reineck 1970c)

ences. Later, when parts of tidal flats are emerged, current direction is controlled by morphological slope. Strong winds may also affect the current direction. The direction of waves on tidal flats is highly variable; waves can come from various directions. The variable direction of waves and currents is expressed in changing directions of current and wave ripples. That is why herringbone cross-bedding in the surface sediments of the tidal flat is produced in only exceptional circumstances.

The tidal flat sediment-body is elongated parallel to the shore line over tens of kilometers and is intersected by tidal channels and river estuaries (Fig. 590). In the bights the tidal flat sediment-bodies are semi-circular, or trumpet-shaped, depending upon the shape of the bight.

Intertidal flat sediments are mostly fine-grained usually mud (silt and clay) and fine sand. Gravels are rare. However, mud pebbles and shells are abundant in channel deposits; occasionally, on tidal flat surfaces undergoing strong erosion, shell concentration may also occur. Most of the intertidal flats show a characteristic pattern of distribution of sediments (Fig. 591). Near the high-water line and watershed, sediments are muddy–muddy intertidal flats (Linke 1939; Stevenson and Emery 1958). The intertidal region near the low-water line is sandy. The muddy zone grades gradually into sandy intertidal flats. The transition part is known as mixed intertidal flats. The mud content also increases near the gullies of intertidal flats (Evans 1958). The mud content sometimes increases in larger channels of the subtidal zone, where deposition takes place mainly by lateral migration.

The tidal flats of the Netherlands are less muddy than the tidal flats in Germany. The tidal flats of Britain are rather sandy. In the Gulf of California tidal flats south of San Felipe are made up of sand. North of San Felipe, the mud content strongly increases toward the Colorado delta, which supplies the mud (Thompson 1968).

The reasons for this characteristic distribution of sediment on tidal flats are because of the energy and partly the transport mechanism. Near the low-water line the wave activity is strongest and active for the longest time as compared to higher parts of the intertidal zone. Thus, the sand is enriched here. Winnowed mud may be deposited in the subtidal zone below the wave base, e. g., on the margins of subtidal channels (Reineck in Dörjes et al. 1969). However, major mud deposition takes place in the muddy intertidal flat near the high-water line. The main reason is low wave and current energy in this part. Moreover, according to Postma (1961), the time of low-current velocity when fine-grained sediment may be deposited is much longer during high tide than during low tide. Besides, during decreasing velocity of flood current fine-grained sediment is deposited, whereas during increasing velocity of ebb current, sediment is eroded. The current velocity needed to erode the sediment of a given size is much higher than the velocity needed to deposit it (Hulström's diagram, Sundborg 1956). This phenomenon in tidal flats is called "scour lag" by van Straaten and Kuenen (1958).

Type of Bedding in Tidal Flat Sediments

None of the bedding types is restricted to tidal flats only. Further, some types of bedding are more abundant in one than in other parts of the tidal flats.

Megaripple bedding is common in the channels (Häntzschel 1938; van Straaten 1950b, 1953a; Reineck 1963a; Klein 1967; Reineck and Singh 1967; Boersma et al. 1968).

On sand flats small-scale cross bedding of current ripple origin is most common. Sometimes this is developed in the form of herringbone cross-bedding in sections cut normal to the ripple crest axis (Fig. 156) and festoon-shaped units in sections parallel to the ripple crest axis. Laminated sand is found in only small amounts. Climbing-ripple lamination is very rare, occurring only near the mouth of gullies. Megacurrent ripples and megaripple bedding is sometimes developed in the intertidal zone of sand bars and shoals (Klein 1970a).

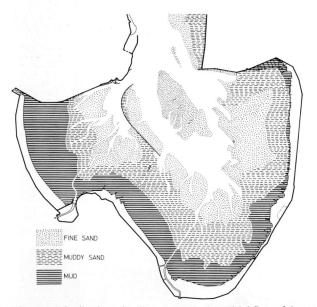

FINE SAND

MUDDY SAND

MUD

Fig. 591. Distribution of sediments on the intertidal flats of the Jade Bay. Mud occupies areas near the high-water line; sand is present near the low-water line. (After Gadow 1970; originally from Linke 1939)

Fig. 592. Lenticular bedding with flat lenses, showing very little bioturbation, subtidal zone, Jade Bay. The water depth is 3.5 m. The scale is in centimeters

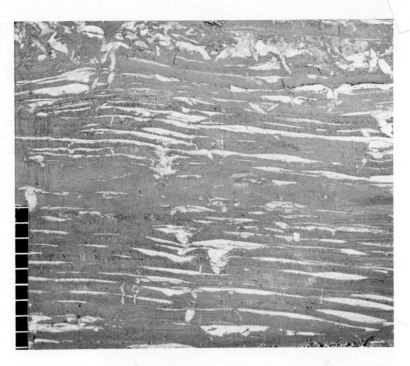

Mixed flat is characterized by flaser bedding (Figs. 184 through 186), wavy bedding, lenticular bedding (Fig. 592), and finely interlayered sand/mud bedding (Fig. 202) (Häntzschel 1963a; Reineck 1960a,b).

These bedding types are especially well-developed in the layers of longitudinal cross-bedding of point bar deposits of gullies (Reineck and Wunderlich 1969). This alternating bedding is mostly related to the alternation of tidal current and slack water phases (see p. 124; Fig. 593).

The sandy layers are deposited during periods of current activity, and mud during slack water periods. Mud layers often show abundant scattered sand grains (Fig. 594). Some sand grains are attached to algal filaments and other organic matter, and are kept in suspension before being deposited with mud aggregates (Wunderlich 1969).

Some of the thin layers (less than 1 mm) show graded bedding, which can be both normal and reverse (Reineck and Wunderlich 1969). Small-scale erosional features are frequent. In a vertical column there are thick bedsets changing in the type of bedding from set to set, as a result of changes in the intensity and direction of wind and wave activity (Fig. 595) (Reineck and Wunderlich 1969).

Some of the sedimentation takes place in shallow morphological depressions; but most sedimentation occurs on the point bars of the gullies and channels (van Straaten 1954a; Reineck 1958a). On the horizontal areas of the flats there is hardly any net deposition.

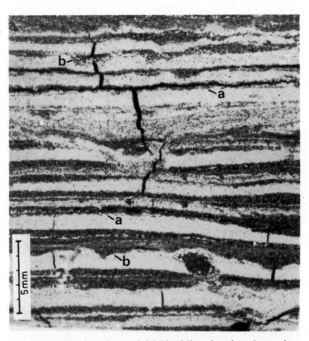

Fig. 593. Thinly interlayered tidal bedding showing alternating sand layers (dark) and mud layers (light). The lower surface of the mud layers is sharp, the upper surface is either somewhat diffused (a) or partly deformed (b) by tracks or imprints. Negative print of a thin section. Sand layers are originally light-coloured, mud layers are originally dark-brown. North Sea tidal flats (Jade Bay). (After Reineck 1978)

Fig. 594. Thin section of a mud layer showing scattered sand grains. North Sea tidal flats. Scale: 1 cm

On mud flats there are mainly thick mud layers deposited with thin sandy intercalations. In the mud flats of more arid climates bedding is highly disturbed due to growth of crystals of gypsum and halite (Fig. 596) from excessive evaporation. Crystal growth, together with mud cracks, produce a chaotic mud evaporite mixture (Thompson 1968).

Salt marsh deposits (supratidal zone and sometimes also intertidal) are characterized by fine-grained sediments interfused by plant roots and possessing uneven, noduled lamination (van Straaten 1954a; Reineck 1962a; Redfield 1967) (Fig. 597). During heavy storms shells are thrown on the salt marshes and become preserved as shell layers over which more sediment with plant growth is deposited. A graphic presentation of the facies model of salt marsh deposits of Netherlands is given by Bouma (1963). This type of salt marsh like all other salt marshes in Europe lies above mean spring high water level.

Sometimes, instead of mud, organic detritus the socalled "coffee ground" may be present (Buller and Green 1976). It is especially significant in the areas where peat is eroded and brought into the sedimentation cycle of the tidal flats. Buller and Green (1976) report this from Scotland. On the tidal flats of Jade Bay, North Sea, locally "coffee ground" is very common.

Bioturbation and Fauna

The North Sea supratidal zone is covered by plants (Fig. 598). The fauna of the tidal flats is characterized by a large number of individuals, which belong to only a limited number of species (Fig. 599). Most parts of the tidal flat surface sediments are highly bioturbated by benthonic organisms (Fig. 595) (van Straaten 1954a; Reineck 1958c; Schäfer 1956, 1962; Thompson 1968; Hertweck 1970).

Bioturbation in mud flats is generally strongest, weaker in mixed flats, and weakest in sand flats. However, this is not always the case. For example, areas of *Arenicola* populations on sand flats are strongly bioturbated; moreover, areas of sand flats and mud flats which are protected in morphological depressions, etc., also show stronger bioturbation than usual.

The distribution of fauna in a gully of a tidal flat is shown in Figs. 600 and 601. The population of gullies differs markedly from the population of tidal flat surfaces (Fig. 599). The banks of gullies and commonly also the gully bottom is muddier in nature than the adjoining surfaces. On the outside older and compacted sediment layers are exposed on which *Corophium* is abundantly populated.

Bioturbation in the sediments which have been deposited very rapidly is negligible; this is especially true with point bar and channel bottom deposits of gullies. Sometimes living horizons of benthonic molluscs are found with escape-traces under them (Reineck 1958c). This is because of sudden rapid rates of sedimentation. In larger gullies and channels escape-traces are rather common (Fig. 255).

Occasionally, fecal pellets are found concentrated in definite layers. In certain climates rolled algal mats are commonly found.

In the intertidal zone, a characteristic spectrum of trace fossils is present. Beside the traces and burrows of the organisms living in the intertidal zone, there are a number of traces made at ebb-tide, e. g., bird tracks, bird feeding traces (Dörjes and Reineck 1979), insect and land animal tracks, and at flood tide. e. g., swimming and resting traces of fishes. Kreft and Michaels (1976) describe traces made by fishes while feeding on the algal mats. Traces of jelly fish are often observed (Häntzschel 1937).

Tidal Channels

The larger tidal channels of a tidal delta and estuary are described on pp. 386 and 392. Tidal channels contribute an important part to the develop-

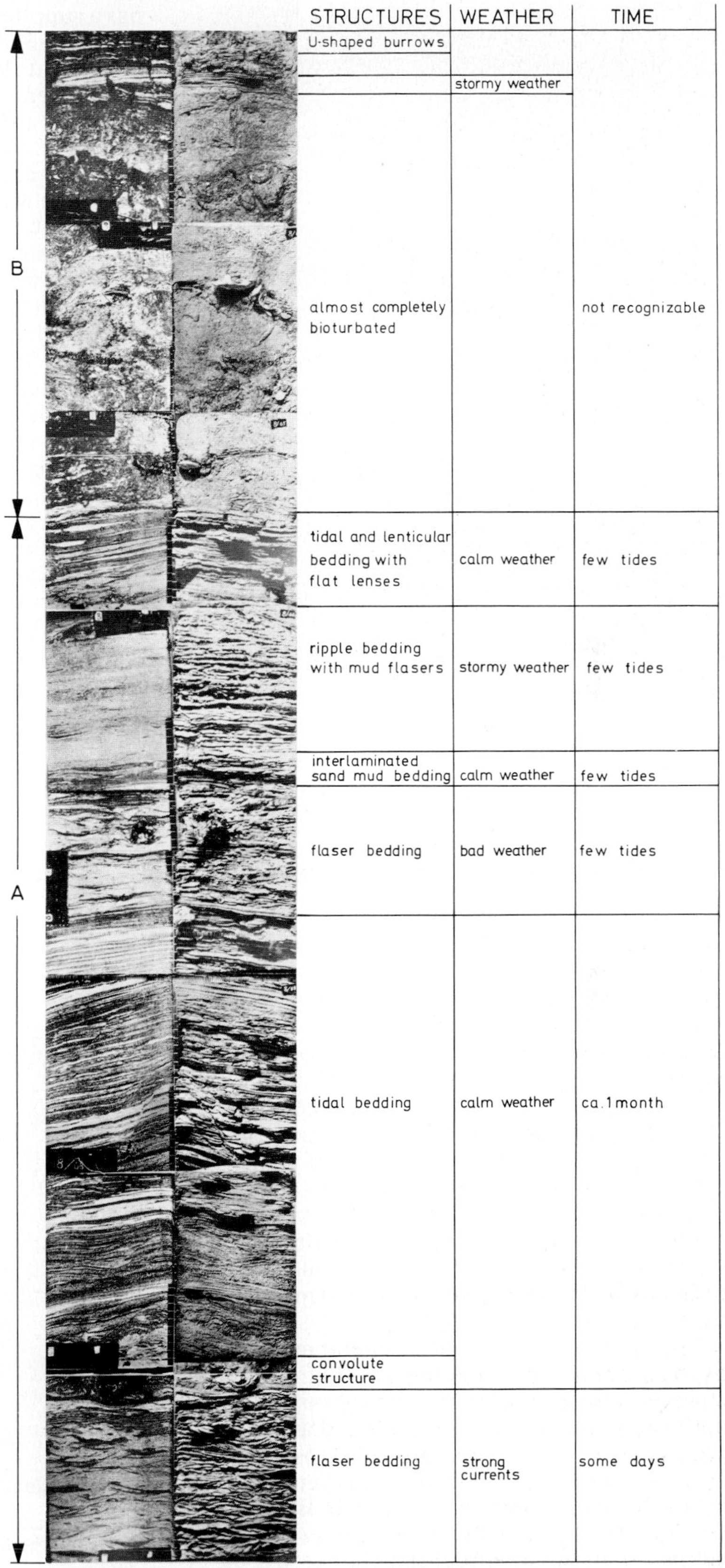

STRUCTURES	WEATHER	TIME
U-shaped burrows		
	stormy weather	
almost completely bioturbated		not recognizable
tidal and lenticular bedding with flat lenses	calm weather	few tides
ripple bedding with mud flasers	stormy weather	few tides
interlaminated sand mud bedding	calm weather	few tides
flaser bedding	bad weather	few tides
tidal bedding	calm weather	ca. 1 month
convolute structure		
flaser bedding	strong currents	some days

Fig. 595. Bed sets of point bar deposits of a meandering channel. A Bed sets of the sides of the channel. Different weather conditions produce different bedding types. Rate of deposition and growth of the sequence is rapid, so that bioturbation is rather low. Time required for the deposition of individual units can be determined. The bed sets are primarily deposited on the inclined point bar surface; thus, they possess a primary dip (longitudinal cross-bedding). B As soon as rapid deposition on the point bar reaches the level surface of the intertidal flat surface, the rate of deposition and growth of sequence is slowed down. The time interval represented by individual sediment units in section B is much longer than in the case of section A. Bioturbation increases to the extent of complete destruction of primary bedding. Numerous erosional unconformities are present, but are obscured because of a high degree of bioturbation. North Sea tidal flats. (Modified after Wunderlich 1970d)

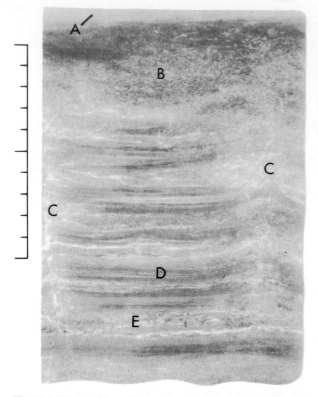

Fig. 596. X-ray radiograph showing gradational change from laminated silt (below) to chaotic mud (above). This change is due to increased desiccation and salt crystallization in the mud flats near the maximum spring tide level. Northwestern Gulf of California. A algal mat (mainly silty clay); B chaotic muds (highly gypsiferous silty clay); C shrinkage cracks containing prominent gypsum concentrations; D interlaminated clayey silt, silt, and gypsum; E mainly clayey silt and gypsum. The scale is in centimeters. (After Thompson 1968)

has a simple, sigmoidal shape in cross-section normal to the channel (Fig. 167), and type B has a prominent flat, lower platform up to 2–3 m wide.

Different types of bedsets in lateral deposits of channels are caused by changing weather conditions (Fig. 595). The major factor is change in wind direction and intensity, causing a corresponding change in wave activity–its intensity and direction. In general, higher wave energy means a rapid lateral deposition in channels. High wave activity causes increased erosion on the surface of tidal flats, and this sediment is deposited on channel slopes, which are protected from the waves (Reineck and Wunderlich 1969).

The lateral deposition is very rapid and water content of sediment is very high. During emergence at low tides fault planes develop. Fault planes are usually concave-upward, or Y-shaped (Fig. 145).

Furthermore, small mud ridges may develop on the steeply inclined channel walls (Figs. 604 and 605), which originate due to erosion of freshly deposited mud (Reineck 1974b). They are 10–20 cm broad, 2–10 cm high and 1 m or more long, occurring usually in groups and sometimes showing bifurcation. Mostly these small mud ridges are erosional forms covered by a few form-concordant layers. From fossil tidal flats of Devonian age Jux (1964) has described similar small mud ridges.

Fig. 597. Small cliff formed in the supratidal salt marsh. Note the irregular bedding planes and shell horizons. (After Reineck 1962)

ment of tidal flats. In North Sea tidal flats the subtidal zone is made up of channels and sand bar sediments (van Straaten 1964a) (Fig. 602).

A comparison of older maps by Reineck (1958a) demonstrated that about 58 % of a tidal flat has been reworked by tidal channels in 68 years (Fig. 603). The reworking is the result of lateral migration of channels. Trusheim (1929 b) gives the rate of migration of channels in muddy sediments as 25 m/year. In sandy sediments rates of migration are rather high–25 to 30 m/year up to 100 m/year (Lüders 1934).

The most common mode of lateral migration is by the process of meandering. In addition straight channels also demonstrate a lateral migration. In both cases. during lateral migration deposition on one side produces longitudinal cross-bedding (Figs. 167 and 168). Bioturbation is very weak, as the rate of sedimentation is very high. Bridges and Leeder (1976) differentiate between two types of points bar in the intertidal flat channels. Type A

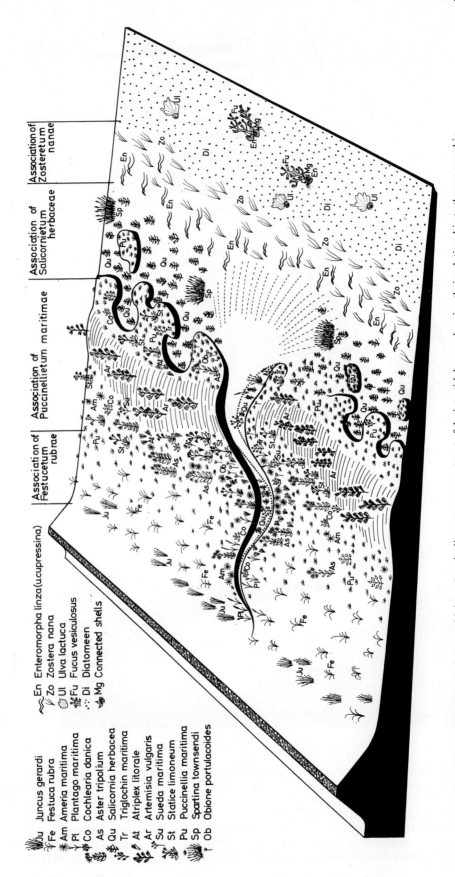

Legend (left):

Ju Juncus gerardi
Fe Festuca rubra
Am Ameria maritima
Pl Plantago maritima
Co Cochlearia danica
As Aster tripolium
Qu Salicornia herbacea
Tr Triglochin maritima
At Atriplex litorale
Ar Artemisia vulgaris
Su Sueda maritima
St Statice limoneum
Pu Puccinellia maritima
Sp Spartina townsendi
Ob Obione portulacoides

En Enteromorpha linza (u.cupressina)
Zo Zostera nana
Ul Ulva lactuca
Fu Fucus vesiculosus
Di Diatomeen
Mg Connected shells

Association of Festucetum rubrae

Association of Puccinellietum maritimae

Association of Salicornietum herbaceae

Association of Zosteretum nanae

Fig. 598. Plant associations of the salt marsh (supratidal zone) and adjacent uppermost part of the intertidal zone, showing their relationship to the geomorphic position. German North Sea tidal flats. (After Dörjes 1970)

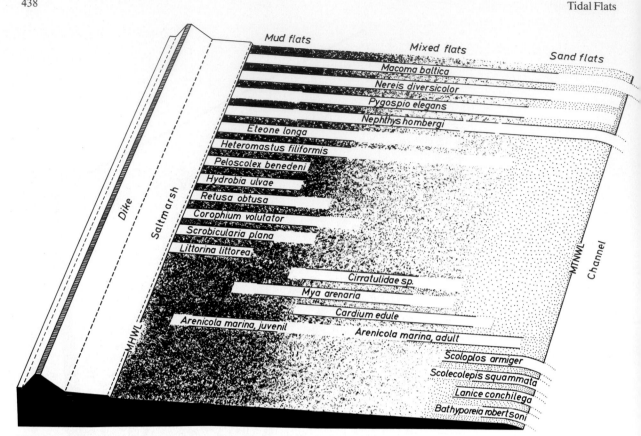

Fig. 599. Population zones of important macrobenthic animals in an intertidal flat located in Jade Bay. The major occurrences are shown by the white area between the black lines; scanty population and sporadic occurrences are shown by the white areas without marked borders. (After Dörjes 1970)

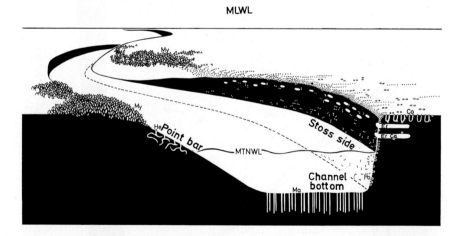

Fig. 600. Distribution of macrobenthic organisms in a point bar, cut side, and bottom of a tidal channel of the intertidal zone. Ca = *Carcinus maenas;* Co = *Corophium volutator;* Er = *Ericheir sinensis;* He = *Heteromastus filiformis;* Ma = *Magelone papillicornis;* My = *Mytilus edulis;* Po = *Polydora ciliata.* (After Dörjes 1970)

In cases where meander loops are cut off or gullies are captured by the adjoining gullies, the inactive parts are filled up by sediments; the beds dip normal to the curvature of the channel.

The channel bottoms of the larger tidal channels are mostly sandy, enriched in shells and mud pebbles. The bottom sediments of smaller channels and gullies of the intertidal zone are mainly muddy in nature with a few sandy intercalations (coarsely interlayered sand mud bedding). Often, lenticular bedding is developed (Fig. 592). Nevertheless, shell layers and mud pebbles are always present. Shells and mud pebbles may be present as channel lag deposits (Fig. 543). Such channel lag deposits

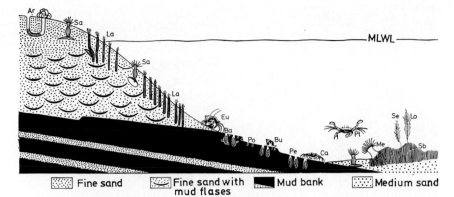

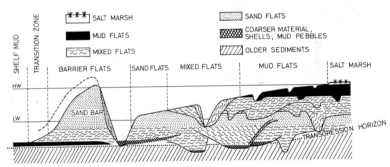

Fig. 601. Sequence of zones of animal population in a large tidal channel. The sandy intertidal flat is followed by a *Lanice* population in the underwater areas of the channel margins. The exposed layers of compact clay exhibit a different spectrum of species than the sandy biotopes. These layers of compact clay are preferentially populated by boring organisms. On the channel bottoms Sabellaria "reefs" may develop. Such Sabellaria "reefs" show specific associated fauna and flora. Ar = *Arenico-* *la marina;* Ba = *Barnea candida;* Bu = *Buccinum undatum;* Ca = *Carcinus maenas;* Eu = *Eupagurus bernhardus;* La = *Lanice conchilega;* Lo = *Laomedea flexuosa;* Me = *Metridium senile;* Pe = *Petricola pholadiformis;* Po = *Polydora ciliata;* Pt = *Portunus hosatus;* Sa = *Sagartia troglodytes;* Sb = *Sabellaria spinulosa;* Se = *Sertularia cupressina.* (After Dörjes 1970)

Fig. 602. A cross-section across tidal flats with tidal channels in the subtidal zone. The open sea is on the left. (Based on van Straaten 1964a. After Reineck 1970b)

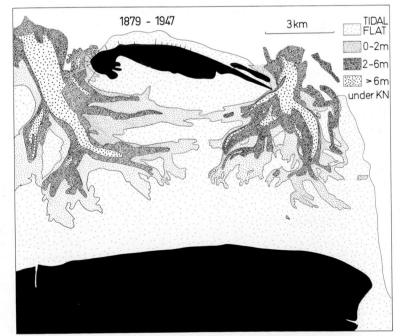

Fig. 603. Reworking of a tidal flat by laterally migrating small and large tidal channels and creeks over a period of 68 years. Lateral migration results in the formation of longitudinal cross-bedding. This illustration is based on the comparison of maps at 1 and 2 year intervals. The depth of reworking has been contoured into 0 to 2 m, 2 to 6 m, and > 6 m. Black in this figure is the mainland and Wangerooge barrier island. (After Reineck 1958a)

Fig. 604. Bifurcating mud ridges on a tidal channel-margin. Mud is flowing down along the troughs. The channel margin is inclined at 14°. Mudflats, Jade Bay. Length of scale = 10 cm. (After Reineck 1974b)

Fig. 605. Cross section of a mud ridge with erosional nucleus and form concordant layers as apron on the outside. X-ray radiograph positive: light grey – mud; dark grey – sand. Burrows are visible as 2 mm wide light-colored bands. Vareler Tief, Jade Bay. Scale in cm. (After Reineck 1974b)

may develop over wider distances than the width of the channel, as a result of channel migration.

Tidal channels are very large relative to their catchment of tidal area compared with those of rivers. Geyl (1976) calls channels created by tidal action "tidal neomorphs". In ancient landforms it is possible to recognize oversized valleys as "tidal palaeomorphs".

Surface Structures

In sand flats and mixed flats small-current ripples and wave ripples–mainly asymmetrical wave ripples–are most abundant. Mostly, depending upon the morphological and hydrographical situation, various types of ripple systems of varied directions are present which are produced at the same time or one after the other in a hierarchical system (Fig. 606a–c). A falling water level produces wave ripples with double crests (Fig. 607). Firstly, in deeper water with larger surface waves larger wave ripples are produced. With a falling water level the surface waves become smaller and their effective depth of action is only up to the existing ripple crests. These smaller waves produce smaller wave ripples on the crests of earlier formed larger ripples. In general, two and sometimes three smaller crests are produced.

Longitudinal ripples are also characteristic of very shallow water, commonly in rather muddy sediments (van Straaten 1951). Another characteristic of a falling water level and an ultimate emergence of sediments is modified ripples. They may be the wave ripples with rounded crests and pointed troughs (Fig. 608). Such ripples are produced when normal wave ripples (pointed crests and rounded troughs) are modified by a falling water level so that the sediment from ripple crests is transferred into ripple troughs. In other cases ripples may be capped-off, thus posses flat crests (Fig. 609). In some cases, a falling water level in the last phase modifies the wave ripples into current ripples. If some water is retained in the troughs at the time of subaerial exposure, the ripple crests show marked erosional markings on the sides (Fig. 610). If the falling water level is irregular (in-phases), during the emergence of sediment surface water le-

vel marks are engraved on the sides of ripple crests (Fig. 88).

During final emergence, ripple crests are sometimes broken through, producing rill marks. Rill marks also develop on unrippled inclined surfaces during emergence (Figs. 97 through 104) (Čepek and Reineck 1970).

Other important features suggesting intermittent subaerial exposure of emergence of a sedimencation surface are raindrop imprints, hail marks, foam marks, adhesion ripples and warts, and

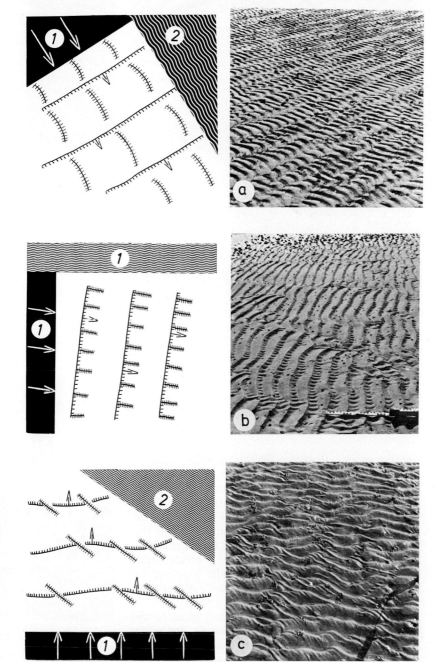

Fig. 606. Interfering ripple systems. The numbers denote the order of the influence. The same numbers mean a simultaneous effect. The arrow points the direction of current and the direction of the migration of current ripples. Wavy lines indicate wave action. Tidal flats, North Sea. a Straight crests of the current ripples (or asymmetrical wave ripples), modified later by the asymmetrical wave ripples. b Ripples with straight crests; probably small current ripples. Simultaneously working waves could produce wave ripples only in the ripple troughs. c Current ripples, and superimposed on them are wave ripples produced later by waves approaching at an angle

Fig. 607. Ripples with double crests, a feature characteristic of falling water level. The larger ripples are formed during the high-water stage. At the lower-water stage smaller waves produce smaller ripples, superimposed on the earlier formed larger ones. Tidal flats, North Sea

Fig. 609. Capped-off wave ripples in the lower and upper margins of the photograph. The capping-off (flattening) of the crests takes place during subaerial emergence. The flattening is caused by "capillary waves", which are produced by strong winds on the water surface. In the central part of the photograph wave ripples are seen to be deformed into lingoid-shaped current ripples by the last film of water. Tidal flats, North Sea

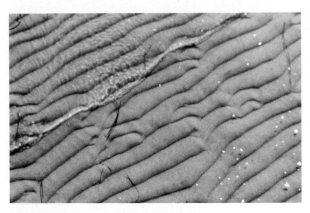

Fig. 608. During subaerial emergence of the depositing surface, the wave ripples are modified by the swash water. The crests become rounded and lowered. Some sand flows down into the troughs. As sand flows down from both sides, the troughs become very pointed grooves

wrinkle marks. Features similar to wrinkle marks are also known in deep-water sediments (flysch deposits). Probably there is more than one way to produce wrinkle marks (Teichert 1970).

Mud cracks are more common in warmer climates. Here, if microbial mats are present, the sides of cracked plates are curled up. Sometimes mudcracked microbial mats can be eroded into thin "tidalflat paper" and brought together by blowing wind (Trusheim 1936, see also clay dunes, pp. 428 and 445; Scott and Hayes 1964).

Shallow erosional depressions are often found on intertidal flats. Such depressions are often sculptured by wave ripples, whereas the surrounding sediment is covered by algal mats (Fig. 611).

Fig. 610. Rests of the ripple crests. In the troughs small wave ripples are developed. The larger ripples have been destroyed by wind-moving water. Tidal flats, North Sea

Fig. 611. Sandy surface protected by thin algal mats, and shallow erosional depression with ripples. Tidal flats, North Sea

sediment frozen in the ice, sediment brought on the ice by wind, sediment thrown on the ice by waves, frozen sediment of tidal flat surface, which can be present in several layers within the drift-ice, and organisms, e.g., molluscs *(Cardium edule)*, can be transported into areas where normally they do not occur. Mud layers which have been transported by the ice are intercalated with vertical ice needles (Reineck 1976a,b; Dionne 1972a).

An excellent compilation of the phenomena of drift-ice is given by Dionne (1976a, ed.). Dionne (1972b) provides terminology on drift ice in English and French.

Drift Ice Action

In areas where in winter freezing temperatures are reached, ice is produced on the tidal flats. The tidal flats develop ice earlier than the shelf zone because of high radiation and cooling effect due to evaporation on the tidal flats. As a result of ice cover on the tidal flats numerous surface markings are produced. Drift ice produces groove marks, especially during the ebb current (Reineck 1956a; Dionne 1969, 1971, 1974; Knight and Dalrymple 1976). The mud is squeezed under the weight of ice blocks (Figs. 114 and 115) (Wunderlich 1973). Dionne (1973) describes monroes (Fig. 612), 5–20 cm-high mud volcanoes, which develop due to weight of ice. Small mud volcanoes described by Knight and Dalrymple (1976) are called twiggies (pers. comm. by R. J. Knight).

Drift ice plays a significant role in the transport of sediment. Large pebbles and blocks, suspended

Horizontal and Vertical Sequences of Tidal Flat Deposits and Intertidal Sandbars

The horizontal sequence of a given environment means inclined, laterally associated depositional regions, which during a progradation may grow one above the other (Reineck 1972). The tidal flat can be regarded as a margin of the sea, located next to the land. Toward the land the tidal flat is associated with continental deposits of coastal plain, marshes, etc. Toward the open sea the tidal flat is associated with coastal sands or shoreface-like deposits. In a prograding sequence these units grow one above the other (Fig. 613). The tidal flat deposits are in themselves differentiated into supratidal (marsh), intertidal (mud flat, mixed flat, and sand flat), and subtidal (channels and sand bars). Both German and Dutch tidal flats show such associations. Sometimes tidal flats show a retrogradational sequence (transgressive sequence) in the lower part (Fig. 614). The retrogradational sequence shows a reverse sequence of the deposits:

Fig. 612. Monroes in a mud tidal flat. Small mud mounds form under the ice-foot as a result of ice pressure which expels air, water or liquefied mud from the underlying strata. These features owe their name to the late movie star, Marilyn Monroe. (After Dionne 1974)

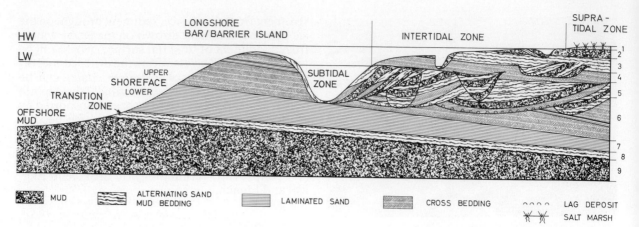

Fig. 613. Schematic cross-section across the tidal flats of the North Sea in a situation of a hypothetical prograding shore line. In reality the North Sea tidal flats are resting on the Pleistocene and Holocene deposits. The section is based on the information of van Straaten (1964a) and additional new data of recent years. 1. *Salt marsh (supratidal zone).* Very fine sand and mud, interbedded shell layers, plant roots, irregular wavy bedding. 2. *Mud flat (intertidal zone).* Mud, occasional very fine sand layers, lenticular bedding with flat lenses, strong bioturbation. 3. *Mixed flat (intertidal zone).* Sandy mud, thinly interlayered sand/mud bedding, lenticular bedding, flaser bedding, shell layers, bioturbation strong to weak. 4. *Sand flat (intertidal zone).* Very fine sand, small-ripple bedding sometimes of herringbone structure, flaser bedding, laminated sand, occasional strong bioturbation. 5. *Subtidal zone.* Medium to coarse sand, mud pebbles, megaripple bedding, small-ripple bedding, laminated sand, weak bioturbation. 6. *Upper shoreface.* Beach bar- and ripple cross-bedding, laminated sand, weak bioturbation. 7. *Lower shoreface.* Laminated sand, bioturbation stronger than in the upper shoreface. 8. *Transition zone.* Alternating sand/mud bedding, i.e., flaser and lenticular bedding, thinly and thickly interlayered sand/mud bedding, moderate bioturbation. 9. *Shelf mud.* Mud with storm silt layers, moderate to strong bioturbation

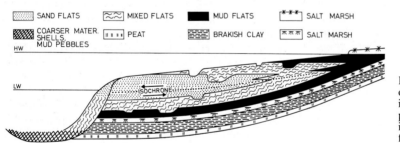

Fig. 614. Fully developed vertical sequences of transgressive (= retrogradational deposits), and regressive (= progradational deposits). Sometimes instead of peat and brakish clay, a transgressive horizon may be found. (After Reineck 1970b)

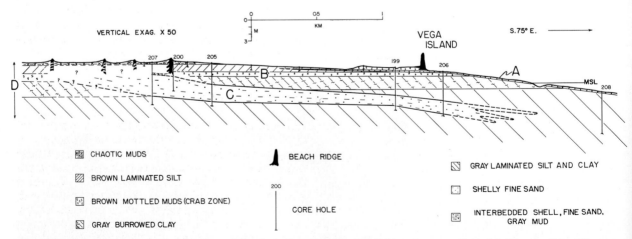

Fig. 615. Cross-section showing a sedimentary sequence in the Vega Island region, northwestern Gulf of California. (After Thompson 1968)

Sand flat deposits
Mixed flat deposits
Mud flat deposits
Brackish or freshwater clay   } or trans-
Peat with spagnum   } gression
Older sediments   } horizon

Thompson (1968) gives a regressive (prograda-tional) sequence of tidal flat deposits of the Color-ado River delta, Gulf of California (Fig. 615):

Chaotic mud
Brown laminated silt
Brown mottled mud (crab zone)
Gray burrowed clay
Gray laminated silt and clay

In many cases of modern and ancient tidal flat deposits, the above-mentioned sequences may not be fully developed. A major controlling factor is availability of appropriate sediments.

The Tidal Flats of Saint-Michel

The tidal flats of Mont Saint-Michel Bay, France belong to the tidal flats of very high tidal range (15.30 m). The clastic sediments contain a high content of biogenic carbonates (> 50 %). The la-teral distribution of the sediments is typically fin-ing-landward type (Fig. 616); however, the sedi-ments of the subtidal zone are coarse-grained and pebbly in nature, related to the high velocities of the tidal currents (Larsonneur 1975). The distribu-tion of benthonic fauna is described by Lang et al. (1973), where a comprehensive list of references on the geological and biological aspects of the area is available. Geomorphology of the French tidal flats is discussed by Verger (1968).

Wind Tidal Flats

The term wind tidal flats is used to describe the ex-tensive coastal sediment areas, which are covered or exposed for several hours and days as a result of water-level changes produced by wind action. Wind flats of the Baltic Sea are discussed by Häntzschel (1935c), Wassmund (1939), and Gripp (1956).

An extensive area of wind tidal flats is located behind Padre Island in the area of Laguna Madre (Fisk 1959; Rusnak 1960; Miller 1975), and Lagu-na Mormona, Gulf of California, Mexico (Horo-dyski et al. 1977). These wind tidal flats are charac-terized by blue-green microbial mats and light olive-gray clay deposits. Because of dessication, gypsum crystallization, and bioturbation the algal mats are distorted (Fig. 617). They break up into polygons, curling up on the margins. Within the clay deposits, thin wind-blown sand layers are

present. For the wind tidal flats of Laguna Madre (Miller 1975) on the basic of C^{14} dating calculated rate of sedimentation to be 1.5–5 cm/100 years.

In the Laguna Madre, the horizontal beds are only a few decimeters thick and represent the final stage of infilling of the lagoon (Fig. 468).

The interstitial water shows a distinct increase in salinity in comparison to the salinity of the lagoon-al water. The surface water of the Laguna Madre flats possesses a salinity of 140 ‰ while the subsur-face water has a maximum value of 242 ‰.

Bacteria play an important role in the decompo-sition of organic matter, production of interstitial gas, Eh, pH values, as well as in the genesis of car-bonates, gypsum, and iron sulfides (Fisk 1959; Miller 1975).

After a long spell of dry periods, the cohesive al-gal layer on the sediment surface dries out and rolls up. The dried mud breaks down into sand-size par-ticles and is blown away by the wind. These sand-size clay particles accumulate in the form of clay dunes near the landward margins of the Laguna Madre flats. During rains, the clay aggregates dis-integrate. The clay dunes are made up of thin-bed-ded clayey sand, land snails, and rodent bones (Fisk 1959; Price and Kornicker 1961; Scott and Hayes 1964). Inactive clay dunes develop zonal soils. Price (1963b) discussed the factors in genesis and the characteristics of clay dunes. Such dunes seem to be indicative of arid and semi-arid cli-mates.

A variant of wind tidal flats in the arid climate are the coastal sabkhas, which represent broad, salt-encrusted supratidal areas, and are only occa-sionally flooded (Dean and Schreiber 1978). Coastal sabkha of Abu Dhabi, Trucial Coast, are discussed by Schneider (1975), and Shearman (1978). The supratidal flats covered with halite-bearing salina near the Colorado delta, North-West Gulf of California, are described by Holser (1966), Phleger (1969), Shearman (1970). The Rann of Kutch is an extensive area of coastal sabk-has, flooded only by storm surges of monsoon peri-ods, where fine-grained gypsiferous silt-clay is de-posited (Glennie and Evans 1976).

Tidal Flats of Taiwan

On the east coast of Taiwan, 3–4 km-broad tidal flats are developed, running over 240 km. A strip of 170 km belongs to the category of exposed tidal flats. On the seaward side of such tidal flats a front bar of medium sand is developed (Fig. 618), which possesses all the characteristics of a foreshore with ridge and runnel system (Reineck and Cheng 1978). The tidal flat shows a fining-landward de-velopment (Figs. 619 and 620). The sand is brought in by rivers (Fig. 621) and is distributed on the tidal flats by the wave action. The fine-grained fraction is transported mostly towards land, while a part of it is probably transported into the offshore area. The coarser fraction builds up the front bar. The sand is made up of black shale particles. Towards the south, in the direction of general sediment

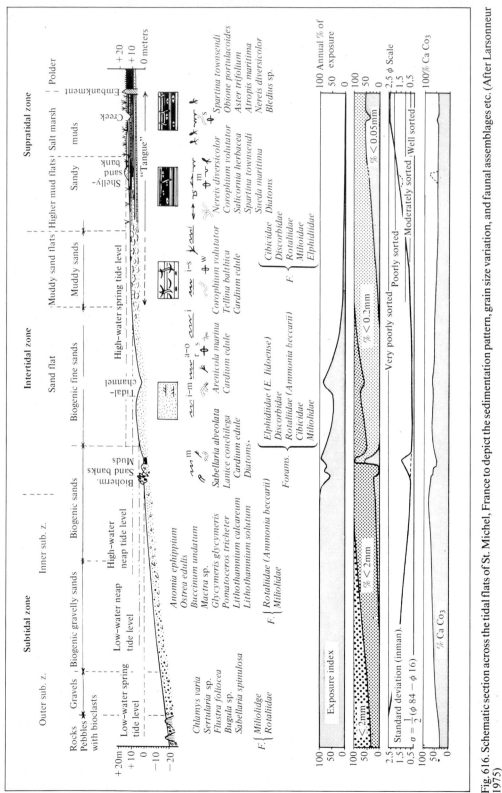

Fig. 616. Schematic section across the tidal flats of St. Michel, France to depict the sedimentation pattern, grain size variation, and faunal assemblages etc. (After Larsonneur 1975)

transport, the sand becomes finer-grained, the size of shale particles decreases, and quartz grains are better rounded. Unlike the North Sea tidal flats, where polychaetes are the main bioturbating animals, the main bioturbation of the Taiwan tidal flats is caused by the crabs, e. g., *Upogebia wuhsienweni* (Dörjes 1978).

The distribution pattern of primary bedding structures, bioturbation structures, median, and content of the < 63 μ fraction is given in Figs. 619 and 620. The content of the < 63 μ fraction increases towards land. Mud content is also high in the sediments of the depression behind the front bar, where the degree of bioturbation is also high. Distribution of macrobenthos and bioturbation was studied by Dörjes (1978), and is depicted in Figs. 622 and 623.

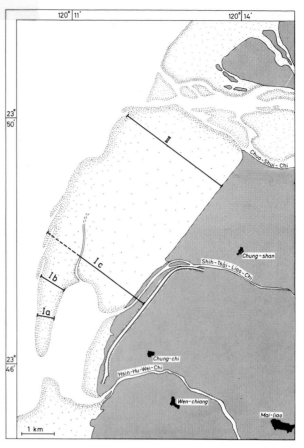

Fig. 618. Location of transects on tidal flats near Mailiao, Taiwan. (After Reineck and Cheng 1978)

Fig. 617. Vertical section showing interlayered clay and algal sediments with minor amounts of sand and sucrosic gypsum. Note also the burrowing, dessication cracks, and distortion of microbial mats. Wind-tidal flats of Laguna Madre, U.S.A. (After Miller 1975)

The Saltmarsh

On the east coast of the U.S.A., south of the ice-limit and north of the mangrove coast, the tidal flats behind the barrier islands are developed as salt marshes. Along the panhandle of Florida, salt marshes may face the open sea (Tanner 1960). Salt marshes are also developed on the west coast of North America (Frey and Basan 1978). Salt marshes of these regions are defined as well-vegetated intertidal flats (Basan and Frey 1977). Distribution and characteristics of salt marshes are discussed by many workers (Teal and Teal 1969; Ranwell 1972; Redfield 1972; Chapman 1974; Chapman 1977).

The ecological conditions in the salt marshes range from marine milieu to terrestrial milieu, and corresponding to it there is a clear zonation of plants (Fig. 624) and animals (Fig. 625) and their lebensspuren.

In a geological record, below the salt marsh sediments a sequence of tidal creeks and estuarine banks is present, where the tidal creek sediments

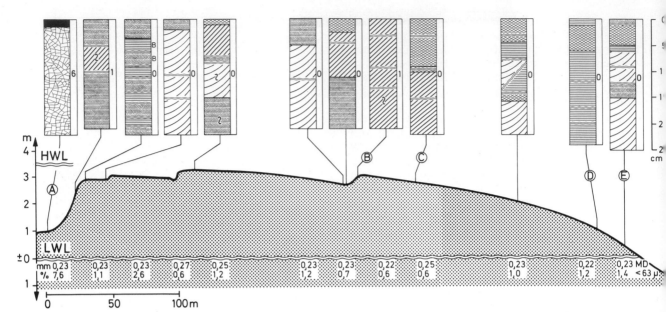

Fig. 619. Profile Ia across the tidal flats near Mailiao (Taiwan). Distribution of bedding types and type and degree of bioturbation, as well as the Md-value of the sediments and the content of < 63 µ fraction in the investigated samples are shown (for location see Fig. 618) cross-section through the seaward situated "swash-ridge" on which three beach bars are developed. On the left hand side a shallow depression is present with steep margin where degree of bioturbation is high. Sea to the right. (After Reineck and Cheng 1978)

are less bioturbated than the salt marsh sediments (Pestrong 1972; Howard and Frey 1975). In the tidal creek sediments laminations, flaser, wavy and lenticular bedding are seen along with slumping and faulting, and mud clasts derived from the banks are present. The sediments of the salt marsh are churned by root patterns. The oysters of the intertidal zone are free of the borings of the sponge *Cliona;* while the subtidal oysters show borings of *Cliona* (Wiedemann 1973). Further, in the salt marsh sediments, tracks of birds, racoons, mink, deer, cattle and marine molluscs are present (Basan and Frey 1977; Edwards and Frey 1977). Greer (1975) gives a progradational sequence from inner shelf to salt marsh deposits.

Freshwater Tidal Flats

In the estuaries, where there is a marked tidal influence, the intertidal flats on the margins are developed which are marine tidal flats near the mouth of the estuary, and gradually becoming brakishwater type, and ultimately upstream grade into freshwater tidal flats. These marine–brakish–freshwater tidal flat transitions have been studied by biologists. The main parameter which influences the distribution pattern of benthonic animals in these areas is the chloride content. Other significant parameters are: grain size, content of organic matter, hydrographical location, rate of reworking, and period of submergence (Carriker 1967). The estuaries are characterized by high seston content, significant deposition, high availability of colloids and organic matter, specialized microbiota, and the displacement of cation content in the longitudinal profile.

Changes in the various parameters in the longitudinal profile along the tidal flats of estuaries cause changes in the competition amongst the benthonic animals, mainly due to disappearance of marine species. Annual fluctuations in the chloride content of the river waters of the estuaries may also cause shifting of the faunal distribution. McLusky (1971) observed such shifting in the distribution of *Corophium volutator.*

The tidal flats of the Weser estuary are characterized by the banks of dense reeds, mainly *Phragmites communis.* With decreasing salinity reedmace, *Typha angustifolia,* and *Glyceria maxima,* and some kinds of club-rush appear. Generally, in the autumn large quantities of leaves and stem are available which are partly thrown on the supratidal areas and partly concentrate as accumulates of swash marks near the high-water line. Coarsergrained organic matter may become concentrated as channel lag in the channels. In other depositional areas, the fine organic debris is deposited on the tidal flats in the form of tidal bedding. Along with the vegetation, microflora and bacteria play a significant role in the biological processes of the brackish water and freshwater tidal areas.

Remane (1958) reports brick-red colored sediments on these tidal flats as a result of the activity

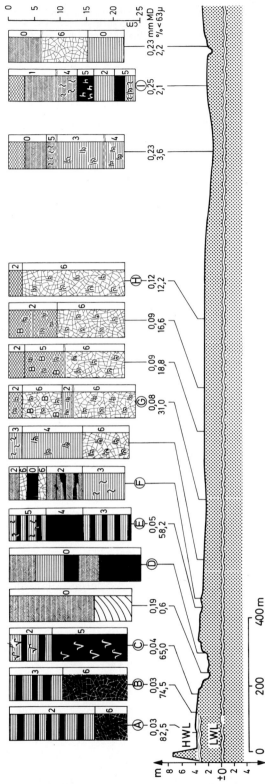

of sulfur bacteria. Diatom layers are quite common. In front of the reeds-belt of the Weser estuary, a 5–25 m broad, and 4 km long continuous band exists which is inhabited by *Vaucheria* of Xanophyceae, and is referred to as *Vaucheria* tidal flats.

Michaelis (1973) studied the distribution of fauna from polyhaline parts (30–18 ‰ salinity) through mesohaline parts (5–18 ‰ salinity) to oligohaline parts (5–0.5 ‰ salinity) of the Weser River tidal flats. These tidal flats are characterized by a fauna of low number of species with poor productivity. From the seaward parts towards freshwater areas, due to decreasing salinity the density of population decreases, size of the animals and shells is reduced, and the fauna itself changes from 26 euryhaline marine species to 13 brackish water species, and from polychaetes to oligochaetes. In the outer seaward parts animals of predatory habits, suspension-feeders, collectors, and substrate-feeders are living. Brackish water areas are inhabited only by the substrate-feeders *(Tubifex costatus)* and collectors *(Streblospio)*. The pattern of sediment distribution in freshwater tidal flats is similar to those of marine tidal flats, namely mainly sand near the low-water line, and mud near the high-water line.

A purely freshwater tidal flat "Fährmannssand" in the Elbe River, near Hamburg, has been investigated by Pfannkuche et al. (1975), Grimm et al. (1976), and Pfannkuche (1977, 1979). The grain size of these tidal flats varies from medium sandy

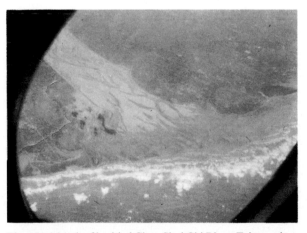

Fig. 621. Mouth of braided Chuo Shui Chi River, Taiwan, during dry period. The river brings sand, which is transported towards south (right hand side) on the tidal flats. This supply of sand is essential for the maintenance of present-day tidal flat areas. There is a strong coast-parallel sediment transport.

Fig. 620. Profile Ic across the tidal flats near Mailiao (Taiwan). Distribution of bedding types and type and degree of bioturbation, as well as the Md-value of the sediments and the content of < 63 μ fraction in the investigated samples are shown (for location see Fig. 618). Profile Ic begins from the embankment. The

seaward situated "swash ridge" is not present in this profile which is shown in profiles Ia (Fig. 619). This profile mainly exhibits the saltmarsh zone (supratidal). (After Reineck and Cheng 1978)

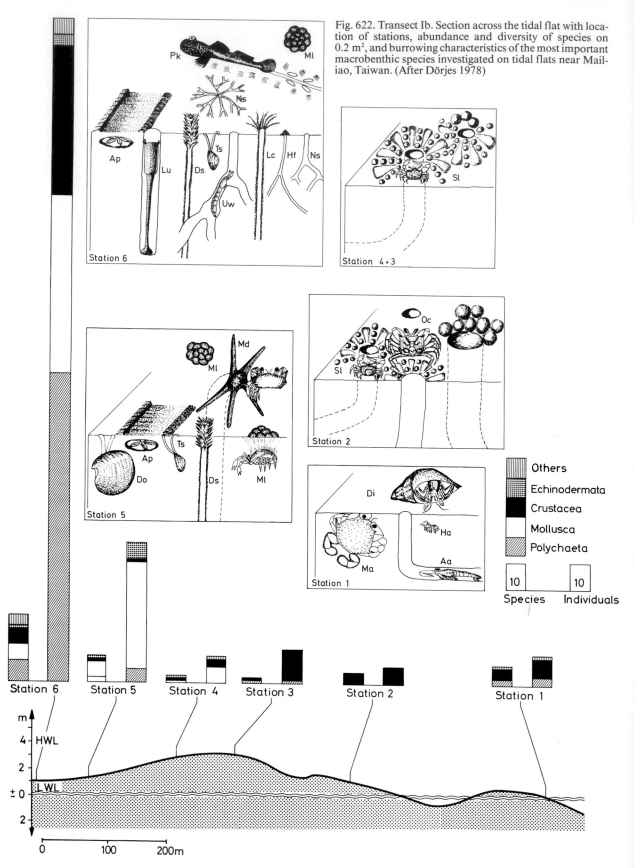

Fig. 622. Transect Ib. Section across the tidal flat with location of stations, abundance and diversity of species on 0.2 m², and burrowing characteristics of the most important macrobenthic species investigated on tidal flats near Mailiao, Taiwan. (After Dörjes 1978)

fine sand (with about 10 % sediment $< 63\,\mu$) to sandy mud (with about 58 % sediment $< 63\,\mu$). The muddy tidal flats are mainly covered by *Scirpus* (Fig. 626). The exposed muddy tidal flats are populated by blue-green algae, dinoflagellates, and diatoms.

The fauna of these tidal flats is made up of eurytopic, freshwater organisms, the saprobic species

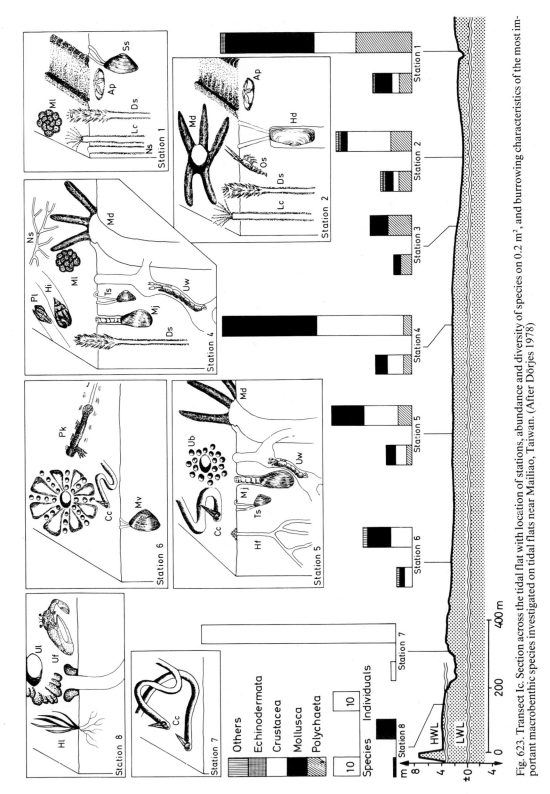

Fig. 623. Transect Ic. Section across the tidal flat with location of stations, abundance and diversity of species on 0.2 m², and burrowing characteristics of the most important macrobenthic species investigated on tidal flats near Mailiao, Taiwan. (After Dörjes 1978)

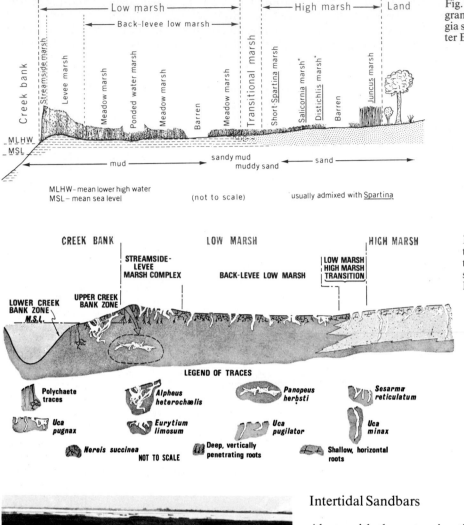

Fig. 624. Classification and diagrammatic distribution of Georgia salt marsh environments. (After Edwards and Frey 1977)

Fig. 625. Schematic illustration of Georgia salt marsh
traces. Dashed line represents mean sea level. (After
Basan and Frey 1977)

Fig. 626. Fresh-water tidal flat "Fährmannssand" in the Elbe
River, near Hamburg, mainly covered by Scirpus.

and marine immigrants. Although much food is
available in these tidal flats, due to high content of
organic matter even in the well-oxygenated sandy
tidal flats only a 5 mm thick oxidized horizon is
present, where only few animal groups can live.

In opposition to the marine tidal flats, in the
freshwater tidal flats collectors and suspension-
feeders are hardly present. From the point of view
of preservation, the burrows of sediment feeders,
i. e., *Paraonis* and Tubificideae are of some significance.

Intertidal Sandbars

Along with the extensive tidal flat surfaces, intertidal sandbars are also significant in the geological
record of tidal sea and estuarine sediments.
Schwartz (1972) gives a useful compilation of papers on bars and spits.

Klein (1970a), Klein and Whaley (1972), and
Dalrymple et al. (1975) describe and discuss the
processes and deposits of sandbars of the intertidal
zone in the Minas Basin, Nova Scotia, an area
characterized by a very high tidal range (ca. 18 m).
Characteristic features of such tide-dominated
sand bodies in the intertidal zone include: sharp
erosional contacts between sets of cross-stratification; rounded upper set boundaries of cross-stratification; unimodal and bimodal distributions of
orientation of cross-stratification; bimodal distributions of set thickness and dip angles of cross-
stratification; orientation of dunes, sand waves,
and cross-stratification in the dominant direction
of flood or ebb tidal flow, basinal topographic
trend and sand body axis; trimodal orientation of
current ripples; oblique or 90° superposition of

smaller current ripples on larger current ripples; double-crested current ripples; superposition of current ripples at 90° or 180° on slip faces and crests of dunes and sand waves; complex organization of internal cross-stratification in sand waves; etch marks on slip faces of dunes and sand waves, and alignment of the long axis of sand bodies parallel to tidal current flow, basinal topographic trend, and basin axis.

The sediments of Girdwood bar, Turnagain Arm, Alaska, exhibit small ripple bedding, oriented in ebb-and-flood directions (Bartsch-Winkler et al. 1975; Bartsch-Winkler and Ovenshine 1975).

Carbonate tidal banks can be very different from their terrigenous clastic counterparts, and are present in very variable developments. Ebanks and Bubb (1975) describe a tidal bank from South Florida, which is made up of coralgal sand and mud on the seaward side, where the shelf-side is made up of muddier molluscan-foraminiferal deposits. Bioturbation structures are abundant. They also provide a useful list of references.

Dörjes and Reineck (1977) describe the characteristics of Mellum bank, North Sea, an intertidal sand bar or shoal, which is exposed to both currents and wave activity. Distribution pattern of fauna and bioturbation structures in this sand bar show strong similarity with the foreshore environments of beaches and shoals. In comparison to the sandy intertidal flats, sediments of this sand bar contain a high proportion of megaripples bedding, low shell content, low population density of endofauna, and complete absence of polychaetes living in their tubes. Contrary to the subtidal environment, the Mellum bank surface shows many small, drainage channels, which may be preserved as erosional surfaces in the sediments. Such shallow channels can be preserved, if their depth is more than the daily sediment-reworking depth. In the normal case, the depth of sediment reworking is less than the depth of such channels; otherwise shallow endobionts cannot live in these channels.

Boersma and Terwindt (1980) investigated the internal structure of megaripples (dunes) and sandwaves (straight-crested megaripples) on an intertidal sand bar (shoal) in the Westerschelde estuary, the Netherlands. This study demonstrates the processes and the internal structures of megaripples, which are produced by currents of changing direction and intensity. On most points of the sand bar, one of the currents (ebb or flood) was stronger than the other, so that only rarely was herringbone structure observed. The clear dominance of one direction over the other can change during a neap-spring tide period. The straight-crested megaripples survive the neap, and retain their spring tide form. On the contrary, the lunate megaripples during neap tide are replaced by small ripples.

While one tidal direction dominates over the other (in general the flood in the area of study), the internal structure is made up of unidirectional cross-bedding, which is composed of a series of tidal bundles (Fig. 627A). Each tidal bundle is produced in response to a single tidal cycle. Within a single tidal bundle, three units are recognized which can be correlated to varied development of lee-side vortex during a tidal cycle.

The first interval is made up of rather diffuse, somewhat crinkled lamination, and is concordant with the earlier pause plane, which represents the time of megaripple stand-still between every flood-tide dune migration. The pause plane can be an erosional surface and corresponds to the reactivation surface and separates the tidal bundles from each other. This first interval of a tidal bundle corresponds to the prevortex acceleration stage. During this stage the flow separation is insignificant, and the vortexes hardly present, so that the avalanching and back-flow is imperfect. The small ripples move in different directions on the megaripple, and move down the slipface producing indistinct crinkled lamination, in which foreset laminae can sometimes be distinguished.

The next (middle) interval of a tidal bundle is made up of thick and distinct foreset laminae (cross-strata) possessing maximum angle of repose. This interval corresponds to the full vortex stage of main flood phase with continuous avalanching along the slipface.

The last interval of a tidal bundle is made up of foreset laminae, made up of poorly sorted material and showing poorly defined laminae. The angle of repose decreases downstream and the foreset laminae become crinkled. The foresets may pass upstream into topsets. This interval marks the deceleration stage of the tidal current, when much suspended sediment is also deposited on the leeface, reducing its angle of slope. The structures of this stage are also termed slackening structures.

The increasing and decreasing velocities that occur during the ebb and flood tides, as well as the overall change in the energy of flow over the neap-spring tide period are also reflected in the internal structure of megaripples. The tidal bundles described above show a lateral sequence reflecting the neap-spring-neap cycle. The main changes going from neap to spring tide are: the increased length and height of the cross-bedding of full vortex stage, the progressive lowering of the lower set boundary which is the result of more intense and prolonged vortex action, and good distinction between reactivation and slackening structures due to well-developed full vortex structures separating them.

In a vertical sequence of this shoal, the thick cross-bedded co-sets are present in the deeper parts of the shoal, while in the shallower parts, where the maximum tidal current velocities are

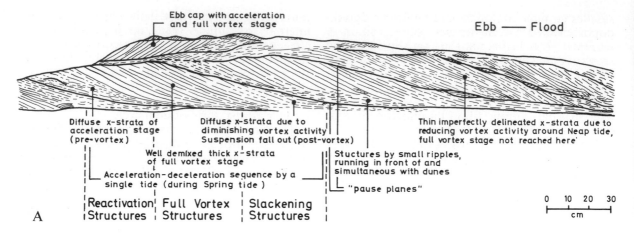

Ebb ——— Flood

Ebb cap with acceleration
and full vortex stage

Diffuse x-strata of
acceleration stage
(pre-vortex)

Diffuse x-strata due to
diminishing vortex activity
Suspension fall out (post-vortex)

Thin imperfectly delineated x-strata due to
reducing vortex activity around Neap tide,
full vortex stage not reached here

Well demixed thick x-strata
of full vortex stage

Acceleration-deceleration sequence by a
single tide (during Spring tide)

Stuctures by small ripples,
running in front of and
simultaneous with dunes

"pause planes"

Reactivation   Full Vortex   Slackening
Structures   Structures   Structures

0   10 20 30
            cm

A

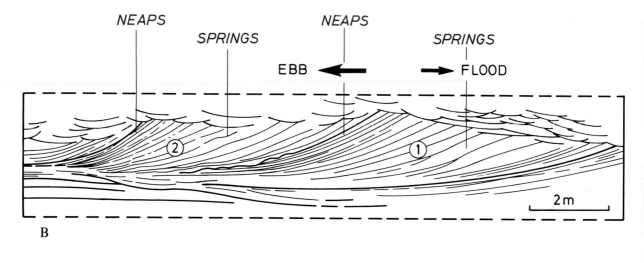

NEAPS   NEAPS

SPRINGS   SPRINGS

EBB ⟵   ⟶ FLOOD

②   ①

2 m

B

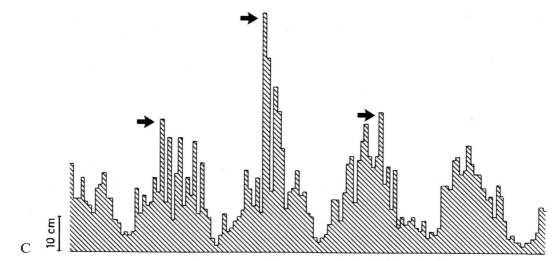

10 cm

C

Fig. 627A. Cross-section through a dune with a succession of tidal bundles containing reactivation (pre-vortex), full vortex, and slackening structures as response of a single-tide. (After Boersma and Terwindt 1980). B. Schematical sketch of tidal bundles reflecting neap-spring tide periods. The neap tide bundles are thinner and less high than the spring tide bundles. (Based on a field sketch of Allen and Homewood 1984.) C. The diagram shows the lateral thickness variation of neap-spring – neap tide cycles. The aperiodic thickness variations (*arrow*) are mainly due to storms. (Modified after Visser 1980)

low, the straight-crested megaripples are only temporarily produced during spring tide. During neap tides and in still higher parts of the shoal only small ripples are developed, where mud drapes are deposited in the troughs.

In the subtidal environment each bundle is produced during a dominant tide current (Visser 1980, Boersma and Terwindt 1980). Sometimes the bundles are separated by thin mud drapes and/or erosional reactivation surfaces. The mud drapes are formed during slackwater periods and respectively after the subordinate tide current phase. Thin sandy layers between the mud drapes are deposited by the subordinate tidal phase. Reactivation surfaces indicate the erosion of mud drapes by the subordinate current. The downcurrent thickening and thinning of the bundles (Fig. 627B) reflect neap-spring cycles (Visser 1980, Allen 1981a,b). A random thickness variation is caused by storms (Fig. 622C).

Diagnostic Features of Tidal Flat Deposits

Subtidal deposits are mainly channel (channel flanks and channel bottoms) and sand bar deposits. Subtidal channels are cut into older underlying sediments. Large-scale erosional unconformities separate older channel deposits from newer channel deposits. Two important characteristics of tidal flat deposits are (de Raaf and Boersma 1971):

1. Vectorial bimodality of cross-stratification.
2. Common joint occurrence at different proportions of large-scale and small-scale structured units in super- or juxtaposition.

The major part of the intertidal zone is made up of longitudinal cross-bedding of small dimensions. Erosional discordances are common. The

deposits often show high population density and strong bioturbation. Horizons of mollusc shells, i. e., *Mya arenaria* with shells in living position, are common, often with escape traces. Bedsets of differing bedding types alternate with each other. Alternating bedding–developed in the form of flaser and lenticular bedding, and finely interlayered sand/mud bedding–is abundant. Surface markings indicating falling water level and subaerial emergence are common on sediment surfaces. Sediments of the supratidal zone are interfused by plant roots; bedding surfaces are uneven and wavy in nature.

Megaripples of tidal deposits often show composite internal structures. Boersma (1969) describes such features as erosional unconformities between the successive cross-stratal bundles that build one large-scale set. Klein (1970a) calls these erosional unconformities "reactivation surfaces", and considers them to be typical parameters for alternating, but unequally strong tidal currents. In such a case, especially the lee-face of a megaripple is partly eroded by the opposite, less strong tidal current. Consequently, the new lee-face migrates laterally producing foreset laminae (Fig. 158).

Klein (1971, 1977a,b) gives an exhaustive list of a clastic tidalite process-response models. Ginsburg (1975) gives useful selected examples of carbonate and terrigenous clastic tidal deposits, both from Recent and ancient rocks. Klein (1977a) describes tidal sedimentary structures from deep-water environments.

Terwindt (1979) recognizes seven lithofacies in the intertidal and subtidal areas of the North Sea, each of which is characterized by grain-size and bedding structures and corresponds to a specific hydrographic conditions (Table 33). The environmental conditions in the inshore areas are rather variable, and consequently the lateral and vertical sequences of tidal deposits exhibit much variabili-

Table 33. Seven main lithofacies of the North Sea tidal sediments with their hydrodynamic interpretation. (After Terwindt 1979)

Lithofacies	Texture	Structures	Hydrodynamic conditions
Strong currents	Coarse-medium sand	X-bedding tabular, festoon, fills	$U_{0.5\,max} > 0.7\,m/sec$
Medium currents	Medium-fine sand, mud drapes	X-lamination tabular, herringbone, flaser	$U_{0.5\,max} > 0.45\,m/sec$
Weak currents	Fine sand, mud	Lenticular, sand/clay alternation	$U_{0.5\,max} < 0.45\,m/sec$ periodically
Strong waves	Coarse-fine sand	Even lamination	$U_m > 0.7\,à\,1.2\,m/sec$
Medium waves	Medium-fine sand, mud drapes	Wave x-lamination flaser bedding	$0.15 < U_m < 0.7\,m/sec$
Weak waves	Fine sand mud layers	Wave x-lamination, lenticular bedding	Periodically $U_m > 0.2$ and $U_{0.5\,max} < 0.45\,m/sec$
Storm deposits	Medium-fine sand	Even lamination	Heavy turbulence

ty. Terwindt (1979) gives a number of microsequences developed in tidal deposits, and concludes that the main characteristics of inshore tidal deposits are:

1. Bidirectional foresets in cross-beds, and sometimes reactivation and slackening structures.
2. Consistent occurrence of tidal bundles and lateral neap-spring tide sequences in megaripples.
3. High frequency of erosional contacts and abrupt facies changes coupled with the absence of a general vertical macro-sequence (in combination with both other features). However, a number of micro-sequences are present.

An important feature, observed predominantly in subtidal environment with a dominant flow component, ebb or flood tide, are tidal bundles described by Boersma and Terwindt (1980) and Visser (1980). Allen (1981a,b), Nio et al. (1983) and Yang and Nio (1985) discuss methods of reconstructing palaeo-hydraulic parameters.

Ancient Tidal Flat Deposits

During the last several years many new tidal flat deposits from the geological record have been described. Reineck (1972) gives a useful bibliography of examples of ancient tidal flat deposits. Klein (1970b) tabulates some 31 ancient intertidal flat deposits. In the following, only a few examples are discussed. Singh (1968, 1969) describes tidal flat deposits from the Precambrium of southern Norway (Telemark Suite). The presence of features such as raindrop imprints and various modified ripple patterns indicate that parts of these deposits are intertidal flat deposits.

Hobday and Eriksson (1977) give many ancient examples of the tidal sea deposits. Button and Vos (1977) describe an interesting Precambrian sequence from South Africa which represents subtidal and intertidal clastic and carbonate sediments of a macrotidal environment. The models reconstructed by them have similarities with the present-day sand tongues (shoals) and channels of Nordergründe, North Sea. Another example is similar to the tidal flats of Taiwan. Both examples are from a non-barred coastline.

Wunderlich (1970b) compares the Devonian tidal deposits (Nellenköpfchenschichten, Germany) with modern tidal flat sediments of the North Sea. There are subtidal as well as intertidal deposits with large channel deposits. Megaripple bedding, flaser- and lenticular bedding, interlayered sand/mud bedding, and tidal bedding are well developed.

MacKenzie (1968, 1972) describes tidal deposits of the Dakota Group (Cretaceous) near Denver, Colorado. The presence of dinosaur tracks, stationary marks of wood, channels of mainly muddy sediments with mud pebbles, capped-off ripples, ripples modified by flowing water during emergence, and erosional depressions with microbial mats sculptured with ripples point to deposition, mainly in an intertidal zone.

All ancient examples of tidal flat deposits show that not all the characteristics are found in a single deposit. A comparison of tidal flat deposits of modern sediments clearly shows a great variation within the tidal flat deposits. In general principle they are very similar; however, in details much variability exists (Reineck 1979).

We agree with de Raaf and Boersma (1971) that the differentiation between subtidal and intertidal deposits, as well as the differentiation between inshore and offshore sediment, is rather problematical in ancient sediments. De Raaf and Boersma (1971) provide a useful discussion on tidal deposits with illustrative descriptions of a few well-known examples from Europe.

Matter (1967) gives an account of Ordovician tidal deposits, Maryland, U.S.A., in carbonate rocks. This example clearly shows that even carbonate rocks of the tidal environment show primary sedimentary structures which are very similar to those of clastic rocks. Matter (1967) distinguishes various subenvironments: Marsh (lumpy limestone), supratidal (laminated dolomite with mud cracks), intertidal (ribboned, and stromatolitic limestone with cut and fill structure and mud cracks).

Klein (1971) suggested a model for determining palaeotidal range, on the basis that the intertidal zone shows sediment-fining towards the high water line, and in a progradational sequence, the intertidal flat sediments develop as fining-upward sequences. Klein (1972) determined palaeotidal ranges for deposits of different geological ages and found that since the Precambrian the paleotidal range has not changed significantly through geological time. However, recognition of high-water and low-water line in most tidal sequences is rather problematic. Thus, this method should be used with the utmost care.

Swett et al. (1971) diagnose the Eriboll sandstone as tidal deposit, where sedimentary structures are planar, herringbone cross-bedding, reactivation surfaces, interference ripples, and both current ripples and micro cross-laminae superimposed at 90° on slip faces of cross strata. Escape structures of burrows are present. The overlying sediments are intertidal and supratidal carbonates.

Fossil-tide generated sandwaves with tidal bundles reflecting neap-spring tide cycles are described by Allen (1981a) from the Lower Cretaceous, Folkstone Beds around the Weald of southeast England, and Allen and Homewood (1984) from the Miocene marine molasse in western Switzerland.

Continental Margin, Slope and Ocean Basin

General Information

Marine waters cover 70.8 % of the earth's surface. And, if we substract the 5.5 % area of the continental shelf, 65.3 % of the earth's surface is occupied by continental slope and rise, and deep-sea parts (data after Dietrich and Kalle 1957). The deep sea can be subdivided into several topographical units. There are *deep-sea grabens* (15 in number), which cut deeper than 6,000 m. Deep-sea grabens are restricted to tectonic active regions near the foot of continental slopes.

The *ocean basins* are usually rather flat, monotonous regions, with average depths of about 4,000 m. The transition from ocean basins to continental margin is usually between 3,000 to 4,000 m water depth. The rather monotonous relief of ocean basins is interrupted at places by volcanic cones. Such features are known as seamounts if they are more than 1 km in height above the average surface of the sea bottom. Smaller features are commonly referred as sea knolls or abyssal hills. Mounts that have once been above water level and have been planned off are called guyots on their submergence. Guyots are submerged mounts with their planned-off upper surface located at water depths of up to 2,000 m. On some of the guyots which sank down rather slowly, corals began inhabitance, making atolls. Other guyots which sank down rather rapidly are covered by a sediment layer of shallow-water carbonates.

Ocean basins are surrounded by the continental margins, which connect them to the continents. In the open ocean itself, oceanic ridges provide another limit to the ocean basins. Ocean ridges are of huge dimensions and are much larger than any of the mountain systems on the continents and are made up of igneous rocks.

As depositional environments continental margins and ocean basins are more important. Mainly pelagic sediment is deposited in the basins, whereas continental margins characteristically show detrital sediments. Submarine canyons play an important role in transporting huge quantities of sediments across continental margins, which are deposited in the form of submarine fans. Many of them show features comparable to turbidite deposits.

Continental margins have gained much importance in petroleum exploration (Hedberg 1970; Walker 1978; Wilde et al. 1978). Ross (1979) discusses many aspects of the oceans, i.e. marine shipping, ocean resources and pollution, and military uses. On the geological and geophysical investigations of the Continental Margins Watkins et al. (1979) provide a collection of 32 papers.

Major Units of Continental Margins

The continental margin is comprised of the continental shelf, the continental slope and the continental rise. The inclination of the continental shelf on average is 7 minutes away from mainland.

According to Shepard (1963), the major break in the inclination of the continental shelf is at an average depth of 132 m.

Many continental shelves of today deviate from this average value; some of them show this break at lesser depths, while others at much greater depth. This break is conventionally considered to occur at a depth of 200 m. The continental slope shows an average inclination of 4°. However, the continental slope is only rarely a plane, inclined surface. In most cases, the continental slope shows well-defined depressions, ridges, step-like features, isolated mounts, and it is sometimes cut across by numerous submarine canyons. Near the lower part of the continental slope the inclination becomes very gentle and grades into the continental rise, which can be 300 to 400 km wide. In front of submarine canyons large deep-sea fans are often present, which extend into oceanic basins. Burk and Drake (1974) and Stanley and Kelling (1978) give an excellent collection of papers dealing with the continental margins, both present-day and ancient.

Some data from Emery (1970) provides an insight into the dimensions of continental margins. Today, continental shelves occupy an area of $27 \times 10^6 \, km^2$, the continental slopes $28 \times 10^6 \, km^2$, and the continental rises $19 \times 10^6 \, km^2$. These figures together are $74 \times 10^6\,\mathrm{km^2}$, which is roughly 15 % of the earths surface, equivalent to the 50 % of the area of continents. The amount of sediment on continental margins is estimated to be $150 \times 10^6\,\mathrm{km^3}$, considering the average thickness of sediment, to be 2 km. The thickness of sediment in ocean basins is considered to be 0.2 km. Thus, the $287 \times 10^6\,\mathrm{km^2}$ area of ocean basins is covered by $57 \times 10^6\,\mathrm{km^3}$ sediment. Although the continental margins of the world amount only to the 21 % of the sea surface, 73 % of the total marine sediments are present on the continental margins. Even the land area shows only $125 \times 10^6\,\mathrm{km^3}$ sediment and sedimentary rocks, against the $150 \times 10^6\,\mathrm{km^3}$ sediments of the continental margins. No doubt, continental margins show the largest accumulation of the sediment in the world.

After one generation's gap, the theory of continental drift of Wegener, Du Toit, and Taylor has been confirmed by new data. This has led to the theory of a spreading sea floor, according to which the separation of North America and South America from Europe and Africa started during early Permian time (Emery et al. 1970). However, the oldest sediments in the continental rise are of Jurassic and lower Cretaceous age. According to Drake et al. (1959), the sediments coming from the land into the sea were trapped into marginal trenches and dams. These sediment traps were filled up during Late Cretaceous, and then the dams were overflowed and the sediments were transported toward the continental rise.

Emery (1970) postulates that such dams were distributed the world over and were of tectonic, organic reef, or diapir origin. On the continental rise of the eastern coast of the United States a horizontal horizon (horizon A) is present below the sediment masses of the continental rise. This probably represents the pelagic sediments of late Cretaceous age. This horizon can be traced near the continental slopes (Fig. 628). One can say with certainty

that during the Mesozoic in the Atlantic no continental rise was present. The stratified sediments of the continental rise lap against the stratified sediments of the continental slope. The continental rises are absent where deep-sea trenches border continental slopes (Fig. 629). In such regions tectonic horsts near the edge of the continental shelves, and deep-sea trenches, have trapped almost all of the sediment brought in from the adjoining land area. On the other hand, where the continental shelf is erosional and no adjacent deep-sea trenches are present, the continental rise is well developed, reaching far into the ocean basin.

In general terms, continents have grown by the successive addition of younger sediments around the older stable shields, as exemplified by Tertiary sediments bordering the continents.

The continental terrace, defined as the sediment and rock underlying the coastal plain, the continental shelf, and continental rise have been seismically explored by Lewis (1974). The development of continental terraces shows three main stages: (1) generation of a strong discontinuity between continent and ocean basin, (2) deposition of sediments, and (3) modification of continental terrace during Quaternary. The genesis of main discontinuity is related either to rifting or to accretion of a large mass of contorted continental terrace and continental rise sediments to a pre-existing edge of a continent (Fig. 631).

On the basis of internal structure, Moore and Curray (1963) distinguish four types of continental margins, to which Curray (1969b) added one more new group (Fig. 630). They distinguish into one erosive and four depositional types of continental terraces. A continental terrace includes the continental shelf and the continental slope. However, all sorts of transitional types of the basic are known to occur in nature.

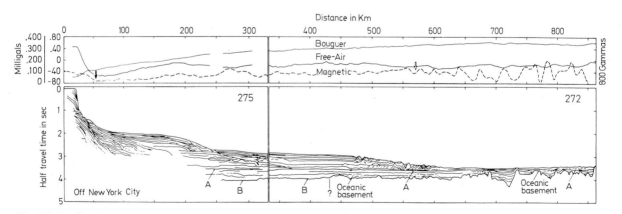

Fig. 628. Profile of continuous seismic reflections, geomagnetics, and gravity measurements off New York City. Note the landward slope of the basement B, flat horizon A, and slumped masses in the continental rise deposits. (After Emery et al. 1970)

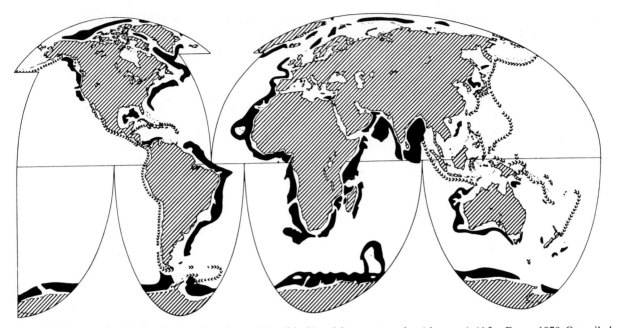

Fig. 629. Map showing the distribution of continental rises (black) and deep-sea trenches (chevrons). (After Emery 1970. Compiled from various sources)

Tectonic or Erosional Type of Continental Margin

The tectonic continental margin represents the shelf and slope regions on which deposition is not continuous. The strata have been folded, faulted, and eroded. An example for such a margin is the narrow continental terrace of southern California. It is made up of a folded and faulted sequences of Tertiary rocks overlaid by thin discontinuous Quaternary sediments. The present-day surface was formed by erosion during the Pleistocene period. Several continental margins of the Pacific Ocean are tectonically active and probably belong to this group.

Depositional Continental Margins

Depositional continental margins are continental terraces on which deposition has been a continuous process. No major interruptions leading to deformation activity are present. Here four subtypes can be distinguished:

1. Dominantly upbuilding. This is the case if deposition is mainly on the continental shelf, and no sediment is deposited on the continental slope (Heezen et al. 1959). The material reaching the continental slope is quickly removed by slumping and turbidity currents. Heezen et al. (1959) proposed this model for the continental terraces of the North Atlantic.

2. Dominantly outbuilding. In opposition to model 1, here deposition is mainly on the continental slope, and no sediment is laid down on the continental shelf. Wave-built terraces of Johnson (1919) belong here, for which it is assumed that shelf break corresponds to the wave base. Permanent deposition is impossible on the continental shelf because of wave activity; thus all sediments by-passing the shelf are deposited on the continental slope. This type is rather rare in ideal form and on a large scale.

3. Both upbuilding and outbuilding. In this case deposition is possible both on the continental shelf and the continental slope. Internally, sedimentary layers are present running roughly parallel to the surface profile of the continental terrace. In other words both topsets and foresets are present. The topsets may represent shelf, beach, lagoonal, delta, or other deposits. The foresets represent deposits of the continental slope. A good example of this type is the continental terrace of the Gulf of Mexico. The continental margins of the west coast of Mexico and Alaska also belong to this type.

4. Entrapment by dam. On the outer margin of several continental shelves dams and ridges have been noted. In the beginning they hindered transport of sediment into deeper parts. Deposition took place in such trenches until they were filled up and sediment overflowed the dams.

The major types "tectonic", "erosional", and "dominantly upbuilding" clearly show that the major structural framework of the continental terraces is of diastrophic origin, whereas tectonic, erosion and deposition processes vary from place to place. In regions where tectonic and erosional processes are more important, e. g., the upper one-third part of the continental slope, older sediments, and volcanic and igneous rocks are seen outcropping.

The above-mentioned major types of internal structure of continental margins are not always ideally developed. Transitional forms are rather common. Moreover, one should not try to see the development of the continental margin, especially the continental shelf, as a continuous process. For example, sea-level fluctuations during the ice age, eustatic sea-level changes, tectonic movements

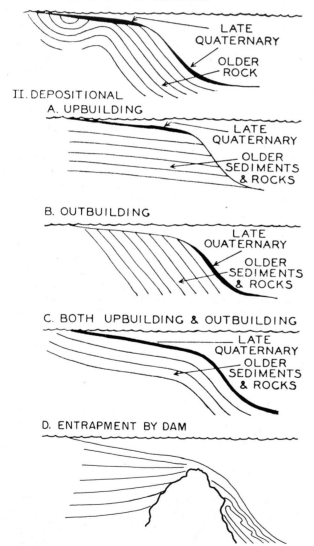

Fig. 630. Different types of internal structure of the continental margin. (After Curray 1969b)

and changing rates of supply of sediment cause changes in the form of transgression and regression.

During *maximum transgression,* which seems to be the case today, wide and flat alluvial plains are formed near the coast. The sediment layer is present on broad shelves, flat deltas are abundantly built up and fine-grained material is distributed over vast areas of the shelf.

During *maximum regression* river mouths are built up near the steep upper continental slope. No broad deltaic deposits are formed, as most of the sediment is deposited in deeper waters of the continental slope and there is no lateral drifting of the sediment. The influence of waves and currents is negligible. Deposition is extremely rapid in restricted areas, resulting in a thick pile of sediment. This pile consists of underconsolidated, unstable sediment deposits, which undergo frequent slumping, leading to turbidity currents. These turbidity currents transport sediment to the continental rise and the ocean basins (Ewing et al. 1958; Moore 1961).

Thus, similar amounts of deposition both on the continental shelf and the continental slope are possible only during some intermediate stages of maximum transgression and maximum regression. Curray (1969b) distinguishes among five major morphological types of continental margins (Fig. 632).

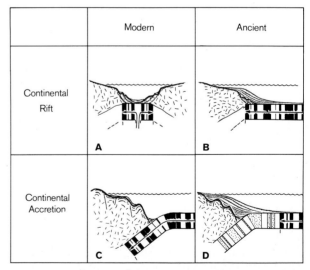

Fig. 631. Diagram sections of continental terraces showing their primary origin. Random hatching – continental crust and vertical bars – ocean crust. Arrows indicate movement of oceanic crust. A, B, D show oceanic crust and continental crust coupled; C shows oceanic crust being thrust beneath continental crust which is being warped along with the most recently deposited sediment. A modern continental rifting some normal faulting; B ancient continental rifting – slow flexure; C modern continental accretion – orogenesis; D ancient continental rifting – slow flexure. (After Lewis 1974)

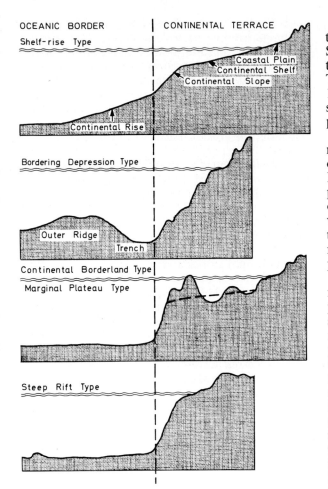

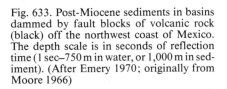

Fig. 632. Scheme showing various morphological types of continental margins. (After Curray 1969b)

1. *Shelf-rise Type.* A broad continental shelf, slope, and rise of all units are well developed. The east coast of the United States is a good example. Internally, dams seem to be present below the continental shelf and the continental rise. The continental margin of the Gulf of Mexico is also of this type.

2. *Bordering-depression Type.* On several continental margins deep-sea trenches are present. Some of these trenches have been filled up with thick deposits of sediments, e.g., the Peru-Chile Trench, which contains sediment more than 1,000 m thick (Scholl et al. 1968). Other trenches show only a little filling from lack of sediment supply, e.g., the Japan Trench.

3. *Marginal Plateaus.* On some continental margins plateau-like structures are present on the continental slope. The most well known is the Blake Plateau. Here, during Palaeocene times depositional rates on the plateau and the adjacent continental shelf were similar. Later, with the development of the Gulf Stream, the rate of deposition slowed down on the plateau. On the outer margin of the plateau a dam structure seems to be present.

4. *Continental Borderlands.* In opposition to marginal plateaus continental borderlands are repeatedly folded and faulted, so that a series of basins and ridges is produced. Some of the ridges reach above sea level as islands. Best-known examples are off southern California and northern Baja California. According to Moore (1960, 1969), these tectonic basins of the continental borderland are filled with sediments. Some of the sediment fill is deformed from continuous tectonism. However, due to high rates of deposition Pleistocene and later sediments are undeformed (Fig. 633). Most of the sediment fill of these basins is turbidite in nature, deposited probably through low-velocity low-density turbidity currents. The basin closest to land is filled up first.

5. *Steep Rift Continental Margins.* Some of the continental margins do not show a well-developed continental rise, although no trench or depression is present at the base of the continental slope. It seems that with time, when enough material accumulates at the base of continental slope, a rise of the shelf-rise type will develop. On the continental slope of such margins older sediments outcrop. A good example of this type is the west coast of the Iberian peninsula.

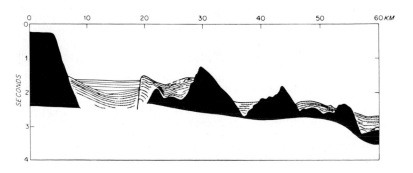

Fig. 633. Post-Miocene sediments in basins dammed by fault blocks of volcanic rock (black) off the northwest coast of Mexico. The depth scale is in seconds of reflection time (1 sec–750 m in water, or 1,000 m in sediment). (After Emery 1970; originally from Moore 1966)

Submarine Canyons and Fan Valleys

Few submarine canyons were known in the nineteenth century; a more complete picture of canyons cutting across the continental slope has been made available only recently with the application of continuous acoustic profiling with echographs. Vaetch and Smith (1939) prepared a detailed hydrographic chart for the east coast of the United States, which has well-defined canyons. Such submarine canyons represent straight or curvilinear valleys, which can be bifurcating in their upper reaches and run from the continental shelf down to the continental slope. In front of such canyons large deep-sea fans and fan valleys are present. Shepard and Emery (1941) published a detailed map with canyons off the coast of California. Scientists started pondering the mode of genesis of canyons, first thinking them to be of subaerial origin.

Shepard and Dill (1966) suggest correctly that many different forms of submarine valleys exist, and that they can be of varied origin. They differentiate 8 types.

1. Submarine Canyons. They are V-shaped in cross-section and possess rather steep walls with exposed rocks. They often possess several branches reaching to the main canyon. The course of the canyons is usually curved and winding. They are the commonest type of submarine valleys and occur extensively on continental slopes (Fig. 634).

2. Fan Valleys. These are valley cutting across the deep-sea fans. They are V- or U-shaped in cross-section. The valley walls do not show rock outcrops, although they are sometimes very steep and more than 180 m deep. The course is winding and there are no tributaries, but frequently many distributaries are present. Along the valley sides low ridges–natural levees–are developed.

3. Shelf Channels. They are relatively shallow channels running across the continental shelves, and are often discontinuous. Shelf channels do not show any direct connection with deeper canyons.

4. Glacial Troughs. They are submarine valleys on the continental shelves off glaciated coasts. They are U-shaped in cross-section, having both tributaries and distributaries. They occasionally reach depths greater than 180 m, with rather large basins.

5. Delta-front Troughs. Such valleys are present in front of some of the large river mouths, e. g., Ganges, Indus, Niger and Mississippi. They are U-shaped, and begin on the continental shelves and slope down continuously across the continental slope and beyond.

6. Slope Gullies. These features are present on rather straight slopes and often in front of deltas. They run across the continental shelves. They are of low relief and often discontinuous.

7. Valleys Resembling Grabens or Rifts. They are restricted to areas of active diastrophism. They are V- or U-shaped in cross-section and are rather straight, without any tributaries.

8. Deep-sea Channels. They are rather broad, trough-shaped valleys in the deep-sea basins, sometimes showing tributaries. They are of very large dimensions. Chough and Hesse (1976) investigated 4,000 km-long meandering Northwest Atlantic Mid-ocean channel of the Labrador Sea. This channel shows natural levees. The sediments of the channel suggest a mass sediment transport by turbidity currents and related processes.

The origin of submarine valleys is rather uncertain even today. Daly (1936) was first to propose the importance of turbidity currents in the genesis of most of the canyons; he was later supported by Kuenen (1937). Shepard and Dill (1966) give a useful summary of various hypotheses and theories of the origin of submarine canyons. There is

Fig. 634. Near true-plan view of canyons towards the top of the continental slope in the Bay of Biscay. In this figure the side-scan sonar beam was directed from the ship (top of sonograph) to look downslope. Sea floor in shadow appears black and where "illuminated" by the soundbeam is light toned. Numerous secondary gullies are seen on slopes facing the sound beam and lie in shadow on slopes facing away. The top of the continental slope is intensely dissected, while further downslope (bottom of sonograph) there are indications of broad, smoother "interfluves" which have been confirmed by a parallel survey track through this area. On the left of the sonograph the canyon trend across the regional slope, perhaps controlled by faulting. Other possible faults are found with a trend that is about normal to the main trend. Towards the right of the sonograph the straight or slightly curved canyon axes lead down the regional slope. (After Belderson and Kenyon 1976)

much evidence that in the upper part of submarine canyons bottom sediment transport (traction transport) and small sand fall are active (Dill 1964). Such sediment movement produces active erosion of canyon walls (Fig. 635). However, the question remains open whether (1) such sediment transport alone can account for erosion and transportation in submarine channels, especially in the fan channels, where the gradient is gentle, or (2) the high-velocity turbidity currents are the major eroding agencies for the submarine channels. Although no direct measurements are available for the occurrence of high-velocity turbidity currents, it seems reasonable to believe that such currents are far too rare and that is why they have not been observed. Even in areas of exceptionally high rates of sedimentation such turbidity currents are observed only at long intervals. For the coast off southern California Gorsline and Emery (1959) state that turbidity currents bring sand into the ocean basin every 400 years.

Shepard (1972) suggests that submarine canyons are largely of marine origin because he has much evidence that active erosion is taking place on the sea floor today.

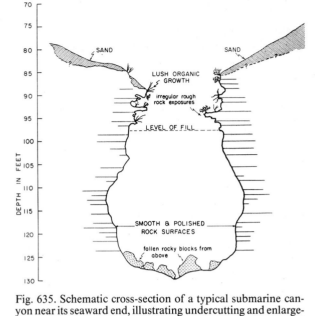

Fig. 635. Schematic cross-section of a typical submarine canyon near its seaward end, illustrating undercutting and enlargement of the lower section that is normally filled with organic sediment. The blocks that fall into these tributaries slowly move into deeper water. "Level of fill" denotes the approximate amount of sediments these features hold before the sediment becomes unstable and moves into deeper water. (After Dill 1964)

Transport System in the Continental Shelf, Margin and Ocean Basins

The occurrence of extensive sandy layers in some ocean basins demanded some transport mechanism other than the slow sinking deposition of pelagic and hemipelagic material from suspension. Daly (1936), Kuenen (1937, 1947b, 1950b) and many later publications proposed the concept of a turbidity current of high density, by which sand and coarser sediments could be transported over long distances, i. e., across the continental slope into the ocean basins. At the same time turbidity currents were suggested to be the major factor in the genesis and development of submarine canyons. In the years since, the concept of high-velocity high-density turbidity currents has been applied to explain numerous ancient examples of extensive sand blankets in flysch sediments.

The study of post-orogenic filling of basins in the continental borderland off the California coast demonstrated that several questions remain unanswered, if high-density turbidity currents were considered to be the only transport mechanism to bring sand into the basins. The green mud layers were interpreted to be a product of particle-by-particle deposition, because these layers contain only deep-water foraminifera and no shallow-water foraminifera. The graded bedding of sand layers did not go beyond the silt of the green mud layer

(Gorsline and Emery 1959; Hand and Emery 1964).

On the other hand, particle-by-particle deposition of green mud layers leaves some other questions unanswered. Why does the bedding of these layers either show an initial dip from the canyon, gully, and distributary channel, or show plane-bedding rather than showing a general dip away from the coastal drainage? Why does part of a single basin rapidly fill, while the other part of the same basin located behind a sill receives almost no sediment? Why are only those basins filled with sediment which are somehow connected to the distributary system of a canyon, valley, or channel?

Moore (1969) points out that these questions remain unanswered if particle-by-particle deposition of hemipelagic type is considered to be the major process.

Moore (1969) discusses various possible modes of transport in the deep-sea region. He places emphasis on a particular concept, i. e., low-density, low-velocity turbidity currents, which is based on both new and old concepts and data.

Moore (1969) listed the following as the four most important transport systems, which produce characteristic types of deep-sea deposits. They are depicted diagramatically in Fig. 636:

Fig. 636. Schematic representation of the routes of transportation of sand (solid arrow) and mud (dotted arrow) from river mouth to deep-sea basin floor. (After Moore 1969)

1. Fluvial mixed transport and separation of sand and lutum
2. Transport and deposition of sand, including longshore transport and canyon and distributary transport
3. Low-velocity, low-density turbidity currents
4. High-velocity, high-density turbidity currents

In order to give a more complete picture of various modes of transport in the deep sea area, the following mechanisms and types of deposits can be added to the above-mentioned ones:

5. Sediment slides of larger dimensions
6. Transport through contour currents

Middleton and Hampton (1976) distinguish four types of sediment gravity flow with their characteristic deposits. Sediment gravity flows denote that the sediments are moved under the influence of gravity, and the movement of sediment grains causes movement of interstitial fluid. In the case of fluid gravity flow, moving water moves the sediment parallel to the bed. The four types of sediment gravity flow are (Fig. 637):

1. Turbidity current flow → turbidite
2. Liquefied sediment flow → turbidite
3. Grain flow → resedimented conglomerate and some fluxo-turbidite
4. Debris flow → pebbly mudstone.

These four types of sediment gravity flow represent end-members, while in nature all the transitions between these flow types may occur.

In accordance with the work of Middleton and Hampton (1976), Walker (1978) put together modes of transport, grain-supporting mechanisms, and the resulting deposits through space and time (Fig. 638). For the initiation of flow, Walker (1978) distinguishes a river in flood or slump and slide. The main long-distance transport takes place as debris flow, high-concentration turbidity currents, or low-concentration turbidity currents. The grain-supporting mechanisms are: fluid turbulence, liquefaction, collision between individual grains (dispersive pressure in grain flow after Bagnold 1954), and matrix strength in debris flows. The resulting deposits are: classical turbidites, massive sandstones, pebbly sandstones and conglomerates, and pebbly mudstones (debris flows). A schematic representation of these deposits is shown in Fig. 649.

In the following a short account of various transport and deposition processes, which operate after the separation of sand and mud in the river mouths, is given. It includes four types of sediment gravity flow; low-density, low-velocity turbidity currents, contour currents, and slice and slumps. These processes are the main agents of transport and deposition in the continental slopes and ocean basins.

Fluvial Mixed Transport and Separation of Sand and Lutum (Mud)

The rivers bring both sand and mud into the sea. The sand is transported mainly as bed load, while minor amounts of sand, silt and clay are transported in suspension. In the sea an effective separation between sand and lutum takes place. The sand is deposited mainly near the river mouth in the form of mouth bars, etc., whereas silt and clay are transported farther offshore in suspension. At last, not very far away from the coast, major amounts of lutum are deposited, and only a small amount is taken farther into deeper water. In the case of large rivers, much lutum can be transported over great distances. For example, in the Mississippi delta, the surface layer of fresh turbid water coming from the river can be seen over 100 km offshore under specific weather and current conditions.

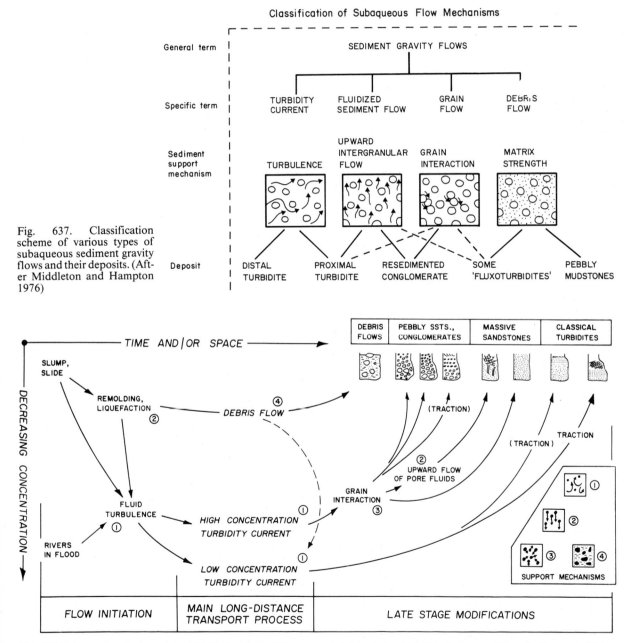

Fig. 637. Classification scheme of various types of subaqueous sediment gravity flows and their deposits. (After Middleton and Hampton 1976)

Fig. 638. Scheme showing processes of initiation, long-distance transport, and deposition for currents transporting sediment into deep water. Grain-supporting mechanisms (insert, lower right) include: *1* fluid turbulence; *2* liquefaction; *3* collision between individual grains (dispersive pressure in grain flow); and *4* matrix strength (as in debris flow). (After Walker 1978)

High-density, High-velocity Turbidity Currents and Turbidite Deposits

Moore (1969) calls high-density, high-velocity turbidity currents marine mixed transport. Such turbidity currents originate as a result of accumulation of a large amount of sediments on the canyon
slopes. On the steep slopes the sediment starts
slumping periodically. Besides, thick sediment deposits in the canyon floor may undergo liquefaction and start moving downslope. A through-mix-
ing of this sediment produces a high-density turbid
layer, which moves down-slope at high speed.
However, such high-density high-velocity turbidity currents have not been observed first hand. Turbidity currents can originate from various processes, e.g., oversteepening of a slope, earthquakes, etc. However, in every case huge amounts
of sediments are put into motion. Such turbidity
currents may initiate active erosion and they are
capable of producing new submarine valleys or
modifying the existing ones. Turbidity currents
flow like rivers toward ocean basins and deposit
large amounts of sediments near submarine valley
mouths in the form of deep-sea fans. Sometimes
they overflow the channels and deposit the sediment on the valley sides in form of levees. Much
sediment is brought into ocean basins and spread
out in the form of an extensive blanket. Coarser
sediments are deposited near the canyon mouth,
while finer-grained sediments are spread into the
ocean basin.

More recently, Middleton (1966a,b, 1967) made
various experiments with the turbidity currents
and density currents. Middleton (1970) provides a
useful summary on the hydrodynamic aspects of
the turbidity currents.

A turbidity current is distinguished into head –
the thickest part, body – which is rather uniform in
thickness throughout its length, and tail – where
flow thins out and becomes dilute. According to
Middleton and Hampton (1976) this flow pattern
of the turbidity current has important conse-
quences:

1. The head is the region of the erosion and formation of scour marks. Probably this process also
   continues when deposition from the body of the
   current behind the head has started.
2. Dense medium (sediment-water mixture) must
   be continuously supplied in the head to compensate for the losses caused by the turbulences. Thus, it is presumed that in the head
   there is a progressive enrichment of the coarse-
   grained material.

Komar (1972) studied the relative thickness of
the head and the body, and concludes that at higher Froude's numbers the head is thicker, while at

lower Froude's numbers (< 0.75) the body is
thicker. Hence, in the deeper waters, near the abyssal plain, the body is thicker and possesses more
significance for overspill out of the channel and
fan sedimentation.

In general terms a turbidity current denotes a
high-density current flowing down a subaqueous
slope, or spreading horizontally because suspended sediment gives it a higher density than the
surrounding clear water. A sedimentary deposit resulting from the deposition of a turbidity current is
called a turbidite deposit, or a turbidite.

Where ever two bodies of water of different den-
sities are in contact with each other, the water body
of higher density tends to flow and spread out below the less dense one. Such density currents are
well known in oceans, because of accompanying
differences in temperature or salinity. Also turbid,
suspension-rich layers of higher density, may flow
down an available slope and spread out below the
clear water of lower density. Forel (1885) applied
this idea to the water of the Rhône River entering
Lake Geneva. Daly (1936) brought this idea to the
attention of geologists to explain the origin of submarine canyons. Kuenen (1937, 1947b, 1950b)
studied the mechanism of turbidity currents experimentally and applied this concept to explain the
erosion of submarine canyons. Later, Kuenen and
Migliorini (1950) and Natland and Kuenen (1951)
applied this concept to explain the origin of flysch
sediments of deep-sea origin.

Kuenen and Humbert (1964), and Bouma and
Brouwer (1964) give detailed bibliographies of the
publications on the turbidity currents and turbidites.

Bouma (1962) made a comprehensive study of
ancient turbidites and developed a turbidite facies
model, now known as the Bouma sequence (Fig.
639). An ideal single turbidite sequence is made up
of five units with specific sedimentary structures:

1. Graded Interval. This is the lowermost part
of a sequence showing a more or less distinct
graded bedding. In well-sorted sediments grading
may be inconspicuous. No other sedimentary
structures are present. It is sandy or gravelly in nature.

2. Lower Interval of Parallel Lamination. This
interval shows predominantly thick parallel laminae of sand. Grading may be present. The contact
to the graded interval is gradual.

3. Interval of Current Ripple Lamination. This
interval is made up of fine sand and silty sediments
showing small-current ripple bedding. The height
of ripples is less than 5 cm and the length is less
than 20 cm. Ripple bedding may be developed in
the form of climbing-ripple lamination. Sometimes convolute laminations are present with
piled-up ripples. The contact with the lower interval of parallel lamination is rather sharp. Within

Fig. 639. Ideal sequence of sedimentary structures in a single turbidite deposit ("the Bouma sequence"). A tentative interpretation of the flow regime conditions during deposition of various units is also added. (After Middleton and Hampton 1976)

Grain Size		Bouma (1962) Divisions	Interpretation
Mud	E	Interturbidite (generally shale)	Pelagic sedimentation or fine grained, low density turbidity current deposition
Sand-Silt	D	Upper parallel laminae	? ? ?
Sand-Silt	C	Ripples, wavy or convoluted laminae	Lower part of Lower Flow Regime
Sand-Silt	B	Plane parallel laminae	Upper Flow Regime Plane Bed
Sand (to granule at base)	A	Massive, graded	? Upper Flow Regime Rapid deposition and Quick bed (?)

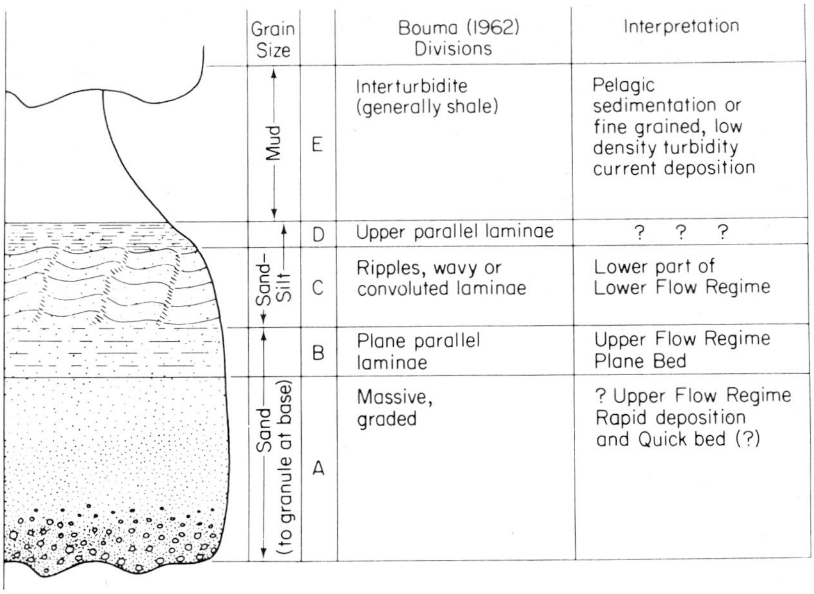

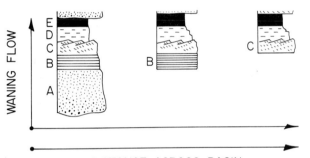

Fig. 640. Schematic diagram depicting interpretation of Bouma sequence (ABCDE) in terms of distance across basin, and waning flow conditions. This suggests that the turbidites beginning with divisions B and C represent deposition from progressively slower flows, which can be due to increasing distance across basin. However, the levees of the proximal turbidite environment may also show sequences beginning with divisions B or C. (After Walker 1978)

the interval of current ripple lamination an indistinct grading is present from bottom to top. Megaripple bedding may be seen very infrequently.

4. Upper Interval of Parallel Lamination. This zone of very fine sandy to silty clay shows a distinct parallel lamination. The contact with the lower zone is distinct.

5. Pelitic Interval. This interval of clayey sediments does not show any distinct sedimentary structures. A decrease in sand content and grain size may be found toward the top. Foraminifera may be present in this interval. The contact with the lower interval is gradational. Sometimes on the top of the pelitic interval a marl or clayey marl may be found, which represents pelagic sediments.

Bouma (1962) pointed out that this complete sequence has been found only in the thicker layers of flysch deposits. Usually the sequence is incomplete, and the topmost part, bottom part, or both are missing. To explain this feature, Bouma hypothizes the development of various intervals as tongue-shaped bodies, whereby the successively finer intervals are of larger areal extent. This is the result of decreasing current velocity and grain size in the direction of the current. Thus, near the source all intervals are represented, whereas in the down current direction lower intervals are missing (Fig. 640). Removal of the upper interval by erosion before next sequence is deposited produces sequences where top or both top and bottom intervals are missing. Middleton and Hampton (1976) provides more recent hydrodynamic interpretation of the Bouma sequence (Fig. 639).

Van der Lingen (1969) worked out a modified facies model of flysch-type sediments, which covers more variations and varieties than the Bouma sequence.

Harms and Fahnestock (1965) and Walker (1965) tried to interpret the Bouma sequence in terms of decreasing flow regime from bottom to top with increasing path across a basin, the turbidity current loses both the energy and the sediment; hence the distal turbidites, as shown in Fig. 640, as well as deposits on the levees, do not show the complete turbidite sequences (Fig. 641). Walker (1965) and Walton (1967) discussed the problem of the absence of megaripple bedding. It appears likely that a combination of factors, fine grain size and

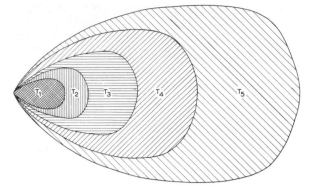

Fig. 641. Hypothetical plan showing geographic distribution of intervals in a flysch profile from T_1 (= A) at the base to T_5 (= E) at the top. (After Bouma 1962)

rapid rate of deposition, inhibit the formation of megaripples. Recently, several examples have been found where a division of mega-ripple cross-bedding is developed within the Bouma sequence. Such cases of large-scale cross-bedding are seen in coarser-grained sediments (Hubert 1966; Thompson and Thomasson 1969).

If one tries to put together various characteristics of turbidite deposits, the following indices seems to be most important. The graded bedding and current ripple lamination beds alternate with pelagic beds. There is a different fauna in the turbidite and adjacent pelagic beds. Besides, the sole markings are abundantly developed on the bottom of the sandy turbidite layers. These markings were produced as a result of the scouring action of turbidity currents on the muddy bottom over which the current flowed. These markings are various kinds of grooves made by current erosion or dragging tools, like pebbles, organic remains, etc. Other characteristic features of turbidite deposits are thick sequences, regular bedding, extensive aerial extension of individual units, and rather uniform current direction shown by the sole markings. Typical shallow-water features are completely absent, i. e., wave ripples, beach features, bioherms, large-scale cross-bedding, well-sorted sand, etc. Features indicating intermittent subaerial exposure, e. g., raindrop imprints, are completely missing.

Furthermore, features made by liquefaction are common in the turbidites (Middleton and Hampton 1976). Convolute bedding in turbidites is related to partial liquefaction. A quickly deposited sand layer (by turbidity current or grain flow) on a soft mud readily produces load casting. Dish and pillar structures are quite common in the turbidites and are related to liquefaction leading to escape of water upwards (Lowe 1975).

Dzulynski and Walton (1965) give a detailed account of various structures and textures of the flysch sediments. We know more about flysch-type sediments from ancient rocks than their modern analogues, especially the sedimentary structures. And in this case, as has also said by van der Lingen (1970), the actualistic principle has been reversed and the past has become the key to the present. The concept of turbidity current has been applied equally to the erosion of submarine canyons and the deposition of sand in deep-sea fans, as well as to the explanation of the genesis of sandy flysch sediments.

A major reason to believe the existence of turbidity currents is because of submarine topography. At the deeper end of the various submarine canyons deep-sea fans are present which gradually grade into the abyssal plain of the oceanic basins. The slope decreases from more than 1° to less than 1° and continues to fall away from the source over distances of 2,000 km. Areas of the deep sea which are supposed to be uneffected by the turbidity currents show extremely uneven topography, with ridges of several hundred meters high protruding upward. Basins and deep-sea trenches often show a plain top surface and in some cases, a spillway is present in the lower part, leading to the adjacent basin.

Long sediment cores taken from the abyssal plain and basins show thick deposits of sand or silt sediments, whereas such strata are absent from the surrounding slopes and ridges protruding out of the plains. Many of these deep-sea sands contain skeletal remains of shallow-water bottom-dwelling organisms and calcareous algae. Each sandy unit shows well-developed graded bedding, coarse-grained near the bottom and finer-grained toward the top and ultimately merging to pelagic sediment layers, which separates it from the next sandy layer. The sandy layer is parallel-laminated and partly current-ripple laminated, suggesting deposition under the influence of currents. Each sandy unit ranges in thickness from several millimeters to 6 m. Individual beds can be traced by echo sounders (subsurface reflections) over tens of kilometers. In some cases, distinctive beds have been traced in cores over areas of almost 10,000 km² (Kuenen 1966b).

Hülsemann and Emery (1961) made detailed corings in Santa Barbara, a basin off the California coast. Cores were made up of mainly bioturbated and nonbioturbated varve-like bedding. Each couplet is made up of a dark-colored, detrital layer (winter) and a light-colored diatom layer (summer). Intercalated were graded silt-clay beds (turbidites) which could be traced over the complete basin. Bouma (1964, 1965) was able to obtain box-core samples from a channel in the outer fan of the La Jolla Canyon which showed graded bed units of turbidite facies showing all five intervals of the ideal Bouma sequence.

Andaman Basin, in the northeastern Indian Ocean, can be regarded as a good example of a modern equivalent of flysch sedimentation. Rodolfo (1969) obtained sandy layers up to 35 cm thick which show graded bedding that is finely laminated. Such layers are found intercalated with the basin-flat clays in deeper parts of the central trough basin flat. These sandy layers are interpreted as a product of density flows initiated in the Martaban Canyons. Basin-flat sediments in adjoining parts are poorly laminated or unlaminated and show animal burrows.

From the 1,218 m deep basin in the south eastern Adriatic Sea van Straaten (1967, 1970) reports 4 Holocene and 4 Pleistocene turbidite layers, which can be traced in cores over a large area of the basin. Turbidite layers are separated by intervals of pelagic mud deposits. The Pleistocene turbidite layers are much thicker than the Holocene turbidite layers. Moreover, Pleistocene turbidite layers are separated by much thinner pelagic mud layers. The turbidite layers show graded bedding, cross-bedding, and horizontal lamination in the coarser-grained lower part, with fine parallel lamination, occasional convolutions in the fine sandy to silty middle part, and isolated ill-defined parallel lamination in the clayey topmost part. Animal burrows were found in only the topmost layer, 5 to 15 cm thick. The clayey deposits of the basin are usually strongly bioturbated.

Aeolian-sand turbidites form a clearly distinguishable "eolomarine" sediment facies. Sarnthein and Walger (1974) and Sarnthein and Diester-Haas (1977) report aeolian-sand turbidites from the Atlantic coast of the West Sahara. They are almost devoid of gradation, mica, and fine fraction. The sands consist predominantly of quartz grains, showing yellowish-red stains and frosted surfaces, both characteristic of desert sands in subtropical latitudes. The turbidite beds reach a thickness of 100 to more than 600 cm. Aeolian-sand turbidites occur where active desert dunes have migrated seaward by dominantly offshore winds during (glacial) periods of low sea level.

Despite the above-mentioned examples (and many more) of modern turbidites, they do not provide suitable and comparable modern analogues to the ancient flysch sediments. For example, thick sandy layers of extensive measurements are not known in modern analogues of the flysch sediments, although this is the most characteristic feature of flysch sediments in ancient rocks. The investigations on the continental slope and the continental rise off Sable Island Bank, Canada and near the Wilmington Canyon off the east coast of the U.S.A. have thrown some light on the problem of turbidite deposits and modern flysch deposits (Stanley and Kelling 1967; Stanley and Silverberg 1969; Stanley 1970). Horn et al. (1971) describe turbidite deposits from the western North Atlantic. Stanley (1970) provides a useful summary of the sedimentation in the continental margin off northeastern North America in respect to flysch-type sedimentation. Most of the ancient flysch-type sequences are deposits of a characteristically unstable tectonic-geographic framework. They have been deposited during advanced phases of orogenic activity in a geosyncline, showing trough and ridge characteristics. On the continental margin off North America turbidite deposits represent only one of the several other types of deposits. Quarternary deposits contain slump deposits, bottom-current deposits and hemipelagic deposits. They are much more abundant than the turbidite deposits. Seismic data of the continental rise indicates that a thick coalescing sediment wedge of turbidite and hemipelagic deposits extend across a former ocean basin.

In many respects sediments of the continental margin off North America are similar to ancient flysch deposits. However, because they lack the characteristic tectonic framework of flysch-type sediments of ancient sequences Stanley (1970) prefers to call the sediments of the continental margin off North America *flyschoid sediments*. The greatest thickness of sediments is near the base of the continental slope, not toward the central part of the ocean basin, as is the case with flysch deposits. However, the characteristics of sediments and their mode of deposition are rather similar to the deposits of the flysch basin. Most of the modern analogues of flysch-type deposits are thus flyschoid sediments. Only the late Miocene to Recent Mediterranean basins can be compared to ancient flysch deposits of the alpine type.

The generation of a high-velocity turbidity current has been assigned to the Grand Banks earthquake of 1929. Calculations based on the time intervals of breaking of cables give a velocity of 55 knots (Heezen and Ewing 1952; Heezen et al. 1954). Thus this could be a good example of a high-velocity turbidity current. However, Shepard and Dill (1966) recalculated the velocity and obtained a value of only 15 knots. There are several other examples of breaking of cables by turbidity currents, where their calculated velocity is many tens of kilometers/hour (Heezen 1963; Heezen and Hollister 1971).

Transport and Deposition of Sand

As mentioned earlier the major part of the sand is deposited near the shore line. This sand is kept actively in motion by coastal currents and longshore currents and main transport takes place in the breaker zone (Ingle 1966). However, for the transport of sand in the region of the shelf the concept of *null point* is important. According to this concept, for given slope and wave characteristics each sand grain has its own null point, where net movement is zero. Landward of the null-point sand grains move onshore and those on the seaward side move slowly offshore (Vernon 1965).

This is important for the entrapment of sand in the canyon heads, which extend up to the null point. Shepard and Emery (1941) observed abrupt depth changes in canyon heads off the California coast. Chamberlain (1964) suggested that major sand transport in canyon heads is mainly because of liquefaction of metastable sands.

Dill (1964), however, considered surface creep and small-scale slumping on slopes up to 10° to be more important factors than the liquefaction. On still steeper slopes sand flows similar to river flows are encountered (Fig. 642). These processes also produce a headward erosion. Shepard and Dill (1966) provide an excellent account of such processes.

The mechanism of grain flow is explained by the dispersive pressure concept of Bagnold (1954c), where an upward supportive stress is imagined to affect the flowing sediment grains. There is a grain-to-grain interaction, which is much stronger than in the fluid turbulence. Deposits of grain flow are characterized by massive bedding, sharp top and bottom bedding surfaces, and presence of reverse grading, as a result of the kinetic sieving effect (Middleton 1970). Stauffer (1967) considers mud pebbles to be typical of grain-flow deposits. According to Rees (1968) grain orientation is parallel to flow, with upflow imbrication. The grain flow probably takes place on rather steep slopes (> 18°).

Another mode of sand transport may be liquefaction, related to upward intergranular flow (Fig. 637) which can occur at very gentle slopes of 3–10° (Middleton 1969). Its importance in nature is not fully understood.

Middleton and Hampton (1976) believe that within a given flow, mechanisms of turbidity current, grain flow and debris flow may operate. Thus, the resulting deposit would exhibit characteristics of different processes (Fig. 643).

Reimnitz (1971) observed in front of a distributary of Rio Balsas that during an exceptionally heavy swell a pulsating longshore current of maximum current velocities up to 2 m/sec is produced, which is turned in the offshore direction under the influence of the river flow. This pulsating flow also causes rhythmic fluctuations with an amplitude of 30 cm and a period of 3 minutes. In a tributary of the Rio Balsas submarine canyon system current pulses reaching values over 1 m/sec were recorded during diving. This pulsating current transports and deposits large amounts of suspended sand down the canyon, which has a slope of 26° (Fig. 644). The current pulsations are in phase with the pulsation of the rip current on the surface.

Similar down canyon currents up to 160 cm/sec as a result of a surf-beat phenomen during storm winds are reported by Inman (1970). Measurements of currents in the submarine canyons have been extensively made by Shepard and his coworkers (Keller and Shepard 1978; Shepard and Marshal 1978). These studies reveal that there are both up- and down-currents in the canyons, and the down-current is mostly stronger. Rarely, near-bottom currents of > 50 cm/sec are recorded. In a few cases, even stronger current velocities have been registered, whereby current meters have been lost (Inman et al. 1976). These high current velocities in

Fig. 642. Submarine sandfall at a water depth of 40 m in a gully down into San Lucas Canyon. (After Shepard and Dill 1966)

Fig. 643. Schematic representation of sequence of sedimentary structures in the hypothetical single-mechanism deposits of the main types of gravity flows. (After Middleton and Hampton 1976)

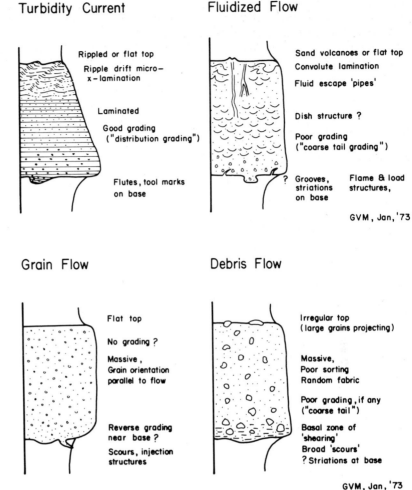

the canyons cause erosion of sand and silt of the canyon floor, and troughs with steep 0.5 m-high walls are produced (Shepard and Marshall 1978). Tidal and internal wave forces are the main cause of the semi-diurnal up-and-down canyon currents.

Transport and Deposition of Sand-Debris Flow and Pebbly Mudstone Deposits

Pebbly mudstone deposits are unsorted admixtures of sand and pebbles in a muddy matrix resembling till deposits. Such deposits have been reported from various regions of continental margins. Gorsline and Emery (1959) report such deposits from the mouths of canyons in the region of the San Pedro and Santa Monica basins off southern California. Stanley (1967) describes them from the continental margin off Nova Scotia, Canada and Stanley and Kelling (1967), from the Wilming-

ton submarine canyon area. Shepard et al. (1969) describe them from La Jolla submarine fan off California. An impressive example of debris flow is present off the Spanish Sahara, in the Lower continental rise, west of the Canary Islands (Embley 1976). Unlike many of the debris flow deposits, which are associated with canyons, this example is not associated with any canyon. The terminus of main debris flow is more than 700 km away from the upper, and 500 km away from the lower part of the slide scar, and covers an area of about 30,000 km^2, with an average thickness of about 20 m. Embley (1976) gives many other examples of debris-flow deposits.

Their ancient analogues are known from various flysch-basins of the Alps, Apennines, Carpathians, and coastal mountains of Oregon and California.

Stanley (1969) discusses the mode of transport and deposition leading to pebbly mudstone deposits. While diving near the mouth of the Paillon River, France, he observed carbonate pebbles, rounded to subrounded, resting on a muddy bot-

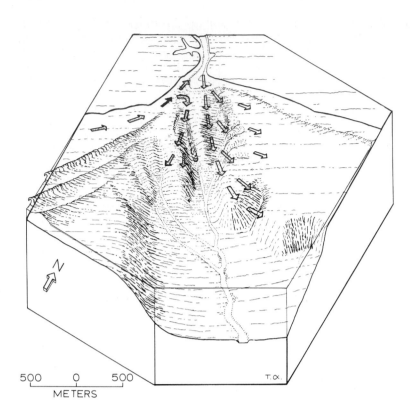

Fig. 644. Schematic block diagram of the head of Canon de la Necesidad showing current patterns. The surf zone wave pattern, longshore current, rip current (dotted arrows), freshwater jets (open arrows), and bottom flow are shown, as observed by divers on June 2, 1967. The black arrow shows the location of the strongest longshore current. (After Reimnitz 1971)

500   0   500

METERS

tom with an angle of slope of 5 to 10° at a water depth of 25 m. These pebbles, brought in from rivers as bed load, moved on the muddy surface some 10 to 30 cm by slight touch. Stanley (1969) believes that increased rates of deposition during flood periods may be enough to trigger failure and mass movement down-slope on such oversteepened metastable slopes. Cores collected offshore often show an unsorted pebble-mud admixture. Bourcart (1964) believes that the pebbles are transferred by slumping and partly by rolling movements.

Thus, pebbly mudstone deposits represent plastic mass-flow downslope movement under the influence of gravity. Such beds are often interbedded with graded beds of turbidite origin. Crowell (1957) emphasizes that genesis of pebbly mudstone deposits requires deposition of pebbles on a muddy surface which partially sinks down and causes a downslope movement.

According to Middleton and Hampton (1973, 1976) a debris flow is made up of a mixture of solids, e. g., sand and boulders, clay minerals and water, which are put into movement by gravitational force. The sand, pebbles, and boulders swim in a debris flow, because water and clay minerals make a fluid with a higher specific gravity than the water alone. This situation helps in the floating of sediment grains in the debris flow.

Sometimes pebbles of pebbly mudstone deposits show a higher degree of good rounding than the pebbles of the associated canyon deposits (Unrug 1963). This suggests that pebbles of such pebbly mudstone deposits remain in the beach environment for a longer time before being moved into deeper water.

Sometimes pebbly mudstone deposits possess large rock-blocks, some of which are derived from exposed bedrocks in canyons and slopes. Such deposits are often designated as *rock-fall deposits*.

The pebbly mudstone deposits often contain broken pieces of silty layers, and the lower surface of deposition is often erosional, showing features like irregular scour and fills, thus indicating erosion during their deposition.

Deposition in Submarine Canyons and Deep-Sea Fan Regions

Sandy sediment is deposited along the axis of submarine canyons. Detailed studies of the sediments of the canyons off southern California and the southern tip of Baja California, Mexico, by Shepard and Einsele (1962), and Bouma (1964, 1965) show that sediment is mainly coarse and fine sand, with mud intercalations and plant remains. Microorganisms and shells are brought in mainly from the continental shelf; some are derived from the erosion of canyon walls. The sandy layers show current-ripple bedding on a small scale. Slump phenomena and bioturbation structures are subordinate.

Outside the axis of the canyons, sediments are much finer-grained and show a much higher degree of bioturbation than the coarser sediments of the canyon axis.

Stanley (1967) describes the sediment of the Gully submarine canyon on the continental margin off Nova Scotia, Canada. Here also coarser sediments are restricted to the axis of the canyon. They are associated with dark muddy layers rich in organic matter, showing a high degree of bioturbation. In the lower parts sediments which move downslope along steep walls merge with those brought down along the canyon axis.

Gravity slumping along the flanks is an important sedimentation process in the case of Gully submarine canyon. This process accounts for such features as contorted mud and sand layers, mud clasts, pockets of sand, small and big sand spheroids. Figure 645 shows mud clasts.

Deposits of gravity-slumping origin are in close association with irregularly stratified coarse-grained sand, especially near the base of canyon walls, near tributary canyons and in the canyon axis. Such sandy units range in thickness from < 1 m to > 16 m and lack mud-layer intercalations, and thus represent the fusion of several rapidly deposited sand units. Such poorly sorted coarse-grained sand deposits often either lack graded bedding or show only poorly developed grading.

Such canyon deposits are also known as *fluxo-turbidites*. Kuenen (1964c) and Dzulynski et al. (1959) applied this term to denote some of the sandy beds in flysch basins which show characteristics that are intermediate to turbidites and gravity slides. Thus, fluxo-turbidites are rather coarse-grained but mud content is little. Bedding is thick. Current markings are poorly developed and scarce. Grading is absent or poorly developed and cannot be followed laterally over large distances. Instead, cross-bedding is more commonly made by the bottom currents. Indications of gravity slumping are found.

However, several geologists have criticized the use of the term fluxo-turbidites, e.g., Walker (1967). Deposits of submarine canyons and other

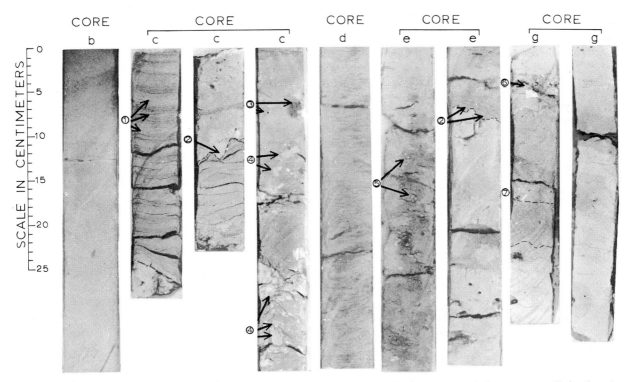

Fig. 645. Selected cores from a submarine gully showing fluxo-turbidites. Because of down-flank slumping in the submarine valley, various features are produced: Mud clasts (core C), pockets of sand and sand spheroids (core C4), and scour-and-fill structures. In few cases such features as well-developed graded bedding (core C1) in silt and sand appear. Parts with slow rates of sediment deposition are marked by bioturbation (mottling). (After Stanley 1967)

deep-sea valleys occur as coarse, shoestring-shaped bodies interfingered with finer-grained deposits showing characteristics intermediate to turbidity slump, and gravity slump deposits (Stanley 1970).

Many ancient examples are known of the submarine canyon and other deep-sea valley deposits from flysch basins. The most well known are from the Maritime Alps (Stanley and Bouma 1964; Stanley 1967) and the Carpathian Mountains (Unrug 1963; Stanley and Unrug 1972). Other examples have been described by Bornhauser (1948), Neev (1960), Whitaker (1962), and Wezel (1968). Stanley and Unrug (1972) give a good account of these deposits.

Deep-sea fans are developed near the points where submarine canyons open into an ocean basin. They are fan-shaped geomorphic features formed as a result of deposition of mainly terrigenous sediment where submarine canyons reach the lower slopes of ocean basins. Deep-sea fans are almost always present at the base of a submarine canyon; however, the size of the fans is highly variable. Sometimes they acquire huge dimensions. Table 34 gives general data about the size of deep-sea fans. One of the largest known deep-sea fans is the Bengal Deep-Sea Fan with a radius of about 3,000 km and a thickness of more than 10 km.

Curray and Moore (1974) discuss that at present a single large turbidity channel is active in the Bengal fan. However, a number of earlier, abandoned channels are present. None of the channels shows bifurcation and braiding. In the upper part, the channels are 10–20 km broad, and several hundred meters deep. The channels are bounded by large natural levees of rather sandy sediments, while the inter-channel fan sediments contain horizontal layered sheet flow strata (Fig. 646).

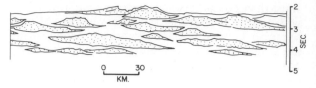

Fig. 646. Transect of Bengal deep-sea fan at 19° N, showing interleaved channel-levee complexes (stippled) and fan sheet-flow turbidites. (After Curray and Moore 1974)

Invariably, deep-sea cones possess a well-developed deep-sea valley or a system of deep-sea valleys with distributaries. These deep-sea valleys are better known as fan valleys. Some of the submarine canyons continue into fan-valleys, become shallower because of the decreased gradient; and in the lower part of the deep-sea fan, fan-valleys are extremely shallow (Nelson et al. 1970). The fan valleys are usually of winding or meandering type. Fan valleys migrate laterally; old valleys are abandoned and new valleys are cut into deep-sea fan sediments. The shape and nature of fan valleys is strongly controlled by the nature of sediment and the rate of deposition. In the upper part of a deep-sea fan, valleys are symmetrical in cross-section. Fan valleys can be more than 300 m deep and 1 or 2 km broad.

Walker (1978) gives an excellent summary of our present-day knowledge on the Recent and ancient deep-water sandstone facies, related to submarine canyon and submarine fans. The anatomy of canyon and fan is depicted in Fig. 647, after Walker (1978), where morphological subdivisions by Normark (1978) are also shown.

The upper fan is characterized by a leveed channel, which is slightly curved. The middle fan is made up of individual supra-fan lobes, which mi-

Table 34. Dimension of representative submarine fans of the world. (From Nelson 1968; modified after Bates et al. 1959)

Associated canyon name	Location		Kilometers		Gradient	Infra-fan relief (m)	Depth (m)		Levees	Apparent number of distributaries
	Latitude	Longitude	Length	Width			Apex	Base		
Mississippi	27°30N	89°W	222	148	1:120	120	1281	2928	Yes	7
Hudson	37°N	70°W	148	148	1:225	60	3843	4758	Yes	4
Congo	6°S	8°E	278	185	1:227	35	2562	3843	Yes	7
Rhône	42°N	4°30E	167	167	1: 90	75	732	2196	Yes	11[b]
Monterey	36°N	123°N	118	115	1:100[a]	75	3111	3843	Yes	7
Coronodo	32°30N	117°22W	9	7	1:100	11	1235	1391		3
Santa Cruz	33°47N	119°45W	7	9	1: 60	8	1647	1958		2
Hueneme	33°57N	119°16W	9	9	1: 70	9	641	824	Yes	3
Redondo	33°55N	118°30W	7	11	1: 48	20	549	759		2
Trinidad	41°05N	125°03W	13	17	1:140	12	2489	3038		2
	32°20N	134°50E	20	24	1: 40	38	4392	4804		3
	34°24N	138°21E	7	6	1: 17	42	2562	2745		2
Astoria	45°53 N	125°30W	167	102	1:200	45	2013	2836	Yes	40

[a] Wilde (1965) states 1:250 overall average for Monterey Fan.
[b] Menard (1965) shows 15–20 on recent charts.

Fig. 647. Schematic block diagram showing depositional relationship of shale grit and associated formations to demonstrate various geomorphic units of deep sea fan valleys. (After Walker 1978)

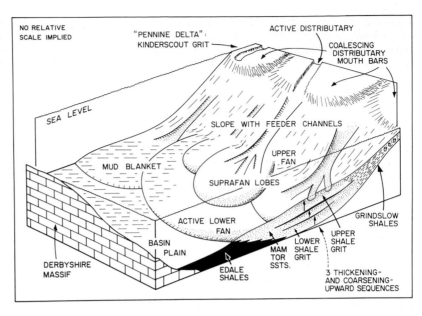

grate laterally. The inner part of the supra-fan is characterized by shallow, non-leveed, braided channels. The outer part of the supra-fan lobe is smooth and merges into the smooth lower fan. In many cases, there is no marked transition to the basin plain.

The development of submarine fans is controlled by a number of parameters, which Maldonado and Stanley (1979) discuss on the examples of the Menorca Fan and the Rosetta Fan in the Mediterranean Sea (Table 35).

Sedimentation in the fan valleys resembles the sedimentation in the submarine canyons and possesses coarser-grained sediments than the adjoining sediments. These coarse-grained sediments are brought in through submarine canyons by turbidity currents, slumping, surface creep, and sand flows and are transported farther into distal parts of the fan valleys. In general, sediments of fan valleys show characteristics similar to the canyon sediments, i.e., fluxo-turbidites. Walls of fan valleys often show steep slopes and undercutting. Muddy sediments of the walls are eroded and huge blocks of muddy sediments are incorporated into deposits of fan valley bottoms.

Most of the fan valleys show well-developed levees on the sides. It is believed that often sediment coming along the fan valleys is so abundant that sediment is spilled over, thus adding finer-grained sediments on the levees and the adjoining fan sur-

Table 35. A comparison of factors influencing Quaternary deposition on the Menorca fan and Rosetta fan sector of the Nile Cone. (After Maldonado and Stanley 1979)

	Menorca fan	Nile Cone
Structural setting	Considerable vertical displacement of balearic rise since the late Miocene. Fault and salt tectonics have affected the Quaternary depositional cover	Gentle, fairly continuous seaward-dipping late Miocene basement; possibly gentle subsidence. Levant platform east of the fan affected by salt tectonics
Sediment supply	Low volumes, consisting in large part of coarse bioclastic sands and clastic carbonates	Large volumes of mostly terrigenous fine-grade material (to 140 million tons of sediments per year, pre-Aswan dam, 1964)
Physical oceanography	Alternations of well mixed to partially ventilated basin conditions during late Quaternary sea floor locally swept by bottom currents	Phases of water mass stratification and stagnant anaerobic (H_2S-$_{rich}$) sea floor alternating with vertically mixed oxygenated basin conditions. Possible current reversal model well developed
Late Quaternary climatic-eustatic oscillations	Sharp increase of sediment input and transfer of sediment supply to deep environments during low sea level stands. Reduction of fluvial supply and shelf edge spill-over processes at times of higher sea level stands	Marked changes affected Nile river headwaters and freshwater discharge, coupled with major shifting of the Nile depocenters across the shelf and margin

face (fan apron). Sometimes sedimentation in the deep-sea fan is in the form of a flash flood, especially when fan valleys are not well developed or when they are absent.

The coarsest sediments, e. g., pebbles, are deposited near the mouth of a submarine canyon, in the upper part of deep-sea fan (Hand and Emery 1964; Stanley and Kelling 1967; Nelson 1968). The sediments of fan valleys are made up of medium to fine sand showing parallel bedding and small-ripple bedding. Graded bedding is uncommon. They constitute finger-like sand bodies along the fan valleys (Fig. 648). Other features found in fan valley sediments are plant remains, load structures, scour-and-fill structures, and slump structures. Sandy layers are covered and intercalated with poorly sorted clayey silt and mud layers. Bouma (1965), von Rad (1968), and Shepard et al. (1969) describe in detail the sedimentary features of fan valley deposits.

Sediment cores from the levees and the terrace-like features of fan valleys show that sand layers in the levee sediments are less common. Sediments of

the fan surface are rather fine-grained and do not have sand layers. Moreover, fan surface sediments show a high degree of bioturbation.

The distribution of sedimentary structures and sequences within a submarine canyon and submarine fan is given by Walker (1978), where information from both Recent and ancient examples has been utilized (Fig. 649).

In the area of basin plain and lower fan, slow hemipelagic sedimentation takes place, interrupted by parallel, thin-bedded turbidites, which become thicker towards the mid-fan region.

In the area of supra-fan lobes, within the braided channel, massive and pebbly sandstone are deposited in the form of channels. As the braided channels shift laterally, there is a tendency for merging of different massive sandstone bodies. In the adjacent parts smaller turbidites are present. Similar occurrences are known from the Recent fan channel, e. g., the Redondo Fan (Haner 1971), the Hatteras Fan (Clearly and Conolly 1974). In such areas gravels and boulders may be sporadically present.

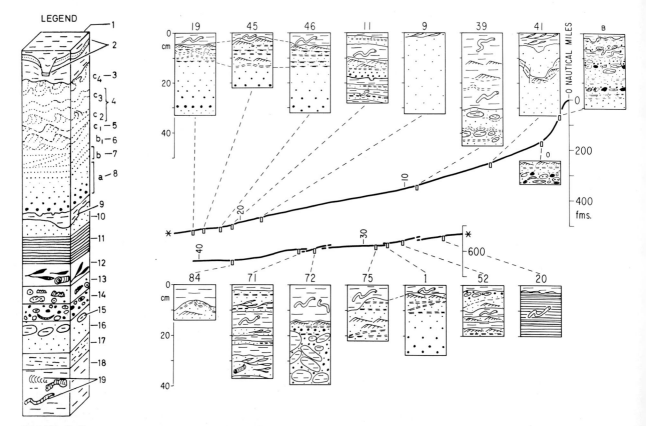

Fig. 648. Sediment structures in some of the characteristic box-core samples from the La Jolla submarine canyon and fan valley, 1 = mud (mainly clayey silt); 2 = slump structure and microfaulting; 3 = (c₄) convolute bedding; 4 = (c₃) wavy bedding; 5 = (c₂) deformed cross-bedding; 6 = (b₁) inclined lamination; 7 = (b) parallel lamination; 8 = (a) massive graded bedding; 9 = erosional channel filled with clayey silt; 10 = ungraded fine sand; 11 = semiconsolidated clay (Pleistocene?); 12 = layer of plant debris (kelp, grass, wood, etc.); 13 = layers of coarse sand, gravel, and pebbles; 14 = layers of gravel and shells; 15 = mud pebbles; 16 = fine to medium sand; 17 = sandy silt; 18 = clayey silt; 19 = bioturbation structures. (After von Rad 1968 and Shepard et al. 1969)

Fig. 649. Schematic model of submarine-fan deposition, relating facies, fan morphology, and depositional environment. D–B indicates disorganized-bed con-

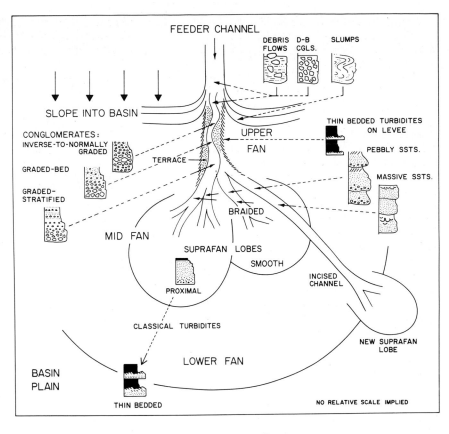

In the upper fan region, the upper-fan channel is the area of deposition of conglomerates. The levees of the upper-fan channel are made up of alternating sand/mud layers, and look similar to the turbidites of the basin plain.

The feeder channel or the submarine canyon can be filled up by slumps and debris flow, if these processes are active. Sometimes the canyons may be filled by muddy sediments, as is the case in a few modern canyons, because of the rise in sea level.

Walker (1978) proposes that these lateral facies can build up a progradational vertical sequence. Mutti and Ghibaudo (1972) and Mutti and Ricci-Lucchi (1972) recognized progradational sequences in the ancient submarine fan deposits of Italy. During progradation of a submarine lower fan sequence, a coarsening upward sequence is produced, as is also the case in a progradational delta sequence. In such a sequence the sandy turbidite layers become coarser-grained and thicker upwards. (In Fig. 650, such a sequence is marked as cu = coarsening upward.)

Above this, sediments of the middle fan are expected. These sediments also show coarsening-upward sequences of the turbidites; though near the top they are replaced by massive or pebbly sandstone (Fig. 650, sequences 3 and 4). Due to lateral shifting of a lobe in the mid-fan region, it is possible that thickening and coarsening-upward sequences are repeated. However, within the abandoned channels, on the contrary, fining-upward sequences are recorded, similar to those in the fluvial deposits (Fig. 650, sequence 5). This is overlain by pebbly to massive sandstone and ultimately by mudstone (Fig. 650, sequences 6 and 7). Similar sequences are also present in the channels following them, and ultimately they are filled up by conglomerates, debris flows, and slumps. Thus, the total profile acquires the character of a coarsening-upward sequence.

Almost all workers agree that well-developed deep-sea fans are present in areas where much sediment is brought into the sea by large rivers. Sediments of deep-sea fans often show a large content of shallow-water micro-organisms, i.e., foraminifera.

Examples of Ancient Deep-Sea Fan and Fan Valley Deposits

Shepard et al. (1969) compares deep-sea fan and fan valley sediment of La Jolla with those of deep-water Pliocene deposits of the Ventura and Santa

FACIES   SEQUENCE   INTERPRETATION

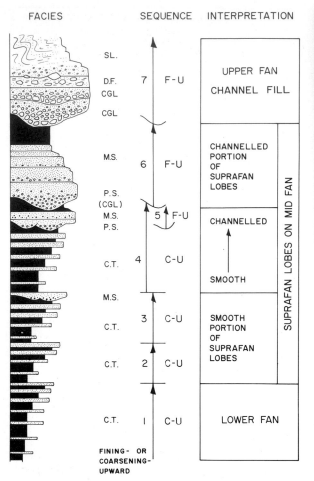

Fig. 650. Hypothetical stratigraphic sequence that could be developed during fan progradation. C–U represents thickening- and coarsening-upward sequence; F–U – thinning- and fining-upward sequence; C.T. – classic turbidites; M.S. – massive sandstones; P.S. – pebbly sandstones; CGL – conglomerate; D.F. – debris flows; SL. – slumps. (After Walker 1978)

Fig. 651. A sequence of graded sand, followed by laminated finer sand, and then by cross-bedded sand. A characteristic of a turbidite facies model from a box-core along a fan valley at a water depth of 1,039 m. (After Shepard et al. 1969)

Paula Creek area of southern California. These rocks have been studied in detail by Crowell et al. (1966), Natland and Kuenen (1951), and Winterer and Durham (1962).

Convolute lamination in the modern deep-sea fan sediments resembles that found in Hall Canyon (Ventura area), Pico formation. The sequence of massive graded sand followed by parallel-bedded sand, small current ripple bedded sand (A-B-C of Bouma sequence) are present in both cases (Fig. 651). Mud balls, common in the outer La Jolla fan, are also present in the ancient example (Fig. 652). Graded sand covered by a layer of plant debris is common in both cases. However, graded bedding seems to be more common in ancient examples than in their modern analogue.

Bartow (1966) investigated a large channel eroded in the silty shales and fine-grained sandstone of the Capistrano formation (Miocene-Pleistocene) in Orange County, California. This channel is filled with thickly bedded coarse-grained sandstone. This channel has been interpreted as a channel cutting across a deep-sea fan. Conrey (1959), Stanley (1969), Stanley and Bouma (1964); von Rad (1968), and Piper and Normark (1971) give some more examples of probable fan valley deposits in ancient sediments. Mutti (1974) describes ancient deep-sea fan deposits from circum-Mediterranean geosynclines.

Mutti and Ricci-Lucchi (1972) give an excellent analysis of ancient turbidites, differentiated into outer fan, mid fan, and inner fan, mainly on the basis of their work in the northern Apennines. Whitaker (1974) made a useful compilation of information on 32 areas of ancient submarine canyon and fan valley deposits, ranging in age from Palaeozoic to Tertiary. Whitaker (1974) gives following criteria as important in recognizing ancient submarine channels: (1) size comparable to modern canyon and fan valleys; (2) comparable geometry (steep to overhanging walls, increase in flatness towards sea); (3) location between shallow sea and deep sea for fan valleys – cutting into the deep-sea deposits; (4) comparable lithology and sedimentological characteristics; (5) comparable indices for quick filling and partial flashing out, smaller channels at the base or within channel fill, compaction effect; (6) fauna of diverse affinity and provenance.

Low-velocity, Low-density Turbidity Currents

Investigation of sediment filling of basins on the continental margin and sediment deposition on the continental rise shows that they are made up of horizontally stratified homogeneous sediments. This leads to the hypothesis that fine-grained material is transported over the continental shelf into the deep sea by some kind of gravity-generated bottom flows. Ewing and Thorndike (1965) argued

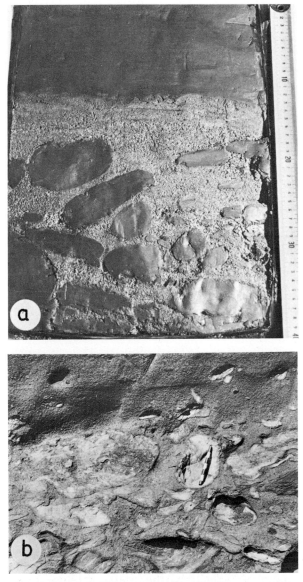

Fig. 652. Mud balls in the sandy sediments. a Box-core sample from disconnected short fan valley at a water depth of 1,085 m. b Miocene sediments at La Jolla, southern California. (After Shepard et al. 1969)

that there should be some transport mechanism for the transport of clay sediments into the deep sea, similar to the transport of sand by turbidity currents. These facts led Moore (1969) to believe in a process of turbid layer transport, generated on the continental shelf.

Fine silt and clay particles are brought in suspension by rivers into the sea and are deposited on the upper part of the continental shelf. Occasionally during a heavy swell, mud on the shelf is resuspended and a turbid layer is produced near the bottom, which moves under the influence of coastal currents and the downslope gravity flow. The net movement of this turbid layer is offshore and at an angle to the coast. The rate of net movement is rather low, i.e., 10 cm/sec. As soon as this turbid layer reaches a submarine canyon or other depression, it is out of the reach of the wave effect. Channelization of the turbid layer in a submarine valley results in an increase in thickness of the turbid layer. Moreover, steep axial slopes of canyons cause an interplay of gravity effects. If enough sediment is available, a kind of low-velocity low-density turbidity current develops, moving under the influence of gravity. Such currents do not possess any erosive power; however, they are capable of transporting large amounts of fine-grained sediment into deeper parts, where it is spread over wide areas and deposited as a thin layer in the basin plains. Graded bedding in such deposits is absent because the layer is so thin and lacks coarser grades. The thin lamination in such clay deposits is due to individual flows coming at long intervals.

Sometimes the turbid layer produced on the shelf is not thick and dense enough to produce low-density low-velocity turbidity currents. As soon as this turbid layer reaches the submarine valleys, where there are no wave movement or coastal currents, the suspended grains are rapidly deposited on valley walls and valley bottoms. Where no submarine valleys are present, the turbid layer flows over the shelf edge and starts depositing slowly on the continental margin.

Ewing and Thorndike (1965) report a 195-to-950-m thick, very dilute turbid zone above the sea bottom in water depths of 2,680 to 4,975 m. These dilute turbid zones are called "nepheloid layers". They are made up of extremely dilute suspensions of clay-sized sediment grains. The genesis of "nepheloid layers" is still unclear. However, Moore (1969) believes that they are not related to the turbid layer originating on the continental shelf.

Stanley (1969) expresses the opinion that during heavy storms much sediment can be taken into suspension near the shelf-break. Such suspended sediment may become low-density turbidity currents and may move across the continental slope (Fig. 653). These currents are capable of transport-

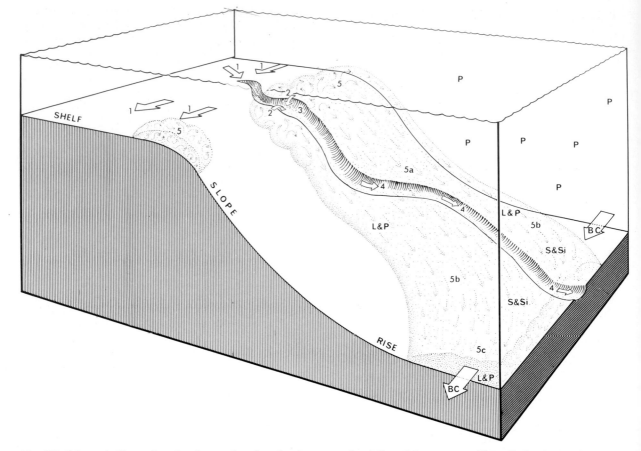

Fig. 653. Schematic illustration showing modern low-density turbidity current flow across the continental slope and rise in the vicinity of Wilmington Canyon. As the outer shelf (5) is stirred, fine-grained sediments are taken into suspension; these move in slope water (5a) onto the continental rise and beyond. Contour currents (arrow, BC) deflect the turbid suspension-rich (clay and pelagic fraction, L and P) downslope moving water mass (5b, 5c). Sand and silt (S and Si) of this and earlier flows are also deflected, but are moved laterally for shorter distances than the suspended fines. At the same time, sand moved by bottom currents near the shelf edge (open arrow 1) is trapped in the canyon heads (2). Slumping down the canyon walls is also prevalent. The canyons function as a funnel (4) in the process of sediment transfer to the continental rise and abyssal plain. (After Stanley 1969)

ing only silt and clay in suspension. At the same time they are capable moving sand-sized sediment on the bottom by traction and producing different kinds of scours, as can be seen in bottom photographs (Stanley 1969).

There have been many studies dealing with the transport of suspension from shelf into continental slope, and much interesting information is available in Swift et al. (1972) and Swift (1976).

Harlett and Kulm (1973) observed bottom currents on the Northern Oregon Continental shelf, with an average velocity of 10 cm/sec. Measurements of the turbidity demonstrated that suspended material is, in general, concentrated as layers at the seasonal thermocline, at the permanent pycnocline, and at the bottom. The seaward movement of the upper two layers can transport the terrigenous and biogenic material into the deeper parts, by-passing the outer continental shelf.

Komar et al. (1974) demonstrated the presence of a bottom turbid layer on the Oregon continental shelf. These turbid layers are restricted to depressions and highs of the sea bottom, and flow around a topographical high. These turbid layers are probably initiated by currents associated with surface and internal tides, wind-driven currents, and by sediment stirring due to surface wave activity. The flow velocity of these layers is highest at about 2 m above the sediment bottom, and attains values of 20 cm/sec. This part also exhibits the highest concentration of suspended material. Komar et al. (1974) considers autosuspension to be the main driving force for this layer. Bagnold (1962) used this concept to explain the turbidity current flow. These turbid layers may be transported from the continental shelf down the continental slope, to the lower part of the continental slope, where the rate of sedimentation is much higher; and also in the submarine canyons, where sedimentation is several times higher than on the adjacent upper continental slope.

Pierce and Siegel (1979) made measurements of suspended material in the southern Argentina shelf, and suggest that about $48-117 \times 10^6$ metric ton/year sediment is transported from the continental shelf into deeper parts. The ultimate source of these sediments is fluvial sediments and cliff erosion. Pierce and Siegel (1979) believe that the fluvial material is at first deposited on the continental shelf, and is several times resuspended and redeposited before it finally leaves the continental shelf.

The measurements made by Drake and Gorsline (1973) on the suspended material in the waters of Hueneme, Redondo, New Port, and La Jolla submarine canyons off southern California exhibit that suspension content is dependent upon the density structure of the water column. The concentration is mostly 5–7 mg/l. This concentration is far below the value on which this layer can flow as an independent layer. However, it seems quite likely that a turbid layer flow may develop if other current-producing factors are operative.

Bottom currents in some canyons exhibit a net down-canyon flow, which is much stronger than can be explained as a turbid-layer flow. By this process 10–15 % of the annual terrigenous supply into the deeper parts is transported through the canyons off southern California (Drake 1974). McCave (1972) discusses the transport and escape of fine-grained sediment from the shelf areas. On the continental shelf there is mud deposition in the form of muddy coastal sediments; nearshore, mid-shelf, and outer-shelf mud belts; and as mud blankets commonly found off major supply points. The concentration of mud in the water shows seasonal fluctuations related to current activity, activity of water movement, and concentration of suspension.

If we try to summarize the transport of sediment on the shelf, most of the workers agree that the movement of sediment across the shelf, beyond the shelf break, takes place by step-wise resuspension and redeposition. However, movement of sediment from shelf break into deeper parts incorporates theoretical difficulties in explaining the movement of sediment through water stratification. To this point, McCave (1972) considers (Fig. 654) that the lutite flow is not capable of breaking through the thermal stratification because of its low density. Also Bouma et al. (1969) and Drake (1971) showed enrichment of the turbid layer at the thermal interfaces; but McCave (1972) picks up the idea of Postma (1969) that such suspensions lose sediment grains by simple downward falling, and a suspension-cascade is developed (Fig. 654). Nevertheless, the matter is not yet fully understood. However, most probably the suspended material is channelized through canyons and brought into the deep sea. Observations made by Baker (1976) also support this.

Baker (1976) made measurements on the nepheloid layer above the Nitinat Deep-Sea fan over a period of four years, and found a characteristic distribution pattern of suspended particulate matter: (1) the levee-type with a bottom nepheloid layer, with a marked separation from the overlying clear water, (2) the fan-valley type with a bottom nepheloid layer of variable thickness and with prominent internal layering. This nepheloid layer dissipates only gradually into the overlying clearer water. Because these nepheloid layers exist over a period of several years, Baker (1976) concludes the existence of some combination of continually operative processes, such as a low-density turbidity current.

Pierce (1976) points out that we still have too little data to make a clear picture of the sediment transport on the continental margin. However, he concludes that there are two important processes which actively bring sediment on the continental slope and rise. The one is down-canyon transport and formation of coalescing fans; the other is boundary currents from Arctic and Antarctic cold

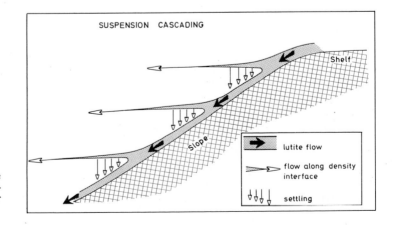

Fig. 654. Schematic representation of lutite flows with cascading. Decrease in width of arrows indicates decrease in concentration. (After McCave 1972)

waters which bring the nepheloid layers from which suspended material is deposited.

Studies in the Corsican trough, NW Mediterranean Sea, indicate that the accumulation of clay and silt in the distal parts is much greater than in the coarser sediments of more proximal setting. This observation led Stanley et al. (1979) to develop a turbid layer – by passing model (Fig. 655):

1. Erosion dominates on the shelf, shallow ridges, and upper slope areas.
2. Erosion and deposition, along with resuspension, are almost in equilibrium in the main slope region.
3. In the deeper parts, e.g., more distal basinal part, high rates of deposition (accumulation) are dominant.

Contour Currents and Contourites

In several places on the continental slope off the east coast of North America older rocks crop out, whereas the continental rise is built-up of a young thick sediment wedge. Such a sediment wedge is 100 to 1,000 km wide and 1 to 10 km thick. In the case of a preferred transport of sediment from the shelf into the ocean basin along submarine canyons only, the faulted continental slope should be visible and near the bottom of the slope only coalescing deep-sea fans and active gravity slide blocks should be present. However, this is not the case.

According to Heezen et al. (1966), the rate of deposition near the continental rise during the post-Pleistocene period is rather high, i.e., 5 to 50 cm/ 1,000 years. Primarily, it was believed that deep-sea depositional basins are deep-sea valleys and have been filled by turbidite deposits. However, intensive coring of the sediments of the continental rise demonstrated that only part of it is made up of turbidite sediments. Thus, other modes of sediment transport and deposition were searched for. On the east coast of North America near Cape Hatteras a direct transport of sediment from the shelf across the slope can be ruled out due to the presence of the northerly flowing Gulf stream, which acts as a barrier. At the same time, Heezen and Hollister (1964) concluded from ripples and other current markings in deep-sea photographs that there are certain types of currents present that follow the contours of the continental slope and are directed southward (Fig. 656). Wüst (1958) also postulated the presence of a southerly current. He found that a relatively strong antarctic bottom current is flowing northward in water depths of 3,000 to 5,000 m on the western flank of the South Atlantic. He thus postulated the presence of a southward-flowing current on the western flank of the North Atlantic. Heezen et al. (1966) provide evidence for such a southward-flowing bottom current by bottom photographs. The current follows the contours of the base of the continental slope, and thus the name *contour current*. The velocity of such currents is 15 to 20 cm/sec. or even more.

In part, the source of the sediment has been determined on the basis of color. According to Schneider et al. (1967), the upper few meters of sediment deposited on the upper part of continental rise off the east coast of North America are made up of grayish, homogeneous, silty clay (Fig. 657), while the lower continental rise is made up of

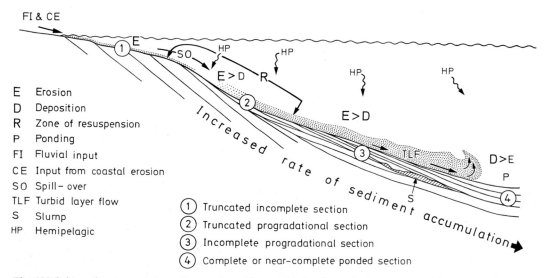

E   Erosion
D   Deposition
R   Zone of resuspension
P   Ponding
FI   Fluvial input
CE   Input from coastal erosion
SO   Spill-over
TLF   Turbid layer flow
S   Slump
HP   Hemipelagic

① Truncated incomplete section
② Truncated progradational section
③ Incomplete progradational section
④ Complete or near-complete ponded section

Fig. 655. Schematic representation of by-passing of fine-grained sediment in deeper areas of deep-see basins through mainly gravitational processes. (After Stanley et al., 1979 in press)

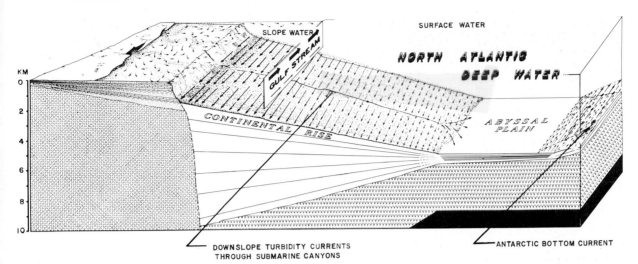

Fig. 656. Diagram illustrating the shaping of the continental rise by geostrophic contour currents. Arrows denote prevailing bottom currents. The continental and oceanic crust is shown by patterns. The mantle is solid black. Sedimentary rocks are shown by conventional symbols. Turbidite deposits are shown by horizontal ruling, and continental rise deposits are shown by open wedges. In addition, the transport directions are supported by underwater photographs on the continental rise of Nova Scotia and on the Western Bermuda Rise. This diagram depicts the principal processes active in shaping a "normal continental rise". (After Heezen et al. 1966)

predominantly thin layers of clean quartz silt (Fig. 658). Up to several hundred clean quartz silt layers may be present in a core 10 m long. Such silty beds are often cross-bedded, with occasional heavy mineral placers; they are very well sorted. The sediments are brown to pinkish-brown in color. Most of the cores also show bands of red to brick-red clay. Similar red-colored clays have been observed in sediments near the Laurentian Channel off Nova Scotia and the outer ridges off the Carolinas and the Bahamas. The red-colored clays found on the contour current deposits of the lower continental

rise are probably derived from the St. Lawrence area and have been transported by contour currents southward as far as Florida. Sediment cores from the abyssal plains show characteristic brown-colored clays intercalated with poorly sorted thick sand and silt layers, many of which show graded bedding.

In many cases along with the bands of red clays and silty layers, clay layers rich in pelagic micro-organism are also present. This suggests that sediment brought by contour currents comes only occasionally and in the form of suspension clouds;

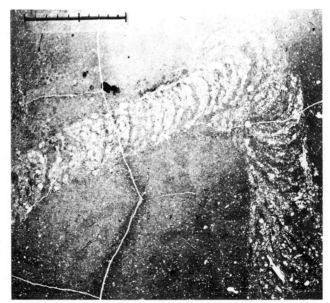

Fig. 657. Thin-section of piston corer sample from HR/V Eastward (Cruise No. E–37–67). This is homogeneous-looking silty clay from the continental slope off the east coast of North America. *Zoophycus*-like sreiten are visible. The water depth is 1,700 m; the sedemintdepth is 222 to 236 cm. Length of the scale = 1 cm. (Sample courtesy of O. Pilkey.) Duke University, North Carolina

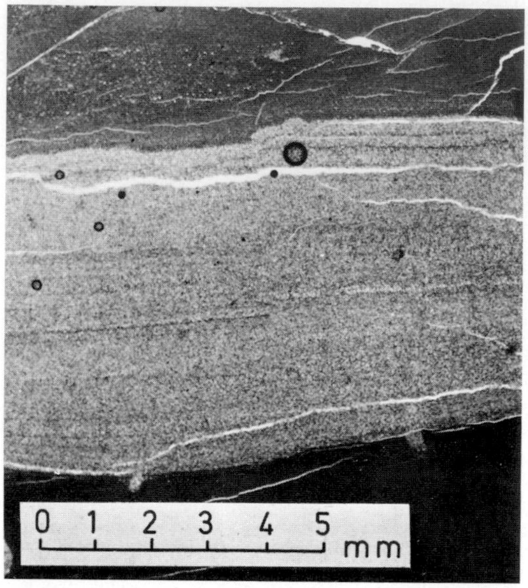

Fig. 658. Thin-section of piston corer sample from HR/V East-ward (Cruise No. E–37–67) showing laminated silty layers intercalated in the red clay. Deposition is by contour currents. The water depth is 5,000 m; the sediment depth is 32 to 36 cm. East coast of North America. (Sample courtesy of O. Pilkey)

but during certain periods the rate of deposition is so low that pelagic micro-organisms become major constituents of the layer (Fig. 659) (Reineck 1974d).

According to Klasik and Pilkey (1975), on the continental rise off the southeastern U.S.A., contour currents bring a significant amount of the sediments, which come from a Northern direction. These sediments belong partly to hemipelagic sediment and partly to bottom sediments resuspended by the bottom-dwelling organisms (Rowe et al. 1974).

Heezen et al. (1966) come to the following conclusion about the contour currents and their deposits. They state that maximum thickness of the sediment is along and near the axes of the geostrophic contour currents and the deposits become thinner with increasing distance from the axes of currents. In certain parts of the continental margins such patterns have been demonstrated by the study of sediment cores, echograms, reflection profiles, etc. Thus, the characteristic downslope thinning sediment wedges, stocked one upon another, make the continental rise, controlled mainly by contour currents.

Table 36. Important characteristics of turbidite and contourite sand or silt. (After Bouma and Hollister 1973)

		Turbidite	Contourite	Conclusions
	Size sorting	Moderate to poorly sorted < 1.50 (Folk)	Well to very well sorted < 0.75 (Folk)	Contourite is better sorted
	Bed thickness	Usually 10–100 cm	Usually < 5 cm	Contourite has thinner bedding
Primary sedimentary structure	Grading	Normal grading ubiquitous, bottom contacts sharp, upper contacts poorly defined	Normal and reverse grading, bottom and top contacts sharp	Contourite tends to be less regularly graded and has sharp upper contacts
Primary sedimentary structure	Cross Laminations	Common, accentuated by concentrations of lutite	Common, accentuated by concentrations of heavy minerals	Contourite contrasts sharply with turbidite in that heavy mineral placers are in the form of small scale stratification
Primary sedimentary structure	Horizontal lamination	Common in upper portion only, accentuated by concentration of lutite	Common throughout, accentuated by concentrations of heavy minerals or foraminifera shells	Contourite contrasts sharply with turbidite in that heavy mineral placers are in the form of small scale stratification
	Massive bedding	Common, particularly in lower portion	Absent	Contourite ubiquitously laminated
	Grain fabric	Little or no preferred grain orientation in massive graded portions	Preferred grain orientation parallel to the bedding plane is ubiquitous throughout bed	Contourite has better grain orientation
Principal constituents of sand and silt beds	Matrix (< 2μ)	10–20%	0–5%	Contourite has less matrix
Principal constituents of sand and silt beds	Micro-fossils	Common and well preserved, sorted by size throughout bed	Rare and usually worn or broken, often size sorted in placers	Contourite shows more evidence of reworking
Principal constituents of sand and silt beds	Plant and skeletal remains	Common and well preserved, sorted by size throughout bed	Rare and usually worn or broken	Contourite shows more evidence of reworking
	Classification (Pettijohn)	Graywacke and subgraywacke	Subgraywacke, Arkose, and orthoquartzite	Contourite is more "mature"

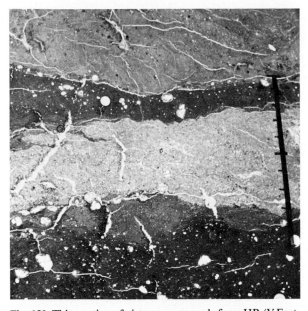

Fig. 659. Thin-section of piston corer sample from HR/V Eastward (Cruise No. E–37–67) showing interlayered gray silty clay with abundant pelagic micro-organisms, and silt and reddish clay without pelagic micro-organisms. The silt and reddish clay are brought by contour currents. Core no. 7872. The water depth is 5,000 m; the sediment depth is 96 to 100 cm. East coast of North America. (Sample courtesy of O. Pilkey)

Large-scale Gravity Slide and Slump Deposits

Many large-scale gravity slides and slump masses have been recorded on seismic profiles of the continental slope and the continental rise (e. g., Moore 1961; Kelling and Stanley 1970). Such slides are capable of transporting huge masses of sediments into deeper parts, i. e., sediment from shallow water and outer shelf parts is moved into deep-sea basins. During slumping there is compression near the toe of a slump-sheet and tension near the head (Fig. 660). Compression results in thrusting and folding, while tension causes faulting and downslope gliding. In the central part the slump-sheet is undeformed. Sometimes the slump-sheet may show rotational movements.

Kenyon et al. (1978) give impressive illustrations of the slump folds on the Landes Marginal Plateau, Bay of Biscay, which they obtained by long-range side-scan sonar. Investigations of Hampton and Bouma (1978) on the shelf break, western Gulf of Alaska, indicate that strong earthquakes and the associated strong accelerations are the main reasons for causing the large-and-small dimensional slides. Yagishita (1977) discusses that the mechanism of submarine sliding is mainly related to the high anomalous pore pressure zone at the base of the sliding block. This high pore pressure causes reduction in the weight of blocks and diminishes the shear strength of the sliding blocks.

Stanley and Silverberg (1969) describe contorted stratification in the cores from slump deposits off Sable Island Bank, southeastern Canada. Gravity slides are common on continental slopes with high rates of deposition. Such conditions are present in front of the Mississippi delta (Shepard 1965), the Rhône delta (Duboul-Razaret 1956), and continental slopes with fluvio-glacial deposition, e. g., off Sable Island Bank, Canada (Stanley

Bouma (1973b) and Bouma and Hollister (1973) give the criteria to differentiate between contourites and distal turbidites Bouma (1972, 1973a) gives examples of ancient contourites from Niesenflysch, Switzerland (Table 36).

Terrigenous clastic sediments, derived from the continents and transported into the deep sea, constitute a significant part of the geosynclinal sediments. Dott and Shaver (1974) give a number of interesting papers dealing with the geosynclinal sedimentation.

Fig. 660. Diagrammatic cross-section on a gentle slope. Unstippled areas are slumped sediments; sparsely stippled areas are sediments disturbed perhaps by instrastratal movement; densely stippled areas are undisturbed sediments. Areas that are cross-hatched overlie a zone of compressional folding and thrusting at the toe of the slump sheet. Areas of vertical cross-hatching overlie a zone of undeformed or slightly deformed beds in the middle of the slump sheet. Areas of solid black overlie an exposed glide plane in a zone of tension at the head of the slump sheet. Large stippled areas overlie disturbed sediments beneath a smooth sea bed. (After Lewis 1971)

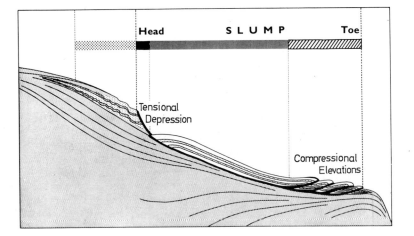

and Silverberg 1969). It seems likely, that during Pleistocene times such gravity slumping used to be much more common, where because of the lowered sea level rivers transported much material directly to the outer part of the shelf and upper part of the continental slope.

Slump-masses can be of huge dimensions. Heezen and Drake (1963) estimate the size of the slump mass produced by Grand Bank earthquake to be 400 m thick, more than 100 km long, and even more in width. Moore et al. (1970) give comparable dimensions for the slump-masses recorded from the continental margin off Brazil (Fig. 661). According to them, the total depth of slumping is 2,500 m, and the horizontal distance moved ranges between 4 and 50 km. The thickness of the slide-slice is 500 m and the width 20 km. The mound of sediment rising above the sediment level of slide block up to a height of 125 m suggests that much contortion must be present in slide slices.

Size of slide sheets and slump-sheets of most of the continental margins of today are comparable. But they are much larger than the submarine slide sheets known from the geological record. It has been suggested that sediment deposits at the foot of the continental slope with abundant slide masses and strongly contorted sediments may be considered modern analogues to the wild-flysch, olistostromes, and chaotic boulder clays. Uchupi (1967) and Stride et al. (1969) provide useful summaries on the subject of large-scale slumping. Lewis (1971) describes the slumping phenomena from the continental slope off North Island, New Zealand, on slopes of 1 to 4°. Lewis (1971) gives a schematic picture of the slumping blocks (Fig. 660).

Moore et al. (1970) list several examples of gravity slide deposits from the geological record. Brown (1938) describes the gravity slide in the Tertiary of Ancon, Ecuador. Other examples are given by Moore (1934), Kindle and Whittington (1958), Kugler (1953), Marchetti (1957), Renz et al. (1955), and Rigby (1958).

Small-scale gravity slumps are also common in the lower continental margin. They are often called olisthonites and mass-flow deposits (Klemme 1958).

Deep-Sea Sediments: Recent and Ancient

Deep-sea sediments are extremely heterogeneous in nature and many different sedimentation processes are involved in their deposition. The term "pelagic sediments" is often used in association with deep-sea sediments. Pelagic sediments are deposits of deeper water that settled from the overlying water by particle by particle deposition in the absence of any major current activity. Pelagic sediments are made up mainly of organic skeletal remains of micro-organisms and are known as *ooze* or *pelagic ooze*. Sediments which are brought in and deposited by bottom-current activity, ice bergs, etc., are known as *terrigenous* sediments, and are much coarser-grained than the pelagic sediments. Sediments brought in and deposited under the influence of gravity are grouped as gravitational deposits, which include rock fall deposits, small- and large-scale slumps and slides, fluxo-turbidites, and turbidites (de Raaf 1968; Stanley 1970). *Hemipelagic* sediments show intermediate characteristics. Locally, chemical and biochemical precipitation of various minerals is an important factor contributing to deep-sea sediments, leading to deposition of Mn, Fe, P, etc.

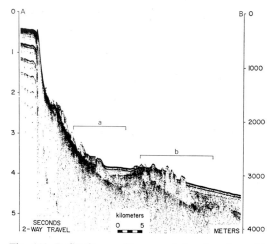

Fig. 661. Reflection profiler record showing slump block (a) and slump sheet (b). The vertical is exaggerated approximately 11 times. Off the coast of Brazil. (After Moore et al. 1970)

Turekian (1968), Keen (1969), Hsu and Jenkyns (1973), and Seibold (1974) give useful information on the deep-sea sediments. Neuman (1968) discusses the oceanic currents; while Heezen (1977) provides useful papers dealing with the abyssal circulation. Lisitzin (1972) gives an exhaustive review of the sediments and sedimentation in the world oceans.

Wind-blown dust (Pratje 1934) and ice-rafted sediments often contribute a large amount to the deep-sea sediments. Mullen et al. (1972) estimate that about 10 % of the sediments in parts of the Arctic Ocean are contributed by wind-blown dust (see also Windom 1975). Windom and Chamberlain (1978) discuss the effect of a dust storm to the North Atlantic Ocean, and they suggest that such a dust storm may account for 25 % of the sedimentation rate in that part of the North Atlantic. Extraterrestrial material, e. g., magnetic spherules, are often incorporated in deep-sea deposits.

Menzies et al. (1973) give a compilation of ecology of the oceanic deep-sea bottoms.

Various attempts have been made to classify the deep-sea sediments. The most important recent attempts are those by Olausson (1960) and Shepard (1963). Emiliani and Milliman (1966), Berger (1974), Dott and Shaver (1974), and Hay (1974) provide a useful review of deep-sea sediments. Table 37 gives the classification of deep-sea sediments after Shepard (1963). Characteristics of the most important type of deep-sea sediments are discussed below. (Some of the deep-sea deposits have been already discussed in connection with depositional and transport processes.) Bernoulli and Jenkyns (1974) give an impressive example of ancient pelagic sedimentary rocks, describing the early evolution of the Tethys.

The distribution of Recent sediments on the ocean floors (Fig. 662) can mainly be related to depth, temperature, fertility of surface waters, and distance from land. These relationships can best be represented in a depth-fertility diagram (Fig. 663), showing the importance of the carbonate compensation depth (CCD). The CCD, below which calcite does not accumulate on the deep sea floor, only red brown clay, radiolarian and diatom oozes may occur. Fluctuations of CCD since the Jurassic is reported by Ramsay (1974).

Table 37. Main types of deep-sea deposits. (After Shepard 1963)

1. Pelagic deposits
a) *Brown clay* (Red clay) (less than 30 % biogenous material)
b) *Diagenetic deposits* (consisting dominantly of minerals crystallized in sea water as authigenic minerals such as phillipsite and manganese nodules).
c) *Biogenous deposits* (more than 30 % material derived from organisms).
  (I) Foraminiferal ooze (having more than 30 % calcareous biogenous largely Foraminifera. For example, globigerina ooze).
  (II) Diatom ooze (having more than 30 % siliceous biogenous, largely diatoms).
  (III) Radiolarian ooze (having more than 30 % siliceous biogenous largely radiolarians).
  (IV) Coral reef debris (derived from slumping around reefs)
  (V) Coral sands
  (VI) Coral muds (white)

II. Terrigenous deposits
a) *Terrigenous muds* (having more than 30 % of silt and sand of definite terrigenous origin)
  (I) Green muds
  (II) Black muds
  (III) Red muds

b) *Turbidites* (derived by turbidity currents from the lands or from submarine highs)
c) *Slide deposits* (carried to deep water by slumping)
d) *Glacial marine* (having a considerable percent of allochthonous particles derived from iceberg transportation).

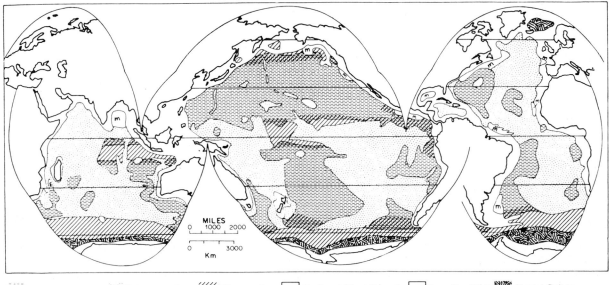

:::::: Clay or no deposit   Calcareous Ooze   ///Siliceous Ooze   □ Shelf and Slope Deposits |m| Deep Sea Muds   ▨ Glacial Debris

Fig. 662. Distribution pattern of Recent sediments on the deep sea floor (After Berger 1974; original data from various sources)

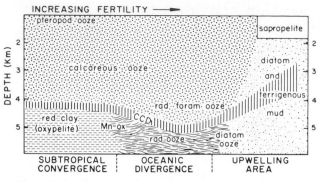

Fig. 663. Depth-fertility diagram showing distribution of major facies and CCD (After Berger 1974)

cles. A 1,500-km-wide band of diatom ooze is present along the 60th parallel south of the Pacific Ocean.

Pteropod Ooze

Pteropod ooze is made up of mainly aragonitic pteropod shells, together with abundant planktonic foraminifera. Pteropod ooze is concentrated at moderate depths of 1,500 to 3,000 m, because at greater depths pteropod shells dissolve easily. Inorganic components are the same as those of red clay. Today, pteropod ooze is present on the slopes of the mid-Atlantic ridge, the slopes of the Bahama Platform, and the slopes of the Bermuda Pedestal.

Brown Clay (Red Clay)

Most of the ocean basins show vast areas covered with red- to brown-colored clay. The median diameter is in the clay size particles ($< 2\mu$). Clay minerals and other resistant mineral residues derived from the land are the main constituents, together with volcanic ash and cosmic spherules. Carbonate content is $< 30\%$. The red clay of the South Pacific is made up mainly of authigenic clay minerals, produced by the *in situ* alteration of volcanic material. The reddish color of the clay is a result of the oxidation state of iron.

Globigerina Ooze

Globigerina ooze is made up mainly of tests of various planktonic foraminifera, especially the *Globigerina*. Besides, platlets of coccoliths, radiolarian shells, diatom parts and pteropods are also abundant. Inorganic constituents are less abundant and are the same as in the red clay. Carbonate content is $> 30\%$.

Radiolarian Ooze

Radiolarian ooze is characterized by the high content of radiolarian shells, usually more than 50 % (cf. Sverdrup et al. 1942). It is widely distributed at a depth of about 4,600 m in a 200-km-wide belt stretching across the equatorial Pacific. Inorganic components are the same as for red clay.

Diatom Ooze

Diatom ooze is made up of 50 % or more diatom frustules. Inorganic components (about 20 %) are coarser-grained, made up mainly of silt-size parti-

Diagenetic (Hydrogenous) Deposits

The report of the extensive occurrence of phillipsite in deep-sea sediments by the Challenger expedition led to the conclusion that many minerals, e.g., phillipsite, are produced *in situ*. Arrhenius (1961) calls such deposits *halmeic deposits*.

Manganese Nodules

Manganese nodules are the most well-known and widespread authigenic deposits. Agassiz (1906) pointed to their great abundance and wide distribution in ocean basins. Manganese nodules occur concentrated in patches, usually a few centimeters across. Nodules show a concentric structure indicating interrupted depositional processes. Locally, they are economically important. Their maximum concentration takes place below the calcium carbonate compensation depth (CCD), where detrital sedimentation is minimal or even erosion takes place (Cronan 1977).

Manganese nodules may be dissolved, and if covered, may again act as source of manganese for the new mangenese nodules (Marchig and Gundlach 1976).

Terrigenous Deposits

Terrigenous deposits are the main sediments on the continental margins and parts of ocean basins. Most important of them are gravitational deposits, e.g., gravity slides and turbidity deposits. These deposits have been dealt in detail under various geomorphic units of the deep sea.

Current Markings and Related Sedimentary Structures in the Deep-Sea Sediments

Development of multiple exposure cameras for use in the deep sea in the late 1940's provided a possibility for observing surface features on the deep-sea bottoms (Ewing et al. 1946; Thorndike 1959; Edgerton 1963). Such surface features as ripple marks, scour marks, etc., have been abundantly observed on the deep-sea bottoms (Menard 1952; Heezen et al. 1959). In the last decade development of various instruments for direct-current measurements in the deep sea, e. g., neutrally buoyant floats, moored current meters, and bottom current meters have provided further information about bottom currents. It has been found that rather strong bottom currents are present in many parts of the deep sea, with current velocities ranging from 4 to 40 cm/sec. These current velocities are strong enough to move and transport the sediment particles present on the deep-sea bottoms. Thus, it is reasonable to think that in the sediment cores from such areas, various current markings should be found preserved, which is also the case. Cross-bedding, etc., in the sediment cores indicates the presence of a current in the past, and the ripples on the present-day sediment surface indicates the presence of contemporary currents. Grain size can provide an approximation about the minimum current velocity.

On the present-day sediment surface a variety of markings have been observed, e. g., ripple marks, scour marks, current lineation. Besides, bare rock surfaces and coarse residual concentration are indirect evidence of current activity.

The ripple marks of deep-sea bottoms can be of asymmetrical, symmetrical, lingoid, lunate, and longitudinal type (Fig. 664). Ripple length varies generally from 10 cm to several meters, and height may be up to 20 cm or even more.

Scour marks around various tools, e. g., nodules, rock pieces, etc., are commonly seen. At places where current velocities are very high, rock pieces are found concentrated in depressions as residual lag.

Distribution of current markings in different geomorphic parts of the deep-sea is highly irregular. They have been most commonly photographed on the sea mounts (Fig. 665) and island slopes. The same is true of the ridges and slopes of the mid-Atlantic rift ridge system (Cousteau 1963). In contrast, the current markings are rather rare on the ocean basin bottoms. A change in the characteristics of the sediment bottom is seen coming from globigerina ooze to red clay of abyssal plain. Fewer bottom-living organisms are observed. The finer grain size, coupled with other ecological factors, also affects the configuration of burrows and trails.

The deepest parts of ocean bottoms covered often with manganese nodules commonly show scour marks (Shipek 1960). Heezen et al. (1959) suggest that nodules have most probably rolled on the bottom. It has been suggested that manganese nodules are formed in areas of moderate current activity.

Evidence of current activity on the continental margin below a depth of 200 m seems to be less common than on the sea mounts, etc., located at the same depth. Most of the photographs of continental slopes show a muddy bottom with abundant animal life, represented by sessile and vagile benthic organisms. Besides, scour marks, current ripple marks and current lineation, outcrops of ancient sediments have been observed (Northrop and Heezen 1951; Jordan 1951; Elmendorf and Heezen 1957). However, evidence of current activity on continental slopes is rather uncommon. Extensive evidence of current activity in the form of asymmetrical ripples is noted from the continental slope of Brest (47° 43′ N, 08° 02′ W) at water depths of 490 m and 1,530 m. Other examples are known from the Grand Banks area, the continental slope off Portugal, the continental slope off eastern South America, etc.

The marginal plateaus often show ripples and scour marks. On the north flank of the Carnegie Ridge, in a water depth of 2,650 m, Lonsdale and Malfait (1974) describe abyssal dunes (megaripples) made in the foraminiferal sand (see also on p. 46).

On the continental rise evidence of current activity is extremely rare; only a few areas have revealed extensive evidence of current activity. Photographs of the bottom sediments of the continental rise show characteristic features of organic activity, e. g., burrow mounds and surface trails. In comparison to the continental slope, rise bottom show lesser sessile organisms; however, burrows and vagile benthic organisms are common.

Stanley and Kelling (1967) describe in detail bottom photographs of a submarine canyon–Wilmington Canyon–off the North American Atlantic coast. They observed abundant bottom fauna and burrowing activity, suggesting lack of coarse sediment transport along the canyon at the present-day. Ripple marks of various types are common and empty shells are concentrated in the ripple troughs. Scour marks are rare. Angular rock slabs are also recorded, suggesting active erosion, slumping, and sliding along the canyon.

Wüst (1955, 1957) deduced regional patterns of bottom currents in the South Atlantic and Heezen and Hollister (1964) found ripple marks on the ocean floor in accordance with current patterns of Wüst (1955, 1957). This Antarctic bottom current (contour current) reaches current velocities of 15 cm/sec and more. The calculated velocities give only the minimal values because of large spacing of the stations. However, Wüst (1958) points out that tidal velocities in the range of 5 to 10 cm/sec are probably superimposed upon the calculated bottom current velocities. The photographs show

an extensive record of ripples, scour marks, current lineation (Fig. 666), tool marks, etc. In the sediment cores of this region small-scale cross-bedding has been recorded. However, the presence of ripple bedding can not be regarded as indicator of a bottom current, because the turbidity current also produces ripple-bedded units.

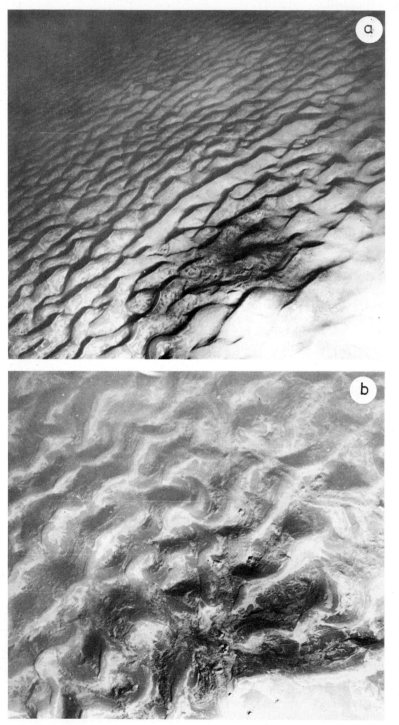

Fig. 664. Lunate- and lingoid-shaped ripples. Well-developed short-crested ripples in sand, high in heavy minerals. White bands probably represent local concentrations of Globigerina sand. The photograph is from the sea floor of the Scotia Sea, southeast of Terra del Fuego. Eltanin 4–13, 4,010 m water depth. In (b) note the animal tracks across the ripple crests, indicating the period of quiescence since the time of ripple formation. (After Heezen and Hollister 1964)

Fig. 665. Well-developed symmetrical ripples (a) and scour marks (b) in the sea mount region 3,218 m water depth. The photograph is from the flank of Bellafontaine Sea mount (100 miles east of Recife, Brazil). (After Heezen and Hollister 1964)

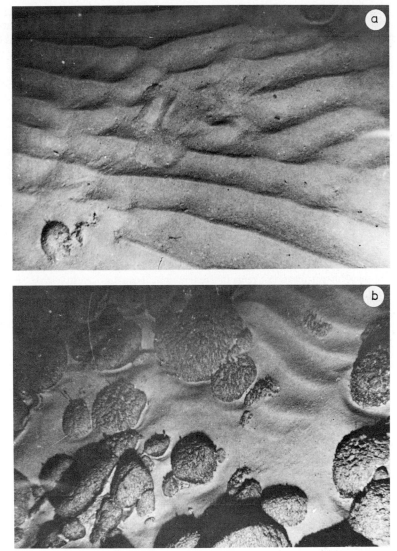

The preceding discussion is based mainly on Heezen and Hollister (1964). Further literature can be obtained in Hersey (ed.) (1967).

Bedding and Bioturbation Structures in the Bottom. Samples of Deep-Sea Sediments of the East African Coast

Distribution of Sediments

A general survey of the sediments of the continental shelf, slope and adjoining deep-sea parts is given by Schott and von Stackelberg (1965) and Müller (1966). They differentiated five sediment facies (Fig. 667): Biogenic carbonate sand, olive-gray mud, foraminiferal sand, globigerina ooze and red deep-sea clay.

The relatively narrow continental shelf between Sakotra Island and Mombasa is covered by biogenic carbonate sand. It is likely that part of this is of Pleistocene age, when due to lowered sea level, coral reef, etc., developed very near to the shelf break.

The upper part of the continental slope is occupied by foraminiferal sand and olive-gray mud. In deeper parts of the continental slope is globigerina ooze, which extends into the deep ocean basin up to a water depth of 4,500 m. The deeper parts of ocean basin itself are covered by red deep-sea clay. These zones run parallel to the coast (cf. Schott and von Stackelberg 1965; Müller 1966). In the follow-

Fig. 666. Current lineations. Vema 14–36, 4,909 m. This is a flat-floored trough south of Reunion Island, Indian Ocean. Note the well-developed current lineations ("crag and tail") due to deposition in the lee of the rocks. (After Heezen and Hollister 1964)

ing, important inorganic sedimentary structures and bioturbation structures of various sediments are described (cf. Reineck 1973).

Red Deep-Sea Clay

Red-colored deep-sea clay occupies regions deeper than 4,500 m. The deep-sea clay of the Somali Basin shows a carbonate content of < 50 % and a rather high $MgCO_3$ content. The grain-size analysis shows sediment to be clayey silt to sandy silt. A bedding is developed due to change in color in units several centimeters thick. The degree of bioturbation is rather high. The depth of bioturbation may be up to 15 cm. The depth of bioturbation denotes up to what depth material of a higher sediment layer is churned and moved by organisms. Typical bioturbation structures for the deep-sea clay are, e.g., almond-seed structures, spreite structures, and high content of fecal pellets. Following are the most important bioturbation structures of red deep-sea clays.

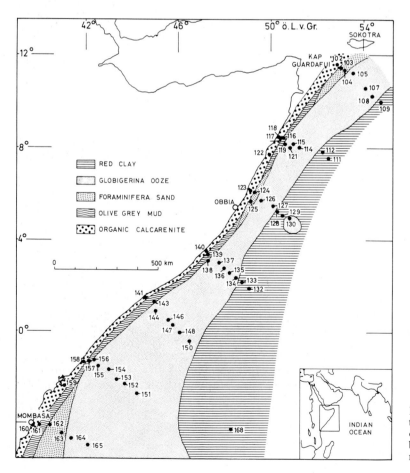

Fig. 667. Distribution of various sedimentation areas off the African East Coast. Based on information by Schott and von Stackelberg (1965); and Müller (1966). (After Reineck 1973)

1. Almond-seed Structure. These structures are biogenic structures, spindle-shaped or ellipsoidal-shaped, 13 to 20 mm in length and 7 to 10 mm in width (Figs. 668 and 669). The outer boundary is more or less even. Internally, they are made up of light-colored irregular ball-and-sausage-like features. The matrix in between shows the same color as the surrounding sediment. Some thin sections have a dark-colored rim, which surrounds the almond-seed structure. The bioturbating organism and the mode of genesis is unknown. It is likely that they are not the stuffed burrows, because sections in various directions show the same feature. Probably they are accumulations of fecal pellets enclosed by a mucus layer.

2. Spreite Structures. Spreites are developed as lateral and upward-directed burrows. In one case burrows with downward-directed spreites are also seen. The organism producing such structures is not known. Figure 670 shows spreite structures.

3. Burrows. In all the samples burrows are abundant. They are 1 to 10 mm in diameter (Fig. 671). The burrows do not have any preferred orientation; they are straight or curved, and often show bifurcation. Two sizes are most abundant, i.e., 1 to 2 mm and 4 to 5 mm in diameter.

4. Deformative Bioturbate Structures. These are ill-defined structures showing no definite forms. They are up to 1 cm broad and several centimeters in depth. They are most commonly visible as irregular pockets of sediments of differing character.

5. Fecal Pellets. Fecal pellets are present mostly concentrated in form of nests (Fig. 672). Single fecal pellets are in general ellipsoidal in shape, 0.7 mm in length, and 0.3 mm thick.

Globigerina Ooze

Globigerina ooze is present in a region of 1,600 to 4,500 m water depth. Grain size analysis shows this to be sandy silt. The carbonate content is > 50 %, which is much higher than that of red deep-sea mud. In thin-sections tests of *Globigerina* are

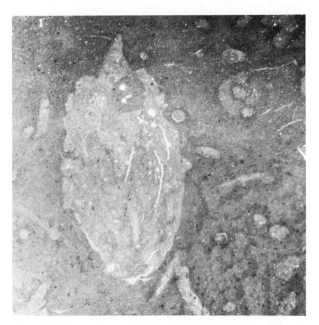

Fig. 669. Thin section of deep-sea clay showing almond-seed structure, probably balls of fecal pellets. The water depth is 5,090 m width of pellet = 1 cm. The box-core sample is from off the East African coast. (After Reineck 1973)

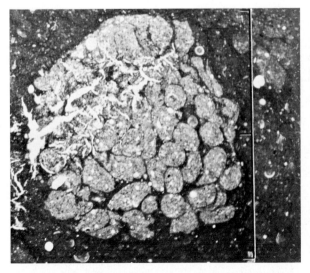

Fig. 668. Thin section of deep-sea clay showing almond-seed structure, probably balls of fecal pellets. The water depth is 5,090 m, length of scale = 1 cm. The box-core sample is from off the East African coast. (After Reineck 1973)

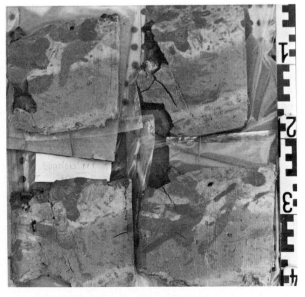

Fig. 670. Section of a box-core sample of deep-sea clay with burrows and spreiten. The water depth is 5,080 m. This sample is taken from the East African coast

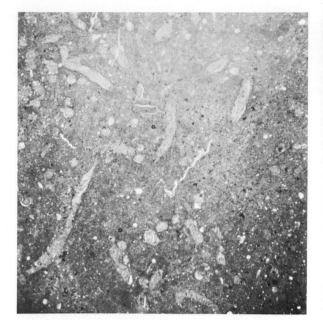

Fig. 671. Stuffed burrows in the deep-sea clay. The water depth is 5,090 m. The box-core sample is from off the East African coast. (After Reineck 1973)

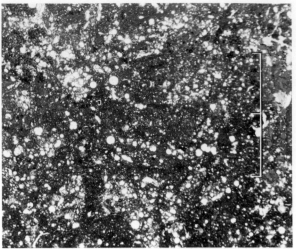

Fig. 673. Thin-section of Globigerina ooze showing typical irregular distribution of microshells. The water depth is 3,803 m, length of scale = 1 cm. The box-core sample is from off the East African coast. (After Reineck 1973)

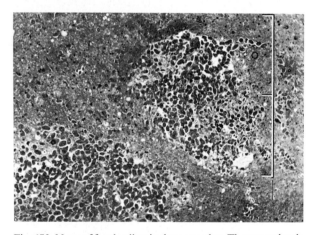

Fig. 672. Nests of fecal pellets in deep-sea clay. The water depth is 4,690 m, length of scale = 1 cm. The box-core sample is from off the East African coast. (After Reineck 1973)

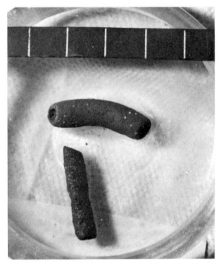

Fig. 674. Part of a burrow with enforced walls, made up of Globigerina ooze. The water depth is 4,355 m. The scale is in centimeters. The box-core sample is from off the East African coast

identifiable. As a result of strong bioturbation the tests of *Globigerina*, other foraminifera, and diatoms are irregularly distributed and concentrated as nests (Fig. 673). Fecal pellets are much less abundant than in deep-sea clay. Characteristic bioturbation structures are *Zoophycos*-type of burrows, and burrows filled with pyrite. Besides, irregularly directed burrows are common, some of them with rather thick walls (Fig. 674).

Zoophycos. *Zoophycos*-type of burrows are restricted to the Globigerina ooze. Only in one case was a *Zoophycos* burrow found in red deep-sea clay.

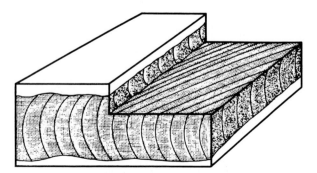

Fig. 675. A laterally displaced *Zoophycos* burrow in spatial view

The *Zoophycos* burrows are the result of the lateral shifting of burrows. In sections cut normal to the bedding planes and the axis of the burrow, lateral displacement of structures is clearly visible; the C-shaped spreites placed one into the other make a broad band which can be followed across a sample (Fig. 675). Mostly *Zoophycos* burrows occur deeper than 20 cm in sediment. They are 5 to 10 mm in thickness. The direction of shifting (displacement) is usually horizontal. If inclined, the angle of inclination is less than 10°. Laterally, the burrow is mostly straight, or sometimes slightly curved (both concave and convex). Within the burrow, tests of foraminifera and fecal pellets are recognizeable. The layers of fecal pellets alternate with the ground mass. Around the burrow a burrow halo is developed. In the burrow halo forams are concentrated together, and the sediment seems to be darker-colored, suggesting that in the burrow halo sediment is pressed together. The burrow halo is easily seen both in thin sections and X-ray radiographs (Fig. 676).

Zoophycos burrow are relatively abundant. In cores they are present at distances ranging from 56 to 3 cm. The usual distance between *Zoophycos* burrows is 20 to 30 cm. *Zoophycos* burrows are well known from ancient sediments. Seilacher (1967) discusses them in detail. Jones (1961) and Taylor (1967) describe similar structures from the lower Cretaceous of Saskatchewan and Alexander Island, respectively. Taylor (1967) describes *Zoophycos* burrows in both horizontal and vertical sections and demonstrates that some of them are built up spirally, e.g., *Zoophycos briantus* (Villa) or *Z. Taonurus* (Sarle) or *Z. Crassus* (Hally). Besides, straight *Zoophycos* burrows running in a single plane are present, as in the case of Recent sediments off the African coast. They are named *Zoophycos laminatus*.

Pyritic Burrows. They are extremely narrow, mostly straight unbranched burrows, running vertically in the sediment and filled with pyrite (Fig. 677). They are present in sediments of water depths of 3,500 to 3,800 m. In the cores they are found in sediment depths of > 1 m. In one case, they were found in sediment depths of 20 cm in 4,540 m water depth. Einsele and Werner (1968) describe si-

Fig. 676. X-ray radiograph of *Zoophycos* burrows in Globigerina ooze. The water depth is 3,900 m. The box-core sample is from off the East African coast. (After Reineck 1973)

Fig. 677. Thin section of Globigerina ooze showing very narrow burrows, partly filled with pyrite. The water depth is approximately 3,500 m. Length of scale = 1 cm. The box-core sample is from off the East African coast. (After Reineck 1973)

milar thread-thin burrows filled with aggregates of pyrite-spherules from the Nile delta. Hesse et al. (1971) report similar burrows from cores of the Strait of Otranto, Mediterranean Sea. Bouma (1968) regards similar features from the Gulf of Mexico as *Mycelien*.

Burrows with Thick Walls. Some of the horizontal burrows of 10 mm outer diameter show walls as thick as 3 mm. In one case a burrow was separated by washing the sediment (Fig. 674).

Nonoriented Burrows. In many samples branched stuffed burrows are visible, showing no preferred orientation. The diameter is 1 to 2 mm. Mostly the stuffed material is darker-colored than the surrounding sediment.

Green-colored Streaks. Occasionally, cores show 1-to-5-mm-thick streaks of green color. These streaks sometimes show a layer of forams in the center, where the color is also most intensive. Upward and downward from the streak green color grades over to other colors.

Graded Bedding. In the deep-sea region off the East African coast no sandy turbidite layers were recorded. However, parts of some cores show graded bedding, where the sandy part was made up exclusively of foraminifera. Mostly such parts are also strongly bioturbated. Only in one case are fine lamination visible, made up of foraminiferal tests.

Foraminiferal Sand

The major component of foraminiferal sand is tests of forams. Larger tests and shells are also visible. The degree of bioturbation is very high. Primary inorganic structures and burrows are not recognizable. Fecal pellets are abundant, sometimes up to 1 mm in length.

Olive-gray Mud

The content of small tests and shells is variable; larger tests and shells are present. Fecal pellets are commonly present. The larger fecal pellets are not so common, but smaller fecal pellets 0.5 mm long are mostly present in nests. The bioturbation structure is mainly deformative. The concentration of shells in layers can be assigned to burrows without strong walls. In some cores bifurcating burrows with a diameter of 2 to 3 mm are present. In the deeper parts of the cores flecks of about 2 cm diameter are seen clearly flattened. Despite the strong bioturbation, some bedding is visible in X-ray radi-

ographs. The thickness of beds varies from 0.5 to 2.0 cm. The bedding is broken by burrows, etc. The bedding planes are poorly defined.

Biogenic Carbonate Sand

The carbonate sand is made up of reef detritus, quartz, mica, and glauconite (Schott and von Stackelberg 1965). Besides, shells of lamellibranchs and gastropods, spines of echinoderms, spines of siliceous sponges and tests of forams are present (Gutmann, personal communication). The degree of bioturbation is strong, partly made by echinoderms.

Continental Slope, Rise and Abyssal Plain of the Gulf of Mexico

General Information

The Gulf of Mexico is one of the oceanic regions that have been studied in certain detail. The history of the Gulf of Mexico and its sediment has been discussed by Uchupi and Emery (1968). In the following, a short account of this region is given. General data about bathymetry are taken from Uchupi and Emery (1968).

The continental shelf in the Gulf of Mexico is variably developed in different regions. The eastern part of the Gulf of Mexico is made up of carbonate sediments, whereas the western part possesses clastic sediments. The boundary is located along the line connecting De Soto Canyon to Campeche Canyon. Off the coast of Florida and Texas the shelf is about 200 km wide, but almost absent or extremely narrow in the area where the Mississippi delta has prograded across the shelf. The continental shelf off Florida is a carbonate platform with minor geomorphic irregularities caused by the growth of algal reefs. The carbonate platform off Yucatan is covered by coral reefs. The shelf region west of the Mississippi trough along the Mexican coast is noncarbonate and shows many isolated hills, most probably caused by salt domes.

The continental slope of the Gulf of Mexico in front of the Mississippi delta is rather smooth and well developed in the shape of a cone, with a slope of 2° near the apex and flattening seaward. This cone, named the Mississippi Cone (Ewing et al. 1958) extends eastward to the Straits of Florida, filling the narrow trough between Florida and the Campeche Escarpments. Westwards it grades into the continental rise and Sigsbee abyssal plain. On each side of the cone is a canyon–De Soto Canyon and the Mississippi Trough; but both seem to be inactive at present. The continental rise in front of the Sigsbee Escarpment is made up of a thick apron of sediments, deposited mainly by turbidity currents. This sediment apron of the continental rise grades into the flat Sigsbee abyssal plain in the south and the Mississippi Cone in the east.

Sedimentary Structures in the Sediments of the Gulf of Mexico

Bouma (1968) made a systematic study of the piston core and box core samples obtained from the deeper parts of the Gulf of Mexico. He studied in detail the sedimentary structures in these samples and found main sedimentary structures, which are described in the following to give a more complete picture of the characteristics of the deep-sea sediments. It covers the region of the continental slope, the rise, the abyssal plain and the Mississippi fan.

Thin Bedding and Lamination. The layers are of variable thickness–some more than 5 mm, others less than 5 mm. They are regular, mostly horizontal parallel varve-like layers (thinly interlayered bedding). Layering is due to minor differences in grain size, and mineralogical and chemical composition of the sediments. Sometimes layers are lenticular in shape (Fig. 678). Occasionally, laminae show irregular, indistinct contact planes. Layers are sometimes composed of rather coarse material. They are solitary layers within finer-grained sediment and are mostly a concentration of tests of foraminifera and pteropods (see p. 489); sometimes, however, they are coarse-grained detrital material. Deposi-

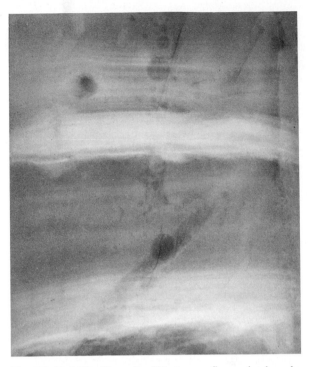

Fig. 679. Turbidite-like units. Silty to very fine-grained sandy sediments showing parallel lamination. The water depth is 3,515 m. The X-ray radiograph of piston-core samples is from the Gulf of Mexico. (After Bouma 1968)

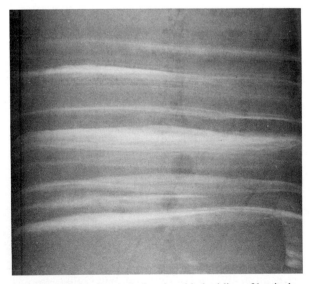

Fig. 678. X-ray radiograph showing thin bedding of lenticular units. The water depth is 3,400 m, the sediment depth is 1,044 to 1,049 cm. Piston-corer samples are from the Gulf of Mexico. (After Bouma 1968)

tion of such layers is thought to be due to some currents or temporary still-stand of the normal sedimentation processes. The process of winnowing can be ruled out, as such bedding is common in deposits of the continental slope and beyond. Lack of any internal structure in these layers excludes the process of bottom traction current.

The presence of such fine layering suggests fluctuations in the transport-deposition process of the Gulf.

Turbidite Layers. In several cores from the Mississippi fan and the deeper central parts of the Gulf of Mexico are bands of silty to very fine sandy sediments, which exhibit foreset bedding overlaid by parallel bedded units. Two types are observed–a distinct and an indistinct type.

The distinct type shows a short basal contact and a normal grading upward, throughout the foreset bedding and parallel bedding (Fig. 679). This corresponds to the incomplete turbidite sequences of types Tc-e (current-ripple lamination, upper parallel lamination, and pelitic interval) and Td-e (upper parallel lamination and pelitic interval) of the Bouma sequences (cf. Bouma, 1962). Unlike the ancient turbidites, foreset bedding of these layers is indistinct, and foreset laminae have lower angles of dip. Such incomplete turbidite layers are thought to represent distal parts of turbic te deposits, or that the intensity of turbidity currents was not large enough to deposit the complete turbidite sequence.

The indistinct type shows rather vague foreset laminae and parallel laminations. However, they are also thought to be deposits of turbidity currents.

Load Structures. Small-scale nodule-like structures are present in a few sediment cores. They are often associated with turbidites with a thin layer of coarser sediment on the top.

Muddy Conglomerate Layers. Few cores from the Florida Escarpment contain calcarenitic muddy conglomerate layers. They are present below the sediment depths of 4 to 6 m. Clayey pebbles are angular to subrounded. They may represent some kind of slump deposit.

Primary Homogeneous Layers. Most cores contain zones in which no bedding structure is visible, even on X-ray radiographs. Such layers are, most probably, the result of a regular rates of deposition without any fluctuation; thus there is no change in the nature of the material deposited. It can be of either pelagic or turbidite nature.

Burrows. Bioturbation structures are rather common, but their nature, density, and size are highly variable. Burrows vary in size from less than 1 mm to more than 2 cm. They may be horizontal, vertical, or even inclined in various planes. Samples of the box-core show more intensive bioturbation. As these samples represent only the topmost part of the sediments, it may be due to rather slow rate of deposition during sub-Recent times.

Mottling. Destruction of sedimentary features leading to mottling can be produced by slumping as well as by bioturbation. Biogenic mottling may be taken as an indicator of a low rate of sedimentation, if the density of burrowing animals can be taken as constant through time. Mottling is common in the upper 20 to 50 cm of the samples.

Mycelia. Radiographs of many cores show thin light colored strings, whose length varies from about 1 cm to several centimeters (Fig. 680). Such strings are straight or slightly curved, and possess sharp bends. The thickness is usually less than 0.5 mm. In cross-section they are round to slightly oval or even band-shaped. It is difficult to suggest their mode of origin. They are found only in the homogeneous parts of the cores.

Shells and Shell Fragments. Complete shells and fragments of molluscs, foraminifera, and pterop-

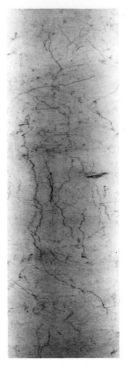

Fig. 680. X-ray radiograph showing thin, dark-colored strings filled with carbonate material. Probably they are *Mycellia*. The water depth is 1,110 m. The piston-corer sample is from the Gulf of Mexico. (After Bouma 1968)

ods are found scattered, or concentrated in bands or nests. Fragmentation of shells seems to be due to carnivores, before their deposition. Large shells and solitary corals, which are sometimes present, may have been transported by algae.

Distribution of Sedimentary Structures over Physiographic Provinces

In general there is little variation in the type and abundance of sedimentary structures within a physiographic province. However, sometimes change from a shallower part to a deeper part is well documented. For example, in the Mississippi fan, indistinct types of turbidites are more abundant in the lower fan region than in the upper part. Thus, the distinct type of turbidites decrease in number with increasing depth and distance from source. Carbonate sediments exhibit only a few recognizable sedimentary structures.

The degree of bioturbation in box-core samples is extremely high. If no change in the density of fauna since late Pleistocene is assumed, the increased bioturbation seems to indicate decrease in the rate of deposition. (See the chapter on adjacent seas on p. 499 for comparison, where increased bioturbation during the Holocene is explained as a result of better circulation of O_2 content in the lower layers of water.)

Sedimentation in the Deeper Parts of the Adjacent Seas

Deposition in the adjacent seas shows a great variability in its characteristics. Moreover, not all kinds of adjacent seas that might have existed in the geological past have their analogues in the Holocene times.

In any case, the study of present-day adjacent seas is of great importance for the environmental analysis of ancient adjacent sea deposits. Seibold (1970) states that, fundamentally, adjacent seas can be distinguished into two kinds—those of humid regions and those of arid regions. In addition many other parameters influence the sedimentation characteristics in the adjacent seas, e.g., nature of swell or sill between adjacent seas and the open ocean, the nature of incoming sediment, etc.

The characteristics of the lagoonal-, coastal sand-, and continental-shelf deposits of the adjacent seas are similar to those of the open sea and have been discussed in respective chapters. However, bedding characteristics of the deeper parts of the adjacent seas show some specific characteristics. The following types of bedding features are usually found in such deposits:

Varve-like thinly interlayered bedding
Alternating bedding of fine sand, silt/clay
Primary homogeneous bedding
Strongly bioturbated layers
Turbidite-like intercalations.

A short description of these bedding types is given in the following.

Varve-like Thinly Interlayered Bedding. Light- and dark-colored, thin layers alternate with each other, as in varves (see p. 127). The alternation may be a result of seasonal variation in the supply of sediment and organisms. Such is the case in the sediments of the Black Sea (Müller and Blaschke 1969a, 1971; Bukry et al. 1970). Here, carbonate-rich light-colored laminae made up of coccoliths

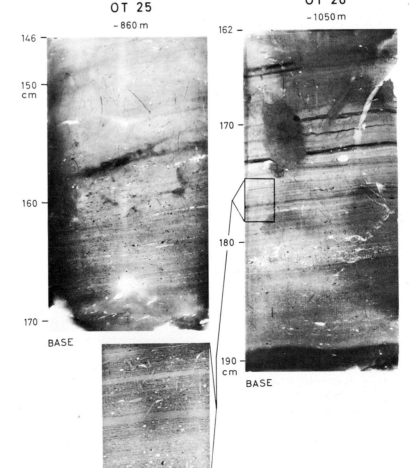

Fig. 681. X-ray radiographs showing basal varved sediment of two box-cores from the Apulian-Ionan Ridge, Mediterranean Sea. Both cores are 12 km away from each other, but contain a characteristic dark-colored (olive gray) varved part at comparable core depths.
Core OT25 (water depth 850 m). 146 to 155 cm: Faint lamination, partly destroyed by bioturbation, pyritized filaments. 185 cm: *Pteropod* layer, mostly *Cavolinia tridentata tridentata* Forkksal. 160 to 168 cm: forminiferal layers, partly pyritized (*Orbulina, Globigerina*). Alternation of dark- and light-colored laminae developed as annual varves. Below 168 cm: pteropod-rich homogeneous mud.
Core OT26 (water depth 1,015 m). 162 to 173 cm: Laminated, pteropod-rich mud, in part disturbed by large burrows. 175.5 to 178.5 cm (thin section; see inset): varved sediments made up of alternating dark laminae rich in clay, organic material, and carbonate; and light-colored laminae made up of quartz-rich sediments containing coccolithophorids. In opposition to the *Globigerina*-pteropod layers below 180 cm, organisms are irregularly distributed. The individual laminae are 0.2 to 0.5 mm thick. Some pteropod shells are filled with framboidal pyrite spherules. Fucoid-like patterns are present throughout the varved zone. 180 to 189 cm: coarse-grained *Globigerina* pteropod layers, 189 to 191 cm: greenish-gray homogeneous mud. (After Hesse et al. 1971)

OT 25
−860 m

OT 26
−1050 m

BASE

BASE

alternate with the carbonate-poor, darker layers, rich in organic matter (Fig. 207). In other cases varve-like features may be a result of difference in grain size, petrological features, or chemical composition. Such an alternation is caused by the fluctuation in the incoming sediment and it is a case of annual varves (Fig. 681) (van Straaten 1967).

Alternating Bedding of Fine Sand, Silt/Clay. Such alternating bedding is made up of horizontal parallel layers; sometimes layers may be lenticular (Fig. 681).

Primary Homogeneous Bedding. These are sections in a profile which do not show any bedding features. They are not the layers homogenized due to bioturbation. Bouma (1968) interprets such layers as a result of longer periods of constant rates of

sedimentation where no fluctuation in the kind and supply of material takes place (Fig. 681).

Strongly Bioturbated Layers. These are layers disturbed by organisms. In most cases, remanents of primary inorganic bedding, or nests of sand, silt and broken shells are encountered. Such layers may show mottled structures or they may be homogenized. Homogenized bioturbated layers may not be differentiated from primary homogeneous layers in narrow cross-section of the cores (Fig. 681).

Turbidite-like Intercalations. Turbidite-like horizons are commonly found in the deposits of adjacent seas. However, in most cases they show only a condensed part of the ideal turbidite sequence (Fig. 682).

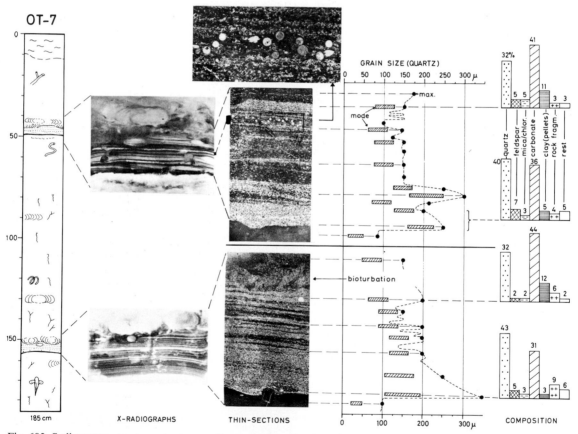

Fig. 682. Sedimentary structures, texture, and composition of two sand layers in the box-core. OT 7. Strait of Otranto, Mediterranean Sea. Water depth 1,050 m.
Upper sand layer: Sharp erosional basal boundary. Basal 5 mm of unlaminated, well-sorted fine sand that is rich in reworked shallow-water foraminifera is overlaid by a cross-bedded (low-angle) unit rich in dark laminae (clay and fecal pellets). The middle part shows thick clay-rich dark laminae and a layer containing large specimens (0.5 to 1 mm) of *Orbulina sp.* The upper boundary of the sand layer is well defined and disturbed by

large burrows. Some isolated burrows (echinoid?) are visible in the parallel-bedded and cross-bedded parts.
Lower sand layer: Basal contact is very sharp (the burrows protrude from the underlying mud layer), the upper contact is gradational as a result of bioturbation. The sand layer is unlaminated and graded near the base, and parallel-bedded in its upper part. Dark laminae are made up of fecal pellets up to 0.5 mm in size. Mostly redeposited organisms, 1 to 2 % ooids, rock fragments, heavy minerals, opaques, etc. (After Hesse et al. 1971)

In general, the above-mentioned bedding features are present in the sediments of the adjacent seas of humid and arid climates (Seibold 1970). In cases where adjacent seas are separated from the open ocean by a swell, the difference in sediment characteristics and organisms becomes apparent (Fig. 683). Table 38 gives characteristics of the adjacent seas of humid and arid regions.

In the adjacent seas of the arid climate water layering is less apparent (Seibold and Vollbrecht 1969). This leads to the increased oxygen content in the near-bottom waters. Corresponding to this factor, benthonic organisms are abundant and degree of bioturbation in the sediments is high. Examples of such occurrences are Holocene sediments of the Persian Gulf (Hartmann et al. 1971), the Adriatic Sea (van Straaten 1967, 1970), and the Strait of Otranto, Mediterranean Sea (Hesse et al. 1971).

In the adjacent seas of humid climates a stable water layering is mostly present, leading to oxygen deficiency and formation of H_2S in the bottom layers. As living conditions for macrobenthonic organisms are unfavorable, there is no bioturbation activity, and fine bedding features are beautifully preserved. Pratje (1939, 1948), Kolp (1966), and Seibold (1965, 1970) describe such features from the Baltic Sea. Further examples are the Black Sea (Zenkovitch 1966; Müller and Blaschke 1969) (Fig. 207), and the Pleistocene sediments in the Adriatic Sea (van Straaten 1967). Degens and Ross (1974) give a collection of papers dealing with the structure, water composition, sediments, biology, and geochemistry of the Black Sea. Laking (1974) compiled a comprehensive bibliography on the Black Sea.

The Gulf of California represents a mixed kind of adjacent sea. Here benthonic organisms are well represented and the sediment is strongly bioturbated; however, a water layering is developed at water depths of 400 to 800 m, and below it the oxygen content is strongly reduced. During winter upwelling water and production of diatoms is so much that the sinking diatom tests produce diatom ooze. Near the central part of the Gulf varve-like, thinly interlayered bedding is developed, consisting of al-

Table 38. Adjacent seas in humid and arid climatic zones are differentiated on the basis of their water circulation patterns. This has its direct implication in the organisms and the sediments. Thus, the following two models of adjacent seas can be contructed. (After Seibold 1970)

Characteristics	Humid region	Arid region
Surface current	Outflowing	Inflowing
Bottom current	Inflowing	Outflowing
Water layering	Clear-cut when winter convection has not reached bottom	Less clear
Bottom water character:		
Salinity	Low	High
Salinity	Low	High
Oxygen content	Low	High
Nutrients (PO_4, SiO_2)	High	Low
Sediment character:		
Carbonate content	Low, higher in stagnant water	High
Organic carbon	High	Low
Metal content	High	Low
Biofacies boundaries	Often sharp	Often vague
Lagoons	Peat	Evaporites

ternating layers of clayey silt and diatoms. They are well preserved in the oxygen deficient zone below 400 to 800 m water depth (Byrne and Emery 1960; van Andel 1964).

Seibold (1970) points to the possibility of enrichment of iron, manganese, and other metals in the sediments of adjacent seas of humid climates.

In adjacent seas turbidite-like intercalations are often found. Thus, some of the deeper parts of the adjacent seas can be regarded as flysch-like sediments.

A summary of various flysch-like basins in the Mediterranean Sea is given by Ryan et al. (1971).

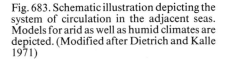
Fig. 683. Schematic illustration depicting the system of circulation in the adjacent seas. Models for arid as well as humid climates are depicted. (Modified after Dietrich and Kalle 1971)

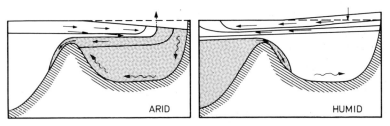

Concluding Remarks

We would like to point out that the intention of this book has been to present all available data on recent sediments and, by doing so, to provide the field geologist with a working tool for reconstructing the ancient environment. Further, we do not hesitate to state that as the reader goes through the book, he will encounter a number of debatable points, at least two of which will strike everyone: (1) Do we have modern analogues or sedimentation models applicable to all ancient environments? (2) Is it always possible to recognize an ancient environment in the light of recent analogues? Though these questions are logical, they are difficult to answer; and if we were asked to answer them categorically, given the present state of our knowledge, we would choose "no". This does not mean that we wish to dishearten those engaged in environmental reconstruction, but we wish to emphasize that at times the nature is not yet properly understood.

This led us to discuss briefly, in the opening chapter of Part II, the limitations involved in recognizing ancient environments and also in constructing environmental models on the basis of data obtained from the study of recent sediments. Further, it is appropriate to mention that this concept of limitation is based not only on recent data but also on inferences drawn from a number of facts about the earth's past history.

Certainly, the present is the key to the past; but the implication of this is limited, as many of the conditions of sedimentation were unique to the geological past and are not present today, or if present, do not function at the same rate and with the same effectiveness.

For instance, before the Devonian, continents were not covered with plants. Under such conditions, entirely different weather and transportation mechanisms would dominate, influencing fluvial deposits, the development of deltas and even the nature and quantity of sediments on sea coasts. We all know today that plant growth strongly influences the development of delta bodies. Thus, in the absence of plant cover before Devonian times, delta growth and the relationship between deltaic plain deposits and coastal sands would be somewhat different.

There were periods in the geological past with much more stable conditions than today. It is generally agreed that conditions in the Jurassic and Cretaceous, for example, were stable in comparison with today's. Therefore, one must infer that the geometry and the extent of deposits in those periods are not analogous to those in present-day environments.

Nevertheless, the principle of uniformitarianism (principle of actuality) is certainly valid in broad terms. Through the comparison of ancient and recent sediments, it has long been established that the physical and chemical laws governing sedimentation have remained constant throughout geological history. The same primary structures have developed in ancient and recent sediments under given hydrodynamic conditions.

It can be stated, on the basis of the study of recent sediments, that when making environmental models, one is handicapped by the uniqueness of the Quaternary period which has affected the pattern of sedimentation in current environments. The major problem is that present-day geomorphology has been formed in response to the Pleistocene ice age and has not yet reached equilibrium, resulting in unique deposits. Moreover, the short period after the Pleistocene ice age produced only a thin layer of deposits; these deposits give almost no information on the development of vertical sequences in which changes with time in a single environment can be registered. This factor has played an important role in the geological past.

A good example of such limitations is continental shelf deposits. Only small parts of today's continental shelf are covered with recent sediments; the major part is occupied by relict and palimpsest sediments. In the geological past we have examples of well-developed thick deposits of continental shelf sediments, which must have been formed under more favourable conditions.

It has been pointed out that climate and geomorphology of both the provenance and basin of deposition are most important in controlling the nature of deposits. In our studies of modern sediments we have not yet sufficiently investigated the effect of these factors on specific environments. Actually, we have studied only a fraction of the earth's surface. Even according to optimistic estimates, less than 20 % of the recent depositional environments have been investigated to obtain data for reconstruction of environments. Very little information is available about transgressive horizons, marine windflats, flysch basins and continental margin deposits. Much work is still needed in the areas of fluvial sediments, desert sediments and shoal of the shallow seas. Moreover, almost no information is available about the control of deposition by sedimentary tectonics, which is certainly the major force controlling the thickness of the deposits.

Another important problem in the reconstruction of depositional environments is the limitations of the principle of uniformitarianism (i.e. the principle of actuality). For example, red beds are abundant in ancient desert deposits, but very scarce in modern deserts. This has been a major point of debate in the interpretation of red bed sequences, leading to the humid versus arid climate controversy.

Recognition of ancient environments is not a mathematical problem; it also involves much hypothetical thinking, which should certainly be based on the understanding of modern sedimentary processes and their environmental relationship. In nature, there is no ideal or characteristic model; each sequence has its own uniqueness. Results should be comparable to our well-studied models, but in no case should models become the controlling force in our thinking so that we ignore the facts. We must continually try to expand and better establish our standard models by incorporating information gained by the study of ancient and recent sediments.

The success of this book will be determined by the extent to which it can provide a basis for interpreting ancient environments. We hope that with the help of supplementary information, existing sedimentation models can soon be refined and better defined. For each environment, several models should be developed which include climate and geomorphological setups.

Future work will require the study of more and more modern environments by actuogeologists assessing all possible aspects of modern sediments. This work could also be helpful to all geologists trying to interpret ancient sediments. During such investigations, processes and features that are potentially preservable should be especially emphasized. When making a model one should consider which part of a sedimentary environment might appear in a sedimentary sequence and is easily recognizable. Almost all depositional environments need to be investigated by a close network of drilling so as to ascertain the geometry of various kinds of deposits. On the other hand, the data obtained by actuogeologists should be properly criticized by geologists working on fossil rocks during environmental reconstructions, and it should be modified or rejected according to specific problems.

Whith each solved scientific problem a new complex of questions arises. Scientific work is sometimes like the fight of Hercules against the many-headed serpent Hydra. Only those will find satisfaction in scientific research to whom the result is only one step on a long long way and who are not disheartened by the ever-appearing complex of questions arising at the end of an investigation.

References

AAPG, ed. (1973 a) Sandstone reservoirs and stratigraphic concepts I. Selected papers repr. from AAPG Bull. and Mem. 18, 190 p. Tulsa, Oklahoma: AAPG

AAPG, ed. (1973 b) Sandstone reservoirs and stratigraphic concepts II. Selected papers repr. from AAPG Bull. and Geometry of Sandstone Bodies, 216 p. Tulsa, Oklahoma: AAPG

AAPG, ed. (1974) Facies and the reconstruction of environments. Selected papers repr. from AAPG Bull., 223 p. Tulsa, Oklahoma: AAPG

Aarseth, I., Bjerkli, K., Björklund, K. R., Böe, D., Holm, J. P., Lorentzen-Styr, T. J., Myhre, L-A., Ugland, E. S., Thiede, J. (1975) Late quaternary sediments from Korsfjorden, Western Norway. Sarsia 58, 43–66

Abdul-Razzak, A. K. (1974) Geochemisch-sedimentpetrographischer Vergleich lakustrischer Sedimente aus verschiedenen Klimabereichen. Chemie der Erde 33, 152–187

Adler, A. A., Lattman, L. H. (1961) Floodplain sediments of Halfmoon Creek, Pennsylvania. Bull. Mineral Ind. Expt Sta., Penna State Univ. 77, 1–11

Agassiz, A. (1906) General report of the expedition steamer Albatross from October, 1904, to March, 1905. Mem. Museum Comp. Zool. 33, 1–77

Agterberg, F. P., Banerjee, I. (1969): Stochastic model for the deposition of varves in glacial Lake Barlow-Objibway, Canada. Can. J. Earth Sci. 6, 625–652

Ahlbrandt, T. S. (1975) Comparison of textures and structures to distinguish eolian environments, Killpecker dune field, Wyoming. The Mountain Geologist 12, 61–73.

Ahlbrandt, T. S., Andrews, S. (1978) Distinctive sedimentary features of cold-climate eolian deposits, North Park, Colorado. Paleogeogr. Palaeoclimatol., Palaeoecol. 25, 327–351.

Ahlbrandt, T. S., Andrews, S., Gwynne, D. T. (1978) Bioturbation in eolian deposits. J. Sediment. Petrol. 48, 839–848

Ahlmann, H. W., son (1935) Contribution to the physics of glaciers. Geogr. J. 86, 97–107

Aigner, T. (1979) Schill-Tempestite im Oberen Muschelkalk (Trias, SW Deutschland). N. Jb. Geol. Paläont. Abh. 157, 326–343

Aigner, Th. (1985) Storm depositional systems, 173 p. Berlin, Heidelberg, New York, Tokyo: Springer

Aigner, Th., Reineck, H.-E. (1982) Proximity trends in modern storm sands from the Helgoland Bight (North Sea) and their implication for basin analysis. Senckenbergiana marit. 14, 183–215

Albrecht, P. (1970) Etude de constituants organiques des séries sédimentaires de Logbaba et de Messel. Transformations diagénétiques. Mém. du serv. de la Carte Geol. d'Alsace, Lorraine 32, 119 p.

Allen, G. P. (1971) Relationship between grain size parameter distribution and current patterns in the Gironde estuary (France). J. Sediment. Petrol. 41, 74–88

Allen, G. P. (1972) Etude des processus sédimentaires dans l'estuaire de la Gironde. Thèse Fac. Sci. Bordeaux, 310 p.

Allen, G. P., Castaing, P. (1973) Suspended sediment transport from the Gironde estuary (France) onto the adjacent continental shelf. Mar. Geol. 14, M47–M53

Allen, G. P., Castaing, P., Klingebiel, A. (1972) Distinction of elementary sand populations in the Gironde Estuary (France) by R-monde factor anlaysis of grain-size data. Sedimentology 19, 21–35.

Allen, J. R. L. (1963) Asymmetrical ripple marks and the origin of water laid cosets of cross-strata. Liverpool Manchester Geol. J. 3, 187–236

Allen, J. R. L. (1964 a) Primary current lineation in the lower Old Red Sandstone (Devonian), Anglo-Welsh Basin. Sedimentology 3, 98–108

Allen, J. R. L. (1964 b) Studies in fluviatile sedimentation: Six cyclothems from the lower Old Red Sandstone, Anglo-Welsh Basin. Sedimentology 3, 163–198

Allen, J. R. L. (1965 a) Sedimentation in the lee of small underwater sand waves: an experimental study. J. Geol. 73, 95–116

Allen, J. R. L. (1965 b) Late Quaternary Niger delta, and adjacent areas: Sedimentary environments and lithofacies. Bull. Am. Assoc. Petroleum Geologists 49, 547–600

Allen, J. R. L. (1965 c) A review of the origin and characteristics of recent alluvial sediments. Sedimentology 5, 89–191

Allen, J. R. L. (1965 d) Coastal geomorphology of eastern Nigeria: Beach-ridge barrier islands and vegetated tidal flats. Geol. Mijnbouw 44, 1–21

Allen, J. R. L. (1965 e) The sedimentology and paleogeography of the Old Red Sandstone of Anglsey, North Wales. Yorks. Geol. Soc. Proc. 35, 139–185

Allen, J. R. L. (1966) On bed forms and palaeocurrents. Sedimentology 6, 153–190

Allen, J. R. L. (1968 a) Current ripples. Their relation to patterns of water and sediment motion, 433 p. Amsterdam: North-Holland Publ. Comp.

Allen, J. R. L. (1968 b) Flute marks and flow separation. Nature 219, 602–604

Allen, J. R. L. (1970) Physical processes of sedimentation. An Introduction, 248 p. London: G. Allen & Unwin

Allen, J. R. L. (1970 a) Studies in fluviatile sedimentation: A comparison of fining-upwards cyclothems with special reference to coarse-member composition and interpretation. J. Sediment. Petrol. 40, 298–323

Allen, J. R. L. (1970 b) A quantitative model of grain size and sedimentary structures in lateral deposits. J. Geol. 7, 129–146

Allen, J. R. L. (1971) Transverse erosional marks of mud and rock: Their physical basis and geological significance. Sediment. Geol. 5, 167–385

Allen, J. R. L. (1971 a) Instanteneous sediment deposition rates deduced from climbing – ripple cross-lamination. J. Geol. Soc. London 127, 553–561

Allen, J. R. L. (1971 b) A theoretical and experimental study of climbing-ripple cross-lamination, with a field application to the Uppsala esker. Geograf. Ann. 53, 157–187

Allen, J. R. L. (1973 a) A classification of climbing-rippel cross-lamination. J. Geol. Soc. London 129, 537–541

Allen, J. R. L. (1973 b) Development of flute mark assemblages, 1. Evolution of pairs of defects. Sediment. Geol. 10, 157–177

Allen, J. R. L. (1973 c) Features of cross-stratified units due to random and other changes in bed forms. Sedimentology 20, 189–202.

Allen, J. R. L. (1973 d) Studies in fluviatile sedimentation implications of pedogenic carbonate units, Lower Old Red Sandstone, Anglo-Welch outcrop. J. Geol. 9, 181–208

Allen, J. R. L. (1975) Development of flute-mark assemblages, 2. Evolution of trios of defects. Sediment Geol. 13, 1–26

Allen, J. R. L. (1977) The possible mechanics of convolute lamination in graded sand beds. J. Geol. Soc. London 134, 19–31

Allen, J. R. L. (1978) Polymodal dune assemblages: an interpretation in terms of dune creation – destruction in periodic flows. Sediment. Geol. 20, 17–28

Allen, J. R. L. (1981 a) Lower Cretaceous tides revealed by cross-bedding with mud drapes. Nature 289, 579–581

Allen, J. R. L. (1981 b) Palaeotidal speeds and ranges estimated from cross-bedding sets with mud drapes. Nature 293, 394–396

Allen, J. R. L. (1982) Sedimentary structures, their character and physical basis. Dev. Sedimentol. 30 A and B, Vol. I, 593 p.; Vol. II, 663 p., Amsterdam, Oxford, New York: Elsevier Sci. Publ. Comp.

Allen, J. R. L. Banks, N. L. (1972) An interpretation and analysis of recumbant-folded deformed cross-bedding sedimentologie 19, 257–283

Allen, J. R. L., Friend, P. F. (1968) Deposition of the Catskill facies, Applachian region: With notes on some other Old Red Sandstone basin. In: Klein, G., De Vries, ed. Late Paleozoic and Mesozoic continental sedimentation, northeastern North America. Geol. Soc. Am. Spec. Papers 106, 21–74

Allen, P. (1959) The wealden environment, Anglo-Paris Basin. Phil. Trans. Roy. Soc. (London), Ser B 242, 283–346

Allen, P. A., Homewood, P. (1984) Evolution and mechanics of a Miocene tidal sandwave. Sedimentol. 31, 63–81

Almeida, F. F. M. De (1953) Botucatu, a Triassic Desert of South America. In: Déserts Actuels et Anciens – Congr. Géol. Intern. Compt. Rend., 19e, Algiers, 1952, 7, 9–24

Amstutz, G. C., Bernard, A. J. (1973) Ores in Sediments. Internat. Union of Geological Sci., Series A, No. 3, 350 p. Berlin, Heidelberg, New York: Springer

Andel, T. H. van (1964) Recent marine sediments of Gulf of California. Marine geology of the Gulf of California. A symposium Memoir. 3, 216–310. Tulsa, Oklahoma: Am. Assoc. Petroleum Geologists

Andel, T. H. Van (1967) The Orinoco Delta. J. Sediment. Petrol. 37, 297–310

Andel, T. H. Van, Curray, J. R. (1960) Regional aspects of modern sedimentation in northern Gulf of Mexico and similar basins, and paleogeographic significance. In: Shepard, F. P., Phleger, F. B., Andel, T. H. Van, eds., Recent sediments, northwest Gulf of Mexico. p. 345–381. Tulsa, Oklahoma: Am. Assoc. Petroleum Geologists

Anderson, A. G. (1953) The characteristics of sediment waves formed by flow in open channels. In: Proceedings of the Third Midwestern Conference Fluid Mechanics p. 379–395

Anderson, B. (1961) The Rufiji Basin, Tanganyika. 7. Soils of the main irrigable areas. Report of the Government Tanganyika. Preliminary Reconnaissance Survey Rufiji Basin, 125 p.

Andrée, K. (1920) Geologie des Meeresbodens, Bd. II: Bodenbeschaffenheit, nutzbare Materialien am Meeresboden. 689 p. Leipzig: Gebr. Borntraeger

Andrews, P. B. (1970) Facies and genesis of a hurricane washover fan, St. Joseph Island, Central Texas Coast. Bur. Econ. Geol., Univ. Texas, Austin, Rep. Invest. 67, 1–147

Andrews, P. B. (1976) Guide to recording of field observations in sedimentary sequences. Rep. N. Z. Geol. Survey 79, 68 p.

Andrews, P. B., Lingen, G. J. Van der (1969) Environmentally siginifical sedimentologic characteristic of beach sands. N. Z. J. Geol. Geophys. 12, 119–137

Anketell, J. M., Dzulynski, S. (1968 a) Patterns of density controlled convolutions involving statistically homogeneous and heterogeneous layers. Rocznik Polsk. Towarz. Geol. (Ann. Soc. Geol. Pol.) 38, 401–409

Anketell, J. M., Dzulynski, S. (1968 b) Transverse deformational patterns in unstable sediments. Rocznik Polsk. Towarz. Geol. (Ann. Soc. Geol. Pol.) 38, 411–416

Anketell, J. M., Cegla, J. Dzulynski, S. (1970) On the deformational structures in systems with reversed density gradients. Roczn. pol. Tow. geol. 41, 3–30

Arkhangelsky, A. D., Strakhov, N. M. (1938) Geological structure and development of the Black Sea. Akad. Nauk SSSR

Armon, J. W. & McCann, S. B. (1979) Morphology and landward sediment transfer in a transgressive barrier island system, Southern Gulf of St. Lawrence, Canada. – Mar. Geol. 31, 333–344

Arnborg, L. (1958) The lower part of the River Angermanälven. 1. Publ. Geogr. Inst. Univ. Uppsala 1, 181–247 Univ.

Arpino, P. (1973) Les lipides de sédiments lacustres éocènes. Sci. Geol. Mem. 39, 107 p.

Arrhenius, G. (1963) Pelagic sediments. In: Hill, M. N., ed., Thesea, vol. 3, p. 655–727. New York – London: Wiley & Sons.

Ashley, G. M. (1975) Rhythmic sedimentation in glacial lake Hitchcock, Massachusetts – Connecticut: In: Topling, A. V., McDonald, B. C., eds., Glaciofluvial and Glaciolocustrine Sedimentation. Soc. Econ. Palaeontologists Mineralogists Spec. Publ. 23: 304–320

Asserto, R. L. A. M., Kendall, C. G. St. C. (1977) Nature, origin and classification of peritidal tepee structures and related breccias. Sedimentology 24, 153–210

Augustinus, P. G. E. F., Riezebos, H. T. (1971) Some sedimentological aspects of the fluvioglacial outwash plain near Soesterberg (The Netherlands) Geol. Mijnbouw 50, 341–348

Augustinus, P. G. E. F., Terwindt, J. H. J. (1972) Sediments of the middle Mause Esker, Scotland. (Personal communication)

Axelsson, V. (1967) The Laiture delta, a study of deltaic morphology and processes. Geogr. Ann. 49, 127 p.

Bärtling, R. (1905) Der Ås am Neuenkircnener See an der mecklenburgisch-lauenburgischen Landesgrenze. Jahrb. Preuß. Geol. Landesanst. 26, 15–25

Bagnold, R. A. (1946) Motion of waves in shallow water. Interaction between waves and sand bottoms. Prov. Roy. Soc. (London), Ser. A 187, 1–18

Bagnold, R. A. (1954 a) The physics of blown sand and desert dunes, 265 p. London: Methuen

Bagnold, R. A. (1954 b) Experiments on a gravity-free dispersion of large solid spheres in a Newtonian fluid under shear. Proc. Roy. Soc. (London), Ser. A 225, 49–63

Bagnold, R. A. (1956) The flow of cohesionless grains in fluids. Phil. Trans. Roy. Soc. Ser. A 249, 235–297

Bagnold, R. A. (1960) Some aspects of river meanders. U. S. Geol. Surv. Profess. Papers 283-E, 135–144

Bagnold, R. A. (1962) Auto-supsension of transported sediment; turbidity currents. Royal Soc. Proc., ser. A 265, 315–319

Bagnold, R. A. (1963) Beach and nearshore processes. In: Hill, M. N., ed., The sea; vol. 3, p. 507–528. New York: Wiley & Sons

Baker, E. T. (1976) Temporal and spatial variability of the bottom nepheloid layer over a deep-sea fan. Marine Geol. 21, 67–79

Baker, S. R., Friedman, G. M. (1973) Sedimentation in an arctic marine environment: Baffin Bay between Greenland and the canadian arctic archipelago. Geol. Surv. Can. 71–23 471–498

Baker, V. R. (1974) Paleohydraulic interpretation of Quaternary alluvium near Golden, Colorado. Quaternary Res. 4, 94–112

Baker, V. R., Ritter, D. F. (1975) Competence of rivers to transport coarse bedload material. Geol. Soc. Am. Bull. 86, 975–978

Baker-Blocker, A., Callender, E., Josephson, P. D. (1975) Trace-element and organic carbon content of surface sediment from Grand Traverse Bay, Lake Michigan. Geol. Soc. Am. Bull. 86, 1358–1362

Baldwin, C. T. (1974) The control of mud crack patterns by small gastropod trails. J. Sediment. Petrol. 44, 695–697

Ball, M. M. (1967) Carbonate sand bodies of Florida and the Bahamas, J. Sediment. Petrol. 37, 556–591

Banerjee, D. M. (1971) Precambrian stromatolitic phosphorites of Udaipur, Rajsthan, India. Geol. Soc. Am. Bull. 82, 2319–2330

Banerjee, I. (1966) Turbidites in a glacial sequence: A study of the Talchier Formation, Raniganj Coalfield, India. J. Geol. 74, 593–606

Banerjee, I. and McDonald, B. C. (1975) Nature of esker sedimentation: In Jopling, A. V. McDonald, B. C., eds., Glaciofluvial and Glaciolacustrine Sedimentation. Soc. Econ. Palaeontologists Mineralogists Spec. Publ. 23, 132–154

Banks, N. L. (1973) Tide-dominated offshore sedimentation, Lower Cambrian, North Norway, Sedimentology 20, 213–228

Banner, F. T., Collins, M. B., Massie, K. S. eds. (1979) The North-West European shelf seas: The sea-bed and the sea in motion. I. Geology and Sedimentology, 300 p. Amsterdam: Elsevier.

Baria, L. R. (1977) Desiccation features and the reconstruction of paleosalinities. J. Sediment. Petrol. 47, 908–914

Barnes, R. A., Barnes, W. C. (1978) Organic compounds in lake sediments. In: Lerman, A., ed., Lakes, Chemistry, Geology, Physics, 127–152. New York: Springer

Barrel, J. (1914) The Upper Devonian delta of the Appalachian geosyncline. Am. J. Sci. 37, 225–253

Bartberger, C. E. (1976) Sediment sources and sedimentation rates, Chincoteague Bay, Maryland and Virginia. J. Sediment. Petrol. 46, 326–336

Barthurst, R. G. C. (1976) Carbonate sediments and their diagenesis. 2nd enlarged ed., 658 p. Amsterdam: Elsevier

Bartow, J. A. (1966) Deep submarine channel in Upper Miocene, Orange County, California. J. Sediment. Petrol. 36, 700–705

Bartsch-Winkler, S., Ovenshine, A. T. (1975) Sedimentological maps of the Girdwood Bar, Turnagain Arm, Alaska for July-August 1974. Miscellaneous Field Studies Map MF-712

Bartsch-Winkler, S., Ovenshine, A. T., Lawson, D. E. (1975) Sedimentological maps of the Girdwood Bar, Turnagain Arm, Alaska for July-August 1973. Miscellaneous Field Studies Map MF-672

Barwis, J. H. (1976) Internal geometry of Kiawah Island beach ridges. In: Hayes, M. O., Kana, T. W., Terrigenous clastic depositional environments. Techn. Rep. No. 11-CRD, 115–125

Basan, P. B., Frey, R. W. (1977) Actual-palaeontology and neoichnology of salt marshes near Sapelo Island, Georgia. In: Crimes, T. P., Harper, J. C., eds., Trace fossils 2. Geol. J. Spec. Issue 2, 41–70, Liverpool: Seel House Press

Basazs, R. J., Klein, G. De V. (1972) Roundness-mineralogical relations of some intertidal sands. J. Sediment. Petrol. 42, 425–433

Bascom, W. N. (1951) The relationship between sand size and beach face slope. Trans. Am. Geophys. Un. 32, 866–874

Bass, N. W., Leatherrock, C., Dillard, W. R., Kennedy, B. E. (1937) Origin and distribution of Barlesville and Burbank shoestring oil sands in parts of Oklahoma and Kansas. Bull. Am. Assoc. Petroleum Geologists 21, 30–66

Baudoin, R. (1949) Observations sur les dépots alvéolaires des sables marins dans la région de Ronce-Les Bains (Charentes-Maritimes). Bull. Soc. Géol. France 19, 189–194

Beal, M. A., Shepard, F. P. (1956) A use of roundness to determine depositional environments. J. Sediment. Petrol. 26, 49–60

Beaumont, P. (1972) Aulluvial fans along the foothills of the Elburg Mountains, Iran. Palaeogeogr. Palaeoclimtol. Paleoecol. 12, 251–274

Beek, J. L. Van, Koster, E. A. (1972) Fluvial and estuarine sediments exposed along the Oude Mass (The Netherlands). Sedimentology 19, 237–256

Beel, D. L., Goodell, H. G. (1967) A comparative study of Glauconite and the associated clay fraction in modern marine sediments. Sedimentology 9, 169–202

Behrens, M. (1977) Zur Stereometrie von Geröllen. Mitt. Geol.-Paläont. Inst. der Univ. Hamburg 47, 1–124

Belderson, R. H., Kenyon, N. H. (1976) Long-range sonar views of submarine canyons. Mar. Geol. 22, M69–M74

Belderson, R. H., Stride, A. H. (1966) Tidal current fashioning of a basal bed. Marine Geol. 4, 237–257

Belderson, R. H., Wilson, J. B. (1973) Iceberg plough marks in the vicinity of the Norwegian Trough. Norsk Geologisk Tidsskr. 53, 323–328

Belderson, R. H., Kenyon, N. H., Stride, A. H., Stubbs, A. R. (1972) Sonographs of the Sea Floor, 185 p. Amsterdam: Elsevier

Belt, E. S. (1965) Stratigraphy and paleogeography of Mabou Group and related middle carboniferous facies. Nova Scotia, Canada. Geol. Soc. Am. Bull. 76, 777–802

Belt, E. S. (1968) Carboniferous continental sedimentation, Atlantic Provinces, Canada, in: Klein, G. de V., ed., Late paleozoic and mesozoic continental sedimentation, northeastern North America. Geol. Soc. Am. Spec. Papers 106, 127–176

Berger, W. H. (1974) Deep-Sea Sedimentation. In: Burk, C. A., Drake, C. L., eds., The geology of continental margins, p. 213–241. New York, Heidelberg, Berlin: Springer

Bernard, A. J. (1973) A review of processes leading to the formation of mineral deposits in sediments. In Amstutz, G. C., Bernard, A. J., eds., Ores in sediments, 350 p. Berlin, Heidelberg, New York: Springer

Bernard, H. A., Major, C. F. (1963) Recent meander belt deposits of the Brazos River: An alluvial "sand" model. Bull. Am. Assoc. Petrol. Geologists 47, 350 (Abstract)

Bernard, H. A., Major, C. F., Parrott, B. S., (1959) The Galveston Barrier Island and environs – a model for predicting reservoir occurrence and trend. Trans. Gulf Coast Assoc. Geol. Soc. 9, 221–224

Bernard, H. A., LeBlanc, R. J., Major, C. F. (1962) Recent and Pleistocene geology of southeast Texas. Geol. Gulf Coast and central Texas and guidebook of excursion. Houston Geol. Soc., 175–224

Berner, R. A. (1971) Principles of chemical sedimentology, 240 p. New York: McGraw Hill

Berner, R. A. (1976) The benthic boundary layer from the viewpoint of a geochemist. In: McCave, I. N., ed., The Benthic Boundary Layer, p. 33–55. New York: Plenum Press

Bernoulli, D., Jenkyns, H. C. (1974) Alpine, Mediterranean, and Central Atlantic Mesozoic facies in relation to the early evolution of the Tethys. In: Dott, R. H., Shaver, R. H., eds., Modern and ancient geosynclinal sedimentation. Soc. Econ. Paleontologists Mineralogists Spec. Publ. 19, 129–160

Berthois, L. (1956) Comportement du "bouchon vaseux" dans l'estuaire de la Loire. Rapport Quatrièmes J. de l'Hydraulique, p. 564–568

Berthois, L. (1978) Estuarine Sedimentation. In: Fairbridge, R. W., Bourgeois, J., eds., The Encyclopedia of Sedimentology. Earth Sci. 6, 288–292. Stroudsburg, Pa.: Dowden, Hutchinson, Ross.

Biedermann, E. W. (1962) Distinction of shoreline environments in New Jersey. J. Sediment. Petrol. 32, 181–200

Bigarella, J. J. (1965) Sand-ridge structures from Paraná Coastal Plain. Marine Geol. 3, 269–278

Bigarella, J. J. (1969) Dune sediments: characteristics, recognition, and importance. Bull. Am. Assoc. Petroleum Geologists 53, 707 (Abstract)

Bigarella, J. J. (1972) Eolian environments: their characteristics, recognition and importance. In Rigby, J. K., Hamblin, W. K., eds., Recognition of Sedimentary Environments. Soc. Econ. Paleontologists Mineralogists Spec. Publ. 16, 12–62

Bigarella, J. J., Salamuni, R. (1961) Early Mesozoic wind patterns as suggested by dune bedding in the Botucatú Sandstone of Brazil and Uruguay. Geol. Soc. Am. Bull. 72, 1089–1106

Bigarella, J. J., Alessi, A. H., Becker, R. D., Duarte, G. M. (1969) Textural characteristics of the coastal dune, sand ridge and beach sediments. Bol. Paranaense Geociencias 27, 15–80

Bigarella, J. J., Becker, R. D., Duarte, G. M. (1969) Coastal dune structures from Paraná (Brazil). Marine Geol. 7, 5–55

Biggs, R. B. (1978) Coastal Bays. In: Davis Jr., R. A., ed., Coastal Sedimentary Environments, p. 69–99. New York, Heidelberg, Berlin: Springer

Birch, G. F. (1977) Surficial sediments on the continental margin off the West coast of South Africa. Mar. Geol. 23, 305–337.

Bird, E. C. F. (1969) Coasts, 246 p. Cambridge, Mass: The M.I.T. Press

Birkeland, P. W. (1974) Pedology, Weathering and Geomorphological Research, 285 p. New York: Oxford Univ. Press.

Birkenmaier, K. (1959) Classification of bedding in flysch and similar graded deposits. Studia Geol. Polon. 3, 1–133

Blatt,H. Middleton, G. V., Murray, R. C. (1972) Origin of sedimentary rocks, 634 p. New Jersey: Prentice Hall.

Blissenbach, E. (1952) Relation of surface angle distribution to particle size distribution on alluvial fans. J. Sediment. Petrol. 22, 25–28

Blissenbach, E. (1954) Geology of alluvial fans in semi-arid regions. Geol. Soc. Am. Bull. 65, 175–190

Blohm, M. (1974) Sedimentpetrographische Untersuchungen am Neusiedler See/Österreich. Inaug. Diss. Univ. Heidelberg.

Bloos, G. (1976) Untersuchungen über Bau und Entstehung der feinkörnigen Sandsteine des Schwarzen Jura α (Hettangium und tiefstes Sinemurium) im schwäbischen Sedimentationsbereich. Arb. Inst. Geol. Paläont. Univ. Stuttgart, N. F. 71, 1–269

Bloos, G. (1982) Shell beds in the lower Lias of south Germany – facies and origin. In: Einsele, G., Seilacher, A., eds., Cyclic and event stratification, p. 223–239. Berlin, Heidelberg, New York: Springer

Bluck, B. J. (1964) Sedimentation of an alluvial fan in southern Nevada. J. Sediment Petrol. 34, 395–400

Bluck, B. J. (1967) Sedimentation of beach gravels: Examples from South Wales. J. Sediment. Petrol. 37, 128–156.

Bluck, B. J. (1971) Sedimentation in the meandering River Endrick. Scott. J. Geol. 7, 93–138.

Bluck, B. J. (1974) Structure and directional properties of some valley sandur deposits in Southern Island. Sedimentology 21, 533–554.

Bluck, B. J., Kelling, G. (1963) Channels from the upper carboniferous coal measures of South Wales. Sedimentology 2, 29–53

Boer, P. L. De (1979) Convolute lamination in mordern sands of the estuary of the Oosterschelde, the Netherlands, formed as a result of entrapped air. Sedimentology 26, 283–294

Boersma, J. R. (1967) Remarkable types of mega cross-stratification in the fluviatile sequence of a subrecent distributary of the Rhine. Amerongen; The Netherlands. Geol. Mijnbouw 46, 217–235

Boersma, J. R. (1969) Internal structure of some tidal mega-ripples on a shoal in the Westerschelde estuary, the Netherlands. Report of a preliminary investigation. Geol. Mijnbouw 48, 409–414

Boersma, J. R. (1970) Distinguishing features of wave-ripple cross-stratification and morphology. Doctoral Thesis, University of Utrecht. 65 p.

Boersma, J. R., Terwindt, J. H. J. (1980) Neap-spring tide sequences of intertidal shoal deposits in the Westerschelde Estuary, The Netherlands. Internat. Assoc. Sedimentologists Spec. Publ. 3, (in press). Oxford, London, Edinburgh, Melbourne: Blackwell

Boersma, J. R., Meene, E. A. van de, Tjalsma, R. C. (1968) Intricated cross-stratification due to interaction of a mega ripple with its lee-side system of backflow ripple (upper-pointbar deposits, lower Rhine). Sedimentology 11, 147–162

Bogárdi, J. L. (1961) Some aspects of the application of the theory of sediment transportation to engineering problems. J. Geophys. Res. 66, 3337–3346

Bogárdi, J. L. (1965) European concepts of sediment transport. J. Hydraulics Div., Am. Soc. Civil Engrs. 91, 29–54

Bokuniewcz, H. J., Gordon, R. B., Kastens, K. A. (1977) Form and migration of sand waves in a large estuary, Long Island sound. Mar. Geol. 24, 185–199

Bonatti, E., Arrhenius, G. (1965) Eolian sedimentation in the Pacific Ocean off northern Mexico. Marine Geol. 3, 337–348

Boothroyd, J. C. (1969) Crane beach. Field trip guidebook coastal environments of northeastern Massachusetts and New Hampshire, Contribution No. 1–CRG, Geology Department, University of Massachusetts, Amherst: Mass. 146–173

Boothroyd, J. C. (1978) Mesotidal inlets and estuaries. In: Davis Jr., R. A., ed., Coastal Sedimentary Environments, p. 287–360; New York, Heidelberg, Berlin: Springer

Bornhauser, M. (1948) A possible ancient submarine canyon in southwestern Louisiana. Bull. Am. Assoc. Petroleum Geologists 32, 2287

Bott, M. H. P., ed. (1976) Sedimentary basins of continental margins and cratons. Tectonophysics 36, 314 p.

Boulton, G. S. (1972) The role of thermal regime in glacial sedimentation. Inst. Brit. Geogr. Spec. Publ. 4, 1–19

Boulton, G. S. (1972) Modern Arctic glaciers as depositional models for former ice sheets. J. Geol. Soc. London 128, 361–393

Boulton, G. S. (1974) Processes and patterns of glacial erosion. In: Coates, D., ed., Glacial Geomorphol., p. 41–87. New York: Binghamton

Boulton, G. S. (1975) Processes and patterns of subglacial sedimentation. In: Wright, A. E., Moseley, F., eds., Ice Ages: Ancient and Modern, p. 7–42. Liverpool: Seel House Press

Boulton, G. S. (1978) Boulder shopes and grainsize distributions of debris as indicators of transport paths trough a glaciar and till genesis. Sedimentology 25, 773–799

Boulton, G. S., Dent, D. L. (1974) The nature and rates of postdepositional changes in recently deposited till from southeast Iceland. Geogr. Annal. 56A, 121–134

Bouma, A. H. (1962) Sedimentology of some flysch deposits, 168 p. Amsterdam: Elsevier

Bouma, A. H. (1963) A graphic presentation of the facies model of salt marsh deposits. Sedimentology 2, 122–129

Bouma, A. H. (1964) Ancient and recent turbidites. Geol. Mijnbouw 43 e, 375–379

Bouma, A. H. (1965) Sedimentary characteristic of samples collected from some submarine canyons. Marine Geol. 3, 291–320

Bouma, A. H. (1968) Distribution of minor structures in Gulf of Mexico sediments. Trans. Gulf Coast Assoc. Geol. Soc. 18, 26–33

Bouma, A. H. (1972) Fossil contourites in Lower Niesenflysch, Switzerland. J. Sediment. Petrol. 42, 917–921.

Bouma, A. H. (1973a) Contourites in Niesenflysch, Switzerland. Eclogae geol. Helv. 66, 315–323.

Bouma, A. H. (1973b) Leveed-channel deposits, turbidites, and contourites in deeper part of Gulf of Mexico. Transactions-Gulf Coast Assoc. Geol. Soc. 23, 368–376.

Bouma, A. H., Brouwer, A., eds. (1964) Turbidites. Developments in sedimentology, vol. 3, 264 p. Amsterdam: Elsevier

Bourcart, J. (1964) Les sables profonds de la Méditerranée occidentale, In: Bouma, A. H., Brouwer, A., eds., Turbidites. P. 145–155. Developments in sedimentology, vol. 3, Amsterdam: Elsevier

Boy, J. A. (1977) Typen und Genese jungpaläozoischer Tetrapodenlagerstätten. Paläontogr. Abt. A. 156, 111–167

Boyd, W. B., Ore, H. T. (1963) Patterned cones in Permo-Triassic Redbeds of Wyoming and adjacent areas. J. Sediment. Petrol. 33, 438–451

Bradacs, L. K., Ernst, W. (1956) Geochemische Korrelationen im Steinkohlenbergbau. Naturwissenschaften 43, 33

Bradley, W. H. (1931) Origin and microfossils of the oil shale of the Green River Formation of Colorado and Utah. U.S. Geol. Surv. Profess. Papers 168, 58 p.

Bradley, W. H. (1964) Geology of Green River Formation and associated Eocene rocks in southwestern Wyoming and adjacent parts of Colorado and Utah. U.S. Geol. Surv. Profess. Papers 496-A, 86 p.

Bramer, H. (1965) Bestimmung der Oberflächenbeschaffenheit von Quarzkörnern mit dem Elektronenmikroskop. Geologie 14, 1114–1117

Braun, E. von (1953) Geologische und sedimentpetrographische Untersuchungen im Hochrheingebiet zwischen Zurach und Eglisau. Eclogae Geol. Helv. 46, 143–170

Bremer, H. (1965) Musterböden in tropisch-subtropischen Gebieten und Frostmusterböden. Z. Geomorphol., N. F. 9, 222–236

Brenner, R. L., Davies, D. K. (1973) Storm-generates coquinoid sandstone: Genesis of high energy marine sediments from the upper Jurassic of Wyoming and Montana. Geol. Soc. Am. Bull. 84, 1685–1698

Brenninkmeyer, B. M. (1976) Sand fountains in the surf zone. In: Davis, Jr. R. A., Ethington, R. L., eds., Beach and nearshore sedimentation. Soc. Econ. Paleontologists Mineralogists Spec. Publ. 24, 69–91

Bridge, J. S. (1975) Computer simulation of sedimentation in meandering streams. Sedimentology 22, 3–43

Bridge, J. S. (1978 a) Origin of horizontal lamination under turbulent boundary layers. Sediment. Geol. 20, 1–16

Bridge, J. S. (1978 b) Palaeohydraulic interpretation using mathematical models of contemporary flow and sedimentation in meandering channels. In: Miall, A. D., ed., Fluvial Sedimentology. Can. Soc. Petrol. Geol., Mem. 5, 723–742

Bridges, E. M. (1970) World Soils, 89 p. Cambridge: Cambridge Univ. Press

Bridges, P. H., Leeder, M. R. (1976) Sedimentary model for intertidal mudflat channels, with examples from the Solway Firth, Scotland. Sedimentology 23, 533–552

Brock, E. J. (1974) Coars sediment morphometry: a comparative study. J. Sediment. Petrol. 44, 663–672

Bromley, R. G. (1967) Marine phosphorites as depth indicators. Marine Geol. 5, 503–509

Brongersma-Sanders, M. (1957) Mass mortality in the sea. Geol. Soc. Am. Mem. 67, 941–1010

Brookfield, M. E. (1977) The origin of bounding surfaces in ancient aeolian sandstone. Sedimentology 24, 303–332

Broussard, M. L. (1975) Deltas, Models for exploration, 555 p. Houston: Houston Geol. Soc.

Brown, C. B. (1938) On a theory of gravitational sliding applied to the Tertiary of Ancon, Ecuador, Geol. Soc. London Quart. J. 94, 359–370

Bruns, E. (1955) Handbuch der Wellen der Meere und Ozeane, 255 p. Berlin: VEB Dtsch. Verl. Wiss.

Bruns, E. (1958) Ozeanologie, Bd. 1, 420 p. Berlin: VEB Dtsch. Verl. Wiss.

Brush, L. M. (1965) Sediment sorting in alluvial channels. In: Middleton, G. V., ed. Primary sedimentary structures and their hydrodynamic interpretation, Special Publication 12, 25–33. Tulsa, Oklahoma: Soc. Econ. Paleont. Mineral.

Bryan, K. (1945) Glacial versus desert origin of loess. Am. J. Sci. 143, 245–248

Bucher, W. H. (1919) On ripples and related sedimentary surface forms and their paleogeographic interpretations. Am. J. Sci. 47, 149–210, 241–269

Bukry, D., King, S. A., Horn, M. K., Manheim, F. T. (1970) Geological significance of coccoliths in fine-grained carbonate bands of post-glacial Black Sea sediments. Nature 226, 156–158

Bull, W. B. (1960) Types of deposition on alluvial fans in western Fresno County. California. Geol. Soc. Am. Bull. 71, 2052 (Abstract)

Bull, W. B. (1962) Relation of textural (CM) patterns to deposition environment of alluvial fan deposits. J. Sediment Petrol. 32, 211–216

Bull, W. B. (1964 a) Alluvial fans and near surface subsidence in Western Fresno County, California. U. S. Geol. Surv. Profess. Papers 437 A, 1–71

Bull, W. B. (1964 b) Geomorphology of segmented alluvial fans in western Fresno County, California. U. S. Geol. Surv. Profess. Papers 352 E, 89–129

Bull, W. B. (1972) Recognition of alluvial-fan deposits in the stratigraphic record. In: Rigby, J. K., Hamblin, W. K., eds., Recognition of ancient sedimentary environments. SEPM. Spec. Publ. 16, 63–83

Buller, A. T. (1974/75) 5. Sediments of the Tay estuary. II formation of ephemeral zones of high suspended sediment concentrations. Proc. Roy. Soc. Edinb. B 75, 67–89

Buller, A. T., Green, C. D. (1976) The role of organic detritus in the formation of distinctive sandy tidal flat sedimentary structures, Tay Estuary, Scotland. Estuarine and Coastal marine Sci. 4, 115–118

Buller, A. T., McManus, T. (1972) Simple metrie sedimentary statistics used to racognize different environment. Sedimentology 18, 1–21

Buller, A. T., McManus, J. (1974/75) Sediments of the Tay Estuary. I. Bottom sediments of the upper and upper middle reaches. Proc. Roy. Soc. Edinb. B. 75, 41–64

Burk, C. A., Drake, C. L., eds. (1974) The geology of continental margins, 1009 p. New York, Heidelberg, Berlin: Springer

Burst, J. F. (1965) Subaqueously formed shrinkage cracks in clay. J. Sediment. Petrol. 35, 348–353

Busch, D. A. (1953) The significance of deltas in subsurface exploration. Tulsa Geol. Soc. Dig. 21, 71–80

Busch, D. A. (1959) Prospecting for stratigraphic traps. Bull. Am. Assoc. Petrol. Geologists 43, 2829–2843

Busch, D. A. (1974) Stratigraphic traps in sandstones – exploration techniques. Am. Assoc. Petrol. Geol. Mem. 21, Tulsa, Oklahoma

Bussche, H. K. J. van den, Houbolt, J. J. H. C. (1964) A corer for sampling shallow-marine sands. Sedimentology 3, 155–159

Busson, G. (1968) Les sables ronds-mats, émoussés-luisants et non-usés observés au microscope électronique à balayage (Stereoscan). Bull. Mus. Natl. Hist. Nat., Paris 40, 850–856

Butrym, J., Cegla, J., Dzulynski, S., Nakonieczny, S. (1964) New interpretation of "Periglacial structures". Folia Quaternaria 17, 1–34

Button, A. Vos, R. G. (1977) Subtidal and intertidal clastic and carbonate sedimentation in a macrotidal environment: An example form the Lower Proterozoic of South Africa. Sediment. Geol. 18, 175–200

Byers, C. W. (1977) Biofacies patterns in euxinic basins: a general model. Soc. Econ. Paleontologists Mineralogists Spec. Publ. 25, 5–17

Byrne, J. V. (1963) Variations in fluvial gravel imbrication. J. Sediment. Petrol. 33, 467–469

Byrne, J. V., Emery, K. O. (1960) Sediments of the Gulf of California. Geol. Soc. Am. Bull. 71, 983–1010

Byrne, J. V., LeRoy, D. O., Riley, C. M. (1959) The chenier plain and its stratigraphy, southwestern Louisiana: Gulf Coast Association Geological Society 9th Annual Meeting, Houston, Trans. 9, 237–260

Cailleux, A. (1942) Les actions éoliens périglaciaires en Europe. Mém. Soc. Géol. France 46, 1–176

Cailleux, A. (1943) Distinction des sables marins et fluviatiles. Bull. Soc. Géol. France 13, 125–138

Cailleux, A. (1945) Distinction des galets marins et fluviatiles. Bull. Soc. Géol. France 15, 375–404

Cailleux, A. (1947) Granulométrie des formations à galets. Sess. Extraord. Soc. Belg. Soc. Géol. 1947, 91–114

Cailleux, A. (1952) Morphoscopische Analyse der Geschiebe und Sandkörner und ihre Bedeutung für die Paläoklimatologie. Geol. Rundschau 40, 11–19

Cailleux, A., Schneider, H. E. (1968) L'usage des sables vue an microscope électronique à balayage. Sci. Progr. 3395, 92–94

Caine, N. (1972) Airphoto analysis of blockfield fabrics in Talus Valley, Tasmania, J. Sediment. Petrol. 42, 33–48

Calvert, S. E. (1964) Factors affecting distribution of laminated diatomaceous sediments in the Gulf of California. Marine Geology of the Gulf of California: A Symposium, Memoir. 3, 311–330

Campbell, C. V. (1966) Truncated wave-ripple laminae. J. Sediment. Petrol. 36, 825–828

Campbell, C. V. (1967) Lamina, laminaset, bed and bedset. Sedimentology 8, 7–26

Campbell, C. V. (1971) Depositional model – upper Cretaceous Gallup beach shoreline, Ship Rock Area, northwestern New Mexico. J. Sediment. Petrol. 41, 395–409

Campbell, C. V. (1976) Reservoir geometry of a fluvial sheet sandstone. Bull. Am. Ass. Petroleum Geol. 60, 1009–1020

Campbell, C. V. (1979) Model for beach shoreline in Gullup Sandstone (Upper Cretaceous) of northwestern New Mexico. New Mexico Bur. Min. Miner. Resources, Circ. 164, 32 p.

Carey, S. W., Ahmad, N., (1961) Glacial marine sedimentation. In: Raasch, G. O., ed., The geology of the Arctic, 2, 865–894 Toronto: University of Toronto Press

Carter, C. H. (1978) A regressive barrier and barrier – protected deposit: Depositional environments and geographic setting of the late Tertiary Cohansey sand. J. Sediment. Petrol. 48, 933–950

Carter, L., Heath, R. A. (1975) Role of mean circulation, tides, and waves in the transport of bottom sediment on the New Zealand continental shelf. N.Z. J. Mar. and Freshwater Res. 9, 423–448

Carter, R. M. (1975) A discussion and classification of subaqueons mass-transport-with particular application to grain flow, slurry flow and fluxoturbidites. Easth. Sci. Rev. 11, 145–177

Cartwright, D. E., Stride, A. H. (1958) Sand transport at the southern end of the North Sea. Dock. Harb. Auth. 38, 323–324

Casshyap, S. M. (1970) Sedimentary cycles and environment of deposition of the Barakar Coal Measures of Lower Gondwana, India. J. Sediment. Petrol. 40, 1302–1317

Casshyap, S. M. (1973) Paleocurrents and paleogeographic reconstruction in the Barakar (Lower Gondurana) sandstones of Peninsular India. J. Sediment. Geol. 9, 283–303

Casshyap, S. M., Quidwai, H. A. (1974) Glacial sedimentation of Late Paleozoic Talchir Diamictite, Peneh Valley Coalfield, Central India. Geol. Soc. Am. Bull. 85, 749–760

Caswy, S. R., Cantrell, R. B. (1941) Davis sand lens, Hardin Field, Liberty County, Texas. In: Levorsen, A. I., ed., Stratigraphic type oil fields, p. 564–570. Tulsa, Oklahoma; Am. Assoc. Petroleum Geologists

Čepek, P., Reineck, H.-E. (1970) Form und Entstehung von Rieselmarken im Watt- und Strandbereich. Senckenbergiana Marit. 2, 3–30

Chamberlain, T. K. (1964) Mass transport of sediment in the heads of Scripps Submarine canyon, California. In: Miller, R. L., ed., Papers in Marine Geology (Shepard commemorative volume) p. 42–64. New York: Macmillan.

Chappell, J. (1967) Recognizing fossil strand lines from grain-size analysis. J.Sediment. Petrol. 37, 157–165

Chow, V. T. (1959) Open-channel hydraulics, 680 p. New York: McGraw-Hill

Chowdhuri, K. R., Reineck, H.-E. (1978) Primary sedimentary structures and their sequence in the shoreface of barrier island Wangerooge (North Sea). Senckenbergiana marit. 10, 15–29

Church, M. (1972) Baffin Island sandurs. Geol. Surv. Canada, Bull. 216, 208 p.

Church, M. (1978) Palaeohydrological reconstructions from a Holocene valley fill. In: Miall, A. D., ed., Fluvial Sedimentology. Can. Soc. Petrol. Geol., Mem. 5, 743–772

Church, M., Gilbert, R. (1975) Proglacial and lacustrine environments. In: Jopling, A. V., McDonald, B. C., eds., Glaciofluvial and Glaciolacustrine sedimentation. Soc. Econ, Paleontologists Mineralogists Spec. Publ. 23, 22–100

Clague, J. J. (1975) Sedimentology and Paleohydrology of Late Wisconsinon outwash, Rocky Mountain trench, southeastern British Columbia. In: Jopling, A. V., McDonald, B. C., eds., Glaciofluvial and Glaciolacustrine Sedimentation. Soc. Econ. Palaeontologists Mineralogists Spec. Publ. 23, 223–237

Cleary, W. J., Conolly, J. R. (1974) Hatteras deep-sea fan. J. Sediment. Petrol. 44, 1140–1154

Clifton, H. E. (1969) Beach lamination: Nature and origin. Marine Geol. 7, 553–559

Clifton, H. E. (1973) Pebble segregation and bed lenticularity in wave-worked versus alluvial gravel. Sedimentology 20, 173–183

Clifton, H. E. (1976) Wave-formed sedimentary structures – A conceptual model. In: Davis, Jr. R. A., Ethington, R. L., eds., Beach and nearshore sedimentation. Soc. Econ. Paleontologists Mineralogists Spec. Publ. 24, 126–146

Clifton, H. E. (1977) Rain-impact ripples. J. Sediment. Petrol. 47, 678–679

Clifton, H. E., Hunter, R. E., Phillips, R. L. (1971) Depositional structures and processes in the non-barred high energy nearshore. J. Sediment. Petrol. 41, 651–670

Cloud, P. E. (1966) Beach cusps: Response to Plateau's Rule? Rep. Sci. 154, 890–891

Coastal Research Group, University of Massachusetts (1969) Field trip guidebook, coastal environments of northeastern Massachusetts and New Hamsphire. Contribution 1–CRG, Geology Department University of Massachusetts. Amherst, Mass, 462 p.

Colburn, I. P. (1968) Grain fabrics in turbidite sandstone beds and their relationship to sole mark trends on the same beds. J. Sediment. Pentrol. 38, 146–158

Colby, B. R. (1964) Discharge of sands and mean-velocily relationships in sand-bed streams. U. S. Geol. Surv. Prof. Paper 462–A, 1–47

Colby, B. R., Hembree, C. H. (1955) Computations of total sediment discharge Niobrare River near Cody, Nebraska. U. S. Geol. Surv. Water Supply Paper 1357, 1–187

Coleman, J. M. (1966) Ecological changes in a massive freshwater clay sequence. Trans. Gulf Coast Assoc. Geol. Soc. 16, 159–174

Coleman, J. M. (1968) Deltaic evolution. In Fairbridge, R., ed., Encyclopedia of geomorphology, p. 255–261. New York: Reinhold.

Coleman, J. M. (1969) Brahmaputra River: Channel processes and sedimentation. Sediment. Geol. 3, 129–239

Coleman, J. M. (1976) Deltas: Processes of deposition and models for exploration, 102 p. Champaign: Continuing Education Publ. Comp., Inc.

Coleman, J. M., Gagliano, S. M. (1965) Sedimentary structures: Mississippi River deltaic plain. In: Middleton, G. V., ed., Primary sedimentary structures and their hydrodynamic interpretation. Soc. Econ. Paleontologists Mineralogists Spec. Publ. 12, 133–148

Coleman, J. M., Gagliano, S. M., Webb, J. E. (1964) Minor sedimentary structures in a prograding distributary. Marine Geol. 1, 240–258

Collinson, J. D. (1978) Alluvial sediments. In: Reading, H. G., ed., Sedimentary Environments and Facies, p. 15–60. Oxford, London, Edinburgh, Melbourne: Blackwell

Colquhoun, D. J. (1969 a) Coastal plain terraces in the Carolinas and Georgia, U. S. A. Quaternary Geol. Climate, Publ. 1701, 150–162

Colquhoun, D. J. (1969 b) Geomorphology of the lower coastal plain of South Carolina. Div. Geol. State Dev. Board MS-15, 1–36

Congrès International de Sédimentologie, 5. (1958) Eclogae Geol. Helv. 51, 485–1172

Conrey, B. L. (1967) Early Pliocene sedimentary history of the Los Angeles Basin, California. Spec. Rep. 93, 1–63 San Francisco: California Division Mines Geology.

Conybeare, C. E. B., Crook, K. A. W. (1968) Manual of sedimentary structures, 327 p. Bureau of Mineral Resources, Geology and Geophysics, Canberra A. C. T., Bull. No. 102.

Cooper, W. S. (1958) Coastal sand dunes of Oregon and Washington. Geol. Soc. Am. Mem. 72, 1–169

Cooper, W. S. (1967) Coastal dunes of California. Geol. Soc. Am. Mem. 104, 1–131

Cornish, V. (1914) Waves of sand and snow, 383 p. London: T. F. Univin.

Correns, C. W. (1950) Zur Geochemie der Diagnese. Geochim. Cosmochim. Acta 1, 49–54

Costello, W. R. (1974) Development of bed configurations in coarse sands. Mass. Inst. Techn. Exper. Sediment. Lab. Rept. 74–2, 120 p.

Costello, W. R., Southard, J. B. (1974) Development of sand bar configurations in coarse sands. Am. Assoc. Petroleum Geologists Ann. Meeting Abstr. 1, 20–21

Cotter, E. (1971) Paleoflow characteristics of a late Cretaceous river in Utah from analysis of sedimentary structures in the Ferron sandstone. J. Sediment. Petrol. 41, 129–138

Cotter, E. (1978) The evolution of fluvial style, with special reference to the central Appalachian Paleozoic. In: Miall, A. D., ed., Fluvial Sedimentology. Can. Soc. Petrol. Geol., Mem. 5, 361–383

Cousteau, J. (1963) The living sea, 325 p. New York: Harper

Crimes, T. P. (1975) The stratigraphical significance of trace fossils. In: Frey, R. W., edy The Study of Trace Fossils, 562 p. New York, Heidelberg, Berlin: Springer

Cronan, D. S. (1977) Deep-sea nodules: Distribution and geochemistry. In: Glasby, G. P., ed., Marine manganese deposits, p. 11–44. New York: Elsevier

Cronan, D. S., Thomas, R. L. (1972) Geochemistry of ferromanganese oxide concretions and associated deposits in Lake Ontario. Geol. Soc. Am. Bull. 83, 1493–1502

Crosby, E. J. (1972) Classification of sedimentary environments. In: Rigby, J. K., Hamblin, W. K., eds., Recognition of ancient sedimentary environments. Soc. Econ. Paleontologists Mineralogists Spec. Publ. 16, 4–11

Crowell, J. C. (1957) Origin of pebbly mudstones. Geol. Soc. Am. Bull. 68, 993–1010

Crowell, J. C., Hope, R. A., Kahle, J. E., Ovenshine, A. T., Samps, R. H. (1966) Deep-water sedimentary structures, Pliocene Pico Formation., Santa Paula Creek, Ventura Basin, California. Calif. Div. Mines Geol. Spec. Rep. 89, 1–40

Cruickshank, M. J. (1970) Mining and mineral recovery 1969. Under sea Technology Handbook. A 11–21, Reprint E7

Curray, J. R. (1956) Dimensional grain orientation studies of recent coastal sands. Bull. Am. Assoc. Petroleum Geologists 40, 2440–2456

Curray, J. R. (1964) Transgressions and regressions: In: Miller, R. L., ed., Papers in marine geology (Shepard Commemorative volume), p. 175–203. New York: Macmillan

Curray, J. R. (1969 a) Estuaries, lagoons, tidal flats and deltas. In: Stanley, D. J., ed., The new concepts of continental margin sedimentation, JC-3, 1–30. Washington: American Geological Institute

Curray, J. R. (1969 b) Shallow structure of the continental margin. In: Stanley, D. J., ed., The new concepts of continental margin sedimentation, JC-12, 1–22. Washington: American Geological Institute

Curray, J. R., Moore, D. G. (1964) Pleistocene deltaic progradation of continental terrace. Costa de Nayarit, Mexico. In: Andel, Tj. H., van, Shor, G. G. Jr., ed., Marine geology of the Gulf of California. Am. Assoc. Petroleum Geol., Tulsa, Oklahoma, p. 193–215

Curray, J. R., Moore, D. G. (1971) Growth of the Bengal deep-sea fan and denudation in the Himalayas. Geol. Soc. Am. Bull. 82, 563–572

Curray, J. R., Moore, D. G. (1974) Sedimentary and Tectonic Processes in the Bengal Deep-Sea Fan and Geosyncline. In: Burk, C. A., Drake, C. L., eds., The Geology of Continental Margins, p. 617–627. New York, Heidelberg, Berlin: Springer

Curray, J. R., Emmel, F. J., Crampton, P. J. S. (1969) Holocene history of strand plain, lagoonal coast, Nayarit, Mexico. In: Castanares, A. A., Phleger, F. B. eds., Coastall Lagoons, a Symposium, p. 63–100. Mexico: Universidad Nacional Autónoma

Curtis, B. F., and Sandstone Reservoir Committee (1961) Characteristic of sandstone reservoirs in United States. In: Peterson, J. A., Osmond, J. C., eds., Geometry of sandstone bod-

ies, p. 208 219. Tulsa, Oklahoma: Amer. Assoc. of Petroleum Geologists

Curtis, D. M. (1970) Miocene deltaic sedimentation, Louisiana Gulf Coast. In: Morgan, J. P., ed., Deltaic sedimentation modern and ancient. Soc. Econ. Paleontologists Mineralogists Spec. Publ. 15, 293–308

Curtis, D. M., ed. (1978) Depositional Environments and Paleoecology: Environmental Models in Ancient Sediments. Reprint Series Number 6, 240 p. Tulsa, Oklahoma

Daboll, J. M. (1969) Holocene sediments of the Parker River estuary Massachusetts. Contr. No. 3-CRG, Dept. of Geol., Univ. Mass, 138 p.

Dalrymple, R. A., Lannan, G. A. (1976) Beach cusps formed by intersecting waves. Geol. Soc. Am. Bull. 87, 57–60

Dalrymple, R. W., Knight, R. J., Middleton, G. V. (1975) Intertidal sand bars in Cobequid Bay (Bay of Fundy). In: Gronin, L. E., ed., Estuarine Research. 2, 293–307. New York, San Francisco, London: Academic Press

Dalrymple, R. W., Knight, R. J., Lambiase, J. J. (1978) Bedforms and their hydraulic stability relationships in a tidal environment, Bay of Fundy, Canada. Nature 275, 100–104

Daly, R. A. (1936) Origin of submarine "canyons". Am. J. Sci. 31, 401–420

Dangeard, L., Vanney, J. R. (1975) Essai de classification des dépôts glacio-marins d'après l'examen des photographies sous-marines. Ann. Intst. Oceanogr. 51, 143–153

Dangeard, L., Migniot, C., Larsonneur, C., Baudet, P. (1964) Figures et structures observées au cours du tassement des vases sous l'eau. Compt. Rend. 258, 5935–5938

D'Anglejan, B., Brisebois, M. (1978) Recent sediments of the St. Lawrence middle estuary. J. Sediment. Petrol. 48, 951–964

Dapples, E. C., Rominger, J. F. (1945) Orientation analysis of finegrained clastic sediments: A report of progress. J. Geol. 53, 246–261

Davidson-Arnott, R. G. D., Greenwood, B. (1976) Facies relationships on a barre coast, Kouchibouguac Bay, New Brunswick, Canada. In: Davis, Jr., R. A., Ethington, R. L., eds., Beach and nearshore sedimentation. Soc. Econ. Paleontologists Mineralogists Spec. Publ. 24, 149–168

Davies, D. K. (1965) Sedimentary structures and subfacies of a Mississippi River point bar. J. Geol. 74, 234–239

Davies, D. K., Ethridge, F. G., Berg, R. R. (1971) Recognition of barrier environments. Bull. Am. Assoc. Petroleum Geologists 55, 550–565

Davis, Jr., R. A., ed. (1978) Coastal Sedimentary Environments, 420 p. New York, Heidelberg, Berlin: Springer

Davis, Jr. R. A., Ethington, R. L., eds. (1976) Beach and nearshore sedimentation. Soc. Econ. Paleontologists Mineralogists Spec. Publ. 24, 187 p.

Davis, Jr. R. A., Fox, W. T., Hayes, M. O. Boothroyd, J. C. (1972) Comparison of ridge and runnel systems in tidal and non-tidal environments. J. Sediment. Petrol. 2, 413–421

Dawdy, D. R. (1961) Depth-discharge relations of alluvial streams—discontinuous rating curves. U. S. Geol. Surv., Water Supply Papers 1498-C, 1–16

Dean, W. E., Ghosh, S. K. (1978) Factors contributing to the formation fo ferromanganese nodules in Oneida Lake, New York. J. Res. U. S. Geol. Surv. 6, 231–240

Dean, W. E., Schreiber, B. C., eds. (1978) Marine Evaporites. Soc. Econ. Paleontologists Mineralogists Short Course No. 4, 188 p.

Deelman, J. C. (1972) On mechanisms causing birdseye structures. Neues Jahrb. Geol. Paläontol. Mh. 1972, 582–595

Deery, J. R., Howard, J. D. (1977) Origin and character of washover fans on the Georgia Coast, U. S. A. Transactions-Gulf Coast Association of Geological Societies. 27 259–271

Degens, E. T. (1968) Geochemie der Sedimente, 282 p. Stuttgart: Enke-Verlag

Degens, E. T., Epstein, S. (1964) Oxygen and carbon isotope ratios in coexisting calcites and dolomites from recent and ancient sediments. Geochim. Cosmochim. Acta 28, 23–44

Degens, E. T., Ross, D. A., eds. (1974) The Black Sea-Geology, chemistry and biology, 633 p. Tulsa, Oklahoma: Am. Assoc. Petroleum Geol.

Desborough, G. A. (1978) A biogenic-chemical stratified lake model for the origin of oil shale of the Green River Formation: An alternative to the Playa-lake model. Geol. Soc. Am. Bull. 89, 961–971

Dette, H. H. (1974) Über Brandungsströmungen im Bereich hoher Reynolds-Zahlen. Mitt. Leichtweiss-Inst. Wasserbau, Techn. Univ. Braunschweig. 41, 1–254

Dette, H. H., Führböter, A. (1975) Naturuntersuchungen an Brandungsstörmungen. Die Küste. 27, 1–7

Dickinson, K. A., Berryhill, H. L., Jr., Holmes, C. W. (1972) Criteria for recognizing ancient barrier islands. In: Rigby, J. K., Hamblin, W. K., eds. Recognition of ancient sedimentary environments. Soc. Econ. Paleontologists Mineralogists Spec. Publ. 16, 192–214

Dickinson, W. R., ed. (1974) Tectonics and sedimentation. Soc. Econ. Paleontologists Mineralogists Spec. Publ. 22, 204 p.

Dickinson, W. R., ed. (1975) Current concepts of depositional systems with applications for petroleum geology, 108 p. Bakersfield: San Joaquin Geol. Soc.

Dickinson, W. R. (1975 a) Deltaic deposits and cyclothems. In: Dickinson, W. R., ed., Current concepts of depositional systems with applications for petroleum geology. Bakersfield, San Jaquin Geol. Short Course, p. 5.1–5.8

Dickinson, W. R. (1975 b) Fluvial sediments of stream valleys and alluvial fans. In: Dickinson, W. R., ed., Current concepts of depositional systems with applications for petroleum geology. Bakersfield, San Jaquin Geol. Soc. Short Course, p. 1.1–1.6.

Didyk, B. M., Simoreit, B. R. T., Brassel, S. C., Eglinton, G. (1978) Organic geochemical indicators of palaeoenvironmental conditions of sedimentation. Nature 272, 216–222

Diessel, C. F. K. (1966) The determination of the direction of transport of fluviatile arenites by orientation analyses of the detrital mica. Sedimentology 7, 167–177

Diester-Haas, L., Schrader, H.-H. (1979) Neogene coastal upwelling history off Northwest and Southwest Africa. Mar. Geol. 29, 39–53

Dietrich, G., Kalle, K. (1957) Allgemeine Meereskunde, 492 p. Berlin: Gebr. Borntraeger

Dijkema, K. S. (1980) Large-scale geomorphologic pattern of the Wadden Sea area. In: Dijkema, K. S., Reineck, H.-E., Wolff, W. J., eds., Geomorphology of the Wadden Sea area, Rep. 1, 72–80. Leiden, Stichting Veth tot Steun aan Waddenonderzeok

Dill, R. F. (1964) Sedimentation and erosion in Scripps submarine canyon head. In: Miller, R. L., ed., Papers in marine geology (Shepard Commemorative volume), p. 23–41. New York: Macmillan

Dillard, W. D., Oak, D. P., Bass, N. W. (1941) Chanute Oil pool. Neosho County, Kansas. In: Levorsen A. I., ed., Stratigraphic type oil fields. p. 57–77. Symp. Am. Assoc. Petroleum Geol. 57

Dillo, H.-G. (1960) Sandwanderungen in Tideflüssen. Mitt. Franzius Inst. Grund-Wasserbau, T. H. Hannover 17, 135–253

Dionne, J.-C. (1969) Tidal flat erosion by ice at la Pocatière, St. Lawrence Estuary. J. Sediment. Petrol. 39, 1174–1181

Dionne, J.-C. (1970) Aspects morpho-sédimentologiques du glaciel, en particulier des côtes du Saint-Laurent, Québec. Can. Rech. For. Laurentides- Rapp. Infor. Q-FX-9, 324 p.

Dionne, J.-C. (1971) Erosion glacielle de la slikke, estuaire du Saint-Laurent. Rev. Géomorph. Dyn. 20, 5–21

Dionne, J.-C. (1972 a) Caracterisation des schorres des régions froides, en particulier de l'estuaire du Saint-Laurent. Z. Geomorph. N. F. 13, 131–162

Dionne, J.-C. (1972 b) Vocabulaire du glaciel (Drift ice terminology) Centre de Recherches Forestières des Laurentides, Région de Québec, Québec. Rapport d-inform. Q-F-X-34, p. 1–47

Dionne, J. C. (1973) Monroes: A type of so-called mud volcanoes in tidal flats. J. Sed. Petrol. 43, 848–856

Dionne, J.-C. (1974) How drift ice shapes the St. Lawrence. Can. Geogr. J. 88, 4–9

Dionne, J.-C., ed. (1976 a) Le glaciel. Rev. Geôgr. Montr. 30, 236 p. Montréal: Presses de l'Universitède

Dionne, J.-C. (1976 b) Le glaciel de la région de la grande Rivière, Quebec subarctique. Rev. Geogr. Montr. 30, 133–153

Dionne, J.-C., Shilts, W. W. (1974) A Pleistocene clastic dike, upper Chaudière valley, Québec, Can. J. Easth Sci. 11, 1594–1605

Dmitrieva, E. V., Erchova, G. I., Orechnikova, E. I. (1962) Structures and textures of sedimentary rocks, 1. Clastic and Argillaceous Rocks. Gosgeoltekhizdat, Moskau

Dobkins, Jr. J. E., Folk, R. L. (1970) Shape development on Tahiti-Nui. J. Sediment. Petrol. 40, 1167–1203

Doeglas, D. J. (1946) Interpretation of the results of mechanical analysis. J. Sediment. Petrol. 16, 19–40

Doeglas, D. J. (1949) Loess, an aeolian product. J. Sediment Petrol. 19, 112–117

Doeglas, D. J. (1954) The origin and destruction of beach ridges. Leidse Geol. Mededel. 20, 34–47

Doeglas, D. J. (1962) The structure of sedimentary depositis of braided rivers. Sedimentology 1, 167–190

Doeglas, D. J. (1968) Grain-size indices, classification and environment. Sedimentology 10, 83–100

Dörjes, J. (1970) Das Watt als Lebensraum. In: Reineck, H.-E., ed. Das Watt, Ablagerungs- und Lebensraum, p. 71–105. Frankfurt: Kramer

Dörjes, J. (1971) Der Golf von Gaeta (Tyrrhenisches Meer) IV. Das Makrobenthos und seine küstenparallele Zonierung. Senckenbergiana marit. 3 203–24

Dörjes, J. (1972) Georgia coastal region, Sapelo Island, U.S.A.: Sedimentology and biology. VII. Distribution and zonation of macrobenthic animals. Senckenbergiana marit. 4, 183–216

Dörjes, J. (1975) Biological aspects. In: Dörjes, J., Hertweck, H., Recent biocoenoses and ichnocoenoses in shallow water marine environments. In: Frey, R. W., ed., The study of trace fossils, p. 460–468. New York, Heidelberg, Berlin: Springer

Dörjes, J. (1976) Primärgefüge, Bioturbation und Makrofauna als Indikatoren des Sandwatzes im Seegebiet vor Norderney (Nordsee). II. Zonierung und Verteilung der Makrofauna. Senckenbergiana marit. 8, 171–188

Dörjes, J. (1978) Sedimentologische und faunistische Untersuchungen an Watten in Taiwan. II. Faunistische und aktuopaläontologische Studien. Senckenbergiana marit. 10, 117–143

Dörjes, J., Hertweck, G. (1975) Recent biocoenoses and ichnocoenoses in shallow-water marine environments. In: Frey, R. W., ed., The study of trace fossils. Chap. 20, 459–491. New York: Springer

Dörjes, J., Gadow, S., Reineck, H.-E., Singh, I. B. (1969) Die Rinnen der Jade (Südliche Nordsee). Sedimente und Makrobenthos. Senckenbergiana marit. 1, 5–62

Dörjes, J., Gadow, S., Reineck, H.-E., Singh, I. B. (1970): Sedimentologie und Makrobenthos der Nordergründe und der Außenjade (Nordsee). Senckenbergiana marit. 2, 31–59

Dörjes, J., Howard, J. D. (1975) Estuaries of the Georgia Coast, U. S. A.: Sedimentology and Biology. IV. Fluvial-marine transition indicators in an estuarine environment, Ogeechee River-Ossabaw Sound. Senckenbergiana marit. 7, 137–179

Dörjes, J., Reineck, H.-E. (1977) Fauna und Fazies einer Sandplate (Mellum Bank, Nordsee). Senckenbergiana marit. 9, 19–45

Dörjes, J., Reineck, H.-E. (1980): Quergegliederte Pendelspuren. Nat. und Mus. 110, 44–47

Dolan, R., McCloy, J. (1964) Selected bibliography on beach features and related nearshore processes. Technical Report 23, Part A, Coastal Studies Institute Louisiana State University, Baton Rouge, Louisiana, 59 p.

Dominik, J., Förstner, U., Mangini, A., Reineck, H.-E. (1978) ^{210}Pb and ^{137}Cs chronology of heavy metal pollution in a sediment core from the German Bight (North Sea). Senckenbergiana marit. 10, 213–227

Donaldson, A. C., Martin, R. H., Kanes, W. H. (1970) Holocene Guadalupe delta of Texas gulf coast. In: Morgan, J. P., ed., Deltaic sedimentation, modern and ancient. Soc. Econ. Paleontologists Mineralogists Spec. Publ. 15, 107–137

Dott, R. H., Jr., Howard, J. K. (1962) Convolute lamination in non-graded sequences. J. Geol. 70, 114–120

Dott, Jr., R. H., Shaver, R. H., eds. (1974) Modern and ancient geosynclinal sedimentation. Soc. Econ. Paleontologists Mineralogists Spec. Publ. 19, 380 p.

Drake, C. L., Ewing, M., Sutton, G. H. (1950) Continental margins and geosynclines. In: Physics and chemistry of the earth, vol. 3, Ahrens, L. H., Press, F., Rankama, K., Runcorn, S. K., eds., The east coast of North America north of Cape Hatteras, p. 110–198

Drake, D. E. (1971) Suspended sediment and thermal stratification in Santa Barbara Channel, California. Deep-Sea Res. 18, 763–769

Drake, D. E. (1974) Distribution and transport of suspended particulate matter in submarine canyons off southern California. In: Gibbs, R. J., ed., Suspended solids in water, p. 133–153. New York: Plenum Press

Drake, D. E., Gorsline, D. S. (1973) Distribution and transport of suspended particulate matter in Hueneme, Redondo, Newport, and La Jolla submarine canyons, California. Geol. Soc. Am. Bull. 84, 3949–3968

Drake, D. E., Kolpack, R. L. Fischer, P. J. (1972) Sediment transport on the Santa Barbara-Oxnard shelf, Santa Barbara Channel, California. In: Swift, D. J. P., Duane, D. B., Pilkey, O. H., eds., Shelf sediment transport: Process and Pattern, p. 307–331. Stroudsbury, Penn.: Dowden, Hurchinson and Ross, Inc.

Drake, L. D. (1974) Till fabric control by clast shape. Geol. Soc. Am. Bull. 85, 247–250

Draper, L. (1967) Wave activity at the sea bed around northwestern Europe. Marine Geol. 5, 133–140

Dryden, L. (1958) Monazite in part of the southern Atlantic coastal plain. U. S. Geol. Surv. Bull. 58, 54 p.

Duboul-Razaret, C. (1956) Contribution à l'étude géologique et sédimentologique du delta du Rhône. Soc. Géol. France. Mem. 71-1, 234 p.

Duff, P. McL. D., Hallam, A., Walton, E. K. (1967) Cyclic sedimentation. Developments in sedimentology, vol. 10, 280 p. Amsterdam: Elsevier

Düing, W. (1965) Strömungsverhältnisse im Golf von Neapel Pubbl. staz. zool. Napoli 34, 256–316

Dunbar, C. O., Rodgers, J. (1957) Principles of stratigraphy. 356 p. New York: Willey & Sons

Dyer, K. R. (1972) Sedimentation in estuaries. In: Barnes, R. S. K., Green, J., eds., The estuarine environment, p. 10–32. London.: Applied Sci. Publ. LTD

Dylik, J., Maarleveld, G. C. (1967) Frost cracks, frost fissures and related polygons. Mededel. Geol. Sticht. 18, 7–22

Dzulynski, S. (1965) New data on experimental production of sedimentary structures. J. Sediment Petrol. 35, 196–212

Dzulynski, S., Kotlarczyk, J. (1962) On load-casted ripples. Ann. Soc. Géol. Pologne 32, 148–159

Dzulynski, S., Sanders, J. E. (1962) Current marks on firm mud bottoms. Trans. Conn. Acad. Arts Sci. 42, 57–96

Dzulynski, S., Simpson, F. (1966) Experiments on interfacial current markings. Geol. Rom. 5, 197–214

Dzulynski, S., Slaczka, A. (1958) Directional structures and sedimentation of the Krosno Beds (Carpathian flysch). Ann. Soc. Géol. Pologne 28, 205–259

Dzulynski, S., Walton, E. K. (1965) Sedimentary features of flysch and greywackes. Developments in sedimentology, vol. 7, 274 p. Amsterdam: Elsevier

Dzulynski, S., Ksiazkiewicz, M., Kuenen, P. H. (1959) Turbidites in flysch of the Polish Carpathian Mountains. Geol. Soc. Am. Bull. 70, 1089–1118

Dzulynski, S., Shideler, G. L., Slaczka, A. (1972) Impact-induced dendritic ridges in soft sediments. Roczn. Pol. Tow. geol. 42, 285–288

Ebanks, W. J., Bubb, J. N. (1975) Holocene carbonate sedimentation, Matecumbe Keys tidal bank, South Florida. J. Sediment. Petrol. 45, 422–439

Edge, E. R. (1934) Faecal pellets of some marine invertebrates. Am. Midland Naturalist 15, 78–84

Edgerton, H. E. (1963) Underwater photography. In: Hill, M. N., ed., The sea, vol. 3, p. 473–475. New York: Wiley & Sons

Edwards, B. D., Perkinds, R. D. (1974) Distribution of microborings within continental margin sediments of the southeastern United States. J. Sediment. Petrol. 44, 1122–1135

Edwards, J. M., Frey, R. W. (1977) Substrate characteristics within a holocene salt marsh Sapelo Island, Georgia. Senckenbergiana marit. 9, 215–259

Edwards, M. B. (1975) Glacial retreat sedimentation in the Smalfjord Formation, Late Pre-Cambrian, North Norway. Sedimentology 22, 75–94

Edwards, M. B. (1979) Late Precambrian glacial loessites from North Norway and Svalbord, J. Sediment. Petrol. 49, 85–92

Edzwald, J. K., O'Melia, C. R. (1975) Clay distributions in recent estuarine sediments. Clays Clay Min. 23, 39–44

Eichhoff, H. J., Reineck, H.-E. (1953) Sekundäre Verfärbungen durch Herauslösen von Hämatit aus Gesteinen. Neues Jahrb. Mineral. Monatsh. 1952, 315–324

Einsele, G. (1963) "Convolute bedding" und ähnliche Sedimentstrukturen im rheinischen Oberdevon und anderen Ablagerungen. Neues Jahrb. Geol. Palaeontol. Abhandl. 116, 162–198

Einsele, G., Werner, F. (1968) Zusammensetzung, Gefüge und mechanische Eigenschaften rezenter Sedimente vom Nildelta, Roten Meer und Golf von Aden. "Meteor" Forsch.-Ergebn. C, 1, 21–42. Berlin: Borntraeger

Einsele, W. (1937) Physikalisch-chemische Betrachtung einiger Probleme des limnischen Mangan- und Eisenkreislaufs. Internat. Verein. Limnol. 8, 69–84

Einstein, H. A. (1950) The bedload function for sediment transport in open channel flows. U. S. Dept. of Agri. Soil Conservation Serv. Techn. Rep. 1026, 70 p.

Eisma, D. (1965) Eolian sorting and roundness of beach and dune sands. Netherlands J. Sea Res. 2, 541–555

Elliott, T. (1974) Interdistributary bay sequences and their genesis. Sedimentology. 21, 611–622

Ellwood, M., Evans, P. D., Wilson, I. G. (1975) Small scale aeolian bedforms. J. Sediment. Petrol. 45, 554–561

Elmendorf, C. H., Heezen, B. C. (1957) Oceanographic information for engineering submarine cable systems. Bell System Tech. J. 36, 1047–1093

Elson, J. A. (1961) The geology of tills. Proceedings of the 14th Canadian Soil Mech. Conf., Nat. Res. Co., Ottawa p. 5–17

Embleton, C., King, C. A. M. (1968) Glacial and periglacial geomorphology, 608 p. London: Edward Arnold Ltd.

Embley, R. W. (1976) New evidence for occurrence of debris flow deposits in the deep sea. Geology. 4, 371–374

Emery, K. O. (1945) Entrapment of air in beach sand. J. Sediment Petrol. 15, 39–49

Emery, K. O. (1950) Contorted strata at Newport Beach, California. J. Sediment Petrol. 20, 111–115

Emery, K. O. (1952) Continental shelf sediments of southern California. Geol. Soc. Am. Bull. 63, 1105–1108

Emery, K. O. (1955) Grain size of marine beach gravels. J. Geol. 63, 39–49

Emery, K. O. (1960a) The sea off southern California, 366 p. New York: Wiley & Sons

Emery, K. O. (1960b) Basin plains and aprons off southern California. J. Geol. 68, 464–479

Emery, K. O. (1968a) Positions of empty Pelecypod valves on the continental shelf. J. Sediment. Petrol. 38, 1264–1269

Emery, K. O. (1968b) Relict sediments on continental shelves of the world. Bull. Am. Assoc. Petrol. Geologists 52, 445–464

Emery, K. O. (1970) Continental margins of the world. Geol. east Atlantic continental margin. 1. Gen. Econ. papers, ICSU/SCOR Working Party 31 Symp. Cambridge, Rept. No. 70/13, 7–29

Emery, K. O., Gale, J. F. (1951) Swash and swash mark. Trans. Am. Geophys. Union 32, 31–36

Emery, K. O., Stevenson, R. E. (1950) Laminated beach sand. J. Sediment. Petrol. 20, 220–223

Emery, K. O., Uchupi, E., Phillips, J. D., Bowin, C. O., Bunce, E. T., Knott, S. T. (1970) Continental rise off eastern North America. Bull. Am. Assoc. Petrol. Geologists 54, 44–108

Emiliani, C., Milliman, J. D. (1966) Deep-sea sediments and their geological record. Earth Sci. Rev. 1, 105–132

Engel, C. G., Sharp, R. P. (1958) Chemical data on desert varnish. Geol. Soc. Am. Bull. 69, 487–518

Engelhardt, W. von, Gaida, K. H. (1963) Concentration changes of pore solutions during the compaction of clay sediments. J. Sediment. Petrol. 33, 919–930

Engelund, F., Fredsøe, J. (1974) Transition from dunes to planebed in alluvial channels. Tech. Univ. Denmark, Lyngby Inst. Hydrodyn. and Hydraul. Eng. Series paper 4, 56 p.

Engelund, F., Hansen, E. (1966) Investigations of flow in alluvial streams. Acta Polytechn. Scand., Civil. Eng. Bldg. Construct. Ser. 35, 1–100

Epstein, S., Buchsbaum, R., Lowenstam, H., Urey, H. C. (1951) Carbonate-water isotopic temperature scale. Geol. Soc. Am. Bull. 62, 417–425

Ernst, W. (1963) Diagnose der Salinitätsfazies mit Hilfe des Bors. Fortschr. Geol. Rheinland Westfalen 10, 253–266

Ernst, W. (1970) Geochemical facies analysis, 152 p. Amsterdam: Elsevier

Ethridge, F. G., Schumm, S. A. (1978) Reconstruction paleochannel morphologic and flow characteristics: methodology, limitations, and assessment. In: Miall, A. D., ed., Fluvial Sedimentology. Can. Soc. Petrol. Geol., Mem. 5, 703–721

Eugster, H. P., Hardte, I. A. (1978) Saline lakes. In: Lerman, A., ed., Physics and chemistry of lakes, p. 238–293. New York, Heidelberg, Berlin: Springer

Evans, G. (1958) Some aspects of recent sedimentation in the Wash. Eclogae Geol. Helv. 51, 508–515

Evans, G. (1965) Intertidal flat sediments and their environments of deposition in The Wash. Quar. J. Geol. Soc. 121, 209–241

Evans, G. (1970) Coastal and nearshore sedimentation: A comparison of clastic and carbonate deposition. Proc. Geologists Ass. (Engl.) 81, 493–508

Evans, G., Bush, P. R. (1969) Some oceanographical and sedimentological observations on a Persian Gulf Lagoon. In: Castanares, A., Phleger, F. B., eds., Coastal Lagoons, a Symposium, p. 155–170. Mexico: Universidad Nacional Autonoma

Evans, G., Schmidt, V., Bush, P., Nelson, H. (1969) Stratigraphy and geologic history of the sabkha Abu Dhabi, Persian Gulf. Sedimentology 12, 145–159

Evans, O. F. (1942) The relation between size of wave-formed ripple marks, depth of water and the size of the generating waves. J. Sediment. Petrol. 12, 31–35

Evans, O. F. (1943) Effect of change of wave size on the size and shape of ripple marks. J. Sediment. Petrol. 13, 35–39

Evans, O. F. (1949) Ripple marks as an aid in determining depositional environment and rock sequence. J. Sediment. Petrol. 19, 82–86

Ewing, J. A. (1973) Wave-induced bottom currents on the outer shelf. Marine Geol. 15, M31–M35

Ewing, M., Ericson, D. B., Heezen, B. C. (1958) Sediments and topography of the Gulf of Mexico. Am. Assoc. Petroleum Geologists. 42, 995–1063

Ewing, M., Heezen, B. C. (1956) Oceanographic research programs of the Lamont Geological Observatory Geograph. Rev. 46, 508–535

Ewing, M., Thorndike, E. M. (1965) Suspended matter in deep ocean water. Science 147, 1291–1294

Ewing, M., Vine, A., Worzel, I. L. (1946) Photography of the ocean bottom. J. Opt. Soc. Am. 36, 307–321

Eyon, G., Walker, R. G. (1974) Facies relationships in Pleistocene outwash gravels, southern Ontario: a model for bar growth in braided rivers. Sedimentology 21, 43–70

Fabricius, F., Schmidt-Thomé, P. (1972) Contribution to recent sedimentation on the shelves of the Southern Adriatic, Ionian, and Syrtis Seas. In: Stanley, D. J., ed., The Mediterranean Sea. Stroudsburg, Pa.: Dowden, Hutchinson & Ross

Fabricius, F., Rad, U. von Hesse, R., Ott, W. (1970) Die Oberflächensedimente der Straße von Otranto (Mittelmeer). Geol. Rundschau 60, 164–192

Fahnestock, R. K. (1963) Morphology and hydrology of a glacial stream. U. S. Geol. Surv. Profess. Papers 422-A, 1–70

Fahnestock, R. K. (1969) Morphology of the Slims River, in Icefield Ranges research project. Sci. Results. Am. Geogr. Soc., Arctic Inst. N. A. Publ. 1, 161–172

Fairbridge, R. W. (1947) Possible causes of intraformational disturbances in the Carboniferous varve rocks of Australia. J. Proc. Roy. Soc. N. S. Wales 81, 99–121

Fairchild, H. L. (1901) Beach structures in Medina sandstone. Am. Geol. 28, 9–14

Falke, H., ed. (1976) The continental Permian in Central, West, and South Europe, 352 p. Dordrecht: Reidel

Felföldy, L., Muskalay, L., Rakoczi, L., Szesztay K. (1969) Origin and movement of sediment in Lake Balaton. Mitt. Internat. Verein. Limnol. 17, 282–291

Ferm, J. C., Horne, J. C. (1979) Carboniferous depositional environments in the Appalachian Region. Collected papers, Ferm, J. C., Horne, J. C., Weisenfluh, G. A., Staub, J. R., eds., 760 p. Columbia, South Carolina

Feth, I. H. (1964) Review and annotated bibliography of ancient lake deposits (Precambrian to Pleistocene) in the western states. U. S. Surv. Bull. 1080, 119 p.

Fettke, C. R. (1941) Musik Mountain oil pool, Mac Kean County, Pennsylvania. In: Levorsen, A. I. ed., Stratigraphic type oil fields. p. 492–506. Tulsa, Oklahoma: Am. Assoc. of Petroleum Geologists.

Field, M. E., Duane, D. B. (1976) Post-Pleistocene history of the United States inner continental shelf: Significance to origin of barrier islands. Geol. Soc. Am. Bull. 87, 691–702

Field, M. E., Duane, D. B. (1977) Post-Pleistocene history of the United States inner continental shelf: Significance to origin of barrier islands: Discussion and reply. Geol. Soc. Am. Bull. 88, 735–736

Field Trip Guidebook (1964) Prepared by the Corpus Christi Geological Society. Depositional environments South-Central Texas Coast, 170 p.

Field Trip Guidebook (1969) Coastal environments of Northeastern Massachusetts and New Hampshire. Contr. No. 1, 462 p.

Finck, P., Kelts, K. (1976) Geophysical investigations into the nature of preholocene sediments of Lake Zürich. Eclogae geol. Helv. 69, 139–148

Fischer, A. G. (1961) Stratigraphic record of transgressing seas in light of sedimentation on the Atlantic coast of New Jersey. Bull. Am. Assoc. Petroleum Geologists 45, 1656–1666

Fisher, R. V. (1971) Features of Coarse-grained, high concentration fluids and their deposits. J. Sediment. Petrol. 41, 916–927

Fisher, W. L., Brown, F. Jr. (1972) Clastic depositional systems – A genetic approach to facies analysis. Bureau of Economic Geology. The Univ. of Texas, Austin, p. 1–211

Fisher, W. L., Brown, L.F., Jr., Scott, A. J., McGowen, J. H. (1969) Delta systems in the exploration for oil and gas: A Research Colloquium. Univ. of Texas Bur. Econ. Geol. Austin

Fisk, H. N. (1944) Geological investigation of the alluvial valley of the lower Mississippi River. Mississippi River Commission, Vicksburg, Miss., 78 p.

Fisk, H. N. (1947) Fine-grained alluvial deposits and their effects on Mississippi River activity. Mississippi River Commission, Vicksburg, Miss., 82 p.

Fisk, H. N. (1955) Sand facies of recent Mississippi delta deposits. Proceedings of the Fourth World Petroleum Congress, Sec. I/C, 3, 377–398

Fisk, H. N. (1959) Padre Island and the Laguna Madre Flats Coastal South Texas. National Academy of Science-National Research Council, Second Coastal Geographical-Conference, p. 103–151

Fisk, H. N. (1960) Recent Mississippi River sedimentation and peat accumulation. Congr. Avan. Etudes Stratigr. Geol. Carbonifère, Compte Rendu, 4 Heerlen, 1958, 1. 187–199.

Fisk, H. N. (1961) Bar-finger sands of the Mississippi delta. In: Peterson, J. A., Osmond, J. C., eds., Geometry of sandstone bodies, p. 29–52. Tulsa, Oklahoma: Am. Assoc. Petroleum Geologists.

Fisk, H. N., McFarland, E., Kolb, C. R., Wilbert, L. J. (1954) Sedimentary framework of the modern Mississippi delta. J. Sediment. Petrol. 24, 76–99

Fitzgerald, D. M. (1976) Ebb-tidal delta of Price Inlet, South Carolina: Geomorphology, physical processes, and associated inlet shoreline changes. In: Hayes, M. O., Kana, T. W., eds., Terrigenous clastic depositional environments. Techn. Report No. 11-CRD. II-143-II-157. Columbia, S. C.: Univ. of South Carolina.

Fitzgerald, D., Bouma, A. H. (1972) Consolidation studies of deltaic sediments. Gulf Coast Assoc. Geol. Soc., Trans. 22, 165–173

Flekken, P. M. (1978) Anwendung organisch geochemischer – kohlenpetrologischer- und isotopengeochemischer Untersuchungsmethoden in der Faziesanalyse und der Kohlenwasserstoffexploration am Beispiel des NE-Randes vom Pariser Becken (Trias-Lias Luxemburg). Doctoral Thesis, Techn. Univ. Aachen, 273 p.

Flemming, B. W., Fricke, A. H. (1983) Beach and nearshore habitats as a function of internal geometry, primary sedimentary structures and grain size. In: McLachlan, A., Erasmus, T., eds., Sandy beaches as ecosystems, p. 115–132. Junk Publ., Den Hague

Flint, R. F. (1957) Glacial and Pleistocene geology, 553 p. New York: Wiley & Sons

Förstner, U. (1976) Lake sediments as indicators of heavy-metal pollution. Naturwiss. 63, 465–470

Förstner, U., Müller, G., Reineck, H.-E. (1968) Sedimente und Sedimentgefüge des Rheindeltas im Bodensee. Neues Jahrb. Mineral. Abhandl. 109, 33–62

Förstner, U., Müller, G. (1974) Schwermetallanreicherung in datierten Sedimentkernen aus dem Bodensee und aus dem Tegernsee. Tschermaks Min. Petr. Mitt. 21, 145–163

Foley, M. G. (1977) Gravel-lens formation in antidune-regime flow – a quantitative hydrodynamic indicator. J. Sediment. Petrol. 47, 738–746

Folk, R. L. (1966) A review of grain-size parameters. Sedimentology 6, 73–93

Folk, R. L. (1968) Bimodal supermature sandstones: Product of the desert floor. International Geological Congress, 23rd session 8, 9–32

Folk, R. L. (1971) Longitudinal dunes of the northwestern edge of the Simpson desert, Northern Territory, Australia, 1. Geomorphology and grain-size relationships. Sedimentology 16, 5–54

Folk, R. L. (1976a) Rollars and ripples in sand, streams and sky: rhythmic alteration of transverse and longitudinal vortices in three orders. Sedimentology 23, 649–669

Folk, R. L. (1976b) Reddening of desert sands: Simpson Desert, N. T., Australia. J. Sediment. Petrol. 46, 604–615

Folk, R. L. (1978) Angularity and silica coatings of Simpson Desert sand grains, Northern Territory, Australia. J. Sediment. Petrol. 48, 611–624

Folk, R. L., Ward, W. (1957) Brazos River bar: A study in the significance of grain size parameters. J. Sediment. Petrol. 27, 3–26

Forel, F.-A. (1885) Les ravins sous-lacustres des fleuves glaciaires. Compt. Rend. 101, 725–728

Forel, F.-A. (1888) Le ravin sous-lacustrine du Rhône. Bull. Soc. Vaudoise Sci. Nat. 23, 96, 85–107

Frakes, L. A., Figueiredo, P. M. De, Fularo, V. (1968) Possible fossil eskers and associated features from the Parana basin, Brasil. J. Sediment. Petrol. 38, 5–12

Franz, E. (1931) Insektenbegräbnis im Meer, Bestimmung der Arten in der Insektendrift bei Wilhelmshaven vom 24. Mai 1931. Senckenbergiana 13, 228–230

Franzen, J. L. (1976) Die Fossilfundstelle Messel. Ihre Bedeutung für die paläontologische Wissenschaft. Naturwiss. 63, 418–425

Franzen, J. L. (1977) Die Entstehung der Fundstelle Messel. Ber. Naturforsch. Ges. Freiburg i. Br., 67, 53–58

Fraser, G. S., Hester, N. C. (1977) Sediments and sedimentary structures of a beach-ridge complex, southwestern shore of Lake Michigan. J. Sediment. Petrol. 47, 1187–1200

Frazier, D. E., Osanik, A. (1961) Point-bar deposits. Old River Locksite, Louisiana. Trans. Gulf Coast Assoc. Geol. Soc. 11, 121–137

Freeman, W. E., Visher, G. S. (1975) Stratigraphic analysis of the Navajo Sandstone. J. Sediment. Petrol. 45, 651–661

Frey, R. W. (1971) Ichnology – the study of fossil and Recent lebensspuren. In: Perkins, B. F., ed., Trace fossils, a field guide to selected localities in Pennsylvanian, Permian, Cretaceous, and Tertiary rocks of Texas, and related papers. Louisiana State Univ., School Geosci., Misc. Pub. 71-1, 91–125

Frey, R. W. (1978) Behavioral and ecological implications of trace fossils. Trace Fossil Concepts, Soc. Econ. Paleontologists Mineralogists Short Course 5, 43–66

Frey, R. W., Basan, P. B. (1978) Coastal salt marshes. In: Davis, Jr. R. A., ed., Coastal sedimentary environments, p. 101–169. New York, Heidelberg, Berlin: Springer

Frey, R. W., Howard, J. D. (1969) A profile of biogenic sedimentary structures in a Holocene barrier island salt marsh complex, Georgia. Trans. Gulf Coast Assoc. Geol. Soc. 19, 427–444

Frey, R. W., Howard, J. D. (1972) Georgia coastal region, Sapelo Island, U. S. A.: Sedimentology and biology. VI. Radiographic study of sedimentary structures made by beach and offshore animals in aquaria. Senckenbergiana marit. 4, 169–182

Frey, R. W., Mayou, T. V. (1971) Decapod burrows in Holocene barrier island beaches and washover fans, Georgia. Senckenbergiana marit. 3, 53–77

Friedman, G. M. (1961) Distinction between dune, beach, and river sands from their textural characteristics. J. Sediment. Petrol. 31, 514–529

Friedman, G. M. (1962) On sorting, sorting coefficients, and the lognormality of the grain-size distribution of sandstones. J. Geol. 70, 737–753

Friedman, G. M. (1967) Dynamic processes and statistical parameters compared for size frequency distribution of beach and river sand. J. Sediment. Petrol. 37, 327–354

Friedman, G. M., Johnson, K. G. (1966) The Devonian Catskill deltaic complex of New York, type example of a "tectonic delta complex", In: Shirley, M. L., ed., Deltas in their geologic framework, p. 171–188. Houston: Houston Geol. Soc.

Friedman, G. M., Johnson, K. G. (1969) The Tully clastic correlatives (upper Devonian) of New York State: A model for recognition of alluvial, dune (?), tidal, nearshore (bar and lagoon), and offshore sedimentary environments in a tectonic delta complex, J. Sediment. Petrol. 39, 451–485

Friedman, G. M., Sanders, J. E. (1974) Positive relief bedforms on modern tidal flat that resemble molds of flutes and grooves; implications for geopetal criteria and for origin and classification of bedforms. J. Sediment. Petrol. 44, 181–189

Friedman, G. M., Sanders, J. E. (1978) Principles of sedimentology, 792 p. New York: Wiley & Sons

Friend, P. F. (1965) Fluviatile sedimentary structures in the Woodbay Series (Devonian) of Spitsbergen. Sedimentology 5, 39–68

Frith, D. W. (1977) A selected bibliography of mangrove literature. Phuket marine biolog. center, Bull. No. 19, 142 p.

Füchtbauer, H., Müller, G. (1970) Sediment-Petrologie. Teil II Sedimente und Sedimentgesteine, 726 p. Stuttgart: E. Schweizerbart'sche Verlagsbuchhandlung

Führböter, A. (1970) Air entrainment and energy dissipation in breakers. Proc. XIIth Coastal Engineering Conference, Vol. 1, Washington D. C.

Führböter, A., Büsching, F., Dette, H. H., Hansen, U. A. (1979) Energieumwandlungen in Brandungszonen. DFG-Forschungsbericht "Sandbewegung im Küstenraum". Abschlußbericht, S. 80–96

Fuller, A. O. (1961) Size characteristics of shallow marine sands from Cape of Good Hope, South Africa. J. Sediment. Petrol. 31, 256–261

Futterer, E. (1974) Untersuchungen zum Einsteuerungsverhalten der Einzelklappen von Cardium echinatum L. und Cardium edule L. im Strömungskanal. N. Jb. Geol. Paläont. Mh. 8, 449–455

Futterer, E. (1976) Rezente Schille: Transport und Einregelung tierischer Hartteile im Strömungskanal. Zbl. Geol. Paläont. Teil II. 5/6, 203–494

Gadow, S. (1969) Gips als Leitmineral für das Liefergebiet Helgoland und für den Transport bei Sturmfluten. Natur Mus. 99, 537–540

Gadow, S. (1970) 1. Sedimente und Chemismus. In: Reineck, H.-E. ed., Das Watt, Ablagerungs- und Lebensraum, p. 23–35. Frankfurt a. M.: W. Kramer

Gadow, S. (1971) Der Golf von Gaeta (Tyrrhenisches Meer). I. Die Sedimente. Senckenbergiana marit. 3, 103–133

Gadow, S., Reineck, H.-E. (1968) Der Abbau einer Magnetit-Strandseife in Italien. Natur Mus. 98, 57–63

Gadow, S., Reineck, H.-E. (1969) Ablandiger Sandtransport bei Sturmfluten. Senckenbergiana marit. 1, 63–78

Galloway, W. E. (1975) Process framework for describing the morphologic and stratigraphic evolution of deltaic depositional systems. In: Broussard, M. L. Deltas, models for exploration, 555 p. Houston, Texas: Houston Geol. Soc.

Game, P. M. (1964) Observations on a dustfall in the eastern Atlantic, February, 1962. J. Sediment. Petrol. 34, 355–359

Gardner, L. R. (1972) Origin of the Mormon Caliche, Clark County, Nevada. Geol. Soc. Am. Bull. 83, 143–156

Garrels, R. M., Christ, C. L. (1965) Solutions, minerals, and equilibria, 450 p. New York: Harper & Row

Garrels, R. M., McKenzie, F. T. (1971) Evolution of Sedimentary Rocks, 397 p. New York: Norton

Geer, D. De (1897) Om rullstensåsarnas bildningssätt. Geol. Foren. Stockholm Forh. 19, 366–388

Geer, G. De (1940) Geochronologia suecica principles. K. Svensk. Vet. Akad. Handl. 18

Gees, R. A. (1969) Surface textures of quartz sand grains from various depositional environments. Beitr. Elektronenmikroskop. Direktabb. Oberfl. 2, 283–297

Geib, K. W. (1950) Neue Erkenntnisse zur Paläogeographie des westlichen Mainzer Beckens. Notizbl. Hess. L-Anst. Bodenf. 5, 101–111

Geib, K. W. (1954) Die geologischen Verhältnisse des Mainzer Beckens unter besonderer Berücksichtigung der Entstehung und der wirtschaftlichen Bedeutung seiner Sedimente. 32. Jahrestagung der Dt. Mineralog. Ges., p. 9–16

Geib, K. W., Atzbach, O. (1969) Über eine Trockentalung im Nahebergland – Eine Richtigstellung von geologischer Seite. Mainz. Naturw. Arch. 8, 140–148

Gellert, J. F. (1937) Strandhörner bei Duhnen-Cuxhaven. Senckenbergiana 19, 7–12

George, T. N. (1976) Charles Lyell: The present is the key to the past. Philsopical J. 13, 3–24

Gessler, J. (1971) Beginning and ceasing of sediment motion. In: Shen, H. W., ed., River Mechanics Water Resources Publ. 1, 22 p. Fort Collins, Colorado

Geyl, W. F. (1976) Tidal neomorphs. Z. Geomorph. N. F. 20, 308–330

Ghibaudo, G., Mutti, E., Rosel, J. (1974) Le Spiagge Fossili delle Arenarie di Aren (Creacico Superiore) nella Valle No-

guera Ribagorzana (Pirenei Centro-Meridionali, Province di Lerrida e Huesca, Spaga). Soc. Geol. Italiana, 13, 497–537

Gibbs, R. J., Matthews, M. D., Link, D. A. (1971) The relationship between sphere size and settling velocity. J. Sediment. Petrol. 41, 7–18

Gienapp, H. (1973) Strömungen während der Sturmflut vom 2. November 1965 in der Deutschen Bucht und ihre Bedeutung für den Sedimenttransport. Senckenbergiana marit. 5, 135–151

Gierloff-Emden, H. G. (1959) Lagunen, Nehrungen, Strandwälle und Flußmündungen im Geschehen tropischer Flachlandküsten. Z. Geomorphol. 3, 29–46

Gierloff-Emden, H. G. (1961) Nehrungen und Lagunen. Petermanns geogr. Mitt. 105, 82–92

Gieteling, G. (1973) Sedimentology of a linear prograding coastline followed by three high-destructive delta-complexes (Cambro-ordovician, Catabrian Moutains, NW Spain). Leidse Geol. Mededel. 49, 125–144

Gilbert, G. K. (1899) Ripple marks and cross-bedding. Geol. Soc. Am. Bull. 10, 135–140

Gilbert, G. K. (1914) The transportation of debris by running water, U. S. Geol. Surv. Profess. Papers 86, 263 p.

Gile, L. H., Peterson, F. F., Grossman, R. B. (1966) Morphological and genetic sequences of carbonate accumulation in desert soils. Soil Sci. 101, 347–360

Gill, W. D., Kuenen, P. H. (1958) Sand volcanoes on slumps in the Carboniferous of County Clare, Ireland. Quart. J. Geol. Soc. London 113, 441–460

Ginsburg, R. N., ed. (1975) Tidal deposits, 428 p. New York, Heidelberg, Berlin: Springer

Glaister, R. P., Nelson, H. W. (1974) Grain size distribution, an aid in facies identification. Bull. Canadian Petrol. Geol., 22, 203–204

Glangeaud, L. (1938) Transport et sédimentation dans l'estuaire et à l'embouchure de la Gironde. Bull. Soc. Géol. France 5, 599–531

Glenn, J. L., Dahl, A. R. (1959) Characteristics and distribution of some Missouri River deposits. Proc. Iowa Acad. Sci. 66, 302–311

Glennie, K. W. (1970) Desert sedimentary environments. Developments in Sedimentology, vol. 14, 222 p. Amsterdam: Elsevier

Glennie, K. W. (1972) Permian Rotliegendes of northwest Europe interpreted in light of modern desert sedimentation studies. Am. Assoc. Petrol. Geol. Bull. 56, 1048–1071

Glennie, K. W., Evans, G. (1976) A reconnaissance of the Recent sediments of the Ranns of Kutch, India. Sedimentology 23, 625–647

Glennie, K. W., Evamy, B. D. (1968) Dikaka: Plants and plant-root structures associated with aeolian sand. Palaeogeogr., Palaeoclimatol., Palaeoecol. 4, 77–87

Göhren, H. (1969) Die Strömungsverhältnisse im Elbmündungsgebiet. Hamburger Küstenforsch. 6, 1–83

Göhren, H. (1975 a) Zur Dynamik und Morphologie der hohen Sandbänke im Wattenmeer zwischen Jade und Eider. Die Küste 27, 28–49

Göhren, H. (1975 b) Dynamics and Morphology of Sand Banks in the Surf Zone of Outer Tidal Flats. Proc. 14th Int. Conf. on Coast. Eng., p. 871–883

Goethe, F. (1958) Anhäufungen unversehrter Muscheln durch Silbermöwen. Natur Volk 88, 181–187

Goldring, R. (1964) Trace-fossils and the sedimentary surface in shallow water marine sediments. In: Straaten, L. M. J. U. van, ed., Developments in sedimentology, vol. 1. Deltaic and shallow water marine deposits, p. 136–143. Amsterdam: Elsevier

Goldring, R., Bridges,, P. (1973) Sublittoral sheet sandstones. J. Sediment. Petrol. 43, 736–747

Goldring, R., Bosence, D. W. J., Blake, T. (1978) Estuarine sedimentation in the Eocene of Southern England. Sedimentology 25, 861–876

Goldschmidt, V. M., Peters, C. L. (1932) Zur Geochemie des Bors. 1. 2. Nachdr. Akad. wiss. Göttingen, Math. Phys. Kl. II 1932, 402–407, 528–545

Goldsmith, V. (1973) Internal geometry and origin of vegetated coastal sand dunes. J. Sediment. Petrol. 43, 1128–1142

Goldstein, S., ed. (1938/52) Modern development in fluid dynamics 1032 p. New York: Dover

Goldthwait, R. P., ed. (1971) Till – A Symposium, 402 p. Columbus: Ohio State Univ. Press

Gole, C. V., Chitale, S. V. (1966) Inland delta building activity of Kosi River. J. Hydraul. Proc. Am. Soc. Civ. Eng. 92, 111–126

Gorsline, D. S., ed. (1962) Proceedings of the first national coastal and shallow water research conference, 897 p. Tallahassee. Florida: Nationnal Science Foundation and Office Naval Research

Gorsline, D. S. (1966) Dynamic characteristics of West Florida Gulf Coast beaches. Marine Geol, 4, 187–206

Gorsline, D. S., Emery, K. O. (1959) Turbidity-current deposits in San Pedro and Santa Monica basins off southern California. Geol. Soc. Am. Bull. 70, 279–290

Goudie, A. (1973) Duricrusts in tropical and subtropical landscapes, 174 p. Oxford: London Univ. Press

Gould, H. R. (1970) The Mississippi delta complex. In: Morgan, J. P., ed., Deltaic sedimentation modern and ancient. Soc. Econ. Paleontologists Mineralogists Spec. Publ. 15, 3–47

Gould, H. R., McFarlan, E., Jr. (1959) Geologic history of the chenier plain, southwestern Louisisana. Trans. Gulf Coast Assoc. Geol. Soc., 9th Ann. Meet., Houston 9, 261–270

Gradzinskii, R., Jerzykiewicz, T. (1974) Dinosaur and mammal-bearing aeolian and associated deposits of the upper Cretaceous in the Gobi Desert (Mangolia). J. Sediment. Geol. 12, 249–278

Graf, W. M. (1971) Hydraulics of Sediment transport, 573 p. New York: McGraw Hill

Graham, J. R. (1975) Analysis of an Upper Palaeozoic transgressive sequence in Southwest County Cork, Eire. Sedimentary Geol. 13, 267–290

Grahmann, R. (1932) Der Lösz in Europa. Ges. Erdkde. Leipzig. Mitt. 1930–31 (1932), 5–24

Grant, U. S. (1948) Influence of the water table on beach aggradation and degradation. J. Marine Res. 7, 655–660

Greensmith, J. T., Tucker, E. V. (1969) The origin of Holocene shell deposits in the chenier plain facies of Essex (Great Britain). Marine Geol. 7, 403–425

Greenwood, B., Davidson-Arnott, R. G. D. (1979) Sedimentation and equilibrium in wave-formed bars: a review and case study. Can. J. Earth Sci. 16, 312–332

Greer, S. A. (1975) Estuaries of the Georgia Coast, U. S. A.: Sedimentology and Biology. III. Sandbody geometry and sedimentary facies at the estuary-marine transition zone, Ossabaw Sound, Georgia: A stratigraphic model. Senckenbergiana marit. 7, 105–135

Gregory, K. J., ed. (1977) River channel changes, 448 p. Chichester: Wiley

Griffiths, J. C. (1951) Size versus sorting in some Caribbean sediments. J. Geol. 59, 211–243

Grim, R. E. (1968) Clay mineralogy, 2nd ed., 596 p. New York: McGraw-Hill

Grimm, R., Pfannkuche, O., Podloucky, R., Wilkens, H. (1976) Zoologische Charakterisierung der Wedeler und Haseldorfer Marsch. Die Heimat 83, 236–247

Grinnell, R. S., Jr. (1974) Vertical orientation of shells on some Florida oyster reefs. J. Sediment. Petrol. 44, 116–122

Gripp, K. (1944) Entstehung und künftige Entwicklung der Deutschen Bucht. Arch. Deut. Seewarte Marineobs. 63, 1–44

Gripp, K. (1956) Das Watt; Begriff, Begrenzung und fossile Vorkommen. Senckenbergiana Lethaea 37, 149–181

Gripp, K. (1961) Über Werden und Vergehen von Barchanen an der Nordsee-Küste Schleswig-Holsteins. Z. Geomorphol. 5, 24–36

Gripp, K. (1963) Wenn die Natur im Sande spielt . . ., 54 p. Hamburg: Verl. Ges. der Freunde des vaterländischen Schul- und Erziehungswesens

Gripp, K. (1964) Erdgeschichte von Schleswig-Holstein, 411 p. Neumünster: Karl Wachholtz

Gripp, K. (1968) Zur jüngsten Erdgeschichte von Hörnum/Sylt und Amrum mit einer Übersicht über die Entstehung der Dünen in Nordfriesland. Die Küste 16, 76–117

Gross, D. L., Lineback, J. A., White, W. A., Ayer, N. J., Collinson, C., Leland, H. U. (1970) Studies of Lake Michigan bottom sediments – Number one: Preliminary stratigraphy of unconsolidated sediments from the southwestern part of Lake Michigan. Environmental Geol. Notes. 30, 1–20

Gross, M. G., Gucluer, S. M. (1964) Recent marine sediments in Saanich Inlet, a stagnant marine basin. Limnol. Oceanog. 9, 359–376

Gross, M. G., Gucluer, S. M., Creager, J. S., Dawson, W. A. (1963) Varved marine sediments in a stagnant fjord. Science 141, 918–919

Groth, P. (1971) Untersuchungen über einige Spurenelemente in Seen. Arch. Hydrobiol. 68, 305–375

Grumbt, E. (1975) Johannes Walther – ein Begründer der modernen Sedimentforschung. Z. geol. Wiss. 3, 1255–1263

Grumbt, E. (1976) Die Sedimentgefüge als wichtige geologische Merkmale. I. Sedimentgefüge und Bildungsmilieu. – Vortrag, Weiterbildungsveranstaltung "Fortschritte Untersuchung sedimentärer Komplexe" Sekt. Geowiss. BA Freiberg, 14 p.

Gubler, Y., Bugnicourt, D., Faber, J., Kubler, B., Nyssen, R. (1966) Essai de nomenclature et caracterisation des principales structures sédimentaires, 291 p. Paris: Technip.

Guilcher, A. (1949) Observations sur les Croissants de Plage. Soc. France, 5 Ser. 19, 15–29

Guilcher, A. (1958) Coastal and submarine morphology, 274 p. London: Methuen & Co. Ltd.

Gustavson, T. C. (1974) Sedimentation on gravel outwash fans, Malaspina glacier foreland, Alaska. J. Sediment. Petrol. 44, 374–389

Gustavson, T. C. (1975 a) Sedimentation and physical limnology in proglacial Malaspina lake southeastern Alaska. In: Jopling, A. V., McDonald, B. C., eds., Glaciofluvial and Glaciolacustrine Sedimentation. Soc. Econ. Palaeontologists Mineralogists Spec. Publ. 23, 249–263

Gustavson, T. C. (1975 b) Bathymetry and sediment distribution in proglacial Malaspina Lake, Alaska. J. Sediment. Petrol. 45, 738–744

Gustavson, T. C. (1978) Bedforms and stratification types of modern gravel meander lobes, Nueces River, Texas. Sedimentology 25, 401–426

Gustavson, T. C., Ashley, G. M., Boathroyd, J. C. (1975) Depositional sequences in Glaciolacustrine deltas. In: Jopling, A. V., McDonald, eds., Glaciofluvial and Glaciolacustrine Sedimentation. Soc. Econ. Palaeontologists Mineralogists Spec. Publ. 23, 264–280

Guy, H. H., Simons, D. B., Richardson, E. V. (1966) Summary of alluvial channel data from flume experiments, 1956–61. U. S. Geol. Surv., Profess. Papers 462-I. 96 p.

Guza, R. T., Inman, D. L. (1975) Edgewaves and beach cusps. J. Geophys. Res. 80, 2997–3012

Hadley, M. L. (1964) Wave-induced bottom currents in the Celtic Sea. Marine Geol. 2, 164–167

Häntzschel, W. (1935 a) Erhaltungsfähige Schleifspuren von Gischt am Nordseestrand. Senckenbergiana 65, 461–465

Häntzschel, W. (1935 b) Rezente Eiskristalle in meerischen Sedimenten und fossile Eiskristall-Spuren. Senckenbergiana 17, 151–177

Häntzschel, W. (1935 c) Ein Fisch (Gobius microps) als Erzeuger von Sternspuren. Nat. u. Volk, 65, 562–568

Häntzschel, W. (1936 a) Die Schichtungs-Formen rezenter Flachmeer-Ablagerungen. Senckenbergiana 18, 316–356

Häntzschel, W. (1936 b) Seeigel-Spülsäume. Natur Volk 66, 293–298

Häntzschel, W. (1937) Erhaltungsfähige Abdrücke von Hydro-Medusen. Nat. u. Volk 67, 141–144

Häntzschel, W. (1938) Bau und Bildung von Groß-Rippeln im Watten-Meer. Senckenbergiana 20, 1–42

Häntzschel, W. (1939) Brandungswälle, Rippeln und Fließfiguren am Strande von Wangeroog. Natur Volk 69, 40–48

Häntzschel, W. (1940) Wattenmeer-Beobachtungen am Ringelwurm *Nereis*. Natur Volk 70, 144–148

Häntzschel, W. (1941) Entgasungs-Krater im Wattenschlick. Natur Volk 71, 312–314

Häntzschel, W. (1962) Trace fossils and problematica. In: Moore, R. C., ed., Treatise on invertebrate paleontology, p. 177–245. New York: University of Kansas Press

Häntzschel, W., El-Baz, F., Amstutz, G. C. (1968) Coprolites, an annotated bibliography Mem. vol. 108, 132 p. Boulder, Colorado: Geological Society of America

Häntzschel, W., Reineck, H.-E. (1968) Faziesuntersuchungen im Hettangium von Helmstedt (Niedersachsen). Mitt. Geol. Staatsinst. Hamburg 37, 5–39

Hagemann, Ir. B. P. (1969) Development of the western part of the Netherlands during the Holocene. Geol. Mijnbouw 48, 373–388

Hails, J. R., Hoyt, J. H. (1968) Barrier development on submerged coasts: Problems of sea-level changes from a study of the Atlantic coastal plain of Georgia. U. S. A. and parts of the east Australian coast. Z. Geomorphol. 7, 24–55

Hails, J. R., Carr, A., eds. (1975) Nearshore Sediment Dynamics and Sedimentation, 316 p. New York: Wiley & Sons

Hall, J. (1843) Geology of New York, Part IV, 683 p. Albany: Carroll & Cook

Hallam, A., ed. (1967) Depth indicators in marine sedimentary environments. Marine Geol., Spec. Issue 5, 329–332

Halstead, L. B., Nada, A. C. (1973) Environment of deposition of the Pinjor Formation, Upper Siwaliks near Chandigarh. Bull. Ind. Geol. Assoc. 6, 63–70

Hamblin, W. K. (1965) Internal structures of "homogeneous" sandstones. Kans. Geol. Surv. Bull. 175, 569–582

Hamelin, L.-E. (1976) La famille du mot "Glaciel". Rev. Géogr. Montr. 30, 233–236

Hamilton, W., Krinsley, D. (1967) Upper Paleozoic glacial deposits of South Africa and southern Australia. Geol. Soc. Am. Bull. 78, 783–800

Hampton, M. A. (1975) Competence of fine-grained debris flow. J. Sediment. Petrol. 45, 834–844

Hampton, M. A., Bouma, A. H. (1977) Slope instability near the shelf break, Western Gulf of Alaska. Marine Geotechnol. 2, 309–331

Hand, B. M., Emery, K. O. (1964) Turbidites and topography of the north end of San Diego Trough, California. J. Geol. 72, 526–542

Hand, B. M., Middleton, G. V., Skipper, K. (1972) Discussion: antidune cross-stratification in a turbidite, sequence, Cloridorme Formation, Gaspé, Quebec. Sedimentology 18, 135–138

Haner, B. E. (1971) Morphology and sediments of Redondo Submarine fan, Southern California. Geol. Soc. Am. Bull. 82, 2413–2432

Happ, S. C. (1940) Significance of texture and density of alluvial depositis in the middle Rio Grande Valley. J. Sediment. Petrol. 10, 3–19

Happ, S. C., Rittenhouse, G., Dobson, G. C. (1940) Some aspects of accelerated stream and valley sedimentation. U. S. Dept. Agr., Techn. Bull. 695, 1–134

Harder, H. (1970) Boron content of sediments as a tool in facies analysis. Sediment. Geol. 4, 153–175

Hardie, L. A., Smoot, J. P., Eugster, H. P. (1978) Saline lakes and their deposits: a sedimentological approach. In: Matter, A., Tucker, M. E., eds., Modern and ancient lake sediments. Spec. Publs. Int. Ass. Sediment. 2, 7–41

Harland, W. B., Herod, K. N., Kinsley, D. H. (1966) The definition and identification of tills and tillites. Earth Sci. Rev. 2, 225–256

Harlett, J. C., Kulm, L. D. (1973) Suspended sediment transport on the Northern Oregon continental shelf. Geol. Soc. Am. Bull. 84, 3815–3826

Harms, J. C. (1969) Hydraulic significance of some sand ripples. Geol. Soc. Am. Bull. 80, 363–396

Harms, J. C. (1975) Stratification and sequence in prograding shoreline deposits. Soc. Econ. Paleontologists Mineralogists Short Course 2, 81–102

Harms, J. C., Fahnestock, R. K. (1965) Stratification, bed forms, and flow phenomena (with an example from the Rio Grande). Soc. Econ. Paleontologists Mineralogists Spec. Publ. 12, 84–115

Harms, J. C., MacKenzie, D. B., McCubbin, D. G. (1963) Stratification in modern sands of the Red River, Louisiana. J. Geol. 71, 566–580

Harris, L. M., Jollymore, P. G. (1974) Iceberg furrow marks on the continental shelf Northeast of Belle Isle, Newfoundland. Can. J. Easth Sci. 11, 43–52

Harris, S. A. (1958) Differentiation of various Egyptian aeolian micro-environments by mechanical composition. J. Sediment Petrol. 28, 164–174

Harrison, P. W. (1957) New technique for three-dimensional fabric analyses of till and englacial debris containing particles from 3 to 40 mm in size. J. Geol. 65, 98–105

Harrison, S. S. (1975) Turbidite origin of glaciolacustrine sediments, Woodcock lake, Pennsylvania. J. Sediment. Petrol. 45, 738–744

Hartmann, M., Lange, H., Seibold, E., Walger, E. (1971) Oberflächensedimente im Persischen Golf und Golf von Oman. I. Geologisch-hydrologischer Rahmen und erste sedimentologische Ergebnisse. "Meteor" Forsch.-Ergebn. C, 4, 1–76, Berlin: Borntraeger

Hartwell, A. D. (1970) Hydrography and Holocene sedimentation of the Merrimach River estuary, Massachusetts. Contr. No. 5-CRG, Dept. of Geol., Univ. Mass., 166 p.

Harvey, J. G. (1966) Large sand waves in the Irish Sea. Marine Geol. 4, 49–55

Hatai, K. (1960) A rill-mark observed along the Noribu Beach, Mono-Gun, Miyagi Prefecture. Saito Ho-on Kai Mus. Res. Bull. 29, 28–31

Haubold, H., Katzung, G. (1978) Palaeoecology and palaeoenvironments of tetrapod footprints from the Rotliegend (Lower Permian) of Central Europe. Palaeogeogr. Palaeoclimatol. Palaeoecol. 23, 307–323

Hay, W. W., ed. (1974) Studies in paleo-oceanography. Soc. Econ. Paleontologists Mineralogists Spec. Publ. 20, 218 p.

Hayes, M. O. (1964) Summary of geological effects of hurricanes *Carla*, 1961 and *Cindy*, 1963 on the south Texas coast. Gulf Coast Assoc. Geol. Soc., Field Trip Guidebook 1964. Depositional environments south-central Texas Coast, p. 128–136

Hayes, M. O. (1967) Hurricanes as geological agents, south Texas coast. Bull. Am. Assoc. Petrol. Geologists 51, 937–942

Hayes, M. O. (1975) Morphology of sand accumulation in estuaries: an introduction to the symposium. In: Cronin, L. E., ed., Estuarine research, vol. 2, p. 3–33. New York, N.Y.: Academic Press

Hayes, M. O. (1976) Morphology of sand accumulation in estuaries: an introduction to the symposium. In: Gronin, L. E., ed., Estuarine Research 2, 3–22. London: Academic Press

Hayes, M. O. (1979) Barrier island morphology as a function of tidal and wave regime. In: Leatherman, S. P., ed., Barrier islands, p. 1–27. New York, N.Y.: Academic Press

Hayes, M. O., Boothroyd, J. C. (1969) Storms as modifying agents in the coastal environment. Field trip guidebook coastal environments of northeastern Massachusetts and New Hampshire, Contr. No. 1 – CRG, Geol. Dep. Univ. Massachusetts, Amherst, Mass., p. 245–265

Hayes, M. O., Kana, T. W., eds. (1976) Terrigenous clastic depositional environments – some modern examples. Techn. Rep. 11-CRD, Part I, 131 p.; Part II, 171 p.

Hedberg, H. D. (1970) Continental margins from viewpoint of the petroleum geologist. Bull. Am. Ass. Petrol. Geol. 54, 154–194

Heezen, B. C. (1963) Turbidity currents. In: Hill, M. N., ed., The sea, vol. 3, p. 742–775. London: Wiley & Sons

Heezen, B. C., ed. (1977) Influence of abyssal circulation on sedimentary accumulations in space and time, 215 p. New York: Elsevier Sci. Publ. Comp

Heezen, B. C., Drake, C. L. (1963) Gravity tectonics, turbidity currents and geosynclinal accumulations in the continental margin of eastern North America. University of Tasmania Symposium P. D1-D10

Heezen, B. C., Ewing, M. (1952) Turbidity currents and submarine slumps, and the Grand Banks earthquake. Am. J. Sci. 250, 849–873

Heezen, B. C., Ericson, D. B., Ewing, M. (1954) Further evidence for a turbidity current following the 1929 Grand Banks earthquake. Deep-Sea Res. 1, 193–202

Heezen, B. C., Hollister, C. D. (1964) Deep-sea current evidence from abyssal sediments. Marine Geol. 1, 141–174

Heezen, B. C., Hollister, C. D., eds. (1971) The Face of the Deep, 659 p. New York: Oxford Univ. Press

Heezen, B. C., Hollister, C. D., Ruddiman, W. F. (1966) Shaping of the continental rise by deep geostrophic contour currents. Science 152, 502–508

Heezen, B. C., Tharp, M., Ewing, M. (1959) The floors of the oceans. 1. The North Atlantic. Geol. Soc. Am. Spec. Papers 65, 122 p.

Heim, A. (1882) Der Bergsturz von Elm. Dt. Geol. Ges. Z. 34, 74–115

Hein, F. J., Walker, R. G. (1977) Bar evolution and development of stratification in the gravelly, braided Kicking Horse River, British Columbia. Can. J. Earth Sci. 14, 562–570

Helley, E. J. (1969) Field measurements of the initiation of large bed particle motion in Blue creak near Klamath, California. U. S. Geol. Survey Prof. Paper 562-G, 19 p.

Hendry, H. E., Stauffer, M. R. (1975) Penecontemporaneous recumbent folds in trough cross-bedding of Pleistocene sands in Saskatchewan, Canada. J. Sediment. Petrol. 45, 932–943

Henningsen, D. (1969) Paläogeographische Ausdeutung vorzeitlicher Ablagerungen, 170 p. Mannheim: Bibliographisches Inst. AG.

Hersey, J. B., ed. (1967) Deep-sea photography, 310 p. Baltimore: Johns Hopkins Press

Hertweck, G. (1970) Die Bewohner des Wattenmeeres in ihren Auswirkungen auf das Sediment. In: Reineck, H.-E., ed., Das Watt, Ablagerungs- und Lebensraum, p. 106–130. Frankfurt a. M.: Kramer

Hertweck, G. (1971 a) Der Golf von Gaeta (Tyrrhenisches Meer). V. Abfolge der Biofaziesbereiche in den Vorstrand- und Schelfsedimenten. Senckenbergiana marit. 3, 247–276

Hertweck, G. (1971 b) The animal community of a muddy environment and the development of biofacies as effected by the life cycle of the characteristic species. In: Crimes, T. P., Harper, J. C., eds., Trace fossils, p. 235–242. Liverpool: Geol. J., spec. Iss.

Hertweck, G. (1972) Georgia coastal region, Sapelo Island, U.S.A.: Sedimentology and biology. V. Distribution and environmental significance of lebensspuren and in-situ skeletal remains. Senckenbergiana marit. 4, 125–167

Hertweck, G. (1975) Ichnology aspects. In: Dörjes, J., Hertweck, G., Recent biocoenoses and ichnocoenoses in shallow-water marine environments. In: Frey, R. W., ed., The study of trace fossils, p. 468–491. New York, Heidelberg, Berlin: Springer

Herward, A. P. (1981) A review of wave-dominated clastic shoreline deposits. Earth Sci. Rev. 17, 223–276

Hess, H. (1904) Die Gletscher, 426 p. Braunschweig: F. Vieweg

Hesse, R., Rad, U. von, Fabricius, F. H. (1971) Holocene sedimentation in the Strait of Otranto between the Adriatic and Ionian Seas (Mediterranean). Marine Geol. 10, 293–355

Hessland, I. (1943) Marine Schalenablagerungen Nord-Bohusläns. Bull. Geol. Inst. Univ. Upsala 31, 1–346

Heward, A. P. (1978) Alluvial fan and lacustrine sediments from the stephanian A and B (La Magdalena, Ciuera-Matallana and Sabero) coalfields, northern Spain. Sedimentology 25, 451–488

High, L. R., Jr., Picard, M. D. (1965) Sedimentary petrology and origin of analcime-rich Popo Agie Member, Chugwater (Triassic) Formation, west-central Wyoming. J. Sediment. Petrol. 35, 49–70

High, L. R., Picard, M. D. (1974) Reliability of cross-stratification types as palaeocurrent indicators in fluvial rocks. J. Sediment. Petrol. 44, 158–168

Hill, G. W., Hunter, R. E. (1976) Interaction of biological and geological processes in the beach and nearshore environments, Northern Padre Island, Texas. In: Davis, Jr. R. A., Ethington, R. L., eds., Beach and nearshore sedimentation. Soc. Econ. Paleontologists Mineralogists Spec. Publ. 24, 169–187

Hill, M. L., Carlson, S. A., Dibblee, T. W. (1958) Stratigraphy of Cuyama Valley-Caliente Range area, California. Bull. Am. Assoc. Petrol. Geologists 42, 2973–3000

Hill, M. N., ed. (1963) The sea. 3. The earth beneath the sea, 963 p. London: Wiley & Sons

Hinze, C., Meischner, D. (1968) Gibt es rezente Rot-Sedimente in der Adria? Marine Geol. 6, 53–71

Hjulström, F. (1935) Studies in the morphological activity of rivers as illustrated by the River Fyris. Geol. Inst. Univ. Upsala, Bull. 25, 221–528

Hjulström, F. (1939) Transportation of detritus by moving water. In: Trask, P. D., ed., Recent Marine Sediments. Am. Assoc. Petrol. Geol. p. 5–31

Hjulström, F. (1952) The geomorphology of the alluvial outwash plains (sandurs) of Iceland and the mechanics of braided rivers. Proceedings of the XVIIth International Geographical Conference, Washington, p. 227–342

Hobday, D. K. (1978) Fluvial deposits of the Ecca and Beaufort Groups in the eastern Karo basin, Southern Africa. In: Miall, A. D., ed., Fluvial Sedimentology. Can. Soc. Petrol. Geol., Mem. 5, 413–429

Hobday, D. K., Banks, N. L. (1971) A coarse-grained pocket beach complex, Tanafjord (Norway). Sedimentology 16, 129–134

Hobday, D. K., Eriksson, K. A., eds. (1977) Tidal Sedimentation. With particular reference to South African examples. Sediment. Geol. 18, 287 p.

Hobday, D. K., Reading, H. G. (1972) Fair weather versus storm processes in shallow marine sand bar sequences in late Pre-cambrian of Finnmark, North Norway. J. Sediment. Petrol. 42, 318–324

Hollenshead, C. T., Pritchard, R. L. (1961) Geometry of producing mesaverde sandstones, San Juan Basin. In: Peterson, J. A., Osmond, J. C., eds., Geometry of sandstone bodies, p. 98–118. Tulsa, Oklahoma: Am. Assoc. Petroleum Geologists.

Hollmann, R. (1968) Über Schalenabschliffe bei Cardium edule aus der Königsbucht bei List auf Sylt. Helgoländer Wiss. Meeresuntersuch. 18, 169–193

Holm, D. A. (1960) Desert geomorphology in the Arabian Peninsula. Science 132 (3427), 1369–1379

Holmes, A. (1965) Principles of physical geology (rev. ed.) 1288 p. London and Edinburgh: Nelson

Holst, N. O. (1876) Om de glaciala rullstensa sarne. Geol. Foren. Stockholm Forh. 3, 97–112

Holtedahl, H. (1975) The geology of the Hardangerfjord, West Norway. Norges geol. undersøkelse 323, 1–87

Homeier, H. (1969) Das Wurster Watt – Eine historisch-morphologische Untersuchung des Küsten- und Wattgebiets von der Weser- bis zur Elbmündung. Forsch.-Stelle Norderney. Jahrb. 1967, 19, 31–120

Homewood, P., ed., (1979) Sedimentation detritique (fluviatile, littorale et marine), 418 p. Documents complementaires, 133 p. Fribourg, Suisse. Inst. Geol. Univ. Fribourg

Homewood, P., Allen, P. A. (1981) Wave-, tide- and current-controlled sandbodies of Miocene Molasse, western Switzerland. Am. Assoc. Petroleum Geologists, Bull. 65, 2534–2545

Hooke, R. Leb. (1967) Processes on arid-region alluvial fans. J. Geol. 75, 438–460

Hooke, R. Leb., Yang, H.-Y., Weiblen, P. W. (1969) Desert varnish: an electron probe study. J. Geol. 77, 275–288

Hopkins, D. (1964) The Foynes Series: Namurian turbidites in western Ireland. Doctoral Thesis, University of Reading, 180 p.

Horie, S., ed. (1974) Paelolimnology of Lake Biwa and the Japanese Pleistocene. Contr. on the Paleolimnol. of Lake Biwa and the Japanese Pleistocene No. 43, 288 p.

Horie, S., ed. (1975) Paleolimnology of Lake Biwa and the Japanese Pleistocene. Contr. on the Paleolimnol. of Lake Biwa and the Japanese Pleistocene No. 85, 577 p.

Horie, S., ed. (1976) Paleolimnology of Lake Biwa and the Japanese Pleistocene. Contr. on the Paleolimnol. of Lake Biwa and the Japanese Pleistocene No. 155, 836 p.

Horie, S., ed. (1977) Paleolimnology of Lake Biwa and the Japanese Pleistocene. Contr. on the Paleolimnol. of Lake Biwa and the Japanese Pleistocene No. 203, 372 p.

Horie, S. (1978) Lacustrine sedimentation. In: Fairbridge, R. W., Bourgeois, J., eds., The Encyclopedia of Sedimentologie, 901 p. Stroudsburg, Pa.: Dowden, Hutchinson, Ross Inc.

Horikawa, K., Shen, H. W. (1960) Sand movement by wind action – on the charaeteristics of sand. Beach Erosion Board Technical Memorandum 119, 51 p.

Horn, D. R., Ewing, M., Horn, B. M., Delach, M. N. (1971) Turbidites of the Hatteras and Sohm abyssal plains, northwestern Atlantic. Marine Geol. 11, 287–323

Horne, R. A., Woernle, C. H. (1972) Iron and manganese profiles in a coastal pond with an anoxic zone. Chem. Geol. 9, 299–304

Horodyski, R. J., Bloeser, B., Haar, S., von der (1977) Laminated algal mats from a coastal lagoon, Laguna Mormona, Baja California, Mexico, J. Sediment. Petrol. 47, 680–696

Houbolt, J. J. H. C. (1968) Recent sediments in the southern bight of the North Sea. Geol. Mijnbouw 47, 245–273

Houbolt, J. J. H. C., Jonker, J. B. M. (1968) Recent sediments in the eastern part of the Lake of Geneve (Lac Léman). Geol. Mijnbouw 47, 131–148

Houten, F. B. van (1961) Climatic significance of red beds. In: Nairn, A. E. M., ed., Problems in palaeoclimatology, p. 89–139. New York: Interscience Publishers, Inc.

Houten, F. B. van (1964a) Origin of red beds – some unsolved problems. In: Nairn, A. E. M., ed., Problems in palaeoclimatology, p. 647–661. New York: Intersci. Pub., Inc.

Houten, F. B. van (1964b) Cyclic lacustrine sedimentation, Upper Triassic Lockatong Formation, central New Jersey and adjacent Pennsylvania. In: Merriam, D. F., ed., Symposium on cyclic sedimentation. Kans. Geol. Surv. Bull. 169 497–531

Houten, F. B. van (1965) Composition of Triassic Lockatong and associated formation of Newark Group, central New Jersey and adjacent Pennsylvania. Am. J. Sci. 263, 825–863

Houten, F. B. van (1968) Iron oxides in red beds. Geol. Soc. Am. Bull. 79, 399–416

Houten, F. B. van (1973) Origin of red beds: a review, 1961–1972. Ann. Rev. Earth and Planetary Sci. 1, 39–61

Howard, J. D. (1966a) Sedimentation of the Panther sandstone tongue Coals of central Utah Bull. 80, Utah Geol. Mineral. Surv. p. 23–33

Howard, J. D. (1966b) Characteristic trace fossils in upper Cretaceous sandstone of the Book Cliffs and Wasatch Plateau. Coals of central Utah Bull. 80. Utah Geol. Mineral. Surv. p. 35–53

Howard, J. D. (1975a) The sedimentological significance of trace fossils. In: Frey, R. W., editor, The study of trace fossils, 562 p. New York, Heidelberg, Berlin: Springer

Howard, J. D. (1975b) Estuaries of the Georgia Coast, U.S.A.: Sedimentology and Biology. IX. Conclusions. Senckenbergiana marit. 7, 297–305

Howard, J. D., Frey, R. W. (1975) Estuaries of the Georgia Coast, U.S.A.: Sedimentology and Biology. II. Regional animal-sediment characteristics of Georgia estuaries. Senckenbergiana marit. 7, 33–103

Howard, J. D., Lohrengel, C. F. (1969) Large non-tectonic deformational structures from upper Cretaceous rocks of Utah. J. Sediment. Petrol. 39, 1032–1039

Howard, J. D., Reineck, H.-E. (1972a) Georgia coastal region, Sapelo Island. U.S.A.: Sedimentology and biology. IV. Physical and biogenic sedimentary structures of the nearshore shelf. Senckenbergiana marit. 4, 81–123

Howard, J. D., Reineck, H.-E. (1972b) Georgia coastal region, Sapelo Island, U.S.A.: Sedimentology and biology. VIII. Conclusions. Senckenbergiana marit. 4, 217–222

Howard, J. D., Reineck, H.-E. (1981) Depositional facies of high-energy beach-to-offshore sequence: Comparision with low-energy sequence. Am. Assoc. Petroleum Geologists, Bull. 65, 807–830

Howard, J. D., Frey, R. W., Reineck, H.-E. (1972) Georgia coastal region, Sapelo Island, U.S.A.: Sedimentology and biology. I. Introduction. Senckenbergiana marit. 4, 3–14

Howard, J. D., Elders, C. A., Heinbokel, J. F. (1975a) Estuaries of the Georgia Coast, U.S.A.: Sedimentology and Biology. V. Animal-sediment relationships in estuarine point bar deposits, Ogeechee River-Ossabaw Sound, Georgia. Senckenbergiana marit. 7, 181–203

Howard, J. D., Remmer, Jr. G. H., Jewitt, J. L. (1975b) Estuaries of the Georgia Coast, U.S.A.: Sedimentology and Biology. VII. Hydrography and sediments of the Duplin River, Sapelo Island, Georgia. Senckenbergiana marit. 7, 237–256

Hoyt, J. H. (1962) High-angle beach stratification, Sapelo Island. Georgia. J. Sediment. Petrol. 32, 309–311

Hoyt, J. H. (1967a) Barrier Island formation. Geol. Soc. Am. Bull. 78, 1125–1136

Hoyt, J. H. (1967b) Genesis of sedimentary deposits along coasts of submergence. 23 International Geology Congress 8, 311–321

Hoyt, J. H. (1969) Chenier versus barrier, genetic and stratigraphic distinction. Bull. Am. Assoc. Petrol. Geologists 53, 299–306

Hoyt, J. H., Henry, V. J. (1963) Rhomboid ripple marks, indicator of current direction and environment. J. Sediment. Petrol. 33, 604–608

Hoyt, J. H., Weimer, R. J. (1965) The origin and significance of Ophiomorpha (Halymenites) in the Cretaceous of the western interior. Sed. late Cretaceous, Tertiary Outcrops, Rock Springs Uplift. 19. Field Conference, p. 203–207

Hoyt, J. H., Weimer, R. J., Henry, V. J., Jr. (1964) Late Pleistocene and recent sedimentation, central Georgia coast, U.S.A. In: Straaten, L.M.J. U. van, ed., Developments in sedimentology, vol. 1. Deltaic and shallow marine deposits, p. 170–176. Amsterdam: Elsevier

Hoyt, J. H., Henry, V. J., Howard, J. D. (1966) Pleistocene and Holocene sediments, Sapelo Island, Georgia and vicinity. Field Trip Southeast Section, Geologists Society of America. Field Trip Guidebook, vol. 1, p. 6–27

Hoyt, J. H., Henry, V. J., Jr., Weimer, R. J. (1968) Age of late Pleistocene shoreline depositis coastal Georgia. "Means of correlation of quaternary successions", Congr. Int. Assoc. Quaternary Res. 8, 381–393

Hsü, K. T. (1975) Catastrophic debris streams (sturzstroms) generated by rockfalls. Geol. Soc. Am. Bull. 86, 129–140

Hsü, K. J., Jenkyns, H. C., eds. (1974) Pelagic sediments on land and under the sea. Internat. Assoc. Sedimentologists Spec. Publ., 448 p. Oxford: Blackwell

Hubbard, D. K., Barwis, J. H. (1976) Discussion of tidal inlets sand deposits: Examples from the South Carolina coast. In: Hayes, M. O., Kana, T. W., eds., Terrigenous clastic depositional environments. Techn. Report No. 11-CRD. II-128-II-142 Columbia, S. C.: Univ. of South Carolina

Hubert, J. F. (1966) Sedimentary history of upper Ordovician geosynclinal rocks, Girvan, Scotland. J. Sediment. Petrol. 36, 677–699

Hülsemann, J. (1955) Großrippeln und Schrägschichtungs-Gefüge im Nordsee-Watt und in der Molasse. Senckenbergiana Lethaea 36, 359–388

Hülsemann, J., Emery, K. O. (1961) Stratification in recent sediments of Santa Barbara Basin as controlled by organisms and water character. J. Geol. 69, 279–290

Humbert, F. L. (1968) Selection and wear of pebbles on gravel beaches. Doctoral theisis. University of Groningen, 144 p.

Hunt, A. R. (1904) The descriptive nomenclature of ripplemark. Geol. Mag. 1, 410–418

Hunt, C. B. (1972) Geology of Soils, 344 p. San Francisco: Freeman

Hunt, C. B., Mabey, D. R. (1966) Stratigraphy and structure, Death Valley, California. U. S. Geol. Surv. Prof. Pap. 494-A, 162 p.

Hunt, C. B., Washburn, A. L. (1966) Patterned ground, U. S. G. S. Prof. Paper 494-B, 104–131

Hunt, L. M., Groves, D. G. (1965) A glossary of ocean science and undersea technology terms, 173 p. Arlington: Compass Publishers. Inc.

Hunt, R. E., Swift, D. J. P., Palmer, H. (1977) Constructional shelf topography, Diamond Shoals, North Carolina. Geol. Soc. Am. Bull. 88, 299–311

Hunter, R. E. (1969) Eolian microridges on modern beaches and a possible ancient example. J. Sediment. Petrol. 39, 1573–1578

Hunter, R. E. (1973) Pseudo-cross-lamination formed by climbing adhesion ripples. J. Sediment. Petrol. 43, 1125–1127

Hunter, R. E. (1977a) Basic types of stratification in small eolian dunes. Sedimentology 24, 361–387

Hunter, R. E. (1977b) Terminology of crosstratified sedimentary layers and climbing-ripple structures. J. Sediment. Petrol. 47, 697–706

Hutchinson, G. E. (1975) A treatise on limnology, Vol. I, Part I and II. New York: John Wiley & Sons

Illing, L. V., Wells, A. V., Taylor, J. C. M. (1965) Penecontemporaneous dolomite in the Persian Gulf. In: Pray, L. C., Murray, R. C., eds., Dolomitization and limestone diagenesis a symposium. Soc. Econ. Palaeontologists Mineralogists Spec. Publ. 13, 89–111

Imbowden, D. M., Lerman, A. (1978) Chemical models of lakes. In: Lerman, A., ed., Lakes. Chemistry, Geology, Geophysics, p. 341–356. New York, Heidelberg, Berlin: Springer

Ingle, J. C., Jr. (1966) The movement of beach sand. Developments in sedimentology, vol. 5, 221 p. Amsterdam: Elsevier

Ingram, R. L. (1954) Terminology for the thickness of stratification and parting units in sedimentary rocks. Geol. Soc. Am. Bull. 65, 937–938

Inman, D. L. (1949) Sorting of sediments in the light of fluid mechanics. J. Sediment. Petrol. 19, 51–70

Inman, D. L. (1952) Measures for describing the size distribution of sediments. J. Sediment. Petrol. 22, 125–145

Inman, D. L. (1957) Wave-generated ripples in nearshore sands. Department Army Corps of Engineers, Beach Erosion Board Technical Mem. 100, 1–65

Inman, D. L. (1970) Strong currents in submarine canyons. Abstr. Trans. Am. Geophys. Union 51, 319

Inman, D. L., Bagnold, R. A. (1963) Beach and nearshore processes In: Hill, M. N., eds. The sea. vol. 3. p. 529–553. New York: Wiley & Sons

Inman, D. L., Filloux, J. (1960) Beach cycles related to tide and local wind wave regime. J. Geol. 68, 225–231

Inman, D. L., Ewing, G. C., Corliss, J. B. (1966) Goastal sand dunes of Guerrero Negro, Baja California, Mexico. Geol. Soc. Am. Bull. 77, 787–802

Inman, D. L., Nordstorm, C. E., Flick, R. E. (1976) Currents in submarine canyons: An air-sea-land interaction. Ann. Rev. Fluid Mechanics 8, 275–310

Irion, G. (1973) Die anatolischen Salzseen, ihr Chemismus und die Entstehung ihrer chemischen Sedimente. Arch. Hydrolbiol. 71, 517–557

Irion, G. (1977) Der eozäne See von Messel. Nat. u. Mus. 107, 213–218

Irwin, J. (1972) Sediments of Lake Pukaki, South Island, New Zealand. N. Z. J. Marine & Freshwater Res. 6, 482–491

Irwin, J. (1974) Morphology and classification. In: Jolly, V. H., Brown, J. M. A., eds., New Zealand lakes, p. 25–57. Auckland: Univ. Press.

Jackson, R. G. (1975) Velocity-bed-form-texture patterns of meander bends in the Lower Wabash River of Illinois and Indiana. Geol. Soc. Am. Bull. 86, 1511–1522

Jackson, R. G. (1976a) Depositional model of point bars in the Lower Wabash River. J. Sediment. Petrol. 46, 579–594

Jackson, R. G. (1976b) Largescale ripples of the Lower Wabash River, Sedimentology 23, 593–623

Jackson, R. G. (1978) Preliminary evaluation of lithofacies models for meandering alluvial streams. In: Miall, A. W., ed., Fluvial Sedimentology Can. Soc. Petrol. Geol. Mem. 5, 543–576

Jahns, R. H. (1947) Geologic features of the Connecticut Valley, Massachusetts, as related to recent floods. U. S. Geol. Surv. Water Supply Papers 996, 1–158

Jaquet, J. M., Vernet, J. P., Thomas, R. L. (1975) Etude granulométrique des sédiments du Lac Léman. Congress Internat. de Sedimentologie, Nice 3, 39–48

Jarke, J. (1948) Die Unterschiede in der Sedimentation vor der Ost- und Westküste Schleswig-Holsteins. Dissertation. University of Kiel, p. 1–125

Jenkins, O. P. (1925) Mechanics of clastic dike intrusion. Eng. Miner. J.-Press 120, 1–12

Jennings, J. N., Conventry, R. J. (1973) Structure and texture of a gravelly barrier island in the Fitzroy estuary, western Australia, and the role of mangroves in the shore dynamics. Marine Geol. 15, 145–167

Johannsson, C. E. (1963) Orientation of pebbles in running water. A laboratory study. Geogr. Ann. XLV, 85–112

Johannsson, C. E. (1965) Structural studies of sedimentary deposits. Geol. Fören. Förhand. 87, 3–61

Johannsson, C. E. (1976) Structural studies of frictional sediments. Geografiska Ann. 58, 201–300

Johnsen, R. (1961) Wechselbeziehungen zwischen der Welle und dem strandnahen Unterwasserhang. Großmodellversuche und eine neue Anschauung über das Sandriffproblem. Veröff. Forsch.-Anstalt Schiffahrt, Wasser Grundbau 9, 1–102

Johnson, D. W. (1919) Shore processes and shoreline development, 584 p. New York: Wiley & Sons

Johnson, G. D., Vondra, C. F. (1972) Siwalik sediments in a portion of the Punjab Re-entrant: the sequence at Haritalyangar, District Bilaspur, H. P. Him. Geol. 2, 118–144

Johnson, H. D. (1978) Shallow siliciclastic seas. In: Reading, H. G., ed., Sedimentary Environments and Facies, p. 207–258. London: Blackwell Sci. Publ.

Johnston, W. A. (1921) Sedimentation of the Fraser River Delta. Canada Geol. Surv. Mem. 125, 1–46

Johnston, W. A. (1922) The character of the stratification of the sediments in the recent delta of Fraser River, British Columbia. Can. J. Geol. 30, 115–129

Jones, B. F., Bowser, C. J. (1978) The mineralogy and related chemistry of lake sediments. In: Lerman, A., ed., Lakes. Chemistry, Geology, Geophysics, p. 179–227. New York, Heidelberg, Berlin: Springer

Jopling, A. V. (1963) Hydraulic studies on the origin of bedding. Sedimentology 2, 115–121

Jopling, A. V. (1965a) Laboratory study of sorting processes in cross-bedded deposits. In: Middleton, G. V., ed., Primary sedimentary structures and their hydrodynamic interpretation. Soc. Econ. Paleontologists Mineralogists Spec. Publ. 12, 53–65

Jopling, A. V. (1965b) Hydraulic factors and the shape of laminae. J. Sediment. Petrol. 35, 777–791

Jopling, A. V. (1966) Some principles and techniques used in reconstructing the hydraulic parameters of a paleo-flow regime. J. Sediment. Petrol. 36, 5–49

Jopling, A. V. (1967) Origin of laminae deposited by the movement of ripples along a streambed: A laboratory study, J. Geol. 75, 287–305

Jopling, A. V., McDonald, B. C., eds. (1975) Glaciofluvial and Glaciolacustrine Sedimentation. Soc. Econ. Palaeontologists Mineralogists Spec. Publ. 23, 320 p.

Jopling, A. B., Richardson, E. V. (1966) Backset bedding developed in shooting flow in laboratory experiments. J. Sediment. Petrol. 36, 821–825

Jopling, A. V., Walker, R. G. (1968) Morphology and origin of ripple-drift cross lamination, with examples from the Pleistocene of Massachusetts. J. Sediment. Petrol. 38, 971–984

Jordan, G. F. (1951) Continental slope off Apalachicola River, Florida. Bull. Am. Assoc. Petrol. Geologists 35, 1978–1993

Jordan, G. F. (1962) Large sand waves in estuaries and in the open sea. Proc. First Nat. Coastal Shallow Water Res. Conf., p. 232–236

Jüngst, H. (1934) Zur geologischen Bedeutung der Synärese. Geol. Rundschau 15, 312–325

Jüngst, H. (1942) Schillkalk ("Schelpkalk") als nationale Industrie. Geol. d. Meere u. Binnengewässer 5, 220–231

Jux, U. (1964) Erosionsformen durch Gezeitenströmungen in den unterdevonischen Bensberger Schichten des Bergischen Landes? N. Jb. Geol. Paläont. Mh. 9, 515–530

Kalterherberg, J. (1956) Über Anlagerungsgefüge in grobklastischen Sedimenten. Neues Jahrb. Geol. Palaeontol. Abhandl. 104, 30–57

Kanes, W. H. (1970) Facies and development of the Colorado River Delta in Texas. In: Morgan, J. P., ed., Deltaic sedimentation, modern and ancient. Soc. Econ. Paleontologists Mineralogists Spec. Publ. 15, 78–106

Karcz, I. (1967) Harrow marks, current aligned sedimentary structures. J. Geol. 75, 113–121

Karcz, I. (1968) Fluviatile obstacle marks from the wadis of the Negev (southern Israel). J. Sediment. Petrol. 38, 1000–1012

Karcz, I. (1972) Sedimentary structures formed by flash floods in southern Israel, Sediment. Geol. 7, 161–182

Karcz, I., Goldberg, M. (1967) Ripple controlled dessication patterns from Wadi Shigma, southern Israel. J. Sediment. Petrol. 37, 1244–1245

Katzung, G. (1971) Zur fluviatilen Gerölleinregelung. Z. für angew. Geologie 17, 39–47

Kaye, C. A., Power, W. R. (1954) A flow cast of very recent date from northeastern Washington. Am. J. Sci. 252, 309–310

Keen, M. J., ed. (1969) An Introduction to Marine Geology, 218 p. New York: Pergamon Press

Keller, G. (1952) Beitrag zur Frage Oser und Kames. Eiszeitalter Gegenwart 2, 127–132

Keller, G. H., Shepard, F. P. (1978) Currents and sedimentary processes in submarine canyons off the Northeast United States. In: Stanley, D. J., Kelling, G., eds., Sedimentation in submarine canyons, fans and trenches. Chap. 2, 15–32. Stroudsburg, Penna: Dowden, Hutchinson and Ross

Kelling, G., Mullin, P. R. (1975) Graded limestones and limestone-quartzite couplets: Possible storm-deposits from the Moroccan carboniferous. Sedimentary Geol. 13, 161–190

Kelling, G., Stanley, D. J. (1970) Morphology and structure of Wilmington and Baltimore submarine canyons, eastern United States. J. Geol. 78, 637–660

Kelling, G., Walton, E. K. (1957) Load cast structures: Their relationship to upper surface structures and their mode of formation. Geol. Mag. 94, 481–490

Kelts, K. R. (1978) Geological and Sedimentological Evolution of Lake Zurich and Lake Zug. Unpubl. Ph. D. Thesis, ETH. (cit. Lambert, A. & Hsü, K. J., 1979)

Kelts, K., Hsü, K. J. (1978) Freshwater carbonate sedimentation. In: Lerman, A., ed., Lakes. Chemistry, Geology, Physics, p. 295–232. New York, Heidelberg, Berlin: Springer

Kemp, A. L. W., Anderson, T. W., Thomas, R. L., Mudrochova, A. (1974) Sedimentation rates and recent sediment history of

lakes Ontario, Erie and Huron. J. Sediment. Petrol. 44, 207–218

Kennedy, J. F. (1961) Stationary waves and antidunes in alluvial channels. W. M. Keck Lab. Hydraulics Water Resources, Rep. KH-R-2, Cal. Inst. Techn. Pasadena, 146 p.

Kennedy, J. F. (1964) The formation of sediment ripples in closed rectangular conduits and in the desert. J. Geophys. Res. 69, 1517–1524

Kennedy, W. J., Juignet, P. (1974) Carbonate banks and slump beds in the upper Cretaeous (Upper Turonian-Sontonian) of Haute Normandie, France. Sedimentology 21, 1–42

Kent, H. C. (1976) Modern coastal sedimentary environments Alabama and Northwest Florida. Geol. Exploration Assoc. Ltd. 1–96. Golden, Colorado

Kenyon, N. H. (1970) Sand ribbons of European tidal seas. Marine Geol. 9, 25–39

Kenyon, N. H., Belderson, R. H., Stride, A. H. (1978) Channels, canyons and slump folds on the continental slope between South-West Ireland and Spain. Oceanol. Acta 1, 369–380

Ketchum, B. H. (1953) Circulation in estuaries. Proc. Third Conf. Coastal Eng., p. 65–76

Kindle, C. H., Whittington, H. B. (1958) Stratigraphy of the Cowhead region, western Newfoundland. Geol. Soc. Am. Bull. 69, 315–342

Kindle, E. M. (1917) Recent and fossil ripple-mark. Museum Bull. Geol. Surv. Canada 25, 121 p.

King, C. A. M. (1972) Beaches and coasts, 570 p. London: Arnold Ltd.

King, C. A. M., Buckley, J. T. (1968) The analysis of stone size and shape in arctic environments. J. Sediment. Petrol. 38, 200–214

Klasik, J. A., Pilkey, O. H. (1975) Processes of sedimentation on the Atlantic continental rise off the southeastern U.S. Mar. Geol. 19, 69–89

Klebelsberg, R. von (1948) Handbuch der Gletscherkunde und Glazialgeologie, vol. 1, 403 p. Vienna: Springer

Klein, G. de V. (1962) Sedimentary structures in the Keuper Marl (Upper Triassic). Geol. Mag. 99, 137–144

Klein, G. de V. (1965) Diverse origins of graded bedding. Geol. Soc. Am. Spec. Papers 82, 1–112

Klein, G. de V. (1967) Paleocurrent analysis in relation to modern marine sediment dispersal patterns. Bull. Am. Assoc. Petrol. Geologists 51, 366–382

Klein, G. de V. (1970a) Depositional and dispersal dynamics of intertidal sand bars. J. Sediment. Petrol. 40, 1095–1127

Klein, G. de V. (1970b) Tidal origin of a Precambrian quartzite the lower fine-grained quartzite (middle Dalradian) of Islay, Scotland, J. Sediment. Petrol. 40, 973–985

Klein, G. de V. (1971) A sedimentary model for determining Paleotidal range. Geol. Soc. Am. Bull. 82, 2585–2592

Klein, G. de V. (1972) Determination of Paleotidal range in clastic sedimentary rocks. 24th Int. Geol. Cong. 6, 397–405

Klein, G. de V. (1975) Sandstone depositional models for exploration for fossil Fuels. Short Course Syllabus, 109 p. Champaign, Ill.: Continuing Educ. Publ. Comp.

Klein, G. de V. (1977a) Tidal circulation model for deposition of clastic sediments in epeiric and mioclinal shelf seas. Sediment. Geol. 18, 1–12

Klein, G. de V. (1977b) Clastic tidal facies, 149 p. Champaign. Ill.: CEPCO

Klein, G. de V., Whaley, M. L. (1972) Hydraulic parameters controlling bedform migration on a intertidal sand body. Geol. Soc. Am. Bull. 83, 3465–3470

Klemme, H. D. (1958) Regional geology of Circum-Mediterranean region. Bull. Am. Assoc. Petrol. Geologists 42, 477–512

Knebel, H. J., Folger, D. W. (1976) Large sand waves on the Atlantic outer continental shelf around Wilmington Canyon, off eastern United States. Mar. Geol. 22, M7-M15

Knight, R. J., Dalrymple, R. W. (1976) Winter conditions in a macrotidal environment, Cobequic Bay, Nova Scotia. Rev. Géogr. Montr. 30, 65–85

Kögler, F. C. (1963) Das Kastenlot. Meyniana 13, 1–7

Köhler, J. (1963) Salz- und Schillgewinnung an Europäischen Küsten. Staatsarbeit, Münster, 132 S.

Königswald, R. von (1930) Die Arten der Einregelung ins Sediment bei den Seesternen und Seelilien des unterdevonischen Bundenbacher Schiefers. Senckenbergiana 12, 338–360

Köster, E. (1964) Granulometrische und morphometrische Meßmethoden an Mineralkörnern, Steinen und sonstigen Stoffen, 336 p. Stuttgart: F. Enke.

Köster, R. (1955) Die Morphologie der Strandwall-Landschaften und die erdgeschichtliche Entwicklung der Küsten Ostwagriens und Fehmarns. Meyniana 4, 51–65

Kolb, C. R., Lopik, J. R., van (1966) Depositional environments of the Mississippi River deltaic plain, southeastern Louisiana. In: Shirley, M. L., ed., Deltas in their geologic framework, p. 17–61. Houston: Houston Geol. Soc.

Koldewijn, B. W. (1958) Sediments of the Paraia-Trinidad shelf, Repts. Orinoco Shelf Exped., vol. 3, 109 p. Den Haag: Mouton & Co

Kolmer, J. R. (1973) A wave tank analysis of the beach foreshore grain size distribution. J. sediment. Petrol. 43, 200–204

Kolp, O. (1958) Sedimentsortierung und Umlagerung am Meeresboden durch Wellenwirkung. Petermanns Geogr. Mitt. 102, 173–178

Kolp, O. (1966) Die Sedimente der westlichen und südlichen Ostsee und ihre Darstellung. Beitr. Meereskde. 17–81, 1–60

Komar, P. D. (1972) Relative significance of head and body spill from a channelized turbidity current. Geol. Soc. Am. Bull. 1151–1156

Komar, P. D. (1973) Observations of beach cusps at Mano Lake, California. Geol. Soc. Am. Bull. 84, 3593–3600

Komar, P. D. (1978) Relative quantities of suspension versus bed-load transport on beaches. J. Sediment. Petrol. 48, 921–932

Komar, P. D., Miller, M. C. (1975) The initiation of oscillatory ripple marks and the development of plane-bed at high-shear stresses under waves. J. Sediment. Petrol. 45, 697–703

Komar, P. D., Neudeck, R. H., Kulm, L. D. (1972) Observations and significance of deep-water oscillatory ripple marks on the Oregon Continental Shelf. In: Swift, D. J. P., Duane, D. B., Pilkey, O. H., eds., Shelf Sediment Transport Process and Pattern, p. 601–619. Stroudsbourg, Pa.: Dowden, Hutchinson & Ross

Komar, P. D., Kulm, L. D., Harlett, J. C. (1974) Observations and analysis of bottom turbid layers on the Oregon continental shelf. J. Geol. 82, 104–111

Kondrat'ev, N. E. (1953) Calculations of wind waves and modification of reservoir banks. Gidrometeoizdat. (cit. Zenkovich, 1967)

Koopmann, G. (1962) Wasserstanderhöhungen in der Deutschen Bucht infolge von Schwingungen und Schwallerscheinungen, angewandt auf die Sturmflut vom 16/17 Februar 1962. Deut. Hydrogr. Z. 15

Korn, J. (1910) Erläuterungen zu Blatt Marienfließ 1 : 25 000. Berlin: Kgl. Preuß. Geol. Landesanstalt

Koster, E. H. (1978) Transverse ribs: their characteristics, origin and paleohydraulic significance. In: Miall, A. D., ed., Fluvial Sedimentology. Can. Soc. Petrol. Geol. Mem. 5, 161–186

Kovacs, A. (1972) Ice scouring marks floor of the arctic shelf. Oil and Gas J. Oct. 23, 92–106

Kowalczyk, G. (1974) Kryoturbationsartige Sedimentstrukturen im Pliozän und Altquartär der südlichen Niederrheinischen Bucht. Eiszeitalter u. Gegenwart 25, 141–156

Kraft, J. C. (1971) Sedimentary facies patterns and geologic history of a Holocene marine transgression. Geol. Soc. Am. Bull. 82, 2131–2158

Kraft, J. C. (1972) A reconnaissance of the geology of the sandy coastal areas of Eastern Greece and the Peloponnese. Techn. Rep. 9, 1–158

Kranck, K. (1973) Flocculation of suspended sediment in the sea. Nature 246, 348–350

Kranck, K. (1975) Sediment deposition from flocculated suspensions. Sedimentology 22, 111–124

Krause, P. G. (1911) Über Oser in Ostpreußen. Jahrb. Preuß. Geol. Landesanstalt 32, 76–91

Krauskopf, K. B. (1967) Introduction to geochemistry, 721 p. New York: McGraw Hill

Kreft, G., Michaelis, H. (1976) Die Meeräsche im niedersächsischen Wattenmeer. Nat. u. Mus. 106, 23–29

Krejci-Graf, K. (1932) Definition der Begriffe Marken, Spuren, Fährten, Bauten, Hieroglyphen und Fucoiden. Senckenbergiana 14, 19–39

Krejci-Graf, K. (1964) Geochemical diagnosis of facies. Proc. Yorkshire Geol. Soc. 34, 469–521

Krejci-Graf, K. (1966) Geochemische Faziesdiagnostik. Freiberger Forsch.-H. C 224, 1–80

Krejci-Graf, K. (1972) Trace metals in sediments, oils, and allied substances. In: Fairbridge, R. W., ed., The Encyclopedia of Geochemistry and Environmental Sciences, p. 1201–1209. Stroudsburg: Dowden, Hutchinson & Ross

Krigström, A. (1962) Geomorphological studies of sandur plains and their braided rivers in Iceland. Geogr. Ann. 44, 328–346

Krinsley, D., Donahue, J. (1968) Environmental interpretation of sand grain surface textures by electron microscopy. Geol. Soc. Am. 79, 743–748

Krinsley, D., Doornkamp, J. (1973) Atlas of Quartz Sand Grain Surface Textures, 91 p. New York: Cambridge Univ. Press

Krinsley, D., Funnell, B. (1965) Environmental history of sand grains from the lower and middle Pleistocene of Norfolk, England. Quart. J. Geol. Soc. London 121, 435–456

Krinsley, D., Margolis, S. (1969) A study of quartz sand grain surface textures with the scanning electron microscope. Trans. N. Y. Acad. Sci., Ser. 2, 31, 457–477

Krinsley, D., Takahashi, T. (1962a) Surface textures of sand grains: An application of electron-microscopy. Science 135, 923–925

Krinsley, D., Takahashi, T. (1962b) Electron microscopic examination of natural and artificial glacial sand grains. Geol. Soc. Am. Spec. Papers 73, 1–175

Kruit, C. (1955) Sediments of the Rhône Delta. Grain size and microfauna. Verhandel. Ned. Geol.-Mijnbouwk. Genoot., Geol. Ser. 15, 357–514

Krumbein, W. C. (1941) Measurement and geological significance of shape, and roundness of particles. J. Sediment. Petrol. 11, 64–72

Krumbein, W. C., Aberdeen, E. (1937) The sediments of Barataria Bay. J. Sediment. Petrol. 7, 3–17

Krumbein, W. C., Garrels, R. M. (1952) Origin and classification of chemical sediments in terms of pH and oxidation-reduction potentials. J. Geol. 60, 1–33

Krumbein, W. C., Pettijohn, F. J. (1938) Manual of sedimentary petrology, 549 p. New York: Appleton-Century-Crofts. Inc.

Krumbein, W. C., Sloss, L. L. (1963) Stratigraphy and sedimentation, 2d ed., 660 p. San Fransisco: Freemann

Krumbein, W. E., Cohen, Y. (1974) Biogene, klastische und evaporitische Sedimentation in einem mesothermen monomiktischen ufernahen See (Golf von Aqaba). Geol. Rundschau. 63, 1035–1065

Krynine, P. D. (1948) The megascopic study and classification of sedimentary rocks. J. Geol. 56, 130–165

Krynine, P. D. (1949) The origin of red beds. Trans. N. Y., Acad. Sci. Ser. 2, 11, 60–68

Ksiazkiewicz, M. (1954) Graded and laminated bedding in the Carpathian flysch. Ann. Soc. Geol. Pologne 22, 399–449

Ksiazkiewicz, M. (1958) Submarine slumping in the Carpathian flysch. Ann. Soc. Geol. Pologne 28, 123–150

Ksiazkiewicz, M. (1977) Trace fossils in the flysch of the Polish Carpathians. Palaeontol. Polonica 36, 208 p.

Kudras, H. R. (1974) Experimental study of nearshore transportation of pebbles with attached algae. Mar. Geol. 16, M9–M12

Kuenen, Ph. H. (1937) Experiments in connection with Daly's hypothesis on the formation of submarine canyons. Leidse Geol. Mededel. 8, 327–335

Kuenen, Ph. H. (1947a) Water-faceted boulders. Am. J. Sci. 245, 779–783

Kuenen. Ph. H. (1947b) Two problems of marine geology: Atolls and canyons. Verhandel. Koninkl. Ned. Akad. Wetenschap., Afdel. Natuurk., Sect. 2, 43, 37–68

Kuenen, Ph. H. (1948) The formation of beach cusps. J. Geol. 56, 36–40

Kuenen, Ph. H. (1949) Slumping in the Carboniferous rocks of Pembrokeshire. Quart. J. Geol. Soc. London (1948) 104, 365–385

Kuenen, Ph. H. (1950a) Marine geology, 568 p. New York: Wiley & Sons

Kuenen, Ph. H. (1950b) Turbidity currents of high density. Intern. Geol. Congr. 18th, London, 1948, Rept. 8, 44–52

Kuenen, Ph. H. (1951a) Turbidity currents as the cause of glacial varves. J. Geol. 59, 507–508

Kuenen, Ph. H. (1951b) Mechanics of varve formation and the aaction of turbidity current. Geol. Foren. Stockholm Forh. 73, 69–84

Kuenen, Ph. H. (1953a) Significant features of graded bedding. Bull. Am. Assoc. Petrol. Geologists 37, 1044–1066

Kuenen, Ph. H. (1953b) Graded bedding, with observations on lower Paleozoic rocks of Britain. Verhandel. Koninkl. Ned. Akad. Wetenschap. Afdel. Natuurk., Sect. 1e, 20, 3, 1–47

Kuenen, Ph. H. (1963) Experimentele sedimentstructuren. Koninkl. Ned. Akad. Wetenschap Amsterdam, Versl. Gw. Verg. Afd. Nat. 72, 65–66

Kuenen, Ph. H. (1964a) Experimental abrasion: 6. Surf action. Sedimentology, 3, 29–43

Kuenen, Ph. H. (1964b) Pivotability studies of sand by a shape-sorter. In: Straaten, L. M. J. U. van, ed., Deltaic and shallow marine deposits. Developments in sedimentology, vol. 1, p. 207–215. Amsterdam: Elsevier

Kuenen, Ph. H. (1964c) Deep-sea sands and ancient turbidites. In: Bouma, A. H., Brouwer, A., eds., Turbidites. Developments in sedimentology, vol. 3, p. 3–33. Amsterdam: Elsevier.

Kuenen, Ph. H. (1965) Value of experiments in geology. Geol. Mijnbouw 44, 22–36

Kuenen, Ph. H. (1966a) Experimental turbidite lamination in a circular flume. J. Geol. 74, 523–545

Kuenen, Ph. H. (1966b) Turbidity currents. In: Fairbridge, R. W., ed. The encyclopedia of oceanography. Encyclopedia of earth sciences series, vol. 1, p. 943–948. New York: Reinhold Publ. Corp.

Kuenen, Ph. H., Humbert, F. L. (1964) Bibliography of turbidity currents and turbidites. In: Bouma, A. H., Brouwer, A., eds., Developments in sedimentology, vol. 3, p. 222–246. Amsterdam: Elsevier

Kuenen, Ph. H., Migliorini, C. I. (1950) Turbidity currents as a cause of graded bedding. J. Geol. 58, 91–127

Kuenen, Ph. H., Prentice, J. E. (1957) Flow markings and loadcasts. Geol. Mag. 94, 173–174

Kues, B. S. and Siemers, C. T. (1977) Control of mudcrack patterns by the infaunal bivalve Pseudocyrena. J. Sediment. Petrol. 47, 844–848

Kühl, H., Mann, H. (1954) Über die Hydrochemie der unteren Ems. Veröff. Inst. Meeresforsch. 3, 126–158

Kühlmorgen-Hille, G. (1963) Quantitative Untersuchungen der Bodenfauna in der Kieler Bucht und ihre jahreszeitlichen Veränderungen. Kieler Meeresforsch. 19, 42–66

Kürsten, M. (1960) Zur Frage der Geröllorientierung in Flußläufen. Geol. Rundschau 49, 498–501

Kugler, H. G. (1953) Jurassic to Recent sedimentation in Trinidad. Assoc. Swisse Geol. Ing. Petrole Bull. 20, 27–60

Kukal, Z. (1970) Geology of Recent sediments, 490 p. Prague: Academia

Kulm, L. D., Roush, R. C., Harlet, J. C., Neudeck, R. H., Chambers, D. M., Runge, E. J. (1975) Oregon continental shelf

sedimentation: Interrelationships of facies distribution and sedimentary processes. J. Geol. 83, 145–175

Kumar, N., Sanders, J. E. (1975) Inlet sequence formed by the migration of Fire Island inlet, Long Island, New York. In: Ginsburg, R. N., Tidal Deposits, p. 75–83. New York, Heidelberg, Berlin: Springer

Kumar, S., Singh, I. B. (1978) Sedimentological study of Gomti River sediments, Uttar Prodesh, India. Example of a river in alluvial plain. Senckenbergiana marit. 10, 145–211

Kuster-Wendenburg, E. M. (1974) Fazielle, biostratonomische und feinstratigraphische Untersuchungen dreier Meeressandvorkommen (Rupelium) im Mainzer Tertiärbecken. Geol. Jb. A 22, 107 p.

Lachenbruch, A. H. (1962) Mechanics of thermal contraction cracks and ice-wedge polygons in permafrost. Geol. Soc. Am. Spec. Papers 70, 1–69

Lafond, E. C. (1966) Upwelling. In: Fairbridge R. W., ed., The Encyclopedia of Oceanography. Encycl. Earth Sci. Ser. 1, 957–959. New York: Reinhold Publ. Corp.

Lagally, M. (1933) Mechanik und Thermodynamik der stationären Gletscher. Gerl. Beitr. Geophys., Suppl. 2

Laking, P. N., (1974) The Black Sea. A Bibliography, 368 p. Woods Hole, Mass.: Woods Hole Oceanogr. Inst.

Lambert, A. M. (1978) Eintrag, Transport und Ablagerung von Feststoffen im Walensee. Eclog. geol. Helv. 71

Lambert, A. M., Hsü, K. J. (1979) Non-annual cycles of varve-like sedimentation in Walensee, Switzerland. Sedimentology 26, 453–461

Lambert, A. M., Kelts, K. R., Marshall, N. F. (1976) Measurement of density underflows from Walensee, Switzerland. Sedimentology 23, 87–105

Lamcke, K. (1940) Natürliche Anreicherungen von Schwermineralien in Küstengebieten. (2). Geol. Meere und Binnengewässer 4, 77–92

Lane, E. W. (1938) Notes on the formation of sand. Am. Geophys. Union Trans. 19, 505–508

Lang, A. W. (1970) Untersuchungen zur morphologischen Entwicklung des südlichen Elbe-Ästuars von 1560 bis 1960. Hamburger Küstenforsch. 12, 1–195

Lang, J., Lucas, G., Mathieu, R. (1973) Le domaine benthique littoral de la baie du Mont-Saint Michel (Manche). Sci. Terre 18, 19–78

Lankford, R. R., Shepard, F. P. (1960) Facies interpretations in Mississippi Delta borings. J. Geol. 68, 408–426

Laporte, L. F. (1968) Ancient environments. In: McAlester, A. L., ed., Foundation of earth science series, 115 p. Englewood Cliffs, New Jersey: Prentice-Hall, Inc.

Laprade, K. E. (1957) Dust-storm sediments of Lubbock area, Texas. Bull. Am. Assoc. Petrol. Geologists 41, 709–726

Larsen, V., Steel, R. J. (1978) The sedimentary history of a debris-flow dominated. Devonian alluvial fan – a study of textural inversion. Sedimentology 25, 37–59

Larsonneur, C. (1975) Tidal deposits, Mont Saint-Michel Bay, France. In: Ginsburg, R. N., ed., Tidal deposits, p. 21–30. New York, Heidelberg, Berlin: Springer

Lattman, L. H. (1960) Cross-section of a flood plain in a moist region of moderate relief. J. Sediment. Petrol. 30, 275–282

Lauff, G. H., ed. (1967) Estuaries. Publ. No. 83, 757 p. Washington: Am. Assoc. Advancem. Sci.

Leatherman, S. P., Williams, A. T. (1977) Lateral textural grading in overwash sediments. Earth Surface Processes 2, 333–341

LeBlanc, R. J. (1972) Geometry of sandstone reservoir bodies. In: Underground waste management and environmental implications. Am. Ass. Petrol. Geol. Mem. 18, 133–190

LeBlanc, R. J. (1975) Significant studies of modern and ancient deltaic sediments. In Broussard, M. L., ed. Deltas, models for exploration, p. 13–86. Houston, Tex.: Houston Geol. Soc.

Leeder, M. R. (1973) Fluviatile fining-upwards cycles and the magnitude of palaeochannels. Geol. Mag. 110, 265–276

Leeder, M. R. (1975) Pedogenic carbonate and flood sediment accretion rates: a quantitative model for alluvial, arid-zone lithofacies. Geol. Mag. 112, 257–270

Leeder, M. R., Bridges, P. H. (1975) Flow seperation in meander bends. Nature 253, 338–339

Legget, R. F., ed. (1976) Glacial till: An Interdisciplinary study. Spec. Publ. R. Soc. Canada 12, 412 p.

Leggett, R. F., Brown, R. J. E., Johnston, G. H. (1966) Alluvial fan formation near Aklavik, Northwest Territories, Canada. Geol. Soc. Am. Bull. 77, 15–30

Leliavsky, S. (1955) An introduction to fluvial hydraulics, 257 p. London: Constable

Lenk-Chevitch, P. (1959) Beach and stream-pebbles. J. Geol. 67, 103–108

Leopold, L. B., Wolman, M. G. (1957) River channel patterns: braided, meandering and straight. U.S. Geol. Surv. Profess. Papers 282-B, 39–85

Leopold, L. B., Wolman, M. G. (1960) River meanders. Geol. Soc. Am. Bull. 71, 569–794

Leopold, L. B., Wolman, M. G., Miller, J. P. (1964) Fluvial process in geomorphology, 522 p. San Francisco, London: Freeman

Lerman, A. (1978) Lakes Chemistry, Geology, Physics, 366 p. New York, Heidelberg, Berlin: Springer

Lever, J. (1958) Quantitative beach research. I. The "left-right-phenomenon" sorting of Lamellibranch valves on sandy beaches. Basteria 22, 21–51

Lever, J., Kessler, A., Overbeeke, A. P. van, Thijssen, R. (1961) Quantitative beach research. II. The "hole effect": A second mode of sorting of lamellibranch valves on sandy beaches. J. Sea Res. 1, 339–358

Lever, J., Bosch, M. van den, Cook, H., Dijk, T. van, Thiadens, S. J., A. J. H., Thijssen, R. (1964) Quantitative beach research. III. An experiment with artificial valves of Donax Vittatus. J. Sea Res. 2, 458–492

Levey, R. A. (1978) Bed-form distribution and internal stratification of coarse-grained point bars upper Congaree River, S. C. In: Miall, A. D., ed., Fluvial Sedimentology. Can. Soc. Petrol. Geol. Mem. 5, 105–127

Lewin, (1976) Initiation of bed forms and meanders in coarse-grained sediment. Geol. Soc. Am. Bull. 87, 281–285

Lewis, K. B. (1971) Slumping on a continental slope included at 1°–4°. Sedimentology 16, 97–110

Lewis, K. B. (1974) The continental terrace. Earth-Sci. Rev. 10, 37–71

Lewis, W. V. (1949) An esker in process of formation: Böverbreen. Jotunheimen. J. Glacial. 1, 314–319

Liboriussen, J. (1975) A study of gravel fabric. Sediment. Geol. 14, 235–251

Linden, W. J. M. van der (1974) The surficial geology of Hamilton Bank and periphery. Geol. Surv. Can. 1, 157–160

Lindström, M. (1979) Storm surge turbation. Sedimentology 26, 115–124

Lingen, G. J. van der (1969) The turbidite problem. New Zealand J. Geol. Geophys. 12, 7–50

Lingen, G. J. van der (1970) The turbidite problem: A reply to Kuenen. New Zealand J. Geol. Geophys. 13, 858–872

Linke, G. (1970) Über die Verhältnisse im Gebiet Neuwerk/Scharhörn. Hamburger Küstenforsch. 17, 17–58

Linke, O. (1939) Die Biota des Jadebusens. Helgolaender Wiss. Meeresuntersuch. 1, 201–348

Lisitzin, A. P., ed. (1972) Sedimentation in the world ocean. Soc. Econ. Paleontologists Mineralogists Spec. Publ. 17, 218 p.

Little-Gadow, S., Reineck, H.-E. (1974) Diskontinuierliche Sedimentation von Sand und Schlick in Wattensedimenten. Senckenbergiana marit. 6, 149–159

Liu, H. K. (1957) Mechanics of sediment-ripple formation. Am. Soc. Civ. Engrs. Proc. HY2, 83, 1–23

Lombard, A. (1956) Géologie Sedimentaire, les Series Marines. 722 p. Liège, Belgien: H. Vaillant-Charmanne, S. A.

Lombard, A. (1959) Lithological Sequences in Marine Series. Geol. Mijnbow 21, 185–190

Lombard, A. (1972) Séries sédimentaires. Genèse – Evolution, 425 p., Paris

Longinov, V. V. (1956) The possibility of direct study of the effect of waves in moving material under natural conditions. Trudy okeonogr. Kom. Akad. Nauk SSSR, 1. (cit. Zenkovitch, 1969)

Longinov, V. V. (1958 a) An attempt to determine the material-moving effect of waves from observations of wave transformation in the coastal zone. Turdy Inst. Okeanol. Akad. Nauk SSSR, 28 (cit. Zenkovitch, 1969)

Longinov, V. V. (1958 b) The dynamics of the breaker zone and of the coastal zone as a whole. Liet. TSR Mokslu Akad. Geol. Geogr. Inst. 7. (cit. Zenkovitch, 1969)

Lonsdale, P., Malfait, B. (1974) Abyssal dunes of foraminiferal sand on the Carnegie Ridge. Geol. Soc. Am. Bull. 85, 1697–1712

Lonsdale, P., Normark, W. R., Newman, W. A. (1972) Sedimentation and erosion on Horizon Guyot. Geol. Soc. Am. Bull. 83, 289–316

Loon, A. J. van, Wiggers, A. J. (1975 a) Composition and grain-size distribution of the Holocene Dutch "sloef" (Almere Member of the Groningen Formation). Sediment. Geol. 13, 237–251

Loon, A. J. van, Wiggers, A. J. (1975 b) Erosional features in the lagoonal Almere Member ("sloef") of the Groningen Formation (Holocene, Central Netherlands). Sediment. Geol. 13, 253–265

Loon, A. J. van, Wiggers, A. J. (1975 c) Holocene lagoonal silts (formerly called "sloef") from the Zuiderzee. Sediment. Geol. 13, 47–55

Loon, A. J. van, Wiggers, A. J. (1976 a) Abrasion as an agent for sand supply in a Holocene lagoon (Almere and Zuiderzee Members, Groningen Formation) in The Netherlands. Sediment. Geol. 15, 293–307

Loon, A. J. van, Wiggers, A. J. (1976 b) Primary and secondary synsedimentary structures in the lagoonal Almere Member (Groningen Formation, Holocene. The Netherlands). Sediment. Geol. 16, 89–97

Loon, A. J. van, Wiggers, A. J. (1976 c) Metasedimentary "graben" and associated structures in the lagoonal Almere Member (Groningen Formation, The Netherlands). Sediment. Geol. 16, 237–254

Loring, D. H., Nota, D. J. G. (1973) Morphology and sediments of the Gulf of St. Lawrence. Bull. Fisheries Res. Board Canada. 182, 1–147

Loring, D. H., Nota, D. J. G., Chesterman, W. D., Wong, H. K. (1970) Sedimentary environments on the Magdalen shelf, southern Gulf of St. Lawrence. Mar. Geol. 8, 337–354

Lowe, D. R. (1975) Water-escape structures in coarse-grained sediments. Sedimentology 22, 157–204

Lowe, D. R. (1976) Subaqueous liquefied and fluidized sediment flows and their deposits. Sedimentology 23, 285–308

Lowe, D. R., Lopiccolo, R. D. (1974) The characteristics and origins of dish and pillar structures. J. Sediment. Petrol. 44, 484–501

Ludwig, G., Vollbrecht, K. (1957) Die allgemeinen Bildungsbedingungen litoraler Schwermineralkonzentrate und ihre Bedeutung für die Auffindung sedimentärer Lagerstätten. Geologie, 6, 233–277

Lüders, K. (1929) Entstehung und Aufbau von Großrücken mit Schillbedeckung in Flut- bzw. Ebbetrichtern der Außenjade. In: Lüders, K., Trusheim, F., eds., Beiträge zur Ablagerung mariner Mollusken in der Flachsee. Senckenbergiana 11, 123–142

Lüders, K. (1930) Entstehung der Gezeitenschichtung auf den Watten im Jadebusen. Senckenbergiana 12, 229–254

Lüders, K. (1934) Über das Wandern der Priele, Abhandl. Naturwiss. Ver., Bremen 29, 19–32

Lüders, K. (1956) Grenzlinien und Gebietsbezeichnungen vor der niedersächsischen Nordseeküste. N. Arch. Niedersachsens; Landeskde., Statist., Landesplanung 8, 276–285

Lüneburg, H. (1961) Zur Sedimentverteilung in der Außenweser zwischen Hoheweg und Rotersand. Veröffentl. Inst. Meeresforsch. Bremerhaven 7, 1–14

Lüneburg, H. (1972) Zum Sedimentcharakter rezenter Tone und Silte in Fjorden der schwedischen Skagerakküste (Bohuslän). Veröff. Inst. Meeresforsch. 14, 45–79

Lützner, H. (1976) Die Sedimentgefüge als wichtige geologische Merkmale. II. Sedimentgefüge in ihrer Bedeutung für die Paläogeographie. Vortrag Weiterbildungsveranstaltung "Fortschritte Untersuchung sedimentärer Komplexe" Sekt. Geowiss. Ba Freiberg 31. 8. 76, 26 p.

Lützner, H., Rentsch, J. (1975) Sedimentation und Metallogenie in einem intermontanen Becken der variskischen Molasse. Z. geol. Wiss. 3, 1473–1490

Lützner, H., Falk, F., Ellenberg, J., Grumbt, E. (1974) Tabellarische Dokumentation klastischer Sedimente. Veröff. Zentralinst. Physik der Erde 20, 153 p.

Lyell, C. (1851) On fossil rain-marks of the Recent, Triassic, and Carboniferous periods. Geol. Soc. London Quart. J. 7, 238–247

Macar, P. (1951) Pseudo-nodules en terrains meubles. Ann. Soc. Géol. Belg. 75, 111–115

Macar, P., Ek, C. (1965) Un curieux phénomène d'érosion Fammennienne: Les "Pains de grès" de Chambralles (Ardenne, Belge). Sedimentology 4, 53–64

MacKenzie, D. B. (1965) Depositional environments of muddy sandstone, western Denver Basin, Colorado. Bull. Am. Assoc. Petrol. Geologists 49, 186–206

MacKenzie, D. B. (1968) Studies of students: Sedimentary features of Alameda Avenue Cut, Denver, Colorado. Mountain Geologists 5, 3–13

MacKenzie, D. B. (1972) Tidal sand flat deposits in Lower Cretaceous Dakota Group near Denver, Colorado. Mountain Geologist 9, 269–277

Mackin, J. H. (1937) Erosional history of the Big Horn basin, Wyoming. Geol. Soc. Am. Bull. 48, 813–894

Majou, T. V., Howard, J. D. (1975) Estuaries of the Georgia Coast, USA: Sedimentology and Biology. VI. Animal-sediment relationships of a salt marsh estuary – Doboy Sound. Senckenbergiana marit. 7, 205–236

Maldonado, A. (1972) El delta del Ebro. Bol. Estratigrafia 1, 475 p.

Maldonado, A., Stanley, D. J. (1979) Depositional patterns and late quaternary evolution of two Mediterranean submarine fans: a comparison. Mar. Geol. 31, 215–250

Manker, J. P., Ponder, R. D. (1978) Quartz grain surface features from fluvial environments of Northeastern Georgia. J. Sediment. Petrol. 48, 1227–1232

Manohar, M. (1955) Mechanics of bottom sediment movement due to wave action. Corps of Engineers, Beach Erosion Board Techn. Mem. 75, 1–121

Mantz, P. A. (1978) Bedforms produced by fine, cohesionless, granular and flakey sediments under subcritical water flows. Sedimentology. 25, 83–103

Marchetti, M. P. (1957) Occurrence of slide and flowage material (olistostromes) in the Tertiary series of Sicily, 20th International Geologists Congress (Mexico), Sect. 5, 1, 209–225

Marchig, V., Gundlach, H. (1976) Zur Geochemie von Manganknollen aus dem Zentralpazifik und ihrer Sedimentunterlage. Geol. Jb. 16, 79–83

Margolis, S. V., Kennett, J. P. (1971) Cenozoic paleoglacial history of Antarctica recorded in subantarctic deep-sea cores. Am. J. Sci. 271, 1–36

Margolis, S., Krinsley, D. H. (1974) Processes of formation and environmental occurrence of microfeatures on detrital quartz grains. Am. J. Sci. 274, 449–464

Martini, I. P. (1975) Sedimentology of a lacustrine barrier system at Wasaga Beach, Ontario, Canada, Sediment. Geol. 14, 169–190

Martini, I. P. (1977) Gravelly flood deposits of Irvine Creek, Ontario, Canada. Sedimentology 24, 603–622

Martini, I. P., Ostler, J. (1973) Ostler lenses: possible environmental indicators in fluvial gravels and conglomerates. J. Sediment. Petrol. 43, 418–422

Martinsson, A. (1965) Aspects of a middle Cambrian thanatotope on Öland. Geol. Foren. Stockholm Forh. 87, 181–230

Martinsson, A. (1970) Toponomy of trace fossils. In: Crimes, T. P., Harper, J. G., eds., Trace fossils. Geol. J. Spec. Issue 3, 323–330

Mason, C. C., Folk, R. L. (1958) Differentiation of beach, dune and aeolian flat environments by size analysis, Mustang Island, Texas. J. Sediment. Petrol. 28, 211–226

Masters, C. D. (1965) Sedimentology of the Mesaverde Group and of the upper part of the Mancos formation, Northwestern Colorado. Dissertation. Yale University, U.S.A. 88 p.

Masters, C. D. (1967) Use of sedimentary structures in determination of depositional environments, Mesaverde formation, William Fork Mountains, Colorado. Bull. Am. Assoc. Petrol. Geologists 51, 2033–2043

Matalucci, R. V., Shelton, J. W., Abdelhady, M. (1969) Grain orientation in Vicksburg loess. J. Sediment. Petrol. 39, 969–979

Mathews, W. H., Shepard, F. P. (1962) Sedimentation of Fraser River Delta, British Columbia. Bull. Am. Assoc. Petrol. Geologists 46, 1416–1442

Matter, A. (1967) Tidal flat deposits in the Ordovician of western Maryland. J. Sediment. Petrol. 37, 601–609

Matter, A., Tucker, M. E. eds. (1978) Modern and Ancient Lake Sediments. Spec. Publ. No 2, 290 p. Internat. Assoc. of Sedimentologists. Oxford, London, Edinburgh, Melbourne: Blackwell Scientific Publications

Matter, A., Süsstrunk, A. E., Hinz, K., Sturm, M. (1971) Ergebnisse reflexionsseismischer Untersuchungen im Thunersee. Eclogae geol. Helv. 64, 505–520

Matter, A., Dessolin, D., Sturm, M., Süsstrunk, A. E. (1973) Reflexionsseismische Untersuchung des Brinezer Sees. Eclogae geol. Helv. 66, 71–82

Mattiat, B. (1969) Eine Methode zur elektronenmikroskopischen Untersuchung des Mikrogefüges in tonigen Sedimenten. Geol. Jahrb. 88, 87–111

McBride, E. F. (1974) Significance of color in red, green, purple, olive brown, and gray beds of Difunta Group, Northeaster Mexico. J. Sediment. Petrol. 44, 760–773

McBride, E. F., Shepherd, R. G., Crawley, R. A. (1975) Origin of parallel, nearhorizontal laminae by migration of bedforms in a small flume. J. Sediment. Petrol. 45, 132–139

McBride, E. F., Yeakel, L. S. (1963) Relationship between parting lineation and rock fabric. J. Sediment. Petrol. 33, 779–782

McCabe, P. J. (1977) Deep distributary channels and giant bedforms in the upper Carboniferous of the Central Pennines, northern England. Sedimentology 24, 271–290

McCabe, P. J., Jones, C. M. (1977) Formation of reactivation surfaces within superimposed deltas and bedforms. J. Sediment. Petrol. 47: 707–715

McCave, I. N. (1970) Deposition of fine-grained suspended sediment from tidal currents. J. Geophys. Res. 75, 4151–4159

McCave, I. N. (1971 a) Wave effectiveness at the sea bed and its relationship to bedforms and deposition of mud. J. Sediment. Petrol. 41, 89–96

McCave, I. N. (1971 b) Sand waves in the North Sea off the coast of Holland. Marine Geol. 10, 199–225

McCave, I. N. (1972) Transport and escape of fine-grained sediment from shelf areas. In: Swift, D. J. P., Duane, D. B., Pilkey, O. H., eds. Shelf sediment transport, p. 225–248. Stroudsburg, Pa.: Dowden, Hutchinson, Ross

McCulloch, D., Moser, F., Briggs, L. (1960) Hydraulic shape of mineral grains. Geol. Soc. Am. Bull. 71, 1925

McDonald, B. C., Banerjee, I. (1971) Sediments and bedforms on a braided outwash plain. Can. J. Earth Sciences 8, 1282–1301

McDonald, B. C., Shilts, W. W. (1975) Interpretation of fults in glaciofluvial sediments. In: A. V. Jopling, McDonald, B. C., eds., Glaciofluvial and glaciolacustrine sedimentation. Soc. Econ. Paleontologists Mineralogists Spec. Publ. 23, 123–131

McGowen, J. H. (1970) Gum Hollow fan delta, Neces Bay, Texas. Rep. Invest. 69, 1–91. Bureau of Econ. Geol. Univ. Texas, Austin

McGowen, J. H., Garner, L. E. (1970) Physiographic features and stratification types of coarse-grained point bars: Modern and ancient examples. Sedimentology 14, 77–111

McGowen, H., Groat, C. G. (1971) Van Horn Sandstone, west Texas: an alluvial fan model for mineral exploration. Rep. Bureau Econ. Geol., Univ. Texas, Austin, Invest. 72, 57 p.

McKee, E. D. (1938) Original structures in Colorado River flood plain deposits of the Grand Canyon. J. Sediment. Petrol. 8, 77–83

McKee, E. D. (1939) Some types of bedding in the Colorado River delta. J. Geol. 47, 64–81

McKee, E. D. (1945) Small-scale structures in the Coconino Sandstone of northern Arizona. J. Geol. 53, 316–320

McKee, E. D. (1953) Report on studies of stratification in modern sediments and in laboratory experiments. Office of Naval Research, Project Nonr. 164(00), Nr. 081 123, 61 p.

McKee, E. D. (1957 a) Flume experiments on the production of stratification and cross-stratification. J. Sediment. Petrol. 27, 129–134

McKee, E. D. (1957 b) Primary structures in some recent sediments. Bull. Am. Assoc. Petrol. Geologists 41, 1704–1747

McKee, E. D. (1965) Experiments on ripple lamination. In: Middleton, G. V., ed. Primary sedimentary structures and their hydrodynamic interpretation. Soc. Econ. Paleontologists Mineralogists Spec. Publ. 12, 66–83

McKee, E. D. (1966 a) Significance of climbing-ripple structure. U. S. Geol. Surv. Profess. Papers 550-D, D 94-D 103

McKee, E. D. (1966 b) Structures of dunes at White Sands National Monument, New Mexico (and a comparison with structures of dunes from other selected areas). Sedimentology 7, 1–69

McKee, E. D., Bigarella, J. J. (1972) Deformational structures in Brazilian Coastal dunes. J. Sediment. Petrol. 42, 670–681

McKee, E. D., Goldberg, M. (1969) Experiments on formation of contorted structures in mud. Geol. Soc. Am. Bull. 80, 231–244

McKee, E. D., Moiola, R. J. (1975) Geometry and growth of the white Sands Dune Field, New Mexico. J. Res. U. S. Geol. Surv. 3, 59 66

McKee, E. D., Sterrett, T. S. (1961) Laboratory experiments on form and structure of longshore bars and beaches. In: Peterson, J. A., Osmond, J. C., eds. Geometry of sandstone bodies, p. 13–28. Tulsa, Oklahoma: Amer. Assoc. of Petroleum Geologists

McKee, E. D., Tibbitts, G. C., Jr. (1964) Primary structures of a seif dune and associated deposits in Libya. J. Sediment. Petrol. 34, 5–17

McKee, E. D., Weir, G. W. (1953) Terminology for stratification and cross stratification in sedimentary rocks. Geol. Soc. Am. Bull. 64, 381–390

McKee, E. D., Reynolds, M. A., Baker, C. H. (1962) Laboratory studies on deformation in unconsolidated sediment. U.S. Geol. Surv. Profess. Papers 450-D, D 151–D 155

McKee, E. D., Crosby, E. J., Berryhill, H. L. (1967) Flood deposits. Bijou Creek, Colordao. June 1965, J. Sediment. Petrol. 37, 829–851

McKee, E. D., Douglass, J. R., Ritterhouse, S. (1971) Deformation of Lee-side laminae in eolian dunes. Geol. Soc. Am. Bull. 82, 359–378

McKenzie, F. T. (1964) Geometry of Bermuda calcareous dune cross-bedding. Science 144, 1449–1450

McLusky, D. S. (1971) Some effects of salinity on the mud-dwelling euryhaline Amphipod Corophium colutator. Vie et Milieu, Supp. 22, (cit. Michaelis 1973)

McManus, D. A. (1975) Modern versus Relict Sediment on the continental shelf. Geol. Soc. Am. Bull. 86, 1154–1160

Meade, R. H. (1969) Landward transport of bottom sediments in estuaries of the atlantic coastal plain. J. Sediment. Petrol. 39, 222–234

Meade, R. H. (1972) Transport and deposition of sediments in estuaries. Geol. Soc. Am., Inc. Memoir 133, 91–120

Medeiros, R. A., Schaller, H., Friedmann, G. M. (1971) Facies Sedimentares. – Ciencia-Technica-Petroleo 5, 123 p.

Melton, F. A. (1940) A tentative classification of sanddunes: Its application to dune history in the southern high plains. J. Geol. 48, 113–114

Melton, M. A. (1965) The geomorphic and paleoclimatic significance of alluvial deposits in southern Arizona. J. Geol. 73, 1–38

Menard, H. W. (1952) Deep ripple marks in the sea. J. Sediment. Petrol. 22, 3–9

Menzies, R. J., George, R. W., Rowe, G. T., eds. (1973) Abyssal environment and ecology of the world oceans, 448 p. New York: Wiley & Sons

Merk, G. P. (1960) Great sanddunes of Colorado. In: Guide to the geology of Colorado. Rocky Mt. Assoc. Geol., p. 127–129

Mero, J. L. (1965) The mineral resources of the sea, 312 p. Amsterdam: Elsevier

Meyer, R. E., ed. (1972) Waves on beaches and resulting sediment transport, 462 p. New York: Academic Press

Miall, A. D. (1976) Palaeocurrent and palaeohydraulic analysis of some vertical profiles through a Cretaceous braided stream deposit, Banks Island, Arctic Canada. Sedimentology 23, 459–483

Miall, A. D. (1977) A review of the braided river depositional environment. Earth Sci. Rev. 13, 1–62

Miall, A. D., ed. (1978 a) Fluvial Sedimentology. Can. Soc. Petrol. Geol. Mem. 5, 859 p.

Miall, A. D. (1978 b) Lithofacies types and vertical profile models in braided river deposits: a summary. In: Miall, A. D., ed., Fluvial Sedimentology. Can. Soc. Petrol. Geol. Mem. 5, 597–604

Michaelis, H. (1973) Untersuchungen über das Makrobenthos der Wesermündung. Forschungsstelle f. Insel u. Küstenschutz, Jahrb. 1972, 24, 170 p.

Middleton, G. V. ed. (1965 a) Primary sedimentary structures and their hydrodynamic interpretation. Soc. Econ. Paleontologists Minealogists Spec. Publ. 12, 265 p.

Middleton, G. V. (1965 b) Antidune cross-bedding in a large flume, J. Sediment. Petrol. 35, 922–927

Middleton, G. V. (1966 a) Experiments on density and turbidity currents. I. Motion of the head. Canadian J. Earth Sci. 3, 523–546

Middleton, G. V. (1966 b) Experiments on density and turbidity currents. II. Uniform flow of density currents. Canadian J. Earth Sci. 3, 627–637

Middleton, G. V. (1967) Experiments on density and turbidity currents. III. Deposition of sediment. Canadian J. Earth Sci. 4, 475–505

Middleton, G. V. (1969) Turbidity currents. In: Stanley, D. J., ed., The new concepts of continental margin sedimentation. Am. Geol. Inst. p. GM-A-1-20.

Middleton, G. V. (1970) Experimental studies related to problems of flysch sedimentation. In: Lajoie, J., ed., Flysch sedimentology in North America. Geol. Assoc., Can. Spec. Pap. 7, 253–272

Middleton, G. V. (1976) Hydraulic interpretation of sand size distributions. J. Geol. 84, 405–426

Middleton, G. V. (1977) Introduction – progress in hydraulic interpretation of sedimentary structures. In: Sedimentary Processes, Hydraulic Interpretation of Primary Sedimentary Structures, Soc. Econ. Paleontologists Mineralogists Repr. Ser. 3, 1–15

Middleton, G. V., Hampton, M. A. (1973) Sediment gravity flows: mechanics of flow and deposition. Soc. Econ. Paleontologists Mineralogists, Pacific Sect., Short Course Turbidites and deep Water Sedimentation, 38 p.

Middleton, G. V., Hampton, M. A. (1976) Subaqueous sediment transport and deposition by sediment gravity flows. In: Stanley, D. J., Swift, D. J. P., eds., Marine sediment transport and environmental management, p. 197–218. New York: Wiley-Intersci. Publ.

Middleton, G. V., Southard, J. B. (1978) Mechanics of sediment movement. Soc. Econ. Paleontologists Mineralogists Short Course 3, 246 p.

Migniot, C. (1968) Etude des Propriétés physiques de differents sediments très fins et de leur comportement sous des actions hydrodynamiques. La Houille Blanche 23-7, 591–620

Mii, H. (1957) Peculiar accumulation of drifted shells. Saito Ho-on kai Mus. Res. Bull. 26, 17–24

Mii, H. (1958) Beach cusps on the Pacific coast of Japan. Sci. Rep. Tokoku Univ. Sendai, Japan 29, 77–107

Miller, J. A. (1975) Facies characteristics of Laguna Madre Wind-Tidal Flats. In: Ginsburg, R. N., ed., Tidal deposits, p. 67–72. New York, Heidelberg, Berlin: Springer

Miller, M. C., McCave, I. N., Komar, P. D. (1977) Threshold of sediment motion under unidirectional currents. Sedimentology 24, 507–527

Miller, R. L., Zeigler, J. M. (1958) A model relating dynamics and sediment pattern in equilibrium in the region of shoaling waves, breaker zone, and foreshore. J. Geol. 66, 417–441

Milliman, J. D., Pilkey, O. H., Ross, D. A. (1972) Sediments of the continental margin off the eastern United States. Geol. Soc. Am. Bull. 83, 1315–1334

Millot, G. (1970) Geology of clays, 429 p. Berlin – Heidelberg – New York: Springer

Mills, H. H. (1977a) Differentiation of glacier environments by sediment characteristics: Athabasca glacier, Alberta, Canada. J. Sediment. Petrol. 47, 728–737

Mills, H. H. (1977b) Textural characteristics of drift from some representative Cordilleran glaciers. Geol. Soc. Amer. Bull. 88, 1135–1143

Mills, H. H. (1978) Some characteristics of glacial sediments of Mount-Rainier, Washington. J. Sediment. Petrol. 48, 1345–1356

Milner, H. B. (1962) Sedimentary petrography, 715 p. London: A. Allen & Unwin Ltd.

Mitchell, A. H. G., Reading, H. G. (1978) Sedimentation and Tectonics. In: Reading, H. G., ed., Sedimentary environments and facies, p. 439–476. Oxford: Blackwell

Moign, A. (1973) Strandflats immergés et emergés du Spitsberg central et nord-occidental, 727 p. Brest, Univ. Bretagne occidentale, These. doct.

Moign, A. (1976) L'action des glaces flottantes sur le littoral et les fonds marins du Spitsberg central et nord-occidental. Rev. Géogr. Montr. 30, 51–64

Moiola, R. J., Weiser, D. (1968) Textural parameters: An evaluation. J. Sediment. Petrol. 38, 45–53

Moldvay, L. (1957) Aeolian sedimentation (in Russian) Acta Geol. Acad. Sci. Hung. 4, 271–320

Moody-Stuart, M. (1966) High and low sinuosity stream deposits with examples form the Devonian of Spitsbergen. J. Sediment. Petrol. 36, 1102–1117

Moore, D. G. (1955) Rate of deposition shown by relative abundance of foraminifera. Bull. Assoc. Petrol. Geologists 39, 765–772

Moore, D. G. (1960) Acoustic reflection studies of the continentals shelf and slope off southern California. Geol. Soc. Am. Bull. 71, 1121–1136

Moore, D. G. (1961) Submarine slumps. J. Sediment. Petrol. 31, 343–357

Moore, D. G. (1966) Structure litho-orogenic units, and postorogenic basin fill by reflection profiling: California continental borderland. Rijksuniv. Groningen, Netherlands, Doctoral Dissertation. 151 p.

Moore, D. G. (1969) Reflection profiling studies of the California continental borderland: Structure and quaternary turbidite basins. Geol. Soc. Am. Spec. Papers 107, 1–142

Moore, D. G., Curray, J. R. (1963) Structural framework of the continental terrace, northwest Gulf of Mexico. J. Geophys. Res. 68, 1725–1747

Moore, D. G., Scruton, P. C. (1957) Minor internal structures of some recent unconsolidated sediments. Bull. Am. Assoc. Petrol. Geologists 41, 2723–2751

Moore, D. G., Curray, J. R., Rait, R. W., Emmel, F. J. (1974) Stratigraphic-seismic section correlations and implications to Bengal Fan history. In: Borch, C. C. von der Sclater, J. G. et al., eds., Initial Rep. Deep Sea Drilling Proj. 22, 403–412

Moore, G. T., Asquith, D. O. (1971) Delta: Term and concept. Geol. Soc. Am. Bull. 82, 2563–2568

Moore, H. B., Kruse, P. (1956) A review of present knowledge of fecal pellets. Marine Laboratory University of Miami, 25 p.

Moore, R. C. (1934) The origin of boulder-bearing Johns Valley Shale in the Quachita Mountains of Arkansas and Oklahoma. Am. J. Sci. 27, 432–454

Moore, R. C. (1949) Meaning of facies. Geol. Soc. Am. Mem. 39, 1–34

Moore, R. C. (1957) Modern methods of paleoecology. Bull. Am. Assoc. Petrol. Geologists 41, 1775–1801

Moore, T. C., Jr., Andel, T. H. van, Blow, W. H., Heath, G. R. (1970) Large submarine slide off northeastern continental margin of Brazil. Bull. Am. Assoc. Petrol. Geologists 54, 125–128

Morgan, J. P., ed. (1970) Deltaic sedimentation modern and ancient. Soc. Econ. Paleontologists Mineralogists, Spec. Publ. 15, 312 p.

Morgan, J. P., Coleman, J. M., Gagliano, S. M. (1968) Mud lumps: Diapiric structures in Mississippi delta sediments. In: Braunstein, J., O'Brien, G. D., eds., Diapirism and diapirs a symposium. Am Assoc. Petrol. Geologists Mem. 8, 145–161

Moss, A. J. (1962) The physical nature of common sandy and pebbly deposits. Part I. Am. J. Sci. 260, 337–373

Moss, A. J. (1963) The physical nature of common sandy and pebbly deposits. Part II. Am. J. Sci. 261, 297–343

Moss, A. J. (1972) Bed-load sediments. Sedimentology 18, 159–219

Müller, A. H. (1951) Grundlagen der Biostratonomie. Abh. Deutschen Akad. Wiss. 3, 1–147. Berlin: Akademie

Müller, A. H. (1970) Aktuopaläontologische Beobachtungen an Quallen der Ostsee und des Schwarzen Meeres. Natur Mus. 7, 321–332

Müller, G. (1963) Die rezenten Sedimente im Obersee des Bodensees. Naturwissenschaften 50, 350

Müller, G. (1964) Methoden der Sediment-Untersuchung, 303 p. Stuttgart: E. Schweizerbart'sche Verlagsbuchhandlung (Nägele u. Obermiller)

Müller, G. (1966a) Die Verteilung von Eisenmonosulfid (FeS · nH_2O) und organischer Substanz in den Bodensedimenten des Bodensees – ein Beitrag zur Frage der Eutrophierung des Bodensees. GWF 107, 364–368

Müller, G. (1966b) Grain size, carbonate content, and carbonate mineralogy of recent sediments of the Indian Ocean off the Eastern Coast of Somalia. Naturwissenschaften 21, 547–550

Müller, G. (1969) Sedimentbildung im Plattensee/Ungarn. Naturwiss. 56, 606–615

Müller, G. (1971) Sediments of Lake Constance. Sedimentology of Parts of Central Europe. Guidebook. Eigth International Sedimentology Congress 1971, p. 237–252

Müller, G., Blaschke, R. (1969a) Zur Entstehung des Tiefsee-Kalkschlammes im Schwarzen Meer. Naturwissenschaften 56, 561–562

Müller, G., Blaschke, R. (1969b) Zur Entstehung des Posidonienschiefers (Lias ε). Naturwissenschaften 56, p. 365

Müller, G., Blaschke, R. (1971) Coccoliths: Important rock-forming elements in bituminous shales of central Europe. Sedimentology 17, 119–124

Müller, G., Gees, R. A. (1970) Distribution and thickness of quaternary sediments in the Lake Constance basin. Sediment. Geol. 4, 81–87

Müller, G. Wagner, F. (1978) Holocene carbonate evolution in Lake Balaton (Hungary) a response to climate and impact of man. In: Matter, A., Tucker, M. E., eds., Modern and ancient lake sediments. Spec. Publs. int. Ass. Sediment. 2, 57–81

Müller, G., Reineck, H.-E., Staesche, W. (1968) Mineralogisch-sedimentpetrographische Untersuchungen an Sedimenten der Deutschen Bucht (südöstliche Nordsee). Senckenbergiana Lethaea 49, 321–345

Müller, G., Irion, G., Förstner, U. (1972) Formation and diagenesis of inorganic Ca-Mg Carbonates in the lacustrine environment. Naturwissenschaften 59, 158–164

Mullen, R. E., Darby, D. A., Clark, D. L. (1972) Significance of atmospheric dust and ice rafting for Arctic ocean sediment. Geol. Soc. Am. Bull. 83, 205–212

Mutti, E. (1974) Examples of ancient deep-sea fan deposits from circum-Mediterranean geosyclines. In: Dott, R. H., Shaver, R. H., eds. Modern and ancient geosynclinal sedimentation. Soc. Econ. Paleontologists Mineralogists Spec. Publ. 19, 92–105

Mutti, E. (1977) Distinctive thin-bedded turbidite facies and related depositional environments in the Eocene Hecho Group (South-Central Pyrenees, Spain). Sedimentology 24, 107–131

Mutti, E., Ghibaudo, G. (1972) Un esempio di torbiditi di conoide sottomarina esterna: le arenarie di San Salvatore (Formazione di Bobbio, Miocene) nell'Appennino di Piacenza. Accad. Sci. Torino Mem., Cl. Sci. Fis., Mat. Nat., Ser. 4a, 16, 40 p.

Mutti, E., Ricci Lucchi, F. (1972) Le torbiditi dell'appennino settentrionale: Introduzuone all'analisi di facies. Mem Soc. Geol. It. 11, 161–199

Nachtigall, K. H. (1962) Über die Regelung von Langquarzen in aquatisch sedimentierten Sanden. Meyniana 12, 9–24

Nagahama, H., Isomi, H., Ono, C. and Sato, S. (1966) Dagger blade structure – A new method for detecting line of depositional current of siltstone. J. Geol. Soc. Japan 72, 531–540

Nagle, J. S. (1967) Wave and current orientation of shells. J. Sediment. Petrol., 37, 1124–1138

Nair, R. R. (1976) Unique mud banks, Kerala, Southwest India. Am. Ass. Petrol. Geol. Bull. 60, 616–621

Nami, M., Leeder, M. R. (1978) Changing channel morphology and magnitude in the Scalby Formation (In Jurassic) of Yorkshire, England. In: Miall, A. D., ed., Fluvial Sedimentology. Can. Soc. Petrol. Geol. Mem. 5, 431–440

Nanz, R. H. (1954) Genesis of Oligocene sandstone reservoirs. Seeligson Field, Jim Wells and Kleberg conties., Texas. Bull. Am. Assoc. Petrol. Geologists 38, 96–117

Nanz, R. H. (1955) Grain orientation in beach sands: A possible mean for predicting reservoir trend. J. Sediment. Petrol. 25, 130 (Abstract)

Nasner, H. (1974) Über das Verhalten von Transportkörpern im Tidegebiet. Mitt Franzius Inst. 40, 1–149

Natland, M. L., Kuenen, P. H. (1951) Sedimentary history of Ventura Basin. California, and the action of turbidity currents. In: Hough, J. L., ed., Turbidity currents and the transportation of coarse sediments to deep water. Soc. Econ. Paleontologists Mineralogists Spec. Publ. 2, 76–107

Neal, J. T., ed. (1975) Playas and dried lakes. Benchmark Papers in Geol. 20. Stroudsburg, Penn.: Dowden, Hutchinson & Ross

Neev, D. (1960) A pre-Neogene erosion channel in the southern coastal plain of Israel. Israel Ministry Devel., Geol. Surv. Bull. 25, Oil Div. Paper 7

Negendank, J. F. W. (1972) Turbidite aus dem Unterrotliegenden des Saar-Nahe-Gebietes. (Ein Beitrag zur Sedimentologie limnischer Ablagerungen. N. Jb. Geol. Paläontol. Mh. 561–572

Neihof, R., Loeb, G. (1975) Dissolved organic matter in seawater and the electric charge of immersed surfaces. J. Mar. Res. 32, No. 1

Nelson, C. H. (1968) Marine geology of Astoria deep-sea fan. Thesis, Oregon State University 287 p

Nelson, C. H., Carlson, P. R., Byrne, J. V., Alpha, T. R. (1970) Development of the Astoria canyon-fan physiography and comparison with similar systems. Marine Geol. 8, 259–291

Neumann, G., ed. (1968) Ocean currents. 352 p. New York: Elsevier Publ. Comp.

Neumann-Mahlkau, P. (1976) Recent sand volcanoes in the sand of a dike under construction. Sedimentology 23, 421–425

Newton, R. S. (1968) Internal structure of wave-formed ripple marks in the nearshore zone. Sedimentology 11, 275–292

Newton, R. S., Werner, F. (1969) Luftbildanalyse und Sedimentgefüge als Hilfsmittel für das Sandtransportproblem im Wattgebiet vor Cuxhaven. Hamburger Küstenforsch. 8, 1–46

Newton, R. S., Werner, F. (1972) Transitional-size ripple marks in Kiel Bay (Baltic Sea). Meyniana 22, 89–94

Niggli, P. (1948) Gesteine und Minerallagerstätten I. Allgemeine Lehre von den Gesteinen und Minerallagerstätten. Mineral. Geotechn. Reihe III. 540 p. Basel: Birkhäuser

Niggli, P. (1952) Gesteine und Minerallagerstätten II, Exogene Gesteine und Minerallagerstätten. Mineral. Geotechn. Reihe, IV, 557 p. Basel: Birkhäuser

Nijman, W., Puigdefabregas, C. (1978) Coarse grained point bar structure in a Molasse-type fluvial system, Eocene Castisent Sandstone Formation, south Pyrenean basin. In: Miall, A. D., ed., Fluvial Sedimentology. Can. Soc. Petrol. Geol. Mem. 5, 487–510

Nilsen, T. H. (1969) Old red sedimentation in the Buelandet-Vaerlandet Devonian District, Western Norway. Sediment. Geol. 3, 35–57

Nilsen, T. H., Bartow, J. A., Stump, E., Link, M. H. (1977) New occurrences of dish structure in the stratigraphic record. J. Sediment. Petrol. 47, 1299–1304

Nio, S.-D. (1976) Marine transgressions as a factor in the formation of sandwave complexes. Geol. Mijnbouw 55, 18–40

Nio, S. D., Siegenthaler, C., Yang, C. S. (1983) Megaripple cross-bedding as a tool for the reconstruction of the palaeohydraulics in a Holocene environment, S. W. Netherlands. Geologie en Mijnbouw 62, 499–510

Niyogi, D. (1966) Lower Gondwana sedimentation in Saharjuri Coalfield, Bihar. India. J. Sediment. Petrol. 36, 960–972

Normark, W. R. (1978) Fan valleys, channels, and depositional lobes on modern submarine fans: Characters for recognition of sandy turbidite environments. Bull. Am. Assoc. Petrol. Geol. 62, 912–931

Normark, W. R., Dickson, F. H. (1976) Sublacustrine fan morphology in Lake Superior. Am. Assoc. Petrol. Geol. Bull. 60, 1021–1036

Northrop, J., Heezen, B. C. (1951) An outcrop of Eocene sediment on the continental slope. J. Geo. 59, 396–399

Nota, D. J. G. (1958) Sediments of the western Guiana shelf. Thesis Utrecht. Mededel. Landbouwhogeschool Wageningen 58, 1–98

Novak, I. D. (1972) Swash-zone competency of gravel-size sediment. Mar. Geol. 13, 335–345

Novak, I. D. (1973) Predicting coarse sediment transport: The Hjulström curve revisited. In: Morisawa, M., ed. Fluvial Geomorphology. Publ. in Geomorphology, 13–25. Bringhampton: New York State Univ.

Nummedal, D. (1974) Recent migration of the Skeidarársandur shoreline, Southeast Iceland. Final Report for Contract no. NG0921-73-6-0258, Naval Ordinance Laboratory, Dept. Geology, Univ. South Carolina, 183 p.

Nummedal, D., Fischer, I. A. (1978) Process-response models for depositional shorelinses: The German and the Georgia Bights. Proc. 16th Coastal Eng. Conf., ASCE/Hamburg

Nye, J. F. (1952) The mechanisms of glacier flow, J. Glaciol. 2, 82–93

Nye, J. F. (1965) The flow of a glacier in a channel of rectangular elliptical or parabolic cross-section. J. Glaciol. 5, 661–690

Oak, D. P., Bass, N. W. (1941) Chanute oil pool, Neoshos County, Kansas – a water flooding operation. In: Levorsen, A. I., ed., Stratigraphic type oil fields. Am. Assoc. Petrol. Geologists Spec. Publ., p. 57–77

Obruchev, V. A. (1945) Loess types and their origin. Am. J. Sci. 243, 256–262

Oertel, G. F. (1975) Post Pleistocene island and inlet adjustment along the Georgia coast. J. Sediment. Petrol. 45, 150–159

Oertel, G. F. (1977) Geomorphic cycles in ebb deltas and related patterns of shore erosion and accretion. J. Sediment. Petrol. 47, 1121–1131

Oertel, G. F., Howard, J. D. (1972) Working model for barrier island development along low energy coast of Georgia. Bull. Am. Assoc. Petrol. Geologists 56, 642–643 (Abstract)

Olausson, E. (1960) Description of sedimentary cores from Central and Western Pacific with the adjacent Indonesian Region. Rep. Swedish Deep-Sea Exp. 1947–48, 6, fasc. 5

Olausson, E., Olsson, I. U. (1969) Varve stratigraphy in a core from the Gulf of Aden. Palaeogeography. Paleoclimatol., Palaeoecol. 6, 87–103

Oomkens, E. (1966) Environmental significance of sand dikes. Sedimentology 7, 145–148

Oomkens, E. (1967) Depositional sequences and sand distribution in a deltaic complex. Geol. Mijnbouw 46, 265–278

Oomkens, E. (1974) Lithofacies relations in the late Quaternary Niger Delta complex. Sedimentol. 21, 195–222

Oomkens, E., Terwindt, H. H. J. (1960) Inshore estuarine sediments in the Haringvliet (Netherlands). Geol. Mijnbouw 39, 701–710

Osgood, R. G., Jr. (1975) The paleontological significance of trace fossils. In: Frey, R. W., The Study of Trace Fossils, 562 p. New York, Heidelberg, Berlin: Springer

Otto, G. H. (1938) The sedimentation unit and its use in field sampling. J. Geol. 46, 569–582

Otvos, E. G., Jr. (1964a) Observation of beach cusp and beach ridge formation on the Long Island sound. J. Sediment. Petrol. 34, 554–560

Otvos, E. G. Jr., (1964b) Observations on rhomboid beach marks. J. Sediment. Petrol. 34, 683–687

Otvos, E. G. Jr. (1970a) Development and migration of Barrier Islands, Northern Gulf of Mexico. Geol. Soc. Am. Bull. 81, 241–246

Otvos, E. G., Jr. (1970b) Development and migration of Barrier Islands, Northern Gulf of Mexico: Reply. Geol. Soc. Am. Bull. 81, 3783–3788

Otvos, E. G. Jr., (1977) Post-pleistocene history of the United States inner continental shelf: Significance to origin of barrier Islands: Discussion and reply. Geol. Soc. Am. Bull. 88, 734–736

Owens, E. H. (1976) The effects of ice on the littoral zone at Richibucto Head Eastern New Brunswick. Rev. Géogr. Montr. 30, 95–104

Parel, H. (1973) Detritus in Lake Tahoe, Structural modification by attached microflora. Science 180

Parel, H. (1974) Bacterial uptake of dissolved organic detritus in relation to detrital aggregation in marine and freshwater systems. Limnol. Oceanogr. 19

Parel, H. (1975) Microbial attachment to particles in marine and freshwater ecosystems. Microbial Ecology 2

Panin, N. (1967) Structure de dépôts de plage sur la côte de la mer noire. Marine Geol. 5, 207–219

Panin, N., Panin, St. (1967) Regressive sand waves on the black sea shore. Marine Geol. 5, 221–226

Papp, A., Bachmayer, F., Tauber, A. F. (1947) Lebensspuren mariner Krebse. Sitz.-Ber. Akad. Wiss. Math.-Nat. Kl. Abt. 1, 115, 281–317

Parkash, B., Goel, R. K., Sinha, P. (1975) Palaeocurrent analysis of the Siwaliks of Panjab, Haryana and Himalchal Pradesh. J. Geol. Soc. India 16, 337–348

Parker, R. H. (1959) Macro-invertebrate assemblages of central Texas coastal bays and Laguna Madre. Bull. Am. Assoc. Petrol. Geologists 43, 2100–2166

Parker, R. H. (1960) Ecology and distributional patterns of marine macroinvertebrates, northern Gulf of Mexico. In: Shepard, F. P., Phleger, F. B., Andel, T. H. van, eds., Recent sediments, northwest Gulf of Mexico, p. 302–381, Tulsa, Oklahoma: Am. Assoc. of Petroleum Geologists

Parker, R. H. (1964) Zoogeography and ecology of macroinvertebrates of Gulf of California and continental slope of Western Mexico. Marine Geol. Gulf California, Symp. Mem. 3, 331–376

Passega, R. (1957) Texture as characteristic of clastic deposition. Geol. Soc. Am. Bull. 41, 1952–1984

Passega, R. (1964) Grain size representation by CM patterns as a geological tool. J. Sediment. Petrol. 34, 830–847

Passega, R., Byramjee, R. (1969) Grain-size image of clastic deposits. Sedimentology 13, 233–252

Paterson, W. S. B. (1969) The Physics of glaciers, 250 p. Oxford: Pergamon Press

Pelletier, B. R., Shearer, J. M. (1972) Sea bottom scouring in the Beaufort Sea of the Arctic Ocean: In: Marine geology and geophysics: Internat. Geol. Congr., 24th, Montreal 1972, Proc., Sed. 8, 251–261

Pemberton, G. S., Risk, M. J., Buckley, D. E. (1976) Supershrimp: Deep bioturbation in the Strait of Canso, Nova Scotia. Science 192, 790–791

Pequegnat, W. E. (1961a) New world for marine biologists. Natural History 70/4, 8–17

Pequegnat, W. E. (1961b) New World for marine biologists. Natural History 70/4, 46–55

Perry, W. J., Roberts, H. G. (1968) Late Precambrian glaciated pavements in the Kimberley region. Western Australia. J. Geol. Soc. Australia 15, 51–56

Pestrong, R. (1972) Tidal-flat sedimentation at Colley Landing, southwest San Francisco Bay. Sedim. Geol. 8, 251–288

Petersen, C. G. Joh. (1914) Appendix to report 21. On the distribution of the animal communities of the sea bottom. Rep. Danish Biol. Stat., 22

Petersen, C. G. Joh. (1918) The sea bottom and its production of fish-food. Rep. of the Danish Biol. Stat., 25, 62 p.

Peterson, G. L. (1968) Flow structures in sandstone dikes. Sed. Geol. 2, 177–190

Peterson, J. A., Osmond, J. C. eds. (1961) Geometry of sandstone bodies. 240 p. Tulsa, Oklahoma: Am. Assoc. Petroleum Geologists.

Pettijohn, F. J. (1957) Sedimentary rocks, 718 p. New York: Harper & Row

Pettijohn, F. J., Potter, P. E. (1964) Atlas and glossary of primary sedimentary structures, 370 p. Berlin, Göttingen, Heidelberg, New York: Springer

Pettijohn, F. J., Potter, P. E., Siever, R. (1972) Sand and sandstone, 618 p. Berlin – Heidelberg – New York: Springer

Pfannkuche, O. (1977) Ökologische und systematische Untersuchungen an naidomorphen Oligochaeten brackiger und limnischer Biotope. Dissertation, 138 S., Hamburg

Pfannkuche, O. (1979) Abundance and lifecycle of littoral marine and brackish water Tubificidae and Naididae (Oligochaeta). In: Naylor, E., Hartnoll, R. G., eds., Cyclic phenomena in marine plants and animals, p. 103–111 New York: Pergamon Press

Pfannkuche, O., Jelinek, H., Hartwig, E. (1975) Zur Fauna eines Süßwasserwattes im Elbe-Ästuar. Arch. Hydrobiol. 76, 475–498

Pharo, C. H., Carmack, E. C. (1979) Sedimentation processes in a short residence-time intermontane Lake, Kamloops Lake, British Columbia. Sedimentology 26, 523–541

Philipp, H. (1912) Über ein rezentes alpines Os und seine Bedeutung für die Bildung der Diluvialen Osar. Z. Deut. Geol. Ges. 64, 68–102

Phleger, F. B. (1964) Foraminiferal ecology and marine geology. Marine Geol. 1, 16–43

Phleger, F. B. (1969) Some general features of coastal lagoons. In: Castanares, A. A., Phelger, F. B., eds., Coastal lagoons, a Symposium, p. 5–26. Mexico: Universidad Nacional Autonoma

Picard, M. D., High, L. R., Jr. (1968) Sedimentary cycles in the Green River Formation (Eocene), Uinta Basin, Utah. J. Sediment. Petrol. 38, 378–383

Picard, M. D., High, L. R., Jr. (1972) Criteria for recognizing lacustrine rocks. In: Rigby, J. K., Hamblin, W. K., eds., Recognition of ancient sedimentary environments. Soc. Econ. Paleontologists Mineralogists Spec. Publ. 16, 108–145

Picard, M. D., High Jr., L. R. (1973) Sedimentary structures of ephemeral streams, 223 p. Amsterdam: Elsevier

Pierce, J. W. (1969) Sediment budget along a barrier island chain. Sediment. Geol. 3, 5–16

Pierce, J. W. (1970) Tidal inlets and washover fans. J. Geol. 78, 230–234

Pierce, J. W. (1976) Suspended sediment transport at the shelf break and over the outer margin. In: Stanley, D. J., Swift, D. J. P., eds., Marine sediment transport and environmental management, p. 437–458. Stroudsburg, Pa: Dowden, Hutchinson, Ross

Pierce, J. W., Colquhoun, D. J. (1970) Holocene evolution of a portion of the North Carolina coast. Geol. Soc. Am. Bull. 81, 3697–3714

Pierce, J. W., Siegel, F. R. (1979) Suspended particulate matter on the Southern Argentine shelf. Marine Geol. 29, 73–91

Piper, D. J. W. (1970) Transport and deposition of Holocene sediment on La Jolla deep sea fan, California. Marine Geol. 8, 211–227

Piper, D. J. W. (1972) Turbidite origin of some laminated mudstones. Geol. Mag. 109, 115–126

Piper, D. J. W., Brisco, C. D. (1975) Deep water continental margin sedimentation. In: Hayes, D. A., Franke, O. A. et al. Initial Report of the Deap Sea Drilling Project. Deep Sea Drilling Project 28, 727–755. Washington: Gov. Printing Off.

Piper, J. W. (1970) Eolian sediments in the basal New Red Sandstone of Arran. Scott. J. Geol. 6, 295–308

Piper, J. W., Normark, W. R. (1971) Re-examination of a Miocene deep-sea fan and fan-valley. Southern California. Geol. Soc. Am. Bull. 82, 1823–1830

Plessmann, W. (1961) Strömungsmarken in klastischen Sedimenten und ihre geologische Auswertung. Geol. Jahrb. 78, 503–566

Plumley, W. J. (1948) Black Hills terrace gravels: A study in sediment transport. J. Geol. 56, 526–577

Pollack, J. M. (1961) Significance of Compositional and textural properties of South Canadian River channel sands, New Mexico, Texas and Oklahoma. J. Sediment. Petrol. 31, 15–37

Porrenga, D. H. (1967) Clay mineralogy and geochemistry of recent marine sediments in tropical areas. Publ. Fysisch-Geografisch Lab. Univ. Amsterdam. No. 9, 145 p.

Porter, J. J. (1962) Electron microscopy of sand surface texture. J. Sediment. Petrol. 32, 124–135

Postma, H. (1961) Transport and accumulation of suspended matter in the Dutch Wadden Sea. Neth. J. Sea Res. 1, 148–190

Postma, H. (1967) Sediment transport and sedimentation in the estuarine environment. In: Lauff, G. H., ed., Estuaries 38, 158–179. Washington: Am. Assoc. Advancem. Sci.

Postma, H. (1969) Suspended matter in the marine environment. Morning Review Lectures of the Second Internat. Oceanogr. Congr. 213–219

Postma, H., Kalle, K. (1955) Die Entstehung von Trübungszonen im Unterlauf der Flüsse, speziell im Hinblick auf die Verhältnisse in der Unterelbe. Deutsche Hydrogr. Z. 8, 137–144

Potter, P. E. (1963) Late paleozoic sandstones of the Illinois basin. Illinois State Geol. Surv. Rept. Invest. 217, 1–92

Potter, P. E., Mast, R. F. (1963) Sedimentary structures, sand shape fabrics, and permeability. I., J. Geol. 71, 441–471

Potter, P. E., Pettijohn, F. J. (1963) Palaeocurrents and basin analysis, 296 p. Berlin – Göttingen – Heidelberg: Springer

Powers, M. C. (1953) A new roundness scale for sedimentary particles. J. Sediment. Petrol. 23, 117–119

Prandtl, L. (1952) Essentials of fluid dynamics, 452 p. London: Blackie & Son

Prandtl, L. (1956) Strömungslehre, 407 p. Braunschweig: Friedrich Vieweg & Sohn

Prandtl, L., Oswatitsch, K., Wieghardt, K. (1969) Führer durch die Strömungslehre, 535 p. Braunschweig: Friedrich Vieweg & Sohn

Pratje, O. (1934) Staubfälle auf dem mittleren Atlantischen Ozean. Centralbl. Min. etc. Jg. 1934 (Abt. B.) 177–182

Pratje, O. (1939) Die Sedimentation in der südlichen Ostsee. Ann. Hydrogr. Maritimen Meteor. 1939, 209–221

Pratje, O. (1948) Die Bodenbedeckung der südlichen und mittleren Ostsee und ihre Bedeutung für die Ausdeutung fossiler Sedimente. Deut. Hydrogr. Z. 1, 45–61

Pratt, C. J. (1973) Bagnold approach and bed-form development. Am. Soc. Civil Eng. Proc. 99, HY1, 121–137

Pratt, C. J., Smith, K. V. H. (1972) Ripple and dune phases in a narrowly graded sand. Am. Soc. Civil Eng. Proc. 98, HY5, 859–874

Price, W. A. (1947) Equilibrium of form and forces in tidal basins of coast of Texas and Louisiana. Bull. Am. Assoc. Petrol. Geologists 31, 1619–1663

Price, W. A. (1955) Environment and formation of the chenier plain. Quaternaria 2, 75–86

Price, W. A. (1963 a) Patterns of flow and channeling in tidal inlets. J. Sediment. Petrol. 33, 279–290

Price, W. A. (1963 b) Physicochemical and environmental factors in clay dune genesis. J. Sediment. Petrol. 33, 766–778

Price, W. A., Kornicker, L. S. (1961) Marine and lagoonal deposits in clay dunes, Gulf Coast, Texas. J. Sediment. Petrol. 31: 245–255

Pritchard, D. W. (1955) Estuarine circulation patterns. Proc. Amer. Soc. Civil Eng. 24, 717 p.

Pritchard, D. W. (1967) Observations of circulation in coastal plain estuaries. In: Lauff, G. H., ed., Estuaries. Publ. 83, 3–5. Washington: Am. Assoc. Advancem. Sci.

Pryor, W. A. (1971) Petrology of the Permian Yellow Sands of northeastern England and their North Sea Basin equivalents. Sediment.Geol. 6, 221–254

Pryor, W. A. (1973) Permeability-porosity patterns and variations in some Holocene sand bodies. Am. Assoc. Petrol. Geol. Bull. 57, 162–189

Pryor, W. A. (1975) Biogenic sedimentation and alteration of argillaceous sediments in shallow marine environments. Geol. Soc. Am. Bull. 86, 1244–1254

Pryor, W. A., Amaral, E. J. (1971) Large-scale cross-stratification in the St. Peter Sandstone. Geol. Soc. Am. Bull. 82, 239–244

Psuty, N. P. (1965) Beach-ridge development in Tabasco, Mexico. Technical Report 24, Coastal Studies Institute Louisiana State University, Baton Rouge, Louisiana. Ann. Assoc. Am. Geogr. 55, 112–124

Psuty, N. P. (1966) The geomorphology of beach ridges in Tabasco, Mexico. Technical Report 30. Coastal Studies Institute. Louisiana State University Baton Rouge, Louisiana, 51 p.

Puigdefabregas, C. (1973) Miocene point-bar deposits in the Ebro Basin, Northern Spain. Sedimentology 20, 133–144

Puigdefabregas, C., Vliet, A. van (1978) Meandering stream deposits from the Tertiary of the Southern Pyrenees. In: Miall, A. D., ed., Fluvial Sedimentology. Can. Soc. Petrol. Geol. Mem. 5, 469–485

Purdy, E. G., Imbrie, J. (1964) Carbonate sediments, Great Bahama Bank. A guidebook for field trip No. 2, Geology Society Am. Convention; Nov. 1964, p. 1–58

Quenstedt, W. (1927) Beiträge zum Kapitel Fossil und Sediment vor und bei der Einbettung. Neues Jahrb. Mineral. etc., Beil. 58 B, 253–432

Quirke, T. T. (1930) Spring pits, sedimentation phenomena. J. Geol. 38, 88–91

Raaf, J. F. M. De (1968) Turbidites et associations sedimentaires apparentees. Koninkl. Ned. Akad. Wetenschap. Proc. Ser. 13, 71, 1–23

Raaf, J. F. M. De, Boersma, J. R. (1971) Tidal deposits and their sedimentary structures (seven examples from Western Europe). Geol. Mijnbouw 50, 479–504

Raaf, de, J. F. M., Boersma, R. J., Gelder, A. van (1977) Wave-generated structures and sequences from a shallow marine succession, Lower Carboniferous, County Cork, Ireland. Sedimentology 24, 451–483

Rad, U. von (1968) Comparison of sedimentation in the Bavarian flysch (Cretaceous) and Recent San Diego Trough (California). J. Sediment. Petrol. 38, 1120–1154

Radczewski, O. E. (1937) Die Mineralfazies der Sedimente des Kapverden-Beckens. Deut. Atlantische Expedition "Meteor" 1925–1927, Wiss. Ergebn. 3, 1–262. Berlin: Borntraeger

Radlowska, C. (1969) On the problematics of eskers. Geogr. Polonica 16, 87–104

Radwanski, A., Friis, H., Larsen, G. (1975) The Miocene Hagenor-Borup sequence at Lillebalt (Denmark): its biogenic structures and depositional environment. Bull. Geol. Soc. Denmark 24, 229–260

Rainwater, E. H. (1966) The geological importance of deltas. In: Shirley, M. L., ed., Deltas and their geologic framework. Houston Geol. Soc., 1–15

Rainwater, E. H. (1975) Petroleum in deltaic sediments. In: Broussard, M. L., ed., Deltas, models for exploration, p. 3–11. Houston: Houston Geol. Soc.

Ramsay, A. T. S. (1974) The distribution of calcium carbonate in deep sea sediments. In: Hay, W. W., ed., Studies in paleo-oceanography. Soc. Econ. Paleontologists Mineralogists Spec. Publ. 20, 58–76

Ranwell, D. S. (1972) Ecology of salt marshes and sand dunes, 258 p. London: Chapman and Hall

Rast, U., Schäfer, A. (1978) Delta Schüttungen in Seen des höheren Unterrotliegenden im Saar-Nahe-Becken. Mainzer Geowiss. Mitt. 6, 121–159

Raudkivi, A. J. (1963) Study of sediment ripple formation. Am. Soc. Civil. Engers. Proc. HYG 89, 15–34

Ray, P. K. (1976) Structure and sedimentological history of the overbank deposits of a Mississippi River point bar. J. Sediment. Petrol. 46, 788–801

Rayleigh, L. (1943) The ultimate shape of pebbles natural and artificial. Proc. Roy. Soc. London, Ser. A 181, 107–118

Rayleigh, L. (1944) Pebbles natural and artificial. Their shape under various conditions of abrasion. Proc. Roy. Soc. London, Ser. A 182, 321–335

Reading, H. G., ed. (1978) Sedimentary Environments and Facies, 576 p. Oxford: Blackwell Sci. Publ.

Reading, H. G., Walker, R. G. (1966) Sedimentation of Eocambrian tillites and associated sediments in Finmark, Northern Norway. Palaeogeogr. Paleoclimatol. Palaeoecol. 2, 177–212

Redfield, A. C. (1967) The ontogeny of a salt marsh estuary. In: Lauff, G. H., ed., Estuaries. Am. Assoc. Advanc. Sci., Washington 83, 108–114

Redfield, A. C. (1972) Development of a New England salt marsh. Ecol. Monogr. 42, 201–237

Reed, W. E., Le Lever, R. E., Moir, G. J. (1975) Depositional environment interpretation from setling velocity (Psi) distributions. Geol. Soc. Am. Bull. 86, 1321–1328

Rees, A. I. (1966) Some flume experiments with a fine silt. Sedimentology 6, 209–240

Rees, A. I. (1968) The production of preferred orientation in a concentrated dispersion of elongated and flattened grains. J. Geol. 76, 457–465

Reeves, C. C., Jr. (1968) Introduction to paleolimnology. Developments in sedimentology, vol. 11, 228 p. Amsterdam: Elsevier

Reeves, C. C., Jr. (1970) Origin, classification and geologic history of caliche on the Southern High Plains, Texas and Eastern New Mexico. J. Geol. 78, 352–362

Reimnitz, E. (1971) Surf-beat origin for pulsating bottom currents in the Rio Balsas Submarine Canyon. Mexico. Geol. Soc. Am. Bull. 82, 81–90

Reimnitz, E., Barnes, P. W. (1974) Sea ice as a geologic agent on the Beaufort sea shelf of Alaska. In: Reed, J. C., Sater, J. E., eds., The coast and shelf of the Beaufort Sea, p. 301–353. Arlington, VA: The Arctic Inst. North Am.

Reimnitz, E., Bruder, K. E. (1972) River discharge into an ice-covered ocean and related sediment dispersal, Beaufort Sea, Coast of Alaska. Geol. Soc. Am. Bull. 83, 861–866

Reimnitz, E., Marshall, N. F. (1965) Effects of the Alaska earthquake and tsunami on Recent deltaic sediments. J. Geophys. Res. 70, 2363–2376

Reimnitz, E., Rodeick, C. A., Wolf, S. C. (1974) Strudel scour: A unique arctic marine geologic phenomenon. J. Sediment. Petrol. 44, 409–420

Reineck, H.-E. (1952) Abdrücke von Schneekristallen. Natur Volk 82, 394–397

Reineck, H.-E. (1954) Fossile Schleifspuren und Abdrücke von Schaum und Blasen. Natur Volk 84, 226–233

Reineck, H.-E. (1955a) Marken, Spuren und Fährten in den Waderner Schichten (ro) bei Martinstein/Nahe. Neues Jahrb. Geol. Paläontol. Abh. 101, 75–90

Reineck, H.-E. (1955b) Schleif-Marken und Liege-Marken von strandendem Gischt. Senckenbergiana Lethaea 35, 357–359

Reineck, H.-E. (1955c) Eisblumen im Watt. Natur Volk 85, 400–402

Reineck, H.-E. (1955d) Haftrippeln und Haftwarzen, Ablagerungsformen von Flugsand. Senckenbergiana Lethaea 36, 347–357

Reineck, H.-E. (1956a) Abschmelzreste von Treibeis an den Ufersäumen des Gezeiten-Meeres. Senckenbergiana Lethaea 37, 299–304

Reineck, H.-E. (1956b) Die Oberflächenspannung als geologischer Faktor in Sedimenten. Senckenbergiana Lethaea 37, 265–287

Reineck, H.-E. (1958a) Longitudinale Schrägschichten im Watt. Geol. Rundschau 37, 73–82

Reineck, H.-E. (1958b) Wühlbau-Gefüge in Abhängigkeit von Sediment-Umlagerungen. Senckenbergiana Lethaea 39, 1–24

Reineck, H.-E. (1960a) Über Zeitlücken in rezenten Flachsee-Sedimenten. Geol. Rundschau 49, 149–161

Reineck, H.-E. (1960b) Über die Entstehung von Linsen- und Flaserschichten. Abh. Deut. Akad. Wiss. 3, 1, 370–374

Reineck, H.-E. (1960c) Über eingeregelte und verschachtelte Röhren des Goldköcher-Wurmes (Pectinaria koreni). Natur Volk 90, 334–338

Reineck, H.-E. (1961) Sedimentbewegungen an Kleinrippeln im Watt. Senckenbergiana Lethaea 42, 51–61

Reineck, H.-E. (1962a) Die Orkanflut vom 16. Februar 1962. Natur Mus. 92, 151–172

Reineck, H.-E. (1962b) Reliefgüsse ungestörter Sandproben. Z. Pflanzenernähr. Dueng., Bodenk. 99, 151–153

Reineck, H.-E. (1963a) Sedimentgefüge im Bereich der südlichen Nordsee. Abh. senckenbergische naturforsch. Ges. 505, 138 p.

Reineck, H.-E. (1963b) Der Kastengreifer. Natur Mus. 83, 102–108

Reineck, H.-E. (1965) Der Knechtsand. 6. Die Materialumlagerung auf Grund von Oberflächen- und Gefügestudien. Jahrb. 1964, Forsch.-Stelle Norderney, p. 169–181

Reineck, H.-E. (1967a) Layered sediments of tidal flats, beaches and shelf bottoms of the North Sea. In: Lauff, G. H., ed., Estuaries. Am. Assoc. Advanc. Sci. Publ. 83, 191–206

Reineck, H.-E. (1967b) Parameter von Schichtung und Bioturbation. Geol. Rundschau 56, 420–438

Reineck, H.-E. (1968 a) Sedimentgefüge im Golf von Neapel. Pubbl. Staz. Zool. Napoli 36, 112–134

Reineck, H.-E. (1968 b) Der Schelf. In: Murawski, H., ed., Vom Erdkern bis zur Magnetosphäre, p. 237–247. Frankfurt: Umschau Verlag

Reineck, H.-E. (1969 a) Die Entstehung von Runzelmarken. Senckenbergiana marit. 1, 165–168

Reineck, H.-E. (1969 b) Zwei Sparkerprofile südöstlich Helgoland. Natur Mus. 99, 9–14

Reineck, H.-E. (1970 a) Reliefguß und projizierbarer Dickschliff. Senckenbergiana marit. 2, 61–66

Reineck, H.-E. (1970 b) Marine Sandkörper, rezent und fossil. Geol. Rundschau 60, 302–321

Reineck, H.-E. (1970 c) Topographie und Geomorphologie. In: Reineck, H.-E., ed., Das Watt, Ablagerungs- und Lebensraum, p. 12–15. Frankfurt a. M.: Kramer

Reineck, H.-E. (1971) Der Küstensand. Natur Mus. 101, 45–60

Reineck, H.-E. (1972) Tidal flats. In: Rigby, J. K., Hamblin, W. K., eds., Recognition of ancient sedimentary environments. Soc. Econ. Paleontologists Minieralogists Spec. Publ. 17, 146–159

Reineck, H.-E. (1973) Schichtung und Wühlgefüge in Grundproben vor der ostafrikanischen Küste. "Meteor" Forsch.-Ergebn. C. 16, 67–81, Berlin: Borntraeger

Reineck, H. E. (1974 a) Schichtgefüge der Ablagerungen im tieferen Seebecken des Bodensees. Senckenbergiana marit. 6, 47–63

Reineck, H.-E. (1974 b) Schlickkämme. Senckenbergiana marit. 6, 145–147

Reineck, H.-E. (1974 c) Schlickrücken auf Sandwattflächen. Senckenbergiana marit. 6, 47–63

Reineck, H.-E. (1974 d) Vergleich dünner Sandlagen verschiedener Ablagerungsbereiche. Geol. Rundschau 63, 1087–1101

Reineck, H.-E. (1974 e) Strömungsrippeln in schlickigen Sedimenten. Senckenbergiana marit. 6, 135–143

Reineck, H.-E. (1976 a) Drift ice action on tidal flats, North Sea. Rev. Geogr. Montr. 30, 197–200

Reineck, H.-E. (1976 b) Primärgefüge, Bioturbation und Makrofauna als Indikatoren des Sandversatzes im Seegebiet vor Norderney (Nordsee). I. Zonierung von Primärgefügen und Bioturbation. Senckenbergiana marit. 8, 155–169

Reineck, H.-E. (1977) Natural indicators of energy level in Recent sediments the application of ichnology to a coastal engineering problem. In: Crimes, T. P., Harper, J. C., eds., Trace fossils 2. Geol. J. Spec. Issue 9, 265–272

Reineck, H.-E. (1978) Geopetale Kriterien im Lupenbereich. Senckenbergiana marit. 10, 31–37

Reineck, H.-E. (1979) Rezente und fossile Algenmatten und Wurzelhorizonte. Nat. u. Mus. 109, 290–296

Reineck, H.-E., Cheng, Y. M. (1978) Sedimentologische und faunistische Untersuchungen an Watten in Taiwan. I. Aktuogeologische Untersuchungen. Senckenbergiana marit. 10, 85–115

Reineck, H.-E., Dörjes, J. (1976) Geologisch-biologische Untersuchungen an Geröllstränden und -vorstränden der Costa Brava, Mittelmeer. Senckenbergiana marit. 8, 111–153

Reineck, H.-E., Rosenboom, W. (1969) Stechkästen zur Entnahme von Watten- und Unterwasserproben. Natur Mus. 99, 45–55

Reineck, H.-E., Singh, I. B. (1967) Primary sedimentary structures in the Recent sediments of the Jade, North Sea. Marine Geol. 5, 227–235

Reineck, H.-E., Singh, I. B. (1971) Der Golf von Gaeta/Tyrrhenisches Meer. 3. Die Gefüge von Vorstrand- und Schelfsedimenten. Senckenbergiana marit. 3, 185–201

Reineck, H.-E., Singh, I. B. (1972) Genesis of laminated sand and graded rhythmites in storm-sand layers of shelf mud. Sedimentology 18, 123–128

Reineck, H.-E., Wunderlich, F. (1967 a) Zeitmessungen an Gezeitenschichten. Natur Mus. 97, 193–197

Reineck, H.-E., Wunderlich, F. (1967 b) A new method to measure rate of deposition of single lamina on tidal flats and shelf bottoms. 7th International Sedimentological Congress (Abstracts.)

Reineck, H.-E., Wunderlich, F. (1968 a) Zur Unterscheidung von asymmetrischen Oszillationsrippeln und Strömungsrippeln. Senckenbergiana Lethaea 49, 321–345

Reineck, H.-E., Wunderlich, F. (1968 b) Classification and origin of flaser and lenticular bedding. Sedimentology 11, 99–104

Reineck, H.-E., Wunderlich, F. (1969) Die Entstehung von Schichten und Schichtbänken im Watt. Senckenbergiana marit. 1, 85–106

Reineck, H.-E., Gutmann, W. F., Hertweck, G. (1967) Das Schlickgebiet südlich Helgoland als Beispiel rezenter Schelfablagerungen. Senckenbergiana Lethaea 48, 219–275

Reineck, H.-E., Dörjes, J., Gadow, S., Hertweck, G. (1968) Sedimentologie, Faunenzonierung und Faziesabfolge vor der Ostküste der inneren Deutschen Bucht. Senckenbergiana Lethaea 49, 261–309

Reineck, H.-E., Singh, I. B., Wunderlich, F. (1971) Einteilung der Rippen und anderer mariner Sandkörper. Senckenbergiana marit. 3, 93–101

Reinson, G. E. (1977) Hydrology and sediments of a temperate estuary – Mallacoota Inlet, Victoria. Dept. Nat. Resources, Bur. Miner. Resources, Geol. and Geophys. Bull. 178, 91 p.

Remane, A. (1958) Ökologie des Brackwassers. In: Remane und Schlieper, Die Biologie des Brackwassers (cit. Michaelis 1973)

Renz, O., Lakeman, R., Meulen, E. van de (1955) Submarine sliding in western Venezuela. Bull. Am. Assoc. Petrol. Geologists 39, 2053–2067

Rhoads, D. C. (1975) The paleoecological and environmental significance of trace fossils. In: Frey, R. W., ed., The study of trace fossils, p. 147–160. New York, Heidelberg, Berlin: Springer

Rhoads, D. C., Stanley, D. J. (1965) Biogenic graded bedding. J. Sediment. Petrol. 35, 956–963

Rhoads, D. C., Young, D. K. (1970) The influence of deposit-feeding organisms on sediment stability and community trophic structure. J. Marine Res. 28, 150–178

Ricci-Lucchi, F. R. (1970) Sedimentografia. Atlante fotografico delle structure primarie del sedimenti, 288 p. Bologna: Zanichelli

Ricci-Lucchi, F. (1978) Ambienti sedimentari e facies. Sedimentologia 3, 405 p. Bologna: Cooperativa Libraria Universitaria Editrice Bologna

Richert, P. (1976) Relationship between diatom biocoenoses and and taphocoenoses in upwelling areas off West Africa. Abstracts 4th Symposium on Recent and Fossil Diatoms, Oslo.

Richter, K. (1932) Die Bewegungsrichtung des Inlandeises, rekonstruiert aus den Kritzen und Längsachsen der Geschiebe. Z. Geschiebeforsch. 8, 62–66

Richter, K. (1959) Bildungsbedingungen pleistozäner Sedimente Niedersachsens, aufgrund morphometrischer Geschiebe- und Geröllanalysen. Z. Deut. Geol. Ges. 110, 400–435

Richter, R. (1922) Flachseebeobachtungen zur Paläontologie und Geologie. III–IV. Senckenbergiana 4, 103–141

Richter, R. (1929) Gründung und Aufgaben der Forschungsstelle für Meeresgeologie "Senckenberg" in Wilhelmshaven. Natur Mus. 59, 1–30

Richter, R. (1935) Marken und Spuren im Hunsrückschiefer. I. Gefließ-Marken. Senckenbergiana 17, 244–263

Richter, R. (1936) Marken und Spuren im Hunsrückschiefer. II. Schichtung und Grundleben. Senckenbergiana 18, 215–244

Richter, R. (1937) Marken und Spuren aus allen Zeiten. I–II. Senckenbergiana 19, 150–169

Richter, R. (1952) Fluidal-Textur in Sediment-Gesteinen und über Sedifluktion überhaupt. Notizbl. Hess. Landesamtes Bodenforsch., Wiesbaden 4, 67–81

Richter-Bernburg, G. (1955) Über salinare Sedimentation. Z. Deutsch. Geol. Ges. 105, 593–645

Riedl, R. J. (1971) How much seawater passes through sandy beaches? Int. Revue ges. Hydrobiol. 56, 923–946

Rigby, J. K. (1958) Mass movements in Permian rocks of trans-Pecos, Texas. J. Sediment. Petrol. 28, 298–315

Rigby, J. K., Hamblin, W. K., eds. (1972) Recognition of ancient sedimentary environments. Soc. Econ. Paleontologists Mineralogists Spec. Publ. 16, 340 p.

Riley, J. P., Chester, R. (1971) Introduction to marine chemistry, 465 p. London: Academic Press

Rizzini, A. (1968) Sedimentological representation of grain sizes. Mem. Soc. Geol. Italiana 7, 65–89

Rodine, J. D., Johnson, A. M. (1976) The abilily of debris, heavily freighted with coarse clastic materials, to flow on gentle slopes. Sedimentology 23, 213–224

Rodolfo, K. S. (1969) Sediments of the Andaman Basin, northeastern Indian Ocean. Marine Geol. 7, 371–402

Röder, H. (1971) Gangsysteme von *Paraonis fulgens* Levinsen 1883 (Polychaeta) in ökologischer, ethologischer und aktuopaläontologischer Sicht. Senckenbergiana marit. 3, 3–51

Röder, H. (1977) The relationship between construction and substratum in mechanically boring Lamellibrachiata (Pholadidae, Teredinidae). Senckenbergiana marit. 9, 105–213

Roep, T. B., Beets, D. J., Ruegg, G. H. J. (1975) Wave-built, structures in subrecent beach barriers of the Netherlands. Proc. 9e Int. Congr. Sedimentol., Nice, 6, 141–145

Roep, T. B., Beets, D. J., Drondert, H., Pagnier, H. P. (1979) A prograding coastal sequence of wave-built structures of Messinian age, Sorbas, Almeria, Spain. Sediment. Geol. 22, 135–163

Röthlisberger, H. (1972) Water pressure in intra-and subglacial channels. J. Geol. 11, 177–203

Rona, P. A. (1972) Exploration methods for the continental shelf: Geology Geophysics, Geochemistry. NOAA Techn. Rep. ERL 238-AOML 8, 1–47

Ronai, A. (1969) The geology of Lake Balaton and surroundings. Mitt. Internat. Verein. Limnol. 17, 275–281

Rosenquist, I. T. (1955) Investigations in the clay-electrolyte-water system. Norweg. Geotechn. Inst. Publ. 9, 1–125

Ross, D. A. (1979) Opportunities and uses of the ocean, 320 p. Berlin, Heidelberg, New York: Springer

Royse, C. F. (1970) A sedimentologic analysis of the Tongue River-Sentinel Butte Interval (Paleocene) of the Williston Basin, western North Dakota. Sediment. Geol. 4, 19–80

Rowe, G. T., Keller, G., Edgerton, H., Stavesinic, N., MacIlvaine, J. (1974) Time lapse photograph of the biological reworking of sediments in Hudson Submarine Canyon. J. Sediment. Petrol. 44, 549–552

Rubi, H. C., Horne, J. C., Deinhart, P. J., in consultation with Ferm, J. C. (1981) Cretaceous rocks of Western North America. A guide to terrigenous clastic rock identification, 100 p. Columbia, S.C.: Research planning Inst. Hayes, M. O.

Ruchin, L. B. (1958) Grundzüge der Lithologie, 806 p. Berlin: Akademie-Verlag

Rücklin, H. (1938) Strömungsmarken im unteren Muschelkalk des Saarlandes. Senckenbergiana 20, 94–114

Ruegg, G. H. J. (1977) Features of middle Pleistocene sandur deposits in the Netherlands. Geol. Mijnbouw 56, 5–24

Runcorn, S. K. (1961) Climatic change through geological time in the light of palaeomagnetic evidence for polar wandering and continental drift. Quart. J. Roy. Meteorol. Soc. 87, 282–313

Rusnak, G. A. (1957) The orientation of sand grains under condition of "unidirectional" fluid flow. 1. Theory and experiment. J. Geol. 65, 384–409

Rusnak, G. A. (1960) Sediments of Laguna Madre, Texas. In: Recent sediments Northwest Gulf of Mexico. A Symposium summarizing the results of work carried on in project 51 of the Am. Petrol. Inst., p. 153–196

Russel, R. D. (1939) Effects of transportation on sedimentary particles. In: Trask, P. D., ed., Recent marine sediments, p. 32–47. Tulsa, Oklahoma: Am. Assoc. Petroleum Geologists

Russel, R. D., Taylor, R. E. (1937) Roundness and shape of Mississippi River sands. J. Geol. 45, 225–267

Russell, R. J. (1954) Alluvial morphology of Anatolian rivers. Ann. Assoc. Am. Geogr. 44, 363–391

Russell, R. J. (1967) River and delta morphology. Technical report No. 52. Coastal Studies Institute, Louisiana State University, Baton Rouge, Louisiana, p. 1–55

Russell, R. J. (1968) Glossary of terms used in fluvial, deltaic, and coastal morphology and processes. Technical Report No. 63, Coastal Studies Institute, Louisiana State University, Baton Rouge, Louisiana, 97 p.

Russell, R. J., Howe, H. V. (1935) Cheniers of southwestern Louisiana. Geogr. Rev. 25, 449–461

Russell, R. J., McIntire, W. G. (1965) Beach cusps. Geol. Soc. Am. Bull. 76, 307–320

Rust, B. R. (1963) The geology of the Whithorn area, Wigtownshire. Doctoral Thesis, University of Edinburgh, 135 p.

Rust, B. R. (1972a) Pebble orientation in fluvial sediments. J. Sediment. Petrol. 42, 384–388

Rust, B. R. (1972b) Structure and process in a braided river. Sedimentology 18, 221–245

Rust, B. R. (1975) Fabric and structure in glaciofluvial gravels. In: Jopling, A. V., McDonald, B. C., eds., Glaciofluvial and Glaciolacustrine Sedimentation. Soc. Econ. Paleontologists Mineralogists Spec. Publ. 23, 238–248

Rust, B. R. (1978a) A classification of alluvial channel systems. In: Miall, A. D., ed., Fluvial Sedimentology. Can. Soc. Petrol. Geol., Mem. 5, 187–198

Rust, B. R. (1978b) Depositional models for braided alluvium. In: Miall, A. D., ed., Fluvial Sedimentology. Can. Soc. Petrol. Geol. Mem. 5, 605–625

Rust, B. R., Romanelli, R. (1975) Late Quaternary subaqueous outwash deposits near Ottawa, Canada. In: Jopling, A. V., McDonald, B. C., eds., Glaciofluvial and Glaciolacustrine Sedimentation. Soc. Econ. Palaeontologists Mineralogists Spec. Publ. 23, 177–192

Ryan, W. B. F., Stanley, D. J., Hersey, J. B., Fahlquist, D. A., Allan, T. D. (1971) 12. The tectonics and geology of the Mediterranean Sea. In: Maxwell, ed., The sea, vol. 4, 387–492. London: Wiley & Sons

Ryer, T. A. (1977) Patterns of Cretaceous shallow-marine sedimentation, Coalville and Rockport areas, Utah. Geol. Soc. Am. Bull. 88, 177–188

Sahu, B. K. (1964) Significance of the size-distribution statistics in the interpretation of depositional environments. Res. Bull., N. S., Punjab Univ. 15, 213–219

Salis, A. von (1884) Hydrotechnische Notizen. II. Die Tiefenmessungen im Bodensee. Schweiz. Bau.-Z., 31. Mai 1884, p. 127.

Salomon, W. (1928) Geologische Beobachtungen des Leonardo da Vinci. Sitz. ber. Heidelberg. Akad. Wiss., Math.-Naturw. Kl., Jg. 1928, 8, 1–13

Samu, G. (1968) Ergebnisse der Sandwanderungsuntersuchungen in der südlichen Nordsee. Mitt.-Bl. Bundesanstalt Wasserbau 26, 13–62

Sanders, J. E. (1965) Primary sedimentary structures formed by turbidity currents and related resedimentation mechanisms. In: Middleton, G. V., ed., Primary sedimentary structures and their hydrodynamic interpretation. Soc. Econ. Paleontologists Mineralogists Spec. Publ. 12, 192–219

Sanderson, D. J., Donovan, R. N. (1974) The vertical packing of shells and stones on some recent beaches. J. Sediment. Petrol. 44, 680–688

Sanford, R. M., Lange, F. W. (1960) Basin study approach to oil evaluation of Parana miogeosyncline, South Brazil. Bull. Am. Assoc. Petrol. Geologists 44, 1316–1370

Sarjeant, W. A. S. (1975) Plant trace fossils. In: Frey, R. W., editor, The study of trace Fossils, p. 163–179. New York, Heidelberg, Berlin: Springer

Sarkar, S. K., Basumallick, S. (1968) Morphology, structure, and evolution of a channel island in the Barakar River, Barakar, West Bengal. J. Sediment. Petrol. 38, 747–754

Sarnthein, M. (1970) Sedimentologische Merkmale für die Untergrenze der Wellenwirkung im Persischen Golf. Geol. Rundschau 59, 649–666

Sarnthein, M., Diester-Haas, L. (1977) Eolian-sand Turbidites. J. Sediment. Petrol. 47, 868–890

Sarnthein, M., Walger, E. (1974) Der äolische Sandstrom aus der W-Sahara zur Atlantikküste. Geol. Rundschau 63, 1065–1087

Saunderson, H. C. (1975) Sedimentology of the Brampton esker and its associated deposits: an emperical test of theory. In Jopling, A. V., McDonald, B. C., eds., Glaciofluvial and Glaciolacustrine Sedimentation. Soc. Econ. Paleontologists Mineralogists Spec. Publ. 23, 155–176

Saunderson, H. C. (1977) The sliding bed facies in esker sands and gravels: a criterion for fullpipe (tunnel) flow? Sedimentology 24, 623–638

Sauramo, M. (1918) Geochronologische Studien über die spätglaziale Zeit in Südfinnland. Bull. Comm. Geol. Finlande 50

Sauramo, M. (1923) Studies on the quaternary varve sediments in southern Finland. Bull. Comm. Geol. Finlande 60, 164 p

Schäfer, A. (1972) Petrographische und stratigraphische Untersuchungen an den rezenten Sedimenten des Untersees/Bodensee. Neues Jahrb. Mineral., Abhandl. 117, 117–142

Schäfer, A. (1973) Zur Entstehung von Seekreide – Untersuchungen am Untersee (Bodensee). N. Jb. Geol. Paläont. Mh. 1973, 216–230

Schäfer, A., Rast, U. (1976) Sedimentation im Rotliegenden des Saar-Nahe-Beckens. Nat. u. Mus. 106, 330–338

Schäfer, A., Stapf, K. R. G. (1978) Permian Saar-Nahe Basin and Recent Lake Constance (Germany): two environments of lacustrine algal carbonates. In: Matter, A., Tucker, M. E., eds., Modern and ancient lake sediments. Spec. Publs. Int. Ass. Sediment. 2, 83–107

Schäfer, W. (1938) Die geologische Bedeutung von Bohr-Organismen in tierischen Hart-Teilen, aufgezeigt an Balaniden-Schill der Innenjade. Senckenbergiana 20, 304–331

Schäfer, W. (1952) Biogene Sedimentation im Gefolge von Bioturbation. Senckenbergiana 33, 1–12

Schäfer, W. (1953) Zur Unterscheidung gleichförmiger Kotpillen meerischer Evertebraten. Senckenbergiana 34, 81–93

Schäfer, W. (1954) "Geführte" Trockenrisse. Natur Volk 84, 14–17

Schäfer, W. (1956) Wirkungen der Benthos-Organismen auf den jungen Schichtverband. Senckenbergiana Lethaea 37, 183–263

Schäfer, W. (1962) Aktuo-Paläontologie nach Studien in der Nordsee, 666 p. Frankfurt a. M.: Kramer

Schäfer, W. (1963) Biozönose und Biofazies im marinen Bereich, 37 p. Aufsätze u. Reden Senckenbergische Naturforsch. Ges. Frankfurt a. M.: Kramer

Schäfer, W. (1972) Ecology and palaeoecology of marine environments. 538 p. Edinburgh: Oliver & Boyd

Schäfer, W. (1973) Der Oberrhein, sterbende Landschaft? Natur Mus. 103, 1–29

Schiemenz, S. (1960) Fazies und Paläogeographie der subalpinen Molasse zwischen Bodensee und Isar. Beih. Geol. Jahrb. 38, 1–119

Schindewolf, O. H. (1928) Über *Volborthella tenuis* SCHM. Palaeontol. Z. 10, 1–68

Schmidt, H. (1935) Die bionomische Einteilung der fossilen Meeresböden. Fortschr. Geol., Palaeontol. 38, 1–154

Schmidt, H. (1944) Ökologie und Erdgeschichte Z. Deut. Geol. Ges. 96, 113–128

Schmidt, H. (1958) Zur Rangordnung der Faziesbegriffe. Mitt. Geol. Ges. Wien 49, 333–345

Schmidt, W., Kolp, O. (1965) Beschreibung und Ergebnisse der Erprobung eines im Auftrage des Instituts für Meereskunde Warnemünde gebauten Vibrationsstechrohr 4700/1. Beitr. Meereskde. 12–14, p. 143–148

Schneider, H. E. (1970) Problems of quartz grain morphoscopy. Sedimentology 14, 325–335

Schneider, H. E., Cailleux, A. (1959) Signification geomorphologique des formes des grains de sable des Etats-Unis. Z. Geomorphol. 3, 114–125

Schneider, J. F. (1975) Recent tidal deposits, Abu Dhabi, UAE, Arabian Gulf. In: Ginsburg, R. N., ed., Tidal Deposits, p. 209–214. New York, Heidelberg, Berlin: Springer

Schneider, E. D., Fox, P. J., Hollister, C. D., Needham, H. D., Heezen, B. C. (1967) Further evidence of contour currents in the western North Atlantic. Earth Planetary Sci. Letters 2, 351–359

Schneiderhöhn, P. (1954) Eine vergleichende Studie über Methoden zur quantitativen Bestimmung von Abrundung und Form an Sandkörnern (im Hinblick auf die Verwendbarkeit an Dünnschliffen). Heidelberger Beitr. Mineral. Petrog. 4, 172–191

Schneidermann, N., Pilkey, O. H., Saunders, C. (1976) Sedimentation on the Puerto Rico insular shelf. J. Sediment. Petrol. 46, 167–173

Schöttle, M., Friedman, G. M. (1971) Fresh water Iron-Manganese Nodules in Lake George, New York. Geol. Soc. Am. Bull. 82, 101–110

Scholl, D. W., Huene, R., Ridlon, J. B. von (1968) Spreading of the ocean floor: Underformed sediments in the Peru-Chile trench. Science 159, 869–871

Schott, W. (1976) Mineral (inorganic) resources of the oceans and ocean floors: A general review. In: Wolf, K. H., ed., Handbook of strata-bound and stratiform ore deposits, 245–294. Amsterdam. Elsevier Sci Publ. Comp.

Schott, W., Stackelberg, U. von (1965) Über rezente Sedimentation im Indischen Ozean, ihre Bedeutung für die Entstehung kohlenwasserstoffhaltiger Sedimente. Erdoel, Kohle, Erdgas, Petrochemie (Jg. 18) 12, 945–950

Schroll, E., Wieden, P. (1971) Eine rezente Bildung von Dolomit im Schlamm des Neusiedler Sees. Tschermaks miner. petrogr. Mitt. 7, 286–289

Schütte, H. (1929) Über Sedimentbildung an der Küste des Norddeutschen Wattenmeeres. Senckenbergiana 11, 345–352

Schubart, D. (1927) Die "Schelpenvisscherij" in den Niederlanden. Mitt. Dt. Seefischereivereins 43, 235–237

Schumm, S. A. (1960) The effect of sediment type on the shape and stratification of some modern fluvial deposits. Am. J. Sci. 258, 177–184

Schumm, S. A. (1961) Effect of sediment characteristics on erosion and deposition in ephemeral stream channels. U. S. Geol. Surv. Profess. Papers 352 C, 31–70

Schumm, S. A. (1963) Sinuosity of alluvial rivers on the Great Plains. Geol. Soc. Am. Bull. 74, 1089–1100

Schumm, S. A. (1968) Speculations concerning paleohydrologic controls of terrestrial sedimentation. Geol. Soc. Am. Bull. 79, 1573–1588

Schumm, S. A. (1972) Fluvial paleochannels. In: Rigby, J. K., Hamblin, W. K., eds., Recognition of ancient sedimentary environments. Soc. Econ. Paleontologists Mineralogists Spec. Publ. 16, 98–107

Schumm, S. A. (1977) The fluvial system, 338 p. New York: Wiley & Sons

Schumm, S. A., Lichty, R. W. (1963) Channel widening and flood plain construction along Cimarron River in southwestern Kansas. U. S. Geol. Surv. Profess. Papers 352 D, 71–88

Schwartz, D. E. (1978) Hydrology and current orientation analysis of a braided - to - meandering transition: The Red River in Oklahoma and Texas, U.S.A. In: Miall, A. D., ed., Fluvial Sedimentology. Can. Soc. Petrol. Geol. Mem. 5, 231–255

Schwartz, M. L., ed., (1973) Barrier Islands, USA. Stroudsburg, Pa.: Dowden, Hutchinson & Ross, Inc.

Schwarz, A. (1932) Gestrandete Quallen. Natur Mus. 62, 282–286

Schwarzacher, W. (1951) Grain orientation in sands and sandstones. J. Sediment. Petrol. 21, 162–172

Schwarzbach, M. (1938) Tierfährten aus eiszeitlichen Bändertonen. Z. Geschiebeforsch. 14, 143–152

Schwarzbach, M. (1940) Das diluviale Klima während des Höchststandes einer Vereisung. Z. Deut. Geol. Ges. 92, 565–582

Schwarzbach, M. (1964) Geologische Tätigkeit der Seen. Geologische Tätigkeit des Eises und die Periglazialgebiete. In: Brinkmann, R., ed., Lehrbuch der allgemeinen Geologie, vol. I, p. 177–207; 207–249. Stuttgart: Ferdinand Enke.

Schwertmann, U. (1971) Transformation of hematite to geothite in soils. Nature 232, 624–625

Schwertmann, U., Fischer, W. R. (1974) Natural "amorphous" ferric hydroxide. Geoderma 10, 237–247

Scott, A. J., Hayes, M. O. (1964) Playa lake and clay dunes. Depositional environments. South-Central Texas Coast. Assoc. Corpus Christi Geol. Soc., University of Texas, P. 38

Scruton, P. C. (1960) Delta building and the deltaic sequence. In: Shepard, F. P., Phleger, F. B., Andel, T. H. van, eds., Recent sediments, northwest Gulf of Mexico, p. 82–102. Tulsa, Oklahoma: Am. Assoc. Petroleum Geologists

Seibold, E. (1955) Rezente Jahresschichtung in der Adria. Neues Jahrb. Geol. Palaeontol., Monatsh. 1955, 11–13

Seibold, E. (1958) Jahreslagen in Sedimenten der mittleren Adria. Geol. Rundschau 47, 100–117

Seibold, E. (1963) Geological investigation of near-shore sand transport. In: Sears, M., ed., Progress in oceanography, vol. 1, p. 1–70. Oxford: Pergamon

Seibold, E. (1970) Nebenmeere im humiden und ariden Klimabereich. Geol. Rundschau 60, 73–105

Seibold, E., (1974) Der Meeresboden. 183 p. Berlin, Heidelberg, New York: Springer

Seibold, E., Vollbrecht, K. (1969) Die Bodengestalt des Persischen Golfes. "Meteor" Forsch.-Ergebnisse, C2, 1–12. Berlin: Boerntraeger

Seilacher, A. (1953 a) Studien zur Palichnologie. Neues Jahrb. Geol. Palaeontol., Abhandl. 96, 421–452

Seilacher, A. (1953 b) Die fossilen Ruhespuren (Cubichnia). Neues Jahrb. Geol. Palaeontol., Abhandl. 98, 87–124

Seilacher, A. (1954) Die geologische Bedeutung fossiler Lebensspuren. Z. Deut. Geol. Ges. 105, 214–227

Seilacher, A. (1958) Zur ökologischen Charakteristik von Flysch und Molasse. Eclogae Geol. Helv. 51, 1062–1078

Seilacher, A. (1959) Fossilien als Strömungsanzeiger. Aus der Heimat 67, 171–177

Seilacher, A. (1960) Strömungsanzeichen im Hunsrückschiefer. Notizbl. Hess. Landesamtes Bodenforsch. 88, 88–106

Seilacher, A. (1963) Umlagerung und Rolltransport von Cephalopoden-Gehäusen. Neues Jahrb. Geol. Palaeontol. Monatsh. 11, 593–615

Seilacher, A. (1964) Biogenic sedimentary structures. In: Imbrie, J., Newell, N., eds., Approaches to paleoecology, p. 296–316, New York: Wiley & Sons

Seilacher, A. (1967) Bathymetry of trace fossils. Marine Geol. 5, 413–428

Seilacher, A. (1970) Begriff und Bedeutung der Fossil-Lagerstätten. Neues Jahrb. Geol. Palaeontol., Monatsh. 1970, 34–39

Seilacher, A. (1973) Biostratinomy: The sedimentology of biologically standardized particles. In: Ginsburg, ed. Evolving concepts in sedimentology, p. 159–177. John Hopkins Univ. Press

Seilacher, A. (1978) Use of trace fossil assemblages for recognizing depositional environments. Trace fossil concepts, Soc. Econ. Paleontologists. Mineralogists Short Course 5, 167–181

Seilacher, A. (1982) Distinctive features of sandy tempestites. In: Einsele, G., Seilacher, A., eds., Cyclic event stratification, p. 333–349. Berlin, Heidelberg, New York: Springer

Seilacher, A., Meischner, D. (1964) Fazies-Analyse im Paläozoikum des Oslo-Gebietes. Geol. Rundschau 54, 596–619

Sellards, E. H. (1923) Geologic and soil studies on the alluvial lands of the Red River Valley. Texas Univ. Bull. 2327, 27–87

Selley, R. C. (1965) Diagnostic characters of fluviatile sediments of the Torridonian Formation (Pre Cambrian) of northwest Scotland. J. Sediment.-Petrol. 35, 366–380

Selley, R. C. (1966) The Miocene rocks of Marada and the Jebel Zelten area, Central Libya. A study of shoreline sedimentation. Petrol. Explor. Soc. Libya, 30 p.

Selley, R. C. (1970) Ancient sedimentary environments, 237 p. London: Chapman & Hall, Ltd.

Sellin, R. H. J. (1969) Flow in channels, 149 p. London: Macmillan

Sengupta, S. (1966) Studies on orientation and imbrication of pebbles with respect to cross-stratification. J. Sediment. Petrol. 36, 362–369

Sengupta, S. (1970) Gondwana sedimentation around Bheemaram (Bhimaram) Pranhita – Godavari Valley, India. J. Sediment. Petrol. 40, 140–170

SEPM Short Course No. 2 (1975) Depositional Environments as interpreted from primary sedimentary structures and stratification sequences. Soc. Econ. Paleontologists Mineralogists, Tulsa, Oklahoma

Serruya, C. (1965) Quelques données nouvelles sur la structure profonde du Lac Léman. Arch. Sci. (Geneva) 18, (1)

Sharp. R. P. (1954) Glacier flow: A review. Geol. Soc. Am. Bull. 65, 821–838

Sharp, R. P. (1958) Malaspina Glacier, Alaska. Geol. Soc. Am. Bull. 69, 617–646

Sharp, R. P. (1960) Glaciers. The Condon Lectures, Oregon State System of Education, Eugene, Oregon, 78 p.

Sharp, R. P. (1963) Wind ripples. J. Geol. 71, 617–636

Sharp, R. P. (1966) Kelso Dunes, Mojave Desert, California. Geol. Soc. Am. Bull. 77, 1045–1074

Sharp, R. P., Nobles, L. H. (1953) Mudflow of 1941 in Wrigthwood, Southern California. Geol. Soc. Am. Bull. 64, 547–560

Shaw, J. (1972) Sedimentation in the ice-contact environment, with examples from Shropshire (England). Sedimentology 18, 23–62

Shaw, J. (1975) Sedimentary successions in Pleistocene ice-marginal lakes. In: Jopling, A. V., McDonald, A. C., eds., Glaciofluvial and Glaciolacustrine Sedimentation. Soc. Econ. Palaeontologists Mineralogists Spec. Publ., 23, 281–303

Shaw, J., Kellerhals, R. (1977) Paleohydraulic interpretation of antidune bedforms with applications to antidunes in gravel. J. Sediment. Petrol. 47, 257–266

Shelton, J. W., Noble, R. L. (1974) Depositional features of braided-meandering streams. Am. Assoc. Petrol. Geol. Bull. 58, 742–749

Shelton, J. W., Burman, H. R., Noble, R. L. (1974) Directional features in braided-meandering stream deposits, Cimarron River, North-Central Oklahoma. J. Sediment. Petrol. 44, 1114–1117

Shepard, F. P. (1950) Longshore-bars and longshore-troughs. Beach erosion board, corps of engineers. Techn. Mem. 15, 121–156

Shepard, F. P. (1956 a) Late Pleistocene and Recent history of the central Texas coast. J. Geol. 64, 56–69

Shepard, F. P. (1956 b) Marginal sediments of the Mississippi delta. Bull. Am. Assoc. Petrol. Geologists 40, 2537–2623

Shepard, F. P. (1963) Submarine geology, 557 p. New York: Harper & Row

Shepard, F. P. (1965) Importance of submarine valleys in funneling sediments to the deepsea. In: Sears, M., ed., Progress in oceanography, vol. 3, p. 321–332. Oxford: Pergamon

Shepard, F. P. (1967) The earth beneath the sea, 242 p. Baltimore: John Hopkins

Shepard, F. P. (1972) Submarine canyons. Earth-Sci. Rev. 8, 1–12

Shepard, F. P. (1976) Coastal classification and changing coastlines. Geoscience and Man 14, 53–64

Shepard, F. P., Dill, R. F. (1966) Submarine canyons and other sea valleys, 381 p. Chicago: Rand McNally

Shepard, F. P., Lankford, R. R. (1959) Sedimentary facies from shallow borings in lower Mississippi Delta. Bull. Am. Assoc. Petrol. Geol. 43, 2051–2067

Shepard, F. P., Einsele, G. (1962) Sedimentation in the San Diego Trough and contributing submarine canyons. Sedimentology 1, 81–133

Shepard, F. P., Emery, K. O. (1941) Submarine topography off the California coast: Canyons and tectonic interpretation. Geol. Soc. Am. Spec. Papers 31, 171 p.

Shepard, F. P., Emery, K. O., LaFond, E. C. (1941) Rip currents: A process of geological importance. J. Geol. 49, 337–369

Shepard, F. P., Inman, D. L. (1950) Nearshore water circulations related to bottom topography and wave refraction. Trans. Am. Geophys. Union 31, 196–212

Shepard, F. P., Marshall, N. F. (1978) Currents in submarine canyons and other sea valleys. In: Stanley, D. J., Kelling, G., eds., Sedimentation in submarine canyons, fans, and trenches. Chap. 1, 3–14. Stroudsburg, Pa.: Dowden, Hutchinson and Ross

Shepard, F. P., Moore, D. G. (1960) Bays of central Texas coast. In: Shepard, F. P., Phleger, F. B., Andel, Tj. H. van, eds., Recent sediments, northwest Gulf of Mexico, p. 117–152. Tulsa, Oklahoma: Am. Assoc. Petroleum Geologists

Shepard, F. P., Phleger, F. B., Andel, Tj. H. van, eds. (1960) Recent sediments, northwest Gulf of Mexico, 394 p. Tulsa, Oklahoma: Am. Assoc. Petroleum Geologists

Shepard, F. P., Young, R. (1961) Distinguishing between beach and dune sands. J. Sediment. Petrol. 31, 196–214

Shepard, F. P., Dill, R. F., Rad, U. von (1969) Physiography and sedimentary processes of La Jolla submarine fan and fan-valley. California. Bull. Am. Assoc. Petrol. Geologists 53, 390–420

Shideler, G. L. (1973) Textural trend analysis of coastal barrier sediments along the Middle Atlantic Bight, North Carolina. Sediment. Geol. 9, 195–220

Shields, A. (1936) Anwendung der Ähnlichkeitsmechanik und der Turbulenzforschung auf die Geschiebebewegung Preuss. Versuchsanstalt für Wasserbau und Schiffbau. Berlin, Mitt. 26, 26 p.

Shinn, E. A. (1968) Practical significance of birdseye structures in carbonate rocks. J. Sediment Petrol. 38, 215–223

Shipek, C. J. (1960) Photographic study of some deep-sea floor environments in the eastern Pacific. Geol. Soc. Am. Bull. 71, 1067–1074

Shirley, M. L., ed. (1966) Deltas in their geologic framework, 251 p. Houston Geological Society.

Short, A. D. (1979) Three-dimensional beach stage model. J. Geol. 87, 553–571

Short, A. D. (1985) Rip-current type, spacing and persistence, Narrabeen Beach, Australia. Mar. Geol. 65, 47–71

Shotton, F. M. (1956) Some aspects of the New Red Desert in Britain. Liverpool Manchester Geol. J. 1, 450–465

Shreve, R. L. (1972) Movement of water in glaciers. J. Glaciology 11, 205–214

Shrock, R. R. (1948) Sequence in layered rocks. 507 p. New York: McGraw-Hill Book Co

Shuyskiy, Y. U. D. (1975) Modern eolian processes on the bay barriers of the Black Sea. Doklady Earth Sci. Sec. 226, 177–179 (Doklady Akademii Nauk SSSR 226, 190–193)

Simons, D. B., Richardson, E. V. (1961) Forms of bed roughness in alluvial channels. Am. Soc. Civ. Engrs., Proc. HY3 87, 87–105

Simons, D. B., Richardson, E. V. (1962) Resistance to flow in alluvial channels. Am. Soc. Civ. Engrs., Trans. 127, 927–953

Simons, D. B., Richardson, E. V., Nordin, C. F. (1965) Sedimentary structures generated by flow in alluvial channels. In: Middleton, G. V., ed., Primary sedimentary structures and their hydrodynamic interpretation. Soc. Econ. Paleontologists Mineralogists. Spec. Publ. 12, 34–52

Sindowski, K.-H. (1957) Die synoptische Methode des Kornkurven-Vergleiches zur Ausdeutung fossiler Sedimentationsräume. Geol. Jahrb. 73, 235–275

Sindowski, K.-H. (1965) Das Eem im ostfriesischen Küstengebiet. Z. Deut. Geol. Ges. 115, 163–166

Sindowski, K.-H., Streif, H. (1974) Die Geschichte der Nordsee am Ende der letzten Eiszeit und im Holozän. In: Woldstedt, P., Duphorn, K., eds., Norddeutschland und angrenzende Gebiete im Eiszeitalter, S. 411–431

Singh, I. B. (1968) Lenticular and lenticular-like bedding in the Precambrian Telemark suite, southern Norway. Norsk Geol. Tidsskr. 48, 165–170

Singh, I. B. (1969) Primary sedimentary structures in Precambrian quartzites of Telemark, southern Norway, and their environmental significance. Norsk Geol. Tidsskr. 49, 1–31

Singh, I. B. (1972) On the bedding in the natural-levee and the point-bar deposits of the Gomti River, Uttar Pradesh, India. Sediment. Geol. 7, 309–317

Singh, I. B. (1973) Depositional environment of the Vindhyan sediments in Son Valley area. In: Recent Researches in Geology (a collection of papers in honour of the sixty-fifth birthday of Professor A. G. Jhingran), p. 146–152. Delhi: Hindustan Publishing Corporation (India)

Singh, I. B. (1974) Mineralogical studies of Gondwana sediments from Korba Coalfield, Madhya Pradesh, India. Part II. Scanning electron microscopy of sand grain surface-features. Geophytology 4, 60–82

Singh, I. B. (1975) A sedimentation model for the Siwalik sediments. Chayanica Geologica 1, 91–98

Singh, I. B. (1976a) Mineralogical evidences for climatic vicissitudes in India during Gondwana times. Geophytol. 6, 174–185

Singh, I. B. (1976b) Depositional environment of the Upper Vindhyan sediments in the Satna-Maihar area, Madhya Pradesh, and its bearing on the evolution of Vindhyan sedimentation basin. J. Palaeontol. Soc. India 19, 48–70

Singh, I. B. (1977) Bedding structures in a channel sand bar of the Ganga River near Allahabad, Uttar Pradesh, India. J. Sediment. Petrol. 47, 747–752

Singh, I. B. (1980a) The Bijaigarh Shale, Vindhyan System (Precambrian), India – An example of a lagoonal deposit. Sediment. Geol. 25 (in press)

Singh, I. B. (1980b) Precambrian sedimentary sequences of India: their peculiarities and comparison with the modern sediments. Precambrian Res. 83–103

Singh, I. B., Kumar, S. (1974) Mega-and giant ripples in the Ganga, Yamuna, and Son rivers, Uttar Pradesh, India. Sediment. Geol. 12, 53–66

Singh, I. B., Wunderlich, F. (1978) On the terms Wrinkle marks (Runzelmarken), Millimetre ripples, and mini-ripples. Senckenbergiana marit. 10, 75–83

Skipper, K. (1971) Antidune cross-stratification in a turbidite sequence, cloridorme Formation, Gaspé, Quebec. Sedimentology 17, 51–68

Smalley, I. J. (1966) The properties of glacial loess and the formation of loess deposits. J. Sediment. Petrol. 36, 669–676

Smalley, I. J., ed., (1975) Loess lithology and Genesis, 429 p. Stroudsburg, Pa.: Dowden, Hutchinson & Ross

Smalley, I. J., Krinsley, D. H., Vitafinzi, C. (1973) Observations on the Kaiserstuhl loess, Geol. Mag. 110, 29–36

Smalley, I. J., Leach, J. A. (1978) The origin and distribution of the loess in the Danube Basin and associated regions of East-Central Europe – A review. Sediment. Geol. 21, 1–26

Smalley, I. J., Vita-Finzi, C. (1968) The formation of fine particles in sandy deserts and the nature of "desert" loess. J. Sediment. Petrol. 38, 766–774

Smith, A. E., Jr. (1966) Modern deltas: Comparison maps. In: Shirley, M. L., ed., Deltas in their geologic framework, p. 233, Houston: Houston Geological Society

Smith, A. J. (1968) Lakes. In: Fairbridge, R. W., ed., The encyclopedia of geomorphology, vol. 3, p. 598–603. New York: Reinhold Book Corporation

Smith, A. E. (1975) Ancient deltas: Comparison maps. In: Broussard, M. L., Deltas, models for exploration, 555 p. Houston, Tex.: Houston Geol. Soc.

Smith, N. D. (1970) The braided stream depositional environment: comparison of the Platte River with some Silurian clastic rocks, North-Central Appalachians. Geol. Soc. Am. Bull. 81, 2993–3014

Smith, N. D. (1971 a) Pseudo-planar stratification produced by very low amplitude sand waves. J. Sediment. Petrol. 41, 69–73

Smith, N. D. (1971 b) Transverse bars and braiding in the Lower Platte River, Nebraska. Geol. Soc. Am. Bull. 82, 3407–3420

Smith, N. D. (1972) Some sedimentological aspects of planar cross-stratification in a sandy braided river. J. Sediment. Petrol. 42, 624–634

Smith, N. D. (1974) Sedimentology and bar formation in the upper Kicking Horse River, a braided outwash stream. J. Geol. 82, 205–224

Sneed, E. D., Folk, R. L. (1958) Pebbles in the lower Colorado River, Texas, a study in particle morphogenesis. J. Geol. 66, 114–150

Solle, G. (1966) Rezente und fossile Wüste. Notizbl. Hess. Landesamtes Bodenforsch. 94, 54–121

Solohub, J. T., Klovan, J. E. (1970) Evaluation of grain-size parameters in lacustrine environments. J. Sediment. Petrol. 40, 81–101

Solowiew, M. M. (1921) Über die Rolle der *Tubifex tubifex* in der Schlammerzeugung. Hydrobiol. Stat. Petersburger Agronom. Inst. 90–101

Sorby, H. C. (1859) On the structures produced by the currents present during the deposition of stratified rocks. The Geologist 2, 137–147

Soutendam, C. J. A. (1967) Some methods to study surface textures of sand grains. Sedimentology 8, 281–290

Southard, J. B. (1971) Representation of bed configurations in depth-velocity-size diagrams. J. Sediment. Petrol. 41, 903–915

Southard, J. B. (1975) Bed configuration. Soc. Econ. Paleontologists Mineralogists Short Course 2, 5–44

Southard, J. B., Boguchwal, L. A. (1973) Flume experiments on the transition from ripples to lower flat bed with increasing sand size. J. Sediment. Petrol. 43, 1114–1121

Southard, J. B., Harms, J. C. (1972) Sequence of bedform and stratification in silts, based on flume experiments (abs.). Am. Assoc. Petrol. Geol. Bull. 56, 654–655

Spearing, D. R. (1974) Summary sheets of sedimentary deposits with bibliographies, MC-8. Geol. Soc. Am., 7 Sheets; Boulder

Spearing, D. R. (1975) Shallow marine sands. In: Depositional Environments as interpreted form primary sedimentary structures and stratification sequences. Soc. Econ. Paleontologists Mineralogists Short. Course 2, 103–132

Spjeldnaes, N. (1964) The Eocambrian glaciation in Norway. Geol. Rundschau 54, 24–45

Spotts, J. H. (1964) Grain orientation and imbrication in Miocene turbidity current sandstones, California. J. Sediment. Petrol. 34, 229–253

Sriramadas, A. (1957) Appositional fabric study of the coastal sedimentaries. East Godawari District, Andhra, India. J. Sediment. Petrol. 27, 447–452

Stackelberg, U. von (1972) Faziesverteilung in Sedimenten des indisch-pakistanischen Kontinentalrandes (Arabisches Meer), "Meteor" Forsch.-Ergebn. C, 9, 1–73. Berlin: Borntraeger

Stäblein, G. (1970) Grobsediment-Analyse als Arbeitsmethode der genetischen Geomorphologie. Würzburger Geogr. Arb. 27, 203 p.

Stanley, D. J. (1967) Comparing patterns of sedimentation in some modern and ancient submarine canyons. Earth Planetary Sci. Letters 3, 371–380

Stanley, D. J. (1969) Sedimentation in slope and base-of-slope environments. In: Stanley, D. J., ed., The new concept of continental margin sedimentation, p. DJS-8 1–25. Washington: American Geological Institute

Stanley, D. J. (1970) Flyschoid sedimentation on the outer Atlantic margin off northeast North America. Geol.Assoc. Canada, Spec. Papers 7, 179–210

Stanley, D. J., Bouma, A. H. (1964) Methodology and paleogeographic interpretation of flysch formations: A summary of studies in the Maritime Alps. In: Bouma, A. H., Brouwer, A., eds., Turbidites. Development in sedimentology, vol. 3, 34–64. Amsterdam: Elsevier

Stanley, D. J., Kelling, G. (1967) Sedimentation patterns in the Wilmington submarine canyons area. Ocean Sci. & Engng. Atlantic Shelf, Trans. Nat. Symp. Marine Technol. Soc. p. 127–142

Stanley, D. J., Kelling, G., eds. (1978) Sedimentation in Submarine Canyons, Fans, and Trenches, 395 p. Stroudsburg, Pa.: Dowden, Hutchinson & Ross

Stanley, D. J., Silverberg, N. (1969) Recent slumping on the continental slope off Sable Island Bank, southeast Canada. Earth Planetary Sci. Letters 6, 123–133

Stanley, D. J., Unrug, R. (1972) Submarine channel deposits, fluxoturbidites and other indicators of slope and base-of-slope environments in modern and ancient basins. In: Rigby, J. K., Hamblin, Wm. K., eds., Recognition of ancient sedimentary environments. Soc. Econ. Paleontologists Mineralogists Spec. Publ. 16, 287–340

Stanley, K. O. (1974) Morphology and hydraulic significance of climbing ripples with superimposed micro-ripple-drift cross-lamination in Lower Quaternary lake silts, Nebraska. J. Sediment. Petrol. 44, 472–483

Stapor Jr., F. W. (1973) Heavy mineral concentrating processes and density/shape/size equilibria in the marine and coastal dune sands of the Apalachicola, Florida region. J. Sediment. Petrol. 43, 396–407

Stapor, F. W., Tanner, W. F. (1975) Hydrodynamic implications of beach, beach ridge and dune grain size studies. J. Sediment. Petrol. 45, 926–931

Starkey, J. (1974) The quantitative analysis of orientation data obtained by the Starkey method of X-ray fabric analysis. Canada, J. Earth Sci. 11, 1507–1516

Stauffer, K. W. (1962) Quantitative petrographic study of Paleozoic carbonate rocks, Caballo Mountains, New Mexico. J. Sediment. Petrol. 32, 357–396

Stauffer, P. H. (1967) Grain-flow deposits and their implications, Santa Ynes mountains, California. J. Sediment. Petrol. 37, 487–508

Steel, R. J. (1974) New Red Sandstone floodplain and piedmont sedimentation in the Hebridean province, Scotland. J. Sediment. Petrol. 44, 336–357

Steel, R. J., Aasheim, S. M. (1978) Alluvial sand deposition in a rapidly subsiding basin (Devonian, Norway). In: Miall, A. D., ed., Fluvial Sedimentology. Can. Soc. Petrol. Ged., Mem. 5, 385–412

Steel, R. J., Maehle, S., Nilsen, H., Roe, S. L., Spinnangr, Å. (1977) Coarsening upward cycles in the alluvium of Hornelen Basin (Devonian), Norway. Sedimentary response to tectonic events. Geol. Soc. Am. Bull. 88, 1124–1134

Steinmetz, R. (1967) Depositional history, primary sedimentary structures, cross bed dips, and grain size of an Arkansas river point bar at Wekiwa. Oklahoma. Rep. F 67–G-3 (By courtesy of Amoco Production Company.)

Steinmetz, R. (1972) Sedimentation of an Arkansas River sand bar in Oklahoma: A coutionary note on dipmeter interpretation. Shale Shaker 16, 32–37

Sternberg, H. (1875) Untersuchungen über Längen- und Querprofil geschiebeführender Flüsse. Z. Bauwesen 25, 483–506

Sternberg, R. W., Larson, L. H. (1976) Frequency of sediment movement on the Washington continental shelf: A note. Mar. Geol. 21, M37–M47

Stevenson, R. E., Emery, K. O. (1958) Marshlands at Newport Bay, California. Allan Hancock Foundation Publ. 20, 1–109

Stoertz, G. E., Ericksen, G. E. (1974) Geology of salars of northern Chile. U. S. Geol. Surv. Prof. Pap. 811, 1–65

Stokes, W. L. (1961) Fluvial and eolian sandstone bodies in Colorado Plateau. In: Peterson, J. A., Osmond, J. C., eds., Geometry of sandstone bodies, p. 151–178. Tulsa, Oklahoma: Am. Assoc. Petroleum Geologists

Stokes, W. L. (1968) Multiple parallel-truncation bedding planes – a feature of wind-deposited sandstone formations. J. Sediment. Petrol. 38, 510–515

Stow, D. A. W., Bowen, A. J. (1978) Origin of lamination in deep-sea, fine-grained sediments. Nature 274, 324–328

Straaten, L. M. J. U. van (1949) Occurrence in Finland of structures due to subaqueous sliding of sediments. Bull. Comm. Geol. Finlande 144, 9–18

Straaten, L. M. J. U. van (1950a) Environment of formation and facies of the Wadden Sea sediments. Koninkl. Ned. Aardrijkskde Genoot. 67, 94–108

Straaten, L. M. J. U. van (1950b) Giant ripples in tidal channels. Koninkl. Ned. Aardrijkskde Genoot. 67, 336–341

Straaten, L. M. J. U. van (1951) Longitudinal ripple marks in mud and sand. J. Sediment. Petrol. 21, 47–54

Straaten, L. M. J. U. van (1952) Biogene textures and the formation of shell beds in the Dutch Wadden Sea. Proc. Koninkl. Ned. Akad. Wetenschap. Amsterdam, Ser. B. 55, 500–516

Straaten, L. M. J. U. van (1953a) Megaripples in the Dutch Wadden Sea and in the Basin of Arcachon (France). Geol. Mijnbouw 15, 1–11

Straaten, L. M. J. U. van (1953b) Rhythmic pattern on Dutch North Sea beaches. Geol. Mijnbouw 15, 31–43

Straaten, L. M. J. U. van (1954a) Composition and structure of Recent marine sediments in the Netherlands. Leidse Geol. Mededel. 19, 1–110

Straaten, L. M. J. U. van (1954b) Sedimentology of Recent tidal flat deposits and the psammites du Condroz (Devonian). Geol. Mijnbouw 16, 25–47

Straaten, L. M. J. U. van (1957) The excavation at Velsen, general introduction. In: The Excavation at Velsen. Verhandl. Koninkl. Ned. Geol. Mijnbouw Gen. Ser. 17, 2, p. 93–99

Straaten, L. M. J. U. van (1959a) Littoral and submarine morphology of the Rhône delta. Proc. 2nd Coastal Geograph. Conf., Baton Rouge (Nat. Acad. Sci. Nat. Rec. Concil) p. 233–264

Straaten, L. M. J. U. van (1959b) Minor structures of some recent littoral and neritic sediments. Geol. Mijnbouw 21, 197–216

Straaten, L. M. J. U. van (1960a) Transport and composition of sediments. Verh. Kon. Ned., Geol. Mijnb. K. Gen. Geol. Serie, D1. XIX, Symposium Ems-Estuarium (Nordsee), p. 279–292

Straaten, L. M. J. U. van (1960b) Some recent advance in the study of deltaic sedimentation. Liverpool Manchester Geol. J. 2, 411–442

Straaten, L. M. J. U. van (1964a) De bodem der Waddenzee. In: Het Waddenboek, Nederl. geol. Ver., p. 75–151

Straaten, L. M. J. U. van, ed. (1964b) Deltaic and shallow marine deposits. Developments in sedimentology, vol. 1, 464 p. Amsterdam: Elsevier

Straaten, L. M. J. U. van (1965) Coastal barrier deposits in south and north Holland in particular in the areas around Scheveningen and Ijmuiden. Mededel. Geol. Sticht. 17, 41–75

Straaten, L. M. J. U. van (1967) Turbidites, ash layers and shell beds in the bathyal zone of the southeastern Adriatic Sea. Rev. Geogr. Phys. Geol. Dynamique 9, 219–240

Straaten, L. M. J. U. van (1970) Holocene and late-Pleistocene sedimentation in the Adriatic Sea. Geol. Rundschau 60, 106–131

Straaten, L. M. J. U. van, Kuenen, Ph. H. (1958) Tidal action as a cause of clay accumulation. J. Sediment. Petrol. 28, 406–413

Strachov, N. M. (1956) Vergleichendes lithologisches Schema authigener Sedimentbildung in den Meeresbecken. Z. angew. Geol. 2, 119–130

Strauch, F. (1966) Sedimentgänge von Tjörnes (Nord-Island) und ihre geologische Bedeutung. Neues Jahrb. Geol. Paläontol. Abh. 124, 259–288

Streif, H.-J. (1978) Der holozäne Meeresspiegelanstieg und die heutige Küstengestalt. In: Reineck, H.-E., Hrsg., Das Watt, pp. 31–38. Frankfurt a. M.: Kramer & Co

Streif-Becker, R. (1952) Probleme der Firnschichtung. Z. Gletscherkd. Glazialgeol. 2, 1–9

Stride, A. H. (1963) Current-swept sea floors near the southern half of Great Britain. Quart. J. Geol. Soc. London 119, 175–199

Stride, A. H. (1970) Shape and size trends for sand waves in a depositional zone of the North Sea. Geol. Mag. 1970, 469–477

Stride, A. H., Curray, J. R., Moore, D. G., Belderson, R. H. (1969) Marine geology of the Atlantic continental margin of Europe. Phil. Trans. Roy. Soc. London, Ser. A 264, 31–75

Strøm, K. M. (1936) Land-locked waters. Hydrography and bottom deposits in badly ventilated Norwegian fjords, with remarks upon sedimentation under anaerobic conditions. Norske Vidensk.-Ak. Mat.-Naturv. Kl. 7

Strøm, K. M. (1939) Land-locked waters and the deposition of black muds. In: Trask, P. D., ed., Recent marine sediments, p. 356–372. Tulsa, Oklahoma: Am. Assoc. Petroleum Geologists

Sturm, M. (1975) Depositional and erosional sedimentary features in a turbidity current controlled basin (Lake Brienz). 9th Intern. Congr. Sedimentology 5, 385–390

Sturm, M. (1976) Die Oberflächensedimente des Brienzer Sees. Eclogae geol. Helv. 69, 111–123

Sturm, M. (1979) Origin and Composition of clastic varves. In: Balkema A. A., ed., Moraines and Varves: Origin, Genesis, Classification, Rotterdam, in press

Sturm, M., Matter, A. (1972) Sedimente und Sedimentationsvorgänge im Thunersee. Eclogae geol. Helv. 65, 563–590

Sturm, M., Matter, A. (1978) Turbidites and varves in Lake Brienz (Switzerland): Deposition of clastic detritus by density currents. In: Matter, A., Tucker, M. E., eds., Modern and ancient lake sediments. Spec. Publ. Int. Ass. Sediment. 2, 147–168

Sundborg, Å. (1956) The river Klarälven; A study of fluvial processes. Geograf. Ann. 38, 125–316

Sundborg, Å (1967) Some aspects on fluvial sediments and fluvial morphology. I General views and graphic methods. Geogr. Ann. 49, 333–343

Surdam, R. C., Stanley, K. O. (1979) Lacustrine sedimentation during the culminating phase of Eocene Lake Gosinte, Wyoming (Green River Formation) Geol. Soc. Am. Bull. 90, 93–110

Sverdrup, H. U., Johnson, M. W., Fleming, R. H. (1942) The oceans, their physics, chemistry and general biology, 1087 p. Englewood Cliffs, N. J.: Prentice-Hall

Swett, K., Klein, de V. G., Smit, D. E. (1971) A Cambrian tidal sand body – The Eriboll Sandstone of Northwest Scotland: An ancient-recent analog. J. Geol. 79, 400–415

Swift, D. J. P. (1976) Coastal Sedimentation. In: Stanley D. J., Swift, D. J. P. eds., Marine sediment transport and environmental management, p. 255–310. New York: Wiley & Sons

Swift, D. J. P., Sanford, R. B., Dill, C. E., Jr., Avignone, N. F. (1971) Textural differentiation on the shore face during erosinal retreat of an unconsolidated coast, Cape Henry to Cape Hatteras, Western North Atlantic shelf. Sedimentology 16, 221–250

Swift, D. J. P., Stanley, D. J., Curray, J. R. (1971) Relict sediments on continental shelves: A reconsideration. J. Geol. 79, 322–346

Swift, D. J. P., Duane, D. B., Pilkey, O. H. (1972) Shelf sediment transport: Process and Pattern, 656 p. Stroudsburg, Pa.: Dowden, Hutchinson & Ross

Swineford, A., Frye, J. C. (1945) A mechanical analysis of wind-blown dust compared with analyses of loess. Am. J. Sci. 243, 249–255

Szupryczynski, J. (1965) Eskers and Kames in the Spitsbergen area. Geogr. Polonica 6, 127–140

Tandon, S. K. (1976) Siwalik sedimentation in a part of the Kumaon Himalaya, India. Sediment. Geol. 16, 131–154

Tanner, W. F. (1959) Sample components obtained by the method of differences. J. Sediment. Petrol. 29, 408–411

Tanner, W. F. (1960) Florida coastal classification. Gulf Coast Assoc. Geol. Soc. Trans. 10, 259–266

Tanner, W. F. (1967) Ripple mark indices and their uses. Sedimentology 9, 89–104

Taylor, G., Woodyer, K. D. (1978) Bank deposition in suspended – load streams. In: Miall, A. D., ed., Fluvial Sedimentology. Canad. Soc. Petrol. Geol., Mem. 5, 257–275

Taylor, G., Crook, A. W., Woodyer, K. D. (1971) Upstream-dipping foreset cross-stratification: origin and implications for paleoslope analyses. J. Sediment. Petrol. 41, 573–581

Teal, J., Teal, M. (1969) Life and death of the salt marsh, 278 p. Boston, Toronto: Little, Brown and Co.

Teichert, C. (1958) Concepts of facies. Bull. Am. Assoc. Petrol. Geologists 42, 2718–2744

Teichert, C. (1970) Runzelmarken (Wrinkle marks). J. Sediment. Petrol. 40, 1056

Teisseyre, A. K. (1975) Pebble fabric in braided stream deposits from Recent and frozen Carboniferous channels (Intrasudetic basin, Central Sudeten). Geol. Sudetica 10, 1–46

Ten Haaf, E. (1959) Graded beds of the Northern Apenines. Doctoral Thesis, University of Groningen, 102 p.

Terlecky, Jr. P. M. (1974) The origin of a late Pleistocene and Holocene marl deposit. J. Sediment. Petrol. 44, 456–465

Termier, H., Termier, G. (1960) Erosion et sedimentation, 412 p. Paris: Masson & Cie

Terwindt, J. H. J. (1971 a) Litho-facies of inshore estuarine and tidal-inlet deposits. Geol. Mijnbouw 50, 515–526

Terwindt, J. H. J. (1971 b) Sand waves in the southern bight of the North Sea. Marine Geol. 10, 51–67

Terwindt, J. H. J. (1973) Sand movement in the in- and offshore tidal area of the S. W. part of the Netherlands. Geol. Mijnbouw 52, 69–77

Terwindt, J. H. J. (1979) Origin and sequences of sedimentary structures in inshore meso-tidal deposits along the North Sea. (unpublished)

Terwindt, J. H. J., Breusers, H. N. C. (1972) Experiments on the origin of flaser, lenticular and sand-clay alternating bedding. Sedimentology 19, 85–98

Terwindt, J. H. J., de Jong, J. D., Wilk, E. van der (1963) Sediment movement and sediment properties in the tidal area of the Lower Rhine (Rotterdam waterway). Geol. Serie 21, 244–258

Tessenow, U. (1975) Akkumulationsprozesse in der Maximaltiefe von Seen durch postsedimentäre Konzentrationswanderung. Verh. Internat. Verein. Limnol. 19, 1251–1262

Tessenow, U., Baynes, Y. (1975) Redox-dependent accumulation of Fe and Mn in a littoral sediment supporting Isoetes lacustrines. Naturwiss. 62, 342

Theakstone, W. H. (1976) Glacial lake sedimentation, Austerdalsisen, Norway. Sedimentology 23, 671–688

Thiede, J. (1978) Pelagic sedimentation in immature ocean basins. In: Ramberg, I. B., Neumann, E. R., eds., Tectonics and geophysics of continental rifts p. 237–248

Thom, B. G., Polach, H. A., Bowman, G. M. (1978) Holocene age structure of coastal sand barriers in New South Wales, Australia. Univ. New South Wales. Fac. Military Stud. Dep. Geogr. Royal Military College, Duntroon act 2600, 1–86

Thomas, R. L., Jaquet, J.-M. (1975) The surficial sediments of Lake Superior. Canada Centre Inland Waters Collected Reprints, 9. Congr. Int. Sedimentologie, Nice, 14 p.

Thompson, A. F., Thomasson, M. R. (1969) Shallow to deep water facies development in the Dimple limestone (Lower Pennsylvanian), Marathon region, Texas. Soc. Econ. Paleontologists Mineralogists Spec. Publ. 14, 57–78

Thompson, D. B. (1969) Dome-shaped aeolian dunes in the Frodsham Member of the so-called "Keuper" Sandstone Formation (Scythian-Anisian: Triassic) at Frodsham, Cheshire (England). Sediment. Geol. 3, 263–289

Thompson, R. W. (1968) Tidal flat sedimentation on the Colorado River delta, northwestern Gulf of California. Geol. Soc. Am. Mem. 107, 1–133

Thompson, W. O. (1937) Original structures of beaches, bars, and dunes. Geol. Soc. Am. Bull. 48, 723–752

Thorndike, E. M. (1959) Deep-sea cameras of the Lamont Observatory. Deep-Sea Res. 3, 234–237

Thorslund, P. (1960) Notes on the geology and stratigraphy. of Västergötland. In: Guide to Excursions Nos. A 23 and C 18, 21. International Geol. Congress 6–11

Thorson, G. (1957) Bottom communities (sublittoral of shallow shelf). Geol. Soc. Am. Mem. 67, 461–534

Tietze, K.-W. (1978) Zur Geometrie von Wellenrippeln in Sanden unterschiedlicher Korngröße. Geol. Rundschau 67, 1016–1033

Tissot, B. P., Deroo, G., Hood, A. (1978) Geochemical studies of the Utah Basin: formation of petroleum from the Green River formation. Geochem. Cosmochem. Acta 42, 1469–1485

Tissot, B. P., Welte, D. H. (1978) Petroleum Formation and Occurrence, 538 p. Berlin, Heidelberg, New York: Springer

Trask, P. D., ed. (1939) Recent marine sediments, 736 p. Tulsa, Oklahoma: Am. Assoc. Petroleum Geologists

Tricart, J. (1970) Geomorphology of cold environments, 320 p. Edinburgh: Macmillan

Troll, C. (1925) Methoden, Ergebnisse und Ausblicke der geochronologischen Eiszeitforschung. Naturwissenschaften 13, 909–919

Truckenbrodt, E. (1968) Strömungsmechanik. Grundlagen und technische Anwendungen, 532 p. Berlin, Heidelberg, New York: Springer

Trusheim, F. (1929 a) Massentod von Insekten. Natur Mus. 59, 54–61

Trusheim, F. (1929 b) Zur Bildungsgeschwindigkeit geschichteter Sedimente im Wattenmeer, besonders solcher mit schräger Parallelschichtung. Senckenbergiana 11, 47–55

Trusheim, F. (1931) Versuche über Transport und Ablagerung von Mollusken. In: Lüders, K., Trusheim, F., eds., Beiträge zur Ablagerung mariner Mollusken in der Flachsee. Senckenbergiana 13, 124–139

Trusheim, F. (1935 a) Naturspiel oder Organismus? Natur Volk 65, 521–524

Trusheim, F. (1935 b) Eine Titaneisenerz-Seife von Wangerooge. Senckenbergiana 17, 62–72

Trusheim, F. (1936) Wattenpapier. Natur Volk 66, 103–106

Turekian, K. K., ed. (1968) Oceans, 120 p. Englewood Cliffs, New Jersey: Prentice-Hall, Inc.

Twenhofel, W. H., ed. (1932) Treatise on sedimentation, 926 p. 2nd. ed. Baltimore: The Williams & Wilkins Co

Twenhofel, W. H. (1950) Principles of sedimentation, 673 p. New York: McGraw-Hill

Uchupi, E. (1967) Slumping on the continental margin southeast of Long Island, New York. Deep-Sea Res. 14, 635–639

Uchupi, E., Emery, K. O. (1968) Structure of continental margin off Gulf Coast of United States. Bull. Am. Assoc. Petrol. Geologists 52, 1162–1193

Ulrich, J. (1972) Untersuchungen zur Pendelbewegung von Tiderippeln im Heppenser Fahrwasser (Innenjade). Die Küste 23, 112–121

Ulrich, J. (1973) Die Verbreitung submariner Riesen- und Großrippeln in der Deutschen Bucht. Erg.-H. Deutsch. Hydrogr. Z. (B.) 14, 1–31

Unrug, R. (1963) Istebna beds – a fluxoturbidity formation in the Carpathian Flysch. Ann. Soc. Geol. Pologne 33, 49–92

Urey, H. C. (1951) Measurement of palaeotemperatures. Geol. Soc. Am. Bull. 62, 399–416

U. S. Waterways Experiment Station (1939) Study of materials in suspension, Mississippi River. Techn. Mem. 122–1, Vicksburg. Louisiana, 27 p.

Valdiya, K. S. (1972) Origin of phosphorite of the Precambrian Gangolihat dolomites of Pithoragarh, Kumaon Himalya, India. Sedimentology 19, 115–128

Valentin, H. (1952) Die Küsten der Erde, 118 p. Berlin: Justus Perthes Gotha

Valeton, I. (1958) Der Glaukonit und seine Begleitminerale aus dem Tertiär von Walsrode. Mitt. Geol. Staatsinst. Hamburg 27, 88–131

Vanney, J.-R., Dangeard, L. (1976) Les dépôts glacio-marins actuels et anciens. Rev. Georgr. Montr. 30, 9–50

Vanoni, V. A. (1974) Factors determining bed forms of alluvial streams. Am. Soc. Civil Engineers Proc. 100, 363–377

Vasari, Y., Koljonen, T., Laasko, K. (1972) A case of manganese precipitate in the Taviharjn Esker, Kuusamo, North East Finland. Bull. Geol. Soc. Finland 44, 133–140

Vassoevic, N. B. (1953) On some structures in the flysch (English summary). Tr. Lvovsk. Geol. Obscesto 3, 17–85

Veatch, A. C., Smith, P. A. (1939) Atlantic submarine valleys off the United States and the Congo submarine valley. Geol. Soc. Am. Spec. Papers 7, 101 p.

Veen, J. van (1936) Onderzoekingen in de Hoofden in verband met de gesteldheid der Nederlandse kust. The Hague: Landsdrukkerij, 252 p.

Veen, J. van (1950) Eb- en vloedschaar-systemen in de Nederlandse getigwateren. T. Koninkl. Ned. Aardigskde Genoot. 66, 303–325

Veenstra, H. J. (1964) Geology of the Hinder Banks, southern North Sea. Hydrographic Newsletter. Publ. Netherl. Hydrograph, 1, 2.

Veizer, J., Demovic, R. (1974) Strontium as a tool in facies analysis. J. Sediment. Petrol. 44, 93–115

Verger, F. (1968) Marais et wadden du littoral français, 541 p. Bordeaux: Biscaye Frères Impr.

Vernet, J.-P., Horn, R. (1971) Etudes sedimentologique et structurale de la patie occidentale du lac Léman par le méthode sismique à rérflexion continue. Eclogae geol. Helv. 64, 291–317

Vernet, J.-P., Thomas, R. L. (1972) Levels of mercury in the sediments of some Swiss lakes including Lake Geneva and the Rhone River. Eclogae geol. Helv. 65, 293–306

Vernet, J.-P., Meybeck, M., Pachoud, A., Scolari, G. (1971) Le Léman: une synthèse bibliographique. Bull. du. B. R. G. M. 2, 47–84

Vernet, J.-P., Thomas, R. L., Jaquet, J.-M., Friedli, R. (1972) Texture of the sediments of the Petit Lac (Western Lake Geneva). Eclogae geol. Helv. 65, 591–610

Vernon, J. W. (1965) Final report on shelf sediment transport system. Los Angeles. University of Southern California. Rep. No. USC Geol. 65-2, 135 p.

Vierke, M. (1937) Die ostpommerschen Bändertone als Zeitmarken und Klimazeugen. Abhandl. Geol.-Palaeont. Inst. Univ. Greifswald 18

Visher, G. S. (1965 a) Fluvial processes as interpreted from Ancient and Recent fluvial deposits. In: Middleton, G. V., ed., Primary sedimentary structures and their hydrodynamic interpretation. Soc. Econ. Paleontologists Mineralogists Spec. Publ. 12, 116–132

Visher, G. S. (1965 b) Use of vertical profile in environmental reconstruction. Bull. Am. Assoc. Petrol. Geologists 49, 41–61

Visher, G. S. (1969) Grain size distributions and depositional processes. J. Sediment. Petrol. 39, 1077–1106

Visser, M. J. (1980) Neap-spring cycles reflected in Holocene subtidal large-scale bedform deposits: a preliminary note. Geology 8, 543–546

Völpel, A., Samu, G. (1966) Reliefänderungen in der Tidestromrinne des Wangerooger Fahrwassers im Verlauf einer Sturmperiode und in der darauf folgenden Periode mit ruhigen Wetterlagen. Mitt.-Bl. Bundesanst. Wasserbau 24, 1–20

Vollbrecht, K. (1953) Zur Quarzachsenregelung sandiger Sedimente. Acta Hydrophys. 1, 61–88

Vollbrecht, K. (1957) Aufbau, Veränderlichkeit und Auflösung von Sandriffen. Geologie 8, 753–796

Waddel, E. (1976) Swash- Groundwater- Beach profile interactions. In: Davis, Jr., R. A., Ethington, R. L., eds., Beach and nearshore sedimentation. Soc. Econ. Paelontologists Mineralogists Spec. Publ. 24, 115–125

Wadell, H. (1932) Volume shape and roundness of rock particles. J. Geol. 40, 443–451

Wadell, H. (1935) Volume, shape and roundness of quartz particles. J. Geol. 43, 250–280

Wagner, G. (1950) Einführung in die Erd- und Landschaftsgeschichte, 664 p. Öhringen: Verlag der Hohenlohe'schen Buchhandlung F. Rau

Wagner, G. (1968) Zur Beziehung zwischen der Besiedlungsdichte von Tubificiden und dem Nahrungsangebot im Sediment. Int. Rev. Ges. Hydrobiol. 53, 715–721

Wagner, G. (1971) FeS-Konkretionen im Bodensee. Internat. Rev. Ges. Hydrobiol. 56, 313–320

Wagner, G. (1972) Stratifikation der Sedimente und Sedimentationsrate im Bodensee. Verh. Internat. Verein. Limnol. 18, 475–481

Walger, E. (1962) Die Korngrößenverteilung von Einzellagen sandiger Sedimente und ihre genetische Bedeutung. Geol. Rundschau 51, 494–507

Walker, C. T., (1972) Paleosalinity. In: Fairbridge, R. W., ed., The Encyclopedia of Geochemistry and Environmental Sciences, p. 885–891. Stroudsburg, Pa.: Dowden, Hutchinson & Ross

Walker, R. G. (1963) Distinctive types of ripple-drift cross-lamination. Sedimentology 2, 173–188

Walker, R. G. (1965) The origin and significance of the internal sedimentary structures of turbidites. Proc. Yorkshire Geol. Soc. 35, 1–32

Walker, R. G. (1967) Turbidite sedimentary structures and their relationship to proximal and distal depositional environments. J. Sediment. Petrol. 37, 25–43

Walker, R. G. (1969) Geometrical analysis of ripple-drift cross-lamination. Canadian J. Earth Sci. 6, 383–391

Walker, R. G. (1978) Deep-water sandstone facies and ancient submarine fans: Models for exploration for stratigraphic traps. Am. Ass. Petrol. Geol. Bull 62, 932–966

Walker, T. R. (1967) Formation of red beds in modern and ancient deserts. Geol. Soc. Am. Bull. 78, 353–368

Walker, T. R. (1974) Formation of red beds in moist tropical climates: a hypothesis. Geol. Soc. Am. Bull. 85, 633–638

Walker, T. R., Harms, J. C. (1972) Eolian origin of Flagstone Beds. Lyons Sandstone (Permian) type area, Boulder county, Colorado. The Mountain Geologist 9, 279–288

Walther, J. (1894) Lithogenesis der Gegenwart. Beobachtungen über die Bildung der Gesteine an der heutigen Erdoberfläche. Dritter Teil einer Einleitung in die Geologie als historische Wissenschaft, p. 535–1055. Jena: Verlag Gustaf Fischer

Walther, J. (1909) Über algonkische Sedimente. Z. Deut. Geol. Ges. 61, 283–305

Walther, J. (1924) Das Gesetz der Wüstenbildung in Gegenwart und Vorzeit, 421 p. Leipzig: 4th ed. Quelle & Meyer.

Walton, E. K. (1967) The sequence of internal structure s in turbidites. Scot. J. Geol. 3, 306–317

Wanless, H. R., et al. (1970) Late paleozoic deltas in the central and eastern United States. In: Morgan, J. P., ed., Deltaic sedimentation modern and ancient. Soc. Econ. Paleontologists Mineralogist Spec. Publ. 15, 215–245

Warme, J. E. (1966) Paleoecological aspects of the Recent ecology of Mugu Lagoon, California. Doctoral dissertation. University of Californ ia. Los Angeles, 380 p.

Warme, J. E. (1967) Graded bedding in the Recent sediments of Mugu Lagoon, California. J. Sediment. Petrol. 37, 540–547

Warme, J. E. (1969 a) Live and dead molluscs in a coastal lagoon, J. Paleontol. 43, 141–150

Warme, J. E. (1969 b) Mugu Lagoon, coastal Southern California: origin, sediments and productivity. In: Castanares, A. A., Phleger, F. B., eds., Coastal lagoons, a Symposium, p. 137–154. Mexico: Universidad nacional Autonoma

Warme, J. E. (1971) Paleoecological aspects of a modern coastal lagoon. Univ. California Publ. Geol. Sci. 87, 1–131,

Berkeley-Los Angeles – London: University of California Press

Warme, J. E. (1977) Traces and significance of marine rock borers. In: Crimes, T. B., Harper, J. C., eds., Trace Fossils. Geol. J., spec. Iss. 3, 515–526

Warme, J. E., Marshall, N. F. (1969) Marine borers in calcareous terrigenous rocks of the Pacific Coast. Am. Zoologist 9, 765–774

Warren, A. (1972) Observations on dunes and bi-modal sands in the Ténéré desert. Sedimentology 19, 37–44

Washburn, (1973) Periglacial processes and environments, 320 p. London: Arnold

Wasmund, E. (1926) Biocönose und Thanatocönose. Arch. Hydrobriol. 17, 1–116

Wasmund, E. (1930) Rieselfelder und Blattfächerabdrücke auf rezentem und fossilem Süßwasser-Flachstrand. Senckenbergiana 12, 139–151

Wasmund, E. (1939) Flachsee-Beobachtungen bei Sturm-Niedrigwasser an der gezeitenschwachen Kieler Förde (Ostsee). Geol. d. Meere u. Binnengewässer 3, 284–309

Wasson, R. J. (1977) Last-glacial alluvial fan sedimentation in the Lower Derwent Valley, Tasmania Sedimentology 24, 781–799

Watkins, J. S., Montadert, L., Dickerson, P. W., eds. (1979) Geological and Geophysical investigations of continental Margins. Am. Ass. Petrol. Geol. Mem. 29, 479 p.

Wedepohl, K. H. (1970) Geochemische Daten von sedimentären Karbonaten und Karbonatgesteinen in ihrem faziellen und petrogenetischen Aussagewert. Verh. Geol. Bundesanstalt Wien 4, 692–705

Weeks, L. G. (1965) World offshore petroleum resources. Bull. Am. Ass. Petrol. Geol. 49, 1680–1693

Weggel, J. R. (1972) An Introduction to oceanic water motions and their relation to sediment transport. In: Swift, D. J. P., Duane, D. B., Pilkey, O. H., eds., Shelf sediment transport: Process and Pattern, p. 1–20. Stroudsburg, Pa.: Dowden, Hutchinson & Ross

Weil, C. B. (1977) Sediments, structural framework, and evolution of Delaware Bay, a transgressive estuarine delta. DEL–SG– 4-77, College of Marine Studies, Univ. of Delaware, 199 p.

Weimer, R. J. (1960) Upper Cretaceous stratigraphy, Rocky Mountain area. Bull. Am. Assoc. Petrol. Geologists 44, 1–20

Weimer, R. J. (1970 a) Rates of deltaic sedimentation and intrabasin deformation, upper Cretaceous of Rocky Mountain region. In: Morgan, J. P., ed., Deltaic sedimentation modern and ancient. Soc. Econ. Paleontologists Mineralogists Spec. Publ. 15, 270–292

Weimer, R. J. (1970 b) Stratigraphy principles and practices in petroleum exploration. Lecture Notes and References. Golden, Col.: School of Mines

Weimer, R. J. (1975) Deltaic and shallow marine sandstones: Sedimentation, tectonics and petroleum occurrences. Am. Ass. Petrol. Geol. Continuing Education Course Note Paper 2, 1–167

Weimer, R. J., Hoyt, J. H. (1964) Burrows of *Callianassa major* Say, geologic indicators of littoral and shallow neritic environments. J. Paleontol. 38, 761–767

Weinkauff, H. E. (1859) Die teritären Ablagerungen im Kreise Kreuznach. Verh. Naturhist. Ver. preuß. Rheinl. Westf. 16

Weir, J. E. (1962) Large ripple marks caused by wind near Coyote Lake (dry). California. Geol. Soc. Am. Spec. Papers 73, 1–72

Weller, J. M. (1960) Stratigraphy principles and practice, 725 p. New York: Harper & Row

Wentworth, C. K. (1936) An analysis of shapes of glacial cobbles, J. Sediment. Petrol. 6, 85–96

Wentworth Jr., C. M. (1967) Dish structure, a primary sedimentary structure and coarse turbidites. Am. Assoc. Petrol. Geol. Bull. 51, 485

Werner, F. (1963) Über den inneren Aufbau von Strandwällen an einem Küstenabschnitt der Eckernförder Bucht. Meyniana 13, 108–121

Werner, F. (1967) Sedimentation und Abrasion am Mittelgrund (Eckernförder Bucht, westliche Ostsee). Meyniana 17, 101–110

Werner, F. (1968) Gefügeanalyse feingeschichteter Schlicksedimente . der Eckernförder Bucht (Westliche Ostsee). Meyniana 18, 79–105

Werner, F., Newton, R. S. (1970) Riesenrippeln im Fehmarnbelt (westliche Ostsee). Meyniana 20, 83–90

Werner, F., Newton, R. S. (1975) The pattern of large-scale bed forms in the Langeland Belt (Baltic Sea). Mar. Geol. 19, 29–59

Werner, F., Arntz, W. E., Tauchergruppe Kiel (1974) Sedimentologie und Ökologie eines ruhenden Riesenrippelfeldes. Meyniana 26, 39–62

West, R. G. (1968) Pleistocene geology and biology, 377 p. London: Longmans Green & Co.

Wetzel, W. (1937) Die koprogenen Beimengungen mariner Sedimente und ihre diagnostische und lithogenetische Bedeutung. Neues Jahrb. Mineral. etc., Beil.-Bd. 78 B, 109–122

Weyl, R. (1969) Magnetitsande der Küste Nicoyas (Costa Rica, Mittelamerika). Neues Jahrb. Geol. Paleontol. Monatsh. 8, 499–511

Wezel, F. C. (1968) Osservazioni sur sedimenti dell'Oligocene-Miocene inferiore della Tunesia settentrionale. Mem. Soc. Geol. Ital. 7, 417–439

Whalley, W. B., Krinsley, D. (1974) A scanning electron microscope study of surface textures of quartz grains from a glacial environment. Sedimentology 21, 87–105

Wheeler, H. E., Mallory, N. S. (1956) Factors in lithography. Bull. Am. Assoc. Petrol. Geologists 40, 2711–2723

Whitaker, J. H., McD. (1962) The geology of the area round Leintwardine, Herefordshire. Quart. J. Geol. Soc. London 118, p. 319

Whitaker, J. H. McD. (1973) "Gutter Casts", a new name for scour-and fill-structures, with examples from the Llandoverian of Ringerike and Malmöya, southern Norway. Norsk Geol. Tidsskr. 53, 403–417

Whitaker, J. H. McD. (1974) Ancient submarine canyons and fan valleys. In: Dott, R. H., Jr., Shaver, R. H., eds., Modern and ancient geosynclinal sedimentation. Soc. Econ. Paleontologists Mineralogists 19, 106–125

White, W. A. (1961) Colloid phenomena in sedimentation of argillaceous rocks. J. Sediment. Petrol. 31, 560–570

Whitehouse, U. G., Jeffery, L. M., Debbrecht, J. D. (1960) Differential settling tendencies of clay minerals in saline waters. Proc. 7th Conf. Clays Cla. Mins, 1–79

Whitney, M. I., Dietrich, R. V. (1973) Ventifact sculpture by windblown dust. Geol. Soc. Am. Bull. 84, 2561–2582

Wiedemann, H. U. (1972) Shell desposits and shell preservation in Anatermary und Tertiary estuarine Sediments in Georgia, U. S. A. Sediment. Geol. 7, 103–125

Wiedemann, H. U. (1973) Reconnaissance of the Ciènaga Grande de Santa Marta, Colombia: Physical Parameters and Geological History. Mitt. Inst. Colombo-Alemán Investt. Cient. 7, 85–119

Wiegel, R. L. (1953) Waves, tides currents and beaches: Glossary of terms and list of standard symbols. Council on Wave Research, Eng. Foundation. Berkeley, California, 113 p.

Wiegel, R. L. (1964) Oceanographic engineering. Prentice Hall, New York

Wilde, P., Normark, W. R., Chase, T. E. (1978) Channel sands and petroleum potential of Monterey Deep. Sea Fan, California. Am. Ass. Petrol. Geol. Bull. 62, 967–983

Williams, A. T. (1973) The problem of beach cusp development. J. Sediment. Petrol. 43, 857–866

Williams, E. (1960) Intra-stratal flow and convolute folding. Geol. Mag. 97, 208–214

Williams, G. (1964) Some aspects of aeolian saltation load. Sedimentology 3, 257–287

Williams, G. E. (1969) Characteristics and origin of a Pre-Cambrian sediment. J. Geol. 77, 183–207

Williams, G. E. (1971) Flood deposits of the sand bed ephemeral streams of central Australia. Sedimentology 17, 1–40

Williams, P. F., Rust, B. R. (1969) The sedimentology of a braided river. J. Sediment. Petrol. 39, 649–679

Wilson, I. G. (1972) Aeolian bedforms – their development and origins. Sedimentology 19, 173–210

Wilson, I. G. (1973) Ergs. Sediment. Geol. 10, 77–106

Windom, H. L. (1975) Eolian contributions to marine sediments. J. Sediment. Petrol. 45, 520–529

Windom, H. L., Chamberlain, C. F. (1978) Dust-storm transport of sediments to the north Atlantic ocean. J. Sediment. Petrol. 48, 385–388

Winkelmolen, A. M. (1969 a) Experimental rollability and natural shape sorting of sand. Thesis. University of Groningen, 141 p.

Winkelmolen, A. M. (1969 b) The rollability apparatus. Sedimentology 13, 291–305

Winkelmolen, A. M., Veenstra, H. J. (1974) Size and shape sorting in a Dutch tidal inlet. Sedimentology 21, 107–126

Winkelmolen, A. M., Knapp, W. van der, Eijpe, R. (1968) An optical method of measuring grain orientation in sediments. Sedimentology 11, 183–196

Winnock, E. (1965) Sismique sur le Léman. Résultats obtenus avec le sparker. Bull. Ver. Schweiz. Petrol. Geol. Ingr. 32, 39–48

Winterer, E. L., Durham, D. L. (1962) Geology of the southeastern Ventura basin, Los Angeles County, California. U. S. Geol. Surv. Profess, Papers 334H, p. 275–366

Woldstedt, P. (1961) Das Eiszeitalter, vol. 1, 374 p. Stuttgart: Enke

Wolf, K. H. (1973) Conceptual models. 2. Fluvial-Alluvial, Glacial, Lacustrine, Desert and Shorezone (Bar-Beach-Dune-Chenier) Sedimentary Environments. Sediment. Geol. 9, 261–281

Wood, W. H. (1970) Rectification of windblown sands. J. Sediment. Petrol. 40, 29–37

Woolsey, J. R., Henry, V. J., Hunt, J. L. (1975) Backshore heavy-mineral concentration on Sapelo Island, Georgia. J. Sediment. Petrol. 45, 280–284

Wright, A. E., Moseley, F. (1975) Ice Ages: Ancient and Modern. Geol. J. Spec. Issue 6, 320 p. Liverpool: Seal House Press

Wright, L. D., Short, A. D. (1983) Morphodynamics of beaches and surf zones in Australia. In: Komar, P. D., ed., Handbook of coastal processes and erosion. CRC Press, Boca Raton, Fla. 35 – 64

Wright, M. D. (1959) The formation of cross-bedding by a meandering or braided stream. J. Sediment. Petrol. 29, 610–615

Wüst, G. (1955) Stromgeschwindigkeiten im Tiefen- und Bodenwasser des Atlantischen Ozeans. Deep-Sea Res. 3 (Suppl.), 373–397

Wüst, G. (1957) Quantitative Untersuchungen zur Statik und Dynamik des Atlantischen Ozeans: Stromgeschwindigkeiten und Strommengen in den Tiefen des Atlantischen Ozeans. Deut. Atlantische Expedition "Meteor", 1925–1927, Wiss. Ergebn. 6, 420 p. Berlin: Borntraeger

Wüst, G. (1958) Über Stromgeschwindigkeiten und Strommengen in der atlantischen Tiefsee. Geol. Rundschau 47, 187–195

Wunderlich, F. (1967 a) Die Entstehung von "convolute bedding" an Platenrändern. Senckenbergiana Lethaea 48, 345–349

Wunderlich, F. (1967 b) Feinblättrige Wechselschichtung und Gezeitenschichtung. Senckenbergiana Lethaea 48, 337–343

Wunderlich, F. (1969) Studien zur Sedimentbewegung. 1. Transportformen und Schichtbildung im Gebiet der Jade. Senckenbergiana marit. 1, 107–146

Wunderlich, F. (1970 a) Schill-Einregelung durch Seevögel auf Mellum. Natur Mus. 100, 175–178

Wunderlich, F. (1970 b) Genesis and environment of the "Nellenköpfchenschichten" (lower Emsian, Rheinian Devon) at locus typicus in comparison with modern coastal environment of the German Bay. J. Sediment. Petrol. 40, 102–130

Wunderlich, F. (1970 c) Korngrößenverschiebung durch Lanice conchilega (Pallas). Senckenbergiana marit. 2, 119–125

Wunderlich, F. (1970 d) Schichtbänke. In: Reineck, H.-E., ed., Das Watt, Ablagerungs- und Lebensraum, p. 48–55. Frankfurt: Kramer

Wunderlich, F. (1971) Der Golf von Gaeta (Tyrrhenisches Meer). II. Strandaufbau und Stranddynamik. Senckenbergiana marit. 3, 135–183

Wunderlich, F. (1972) Georgia Coastal region, Sapelo Island, U. S. A. Sedimentology and biology. III. Beach dynamics and beach development. Senckenbergiana marit. 4, 47–79

Wunderlich, F. (1973 a) Sekundäre Schichtdeformationen unter Eisauflast. Senckenbergiana marit. 5, 153–159

Wunderlich, F. (1973 b) Backset bedding" durch Rhomboederrippeln. Senckenbergiana marit. 5, 161–164

Wunderlich, F. (1978) Deposition of mud in the giant ripples of inner Jade, German Bight, North Sea. Senckenbergiana marit. 10, 257–267

Wunderlich, F. (1979) Die Insel Mellum (südliche Nordsee) Dynamische Prozesse und Sedimentgefüge. I. Südwatt, Übergangszone und Hochfläche. Senckenbergiana marit. 11, 59–113

Wurster, P. (1964) Geologie des Schilfsandsteins. Mitt. Geol. Staatsinst. Hamburg 33, 1–140

Yagishita, K. (1977) Possible mechanism of submarine sliding and its associated minor slump fold. Earth Sci. 31, 179–192

Yalin, M. S. (1964) Geometrical properties of sand waves. Am. Soc. Civil Engrs. Proc. HY 5 90, 105–119

Yalin, M. S. (1972) Mechanics of Sediment Transport, 290 p. Oxford: Pergamon Press

Yang, C.-S., Nio, S.-D. (1985) The estimation of paleohydrodynamic processes from subtidal deposits using time series analysis methods. Sedimentol. 32, 41 – 57

Young, R. G. (1955) Sedimentary facies and intertonguing in the Upper Cretaceous of the Book Cliffs, Utah and Colorado. Geol. Soc. Am. Bull. 66, 177–202

Young, R. G. (1957) Late Cretaceous cyclic deposits. Book Cliffs, Eastern Utah. Bull. Am. Assoc. Petrol. Geologists 41, 1760–1774

Zabawa, C. F. (1978) Microstructure of agglomerated suspended sediments in Northern Chesapeake Bay Estuary. Science 202, 49–51

Zahner, R. (1967) Experimente zur Analyse biologischer, chemischer und physikalischer Vorgänge in der Wasser-Sediment-Grenzschicht stehender und langsam strömender Gewässer. I. Beschreibung der Versuchsanlage mit vorläufigen Ergebnissen über das Verhalten der Tubificiden im Wahlversuchen. Int. Rev. Ges. Hydrobiol.. 53, 627–645

Zahner, R. (1968) Biologische Abbauvorgänge im Bodensediment von Seen. Wasser- und Abwasser-Forsch. 4/68, 1–5

Zenkovitch, V. P. (1966) Black Sea, p. 145–151. In: Fairbridge, R. W., ed., The encyclopedia of oceanography. New York: Reinhold

Zenkovitch, V. P. (1967) Processes of coastal development, 738 p. Edinburgh: Oliver & Boyd

Zenkovitch, V. P. (1969) Origin of barrier beaches and lagoon coast. In: Castanares, A. A., Phleger, F. B., eds., Coastal lagoons, a Symposium, p. 27–38. Mexico: Universidad Nacional Autonoma

Zenkovich, V. P. (1970) Nature of the USSR marine shelves and coasts. Quaternaria 12, 71–77

Zeuner, F. E. (1959) The Pleistocene period, 447 p. 2nd. ed. London: Hutchinson

Zimdars, J. (1958) Über Korn-Oberflächen von Sanden. Eine kritische Betrachtung der morphoscopischen Quarzkornanalyse. Dissertation University of Tübingen, 92 p.

Zimmerle, W., Bonham, L. C. (1962) Rapid methods for dimensional grain orientation measurements. J. Sediment. Petrol. 32, 751–763

Zingg, T. (1935) Beitrag zur Schotteranalyse. Schweiz. Mineral. Petrogr. Mitt. 15, 39–140

Subject Index